INORGANIC ELECTRONIC STRUCTURE AND SPECTROSCOPY

Volume I: Methdology

Edited by

Edward I. Solomon

Monroe E. Spaght Professor of Chemistry
Stanford University
Stanford, California

A.B.P. Lever

Distinguished Professor of Chemistry
Faculty of Pure and Applied Science
York University
Toronto, Ontario, Canada

WILEY-INTERSCIENCE

A JOHN WILEY & SONS, INC., PUBLICATION

Published by John Wiley & Sons, Inc., Hoboken, New Jersey.
Published simultaneously in Canada.

Library of Congress Cataloging-in-Publication Data is available.

ISBN-13 978-0-471-97124-5
ISBN-10 0-471-97124-3

Printed in the United States of America.

10 9 8 7 6 5 4 3 2 1

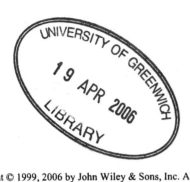

CONTENTS

PREFACE

Spectroscopy has had a huge impact in Inorganic Chemistry and indeed most researchers in the field make significant use of a combination of spectroscopic methods and bonding concepts to understand molecular properties on a fundamental level. The sophistication of the instrumentation available, our understanding of the spectroscopic parameters obtained in terms of electronic structure, and our insight into the contribution of electronic structure to physical properties and chemical reactivity have all made great advances in recent years. Unfortunately, most of the key texts in inorganic spectroscopy and electronic structure that were published before 1970 are out of print and often no longer available. The purpose of these volumes is to provide an up to date text covering the key spectroscopic and electronic structure methods and to develop their application and define their impact on a number of important topics in Inorganic Chemistry. Each chapter is written by an internationally recognized leader in that field, with the level of presentation assuming an advanced undergraduate background with a basic understanding of quantum chemistry (at the level of Pauling and Wilson) and group theory (at the level of Cotton). Each chapter begins with the basics, gives examples and evolves into timely topics in the field.

Volume I presents the different spectroscopic methods and levels of electronic structure calculations for transition metal systems. We start with a summary presentation of the key concepts in ligand field theory and its applications to the physical properties of metal complexes. This is a topic where many of the original texts are "out of print". This chapter serves as a foundation for many of the spectroscopic methods which follow. Ground state methods: electron paramagnetic resonance and Mössbauer; excited state methods: polarized absorption, resonance Raman, luminescence, laser spectroscopy; high energy methods: photoelectron and X-ray absorption spectroscopy. This volume concludes with chapters on the modern "computational approaches" such as "ab initio" and "density functional theory" utilized in the calculation of the electronic structure of transition metal complexes.

Volume II contains two parts. The first presents key topics in Inorganic Chemistry where electronic structure and spectroscopy have had a huge impact: bioinorganic chemistry, electron transfer, mixed valence compounds, electrochemistry and photochemistry and photophysics.

The final section of these volumes presents case studies of major classes of inorganic compounds: metal-metal bonds, metal carbonyls and metallocenes, metal nitrosyls, heme sites, spin crossover compounds and magnetic materials.

This presentation of the methods, applications and case studies in inorganic electronic structure and spectroscopy should provide modern researchers with a strong foundation in physical-inorganic chemistry. We thank all the contributors for their excellent and timely contributions to the field.

EDWARD I. SOLOMON
Monroe E. Spaght Professor of Chemistry
Stanford University
Stanford, CA

A.B.P. (BARRY) LEVER
Distinguished Research Professor of Chemistry
York University
Toronto, Ontario, Canada

CONTRIBUTORS, VOLUME I

G.M. BANCROFT, Department of Chemistry, University of Western Ontario, London, Ontario M6A 5B7, Canada E-mail: scigmb@uwoadmin.uwo.ca

A. BENCINI, Dipartimento di Chimica, Università Degli Studi di Firenze, Via Marigliano 75/77, 50144 Firenze, Italy E-mail: sandro@chim1.unifi.it; bencini@dada.it

T.C. BRUNOLD, Department of Chemistry, Stanford University, Stanford, CA 94305 E-mail: tbrunold@chem.stanford.edu

D.A. CASE, Department of Molecular Biology, MB1, The Scripps Research Institute, 10666 N. Torrey Pines Rd., La Jolla, CA 92037 E-mail: case@scripps.edu

R. CZERNUSZEWICZ, Department of Chemistry, University of Houston, Houston, TX 77204 E-mail: roman@uh.edu

J. ENSLING, Fachbereich Chemie, Institut für Anorganische Chemie, Johannes-Gutenberg-Universität Mainz, Staudinger Weg 9, 6500 Mainz, Germany E-mail: ensling@iacgu7.chemie.uni-mainz.de

D. GATTESCHI, Dipartimento di Chimica, Università Degli Studi di Firenze, Via Maragliano 75/77, 50144 Firenze, Italy E-mail: gattesch@chim1.unifi.it

H.U. GÜDEL, Institut für Anorganische und Physikalische Chemie, Universitat Bern, Freiestrasse 3, CH-3000 Bern 9, Switzerland E-mail: Guedel@iac.unibe.ch

P. GÜTLICH, Fachbereich Chemie, Institut für Anorganische Chemie, Johannes-Gutenberg-Universität Mainz, Staudinger Weg 9, 6500 Mainz, Germany E-mail: p.guetlich@uni-mainz.de

B. HEDMAN, Stanford Synchrotron Radiation Laboratory, Stanford University, Stanford, CA 94305-5080 E-mail: hedman@ssrl.slac.stanford.edu

M.A. HITCHMAN, Department of Chemistry, The University of Tasmania, Box 252C, G.P.O. Hobart, Tasmania 7001, Australia E-mail: Michael.Hitchman@utas.edu.au

K.O. HODGSON, Department of Chemistry, Stanford University, Stanford, CA 94305-5080 E-mail: hodgson@ssrl.slac.stanford.edu

Y.F. HU, Department of Chemistry, University of Western Ontario, London, Ontario M6A 5B7, Canada E-mail: yhu@julian.uwo.ca

E. KRAUSZ, Research School of Chemistry, The Australian National University, G.P.O. Box 4, Canberra A.C.T. 2601, Australia

A.B.P. LEVER, Department of Chemistry, York University, 4700 Keele St., Toronto, Ontario M3J 1P3, Canada E-mail: blever@yorku.ca

J. LI, Department of Molecular Biology, MB1, The Scripps Research Institute, 10666 N. Torrey Pines Road, La Jolla, CA 92037

C.H. MARTIN, Department of Chemistry, University of Florida, Gainesville, FL 32611

L. NOODLEMAN, Department of Molecular Biology, MB1, The Scripps Research Institute, 10666 N. Torrey Pines Road, La Jolla, CA 92037
E-mail: lou@scripps.edu

H. RIESEN, Research School of Chemistry, The Australian National University, G.P.O. Box 4, Canberra, A.C.T. 2601, Australia

M.J. RILEY, Department of Chemistry, University of Queensland, St. Lucia, Queensland 4072, Australia E-mail: riley@chemistry.uq.edu.au

E.I. SOLOMON, Department of Chemistry, Stanford University, S.G. Mudd Building, Stanford, CA 94305 E-mail: solomon@chem.stanford.edu

T.G. SPIRO, Department of Chemistry, Princeton University, Princeton, NJ 08544
E-mail: spiro@chemvax.princeton.edu

M.C. ZERNER, Department of Chemistry, University of Florida, Gainesville, FL 32611 E-mail: zerner@qtp.ufl.edu

H.H. ZHANG, Department of Chemistry, Stanford University, Stanford, CA 94305-5080

CONTENTS, VOLUME II

CONTRIBUTORS, VOLUME II

P. DAY, The Royal Institution, 21 Albemarle Street, London W1X 4BS, UK E-mail: pday@ri.ac.uk

E.S. DODSWORTH, Department of Chemistry, York University, 4700 Keele St., Toronto, Ontario M3J 1P3, Canada E-mail: dod@yorku.ca

J.F. ENDICOTT, Department of Chemistry, Wayne State University, Detroit, Michigan 48202 E-mail: jfe@fourcroy.chem.wayne.edu

J.H. ENEMARK, Department of Chemistry, University of Arizona, Tucson, AZ 85721-0041 E-mail: jenemark@u.arizona.edu

H.B. GRAY, Division of Chemistry and Chemical Engineering, California Institute of Technology, Pasadena, CA 91125 E-mail: hgcm@cco.caltech.edu

N.E. GRUHN, Department of Chemistry, University of Arizona, P.O. Box 210041, Tucson, AZ 85721

P. GÜTLICH, Fachbereich Chemie, Institut für Anorganische Chemie, Johannes-Gutenberg-Universität Mainz, Staudinger Weg 9, D-55099 Mainz, Germany E-mail: p.guetlich@uni-mainz.de

M.A. HANSON, Department of Chemistry, Stanford University, S.G. Mudd Building, Stanford, California 94305-5080 E-mail: melissa@chem.stanford.edu

A. HAUSER, Département de Chimie Physique, Université de Gèneve, 30, Quai Ernest-Ansermet, CH-1211 Genève 4, Switzerland

M.D. HOPKINS, Department of Chemistry, University of Pittsburgh, Pittsburgh, PA 15260

A.B.P. LEVER, Department of Chemistry, York University, 4700 Keele St., Toronto, Ontario M3J 1P3, Canada E-mail: blever@yorku.ca

D.L. LICHTENBERGER, Department of Chemistry, University of Arizona, P.O. Box 210041, Tucson, AZ 85721-0041 E-mail: dlichten@arizona.edu

G.H. LOEW, Molecular Research Institute, 845 Page Mill Rd., Palo Alto, CA 94304 E-mail: loew@montara.molres.org

V.M. MISKOWSKI, Department of Chemistry, University of Pittsburgh, Pittsburgh, PA 15260

D.E. RICHARDSON, Department of Chemistry, University of Florida, Gainesville, FL 32611-7200 E-mail: der@chem.ufl.edu

P.N. SCHATZ, Department of Chemistry, University of Virginia, Charlottesville, VA 22901 E-mail: pns@virginia.edu

E.I. SOLOMON, Department of Chemistry, Stanford University, S.G. Mudd Building, Stanford, CA 94305-5080 E-mail: solomon@chem.stanford.edu

H. SPIERING, Fachbereich Chemie, Institut für Anorganische Chemie, Johannes-Gutenberg-Universitat Mainz, Staudinger Weg 9, 6500 Mainz, Germany

B.L. WESTCOTT, Department of Chemistry, Central Connecticut State University, New Britain, CT 06050 E-mail: westcottb@ccsu.edu

J.R. WINKLER, Division of Chemistry and Chemical Engineering, California Institute of Technology, Pasadena, CA 91125

1 Ligand Field Theory and the Properties of Transition Metal Complexes

A. B. (BARRY) P. LEVER
Department of Chemistry
York University
Toronto, Ontario
Canada, M3J 1P3
E-mail: blever@yorku.ca

EDWARD I. SOLOMON
Department of Chemistry
Stanford University
Stanford, CA 94305-5080, USA
E-mail: solomon@chem.stanford.edu

Inorganic Electronic Structure and Spectroscopy, Volume I: Methodology.
Edited by E. I. Solomon and A. B. P. Lever.
ISBN 0-471-15406-7.

1 GENERAL INTRODUCTION

From the early days of studying transition metal complexes, researchers have been intrigued by their intense colors and unusual magnetic behavior. In recent years the spectroscopic methodology available to rigorously study these and other properties has become extremely sophisticated and covers more than 10 orders of magnitude in photon energy. Different energy regions provide complementary insight into the electronic structure properties of a transition metal complex; and by utilizing the appropriate combination of methods one can address and solve essentially any spectroscopic problem. This can be of importance on a fundamental level (high resolution spectroscopy at 2 K on single crystals of high symmetry complexes) or in application to problems of wide general interest to the scientific community (Bioinorganic Chemistry, Materials Science, Environmental Chemistry, etc.). Ligand field theory provides a basis for understanding many of the wide range of spectroscopic observables in terms of the electronic structures of transition metal complexes.

The roots of ligand field theory are in group theory and geometry and indeed the parameters of the model to be developed in Section 2 are best viewed as defined by the symmetry of the complex. Often real systems have little or no symmetry; however, significant insight can still be achieved by viewing these complexes as low symmetry perturbations on a higher "effective symmetry" system. In 1929 Hans Bethe described the energy splitting of a many electron metal ion in a crystal lattice site and developed much of the formalism of the method.[1] This is a pure electrostatic model (negative point charge ligands having a repulsive interaction with electrons on the metal ion) and allows for no overlap (i.e., no covalent bonds). It was eventually recognized that a quantitative evaluation of the key parameter of crystal field theory, 10 Dq (see Section 2), gives a value that is an order of magnitude smaller than is experimentally observed and does not predict correct trends in 10 Dq with ligand variation. Van Vleck demonstrated that the problem with crystal field theory is its neglect of overlap with the ligand valence orbitals.[2]

Covalency is included in crystal field theory by recognizing that the parameters of the model are dictated by symmetry and by the interelectronic repulsion which is well defined in the free ion. These can then be adjusted to fit the experiment and the resultant parameters can be interpreted in terms of bonding interactions with the ligands. This is known as ligand field theory which has great utility in describing a large number of experimental observables in terms of a few experimental parameters.

Ligand field theory (and the related angular overlap model) provides much of the basis for our understanding of the magnetic, EPR and $d \rightarrow d$ spectroscopic properties of a transition metal complex. However, when one considers higher energy states, more covalent ligands, and the effects of covalency on the parameters of any spectroscopic region, one must utilize the most general description of bonding, molecular orbital theory.

In Section 2, we develop the different models for describing the energy splittings of the d orbitals of the metal ion in any ligand environment: crystal field theory, ligand field theory, the angular overlap model, and molecular orbital theory. In Section 3, we allow for more than one negative electron or positive hole in these d orbitals. One of the powers of ligand field theory is its ability to incorporate the experimental electron repulsion parameters of atomic theory into the bonding description of transition metal complexes. This is developed from both the weak field and strong field approaches, the latter being used to generate the ligand field energy level diagrams for transition metal complexes. In Section 4 we consider how these diagrams and the more general molecular orbital description of bonding relate to key physical properties of metal complexes: geometry, magnetism, spectroscopy and covalency. This chapter provides the basis for understanding the spectroscopic methods developed in detail in the following chapters. The modern molecular orbital methods for describing the electronic structure of transition metal complexes are presented in Chapters 10 and 11.

2 ONE-ELECTRON ORBITAL ENERGIES IN TRANSITION METAL COMPLEXES

2.1 Crystal and Ligand Field Theory

A metal ion in the gas phase, that is, a free ion, has five degenerate *d* orbitals which are defined in Eqn. (1). R_{nl} is the radial function of the electron with quantum numbers n and l, and the remaining component of each orbital is a spherical harmonic (Y_l^m) which is dependent on the quantum number m_l and is described using the polar coordinates of the electron as defined in Figure 1.

$$
\begin{aligned}
d(m_l = 0) &= R_{nl}\sqrt{(1/2\pi)}\sqrt{(5/8)}(3\cos^2\theta - 1) \\
d(m_l = \pm 1) &= R_{nl}\sqrt{(1/2\pi)}\sqrt{(15/4)}\cos\theta\sin\theta\cdot \mathrm{e}^{\pm i\phi} \\
d(m_l = \pm 2) &= R_{nl}\sqrt{(1/2\pi)}\sqrt{(15/16)}\sin^2\theta\cdot \mathrm{e}^{\pm 2i\phi}
\end{aligned}
\tag{1}
$$

From group theory, when the metal ion is placed in a ligand environment of octahedral (O_h) or lower symmetry, these *d* orbitals must split in energy as the highest dimension irreducible representation in O_h symmetry is three (T_1 or T_2). It is the goal of this section to determine this energy splitting quantitatively. In the crystal field model one treats the ligands as negative point charges and evaluates their repulsive interactions with the electron in the *d* orbitals assuming that there is no overlap of the ligand with the metal *d* orbitals.

The energy of an orbital Φ_i is given by the matrix element:

$$E = \langle \Phi_i | \boldsymbol{V} | \Phi_i \rangle \tag{2}$$

where, in crystal field theory, $\boldsymbol{V}$ is a Hamiltonian operator describing the symmetry (geometry) of the environment.

In our case the Φ_i are fivefold degenerate in the free ion and one must solve a secular determinant to extract the energies of the set of wavefunctions in the crystal field. Assuming a general crystal field $\boldsymbol{V}$ perturbing the *d* orbitals, the general secular

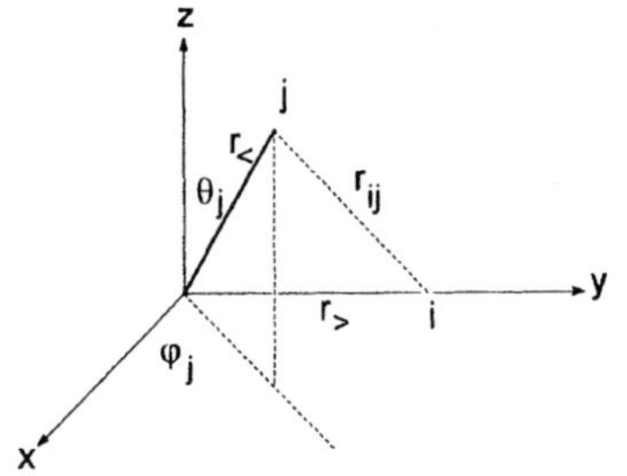

Figure 1. Polar coordinates for electron j and ligand i in the point charge model.

determinant shown in Eqn. (3), is obtained.

m_l	2	1	0	−1	−2
2	$\langle 2\vert\boldsymbol{V}\vert 2\rangle - E$	$\langle 2\vert\boldsymbol{V}\vert 1\rangle$	$\langle 2\vert\boldsymbol{V}\vert 0\rangle$	$\langle 2\vert\boldsymbol{V}\vert -1\rangle$	$\langle 2\vert\boldsymbol{V}\vert -2\rangle$
1	$\langle 1\vert\boldsymbol{V}\vert 2\rangle$	$\langle 1\vert\boldsymbol{V}\vert 1\rangle - E$	$\langle 1\vert\boldsymbol{V}\vert 0\rangle$	$\langle 1\vert\boldsymbol{V}\vert -1\rangle$	$\langle 1\vert\boldsymbol{V}\vert -2\rangle$
0	$\langle 0\vert\boldsymbol{V}\vert 2\rangle$	$\langle 0\vert\boldsymbol{V}\vert 1\rangle$	$\langle 0\vert\boldsymbol{V}\vert 0\rangle - E$	$\langle 0\vert\boldsymbol{V}\vert -1\rangle$	$\langle 0\vert\boldsymbol{V}\vert -2\rangle$
−1	$\langle -1\vert\boldsymbol{V}\vert 2\rangle$	$\langle -1\vert\boldsymbol{V}\vert 1\rangle$	$\langle -1\vert\boldsymbol{V}\vert 0\rangle$	$\langle -1\vert\boldsymbol{V}\vert -1\rangle - E$	$\langle -1\vert\boldsymbol{V}\vert -2\rangle$
−2	$\langle -2\vert\boldsymbol{V}\vert 2\rangle$	$\langle -2\vert\boldsymbol{V}\vert 1\rangle$	$\langle -2\vert\boldsymbol{V}\vert 0\rangle$	$\langle -2\vert\boldsymbol{V}\vert -1\rangle$	$\langle -2\vert\boldsymbol{V}\vert -2\rangle - E$

$= 0$ (3)

The magnitudes of these elements must be evaluated numerically and then the determinant is solved to extract the energies of the five d orbitals in the field described by $\boldsymbol{V}$. Except in situations of extremely low symmetry, we can expect that many, if not most, of the off-diagonal elements in determinant (3) will be zero. Fortunately, it is possible to determine that an element is zero without the need to actually evaluate it, as will be demonstrated below. Once all the numerical magnitudes are available, the determinant is diagonalized. This involves matrix manipulation such that all the off-diagonal elements are zero. The numbers remaining on the diagonal of the matrix are then the five (in this case) solutions of the determinant. They represent the energies of the 5 d orbitals in the field $\boldsymbol{V}$. This secular determinant is simply the matrix form of a set of five simultaneous equations in five unknowns. Note that the sum of the diagonal elements of this determinant equals the sum of the energies of the roots and the wavefunctions which diagonalize the matrix are the appropriate one-electron d orbitals, that is, they are the symmetry-adapted wavefunctions for the ligand environment of the complex (e.g., see development in Eqns. (27–29)).

2.1.1 ***Construction of a Crystal Field Hamiltonian*** The potential associated with an electron, j, at a distance r_{ij} from a negatively charged ligand, i, of charge Z_i is given by (Figure 1):

$$V_i = Z_i\,\mathrm{e}/r_{ij} \tag{4}$$

By summing this potential over all ligands, the total energy of each electron in the field can be derived. The potential, $1/r_{ij}$, experienced by an electron in a crystal field, can be expanded in terms of spherical harmonics $Y_l^m(\theta, \phi)$ centered at the origin. Since the d orbitals (in (1)) are also described in terms of spherical harmonics centered at the metal nucleus as origin, the harmonics describing $1/r_{ij}$ can also be centered at the metal nucleus. The potential then takes the form:

$$\sum_i \frac{Z_i e}{r_i - r_j} = \sum_{k=0}^{\infty}\sum_{q=-k}^{k} (4\pi/(2k+1)) \\ \times \sum_i Z_i e Y_k^{q^*}(\theta_i, \Phi_i)\cdot\left(r_<^k/\left(r_>^{k+1}\right)\right)\cdot Y_k^q(\theta_j, \Phi_j) \tag{5}$$

where $Y_k^q(\theta_i, \phi_i)$ refers to the ligand, and $Y_k^q(\theta_j, \phi_j)$ refers to the electron. The distances $r_<$ and $r_>$ are the shorter and longer of the radial vectors connecting the origin to the electron and to the charge (ligand). Usually it is assumed that the electron does not move far from the metal nucleus so that $r_<$ refers to the electron–nucleus distance, and $r_>$ the metal–ligand bond.

The first spherical harmonic in (5) describes the angular coordinates of the ligand (charge) and is defined by their polar coordinates $(r_>, \theta_i, \phi_i)$. The second spherical harmonic in (5) describes the angular coordinates of the electron, defined by its polar coordinates $(r_<, \theta_j, \phi_j)$. We decide which Y_k^q functions to use for a particular stereochemistry by using group theory. Since the geometry of a molecule is intrinsic and independent of its orientation, in that it is not changed by any symmetry operation such as rotation or reflection, the Hamiltonian operator must be totally symmetric, such that whatever geometry is described by these harmonics, their linear combination must transform as the totally symmetric representation in the molecular point group.

The group theoretical transformation properties of the Y_l^m functions are determined by their L value, which is their angular momentum quantum number (*vide infra*, Eqn. (80)). In an octahedral environment, angular momentum transforms as:

$$L = 0 \rightarrow A_{1g}; \quad L = 1 \rightarrow T_{1u}; \quad L = 2 \rightarrow E_g + T_{2g};$$
$$L = 3 \rightarrow A_{2u} + T_{2u} + T_{1u}; \quad L = 4 \rightarrow A_{1g} + E_g + T_{1g} + T_{2g} \tag{6}$$

Thus in an octahedral environment only harmonics with $L = 0$ and 4 span A_{1g} and can then contribute to the Hamiltonian. For d orbitals we need only consider $L \leqslant 4$. We shall see that matrix element (2) can be construed as the coupling of angular momenta in a vector sense. A property of such coupling is that the three components must be capable of forming a triangle. Thus to couple two d functions, each of $l = 2$, we cannot use harmonics greater than $L = 4$ since otherwise we would not be able to form this triangle. Similarly to couple f functions ($l = 3$) together we can use harmonics up to $L = 6$. For lower point groups, one should take the representations listed in (6) and determine which representations are spanned by these functions in the lower group. For example, in D_{4h} we still exclude $L = 1, 3$ since they are odd in any group with a center of symmetry. However $L = 2$ functions now contribute since $E_g(O_h)$ correlates with $A_{1g} + B_{1g}$ in D_{4h}.

As an example, consider the generation of the Hamiltonian for a square planar (D_{4h}) complex using, in Eqn. (5), the $k = 2, 4$ harmonics as noted above ($k = 0$ can be ignored since it will affect all the d orbitals to the same extent). In the usual polar coordinate representation (Figure 1), the four ligands lie in the xy plane at 90° to the z axis (i.e., $\theta = 90°$). The ϕ values (angle with the x axis when the ligand is projected into the xy plane) for the four ligands, 1–4, will be 0°, 90°, 180° and 270°, respectively. A completely general Hamiltonian using $L = 2$ and 4 harmonics can be written

$$V = \sum_i Z_i e/r_{ij} = c_2^0 \cdot R_2(r) \cdot Y_2^0 + c_2^2 \cdot R_2(r)\left(Y_2^2 + Y_2^{-2}\right)$$
$$+ c_4^0 \cdot R_4(r) \cdot Y_4^0 + c_4^2 \cdot R_4(r)\left(Y_4^2 + Y_4^{-2}\right) + c_4^4 \cdot R_4(r)\left(Y_4^4 + Y^{-4}\right) \tag{7a}$$

where

$$c_k^q = \left(\frac{4\pi}{2k+1}\right) Z_i e \sum_i Y_k^q(\theta, \phi_i)$$

$$R_k(r) = r_<^k / r_>^{k+1} \tag{7b}$$

with the equivalent contributions from the components (−2, +2) and (−4, +4) grouped together. Note that the $c_4^{\pm 1}$ and $c_4^{\pm 3}$ terms are omitted since they do not contribute to fourfold group geometries. They may be needed for threefold axis groups, or if wavefunctions quantized along a threefold axis were required.[3]

The c_k^q values are evaluated as follows, using Y_k^q harmonics from Appendix 1, by summation over the ligands

$$\begin{aligned} c_2^0 &= \sum_{i=0}^{4} Ze(4\pi/5)\sqrt{(5/8)}\,\sqrt{(1/2\pi)}(3\cos^2\theta - 1) \\ &= Ze(4\pi/5)\sqrt{(5/8)}\,\sqrt{(1/2\pi)}(-4) \\ &= -4Ze\sqrt{(\pi/5)} \end{aligned} \tag{8}$$

where the (−4) is the sum of $(3\cos^2\theta - 1)$ over the four ligands for which $\cos(90) = 0$.

The remaining components can be obtained from Table 1.
The $c_2^{\pm 2}$ and $c_4^{\pm 2}$ sum to zero in this geometry. Considering c_4^0 and $c_4^{\pm 4}$, we find:

$$\begin{aligned} c_4^0 &= \sum_{i=0}^{4} \mathrm{Ze}(4\pi/9)\sqrt{(9/128)}\,\sqrt{(1/2\pi)}(35\cos^4\theta_i - 30\cos^2\theta_i + 3) \\ &= \mathrm{Ze}(4\pi/9)\sqrt{(9/128)}\,\sqrt{(1/2\pi)}(+12) = \mathrm{Ze}(\sqrt{\pi}\,) \\ c_4^4 &= \sum_{i=0}^{4} \mathrm{Ze}(4\pi/9)\sqrt{(315/256)}\,\sqrt{(1/2\pi)}\sin^4\theta_i e^{\pm 4i\phi} \\ &= \mathrm{Ze}(4\pi/9)\sqrt{(315/256)}\sqrt{(1/2\pi)}(+4) = \mathrm{Ze}(\sqrt{\pi}\,)\sqrt{35/18} \end{aligned} \tag{9}$$

Table 1

			$c_2^{\pm 2}$	c_4^0	$c_4^{\pm 2}$	$c_4^{\pm 4}$
Ligand	θ_i	ϕ_i	$\sin^2\theta\, e^{\pm 2i\phi}$	$35\cos^4\theta_i - 30\cos^2\theta_i + 3$	$\sin^2\theta(7\cos^2\theta - 1)\, e^{\pm 4i\phi}$	$\sin^4\theta\, e^{\pm 4i\phi}$
1	90	0	1	3	1	1
2	90	90	−1	3	−1	1
3	90	180	1	3	1	1
4	90	270	−1	3	−1	1
Total			0	12	0	4

Collecting the terms and incorporating into Eqn. (7) generates the Hamiltonian suitable for a square planar D_{4h} complex, ML_4:

$$V_{sq.pl.} = \mathrm{Ze}(\sqrt{\pi})\big[-4(1/\sqrt{5})(r^2/a^3)Y_2^0 + (r^4/a^5)\big(Y_4^0 + \sqrt{(35/18)}(Y_4^4 + Y_4^{-4})\big)\big] \qquad (10)$$

where the in-plane metal–ligand distance, $r_>$ is a.

The inclusion of two additional equivalent ligands, L, along the z axis, to the square plane (i.e., at $(\theta, \phi) = (0, 0)$ and $(180, 0)$-generates the Hamiltonian for an octahedral complex. The reader should prove that the addition of these two equivalent ligands will indeed cause the Y_2^0 terms to disappear. The addition of two ligands, X, $(\neq L)$ generates the tetragonal complex ML_4X_2, also of D_{4h} symmetry. This is most simply treated in this formalism by assuming that these ligands lie at a distance "b" instead of "a". It does not matter if, in reality, these are chemically different ligands; the methodology will simply generate semiempirical parameters which will reflect the difference in "field" generated by L and by X.

$$\begin{aligned} V_{D_{4h}} &= V_{O_h} + V_{\mathrm{tetrag}} \\ V(D_{4h}) &= \sqrt{(49/18)}\sqrt{(2\pi)}(\mathrm{Ze}r^4/a^5)\big[Y_4^0 + \sqrt{5/14}(Y_4^4 + Y_4^{-4})\big] \\ &\quad - 2\mathrm{Ze}\sqrt{(2\pi)}\big[\sqrt{(2/5)}\big((r^2/a^3) - (r^2/b^3)\big)Y_2^0\big] \\ &\quad - 2\mathrm{Ze}\sqrt{(2\pi)}\big[\sqrt{(2/9)}\big((r^4/a^5) - (r^4/b^5)\big)Y_4^0\big] \end{aligned} \qquad (11)$$

This Hamiltonian will give (10) if the "b" terms are omitted and generate the O_h Hamiltonian if "b" = "a". The first line in (11) is the octahedral Hamiltonian.

In this fashion, crystal field Hamiltonians for any geometry can be constructed by inserting the angular coordinates of the ligands into the general equation (5) and summing the results. The actual c_k^q (or Y_k^q) which are required are most easily derived by noting which of the $L = 1, 2, 3, 4$ terms span A_1 (or A_{1g}) in the molecular symmetry group concerned. If the molecule does not have a center of symmetry, then harmonics with $L = 1$ and 3 may also contribute. These then are added to the general Hamiltonian (7).

2.1.2 ***Evaluation of One-Electron Crystal Field Matrix Elements*** Several rather different procedures exist to derive the matrix elements of the crystal field Hamiltonian.

2.1.2.1 *Manual Integration* To develop methods to evaluate the magnitudes of the matrix elements of these Hamiltonians consider first that a d orbital may be expressed in terms of spherical harmonics according to its m_l value as given in Eqn. (1).

Suppose we evaluate the energy of the $m_l = 0$ orbital, which corresponds with d_{z^2}, in an octahedral field. If the octahedral Hamiltonian is broken down into the Y_4^0 and $Y_4^{\pm 4}$ components, then the Y_4^0 component may be written $\langle lm|V|lm\rangle =$

$\langle 20|Y_4^0|20\rangle$ and expanded as:

$$
\begin{aligned}
&\langle 20|Y_4^0|20\rangle \\
&= \int R_{nl}\surd(1/2\pi)\surd(5/8)\surd(49/18)\surd(2\pi)\big(\mathrm{Ze}^2r^4/a^5\big)R_{nl}\surd(1/2\pi)\surd(5/8)r^2\,\mathrm{d}r \\
&\quad \times\surd(1/2\pi)\surd(9/128)\int(35\cos^4\theta-30\cos^2\theta+3)(3\cos^2\theta-1)^2\sin\theta\,\mathrm{d}\theta \\
&= \alpha_4(5/8)\surd(49/18)\surd(9/128)(1/\surd 2\pi)\int(35\cos^4\theta-30\cos^2\theta+3) \\
&\quad \times(3\cos^2\theta-1)^2\sin\theta\,\mathrm{d}\theta \qquad (12)
\end{aligned}
$$

where, for $3d$ electrons:

$$
\begin{aligned}
\alpha_2 &= \mathrm{Ze}^2\int R_{3d}\big(r^2/a^3\big)R_{3d}\cdot r^2\,\mathrm{d}r = \mathrm{Ze}^2\big(\overline{r^2}/\overline{a^3}\big) \\
\alpha_4 &= \mathrm{Ze}^2\int R_{3d}\big(r^4/a^5\big)R_{3d}\cdot r^2\,\mathrm{d}r = \mathrm{Ze}^2\big(\overline{r^4}/\overline{a^5}\big) \qquad (13)
\end{aligned}
$$

Two alternate general procedures exist to evaluate the magnitudes of these matrix elements $\langle m_l|V|m_l'\rangle$ in (12). The right-hand side of (12) signifies integration over all space of a triple product of functions. This can be handled in a standard mathematical way but becomes extremely cumbersome for complicated Hamiltonians with many components. We do not develop this classical procedure here.

Before considering an efficient procedure to evaluate these integrals, note that the magnitudes of these d matrix elements are expressed in terms of radial integrals, α_n, usually called, for example for fourfold groups, Ds, Dt, and Dq, where:

$$
\begin{aligned}
Dq &= (1/6)\,\mathrm{Ze}^2\big(\overline{r^4}/\overline{a^5}\big) = (1/6)\alpha_4(a) \\
Ds &= (2/7)\,\mathrm{Ze}^2\big[\big(\overline{r^2}/\overline{a^3} - \overline{r^2}/\overline{b^3}\big)\big] = (2/7)[\alpha_2(a)-\alpha_2(b)] \\
Dt &= (2/21)\,\mathrm{Ze}^2\big[\big(\overline{r^4}/\overline{a^5} - \overline{r^4}/\overline{b^5}\big)\big] = (2/21)[\alpha_4(a)-\alpha_4(b)] \\
Dt &= (4/7)[Dq(Eq)-Dq(Ax)] \qquad (14)
\end{aligned}
$$

where $Dq(Eq)$ and $Dq(Ax)$ are the equatorial and axial crystal field strengths. These radial integrals are treated as empirical parameters and evaluated by solving the electronic spectrum (see Section 2.1.4.).

The left-hand side of Eqn. (12), written in the Dirac formulation in momentum space, signifies the coupling of two angular momenta to give a resultant angular momentum. This more modern approach is especially adaptable to computer evaluation, and will be developed here.

2.1.2.2 *Tensor Methods and the Wigner–Eckart Theorem* [4–11] Given that both the operators and the wavefunctions can each be expressed in terms of spherical harmonics, which are themselves measures of angular momentum, the fundamental element which needs evaluation is of the form $\langle Y_a|Y_b|Y_c\rangle$. Scalars (simple numbers with no direction) and vectors (magnitude and direction) are well known. These are subsets of the more general *tensor* which may be written $C_k{}^q$ where k is the rank of the tensor and q is its component. For example k and q may relate to the quantum numbers L and M_L. Scalars are tensors of rank zero and vectors are tensors of rank 1. A $C_k{}^q$ tensor has $(2k+1)$ components in parallel fashion to the $(2L+1)$ values for M_L.

These tensors are related to the spherical harmonics via:

$$C_k{}^q = (4\pi/(2k+1))^{1/2} Y_k^q \tag{15}$$

Some common $C_k{}^q$ are:

$$C_0{}^0 = 1; \quad C_1{}^0 = \cos\theta; \quad C_1{}^{\pm 1} = \mp(1/2)^{1/2}\sin\theta\, e^{\pm i\phi}$$
$$C_2{}^0 = \frac{1}{2}(3\cos^2\theta - 1); \quad C_2{}^{\pm 1} = \mp\left(\frac{3}{2}\right)^{1/2}\cos\theta\sin\theta\, e^{\pm i\phi}$$
$$C_2{}^{\pm 2} = \left(\frac{3}{8}\right)^{1/2}\sin^2\theta\, e^{\pm 2i\phi} \tag{16}$$

Substituting Eqns. (14) and (15) into our D_{4h} Hamiltonian (11), as an example, yields the rather simple form:

$$V_{D4h} = 21Dq\left[C_4{}^0 + \surd(5/14)\left(C_4{}^4 + C_4{}^{-4}\right)\right] - 21DtC_4{}^0 - 7DsC_2{}^0 \tag{17}$$

Any other ligand field Hamiltonian can be similarly constructed using Dq, Ds, and Dt or α_2, α_4 etc. as appropriate. The one-electron matrix elements (2) of these ligand field operators then collapse to a series of sums of $\langle d_i|C_k{}^q|d_j\rangle$ involving the tensor operators $C_k{}^q$. Now we explore how they are evaluated.

Two one-electron functions $|j, m_j\rangle$ (where j may be l, s, or j) can be coupled by a general tensor operator, $C_k{}^q$, via:

$$\left\langle jm|C_k{}^q|j'm'\right\rangle = (-1)^{j-m}\begin{pmatrix} j & k & j' \\ -m & q & m' \end{pmatrix}\langle j\|C_k\|j'\rangle \tag{18}$$

where $\langle j\|C_k\|j'\rangle$ is a reduced matrix element of the tensor operator, obviously independent of component m, and the term in parentheses which is a Wigner 3-j symbol, contains all component dependence. These are available in a tabulation (see Appendix 2) or readily calculated. The beauty of the Wigner–Eckart theorem[12] is that it separates those aspects relating to the symmetry of the molecule (the 3-j symbol) and the physical properties of the system contained within the reduced matrix element. The procedure is also elegant because one can solve for the reduced matrix element using a simple case and then use the same value for much more complex situations.[10,11]

This expression (18) is an extremely powerful procedure for evaluating integrals arising from the coupling of any wavefunctions by any operator capable of being written in angular momentum terms. The reader should look closely at how the 3-j symbol is written in terms of the various components of the matrix element on the left. The numerical values of these 3-j symbols have all been evaluated and are available in various sources;[5,7,10,11,13] see Appendix 2.

The reader should exercise caution in using these 3-j symbols especially with respect to their sign. They have fundamental properties which should be clearly understood before they are routinely employed. These properties include:

i) They vanish (are equal to zero) if j, k, and j' do not form a triangle in their number space. This is the aforementioned triangle rule; this may also be written that $j + j' \geqslant k \geqslant |j - j'|$. Thus if say, $j = 2$, $j' = 2$, then k can only be equal to $0, 1, 2, 3, 4$.
ii) The sum of the components in the bottom row must equal zero.
iii) If all three components in the bottom row are multiplied by -1, then the 3-j symbol changes sign if the sum of the top row, $j + k + j'$ is odd.
iv) if the columns are permuted, then the 3-j symbol is invariant if $j+k+j'$ is even; otherwise it changes sign under the influence of an odd number of permutations.

Matrix elements of the $C_k{}^q$ operator (in Eqn. (17) for example) can be evaluated in this formalism ($j = j, l$, or s) simply as a product of the function shown in (18) and the component independent reduced matrix element which can be evaluated using:

$$\langle j \| C_k \| j' \rangle = (-1)^j \left[(2j+1)(2j'+1)\right]^{1/2} \begin{pmatrix} j & k & j' \\ 0 & 0 & 0 \end{pmatrix} \tag{19}$$

Thus, knowing the values of the 3-j symbols, and the reduced matrix elements, the evaluation of $\langle jm|C_k{}^q|j'm'\rangle$ becomes straightforward and rapid.

$$\begin{aligned} &\langle jm|C_k{}^q|j'm'\rangle \\ &= (-1)^{-m}\left[(2j+1)(2j'+1)\right]^{1/2} \begin{pmatrix} j & k & j' \\ 0 & 0 & 0 \end{pmatrix} \begin{pmatrix} j & k & j' \\ -m & q & m' \end{pmatrix} \end{aligned} \tag{20}$$

Thus, for example, to evaluate $\langle 22|C_2{}^0|22\rangle = \langle 2-2|C_2{}^0|2-2\rangle$:

$$\langle 22|C_2{}^0|22\rangle = (-1)^{-2}[(2^*2+1)] \begin{pmatrix} 2 & 2 & 2 \\ 0 & 0 & 0 \end{pmatrix} \begin{pmatrix} 2 & 2 & 2 \\ -2 & 0 & 2 \end{pmatrix} \tag{21}$$

$$\langle 22|C_2{}^0|22\rangle = +1 \times 5 \times \left(-\surd(2/35)\right) \times \surd(2/35) = -2/7 \tag{22}$$

For ease of text expression in the following discussion, we use a condensed form of the 3-j symbol, namely $(j_1 j_2 j_3/m_1 m_2 m_3)$ (i.e., for the right-hand 3-j symbol in Eqn. (21), $(222/-202)$. Similarly for the nonzero $C_4{}^0$ matrix elements:

$$\langle 20|C_4{}^0|20\rangle = (-1)^0 \times 5 \times \surd(2/35) \times \surd(2/35) = 2/7$$

$$\langle 21|C_4{}^0|21\rangle = \langle 2-1|C_4{}^0|2-1\rangle = (-1)^{-1} \times 5 \times \surd(2/35) \times \surd(8/315) = -4/21$$

$$\langle 22|C_4{}^0|22\rangle = \langle 2-2|C_4{}^0|2-2\rangle = (-1)^{-2} \times 5 \times \surd(2/35) \times \surd(1/630) = 1/21$$
$$\langle 22|C_4{}^4|2-2\rangle = \langle 2-2|C_4{}^{-4}|22\rangle = (-1)^{-2} \times 5 \times \surd(2/35) \times (1/3)$$
$$= \surd(10/63) \tag{23}$$

The Wigner–Eckart theorem and this tensor methodology provide a rapid procedure for evaluating the matrix elements of crystal field operators.

The octahedral Hamiltonian is simply (from (17), $Ds = Dt = 0$):

$$V_{\text{oct}} = 21Dq\left[C_4{}^0 + \surd(5/14)\left(C_4{}^4 + C_4{}^{-4}\right)\right] \tag{24}$$

Then:

$$\begin{aligned}\langle 20|V_{\text{oct}}|20\rangle &= 21Dq\left[\langle 0|C_4{}^0|0\rangle + \surd(5/14)\langle 0|C_4{}^4 + C_4{}^{-4}|0\rangle\right]\\ &= 21Dq\left[(2/7) + \surd(5/14)(0)\right] = 6Dq\\ \langle 22|V_{\text{oct}}|22\rangle &= \langle -2|V_{\text{oct}}|-2\rangle = 21Dq\left[(1/21) + \surd(5/14)(0)\right] = Dq\\ \langle 21|V_{\text{oct}}|21\rangle &= \langle -1|V_{\text{oct}}|-1\rangle = 21Dq\left[(-4/21) + \surd(5/14)(0)\right] = -4Dq\end{aligned}$$

and finally:

$$\langle 22|V_{\text{oct}}|2-2\rangle = 21Dq\left[0 + \surd(5/14)(\surd(10/63))\right] = 5Dq \tag{25}$$

yielding the secular determinant (26).

V_{oct}	$\|2, 2\rangle$	$\|2, 1\rangle$	$\|2, 0\rangle$	$\|2, -1\rangle$	$\|2, -2\rangle$
$\|2, 2\rangle$	$Dq - E$	0	0	0	$5Dq$
$\|2, 1\rangle$	0	$-4Dq - E$	0	0	0
$\|2, 0\rangle$	0	0	$6Dq - E$	0	0
$\|2, -1\rangle$	0	0	0	$-4Dq - E$	0
$\|2, -2\rangle$	$5Dq$	0	0	0	$Dq - E$

$= 0$ (26)

The solution to this secular determinant generates the well-known octahedral splitting energies $+6Dq$ (twofold degenerate e_g set) and $-4Dq$ (threefold degenerate t_{2g} set).

If an element of the form $\langle\theta_i|V|\theta_j\rangle$ is nonzero, this implies that the basis orbital θ_i is mixed with the basis orbital θ_j by the operator $\boldsymbol{V}$. Thus the symmetry adapted wavefunction (i.e., a wavefunction which transforms properly as one of the irreducible representations of the point group concerned) will be a linear combination of θ_i and θ_j. To obtain the correct form of this wavefunction, or generally of all the wavefunctions described by the secular determinant, we utilize the energies of the wavefunctions to solve the set of simultaneous equations which the matrix represents. Thus for secular determinant (26), we can write, from the top line of the determinant:

$$(Dq - E)x + 5Dqy = 0 \tag{27}$$

where x is the coefficient required for $|2, 2\rangle$ and y is the coefficient required for $|2, -2\rangle$. Further, because of normalization, $x^2 + y^2 = 1$. Inserting the solution $E = +6Dq$ into (27) and solving yields $x = y = 1/\sqrt{2}$ and therefore the wavefunction with the energy $+ 6Dq$ is of the form:

$$1/\sqrt{2}[|2, 2\rangle + |2, -2\rangle] \quad \text{(which corresponds with the real } d(x^2 - y^2) \text{ orbital)} \tag{28}$$

Similarly, inserting $E = -4Dq$, then $x = -y = 1/\sqrt{2}$ and the wavefunction is:

$$1/\sqrt{2}[|2, 2\rangle - |2, -2\rangle] \quad \text{(which corresponds with the real } d(xy) \text{ orbital)} \tag{29}$$

The remaining three orbitals, $|1\rangle$, $|0\rangle$, and $|-1\rangle$ are already eigenstates of the octahedral operator, which is to say they are diagonal with no off-diagonal elements between themselves or with any other d orbital. The orbital $|0\rangle$ corresponds with $d(z^2)$ lying at $+6Dq$, and is degenerate with the $d(x^2 - y^2)$ orbital forming the e_g set of orbitals in the octahedral field. The orbitals $|1\rangle$ and $|-1\rangle$ are degenerate at $-4Dq$ and thus one can take their in and out-of-phase normalized combinations to form the real $d(yz)$ and $d(xz)$ orbitals respectively. Together with $d(xy)$ these form the threefold degenerate t_{2g} set of orbitals in the octahedral field. The e_g and t_{2g} orbitals are said to be symmetry-adapted to the octahedral field.

It should now be evident that if we had used the set of 5 symmetry-adapted wavefunctions themselves, instead of the $|m_l\rangle$ basis set, the resulting secular determinant would already have been diagonal, that is, all the off-diagonal elements would be zero because the field $\boldsymbol{V}$ is totally symmetric (i.e., transforms as the totally symmetric representation of the group) and they therefore would not mix wavefunctions with different symmetry representations. This can be seen by noting that:

$$\begin{aligned}
&\big\langle 1/\sqrt{2}(|2, 2\rangle - |2, -2\rangle)|\boldsymbol{V}|1/\sqrt{2}(|2, 2\rangle - |2, -2\rangle)\big\rangle \\
&\quad = 1/2\big\{\langle 2|\boldsymbol{V}|2\rangle + \langle -2|\boldsymbol{V}|-2\rangle - 2\langle 2|\boldsymbol{V}|-2\rangle\big\} \\
&\quad = 1/2\{1Dq + 1Dq + 2\,\mathrm{x}5Dq\} = +6Dq
\end{aligned} \tag{30}$$

and similarly:

$$\big\langle 1/\sqrt{2}(|2, 2\rangle + |2, -2\rangle)|\boldsymbol{V}|1/\sqrt{2}(|2, 2\rangle + |2, -2\rangle)\big\rangle = -4Dq$$

and:

$$\big\langle 1/\sqrt{2}(|2, 2\rangle - |2, -2\rangle)|\boldsymbol{V}|1/\sqrt{2}(|2, 2\rangle + |2, -2\rangle)\big\rangle = 0 \tag{31}$$

In the octahedral case (determinant in (26)) the diagonal energies of both $|2, 2\rangle$ and $|2, -2\rangle$ (θ_1 and θ_2) are the same which then requires an equal mix $(1/\sqrt{2})$ of each function leading to the in and out-of-phase coupling represented in Eqns. (28,29). Where the diagonal energies are different, a non-equal mix will occur, leading in the two-orbital mixing situation, to the two general wavefunctions:

$$\alpha\theta_1 + \sqrt{(1-\alpha^2)}\theta_2 \quad \text{and} \quad \alpha\theta_2 - \sqrt{(1-\alpha^2)}\theta_1 \tag{32}$$

Finally, we evaluate one of the elements for the tetragonal D_{4h} Hamiltonian (17) [see (21–23)].

$$\begin{aligned}\langle 20|V_{D4h}|20\rangle &= 21Dq[\langle 0|C_4{}^0|0\rangle + \sqrt{(5/14)}[\langle 0|C_4{}^4|0\rangle + \langle 0|C_4{}^{-4}|0\rangle]] \\ &\quad - 21Dt\langle 0|C_4{}^0|0\rangle - 7Ds\langle 0|C_2{}^0|0\rangle \\ &= 21Dq[(2/7) + \sqrt{(5/14)}(0)] - 21Dt(2/7) - 7Ds(2/7)\end{aligned}$$

Hence:

$$\langle 20|V_{D4h}|20\rangle = 6Dq - 2Ds - 6Dt$$

The other terms can be obtained similarly:

$$\begin{aligned}&\langle 22|V_{D4h}|22\rangle = \langle 2-2|V_{D4h}|2-2\rangle = Dq + 2Ds - Dt \\ &\langle 21|V_{D4h}|21\rangle = \langle 2-1|V_{D4h}|2-1\rangle = -4Dq - Ds + 4Dt \\ &\langle 20|V_{D4h}|20\rangle = 6Dq - 2Ds - 6Dt \\ &\langle 22|V_{D4h}|2-2\rangle = 5Dq\end{aligned} \tag{33}$$

The absence of any Ds or Dt contribution in this last off-diagonal element is seen to be due to the inability of a $C_2{}^0$ or $C_4{}^0$ operator to couple $|2, 2\rangle$ with $|2, -2\rangle$ (sum of the bottom row of 3-j symbols must equal zero).

For the real d-orbitals (Eqn. (34)):

$$\begin{aligned}&d(x^2-y^2) = (1/\sqrt{2})[|2,2\rangle + |2,-2\rangle] \\ &d(z^2) = |2,0\rangle \\ &d(xy) = (1/i\sqrt{2})[|2,2\rangle - |2,-2\rangle] \\ &d(xz) = -(1/\sqrt{2})[|2,1\rangle - |2,-1\rangle] \\ &d(yz) = -(1/i\sqrt{2})[|2,1\rangle + |2,-1\rangle]\end{aligned} \tag{34}$$

these energies are (from (33) and (34)):

$$\begin{aligned}&\langle x^2-y^2|V_{D4h}|x^2-y^2\rangle = 6Dq + 2Ds - Dt \\ &\langle z^2|V_{D4h}|z^2\rangle = 6Dq - 2Ds - 6Dt \\ &\langle xy|V_{D4h}|xy\rangle = -4Dq + 2Ds - Dt \\ &\langle xz|V_{D4h}|xz\rangle = \langle yz|V_{D4h}|yz\rangle = -4Dq - Ds + 4Dt\end{aligned} \tag{35}$$

These methods are totally general for generating the d^1 matrix elements of any crystal field Hamiltonian. The energy splitting of the d-orbitals for octahedral, tetrago-

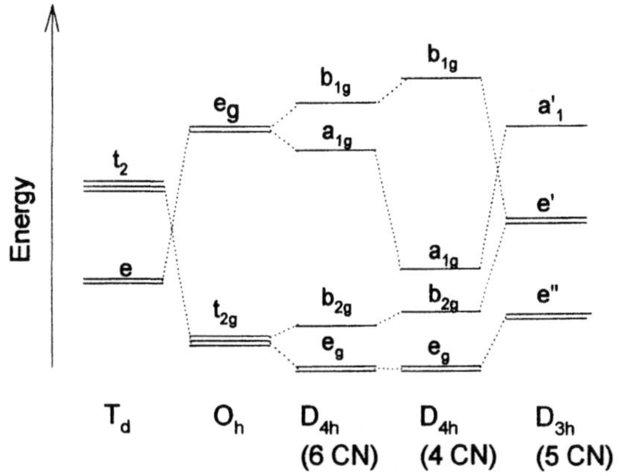

Figure 2. A d-orbital splitting pattern for tetrahedral (T_d), octahedral (O_h), tetragonal six-coordinate (D_{4h}, 6 CN), square (D_{4h}, 4 CN) and trigonal bipyramidal (D_{3h}, 5 CN) fields. The relative energy of the a_{1g} ($d(z^2)$), orbital in a square species is arbitrarily shown. It may lie below the b_{2g} or e_g levels depending on the molecule concerned.

nal, square planar, and tetrahedral fields obtained in this manner are given in Figure 2.

2.1.3 ***Companion and Komarynsky Method*** There may well be instances where one is interested in the d orbital energies and wavefunctions, but not in the actual form of the crystal field Hamiltonian, which, in the case of a low symmetry species, may be quite complicated. Companion and Komarynsky[14] (CK) published a formalism which yields these energies, based upon the polar coordinates of the ligands, without independently deriving the actual Hamiltonian. Tables 2 and 3 give the ligand position functions D_{lm} and G_{lm} and H_{ab} integrals in terms of these ligand position functions. H_{ab} $(a, b$ 1–5) are a shorthand notation for entries in the 5×5 real d orbital crystal field matrix.

$$\langle x^2 - y^2|V|x^2 - y^2\rangle = H_{11}$$
$$\langle xz|V|xz\rangle = H_{22}$$
$$\langle z^2|V|z^2\rangle = H_{33}$$
$$\langle yz|V|yz\rangle = H_{44}$$
$$\langle xy|V|xy\rangle = H_{55}$$

and off-diagonal elements such as:

$$\langle x^2 - y^2|V|z^2\rangle = H_{13} \quad \text{etc.} \tag{36}$$

Four steps are followed to generate the required energies:

i) identify the polar coordinates of the ligand positions in terms of the angle θ_i and Φ_i, in Figure 1.

Table 2 Ligand Position Function, D_m and G_m[14]

$D^i_{00} = \alpha^i_0$
$D^i_{20} = \alpha^i_2(\cos^2\theta_i - 1)$
$D^i_{40} = \alpha^i_4(35/3\cos^4\theta_i - 10\cos^2\theta_i + 1)$
$D^i_{21} = \alpha^i_2 \sin\theta_i \cos\theta_i \cos\phi_i$
$D^i_{22} = \alpha^i_2 \sin^2\theta_i \cos 2\phi_i$
$D^i_{41} = \alpha^i_4 \sin\theta_i \cos\theta_i (7/3\cos^2\theta_i - 1)\cos\phi_i$
$D^i_{42} = \alpha^i_4 \sin\theta_i (7\cos^2\theta_i - 1)\cos 2\phi_i$
$D^i_{43} = \alpha^i_4 \sin^3\theta_i \cos\theta_i \cos 3\phi_i$
$D^i_{44} = \alpha^i_4 \sin^4\theta_i \cos 4\phi_i$
$G^i_{21} = \alpha^i_2 \sin^4\theta_i \cos\theta_i \sin\phi_i$
$G^i_{22} = \alpha^i_2 \sin^2\theta_i \sin 2\phi_i$
$G^i_{41} = \alpha^i_4 \sin^2\theta_i \cos\theta_i (7/3\cos^2\theta_i - 1)\sin\phi_i$
$G^i_{42} = \alpha^i_4 \sin^2\theta_i (7\cos^2\theta_i - 1)\sin 2\phi_i$
$G^i_{43} = \alpha^i_4 \sin^3\theta_i \cos\theta_i \sin 3\phi_i$
$G^i_{44} = \alpha^i_4 \sin^4\theta_i \sin 4\phi_i$

ii) calculate the ligand position functions D^i_{lm} and G^i_{lm}, for each ligand, and sum over all ligands to generate D_{lm} and G_{lm}. These will be associated with a set of α^i_n terms for each non-equivalent ligand in the complex. These functions relate directly to the Y^m_l functions in Eqn. (5). Where d orbitals are concerned, only functions with $l = 0, 2,$ or 4 are required (odd Y^m_l cannot couple even d orbitals). We ignore the $l = 0$ function which only raises all d orbitals by the same degree, and hence need consider only terms in the radial parameter α_2 and α_4;

iii) use these D_{lm} and G_{lm} values to generate the H_{ab} secular determinant using Tables 2 and 3; and

iv) solve this determinant for the desired energies.

For example, for a trigonal bipyramid, distinguishing the axial and equatorial ligands using $\alpha_n(Ax)$ and $\alpha_n(Eq)$, then $\theta_i = 90°$ for the equatorial ligands and $\theta_i = 0°$ for the axial ligands, summing over all five ligands to generate:

$$\begin{aligned} D_{20} &= 4\alpha_2(Ax) - 3\alpha_2(Eq) \\ D_{40} &= (16/3)\alpha_4(Ax) + 3\alpha_4(Eq) \end{aligned} \tag{37}$$

Using $\Phi_i = 0°$ for the axial ligands, and $\Phi_i = 0$, 120, and 240° for the three equatorial ligands, all other D_{lm} and G_{lm} values are seen to sum to zero. Then, neglecting terms in D_{22}, D_{42} and D_{44}, which all sum to zero, from Table 3:

$$\begin{aligned} H_{11} &= H_{55}\left(E(x^2 - y^2, xy)\right) \\ &= -(1/7)D_{20} + (1/56)D_{40} \end{aligned}$$

Table 3 H_{ab} **Integrals in Terms of the Ligand Position Functions,** D_m **and** G_m[14]

$H_{11} = D_{00} - 1/7D_{20} + 1/56D_{40} + 5/24D_{44}$
$H_{22} = D_{00} + 1/14D_{20} - 1/14D_{40} + 3/14D_{22} + 5/42D_{42}$
$H_{33} = D_{00} + 1/7D_{20} + 3/28D_{40}$
$H_{44} = D_{00} + 1/14D_{20} - 1/14D_{40} - 3/14D_{22} - 5/42D_{42}$
$H_{55} = D_{00} - 1/7D_{20} + 1/56D_{40} - 5/24D_{44}$
$H_{12} = 3/7D_{21} - 5/28D_{41} + 5/12D_{43}$
$H_{13} = -\sqrt{3}/7D_{22} + 5\sqrt{3}/84D_{42}$
$H_{14} = -3/7G_{21} + 5/28G_{41} + 5/12G_{43}$
$H_{15} = 5/24G_{44}$
$H_{23} = \sqrt{3}/7D_{21} + 5\sqrt{3}/14D_{41}$
$H_{24} = 3/14G_{22} + 5/42G_{42}$
$H_{25} = 3/7G_{21} - 5/28G_{41} + 5/12G_{43}$
$H_{34} = \sqrt{3}/7G_{21} + 5\sqrt{3}/14G_{41}$
$H_{35} = -\sqrt{3}/7G_{22} + 5\sqrt{3}/84G_{42}$
$H_{45} = 3/7D_{21} - 5/28D_{41} - 5/12D_{43}$

$$
\begin{aligned}
&= (3/7)\alpha_2(Eq) - (4/7)\alpha_2(Ax) + (3/56)\alpha_4(Eq) + (2/21)\alpha_4(Ax)\\
H_{22} &= H_{44}\left(E(xz, yz)\right)\\
&= (1/14)D_{20} - (1/14)D_{40}\\
&= -(3/14)\alpha_2(Eq) + (2/7)\alpha_2(Ax) - (3/14)\alpha_4(Eq) - (8/21)\alpha_4(Ax)\\
H_{33} &= E\left(z^2\right)\\
&= (1/7)D_{20} + (3/28)D_{40}\\
&= (4/7)\alpha_2(Ax) - (3/7)\alpha_2(Eq) + (9/28)\alpha_4(Eq) + (4/7)\alpha_4(Ax) \qquad (38)
\end{aligned}
$$

Since there are no nonzero off-diagonal H_{ab} elements, in this example, the H_{nn} energies accord directly with the real d orbital energies also included in Figure 2. The *CK* method provides a facile procedure to generate these one-electron d-orbital energies and associated wavefunctions through diagonalization of the secular determinant from Table 3, for any stereochemistry. It would be relatively easy to extend the tables of functions to make this also useful for other electron configurations.

2.1.4 ***Ligand Field Theory and the Spectrochemical Series*** While crystal field theory has made a critical contribution to inorganic spectroscopy through its ability to correctly predict the d-orbital splitting diagrams and rationalize and assign the observed spectroscopy in terms of parameters such as *Dq*, *Ds*, *Dt*, and α_n, it fails to provide reliable information about the nature of the metal–ligand chemical bond since it ignores covalency and treats the ligand as a point charge. Its success arises because it is firmly grounded in symmetry.

The approach generally taken is to obtain, experimentally, numerical values for these parameters from fits to spectra; this is known as ligand field theory. Innumerable attempts have been made to extract chemical information from the magnitudes of these parameters. The most important parameter, *Dq*, varies with the bonding interaction of the ligand with the metal center. Good π-acceptors have high *Dq* values, while good π-donors have low *Dq* values. This trend in *Dq* is known as the Spectrochemical Series:

$$\begin{aligned} &I^- < Br^- < Cl^- < S^{2-} < {}^*SCN^- < N_3^- < F^- < dtc^- < urea < OH^- < OAc^- \\ &< oxalate^{2-} \approx malonate^{2-} \approx O^{2-} < H_2O < SCN^{*-} < EDTA < pyridine \\ &\approx NH_3 < en \approx tren \approx [SO_3]^{2-} < [NO_2]- \approx 2,2' - bpy \approx o\text{-}phen < H \\ &\approx CH_3^- \approx ph\text{-} < CN^- < constrained\ phosphites \approx CO \end{aligned} \quad (39)$$

Its origin in molecular orbital theory is developed in Section 2.3.1. The *Dt* parameter is a derivative parameter in that it is a measure of the difference between axial and equatorial *Dq* values (Eqn. (14)), while the *Ds* parameter has not been reliably shown to have chemical significance.

Thus the next step is to recognize that there is covalency in the metal–ligand bond. This can be done through employment of a variety of molecular orbital methodologies which will be addressed in Section 2.3. First, however, we present an early, but still widely used attempt to include covalency—namely the Angular Overlap Model (AOM).

2.2 Angular Overlap Model

2.2.1 ***General Approach*** This theory[15–17] is also well founded in symmetry, resting on earlier work by McClure,[18] and Yamatera[19] who, independently, began the idea of a two-dimensional spectrochemical series by factoring the effects of a ligand field into σ- and π-components. The AOM theory recognizes that the *d* orbitals can be perturbed in a σ, π, or δ-fashion by the ligand orbitals and generates a series of parameters distinguished by these bonding classifications. The connection between the magnitudes of these parameters and the actual σ- or π-bond strengths (free energies) of the chemical bond remains dubious, but their relative values do seem to agree with intuitive expectations concerning the ordering of a metal–ligand bond in terms of the extent of σ- or π-bonding. The numbers thus obtained are therefore more easily linked to chemical information than are the ligand field parameters. These AOM parameters are still semiempirical and are indeed linked to the ligand field parameters by simple mathematical manipulation. More sophisticated MO methods must be employed to obtain rigorous information or insight (vide infra).

Consider an $M{-}L_z$ bond fragment lying along the *z* axis. The AOM theory defines a σ-bonding interaction labeled $e_\sigma(L)$ for interaction of the lone pair σ-electrons on the ligand with a $d\sigma$ orbital lying along the *z* axis and $e_\pi(L)$ for interaction of (equivalent) ligand p_x and p_y $(p\pi)$ orbitals lying along the ligand local *x* and *y* axes, such that:

$$E(z^2) = e_\sigma(L); \quad E(x^2 - y^2) = E(xy) = 0; \quad E(xz) = E(yz) = e_\pi(L) \quad (40)$$

Technically these definitions are considered relative to the δ-interaction being regarded as zero. Thus each of the five d orbitals is perturbed by its specific interaction with ligand σ- and π-symmetry orbitals. Based upon the electron density in each orbital along each coordinate axis, if a σ-interaction with the $d(z^2)$ z axis lobe is defined as $1 \times e_\sigma(L)$, then a σ-interaction with the doughnut of charge in the xy plane of a $d(z^2)$ orbital is $(1/4)e_\sigma(L)$ and the interaction with each of the four lobes of a $d(x^2 - y^2)$ is $(3/4)e_\sigma(L)$. Then, for example, the specific M–L_x fragment, being M–L along the x axis, yields the contributions:

$$E(z^2) = (1/4)e_\sigma(L); \quad E(x^2 - y^2) = (3/4)e_\sigma(L);$$
$$E(xz) = E(xy) = e_\pi(L); \quad E(yz) = 0 \tag{41}$$

To obtain the energies for a species, such as ML_6, one simply sums all the perturbations along each axis, to obtain:

$$E(z^2) = E(x^2 - y^2) = 3e_\sigma(L); \quad E(xz) = E(xy) = E(yz) = 4e_\pi(L) \tag{42}$$

This generates the expected splitting into t_{2g} and e_g with an energy separation of $10Dq = 3e_\sigma(L) - 4e_\pi(L)$. In analogous fashion, if we consider a *trans*-ML_4Z_2 (Z on z axis) species and assume that both L and Z possess lone pairs of π-symmetry along their local x, y axis (each M–L, M–Z bond uses the ligand local z axis), then one can again simply sum the energy contributions from each ligand. This is easily constructed when realizing, for example, that the $d(xz)$ orbital interacts with a π-lone pair on each of two L ligands and with a π-lone pair on each of two Z ligands, thus:

$$E(z^2) = e_\sigma(L) + 2e_\sigma(Z); \quad E(x^2 - y^2) = 3e_\sigma(L);$$
$$E(xz) = E(yz) = 2e_\pi(Z) + 2e_\pi(L); \quad E(xy) = 4e_\pi(L) \tag{43}$$

2.2.2 *More Rigorous Approach* The above analysis is trivial when the ligands occupy the regular axis positions of an octahedron. For almost all other systems and especially for low symmetry systems, we should look more closely into the theoretical basis for this procedure.

Following the Wolfsberg–Helmholz approximation (see Section 2.3), the interaction of a d orbital, initial energy H_M, with a ligand orbital, initial energy H_L, and overlap S_{ML} yields the final energies E_d and E_l according to:[20]

$$E_d - H_M = (H_M + H_L)^2 \cdot (S_{ML})^2/(H_M - H_L) = e_\sigma(L)$$
$$E_l - H_L = -(H_M + H_L)^2 \cdot (S_{ML})^2/(H_M - H_L) = -e_\sigma(L) \tag{44}$$

Thus the metal orbital is destabilized by a σ-antibonding interaction, and the ligand orbital is correspondingly stabilized; $e_\sigma(L)$ is always positive. The AOM parameter is then proportional to the square of the overlap between metal and ligand orbital. The overlap contribution may be factored into a radial part which depends upon the metal, the ligand, and their distance apart; and an angular contribution which is independent of both metal and ligand but is dependent upon the geometry of the molecule. This factorization may be written, for example for a $d(z^2)$ orbital as:

$$E(z^2) = e_\sigma(L)F(z^2, p_z)^2 \tag{45}$$

governing the interaction of a p_z orbital on the ligand with the metal $d(z^2)$ orbital. The angular overlap functions F, such as $F(z^2, p_z)^2$, are defined always as unity for maximum overlap at a given M–L distance. They are unity in the examples described above in Eqns. (40–42). In a regular octahedral molecule, a p_x or a p_y orbital on the ligand (with respect to local ligand x, y, z axes) would have zero overlap with the metal $d(z^2)$ orbital, since the metal orbital has σ-symmetry, while the ligand orbitals have π-symmetry. Note that $F(z^2, p_x)$ and $F(z^2, p_y)$ are identically zero and hence there is no contribution from $e_\pi(L)$ to the energy of $d(z^2)$. However this is not necessarily the case in other geometries or in distorted structures. There may be a mismatch, for example, between the p_z orbital and the $d(z^2)$ orbital such that $F(z^2, p_z)$ is less than unity. It is therefore good practice to employ a general procedure, usable in any geometry, to evaluate these angular functions, F.

In the general case where mismatch may occur, the energy of a d orbital is given by the following expression summed over all the ligands:

$$E(d) = e_\sigma(L)F(d, \sigma)^2 + e_\pi(L)F(d, \pi)^2 + e_\delta(L)F(d, \delta)^2 \tag{46}$$

where the δ-component is usually treated as a zero for the system.

Consider that the ligand is placed on the surface of a unit sphere along the z coordinate of the central metal. We can then move the ligand to another position on the surface of the sphere corresponding to its position in the particular molecular geometry whose orbital AOM energies we desire, and described using the (θ, ϕ) coordinates of Figure 1. The overlap of the ligand (L_i) orbitals centered at the x, y, and z axes *of the ligand* with the *original* metal orbitals is then computed. This is equivalent to creating a new rotated axis system, x', y', z' and describing the original metal orbitals of the x, y, z coordinate system in terms of the primed system,[21] that is the original x, y, z coordinate system undergoes an orthogonal transformation to a new primed system such that the $d(z'^2)$ orbital now points along the M–L_i bond vector. For example, if the ligand is moved to coincide with the x axis of the original axis system, x becomes z' and the original $d(yz)$ orbital now appears as $-d(x'y')$, which can be determined by inputting the appropriate angles into Table 4.

Thus, in general, a d orbital in the x, y, z scheme becomes in the x', y', z' scheme:

$$\begin{array}{cccccc} & z'^2 & x'z' & y'z' & x'y' & x'^2 - y'^2 \\ d_i & F_{i1} & F_{i2} & F_{i3} & F_{i4} & F_{i5} \end{array} \tag{47}$$

where the F_i functions listed in Table 4 are the angular factors, and the energy of this d_i orbital (diagonal energy, no mixing) would be:

$$E(d_i) = \langle d_i | V_{AOM} | d_i \rangle = (F_{i1})^2 e_\sigma(L) + (F_{i2})^2 e_{\pi x}(L) + (F_{i3})^2 e_{\pi y}(L) + (F_{i4})^2 e_\delta(L) + (F_{i5})^2 e_\delta(L) \tag{48}$$

Table 4 Angular Overlap Transformation Matrix for d Orbitals

	z'^2	$y'z'$	$x'z'$
z^2	$(1/4)(1+3\cos 2\theta)$	0	$-(\surd 3/2)\sin 2\theta$
yz	$(\surd 3/2)\sin\Phi\sin 2\theta$	$\cos\Phi\cos\theta$	$\sin\phi\cos 2\theta$
xz	$(\surd 3/2)\cos\Phi\sin 2\theta$	$-\sin\Phi\cos\theta$	$\cos\Phi\cos 2\theta$
xy	$(\surd 3/4)\sin 2\Phi(1-\cos 2\theta)$	$\cos 2\Phi\sin\theta$	$(1/2)\sin 2\Phi\sin 2\theta$
x^2-y^2	$(\surd 3/4)\cos 2\Phi(1-\cos 2\theta)$	$-\sin 2\Phi\sin\theta$	$(1/2)\cos 2\Phi\sin 2\theta$

	$x'y'$	$x'^2-y'^2$
z^2	0	$(\surd 3/4)(1-\cos 2\theta)$
yz	$-\cos\Phi\sin\theta$	$-(1/2)\sin\Phi\sin 2\theta$
xz	$\sin\Phi\sin\theta$	$-(1/2)\cos\Phi\sin 2\theta$
xy	$\cos 2\Phi\cos\theta$	$(1/4)\sin 2\Phi(3+\cos 2\theta)$
x^2-y^2	$-\sin 2\Phi\cos\theta$	$(1/4)\cos 2\Phi(3+\cos 2\theta)$

As an example, we compute the one-electron d orbital energies of a *trans*-MPy_4Cl_2 species. Given the distinction between πp_x and πp_y orbitals we must define the relative orientations of the pyridine ligands. Thus we assume that the pyridine ligands lie in the metal xy plane with their molecular planes $(x'z')$ perpendicular to this metal xy plane. Their πp_x orbitals therefore form the π-bonding framework in the pyridine plane and these will only interact with the metal $d(xy)$ orbital. The chloride ligands occupy the z axis of the metal system.

It is useful to check the symmetry of this species to see if there will be any off-diagonal contributions. The point group symmetry is D_{4h} and therefore all the d-orbitals are unique—they all belong to different irreducible representations and therefore there are no off-diagonal elements. The computed diagonal energies will be our desired one-electron energies. Labeling the pyridine ligands L_{1-4} lying along $+x$, $-x$, $+y$, $-y$, respectively, and the chloride ligands as $L_{5,6}$ lying along $+z$ and $-z$, respectively, then:

$$\begin{array}{ccccccc} & L_1 & L_2 & L_3 & L_4 & L_5 & L_6 \\ \theta^\circ & 90 & 90 & 90 & 90 & 0 & 180 \\ \Phi^\circ & 0 & 180 & 90 & 270 & 0 & 0 \end{array} \tag{49}$$

For each metal orbital, read across a row of Table 4 for each ligand. Consider $d(z^2)$:

$$\begin{array}{cccccc} & z'^2 & x'z' & y'z' & x'y' & x'^2-y'^2 \\ L1 & -0.5 & 0 & 0 & 0 & \surd 3/2 \\ L2 & -0.5 & 0 & 0 & 0 & \surd 3/2 \\ L3 & -0.5 & 0 & 0 & 0 & \surd 3/2 \\ L4 & -0.5 & 0 & 0 & 0 & \surd 3/2 \\ L5 & 1 & 0 & 0 & 0 & 0 \\ L6 & 1 & 0 & 0 & 0 & 0 \end{array} \tag{50}$$

We sum the squares, but ignore the δ-terms, and multiply by the appropriate $e_\sigma(L)$ parameter according to the ligand concerned, to generate the $d(z^2)$ energy:

$$E(z^2) = (1/4+1/4+1/4+1/4)e_\sigma(\text{Py}) + (1+1)e_\sigma(\text{Cl}) = e_\sigma(\text{Py}) + 2e_\sigma(\text{Cl}) \quad (51)$$

Similarly, for the $d(xz)$ orbital:

$$\begin{array}{cccccc} & z'^2 & x'z & y'z' & x'y' & x'^2-y'^2 \\ L_1 & 0 & 0 & -1 & 0 & 0 \\ L_2 & 0 & 0 & 1 & 0 & 0 \\ L_3 & 0 & 0 & 0 & 1 & 0 \\ L_4 & 0 & 0 & 0 & -1 & 0 \\ L_5 & 0 & 0 & 1 & 0 & 0 \\ L_6 & 0 & 0 & 1 & 0 & 0 \end{array} \quad (52)$$

In this case, summing the squares and multiplying by the appropriate $e_\pi(L)$ parameter yields:

$$E(xz) = 2e_\pi(\text{Cl}) + 2e_{\pi y}(\text{Py}) \quad (53)$$

However, we must be alert to the fact that with the pyridine orientation we have defined, the p_y orbital is intimately involved with the σ-framework of the pyridine ligand and is greatly stabilized thereby. It is probable then that the $e_{\pi y}(\text{Py})$ parameter is vanishingly small due to poor overlap and energy match. Similar analysis will yield the other orbital energies:

$$\begin{aligned} E(yz) &= 2e_\pi(\text{Cl}) + 2e_{\pi y}(\text{Py}) \\ E(xy) &= 4e_{\pi x}(\text{Py}) \\ E(x^2-y^2) &= 3e_\sigma(\text{Py}) \end{aligned} \quad (54)$$

The parameter $e_\pi(\text{Cl})$ can be assumed to be positive since the chloride ion is a π-donor, and $e_x(\text{Py})$ is likely negative since pyridine has empty π^*-orbitals (see Section 2.3.1. and assuming $e_{\pi y}(\text{Py}) = 0$); similarly $e_\sigma(\text{Py}) > e_\sigma(\text{Cl})$ on the basis of their relative base strengths, thus the d orbital energy sequence will be:

$$d(x^2-y^2) > d(z^2) \gg d(xz) = d(yz) > d(xy) \quad (55)$$

with the energies cited in Eqns. (51, 53, 54). For a d^1 (or d^9) species, the problem is solved and the spectroscopically observed energy differences between the various d–d electronic transitions correspond to the energy differences between the d orbitals cited above.

Very often however, one is forced to make a further assumption since, using the example above, we have 4 unknowns but only 3 energy differences. Thus, the splitting of the e_g set in this molecule is observable directly from the spectrum and is equal to $2(e_\sigma(\text{Py}) - e_\sigma(\text{Cl}))$ $(= 2d_\sigma)$.[18,22,23] The splitting of the t_{2g} set, $(2e_\pi(\text{Cl}) - 4e_{\pi x}(\text{Py}))$ can also be derived, but the individual values of $e_\sigma(\text{Py})$ and $e_\sigma(\text{Cl})$ and the $e_\pi(L)$ parameters cannot be uniquely evaluated.

A common practice is to study amine complexes and to assume, reasonably, that the e_π(amine) parameter is zero. Under these circumstances, the $e_\pi(L)$ parameters can be evaluated. Within related series of complexes one might then assume that these parameters are transferable[17] and hence can be used to derive the $e(L)$ values. Evidence has been presented that the $e_\sigma(L)$ values are roughly proportional to r^5 (the metal-ligand distance) as they should theoretically be; this then allows parameters to be scaled and transferred even when bond distances change slightly. Another assumption is that for related species, the $e_\pi(L)/e_\sigma(L)$ ratio for a given ligand remains constant.[24–27]

Consider a further example, a planar *trans*-$CuCl_2(Py)_2$ species, where based upon a related study of the lutidine analogs[28–30] the molecular planes of the two pyridine units are again treated as perpendicular to the molecular plane of the complex. Further, the Cu–Cl bonds contain the y axis, the Cu–N bonds the x axis and z is perpendicular to the molecular plane. This molecule has D_{2h} symmetry and $d(z^2)$ and $d(x^2 - y^2)$ each transform as a_g, while the $d(xy)$, $d(xz)$, and $d(yz)$ orbitals transform as b_{1g}, b_{2g} and b_{3g}, respectively. The diagonal matrix elements $\langle d_i|V_{AOM}|d_i\rangle$ are derivable without recourse to solving the whole AOM matrix:

$$\begin{aligned}
\langle z^2|V_{AOM}|z^2\rangle &= (0.5)[e_\sigma(\text{Cl}) + e_\sigma(\text{N})] \\
\langle x^2 - y^2|V_{AOM}|x^2 - y^2\rangle &= (1.5)[e_\sigma(\text{Cl}) + e_\sigma(\text{N})] \\
\langle yz|V_{AOM}|yz\rangle &= 2e_\pi(\text{Cl}) \\
\langle xz|V_{AOM}|xz\rangle &= 0 \\
\langle xy|V_{AOM}|xy\rangle &= 2e_\pi(\text{Cl}) + 2e_\pi(\text{N})
\end{aligned} \tag{56}$$

However, in this case there are two d orbitals of the same symmetry, thus the element $\langle z^2|V_{AOM}|x^2 - y^2\rangle$ is nonzero. The necessary F angular factor is derived by a simple extension of the previous procedure. Eqn. (48) is modified to Eqn. (57) (excluding the δ-functions):

$$\langle d_i|V_{AOM}|d_j\rangle = (F_{i1}, F_{j1})e_\sigma(L) + (F_{i2}, F_{j2})e_{\pi x}(L) + (F_{i3}, F_{j3})e_{\pi y}(L) \tag{57}$$

where again the sum is over all ligands. The relevant lines in the matrix for Py N_1 located at $\theta = 0°$, $\phi = 0°$

	$z'2$	xz'	yz'
z^2	-0.5	0	0
$x^2 - y^2$	$\sqrt{3}/2$	0	0

(58)

Thus the product (from 58) is $-\sqrt{3}/4e_\sigma(N_1)$. Considering the other 3 ligands with their appropriate θ, ϕ values leads to $\sqrt{3}/4e_\sigma(\text{Cl})$, $\sqrt{3}/4e_\sigma(\text{Cl}_2)$, $-\sqrt{3}/4e_\sigma(N_2)$. Thus summing yields:

$$\langle z^2|V|x^2 - y^2\rangle = \sqrt{3}/2[e_\sigma(\text{Cl}) - e_\sigma(\text{N})] \tag{59}$$

Hence the desired energies of the $d(z^2)$ and $d(x^2 - y^2)$ orbitals are the roots of the secular determinant (60).

$$\begin{array}{|c|c|c|} \hline & z^2 & x^2 - y^2 \\ \hline z^2 & (0.5)[e_\sigma(\mathrm{Cl}) + e_\sigma(\mathrm{N})] - E & \sqrt{3}/2[e_\sigma(\mathrm{Cl}) - e_\sigma(\mathrm{N})] \\ \hline x^2 - y^2 & \sqrt{3}/2[e_\sigma(\mathrm{Cl}) - e_\sigma(\mathrm{N})] & (1.5)[e_\sigma(\mathrm{Cl}) + e_\sigma(\mathrm{N})] - E \\ \hline \end{array} = 0 \qquad (60)$$

There are various other subtleties associated with the orbital angular overlap model which we do not have space to deal with here. With bidentate ligands, such as bipyridine, the ligand π-orbitals extend over the polyatomic ligand and it is necessary to ensure that the d orbital matrix used is re-hybridized to take account of the full molecular symmetry. In groups where a d orbital transforms as a_1, mixing with the $s(a_1)$ orbital may take place (d–p mixing is also possible in non-centrosymmetric complexes).[31,32] So-called *sd* mixing has been introduced as a means of improving the fits, especially in square planar systems.[24–27,33] A ligand p orbital may be oriented so that its valency is "mis-directed." Dealing with this phenomenon requires extra angular dependence to be accounted for.[26,30,33,34] Other variants of the angular overlap model have been introduced of which the *Cellular Ligand Field* is the most developed. Among its suppositions is that one can assign a value for e_σ(void), being the effect on the d-energy levels of the absence of a ligand.[35]

2.3 Molecular Orbital Theory

In this section, we develop the procedure employed in the systematic construction of a one-electron molecular orbital energy level diagram for a transition metal complex. We expand our focus to include all the orbitals on the metal and ligands and emphasize the use of group theory and qualitative bonding concepts at the extended Hückel level. We specifically consider the ML_6 complex in the O_h group where M is usually a divalent or trivalent transition metal ion and L is a monatomic and anionic ligand (F^-, Cl^-, O^{2-}, S^{2-}, etc).

The first step is to estimate the relative energies of the valence orbitals on the metal and ligands. Here we assume that the core orbitals do not contribute significantly to bonding and the valence orbitals include the $3d$, $4s$, and $4p$ orbitals for a metal ion of the first transition series and the ns and np levels of the ligands where n is the principal quantum of their valence orbitals. The appropriate orbital energies are the valence state ionization energies (VSIE'S) which are dependent on the charge and configuration of the atom in the molecule and can be determined from calculations and experiment (e.g., photoelectron spectroscopy, see Chapter 8). A reasonable initial energy level diagram for a transition metal complex is given in Figure 3. The ligand valence orbitals are at deeper binding energy relative to the metal ion valence orbitals since the ligands are more electronegative; and thus the electron density should be distributed toward the ligands in the bonding molecular orbitals (*vide infra*).

The next step is to introduce the symmetry of the complex (O_h) into the molecular orbital energy level diagram. Determination of the transformation properties of the

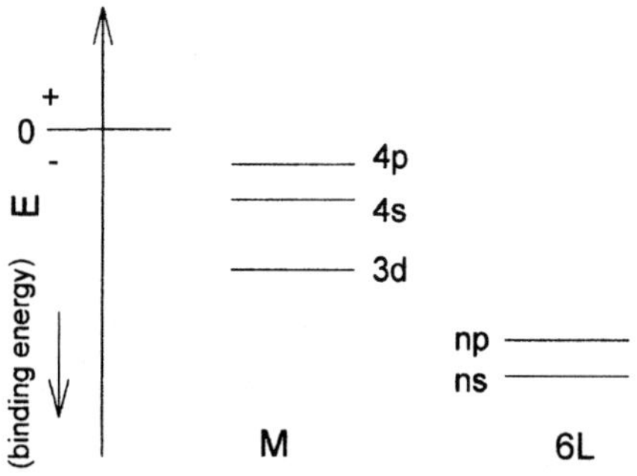

Figure 3. A sketch of relative valence state ionization energies (VSIE) for M and L atoms in an ML_6 species.

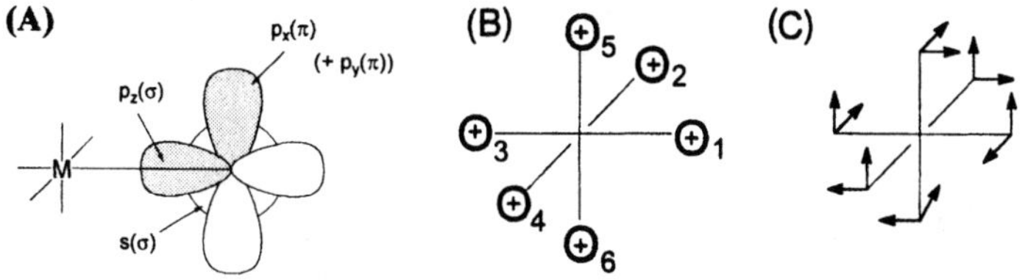

Figure 4. (A) σ- and π-orbitals on ligand L in the fragment M–L; (B) a set of equivalent s-σ orbitals on each ligand atom in the species ML_6; (C) a set of vectors representing the p-π orbitals ($p_{x,y}$) on each ligand in the species ML_6.

central metal is straightforward. The metal ion resides at the "point" of the molecular symmetry group which contains all symmetry elements. Therefore the metal centered orbitals transform directly as basis functions for the octahedral point group. From the O_h character table,[36] the $4p(x, y, z)$ orbitals have t_{1u} symmetry, the $4s$ is totally symmetric and transforms as a_{1g}; the fivefold degeneracy of the $3d$ orbitals must split in energy in O_h into $e_g(x^2 - y^2, z^2)$ and $t_{2g}(xz, yz, xy)$ sets.

Since the ns, and np, valence orbitals on the six ligands interchange with one another under the symmetry operations of the O_h point group, they generate a set of symmetry adapted linear combinations (SALC) whose nature can be derived through the following procedures. First, these orbitals are divided into equivalent sets (i.e., those which are rotated into one another by the symmetry operations of the point group) and their overlap properties are defined with respect to nodal planes along the specific metal–ligand bonds (σ, π, δ corresponding to 0, 1 and 2 nodal planes, respectively). From Figure 4A, the s orbitals form an equivalent set with σ overlap, the p_z orbitals form a separate equivalent set with σ symmetry and the p_x and p_y transform together as an equivalent set with π symmetry (i.e., one nodal plane along the M–L axis). By way of further explanation, there are no symmetry operations which would rotate a p_x or p_y orbital on one ligand into the p_z orbital of the same or another ligand.

Each equivalent set (Figure 4B, C) is then used as a set of basis functions to generate a reducible representation of the molecular symmetry group, a SALC. The

Table 5 Transformation Properties of Some Ligand Basis Sets

R \ L	1	2	3	4	5	6	Contribution to Reducible Character[a]
$E(\sigma L)$	σ_1	σ_2	σ_3	σ_4	σ_5	σ_6	+6
$C_4(\sigma L)$	σ_1	σ_2	σ_4	σ_5	σ_6	σ_3	+2
C_4(in-plane $p_\pi L$)[b]			$-p_{\pi 4}$	$-p_{\pi 5}$	$p_{\pi 6}$	$p_{\pi 3}$	0
$\sigma_v(yz)$(in-plane $p_\pi L$)[b]			$-p_{\pi 5}$	$-p_{\pi 4}$	$-p_{\pi 3}$	$-p_{\pi 6}$	−2

[a] see text;
[b] the sign of orbitals which rotate into another ligand orbital is arbitrary depending on the initial choice of a positive direction.

Table 6 Representations of Ligand *s* and *p* Orbital SALCs in O_h Symmetry

O_h	E	$8C_3$	$6C_2$	$6C_4$	$3C_2$	i	$6S_4$	$8S_6$	$3\sigma_h$	$6\sigma_d$	Red
$\Gamma s\sigma$	6	0	0	2	2	0	0	0	4	2	$a_{1g} + e_g + t_{1u}$
Γp_z	6	0	0	2	2	0	0	0	4	2	$a_{1g} + e_g + t_{1u}$
$\Gamma p_{x,y}\pi$	12	0	0	0	−4	0	0	0	0	0	$t_{1g} + t_{2g} + t_{1u} + t_{2u}$

symmetry operations of the group are performed on each equivalent set of basis functions. The character for a particular operation, R, is obtained by summing 1 for every basis function which is unchanged, −1 for every function which changes sign, and 0 for every function which rotates into a different function (on the same or on a different ligand atom). Thus the full transformation matrix does not have to be generated for each symmetry operation.

For example, label, in an octahedron, a set of ligands 1–6, with 1 and 2 on the $C_4(z)$ axis, 3 and 5 on the x axis, and 4 and 6 on the y axis, the axes being defined with respect to the central metal ion. Table 5 shows the transformation behavior of the σL_i set under identity, E, and $C_4(z)$ and the in-plane (xy) $p\pi(L_i)$ (a subset of the $p - \pi$ equivalent orbitals) under $C_4(z)$ and $\sigma_v(yz)$.

All operations in the same class will have the same character. The complete reducible representation generated by the 6 σ orbitals in Figure 4B is given in Table 6.

$$a_i = \frac{1}{h}\sum_R X_i(R)X_{\text{red}}(R) \tag{61}$$

Reducible representations are reduced to a sum of irreducible representations using Eqn. (61) where a_i is the number of times the ith irreducible representation is contained in the reducible representation Γ_{red}, h is the order of the group (i.e., number of symmetry operations), R is a symmetry operation, $X_i(R)$ is the character of the ith irreducible representation for symmetry operation R and $X_{\text{red}}(R)$ is the character of the reducible representation for symmetry operation R. Reduction of both $\Gamma_{s\sigma}$ in Table 6 and the reducible representation generated by the 6 p_z orbitals

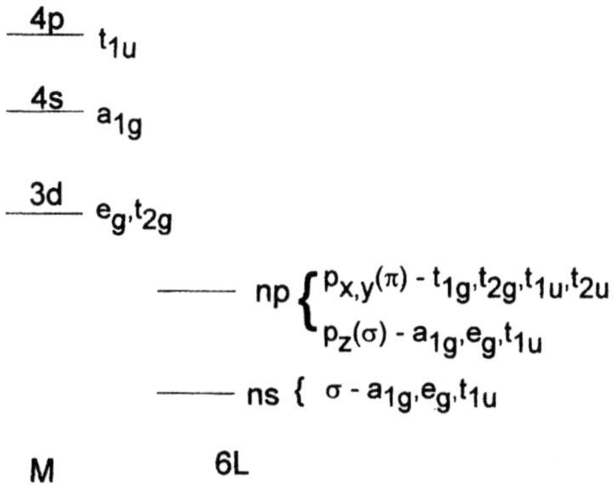

Figure 5. Symmetry representations of metal centered (M) and ligand centered ($6L$) symmetry adapted linear combinations (SALCs) in octahedral symmetry.

having σ symmetry gives $a_{1g} + e_g + t_{1u}$. Finally, the 12 p_x, p_y π-basis functions outlined in Figure 4C give the $\Gamma_{px,y,\pi}$ reducible representation in Table 6, which reduces $\Gamma_{px,y,\pi} = t_{1g} + t_{2g} + t_{1u} + t_{2u}$ using Eqn. (61).

Thus we have determined the symmetry of the metal orbitals and the valence orbitals of the six ligands as shown in Figure 5.

There is a set of wavefunctions corresponding to each ligand irreducible representation (n-fold degenerate wavefunctions for an n-dimensional irreducible representation) which is a set of symmetry adapted linear combinations of equivalent orbitals on the six ligands, $\Psi_{\mathrm{L_{SALC}}}$ obtained using projection operator methods as described.[36] Here we need only utilize their symmetry but the $\Psi_{\mathrm{L_{SALC}}}$ are required for quantitative calculations of overlap and bonding. Very often the SALC wavefunctions can be deduced by consideration of the metal based orbital with which they overlap and with which they necessarily have the same symmetry. For example, in a square complex, the $d(xy)$ orbital transforms as b_{2g} and this will potentially interact with the in-plane π-SALC of b_{2g} symmetry. Using the labeling scheme noted above for in-plane ligands 3, 4, 5, and 6, the necessary SALC must be:

$$(1/2)\big[p_{\pi y}(3) - p_{\pi x}(4) + p_{\pi y}(5) - p_{\pi x}(6)\big] \tag{62}$$

in order to "map" onto the in-plane $d(xy)$ orbital.

From Figure 5 we are interested in bonding interactions between a metal valence orbital e.g., Ψ_{Md} lying at the VSIE of a metal 3d orbital, H_{Md}, and a ligand valence orbital $\Psi_{\mathrm{L_{SALC}}}$ at the VSIE of the six ligand equivalent orbitals, $H_{\mathrm{L_{SALC}}}$. We focus on a specific metal–ligand bonding interaction in Figure 6A. From molecular orbital theory, we assume a good wavefunction is a linear combination of these metal and ligand valence orbitals $\Psi_{\mathrm{MO}} = c_1\Psi_{\mathrm{Md}} + c_2\Psi_{\mathrm{L_{SALC}}}$ and evaluate $E = \langle\Psi_{\mathrm{MO}}|\mathcal{H}_{\mathrm{ML_6}}|\Psi_{\mathrm{MO}}\rangle$ where $\mathcal{H}_{\mathrm{ML_6}}$ is the one-electron molecular Hamiltonian. In the Hückel treatment, these integrals are approximated by experimental values. The variation principle (i.e., minimization of the energy with respect to the coefficients c_1 and c_2) leads to the secular determinant given in Eqn. (63) where $H_{\mathrm{Md}} \equiv \langle\Psi_{\mathrm{Md}}|\mathcal{H}_{\mathrm{ML_6}}|\Psi_{\mathrm{Md}}\rangle$ is the VSIE of a metal 3d electron, $H_{\mathrm{L_{SALC}}} \equiv \langle\Psi_{\mathrm{L_{SALC}}}|\mathcal{H}_{\mathrm{ML_6}}|\Psi_{\mathrm{L_{SALC}}}\rangle$ is the VSIE of the ligand

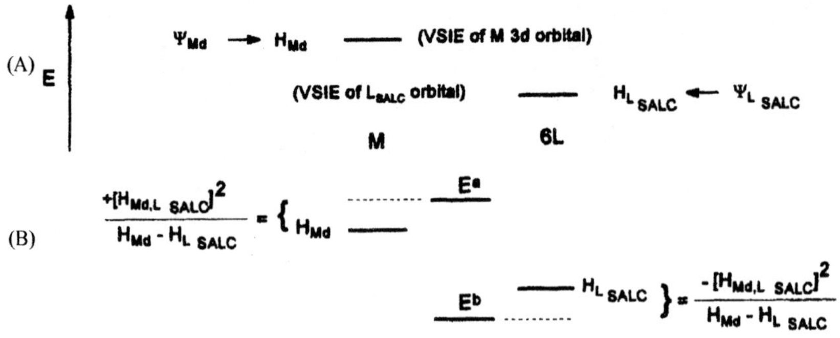

Figure 6. (A) Definition of the valence state ionization energies (VSIE) of the metal center, M, and the ligand set, $6L$, prior to the bonding interaction. (B) Formation of bonding and anti-bonding orbitals displaced from H_{Md} and $H_{\text{L}_{\text{SALC}}}$ by the bonding/anti-bonding interaction.

SALC and $H_{\text{Md,L}_{\text{SALC}}}$ is $\langle \Psi_{\text{Md}} | \mathcal{H}_{\text{ML}_6} | \Psi_{\text{L}_{\text{SALC}}} \rangle$ the resonance integral which leads to bonding. In Eqn. (63), we assume that the dominant overlap integral combination to bonding ($S_{\text{Md,L}_{\text{SALC}}} \equiv \langle \Psi_{\text{Md}} | \Psi_{\text{L}_{\text{SALC}}} \rangle$) is contained in the resonance integral.

$$\delta E / \delta C_i = 0 \rightarrow \begin{vmatrix} H_{\text{Md}} - E & H_{\text{Md,L}_{\text{SALC}}} \\ H_{\text{Md,L}_{\text{SALC}}} & H_{\text{L}_{\text{SALC}}} - E \end{vmatrix} = 0 \tag{63}$$

When the energy difference between valence orbitals ($|H_{\text{Md}} - H_{\text{L}_{\text{SALC}}}|$) is much greater than the interaction between orbitals ($H_{\text{Md,L}_{\text{SALC}}}$) one obtains the two approximate solutions for Eqn. (63) given in Eqn. (64). The bonding solution, Eqn. (64a), involves the ligand SALC being stabilized in energy by an amount $-(H_{\text{Md,L}_{\text{SALC}}})^2/(H_{\text{Md}} - H_{\text{L}_{\text{SALC}}})$. The bonding wavefunction Ψ^{b} is dominantly on the ligands but is now mixed with a small amount of Ψ_{Md} (the mixing coefficient $\lambda \approx -(H_{\text{Md,L}_{\text{SALC}}})/(H_{\text{Md}} - H_{\text{L}_{\text{SALC}}})$. For the antibonding solution, Eqn. (64b) the metal $3d$ orbital is raised in energy by approximately the same amount as the bonding orbital is lowered; this wavefunction is dominantly on the metal but is mixed with a small amount, λ, of $\Psi_{\text{L}_{\text{SALC}}}$. These results are summarized in Figure 6B.

$$\begin{aligned} E^{\text{b}} &= H_{\text{L}_{\text{SALC}}} - \left[H_{\text{Md,L}_{\text{SALC}}}\right]^2 / \left[H_{\text{Md}} - H_{\text{L}_{\text{SALC}}}\right] \\ \Psi^{\text{b}} &= N\left(\Psi_{\text{L}_{\text{SALC}}} + \lambda \Psi_{\text{Md}}\right) \\ N^2&\left(1 + \lambda^2 + 2\lambda S_{\text{Md,L}_{\text{SALC}}}\right) = 1 \end{aligned} \tag{64a}$$

$$\begin{aligned} E^* &= H_{\text{Md}} + \left[H_{\text{Md,L}_{\text{SALC}}}\right]^2 / \left[H_{\text{Md}} - H_{\text{L}_{\text{SALC}}}\right] \\ \Psi^* &= N^*\left(\Psi_{\text{Md}} - \lambda \Psi_{\text{L}_{\text{SALC}}}\right) \\ N^{*2}&\left(1 + \lambda^2 - 2\lambda S_{\text{Md,L}_{\text{SALC}}}\right) = 1 \end{aligned} \tag{64b}$$

From Figure 6B, the requirement for bonding is that the resonance integral $H_{\text{Md,L}_{\text{SALC}}}$ be nonzero. From group theory, this will only be the case when the direct product of the symmetries $\Gamma_{Md} \times \Gamma \mathcal{H}_{\text{ML6}} \times \Gamma_{\text{L}_{\text{SALC}}}$ contains the totally symmetric irreducible representation, A_{1g} in O_h. Since the Hamiltonian represents the total energy of the complex, it must be invariant to all symmetry operations of the molecular symmetry group, in other words, $\Gamma\mathcal{H}_{ML_6} = A_{1g}$. Therefore, for there to be a bonding-antibonding interaction between two valence orbitals Ψ_{Md} and $\Psi_{L_{SALC}}$ the orbitals must transform as the same irreducible representation in the molecular symmetry group.

$$H_{Md,L_{SALC}} = k\left(\tfrac{1}{2}\left(H_{Md} + H_{L_{SALC}}\right)\right)S_{Md,L_{SALC}} \tag{65}$$

From Mulliken,[37,38] Wolfsberg and Helmholtz,[20] the resonance integral is often approximated by Eqn. (65) where k is determined from experiment and is approximately 2. Eqns. (64) and (65) show that the strength of the bonding/antibonding interaction between two orbitals increases as the overlap between these orbitals increases. Due to the directional properties of the orbitals, $S_\sigma > S_\pi > S_\delta$. Finally from Figure 6 the magnitude of the bonding interaction increases as the energy separation between the valence orbitals decreases $(\alpha 1/(H_{Md} - H_{L_{SALC}}))$. Therefore bonding is dominated by the overlap of valence orbitals on the metal and ligands which are close in energy and have the same symmetry, with the extent of overlap dominating the interaction energies.

The bonding interaction between the ligand np and the metal $3d$ is expected to be large (i.e., $H_{Md} - H_{L_{SALC}}$ is small). One ligand p_Z–σ SALC has e_g symmetry and can undergo a σ bonding/antibonding interaction with the metal $3d$ orbitals having e_g symmetry (i.e., $d(x^2 - y^2)$, $d(z^2)$). The overlap is large and the bonding/antibonding splitting will also be large. Further, one ligand $p_{x,y}$–π SALC transforms as t_{2g} and can therefore undergo a bonding/antibonding interaction with the d orbitals of the same symmetry (Figure 7). The energy splittings are however reduced from those of the e_g sets because π rather than σ overlap is involved.

Extending this procedure to include bonding interactions with the metal $4s$ and $4p$ levels leads to the energy level diagram in Figure 7 where the ligand t_{1g} and t_{2u} levels do not have the appropriate symmetry to interact with any of the metal valence orbitals and are thus nonbonding (nb). Note that the ligand ns/np energy splitting in anionic ligands is large and often only a small amount of ligand s orbital is mixed into the ligand p_z hybrid orbitals which dominate the σ-bonding interactions with the metal ion. Finally, we again emphasize that the above treatment presents the qualitative aspects of molecular orbital theory which is quantitatively developed for different levels of calculation in Chapters 10 and 11.

2.3.1 ***Types of Ligand–Metal Bonds*** The energy level diagram in Figure 7 has been derived for single atom anionic ligands which each contribute three p orbitals (p–σ $p_{x,y}$–π) for bonding with the metal ion. This produces a splitting of the e_g and t_{2g} sets of molecular orbitals derived from the metal d orbitals (defined as Δ_{o_h} in molecular orbital theory and 10 Dq in ligand field theory) which is due to the difference between the σ and π antibonding interactions with the ligands. Each of the six ligands has eight valence electrons for a total of 48 electrons which fill the 24 ligand-derived bonding and nonbonding molecular orbitals. Additional electrons

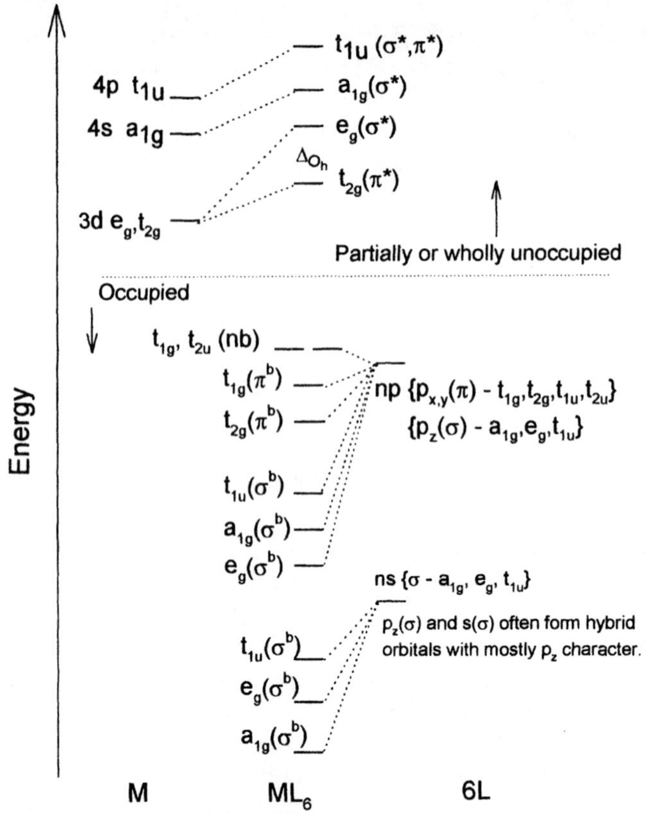

Figure 7. A molecular orbital diagram for an octahedral ML_6 species of a $3d$ transition metal involving a σ- and π-donor ligand (e.g., a halide ion) and derived from general bonding theory.

in d orbitals donated by the metal occupy the $t_{2g}(\pi^*)$ and $e_g(\sigma^*)$ molecular orbitals. From Eqn. (64a), the σ and π bonding molecular orbitals will have dominantly ligand character ($\Psi(\sigma^b$ or $\pi^b) = N(\Psi_{\mathrm{L_{SALC}}} + \Psi\phi_{\mathrm{Md}})$). For $\lambda = 0$, the electrons in these orbitals are localized on the anionic ligands. When there is bonding, $\lambda > 0$ and a portion of this electron density is *donated* to the metal ion. Thus monatomic anionic ligands form σ-donor and π-donor bonds to metal ions and produce a splitting of the d orbitals Δ_{O_h}, which derives from the difference between σ and π antibonding interactions (see also the AOM treatment above). This is summarized in Figure 8A.

For all other ligand types, the coordinating heteroatom is part of a molecular framework and its p_x and p_y orbitals participate in bonding interactions within the ligand. These covalent intraligand bonds will generally be stronger than those to the metal ion. For ligands such as amines, intraligand bonding leaves only one lone pair in an sp^3 type hybrid orbital for bonding to the metal ion. The six doubly occupied sp^3 hybrid orbitals (one on each NH_3 type ligand) transform as $a_{1g} + e_g + t_{1u}$. Each hybrid orbital is oriented along a ligand–metal bond, Figure 8B, and thus has σ overlap with the metal ion. These ligands can only have σ-donor bonding interactions

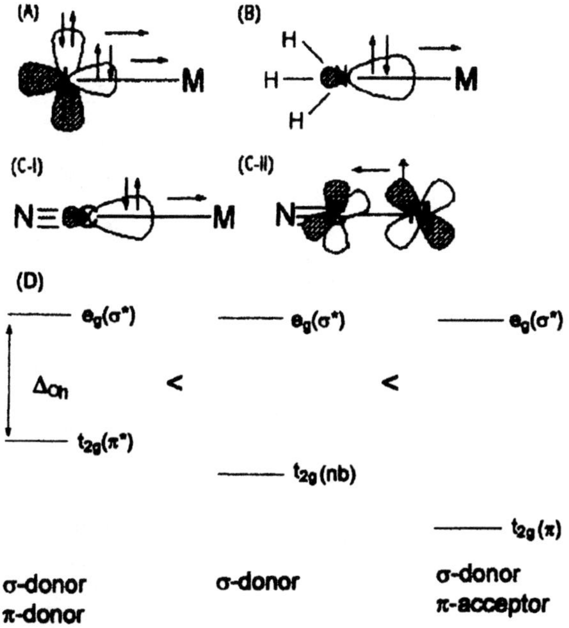

Figure 8. Sketches of bonding with A) a σ-donor and π-donor ligand, B) a σ-donor ligand, C) a σ-donor and π-acceptor ligand (the cyanide ion), i) σ-donation from the $5\sigma^*$ HOMO and ii) π-back-donation in the $2\pi^*$ LUMO and D) showing the trend in Δ_{O_h} over the series.

with the metal ion resulting in the molecular orbital energy level diagram in Figure 9A. This produces a splitting of the t_{2g} and e_g set of d orbitals which derives from the difference between σ-antibonding and nonbonding interactions with the metal ion.

In the final class of ligands, CN^-, CO, py, etc., the coordinating atom is involved in π-bonding interactions within the ligand. This produces ligand π^* orbitals which are higher in energy than the metal d orbitals and unoccupied, but still low enough in energy to participate in bonding. As diagrammed in Figure 8C, for CN^-, these ligands contribute one occupied sp_z type hybrid orbital which is involved in σ-donor interaction with the metal ion and an unoccupied p_x, p_y set which has π symmetry. For six CN^- ligands, in an O_h complex, the six ligand sp_z–σ orbitals transform as $a_{1g} + e_g + t_{1u}$ and the twelve p_x, p_y–π orbitals transform as $t_{1g} + t_{2g} + t_{1u} + t_{2u}$ and are π^* with respect to the cyanide groups. The resulting molecular orbital energy level diagram is shown in Figure 9B. The new feature here is that since the π^* ligand set is higher in energy than the metal ion d orbitals, the latter are stabilized in energy due to bonding and Δ_{o_h} is now the difference between σ-antibonding and π -bonding interactions with the ligands.

Importantly, the wavefunction for the metal d orbital $t_{2g}(\pi^b)$ molecular orbital is $(N(\Psi_{\mathrm{Md}}(t_{2g}\pi) + \lambda\Psi_{\mathrm{L_{SALC}}}(t_{2g}\pi))$. For $\lambda = 0$, electrons in the d orbitals are localized on the metal but when there is bonding, $\lambda > 0$ and the electrons are delocalized onto the ligand. Thus these are classified as σ-donor and π-*acceptor* ligands.

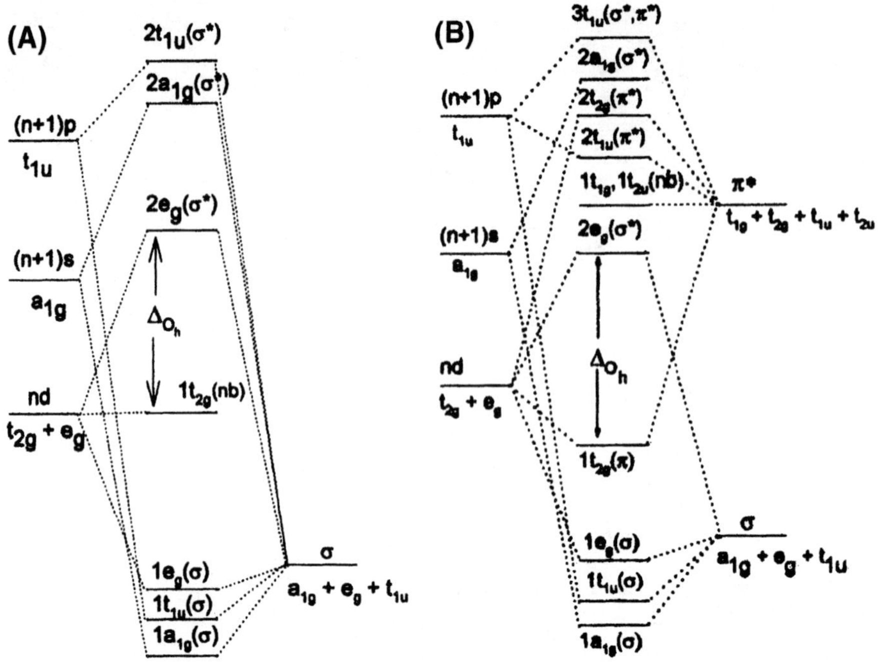

Figure 9. (A) A complete molecular orbital diagram for an octahedral ML_6 species of a $3d$ transition metal involving a σ-donor ligand (e.g., ammonia). (B) A complete molecular orbital diagram for an octahedral ML_6 species of a $3d$ transition metal involving a σ-donor and π-acceptor ligand (e.g., a cyanide ion).

From Figure 8(D), it is seen that there is a general trend in Δ_{oh}. The e_g/t_{2g} splitting is smallest for σ-donor, π-donor ligands; increases for pure σ-donor ligands; and is generally largest for σ-donor, π-acceptor ligands. As described in Section 2.1.4., this trend is observed experimentally through inorganic spectroscopy and is termed the spectrochemical series (see Eqn. (39)).

2.4 Spin–Orbit Coupling for d^1 Species

Since both spin motion and orbital motion generate a magnetic field, they will couple together such that the l and s quantum numbers lose their significance, to be replaced by the total angular momentum quantum number j. The extent of such coupling is proportional to the spin–orbit coupling coefficient, λ (units of $\hbar^2$). For the first row transition elements it is not very large and does not seriously influence the observed electronic spectra except where high resolution data, especially single crystal data, are being analyzed. For the heavier transition metal ions, this coupling process is much more important and indeed is dominant in the Lanthanides and Actinides. However, even with the first row ions, spin–orbit coupling must be considered in magnetic susceptibility or electron paramagnetic resonance applications.

Table 7 Free Ion Single Electron Spin–Orbit Coupling Coefficients, ζ_{3d} (cm^{-1})[a]

M \ Z^+	0	1+	2+	3+	4+	5+	6+
Ti	70	90	123	155			
Zr		{300}	{400}	{500}			
Hf							
V	95	135	170	210	250		
Nb		{420 }	{610}	{800}			
Ta				{1400}			
Cr	135	185	230	275	355	380	
Mo			{670}	800	{850}	{900}	
W			{1500}	{1800}	{2300}	{2700}	
Mn	190	255	300	355	415	475	540
Tc			{950}	{1200}	{1300}	{1500}	{1700}
Re			{2100}	{2500}	{3300}	{3700}	{4200}
Fe	275	335	400	460	520	590	665
Ru				{1250}	{1400}	{1500}	{1700}
Os				{3000}	{4000}	{4500}	{5000}
Co	390	455	515	580	650	715	790
Rh					{1700}	{1850}	{2100}
Ir					{5000}	{5500}	{6000}
Ni		565	630	705	790	865	950
Pd		{1300}	{1600}				
Pt		{3400}					
Cu			830	890	960	1030	1130
Ag			{1800}				
Au			{5000}				

[a] Values in parentheses are estimates, Adapted from ref. 62

With d^1, we have 10 basis states (each d orbital is multiplied by a spin function, α or β with $m_s = \pm 1/2$), with m_j (vector sum of m_l and m_s) running from $5/2$ to $-5/2$. To obtain the spin–orbit energies in a ligand field, V, we need to solve the matrix with elements $\langle d|V + \xi(r)\boldsymbol{l} \cdot \boldsymbol{s}|d'\rangle$ where d and d' are the 10 basis states called spinors.

The one-electron spin–orbit coupling operator $\xi(r)\boldsymbol{l} \cdot \boldsymbol{s}$ has a radial part, $\xi(r)$, operating on the radial part of the wavefunctions, and an orbital part, $\boldsymbol{l}$, and a spin part, $\boldsymbol{s}$, operating on the orbital and spin motion respectively. Since it is a one-electron operator, a multi-electron wavefunction will be factored into a sum of one-electron functions (*vide infra*). The radial part $\langle \mathrm{R}(r)|\xi(r)|\mathrm{R}'(r)\rangle$ is factored out as $\xi/\hbar^2$ (see Table 7 for the one-electron radial integrals of spin–orbit coupling for the transition metal ions) and we are left with deriving the expectation value of the $\boldsymbol{l} \cdot \boldsymbol{s}$ part of the operator.

We express the element $\langle d|\boldsymbol{l} \cdot \boldsymbol{s}|d'\rangle$ in terms of the relevant spin and orbital components as $\langle m_l, m_s|\boldsymbol{l} \cdot \boldsymbol{s}|m_l', m_s'\rangle$. Since $\boldsymbol{l} \cdot \boldsymbol{s} = \mathrm{L_x S_x} + \mathrm{L_y S_y} + \mathrm{L_z S_z} =$

$(1/2)[L_+S_- + L_-S_+] + L_zS_z$ (see also Section 3.2.2.) then for this term to be nonzero, $m_l + m_s = m_l' + m_s'$. Further, there are two solutions: (Diagonal element) $m_s = m_s'$, and $m_l = m_l'$ then $\langle m_l, m_s|\xi(r)\boldsymbol{l}\cdot\boldsymbol{s}|m_l', m_s'\rangle = m_l m_s \xi/\hbar^2$ (Off-diagonal element) $m_s' = m_s \pm 1$, $m_l' = m_l \mp 1$, then

$$\langle m_l, m_s|\xi(r)\boldsymbol{l}\cdot\boldsymbol{s}|m_l', m_s'\rangle = 0.5\sqrt{[(\boldsymbol{l} - m + \tfrac{1}{2})(\boldsymbol{l} + m + \tfrac{1}{2})]}\xi/\hbar^2$$
$$\text{where } m = m_l + m_s = m_l' + m_s'.$$

So,

$$\langle 2^+|\xi(r)\boldsymbol{l}\cdot\boldsymbol{s}|2^+\rangle = m_l m_s \xi/\hbar^2 = (2)\times(+1/2)\tfrac{1}{2} = 1\xi/\hbar^2$$

and

$$\langle 1^+|\xi(r)\boldsymbol{l}\cdot\boldsymbol{s}|2^-\rangle = 0.5\sqrt{[(2 - 1\tfrac{1}{2} + \tfrac{1}{2})(2 + 1\tfrac{1}{2} + \tfrac{1}{2})]} = 1\xi\hbar^2 \tag{66}$$

The analysis above provides the spin–orbit coupling for electrons in d orbitals expressed as complex functions. Often we are interested in evaluating spin–orbit coupling for electrons in the real d orbitals, for example to include spin–orbit coupling in the secular determinants obtained in Section 2.1.3. The effect of the spin–orbit interaction is to mix the real d orbital spinors (i.e., produce nonzero off-diagonal elements) as given by Eqn. (67).[3]

$$
\begin{aligned}
l\cdot s(\overset{+}{z^2}) &= -\frac{\sqrt{3}}{2}(\overset{-}{xz}) - \frac{i\sqrt{3}}{2}(\overset{-}{yz})\\
l\cdot s(\overset{-}{z^2}) &= \frac{\sqrt{3}}{2}(\overset{+}{xz}) - \frac{i\sqrt{3}}{2}(\overset{+}{yz})\\
l\cdot s(\overset{+}{x^2-y^2}) &= i(\overset{+}{xy}) - \frac{i}{2}(\overset{-}{yz}) + \frac{1}{2}(\overset{-}{xz})\\
l\cdot s(\overset{-}{x^2-y^2}) &= -i(\overset{-}{xy}) - \frac{i}{2}(\overset{+}{yz}) - \frac{1}{2}(\overset{+}{xz})\\
l\cdot s(\overset{+}{xy}) &= -i(\overset{+}{x^2-y^2}) + \frac{1}{2}(\overset{-}{yz}) + \frac{i}{2}(\overset{-}{xz})\\
l\cdot s(\overset{-}{xy}) &= i(\overset{-}{x^2-y^2}) - \frac{1}{2}(\overset{+}{yz}) + \frac{i}{2}(\overset{+}{xz})\\
l\cdot s(\overset{+}{xz}) &= \frac{i}{2}(\overset{+}{yz}) - \frac{1}{2}(\overset{-}{x^2-y^2}) - \frac{i}{2}(\overset{-}{xy}) + \frac{\sqrt{3}}{2}(\overset{-}{z^2})\\
l\cdot s(\overset{-}{xz}) &= -\frac{i}{2}(\overset{-}{yz}) + \frac{1}{2}(\overset{+}{x^2-y^2}) - \frac{i}{2}(\overset{+}{xy}) - \frac{\sqrt{3}}{2}(\overset{+}{z^2})\\
l\cdot s(\overset{+}{yz}) &= -\frac{i}{2}(\overset{+}{xz}) + \frac{i}{2}(\overset{-}{x^2-y^2}) - \frac{1}{2}(\overset{-}{xy}) + \frac{i\sqrt{3}}{2}(\overset{-}{z^2})\\
l\cdot s(\overset{-}{yz}) &= \frac{i}{2}(\overset{-}{xz}) + \frac{i}{2}(\overset{+}{x^2-y^2}) + \frac{1}{2}(\overset{+}{xy}) + \frac{i\sqrt{3}}{2}(\overset{+}{z^2})
\end{aligned}
\tag{67}
$$

As an example, the $\langle x^2 - y^{2+}|\xi(r)\boldsymbol{l}\cdot\boldsymbol{s}|xy^+\rangle$ matrix element is evaluated as:

$$\begin{aligned}\langle x^2 - y^{2+}|\xi(r)\boldsymbol{l}\cdot\boldsymbol{s}|xy^+\rangle &= \langle x^2 - y^{2+}|\mathrm{L}_z\,\mathrm{S}_z|xy^+\rangle \\ &= \langle x^2 - y^2|\mathrm{L}_z|xy\rangle\langle +\tfrac{1}{2}|\mathrm{S}_z| + \tfrac{1}{2}\rangle \\ &= \langle(1/\sqrt{2}(2 + -2)|\mathrm{L}_z|(1/i\sqrt{2})(2 - -2)\rangle(1/2) \\ &= -i \end{aligned} \quad (68)$$

The $10 \times 10\langle d|V + \xi(r)\boldsymbol{l}\cdot\boldsymbol{s}|d'\rangle$ matrix can be solved to generate the energies of the d orbitals in a ligand field, inclusive of spin–orbit coupling. The spin–orbit interaction is localized on the metal nucleus; thus covalency which delocalizes the electrons onto the ligands has the effect of a reduction in the magnitude of the spin–orbit coupling relative to the free ion value in Table 7.

3 MANY-ELECTRON LIGAND FIELD THEORY

3.1 Many-Electron Atomic Theory

3.1.1 ***Wavefunctions and Energies*** For a d^n, $1 < n < 9$ free ion, the electrons can be distributed in different combinations of atomic orbitals which have differences in electron-electron repulsion. This produces a number of different states at different energies. For the first and second transition metal series these states are characterized by their many-electron orbital and spin angular momentum and designated by the atomic term symbols, ${}^{2S+1}L$. L is the total orbital angular momentum which can be equal to $0, 1, 2, 3, \ldots$ (designated as $S, P, D, F, \ldots$, respectively). For a given L there are $(2L + 1)$ M_L values $(L, L - 1, \ldots, -L)$. S is the total spin angular momentum which can have half integer or integer (including zero) values depending on whether there are an odd or even number of electrons. For a given S, M_S equals $S, S - 1, \ldots, -S$. This $(2S + 1)$ spin degeneracy is the spin multiplicity of the state.

The atomic term symbols for a given d^n configuration are obtained as follows:

i) determine all microstates (Pauli allowed combinations of m_l, m_s for n electrons) which are designated as shown;

$$\left(m_1^{m_{s1}}, m_2^{m_{s2}}, \ldots, m_n^{m_{sn}}\right)$$

ii) for each microstate obtain $M_L = \Sigma m_l$; $M_S = \Sigma m_s$;

iii) use these microstates to generate the ${}^{2S+1}L$ states where $(2L + 1)(2S + 1)$ microstates are associated with each state. These are usually obtained from a microstate diagram.

Table 8 lists all the microstates corresponding to the d^2 configuration in terms of their total M_L and M_S values. The microstate with the highest M_L or M_S value is chosen. For example, the $(2^+, 1^+)$ microstate is chosen for $M_L = 3$ and $M_S = 1$. Since the highest M_L value is L and the highest M_S is S, this microstate corresponds

Table 8 Two-*d*-Electron Microstate Diagram

M_L \ M_S	1	0	−1
4		$(2^+, 2^-)$	
3	$(2^+, 1^+)$	$(2^+, 1^-)$ $(2^-, 1^+)$	$(2^-, 1^-)$
2	$(2^+, 0^+)$	$(2^+, 0^-)$, $(2^-, 0^+)$ $(1^+, 1^-)$	$(2^-, 0^-)$
1	$(1^+, 0^+)$ $(2^+, -1^+)$	$(1^+, 0^-)$, $(1^-, 0^+)$ $(2^+, -1^-)$ $(2^-, -1^+)$	$(1^-, 0^-)$ $(2^-, -1^-)$
0	$(2^+, -2^+)$ $(1^+, -1^+)$	$(2^+, -2^-)$, $(2^-, -2^+)$ $(1^+, -1^-)(1^-, -1^+)$ $(0^+, 0^-)$	$(2^-, -2^-)$ $(1^-, -1^-)$
−1		Mirrors the above	

to a ^3F state with $M_L = 3$ and $M_S = 1$. The logic here is that this microstate cannot correspond to a component of a state of higher L, such as $L = 4$ (which would also contain $M_L = 3$) since there is no spin triplet microstate with $M_L = 4$ (or higher value). This ^3F state, however, contributes $M_L = 3, 2, 1, 0, -1, -2, -3$ combined with $M_S = 1, 0, -1$ (i.e., 21 microstates) to the diagram. The elimination of the 21 microstates of ^3F from Table 8 leads to a new diagram containing the remaining 24 microstates. The next step is to again identify the term designation for the microstate with the highest remaining M_L (it will be $(2^+, 2^-)$ belonging to ^1G; then remove its 9 microstates) and the process is repeated until all the microstates are eliminated. Note that for the purpose of identifying the terms belonging to the d^n configuration, it does not matter which of the several microstates of given M_L and M_S are eliminated.

The d^2 configuration thus generates ^3F, ^1G, ^1D, ^3P, ^1S states with different interelectronic repulsion energies. Qualitatively, these energies are ordered by Hunds' rules which are based on keeping the electrons apart to minimize electron–electron repulsion. These rules are: 1) the highest spin state is lowest in energy; and 2) for a given S, the highest L is usually lowest in energy. These qualitative rules give the correct ground state (^3F for d^2) but often do not order the other free ion terms correctly.

Indeed, the ground state of any electron configuration can be rapidly derived by identifying the maximum M_L and maximum M_S that it can generate. For example, d^3, the maximum M_L and M_S which can be generated simultaneously are $M_L = 3$ and $M_S = 3/2$ in the microstate $(2^+, 1^+, 0^+)$, hence the ground state is ^4F.

To quantitatively obtain the energies of the excited terms, many electron wavefunctions are required from which electron repulsion (e^2/r_{12}) must be evaluated as a perturbation on these wavefunctions. Many-electron wavefunctions are commonly written in the form of Slater determinants which are products of one-electron functions (Ψ_1, Ψ_2). The wavefunctions are antisymmetric with respect to the exchange of electrons. They are a shorthand notation for:

$$\psi = 1/\sqrt{2}\begin{vmatrix} \Psi_1(1) & \Psi_1(2) \\ \Psi_2(1) & \Psi_2(2) \end{vmatrix}$$

which expands to:

$$\psi = (1/\sqrt{2})[\Psi_1(1)\Psi_2(2) - \Psi_1(2)\Psi_2(1)] \tag{69}$$

where electrons (1) and (2) (which are indistinguishable) may each occupy the one-electron functions ψ_1 or ψ_2. Then, to obtain a d^2 basis state we can write two d^1 bases together, such as combining 2^+ and 1^+ to make $(2^+, 1^+)$. This *Slater determinant* (written in round brackets) expands to:

$$(1/\sqrt{2})[2^+(1)1^+(2) - 1^+(1)2^+(2)] \tag{70}$$

where the numerals in parenthesis (1)(2) are dummy labels to "distinguish" the electrons. Note that since the determinant is antisymmetric with respect to exchange of the columns, $(\Psi_1, \Psi_2) = -(\Psi_2, \Psi_1)$. For more than two electrons, Slater determinants are generated in the following fashion using the three-electron case for demonstration. Consider the determinant $(2^+, 1^-, 0^+)$. First note the standard order, which is decreasing m_l, and for given m_l, then $m_s = +1/2$ before m_s = -1/2. This is expanded as:

$$\begin{aligned}&(1/\sqrt{3!})\left[2^+1^-0^+ + 1^-0^+2^+ + 0^+2^+1^-\right]\\ &\quad - (1/\sqrt{3!})\left[1^-2^+0^+ + 0^+1^-2^+ + 2^+0^+1^-\right]\end{aligned} \tag{71a}$$

or:

$$(1/\sqrt{6})\left[2^+1^-0^+ + 1^-0^+2^+ + 0^+2^+1^- - 1^-2^+0^+ - 0^+1^-2^+ - 2^+0^+1^-\right] \tag{71b}$$

The first bracket of three entries is obtained from the standard order by two permutations, and hence is positive, while the second set of functions is obtained by an odd number of permutations from the standard order, and therefore is negative. Higher number determinants are obtained in a parallel fashion. Note that if were we to contravene the Pauli exclusion principle and place two identical functions in one of these Slater determinants, their expanded form would vanish, that is the first and second terms would cancel out. We now explore how the interelectronic repulsion energies of these Slater determinants are derived.

3.1.2 ***Electron Repulsion Integrals and the Nephelauxetic Series*** Remembering that $|2, 1\rangle = R_{3d}(r_1)Y_2^1(\theta_1, \phi_1)$ etc., and:

$$\frac{1}{r_{12}} = \sum_{l=0}^{\infty}\sum_{m=-l}^{l} \frac{4\pi}{2l+1}\frac{r_<^l}{r_>^{l+1}} Y_l^m(\theta_1, \Phi_1)Y_l^m(\theta_2, \Phi_2) \tag{72}$$

We need to evaluate interelectron repulsion integrals of the general form $\langle ab|e^2/r_{12}|cd\rangle$ using (72) where ab and cd are products of one-electron functions (not Slater determinants). The evaluation of these integrals is achieved by noting first that, similar to Eqns. (15,16), the $Y_l^m(\theta_i, \phi_i)$ can be expressed in terms of normalized Legendre functions.[3,39]

$$Y_l^m(\theta, \phi) = P_l^m(\cos\theta)\frac{1}{\sqrt{2\pi}}e^{im\phi} \tag{73}$$

We now define $c^k(l, m_l, l'm_l')$ and $R^k(abcd)$ as follows:

$$\begin{aligned} c^k(lm_l, l'm_l') &= \sqrt{2/(2k+1)}\int_0^{\pi} P_k^{m_l-m_l'}\cos\theta\, P_l^m\cos\theta\, P_{l'}^{m'}\cos\theta\sin\theta d\theta \\ R^k(abcd) &= e^2\int_0^{\infty}\int_0^{\infty}\frac{r_<^k}{r_>^{k+1}}R_a^1R_b^2R_c^1R_d^2r_1^2dr_1r_2^2dr_2 \end{aligned} \tag{74}$$

where R_a^1 etc. are the radial components of each orbital. It can then be shown[39] that:

$$\begin{aligned} \langle ab|e^2/r_{12}|cd\rangle &= \delta(m_s^a, m_s^c)\delta(m_s^b, m_s^d)\cdot\delta(m_l^a + m_l^b, m_l^c + m_l^d) \\ &\quad\times\sum_{k=0}^{\infty}c^k(l^am_l^al^cm_l^c)c^k(l^dm_l^dl^bm_l^b)\cdot R^k(abcd) \end{aligned} \tag{75}$$

Diagonal matrix elements ($ab = cd$) give rise to two special integrals termed the Coulomb, J, and exchange, K, integral, according to:

$$\begin{aligned} J(ab) &= \langle a(1)b(2)|e^2/r_{12}|a(1)b(2)\rangle = \langle ab|e^2/r_{12}|ab\rangle \\ K(ab) &= \langle a(1)b(2)|e^2/r_{12}|b(1)a(2)\rangle = \langle ab|e^2/r_{12}|ba\rangle \end{aligned} \tag{76}$$

The integrals J and K are always positive. The Kronecker delta functions reveal that for these elements not to vanish, the spins of a and c, and of b and d, should each be the same; and the sums of the m_l values of a and b, and of c and d, should be equal. Table 9 lists values of

$$c^k(lm_l, l'm_l')$$

and also contains

$$a^k(l^am_l^a, l^bm_l^b) = c^k(l^am_l^a, l^am_l^a)c^k(l^bm_l^b, l^bm_l^b)$$

which are useful for J integrals where $ab = cd$. For equivalent electrons $F^k = R^k(abcd)$. And further, $F_2 = F^2/49$ and $F_4 = F^4/441$ (with $F_o = F^0$) which are used to simplify the expressions since the denominators 49 and 441 occur repeatedly (Table 9). The radial integrals are treated as empirical parameters. According to the triangular rule for two d ($l = 2$) electrons only the $l = 0, 2, 4$ terms need be considered in the infinite sum expressed in Eqn. (75). The problem then is to break down a wavefunction into sums of two electron products and evaluate.

Table 9 The $c^k(lm_l, l'm'_l)$ and $a^k(lm_l, l'm'_l)$ Electron Repulsion Parameters for $l, l' = 2$

$c^k(lm_l, l'm'_l)$ for $l, l' = 2,2$

m_l	m'_l	$k = 0$	$k = 2$	$k = 4$
±2	±2	+1	$-\frac{\sqrt{4}}{7}$	$+\frac{1}{21}$
±2	±1	0	$+\frac{\sqrt{6}}{7}$	$-\frac{\sqrt{5}}{21}$
±2	0	0	$-\frac{\sqrt{4}}{7}$	$+\frac{\sqrt{15}}{21}$
±1	±1	+1	$+\frac{1}{7}$	$-\frac{\sqrt{16}}{21}$
±1	0	0	$+\frac{1}{7}$	$+\frac{\sqrt{30}}{21}$
0	0	+1	$+\frac{\sqrt{4}}{7}$	$+\frac{\sqrt{36}}{21}$
±2	∓2	0	0	$+\frac{\sqrt{70}}{21}$
±2	∓1	0	0	$-\frac{\sqrt{35}}{21}$
±1	∓1	0	$-\frac{\sqrt{6}}{7}$	$-\frac{\sqrt{40}}{21}$

$a^k(lm_l, l'm'_l) = a^k(l'm'_l, lm_l)$

$\|m_l\|$	$\|m'_l\|$	$k = 2$	$k = 4$
2	2	4/49	1/441
2	1	−2/49	−4/441
2	0	−4/49	6/441
1	1	1/49	16/441
1	0	2/49	−24/441
0	0	4/49	36/441

$c^k(l'm'_l, lm_l) = (-1)^{m_l - m_{l'}} c^k(lm_l, l'm'_l)$.

For $k = 0$, $a^0(l, m_1, l', m'_1) = F_0$ for all arguments.

Values for other configurations have been determined.[39]

We illustrate this with the $(\Psi(^3F, L, M_L))\Psi(^3F, 3, 1)$ term discussed above. The spin functions can be integrated out (the Kronecker delta functions noted above are unity in this case) and the Slater determinant expanded as:

$$\begin{aligned}\Psi(^3F, 3, 1) &= 1/\sqrt{2}\big[2^+(1)1^+(2) - 1^+(1)2^+(2)\big]\\ &= 1/\sqrt{2}\big[2(1)1(2) - 1(1)2(2)\big]\alpha(1)\alpha(2)\end{aligned}$$

The energy can then be derived via:

$$= 2E_{3d} + 1/2\big\langle 2(1)1(2) - 1(1)2(2)\big|e^2/r_{12}\big|2(1)1(2) - 1(1)2(2)\big\rangle$$

$$\begin{aligned}
&= 2E_{3d} + 1/2[\langle 2(1)1(2)|e^2/r_{12}|2(1)1(2)\rangle + \langle 1(1)2(2)|e^2/r_{12}|1(1)2(2)\rangle \\
&\quad - 2\langle 2(1)1(2)|e^2/r_{12}|1(1)2(2)\rangle] \\
&= 2E_{3d} + 1/2[\langle 21|e^2/r_{12}|21\rangle + \langle 12|e^2/r_{12}|12\rangle - 2\langle 21|e^2/r_{12}|12\rangle] \\
&= 2E_{3d} + J(2,1) - K(2,1) \qquad (77)
\end{aligned}$$

This can be evaluated from the data in Table 9. Thus for $J(2,1)$ we identify $a^k(2,1) = F_0 - 2F_2 - 4F_4$ while for $K(2,1)$ we need $c^k(a,c) \cdot c^k(d,b) = c^k(2,1) \cdot c^k(2,1) = 6F_2 + 5F_4$ and therefore $E_{3_F} = 2E_{3d} + F_0 - 8F_2 - 9F_4$. Note that

$$c^k(l^a m_l^a, l^b m_l^b) = (-1)^{m_l^a - m_l^b} c^k(l^b m_l^b, l^a m_l^a)$$

This procedure can be generalized such that the energy of any n-electron Slater determinant is given by Eqn. (78).[3,40]

$$E(\Psi) = \sum_n h_a + \sum_{m<n} J_{m,n} - \sum_{m<n} K_{m,n} \cdot \delta(s_m, s_n) \qquad (78)$$

where the first term is the sum of the one-electron energies including the kinetic energy and nuclear attraction; $J_{m,n}$ is the Coulombic interaction between electrons m and n; and K is the exchange interaction between m and n which exists only if the spins of n and m are the same (as expressed by the Kronecker delta term).

We have seen then that the problem of defining an energy for a state is broken down into two components, the derivation of the one-electron energies which for the free ion are usually the shielded d orbitals; and the derivation of the interelectronic repulsion, J and K parameters (Eqn. (78), see also Section 3.3. Eqns. (105–110) below, for more detail). Expressions for the energies of the free ion states for the d^2 to d^5 configurations in terms of the F_l radial integrals of electron repulsion are given in Table 10. From the electron/hole formalism $d^{(10-n \text{ holes})}$ is equal to d^{ne^-} therefore, the d^2–d^5 electron repulsion energy expressions are also appropriate for the d^8–d^5 configurations. Finally Figure 10 shows the dependence of the F_l integrals on distance from the nucleus. These integrals are outer orbital properties which can be strongly affected by interactions with the ligands, with the F_4 parameter being least affected by ligand overlap.[41]

An alternative set of radial integral parameters has been given by Racah, and has the advantage of having the energy state splitting with the same spin determined only by the Racah B parameter—specifically $A = F_0 - 49F_4$, $B = F_2 - 5F_4$, and $C = 35F_4$. The F_0 or A parameter affects the average configuration energy but not the energy splitting term. Therefore only two parameters F_2, F_4 or B, C are required to describe the energies splitting of all the many electron states of the free ion. Values for these parameters have been determined experimentally and are given in terms of the Racah parameters in Table 11. From theory using Slater radial functions in Eqn. (74), the C/B ratios should equal 4.0; this is reasonably consistent with the experimental ratios in Table 11. Experimental and theoretical estimates for the electron repulsion integrals over a series of d^2 metal ions are given for F_2 in Figure 11.

Table 10 Free Ion Term Energies for d^n (n = 2–5) Species as a Function of the Slater Condon and Shortley F_0, F_2 and F_4 Parameters.[3] These Expressions Can Also be Used for $d^{6,7,8}$ Except that the Coefficient of the Parameter F_0 Must be Modified for the Number of Electron Pairs

d^2:	$E(^1S) = F_0 + 14F_2 + 126F_4$
	$E(^1G) = F_0 + 4F_2 + F_4$
	$E(^3P) = F_0 + 7F_2 - 84F_4$
	$E(^1D) = F_0 - 3F_2 + 36F_4$
	$E(^3F) = F_0 - 8F_2 - 9F_3$
d^3:	$E(^2P) = 3F_0 - 6F_2 - 12F_4$
	$E(^2_{\pm}D) = 3F_0 + 5F_2 + 3F_4 \pm \sqrt{193F_2^2 - 1650F_2F_4 + 8325F_4^2}$
	$E(^2F) = 3F_0 + 9F_2 - 87F_4$
	$E(^2G) = 3F_0 - 11F_2 + 13F_4$
	$E(^2H) = 3F_0 - 6F_2 - 12F_4$
	$E(^4P) = 3F_0 - 147F_4$
	$E(^4F) = 3F_0 - 15F_2 - 72F_4$
d^4:	$E(^1_{\pm}S) = 6F_0 + 10F_2 + 6\ F_4 \pm 1/2\sqrt{3088F_2^2 - 26{,}400F_2F_4 + 133{,}200F_4^2}$
	$E(^1_{\pm}D) = 6F_0 + 9F_2 - 76.5F_4 \pm 1/2\sqrt{1296F_2^2 - 10{,}440F_2F_4 + 30{,}825F_4^2}$
	$E(^2F) = 6F_0 - 48\ F_4$
	$E(^1_{\pm}G) = 6F_0 - 5F_2 - 6.5F_4 \pm 1/2\sqrt{708F_2^2 - 7500F_2F_4 + 30{,}825F_4^2}$
	$E(^1I) = 6F_0 - 15F_2 - 9\ F_4$
	$E(^3_{\pm}P) = 6F_0 - 5F_2 - 76.5F_4 \pm 1/2\sqrt{912F_2^2 - 9960F_2F_4 + 38{,}025F_4^2}$
	$E(^3D) = 6F_0 - 5F_2 - 129\ F_4$
	$E(^3_{\pm}F) = 6F_0 - 5F_2 - 76.5F_4 \pm 1/2\sqrt{612F_2^2 - 4860F_2F_4 + 20{,}025F_4^2}$
	$E(^3G) = 6F_0 - 12F_2 - 94\ F_4$
	$E(^3H) = 6F_0 - 17F_2 - 69\ F_4$
	$E(^5D) = 6F_0 - 21F_2 - 189\ F_4$
d^5:	$E(^2S) = 10F_0 - 3F_2 - 195F_4$
	$E(^2P) = 10F_0 + 20F_2 - 240F_4$
	$E(^2_{\pm}D) = 10F_0 - 3F_2 - 90F_4 \pm \sqrt{513F_2^2 - 4500F_2F_4 + 20{,}700F_4^2}$
	$E(^2D) = 10F_0 - 4F_2 - 120F_4$
	$E(^2_{\pm}F) = 10F_0 - 9F_2 - 165F_4$
	$E(^2F) = 10F_0 - 25F_2 - 15F_4$
	$E(^2_{\pm}G) = 10F_0 - 3F_2 - 155F_4$
	$E(^2G) = 10F_0 - 13F_2 - 145F_4$
	$E(^2H) = 10F_0 - 22F_2 - 30F_4$
	$E(^2I) = 10F_0 - 24F_2 - 90F_4$
	$E(^4P) = 10F_0 - 28F_2 - 105F_4$
	$E(^4D) = 10F_0 - 18F_2 - 225F_4$
	$E(^4F) = 10F_0 - 13F_2 - 180F_4$
	$E(^4G) = 10F_0 - 25F_2 - 190F_4$
	$E(^4S) = 10F_0 - 35F_2 - 315F_4$

Table 10 ***(Continued)*** **Theoretical Expressions for the Average Energies of the Configuration d^n $D = -(14/9)F_2 - 14F_4$**

Configuration	Average energy, E_{av}
d^1	E_{3d}
d^2	$2E_{3d} + F_0 - D$
d^3	$3E_{3d} + 3F_0 - 3D$
d^4	$4E_{3d} + 6F_0 - 6D$
d^5	$5E_{3d} + 10F_0 - 10D$
d^6	$6E_{3d} + 15F_0 - 15D$
d^7	$7E_{3d} + 21F_0 - 21D$
d^8	$8E_{3d} + 28F_0 - 28D$
d^9	$9E_{3d} + 36F_0 - 36D$

Note: the degeneracy (orbital × spin) weighted average energy of d^n configuration increases as n increases due to electron repulsion.

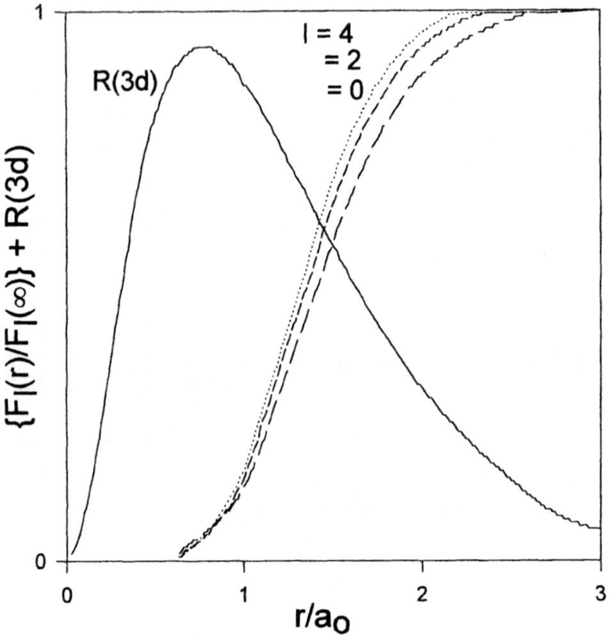

Figure 10. The Slater radial function, R_{3d}, for Ni^{2+}, and the ratio $F_l(r)/F_l(\infty)$ as a function of nucleus-electron distance, r, in atomic units (a.u.) (adapted from[41]).

Electron-electron repulsion increases as Z_{eff} increases. This is a direct demonstration that orbital contraction (associated with an increase in effective nuclear charge) increases electron repulsion.

When the metal ion is placed in a ligand field the values for the electron repulsion parameters are generally found to decrease. Normally the C/B ratio is fixed at the experimental value in Table 11 and B is allowed to vary to fit spectroscopic data (as described in Section 3.3.1.). The parameter $\beta = B_{complex}/B_{free\ ion}$ decreases from 1

Table 11 Free Ion Values of *B* and *C* for Some Gaseous Ions as a Function of Electron Configuration (cm^{-1})[43]

M^{2+}	B	C	C/B	M^{3+}	B	C	C/B
Ti^{2+}	694	2910	4.19	V^{3+}	861	3814	4.43
V	755	3257	4.31	Cr	918	4133	4.50
Cr	810	3565	4.40	Mn	965	4450	4.61
Mn	860	3850	4.48	Fe	1015	4800	4.73
Fe	917	4040	4.41	Co	1065	5120	4.81
Co	971	4497	4.63	Ni	1115	5450	4.89
Ni	1030	4850	4.71				

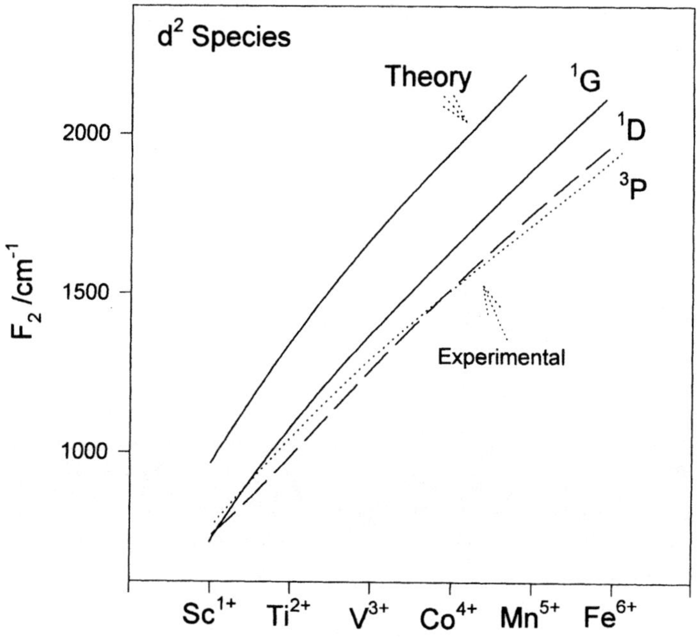

Figure 11. Empirical values of F_2 for various terms of $3d^{2,8}$, as a function of metal ion, compared with theory (adapted from[41]).

as the covalency of the ligand–metal bond increases. The variation in B or β with ligand gives rise to the Nephelauxetic series which reflects the fact that as the electron cloud expands, the electron–electron repulsion decreases. This series may be written with decreasing β from left to right:

$$F^- > H_2O > \text{urea} > NH_3 > \text{en} \approx ox^{2-} > SCN^{*-} > Cl^- \approx CN^- > Br^- > S^{2-} \approx I^- \tag{79}$$

There are considered to be two contributions to the nephelauxetic reduction in electron repulsion. The first is central field covalency which is the dominant contribution. This relates to the fact that the ligands are generally good donors (see Section 2.3.1.) which lower the effective nuclear charge on the metal ion relative to its free ion value. From the above discussion, F_2 and F_4 are proportional to Z_{eff}. The second contribution is called symmetry restricted covalency. Covalency leads to the delocalization of the d electrons onto the ligands, where the σ delocalization is generally greater than π delocalization. This leads to differential radial expansion and is often introduced as differences in β parameters depending upon which orbitals are occupied ($\beta_{5,5}$ (for two electrons in the t_{2g} orbitals) $>$ $\beta_{3,5}$ (one electron in e_g and one in t_{2g}) $>$ $\beta_{3,3}$ (two electrons in the e_g orbitals)). The C/B ratio can also change (usually increases) with covalency due to the relatively larger outer orbital sensitivity of F_2 than F_4 (Figure 10).

The real power of ligand field theory is in its incorporation of experimental estimates of the electron repulsion parameters for coordination compounds. To these we must add the ligand field effects on the d orbital energies as described in Section 2.

To solve the ligand field energies of many-electron states, we need to be able to construct proper many-electron wavefunctions in terms of the one-electron functions and expand energies in terms of $\langle d_i|V|d_j\rangle$. We begin with the weak field approximation which starts with the free ions about whose wavefunctions we know a great deal. Later we consider the strong field approximation which begins with the real d orbitals.

3.2 Weak Field Approach

3.2.1 ***Symmetry of Ligand Field States*** We first consider how the degeneracies of the free ion terms of a d^n configuration are removed by a ligand field, that is, we determine the symmetries of a free ion term in a specific ligand geometry. Thus we recall that the 45 microstates (spinors) of a d^2 configuration yield the spectroscopic terms ^1G, ^1D, ^1S and ^3F and ^3P and that all components of a given degenerate term necessarily have the same interelectronic repulsion energy.[4,39] Writing the electronic repulsion energy of the ground state ^3F term as zero, the energy of ^3P is $15B$. The energies of the spin singlet terms also incorporate the Racah parameter, C. In the weak field approximation, we impose a ligand field upon the free ion terms and consider how their degeneracy is lifted by this field, to generate a set of new energy levels defined by the ligand field parameters and B (and C parameter if low spin states are involved).

To define the group theoretical representations spanned by a particular free ion term in a given point group, it is usually only necessary to ascertain how the free ion functions are affected by the rotations of the group concerned. The transformation properties under rotation of the set of spherical harmonics associated with the orbital part of a free ion term symbol and defined by its L value, for any crystal field symmetry, have been derived by Hans Bethe:[1]

$$
\begin{aligned}
X(\text{rot. by } \alpha) &= \sin\big((L+1/2)\alpha\big)/\sin(\alpha/2) \\
X(E) &= 2L+1
\end{aligned}
\tag{80}
$$

Table 12 Group Theoretical Representations of Free Ion Terms, for d^n Configurations, in some Common Point Groups.[a] Spin is Unaffected and is not Included Here

Free Ion terms	Octahedral[b] O_h	Tetragonal D_{4h}	Trigonal D_{3d}
G	$A_{1g} + E_g + T_{1g} + T_{2g}$	$A_{1g} + 3E_g + A_{2g} + B_{2g}$	$2A_{1g} + 3E_g + A_{2g}$
D	$E_g + T_{2g}$	$A_{1g} + B_{1g} + E_g + B_{2g}$	$2E_g + A_{1g}$
F	$A_{2g} + T_{1g} + T_{2g}$	$B_{1g} + 2E_g + A_{2g} + B_{2g}$	$A_{1g} + 2Eg + 2A_{2g}$
P	T_{1g}	$E_g + A_{2g}$	$A_{2g} + E_g$
S	A_{1g}	A_{1g}	A_{1g}

[a] Correlation tables to determine representations in any other group are available in many sources including refs. 4,36,47

[b] Also applies to tetrahedral, T_d, with omission of the g subscript.

By determining the character for each rotation in the point group, a reducible representation is obtained. By determining the set of irreducible representations comprised within this reducible representation (see Eqn. (61)) one obtains the required group theoretical representations spanned by the particular free ion term in the given point group. These ligand field splittings of the free ion states are listed in Table 12. Note that placing a metal ion into the ligand field of a complex does not affect the spin of the state.

3.2.2 ***Shift Operators, Wavefunctions and Energies*** The symbol $\psi|L, M_L, S, M_S\rangle$ provides a simple way to name the individual microstates of a spectroscopic term. Thus the 21 microstates in ^3F comprise $\psi|3, M_L, 1, M_S\rangle$ with M_L running from $3, 2, \ldots, -3$ and M_S having values $1, 0, -1$. Note the use of capital letters because we are discussing the total many-electron wavefunction. We now want to write these states in terms of one-electron wavefunctions. The Slater determinant wavefunctions attributable to each free ion component wavefunction (e.g., the 21 functions of ^3F) can be readily obtained through application of angular momentum operators, termed the shift, ladder, or raising and lowering operators. The one-electron operators L_z and S_z project out the m_l and m_s values, respectively, of a wavefunction on which they operate, and leave the wavefunction unchanged.

Thus:

$$L_z|2^+\rangle = 2\hbar|2^+\rangle; \quad S_z|2^+\rangle = (1/2)\hbar|2^+\rangle$$

and

$$L_z(2^+, 1^+) = 3\hbar(2^+, 1^+); \quad S_z(2^-, 1^-) = -1\hbar(2^-, 1^-) \tag{81}$$

where the units of angular momentum are $\hbar = h/2\pi$. However, the corresponding L_x, L_y, S_x and S_y operators behave in a rather different fashion and cause the wavefunctions upon which they operate to change. This is the basis of the method used to associate Slater determinants with free ion terms.

Thus:[3]

$$(L_x \pm iL_y)\Phi(m_{l_1}, m_{l_2}, \ldots, m_{l_i}, \ldots, m_{l_n}) = L_{\pm}\Phi(m_{l_1}, m_{l_2}, \ldots, m_{l_i}, \ldots, m_{l_n})$$

$$= \hbar \sum_{i=1}^{n} \sqrt{l_i \pm m_{l_i} + 1)(l_i \mp m_{l_i})}\Phi(m_{l_1}, m_{l_2}, \ldots, m_{l_i} \pm 1, \ldots, m_{l_n}) \quad (82a)$$

$$(S_x \pm iS_y)\Phi(m_{s_1}, m_{s_2}, \ldots, m_{s_i}, \ldots, m_{s_n}) = S_{\pm}\Phi(m_{s_1}, m_{s_2}, \ldots, m_{s_i}, \ldots, m_{s_n})$$

$$= \hbar \sum_{i=1}^{n} \sqrt{(s \pm m_{s_i} + 1)(s \mp m_{s_i})}\Phi(m_{s_1}, m_{s_2}, \ldots, m_{s_i} \pm 1, \ldots, m_{s_n}) \quad (82b)$$

$$(L_x \pm L_y)\Psi|L, M_L, S, M_s\rangle = L_{\pm}\Psi|L, M_L, S, M_s\rangle$$

$$= \hbar\sqrt{(L \pm M_L + 1)(L \mp M_L)}\Psi|L, M_{L\pm 1}, S, M_S\rangle \quad (82c)$$

$$(S_x \pm S_y)\Psi|L, M_L, S, M_s\rangle = S_{\pm}\Psi|L, M_L, S, M_s\rangle$$

$$= \hbar\sqrt{(S \pm M_S + 1)(S \mp M_S)}\Psi|L, M_L, S, M_{S\pm 1}\rangle \quad (82d)$$

Equations (82, a–d) represent the expected behavior of 4 operators, written in short form as, L_+, L_-, S_+, and S_-, towards wavefunctions defined either by their one electron components, $\Phi(m_l, m_s)$ (Slater determinant) or their total angular momentum $\psi|L, M_L, S, M_s\rangle$. Thus when, say, L_+ operates on a wavefunction ψ with angular momentum eigenvalue $M_L\hbar$ it yields a new wavefunction ψ' with an L_z eigenvalue of $(M_L + 1)\hbar$, and similarly for the other operators. The other angular momentum components are unchanged by the operation. These operators provide a means by which, if the Slater determinant is known for one $\psi|L, M_L, S, M_s\rangle$, then all other $\psi|L, M'_L, S, M_s\rangle$ can be derived.

For example, we know that the highest (largest M_L, M_S) microstate of d^2 is (from Table 8):

$$\psi|L, M_L, S, M_s\rangle = \psi|3, 3, 1, 1\rangle = (2^+, 1^+) \quad (83)$$

Then, to obtain $\psi|3, 2, 1, 1\rangle$ we need only to operate on both sides of Eqn. (83) with L_-. (We will drop the $\hbar$ units for simplicity, as they will cancel.)

$$L_-\psi|3, 3, 1, 1\rangle = L_-(2^+, 1^+) \quad (84)$$

From Eqn. (82a,c), operate on each electron of the right-hand side, in turn and use $l = 2$ (d-electron) and $m_l = 2$ and then 1:

$$\sqrt{6}\psi|3, 2, 1, 1\rangle = \sqrt{4}(1^+, 1^+) + \sqrt{6}(2^+, 0^+) = \sqrt{6}(2^+, 0^+)$$

$$\psi(3, 2, 1, 1) = (2^+, 0^+) \quad (85)$$

Table 13 The set of ^3F, ^3P, ^1G, ^1D, and ^1S d^2 Wavefunctions in terms of Linear Combinations of Two-Electron Slater Determinants

^3F

$$|3, 3, 1, 1\rangle = -(2^+, 1^+)$$
$$|3, 2, 1, 1\rangle = (2^+, 0^+)$$
$$|3, 1, 1, 1\rangle = -\left[\sqrt{(3/5)}(2^+, -1^+) + \sqrt{(2/5)}(1^+, 0^+)\right]$$
$$|3, 0, 1, 1\rangle = \sqrt{(1/5)}(2^+, -2^+) + \sqrt{(4/5)}(1^+, -1^+)$$
$$|3, -1, 1, 1\rangle = \sqrt{(2/5)}(0^+, -1^+) + \sqrt{(3/5)}(1^+, -2^+)$$
$$|3, -2, 1, 1\rangle = (0^+, -2^+)$$
$$|3, -3, 1, 1\rangle = (-1^+, -2^+)$$

^3P

$$|1, 1, 1, 1\rangle = -\left[\sqrt{(2/5)}(2^+, -1^+) - \sqrt{(3/5)}(1^+, 0^+)\right]$$
$$|1, 0, 1, 1\rangle = \sqrt{(4/5)}(2^+, -2^+) - \sqrt{(1/5)}(1^+, -1^+)$$
$$|1, -1, 1, 1\rangle = \sqrt{(2/5)}(1^+, -2^+) - \sqrt{(3/5)}(0^+, -1^+)$$

^1G

$$|4, 4, 0, 0\rangle = (2^+, 2^-)$$
$$|4, 3, 0, 0\rangle = -(1/\sqrt{2})\left[(2^+, 1^-) - (2^-, 1^+)\right]$$
$$|4, 2, 0, 0\rangle = \sqrt{(3/14)}\left[(2^+, 0^-) - (2^-, 0^+)\right] + \sqrt{(4/7)}(1^+, 1^-)$$
$$|4, 1, 0, 0\rangle = -\sqrt{(1/14)}\left[(2^+, -1^-) - (2^-, -1^+)\right] - \sqrt{(3/7)}\left[(1^+, 0^-) - (1^-, 0^+)\right]$$
$$|4, 0, 0, 0\rangle = \sqrt{(1/70)}\left[(2^+, -2^-) - (2^-, -2^+)\right] + \sqrt{(8/35)}\left[(1^+, -1^-) - (1^-, -1^+)\right] + \sqrt{(18/35)}(0^+, 0^-)$$
$$|4, -1, 0, 0\rangle = \sqrt{(1/14)}\left[(1^+, -2^-) - (1^-, -2^+)\right] + \sqrt{(3/7)}\left[(0^-, -1^+) - (0^+, -1^-)\right]$$
$$|4, -2, 0, 0\rangle = \sqrt{(3/14)}\left[(0^+, -2^-) - (0^-, -2^+)\right] + \sqrt{(4/7)}(-1^+, -1^-)$$
$$|4, -3, 0, 0\rangle = \sqrt{(1/\sqrt{2})}\left[(-1^+, -2^-) - (-1^-, -2^+)\right]$$
$$|4, -4, 0, 0\rangle = (-2^+, -2^-)$$

^1D

$$|2, 2, 0, 0\rangle = \sqrt{(2/7)}\left[(2^+, 0^-) - (2^-, 0^+)\right] - \sqrt{(3/7)}(1^+, 1^-)$$
$$|2, 1, 0, 0\rangle = -\sqrt{(3/7)}\left[(2^+, -1^-) - (2^-, -1^+)\right] - \sqrt{(1/14)}\left[(1^-, 0^+) - (1^+, 0^-)\right]$$
$$|2, 0, 0, 0\rangle = \sqrt{(2/7)}\left[(2^+, -2^-) - (2^-, -2^+)\right] - \sqrt{(2/7)}(0^+, 0^-) + \sqrt{(1/14)}\left[(1^+, -1^-) - (1^-, -1^+)\right]$$
$$|2, -1, 0, 0\rangle = \sqrt{(3/7)}\left[(1^+, -2^-) - (1^-, -2^+)\right] + \sqrt{(1/14)}\left[(0^-, -1^+) - (0^+, -1^-)\right]$$
$$|2, -2, 0, 0\rangle = \sqrt{(2/7)}\left[(0^+, -2^-) - (0^-, -2^+)\right] - \sqrt{(3/7)}(-1^+, -1^-)$$

^1S

$$|0, 0, 0, 0\rangle = \sqrt{(1/5)}\left[(2^+, -2^-) - (2^-, -2^+) - (1^+, -1^-) + (1^-, -1^+) + (0^+, 0^-)\right]$$

where the first term on the right vanishes because of the Pauli exclusion principle. Since there is only one microstate with $M_L = 2$, $M_s = 1$, this result could have been obtained directly by inspection of Table 8. Continue to use L_-:

$$L_-\psi|3, 2, 1, 1\rangle = L_-(2^+, 0^+)$$
$$\sqrt{10}\psi|3, 1, 1, 1\rangle = \sqrt{4}(1^+, 0^+) + \sqrt{6}(2^+, -1^+)$$
$$\psi|3, 1, 1, 1\rangle = \sqrt{(2/5)}(1^+, 0^+) + \sqrt{(3/5)}(2^+, -1^+) \qquad (86)$$

Note this expression is automatically normalized (the sum of the squares of the contribution of each determinant to $\psi|3, 1, 1, 1\rangle$ is unity) by this procedure. As a further example consider the S_- operator:

$$\begin{aligned} S_-\psi|3, 2, 1, 1\rangle &= S_-(2^+, 0^+) \\ \sqrt{2}\psi|3, 2, 1, 0\rangle &= (2^-, 0^+) + (2^+, 0^-) \\ \psi|3, 2, 1, 0\rangle &= (1/\sqrt{2})[(2^-, 0^+) + (2^+, 0^-)] \end{aligned} \tag{87}$$

The sum of the M_L and M_S values for all the Slater determinant wavefunctions in the linear combination sum to the values in the $\psi|L, M_L, S, M_s\rangle$ function. In this fashion, the shift operators may be used to derive all the required functions of this or any d^n configuration. The results for d^2 are shown in Table 13.

To obtain the ligand field energies of any of these wavefunctions, the appropriate secular determinant must be derived. Since the ligand field operator has no effect upon spin, the ligand field energies of $\psi|3, 1, 1, 1\rangle$, $\psi|3, 1, 1, 0\rangle$ and $\psi|3, 1, 1, -1\rangle$ are all the same and so only one of each set of three need be evaluated; for simplicity we choose the $M_S = 1$ component. Thus, in the case of ^3F, rather than dealing with 21 microstates, only 7 have to be derived. The problem then is to create a 7×7 matrix (or in general a $(2L+1) \times (2L+1)$ matrix) evaluating the expected values of all the diagonal and off-diagonal elements of the general form

$$\langle \psi(L, M_L, S, M_S)|V|\psi(L, M_L', S, M_s)\rangle$$

Where there is more than one term with the same spin, such as the ^nF and ^nP terms of d^2, d^3, d^7 and d^8, the matrix must be expanded (to 10×10 in these examples) to include the cross terms $\langle L, M_L, S, M_S|V|L', M_L', S, M_s\rangle$. For example for the high spin states of d^2, in addition to evaluating the 7 ($M_s = 1$) components of $^3F\langle\psi(3, M_L, 1, 1)|V|\psi(3, M_L', 1, 1)\rangle$, the 3 ($M_s = 1$) components of $^3P\langle\psi(1, M_L, 1, 1)|V|\psi(1, M_L', 1, 1)\rangle$ must be found as must the $^3F/^3P$ cross terms. Since from Section 2 we know how to derive the values of the matrix elements $\langle d_i|V|d_j\rangle$ for any ligand field, then values for these $\langle L, M_L, S, M_S|V|L', M_L', S, M_s\rangle$ matrix elements can be derived by factoring them into sums of one-electron $\langle d_i|V|d_j\rangle$ elements. As an example (from Table 13):

$$\begin{aligned} &\langle\psi(3, 1, 1, 1)|V|\psi(3, 1, 1, 1)\rangle \\ &\quad = \langle\sqrt{(2/5)}(1^+, 0^+) + \sqrt{(3/5)}(2^+, -1^+)|V|\sqrt{(2/5)}(1^+, 0^+) \\ &\qquad + \sqrt{(3/5)}(2^+, -1^+)\rangle \\ &\quad = (2/5)\langle(1^+, 0^+)|V|(1^+, 0^+)\rangle + (3/5)\langle(2^+, -1^+)|V|(2^+, -1^+)\rangle \\ &\qquad + 2\sqrt{6}/5\langle(1^+, 0^+)|V|(2^+, -1^+)\rangle \end{aligned} \tag{88}$$

The many-electron matrix element must be expanded as a sum of one-electron elements according to the general expansion for one-electron operators:

$$\begin{aligned} \langle abc|V|abc\rangle &= \langle a|V|a\rangle\langle b|b\rangle\langle c|c\rangle + \langle b|V|b\rangle\langle a|a\rangle\langle c|c\rangle + \langle c|V|c\rangle\langle a|a\rangle\langle b|b\rangle \\ &= \langle a|V|a\rangle + \langle b|V|b\rangle + \langle c|V|c\rangle \end{aligned} \tag{89}$$

Hence one sequentially integrates out the orbitals not being operated upon as overlap terms which are zero or unity depending upon whether the orbitals are different or identical, respectively. The corresponding spin functions will also be integrated out, viz:

$$\langle a^+b^-c^+|V|a^+b^-c^+\rangle = \langle abc|V|abc\rangle\langle +|+\rangle\langle -|-\rangle\langle +|+\rangle$$

as Kronecker delta functions. Then it follows that:

$$\begin{aligned}\langle abc|V|abd\rangle &= \langle a|V|a\rangle\langle b|b\rangle\langle c|d\rangle + \langle b|V|b\rangle\langle a|a\rangle\langle c|d\rangle + \langle c|V|d\rangle\langle a|a\rangle\langle b|b\rangle \\ &= \langle c|V|d\rangle \end{aligned} \tag{90}$$

The orbitals should be ordered so that they can be integrated out in turn. Usually one is dealing with Slater determinants so that $\langle(acb)|V|(abd)\rangle = -1\langle(abc)|V|(abd)\rangle$. It follows that any many-electron element of a one-electron operator, where the two wavefunctions concerned differ by more than one spin-orbital will vanish identically. In expanding these one-electron elements in this straightforward mathematical fashion we can temporarily set aside the notion that we are actually dealing with Slater determinants, aside from recalling their anti-symmetric nature. If we were to expand each of the kets into their full form according to Eqn. (69), we would double the total number of terms and multiply by (1/2)—hence the result would be the same.

Then, returning to Eqn. (88), and expanding the last line:

$$\begin{aligned}&\langle\psi(3,1,1,1)|V|\psi(3,1,1,1)\rangle \\ &\quad = (2/5)\big[\langle 1|V|1\rangle\langle 0|0\rangle\langle +|+\rangle + \langle 0|V|0\rangle\langle 1|1\rangle\langle +|+\rangle\big] \\ &\qquad + (3/5)\big[\langle 2|V|2\rangle\langle -1|-1\rangle\langle +|+\rangle + \langle -1|V|-1\rangle\langle 2|2\rangle\langle +|+\rangle\big] \\ &\qquad + 2\sqrt{6}/5\big[\langle 1|V|2\rangle\langle 0|-1\rangle\langle +|+\rangle + \langle 0|V|-1\rangle\langle 1|2\rangle\langle +|+\rangle\big] \\ &\quad = (2/5)\big[\langle 1|V|1\rangle + \langle 0|V|0\rangle\big] + (3/5)\big[\langle 2|V|2\rangle + \langle -1|V|-1\rangle\big] \end{aligned} \tag{91}$$

where the overlap terms have been identified as unity or zero. Note that both of the ligand field matrix elements in the cross term vanish because the overlap terms are zero. These individual terms may now be evaluated through the use of any of the ligand field models previously introduced, such as writing them in terms of the crystal field radial integrals, ligand field Dq, Ds, Dt parameters, or AOM parameters etc. The complete matrices can be obtained, for example, for d^2, by expanding out these elements for all the wavefunctions in Table 13. For the D_{4h} Hamiltonian (11, 17), and the results in Eqn. (33), the last line of Eqn. (91) expands to:

$$\begin{aligned}&\langle\psi(3,1,1,1)|V|\psi(3,1,1,1)\rangle \\ &\quad = (2/5)\big[(-4Dq - Ds + 4Dt) + (6Dq - 2Ds - 6Dt)\big] \\ &\qquad + (3/5)\big[(Dq + 2Ds - Dt) + (-4Dq - Ds + 4Dt)\big] \\ &\langle\psi(3,1,1,1)|V|\psi(3,1,1,1)\rangle = -Dq - (3/5)Ds + Dt \end{aligned} \tag{92}$$

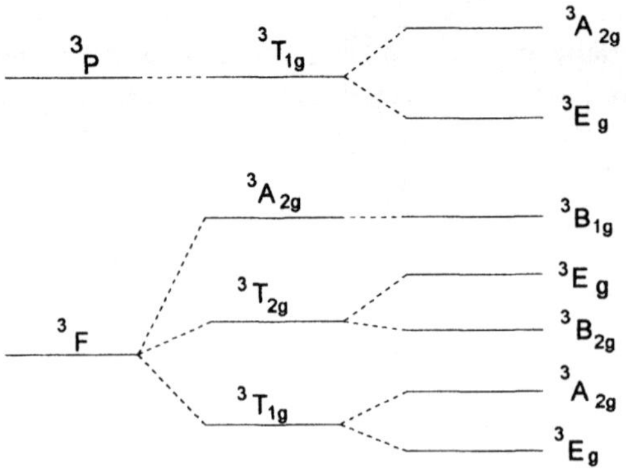

Figure 12. A state splitting diagram for d^2 showing (left) the free ion high spin spectroscopic states, (middle) splitting in an octahedral (O_h) field and (right) splitting in a tetragonal (D_{4h}) field. The sequence of representations in the D_{4h} case is arbitrary and will depend upon the σ, π-bonding characteristics of the complex concerned.

In a similar fashion, the ligand field energies for all the components of the ${}^3P$ and ${}^3F$ manifolds, including cross terms, can be derived and the solution to the resulting 10×10 secular determinant would generate the ligand field energies of these high spin terms, in this case in the D_{4h} environment.

In dealing with the weak field approach, the inclusion of the interelectronic repulsion energies is trivial. All components of a given spectroscopic term have the same interelectronic repulsion energy. The ${}^3P$ term has the energy $15B$ relative to ${}^3F$ as zero. To include interelectronic repulsion in our analysis then only requires that we add $15B$ to all the (three) diagonal elements of the ${}^3P|1, M_L, 1, 1\rangle$ part of our 10×10 matrix. The complete matrix is shown below (Table 14).

From Table 12 the ${}^3P$ splits into ${}^3A_2 + {}^3E$ states in a D_{4h} complex and the ${}^3F$ into ${}^3A_2 + {}^3B_1 + {}^3B_2 + 2{}^3E$ states. Diagonalization of the matrix in Table 14 gives the energy splitting of these states in an O_h ($Ds = Dt = 0$) and D_{4h} complex shown in Figure 12.

The above procedure to obtain the energies of the high spin d^2 states in a D_{4h} field is cumbersome because first we must determine how the $\psi|L, M_L, S, M_S\rangle$ kets are broken down into the Slater determinants and then subsequently compute their energies. These two steps can be usefully combined by using the Wigner–Eckart theorem.

3.2.3 *Many Electron Wavefunctions, Energies and the Wigner–Eckart Theorem*

3.2.3.1 *Wavefunctions* The 3-j symbol has the useful property of being able to uncouple many-electron states (labeled with capital letters) into one-electron states (labeled with small letters) or indeed to do the reverse of coupling one-electron states into many-electron states, via:[5,10,13]

$$|j_1 j_2 J M\rangle = \sum_{m_1, m_2} |j_1 j_2 m_1 m_2\rangle\langle j_1 j_2 m_1 m_2 | j_1 j_2 J M\rangle \tag{93a}$$

Table 14 Matrix Elements $\langle (L, M_L, 1, 1)|V|(L', M'_L, 1, 1)\rangle$ for the D_{4h} Hamiltonian (Eqn. (11))

	$\|3311\rangle$	$\|3211\rangle$	$\|3111\rangle$	$\|3011\rangle$	$\|3{-}111\rangle$	$\|3{-}211\rangle$	$\|3{-}311\rangle$	$\|1111\rangle$	$\|1011\rangle$	$\|1{-}111\rangle$
$\|3311\rangle$	$-3Dq + Ds$ $+3Dt - E$				$-\sqrt{15}Dq$					$-\sqrt{10}Dq$
$\|3211\rangle$		$7Dq - 7Dt$ $-E$				$5Dq$				
$\|3111\rangle$			$-Dq - 0.6Ds$ $+Dt - E$				$\sqrt{15}Dq$	$\sqrt{6}(Dt - Dq$ $+(4/5)Ds)$		
$\|3011\rangle$				$-6Dq - 0.8Ds$ $+6Dt - E$					$4Dq + 2.4Ds$ $-4Dt$	
$\|3{-}111\rangle$	$-\sqrt{15}Dq$				$-Dq - 0.6Ds$ $+Dt - E$					$-\sqrt{6}Dq + 4\sqrt{6}/5Ds$ $+\sqrt{6}Dt$
$\|3{-}211\rangle$		$5Dq$				$7Dq - 7Dt$ $-E$				
$\|3{-}311\rangle$			$\sqrt{15}Dq$				$-3Dq + Ds$ $+3Dt - E$	$-\sqrt{10}Dq$		
$\|1111\rangle$			$\sqrt{6}(Dt - Dq$ $+(4/5)Ds)$				$-\sqrt{10}Dq$	$-1.4Ds + 15B$ $-E$		
$\|1011\rangle$				$4Dq + 2.4Ds$ $-4Dt$					$2.8Ds + 15B$ $-E$	
$\|1{-}111\rangle$	$-\sqrt{10}Dq$				$-\sqrt{6}Dq + 4\sqrt{6}/5Ds$ $+\sqrt{6}Dt$					$-1.4Ds + 15B$ $-E$

$$|j_1 j_2 m_1 m_2\rangle = \sum_{J,M} |j_1 j_2 J M\rangle\langle j_1 j_2 J M | j_1 j_2 m_1 m_2\rangle \tag{93b}$$

Eqn. (93a) shows which one-electron kets $|j_1 j_2 m_1 m_2\rangle$ contribute to a many electron ket defined by $|JM\rangle$, while Eqn. (93b) shows how the one-electron ket $|j_1 j_2 m_1 m_2\rangle$ can be written as a linear combination of many-electron kets $|JM\rangle$. Once again the "J" and "j" symbols can represent J, L, or S, or j, l, or s, respectively. Note these kets are not Slater determinants. The term $\langle jjJM|jjmm\rangle$ is called a "vector coupling coefficient" or "Clebsch–Gordan coefficient" and is tabulated in many places. Its numerical value is obtained through Eqn. (94) where the Wigner 3-j symbols are given in Appendix 2.

$$\langle j_1 j_2 m_1 m_2 | j_1 j_2 J M\rangle = (-1)^{j_2-j_1-M}(2J+1)^{1/2}\begin{pmatrix} j_1 & j_2 & J \\ m_1 & m_2 & -M \end{pmatrix} \tag{94}$$

An example will suffice to show how these expressions Eqns. (93, 94) are employed. Returning to our $d^2\ {}^3F$ term, consider again the component $\psi|L, M_L, S, M_S\rangle = |3111\rangle$. Using Eqn. (93a) we need to sum over all values of the individual electron m_l values, m_1 and m_2 which yield, in this case, $M_L = 1$. We can achieve a total M_L of 1 from the two-electron Slater determinants $(1^+, 0^+)$ and $(2^+, -1^+)$. Evaluation of Eqn. (93) will show us just which combination of these functions comprises $|3, 1, 1, 1\rangle$; thus we expand Eqns. (93, 94) over the sum of $m_1 m_2 = 1, 0$ and $m_1 m_2 = 2, -1$. In this case j_1 and j_2 are l_1 and $l_2 = 2$ (d electron), $J = L = 3$ and $M_J = M_L = 1$, that is, $|j_1 j_2 J M\rangle = |2231\rangle$

$$\begin{aligned}
|2231\rangle &= (-1)^{2-2-1}\sqrt{7}\left[\begin{pmatrix} 2 & 2 & 3 \\ 1 & 0 & -1 \end{pmatrix}|2,1\rangle|2,0\rangle + \begin{pmatrix} 2 & 2 & 3 \\ 0 & 1 & -1 \end{pmatrix}|2,0\rangle|2,1\rangle\right. \\
&\quad \left. + \begin{pmatrix} 2 & 2 & 3 \\ 2 & -1 & -1 \end{pmatrix}|2,2\rangle|2,-1\rangle + \begin{pmatrix} 2 & 2 & 3 \\ -1 & 2 & -1 \end{pmatrix}|2,-1\rangle|2,2\rangle\right] \\
|2231\rangle &= -\sqrt{7}\big[\sqrt{(1/35)}|2,1\rangle|2,0\rangle - \sqrt{(1/35)}|2,0\rangle|2,1\rangle + \sqrt{(3/70)}|2,2\rangle|2,-1\rangle \\
&\quad - \sqrt{(3/70)}|2,-1\rangle|2,2\rangle\big] \\
&= -\big[\sqrt{(2/5)}(1^+,0^+) + \sqrt{(3/5)}(2^+,-1^+)\big]
\end{aligned} \tag{95}$$

where in the last line, the Slater determinant version of the previous line has been inserted—recall the $1/\sqrt{2}$ factor. This result agrees with Eqn. (86) except that the negative of the function is realized here. Pay special attention to the phase generated by the (-1) factors. The expansion Eqn. (94) is readily generalized into dealing with as many electrons as needed. Thus we can write any $|L, M_L, S, M_S\rangle$ microstate in terms of one-electron functions by proceeding through this analysis. The results for the $M_S = 1$ components of all the terms of the d^2 configuration have been given in Table 13.

Table 15 Unitary Matrix Elements $[\langle d^2\,{}^nL\|U_k\|d^2\,{}^nL\rangle]^2$

		U_2 Matrix Elements			
^3P	^3P	*21/25	^1D	^1D	*9/49
	^3F	24/25		^{l}G	144/245
^3F	^3F	6/25	^{l}G	^{l}G	198/49
^1S	^1D	4/5			
		U_3 Matrix Elements			
^3P	^3F	6/5	^1D	^1D	*64/49
^3F	^3F	*3/5		^1G	90/49
			^1G	^1G	99/49
		U_4 Matrix Elements			
^3P	^3F	*2/5	^1D	^1D	16/49
^3F	^3F	*11/5		^1G	110/49
^1S	^1G	4/5	^1G	^1G	143/245

Unitary elements are listed as their square. An asterisk indicates that the negative root should be used.[10]

3.2.3.2 *Many-Electron Energies* We are now interested in evaluating the crystal field or some other tensor operator, $O_k{}^q$ over the many electron wavefunctions $|l^n LM_L\rangle$ where n is the number of electrons in orbitals of angular momentum l ($l = 2$ for d electrons) giving a many-electron state characterized by L, M_L. From the Wigner–Eckart theorem (Eqn. (19)) applied to a many-electron matrix element:

$$\left\langle l^n LM_L \middle| O_k{}^q \middle| l^n L'M_L' \right\rangle = (-1)^{L-M} \begin{pmatrix} L & k & L' \\ -M & q & M_L' \end{pmatrix} \left\langle l^n L \middle\| O_k \middle\| l^n L' \right\rangle \tag{96}$$

We need to evaluate the many-electron reduced matrix element on the right-hand side of Eqn. (96). A many-electron operator and its reduced matrix element can be related to one-electron operators and reduced matrix elements:[42]

$$\langle \| O_k \| \rangle = \left\langle \left\| \sum_i o_k(i) \right\| \right\rangle = A \left\langle \left\| \sum_i u_k(i) \right\| \right\rangle = A\langle \| U_k \| \rangle \tag{97}$$

which in turn relate this to the unit tensor operator u_k. Different operators of the same rank are equal to a multiple, A, of the unit tensor operator. Values for the reduced matrix elements of the unit tensor operator for the different p^n, d^n and f^n configurations are tabulated[10] with values for the $\langle d^2, L\|U_k\|L', d^2\rangle$ reduced matrix element being reproduced in Table 15. The value of the proportionality coefficient A is obtained by evaluating a single one-electron matrix element of the one-electron operator, that is, $A = \langle l \| O_k(i) \| l' \rangle$. For our ligand C_k^q operators in Eqn. (15), A is given by Eqn. (98):

$$A = (-1)^l \left((2l+1)(2l'+1)\right)^{1/2} \begin{pmatrix} l & k & l' \\ 0 & 0 & 0 \end{pmatrix} \tag{98}$$

Thus the final expression for the many-electron ligand field matrix element (Eqn. (96)) in tensor operator form is:

$$\begin{aligned} \langle l^n L M_L | C_k{}^q | l^n L' M_L' \rangle &= (-1)^{L-M+l}(2l+1) \begin{pmatrix} l & k & l \\ 0 & 0 & 0 \end{pmatrix} \begin{pmatrix} L & k & L' \\ -M & q & M' \end{pmatrix} \\ &\quad \times \langle l^n L \| U_k \| l^n L' \rangle \end{aligned} \tag{99}$$

Consideration of this expression shows that we can go directly to the energies of any term component without the need first to deconstruct it into one-electron functions. This expression is cast in terms of the C_q^k operators but may be re-expressed in terms of AOM operators.[29]

We use Eqn. (99) to directly calculate the off-diagonal element $\langle 3111 | V_{D4h} | 1111 \rangle$ in Table 14. We need the various $\langle d^2 L M_L | C_k{}^q | d^2 L' M_L | ' \rangle$ elements contributing to Hamiltonian (17).

Thus:

$$\begin{aligned} \langle 31 | C_2{}^0 | 11 \rangle &= (-1)^{3-1+2}(5) \begin{pmatrix} 2 & 2 & 2 \\ 0 & 0 & 0 \end{pmatrix} \begin{pmatrix} 3 & 2 & 1 \\ -1 & 0 & 1 \end{pmatrix} \langle d^2\,{}^3F \| U_2 \| d^2\,{}^3P \rangle \\ \langle 31 | C_2{}^0 | 11 \rangle &= (5) \times -\surd(2/35) \times \surd(2/35) \times \surd(24/25) = -(4/35)\surd 6 \\ \langle 31 | C_4{}^0 | 11 \rangle &= (-1)^{3-1+2}(5) \begin{pmatrix} 2 & 4 & 2 \\ 0 & 0 & 0 \end{pmatrix} \begin{pmatrix} 3 & 4 & 1 \\ -1 & 0 & 1 \end{pmatrix} \langle d^2\,{}^3F \| U_4 \| d^2\,{}^3P \rangle \\ \langle 31 | C_4{}^0 | 11 \rangle &= (5) \times \surd(2/35) \times \surd(1/42) \times -\surd(2/5) = -\surd(2/147) \end{aligned} \tag{100}$$

The $C_4{}^4$ operator cannot couple these components.

Hence:

$$\begin{aligned} \langle 31 | V_{D_{4h}} | 11 \rangle &= 21Dq \times -\surd(2/147) - 21Dt \times -\surd(2/147) - 7Ds \times -(4/35)\surd 6 \\ &= -\surd 6 Dq + \surd 6 Dt + (4/5)\surd 6 Ds \\ &= \surd 6(Dt - Dq + (4/5)Ds) \end{aligned} \tag{101}$$

obtained readily in a single step without requiring explicit knowledge of the d^2 wave-functions.

Similarly,

$$\begin{aligned} \langle 33 | V_{D_{4h}} | 1-1 \rangle &= 21Dq\left[0 + \surd(5/14)\langle d^2 33 | C_4{}^4 | d^2 1-1 \rangle\right] \\ \langle 33 | V_{D_{4h}} | 1-1 \rangle &= 21Dq \times \surd(5/14)(-1)^2(5) \begin{pmatrix} 2 & 4 & 2 \\ 0 & 0 & 0 \end{pmatrix} \begin{pmatrix} 3 & 4 & 1 \\ -3 & 4 & -1 \end{pmatrix} \\ &\quad \times (-\surd(2/5)) \end{aligned}$$

$$\langle 33|V_{D_{4h}}|1-1\rangle = 21Dq \times \surd(5/14)(1)(5) \times \surd(2/35) \times \surd(1/9) \times (-\surd(2/5))$$
$$= -\surd 10Dq \quad (102)$$

These matrix elements have been included in Table 14. This has been a brief introduction to the Wigner–Eckart theorem and vector coupling coefficients. It can be extended to a wide variety of quantum-mechanical problems including the solution of spin–orbit coupling matrices. It may also be recast in a symmetry based format such that the Clebsch–Gordan coefficients have a form such as

$$\langle \Gamma_1\gamma_1, \Gamma_2\gamma_2|\Gamma\gamma\rangle$$

describing the coupling of wavefunctions represented by their group theoretical labels.[7,40]

3.3 Strong Field Approach

There is perhaps a greater fluency in using real d orbitals (Eqn. (34)) since one can readily imagine them and write them down. In the strong field approach one starts with a one-electron splitting diagram and adds the appropriate number of electrons. Again using our D_{4h} example, we would add, as an example, two electrons to the lowest available levels as shown in Figure 2.

In the weak field approach, the inclusion of interelectronic repulsion energy is essentially trivial, but the generation of ligand field energies is more complex; in the strong field approach, the reverse is true, the derivation of ligand field energies becomes trivial, but the interelectronic contributions are more burdensome to derive.

In the D_{4h} example shown in Figure 2, the total ligand field energy is simply obtained by adding the individual electron occupation energies; for the lowest energy ^3E_g in D_{4h}, this is $E(b_{2g}) + E(e_g)$. Any model may be used so that these energies could be expressed for example in AOM terms or using the ligand field radial parameters. Thus, for this configuration $(b_{2g})^1(e_g)^1$

$$E\left((b_{2g})^1(e_g)^1\right) = -8Dq + Ds + 3Dt \quad (103)$$

(one-electron energies only, from (35)). To this must be added the effects of electron repulsion:

$$E\left((b_{2g})^1(e_g)^1\right) = -8Dq + Ds + 3Dt + \langle d^{2\,3}E_g|e^2/r_{ij}|d^{2\,3}E_g\rangle \quad (104)$$

(in absence of mixing with other ^3E states).

The many-electron excited states are derived by sequentially exciting one or both electrons from the ground state, $^3A_{2g}$, to higher levels, generating states whose symmetries are the direct products of the representations of the occupied orbitals (or obtained using descent in symmetry if there are more than one electron or hole in a degenerate set of orbitals[4,36]). Their ligand field contribution to the energy is also trivially obtained by summing the one-electron energies of the occupied orbitals.

To obtain the interelectronic repulsion energy, matrix elements of the type $\langle d^{2\,3}E_g|e^2/r_{ij}|d^{2\,3}E_g\rangle$ in Eqn. (104) need to be evaluated. The first step in the evaluation of such elements is to construct wavefunction symmetry adapted to the group concerned. This can be done formally using an appropriate projection operator to project out the required functions from the bases (d_i, d_j).[36] However, in most cases, especially in the lower point groups, the required functions can be obtained by inspection.

Where a state symmetry only occurs once, its wavefunction is trivially the product of the one-electron wavefunctions involved. For example, the excited state $(b_{1g})^1(b_{2g})^1$ transforming as ${}^3A_{2g}$ is simply the Slater determinant (x^2-y^{2+}, xy^+).

Before attempting to derive the D_{4h} state energies, consider a general Slater determinant (a^+, b^+) whose interelectronic repulsion energy is given by the element

$$\langle (a^+, b^+)|e^2/r_{ij}|(a^+, b^+)\rangle$$

The determinant is expanded giving the matrix element (analogous to Eqns. (77, 78)):

$$\begin{aligned}
&\langle (a^+, b^+)|e^2/r_{ij}|(a^+, b^+)\rangle \\
&\quad = \tfrac{1}{2}\langle a^+(1)b^+(2) - b^+(1)a^+(2)|e^2/r_{ij}|a^+(1)b^+(2) - b^+(1)a^+(2)\rangle \\
&\quad = \tfrac{1}{2}[\langle a^+(1)b^+(2)|e^2/r_{ij}|a^+(1)b^+(2)\rangle + \langle b^+(1)a^+(2)|e^2/r_{ij}|b^+(1)a^+(2)\rangle \\
&\qquad - 2\langle a^+(1)b^+(2)|e^2/r_{ij}|b^+(1)a^+(2)\rangle] = J(a,b) - K(a,b) \qquad (105)
\end{aligned}$$

and for a determinant (a^+, b^-) where the spins are opposed:

$$\begin{aligned}
&\langle (a^+, b^-)|e^2/r_{ij}|(a^+, b^-)\rangle \\
&\quad = \tfrac{1}{2}\langle a^+(1)b^-(2) - b^-(1)a^+(2)|e^2/r_{ij}|a^+(1)b^-(2) - b^-(1)a^+(2)\rangle \\
&\quad = \tfrac{1}{2}[\langle a^+(1)b^-(2)|e^2/r_{ij}|a^+(1)b^-(2)\rangle + \langle b^-(1)a^+(2)|e^2/r_{ij}|b^-(1)a^+(2)\rangle \\
&\qquad - 2\langle a^+(1)b^-(2)|e^2/r_{ij}|b^-(1)a^+(2)\rangle] = J(a,b) \qquad (106)
\end{aligned}$$

where the Coulomb (J) and exchange (K) integrals were defined in Eqn. (76).

Suppose we have a ground state (a^+, a^-) and excite to the spin triplet (a^+, b^+) and to the corresponding spin-singlet. The one-electron energies, which contain the kinetic energy and the potential energy in the field of the nucleus, but exclude interelectronic repulsion, are written as h_a and h_b, then;

$$E(a^+, a^-) = 2h_a + J(a,a) \quad \text{(no } K \text{ term because of opposite spin)} \qquad (107)$$

$$E(a^+, b^+) = h_a + h_b + J(a,b) - K(a,b) \qquad (108)$$

The spin-singlet excited state cannot be written as (a^+, b^-) since electrons are indistinguishable and (a^-, b^+) would be equivalent. This wavefunction must therefore be written as: $(1/\sqrt{2})[(a^+, b^-) - (a^-, b^+)]$. The corresponding interelectronic repulsion element is

$$\langle (1/\sqrt{2})\big[(a^+, b^-) - (a^-, b^+)\big] \big| e^2/r_{ij} \big| (1/\sqrt{2})\big[(a^+, b^-) - (a^-, b^+)\big] \rangle \quad (109)$$

which can be expanded remembering that these are Slater determinants, as:

$$\tfrac{1}{2}\big[\langle a^+(1)b^-(2) \big| e^2/r_{ij} \big| a^+(1)b^-(2) \rangle + \langle a^-(1)b^+(2) \big| e^2/r_{ij} \big| a^-(1)b^+(2) \rangle \\ +2\langle a^-(1)b^+(2) \big| e^2/r_{ij} \big| b^-(1)a^+(2) \rangle\big] \quad (110)$$

where in the third term, the rightmost function has been interchanged so that the spins can be integrated out as unity (causing the element to be multiplied by -1). The full expansion in Eqn. (110) would yield 8 terms since each Slater determinant is two terms; we have simplified the expansion by collecting equivalent terms together.

From the previous discussion, the first and second terms of Eqn. (110) are simply $J(a, b)$; the third term is already in the K format—thus:

$$E\{(1/\sqrt{2})\big[(a^+, b^-) - (a^-, b^+)\big]\} = h_a + h_b + J(a, b) + K(a, b) \quad (111)$$

Note that the difference in energy between the spin singlet (111) and triplet (108) states of a given two-electron configuration is $2K(a, b)$, which is always true. The spin singlet wavefunction in Eqn. (109) has a negative sign between the two components because if it had been positive, the function would be the $M_s = 0$ component of the $(a)^1(b)^1$ triplet.

The $d^2 D_{4h}$ system generates 3 ^3E_g, 2 $^3A_{2g}$, $^3B_{1g}$, and $^3B_{2g}$ states (Table 12). To obtain the complete energies of the ^3E_g and $^3A_{2g}$ states, one needs to solve a 3×3 ^3E_g matrix and the 2×2 $^3A_{2g}$ matrix, for which we need the relevant wavefunctions. One $^3A_{2g}$ state arises from the configuration (xy^+), $(x^2 - y^{2+})$ (direct product $b_{2g} \times b_{1g}$). The other arises from the configuration (e^2) yielding $^3A_{2g} + {}^1E_g + {}^1A_{1g}$ and the spin triplet component is clearly given by (xz^+, yz^+) this being the only way to write $(e)^2$ to generate $S = 1$, $M_s = 1$. Using the corresponding Slater determinants, the diagonal elements of the $^3A_{2g}$ matrix are easily evaluated (see [35] for one-electron energies):

$$\begin{aligned} \langle (xz^+, yz^+) \big| e^2/r_{ij} \big| (xz^+, yz^+) \rangle &= J(xz, yz) - K(xz, yz) = A - 5B \\ E(^3A_{2g}, e_g^2) &= -8Dq - 2Ds + 8Dt + A - 5B \end{aligned} \quad (112)$$

$$\begin{aligned} &\langle (xy^+, x^2 - y^{2+}) \big| e^2/r_{ij} \big| (xy^+, x^2 - y^{2+}) \rangle \\ &\quad = J(xy, x^2 - y^2) - K(xy, x^2 - y^2) = A + 4B \\ &E(^3A_{2g}, b_{2g}^1, b_{1g}^1) = 2Dq + 4Ds - 2Dt + A + 4B \end{aligned} \quad (113)$$

Table 16 Electron Repulsion Integrals for Pairs of d Electrons in Real Orbitals as a Function of the Racah A, B, and C parameters. *Part I* Coulomb (J) and Exchange (K) Integrals. *Part II* Matrix Elements $\langle ab|e^2/r_{ij}|cd\rangle$

Part I

$$J(xy, xy) = J(xz, xz) = J(yz, yz) = J(x^2-y^2, x^2-y^2) = J(z^2, z^2) = A + 4B + 3C$$
$$J(xy, yz) = J(xy, xz) = J(xz, yz) = J(x^2-y^2, xz) = J(x^2-y^2, yz) = A - 2B + C$$
$$J(z^2, xz) = J(z^2, yz) = A + 2B + C$$
$$J(z^2, xy) = J(z^2, x^2-y^2) = A - 4B + C$$
$$J(x^2-y^2, xy) = A + 4B + C$$
$$K(xy, xz) = K(xy, yz) = K(xz, yz) = K(x^2-y^2, yz) = K(x^2-y^2, xz) = 3B + C$$
$$K(z^2, x^2-y^2) = K(z^2, xy) = 4B + C$$
$$K(z^2, yz) = K(z^2, xz) = B + C$$
$$K(x^2-y^2, xy) = C$$

Part II

a	b	c	d	$\langle ab\|e^2/r_{ij}\|cd\rangle$
xz	z^2	xz	x^2-y^2	$-2\sqrt{3}B$
yz	z^2	yz	x^2-y^2	$2\sqrt{3}B$
yz	yz	z^2	x^2-y^2	$-\sqrt{3}B$
xz	xz	z^2	x^2-y^2	$\sqrt{3}B$
z^2	xy	xz	yz	$\sqrt{3}B$
z^2	xz	xy	yz	$-2\sqrt{3}B$
z^2	xy	yz	xz	$\sqrt{3}B$
x^2-y^2	xy	xz	yz	$3B$
x^2-y^2	xy	yz	xz	$-3B$

NB: $\langle ab\|e^2/r_{ij}\|cd\rangle = \langle ba\|e^2/r_{ij}\|dc\rangle = \langle cd\|e^2/r_{ij}\|ab\rangle\langle ad\|e^2/r_{ij}\|cb\rangle$ obtained by interchanging a and c, b and d, or permutation of ab to ba and cd to dc, or interchange of ab with cd. All combinations not shown above and not obtainable by a permutation of the sort described, are zero.

Table 16 lists the values of these J and K integrals involving the real d orbitals, in terms of the Racah parameters A, B, and C. The Racah A is common to all the integrals and serves only to shift all d configurations by the same amount—it can then usually be ignored.

The off-diagonal element is $\langle(xz^+, yz^+)|e^2/r_{ij}|(xy^+, x^2-y^{2+})\rangle$ which will need to be expanded, taking account of the Slater determinant expansion, as: (we drop the spin designation for simplicity since they are all the same and will be integrated out as unity):

$$\begin{aligned}&\big\langle(xz^+, yz^+)\big|e^2/r_{ij}\big|(xy^+, x^2-y^{2+})\big\rangle\\&\quad= (1/2)\big\langle xz(1)yz(2) - yz(1)xz(2)\big|e^2/r_{ij}\big|xy(1)x^2-y^2(2)\\&\qquad - x^2-y^2(1)xy(2)\big\rangle\end{aligned} \tag{114}$$

which expands to four terms (see Table 16):

$$\big\langle xz(1)yz(2)\big|e^2/r_{ij}\big|xy(1)x^2-y^2(2)\big\rangle = -3B$$
$$\big\langle yz(1)xz(2)\big|e^2/r_{ij}\big|x^2-y^2(1)xy(2)\big\rangle = -3B$$

$$-\langle yz(1)xz(2)|e^2/r_{ij}|xy(1)x^2-y^2(2)\rangle = -(+3B)$$
$$-\langle xz(1)yz(2)|e^2/r_{ij}|x^2-y^2(1)xy(2)\rangle = -(+3B) \quad (115)$$

Certain permutations of orbital functions give rise to the same integral value so that they can be recognized in Table 16. Specifically:

$$\langle ab|e^2/r_{ij}|cd\rangle = \langle ba|e^2/r_{ij}|dc\rangle = \langle ad|e^2/r_{ij}|cb\rangle = \langle cb|e^2/r_{ij}|ad\rangle$$

Using these relationships, and Table 16, the values shown in Eqn. (115) were derived. Thus:

$$\langle (xz^+, yz^+)|e^2/r_{ij}|(xy^+, x^2-y^{2+})\rangle = -6B \quad (116)$$

and the complete ${}^3A_{2g}$ in D_{4h}, secular determinant is:

	${}^3A_{2g}(xz^+, yz^+)$	${}^3A_{2g}(xy^+, x^2-y^{2+})$	
${}^3A_{2g}(xz^+, yz^+)$	$-8Dq-2Ds+8Dt+A-5B-E$	$-6B$	$=0$
${}^3A_{2g}(xy^+, x^2-y^{2+})$	$-6B$	$2Dq+4Ds-2Dt+A+4B-E$	

(117)

Solving this determinant provides the energies of these two ${}^3A_{2g}$ wavefunctions as a function of the various ligand field radial parameters and B. Note that it is useful to check that the anticipated boundary conditions are met by the determinant as an indicator, at least, that it is correct. In this case we can look at the predicted limiting weak and strong field energies generated by this determinant. In the weak field, $Dq = Ds = Dt = 0$ and the derived energies are $-8B$ and $+7B$ which are indeed the correct energies for the ${}^3F$ and ${}^3P$ free ion terms from which two ${}^3A_{2g}$ wavefunctions derive. In the strong field, $B = 0$, and the roots are the diagonal elements, which are, of course, correct.

In a similar manner the 3×3 ${}^3E_g$ secular determinant can be obtained. It is provided here (121) and the interested reader may derive it. The three ${}^3E_g$ terms arise according to Table 12, or from the configurations $(b_{1g})(e_g)$, $(b_{2g})(e_g)$, and $(a_{1g})(e_g)$. The ground state ${}^3E_g$ is, of course, a doubly degenerate pair of functions. One electron resides in $d(xy)$, while the other may with equal probability lie in either $d(xz)$ or $d(yz)$. An acceptable wavefunction would be:

$$\psi({}^3E_g(b_{2g})(e_g)) = (1/\sqrt{2})[(xy^+, xz^+) + (xy^+, yz^+)] \quad (118)$$

whose energy, in the absence of coupling to other ${}^3E_g$ terms, is given by:

$$E({}^3E_g(b_{2g})(e_g)) = \langle (1/\sqrt{2})[(xy^+, xz^+) + (xy^+, yz^+)]|V + e^2/r_{ij}|(1/\sqrt{2})[(xy^+, xz^+) + (xy^+, yz^+)]\rangle \quad (119)$$

Using the one-electron energies from Eqn. (35), the one-electron energy of this state is $-8Dq + Ds + 3Dt$. The interelectronic repulsion can be expanded:

$$\begin{aligned}
&\langle(1/\sqrt{2})[(xy^+, xz^+) + (xy^+, yz^+)]|e^2/r_{ij}|(1/\sqrt{2})[(xy^+, xz^+) + (xy^+, yz^+)]\rangle \\
&\quad = 1/2[\langle(xy^+, xz^+)|e^2/r_{ij}|(xy^+, xz^+)\rangle + \langle(xy^+, yz^+)]|e^2/r_{ij}|(xy^+, yz^+)\rangle \\
&\qquad + 2\langle(xy^+, yz^+)|e^2/r_{ij}|(xy^+, xz^+)\rangle] \\
&\quad = 1/2[J(xy, xz) - K(xy, xz) + J(xy, yz) - K(xy, yz) \\
&\qquad +2\langle(xy^+, yz^+)|e^2/r_{ij}|(xy^+, xz^+)\rangle]
\end{aligned} \tag{120}$$

The last term should now be expanded; however it will expand to a sum of terms involving only $d(xy)$, $d(xz)$, and $d(yz)$. No such element appears in Table 16 (part II) and thus this cross term is zero. The J and K terms in Eqn. (120) sum to $-5B$ (from Table 16) and therefore the first order energy of ${}^3E_g(a)$ is $-8Dq + Ds + 3Dt - 5B$. This then is the diagonal energy of the ${}^3E_g(a)$ term in determinant (121). Evaluation of the other terms is left to the reader.

	${}^3E_g((b_{2g})(e_g))$	${}^3E_g((a_{1g})(e_g))$	${}^3E_g((b_{1g})(e_g))$	
${}^3E_g((b_{2g})(e_g))$	$-8Dq + Ds + 3Dt - 5B - E$	$-3\sqrt{3}B$	$3B$	
${}^3E_g((a_{1g})(e_g))$	$-3\sqrt{3}B$	$2Dq - 3Ds - 2Dt + B - E$	$-3\sqrt{3}B$	$= 0$ (121)
${}^3E_g((b_{1g})(e_g))$	$3B$	$-3\sqrt{3}B$	$2Dq + Ds + 3Dt - 5B - E$	

3.3.1 *Tanabe–Sugano Matrices and Diagrams* Thus using the Wigner–Eckart theorem or expanding the wavefunctions and energies by hand, it is possible to obtain a complete set of configuration interaction (CI) matrices (one for each symmetry and spin) for each d^n configuration for a given molecular structure. Tanabe and Sugano accomplished this, for the octahedral case, some 40 years ago using the strong crystal field formalism;[43] the d^2–d^5 Tanabe–Sugano matrices are given in Appendix 3. Note that these matrices can also be used to treat the d^8–d^5 configurations in an octahedral crystal field by changing the sign of 10 Dq. This is the electron-hole formalism where there is no change in the electrostatic repulsive interaction for two positive holes relative to two negative electrons, but the positive hole can now be considered to undergo an attractive interaction with the negatively charged ligands. These matrices are also appropriate for tetrahedral complexes since in the crystal field limit, Eqn. (5) gives 10 $Dq(T_d) = -4/9$ 10 $Dq(O_h)$. Thus the same matrix corresponds to the d^n in T_d and $d^{10-n\text{(holes)}}$ in O_h cases.

For a given d^n in O_h (or T_d) case, one fixes the C/B ratio to that observed experimentally (Table 11) and calculates the energy of the states in units of B as a function of Dq/B with the ground state energy always defined as zero. This leads to the Tanabe-Sugano diagrams which are given in Figures 13 to 16. The left-hand

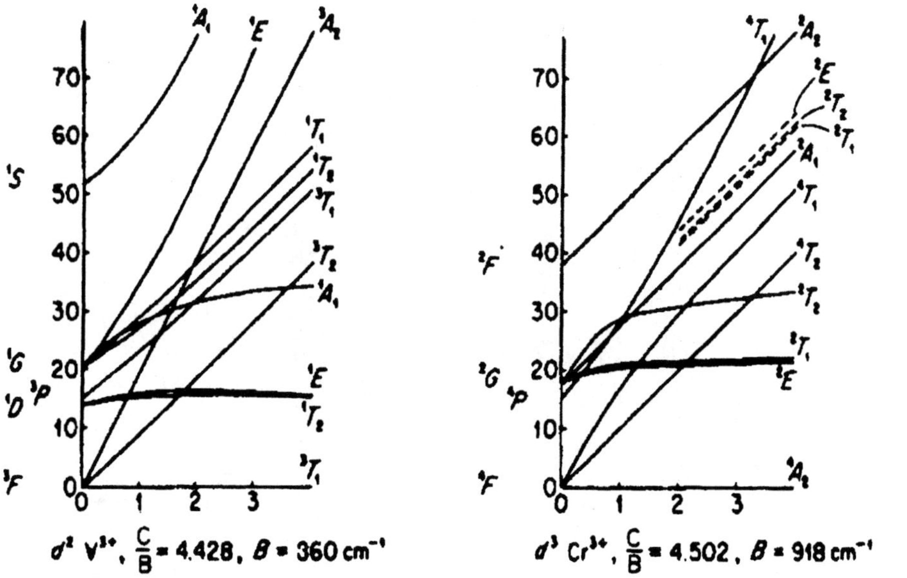

Figure 13. Energy level (Tanabe–Sugano) diagrams showing the effect of an octahedral field on the energies of the free ion spectroscopic states of a d^2 (left) and d^3 (right) species, plotted as a function of E/B versus Dq/B.

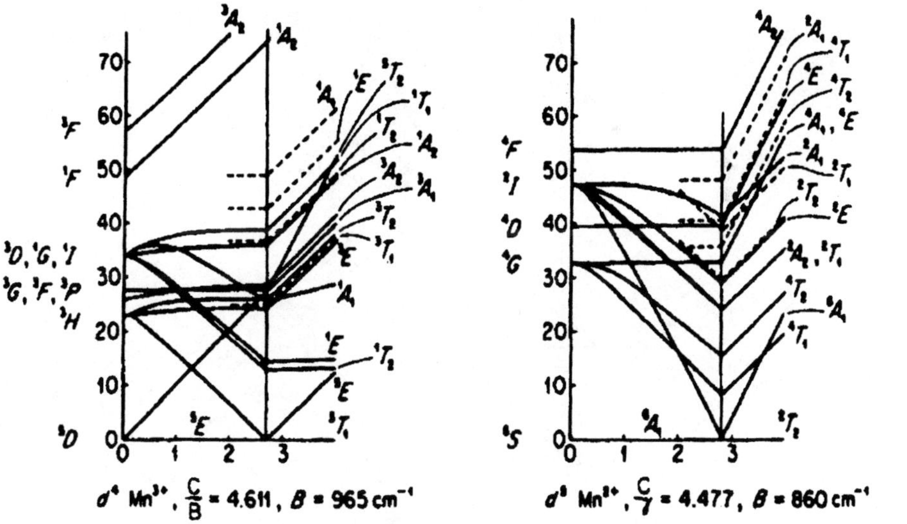

Figure 14. Energy level (Tanabe–Sugano) diagrams showing the effect of an octahedral field on the energies of the free ion spectroscopic states of a d^4 (left) and d^5 (right) species, plotted as a function of E/B versus Dq/B.

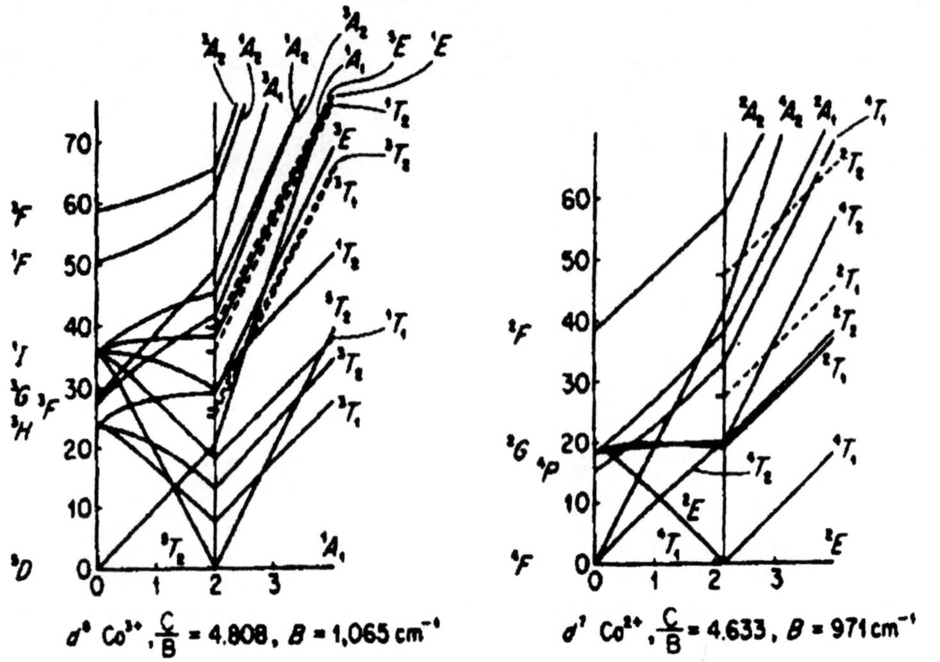

Figure 15. Energy level (Tanabe–Sugano) diagrams showing the effect of an octahedral field on the energies of the free ion spectroscopic states of a d^6 (left) and d^7 (right) species, plotted as a function of E/B versus Dq/B.

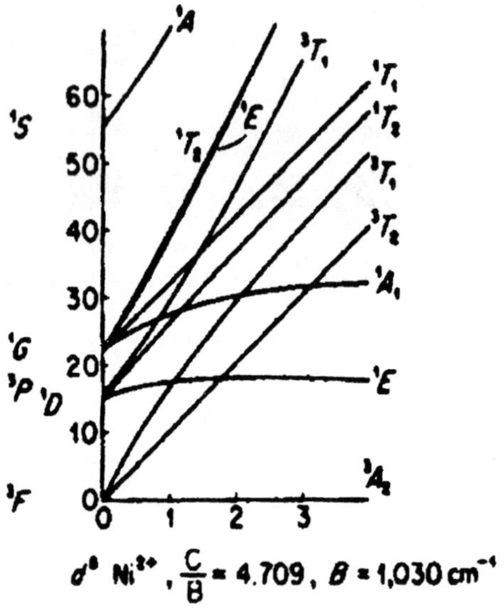

Figure 16. Energy level (Tanabe–Sugano) diagram showing the effect of an octahedral field on the energies of the free ion spectroscopic states of a d^8 species, plotted as a function of E/B versus Dq/B.

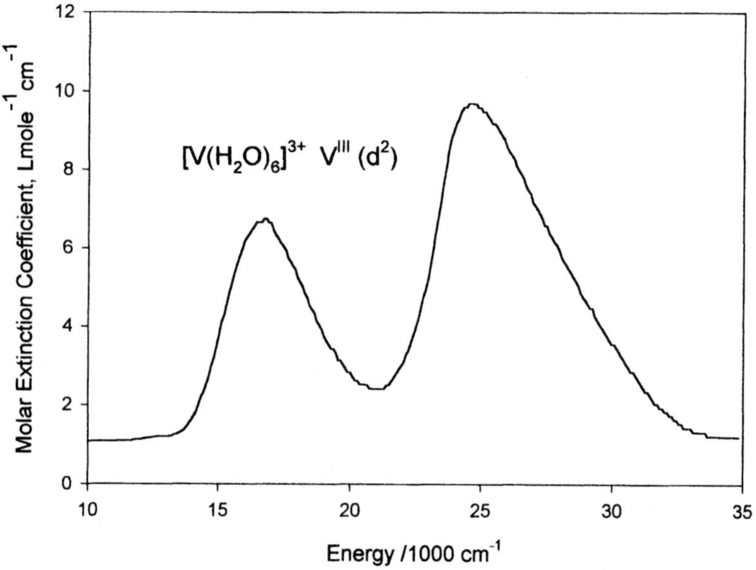

Figure 17. The absorption spectrum of the octahedral d^2 $[V(H_2O)_6]^{3+}$ ion obtained from a solution of $[V(H_2O)_6](ClO_4)_3$ in aqueous perchloric acid [adapted from ref. 59].

side of each diagram corresponds to a weak ligand field which gives, in the limit, the energy order of the free ion atomic term symbols (in units of B) at $Dq = 0$. The right-hand side approaches the strong field limit ($Dq/B = 4$) which separates the molecular states into the strong field $(t_{2g})^m(e_g)^n$ configurations from which they derive, with electron-electron repulsion as a perturbation.

In these diagrams, the two parameters Dq and B quantitatively predict a large number of geometric and electronic structure properties of a transition metal complex. Since energies are given in units of B, this parameter can be reduced from its free ion value to account for the Nephelauxetic effect. The Tanabe–Sugano diagram with a value of Dq/B appropriate for a specific metal complex, gives the ground state which defines the magnetism and structural properties such as the Jahn–Teller effect[44] for the molecule. The Tanabe–Sugano diagram also quantitatively predicts the energy of the many ligand field excited states, while the nature of each state (symmetry, spin, and many-electron wavefunction) gives its spectroscopic properties (i.e., intensity, bandshape, and excited state spin Hamiltonian).

The ligand field parameters are usually obtained experimentally from the absorption spectrum. Continuing with the d^2 example, from Figure 17 hexaaquo vanadium(III) has its two lowest energy ligand field transitions with $\varepsilon \sim 10\ M^{-1}cm^{-1}$ at $\sim 17{,}200$ and $\sim 26{,}000\ cm^{-1}$ giving $E_2/E_1 \sim 1.5$. From the d^2 Tanabe–Sugano diagram in Figure 13, the ground state is $^3T_{1g}$ for any value of Dq/B and one expects to observe ($\varepsilon > 1\ M^{-1}cm^{-1}$) only spin-allowed ($\Delta S = 0$) transitions to excited states. The experimental ratio of 1.5 for transitions to the two lowest energy triplet states occurs in the Tanabe–Sugano diagram at a Dq/B of 2.8. This allows the assignment of the ligand field transitions in Figure 17 as the $^3T_{2g} \leftarrow {}^3T^a_{1g}$ and $^3T^b_{1g} \leftarrow {}^3T^a_{1g}$, respectively. For a Dq/B of 2.8, the $^3T_{2g} \leftarrow {}^3T^a_{1g}$ transition energy is given by the d^2 Tanabe–Sugano diagram as $25.9B$. Since the experimental energy of this transition is 17,200 cm^{-1} this gives a B value of 665 cm^{-1} which is reduced from the free ion value of 860 cm^{-1} ($\beta = 0.77$). Further, $Dq = 2.8B = 1860$ cm^{-1} which provides a reference point in the spectrochemical series for a trivalent metal ion of the first transition metal series with water ligands. A more quantitative estimate of these parameters which also allows C to vary (if spin-forbidden excited states with $\varepsilon < 1$ M^{-1}cm^{-1} are observed) is obtained directly from application of the Tanabe–Sugano matrices in Appendix 3 to fit the observed spectral data.

For the d^4–d^7 Tanabe–Sugano diagrams there is an additional feature to note which is an apparent discontinuity that occurs between Dq/B values of 2.0 and 3.0 depending on the d^n configuration. For these cases the weak field side of the diagram has a ground state governed by the interelectron repulsion of the free ion. The ground state is the Hunds' rule atomic term symbol (highest total spin) with its orbital degeneracy split by the perturbation due to the O_h ligand field as given by Eqn. (80). The right-hand side of the diagram gives the strong field ground state which is determined by the t_{2g}/e_g splitting of the d orbitals due to the strong ligand field. For d^4–d^7, this 10 Dq splitting can overcome electron-electron repulsion which results in spin pairing and a low spin ground state. The apparent discontinuity in the Tanabe–Sugano diagrams occurs when the low spin state (an excited state on the weak field side of the diagram) crosses over to become the ground state on the strong field side of the diagram, since the diagrams are constructed such that the ground state is always at zero energy.

The crossing point gives the strength of the ligand field required to overcome the interelectron repulsion. This is referred to as the spin pairing energy. From the Tanabe–Sugano diagrams, the ratios of Dq to B to cause spin pairing are d^4: $Dq/B = 2.7$; d^5: $Dq/B = 2.8$; d^6: $Dq/B = 2.0$, and d^7: $Dq/B = 2.2$. Thus, it is generally easier to obtain spin paired ground states in d^6 and d^7 complexes.

3.3.2 ***Other Transitions*** Finally, while there is a huge amount of information in the Tanabe–Sugano diagrams, these only give the ligand field states deriving from electrons in d orbitals on the metal ion. From the more general molecular orbital description of bonding (Figure 7, Section 2.3) there are also ligand valence orbitals in the energy vicinity of these metal d orbitals which mix due to covalent interactions. For π donor ligands this involves occupied ligand orbitals which produces ligand to metal charge transfer (LMCT) transitions at fairly low energy. For π acceptor ligands, unoccupied ligand orbitals are close in energy to the metal d orbitals which, when occupied, produce low lying metal to ligand charge transfer (MLCT) excited states. The energy of these states is obtained as developed above; it is the sum of one-electron orbital energies plus electron-electron repulsion. For LMCT transitions this involves coupling of a hole on the ligand to a d^{n+1} metal ion, while for MLCT states, the d^{n-1} metal ion parent states are coupled to the electron in the π^* orbital on the ligand. The electron repulsion within the metal ion $d^{n\pm1}$ final state often dominates the energy splitting of the charge transfer states since the interaction between electrons on the metal and on the ligand is small. To still higher energy are transitions from electrons in core orbitals on the metal (K, L, M) or ligand (K, L) to the valence $3d$, $4s$, and $4p$ orbitals on the metal ion. These edge transitions are the focus of X-ray absorption spectroscopy as described in Chapter 9.

3.4 Spin–Orbit Coupling for Many-Electron States

The aforementioned techniques show how the matrices for ligand field energies and electron repulsion can be derived. For heavier elements where spin–orbit coupling is important, or for problems involving magnetism and EPR spectroscopy, it is necessary to investigate how the many-electron wavefunctions are perturbed by the spin–orbit coupling operator.

When one is interested in the spin–orbit splitting of a many-electron atomic term, the spin–orbit operator can be written in terms of the total orbital and spin angular momentum of the state as given in Eqn. (122) where λ is a radial integral for the specific many-electron state.

$$\mathcal{H}_{\mathrm{SO}} = \sum_i \zeta(r_i) l_i \cdot s_i = \lambda L \cdot S \tag{122}$$

For ground states, $\lambda = \pm\zeta_{3d}/2S$ where the plus sign is for a less than half filled configuration and the minus sign is for a more than half filled configuration and ζ_{3d} is the one-electron radial integral of spin–orbit coupling given in Table 7. Operating on an atomic term symbol with Eqn. (122) gives the Landé formula for the spin–orbit splitting of the term into substates defined by their total angular momentum $J = |L + S|, |L + S - 1|, \ldots, |L - S|$, where there are $(2J + 1)$ M_J values for a given J state:

$$E_J = \frac{\lambda}{2}[J(J+1) - L(L+1) - S(S+1)] \tag{123}$$

This leads to Hunds' third rule: for a less than half filled d^n-configuration the state with the smallest J is lowest in energy, while for a more than half filled configuration the splitting is inverted.

Thus, spin–orbit coupling requires orbital angular momentum which is proportional to the orbital degeneracy of a state. Upon placing a metal ion in a ligand field some of the orbital degeneracy of the atomic term can be removed and thus the orbital angular momentum can be partially quenched.

The $\boldsymbol{l}$ part of the $\boldsymbol{l} \cdot \boldsymbol{s}$ operator transforms as a rotation. Thus $\langle\Gamma|\boldsymbol{l}\cdot\boldsymbol{s}|\Gamma'\rangle$ is nonzero only if the direct product $\Gamma \times \Gamma'$ contains a rotational representation of the group concerned. In O_h symmetry, $R_{x,y,z}$ transforms as T_{1g}. Therefore only T_{1g} or T_{2g} states in O_h symmetry are split by spin–orbit coupling. Other states such as ${}^nE_g$ may be shifted but not split under in-state spin–orbit coupling. In D_{4h} symmetry, the rotations span $A_{2g} + E_g$ so that in this point group ${}^nE_g$ states may split. Orbital singlet states, such as ${}^nA_{2g}$, cannot split to first order in spin–orbit coupling. Second order spin–orbit coupling can lead to zero-field splitting of these spinor states (see Eqn. (138) below).

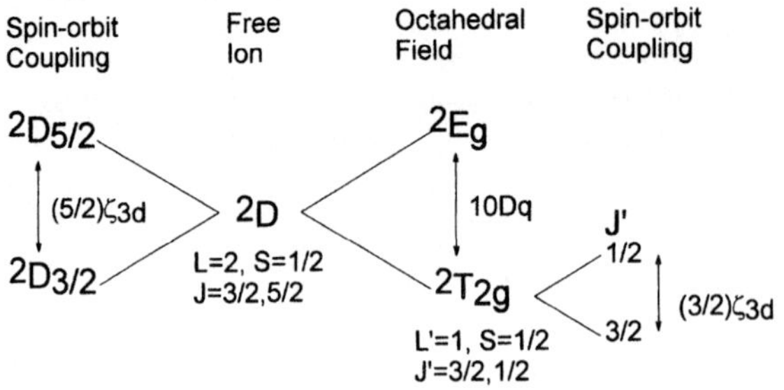

Figure 18. (Left) Spin–orbit splitting of the d^1, ^2D state; (Right) O_h crystal field splitting of the ^2D state showing the spin–orbit splitting of the $^2T_{2g}$ term.

To quantitatively obtain the spin–orbit coupling of T states in cubic symmetry groups it is useful to treat these threefold degenerate states as having an effective orbital angular momentum $L' = 1$, but where the effect of the orbital angular momentum operator is of opposite sign.[7] This leads to the modified Landé formula given in Eqn. (124), where J' is the effective total angular momentum $= |L' + S|, \ldots, |L' - S|$.

$$E_{J'} = -\frac{\lambda}{2}\left[J'(J'+1) - L'(L'+1) - S(S+1)\right] \tag{124}$$

The octahedral d^1 example is given in Figure 18 where $\lambda = \xi_{3d}/2(1/2) = \xi_{3d}$.

Eqn. (124) provides a shortcut to deal with T functions. However, in general, we would need to evaluate the expectation values of spin–orbit matrix elements involving Slater determinants. Using the guidelines expressed in Eqns. (89–90) a typical spin–orbit coupling matrix element of a two-electron Slater determinant will expand to:

$$\begin{aligned}&\left\langle (1^+, 0^+)|\xi(r)\boldsymbol{l}\cdot\boldsymbol{s}|(2^-, 0^+)\right\rangle \\ &= \left\langle 1^+|\xi(r)\boldsymbol{l}\cdot\boldsymbol{s}|2^-\right\rangle\langle 0^+|0^+\rangle + \left\langle 0^+|\xi(r)\boldsymbol{l}\cdot\boldsymbol{s}|0^+\right\rangle\langle 1^+|2^-\rangle\end{aligned} \tag{125}$$

The second term on the right vanishes because its overlap integral vanishes. We are left with a possible sum of one-electron elements which is treated as described in Section 2.4. Note that the spin–orbit coupling operator can couple states of different spin. It is therefore necessary in principle to solve the matrix for all possible states. So for a d^2 species one solves a 45×45 matrix, for d^3, a 120×120, etc. However these large matrices can be simplified by using the double groups and the symmetry properties of the microstates involved.[7]

4 RELATION OF LIGAND FIELD THEORY TO PHYSICAL PROPERTIES

To this point in the chapter we have developed the methodology utilized in describing the electronic structure of transition metal complexes. The combination of d orbital

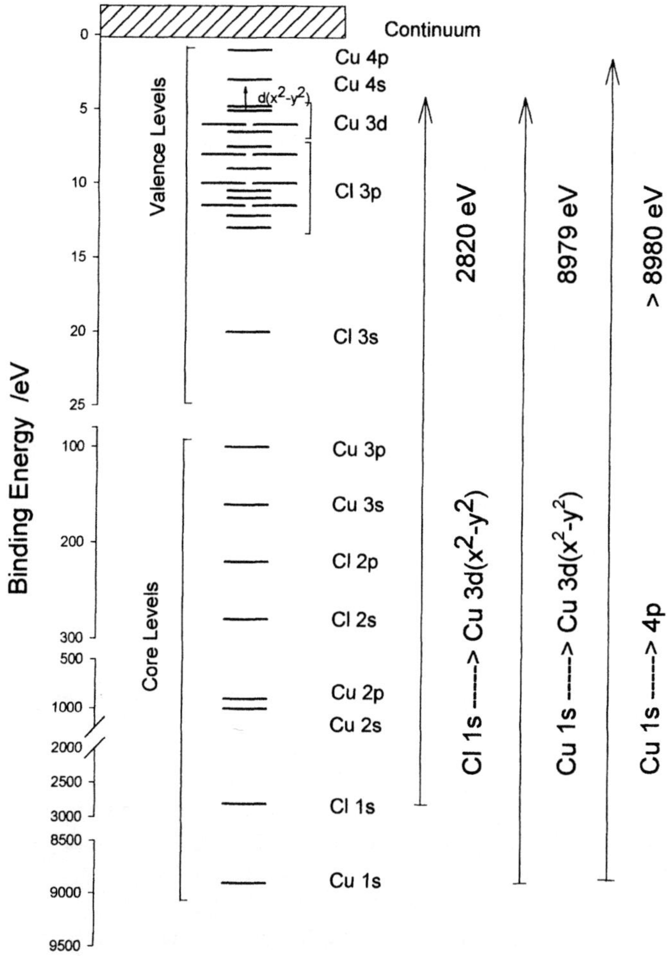

Figure 19. A complete energy level diagram for the d^9 D_{4h} $[CuCl_4]^{2-}$ ion. All core and valency levels deeper in energy than the $3d(x^2-y^2)$ orbital are filled. Vertical arrows represent the K-edge spectra of chloride and copper (adapted from [60]).

splitting due to the ligand field and interelectronic repulsion produces the ground and ligand field excited states of a metal complex. From the more general molecular orbital description of bonding we have seen that π donor or acceptor ligands can result in additional low-lying valence orbitals which produce ligand-to-metal or metal-to-ligand charge transfer excitations and a number of additional d^{n+1} or d^{n-1} excited states. To higher energy are excitations from the ligand or metal core orbitals into the n, d, and $(n+1)[s$ and $p]$ orbitals of the metal leading to X-ray pre-edge and edge transitions.[45,46] A representative complete energy level diagram for the $D_{4h} - [CuCl_4]^{2-}$ (d^9) complex is given in Figure 19.

The nature of the ground state (i.e., spin, symmetry, and many-electron wavefunction) dictates the geometry and magnetic properties of a metal complex (when magnetically dilute). Transitions from the ground state to the different excited states (*dd*, CT, edges, etc.) define the spectroscopic properties of molecules. Spectroscopy provides experimental insight into the covalency of the ligand–metal bond. The relations of electronic structure to these physical properties are briefly summarized below.

4.1 Geometry

In all the above electronic structure considerations, we have assumed that the transition metal complex has a fixed geometric structure, usually the one defined by crystallography. However, molecules vibrate and for an ML_x (nonlinear) complex with N atoms there will be $3N - 6$ normal modes of vibration, Q_i. These are listed for many structural types in Table 17. Including electronic-nuclear coupling, the Hamiltonian for the total energy of an electronic state is given by a Taylor expansion in Q_i ($i = 3N - 6$) around the equilibrium geometry (Eqn. (126)), where the first term is the ligand field electronic Hamiltonian developed in Sections 2 and 3 and the third term is the potential energy associated with nuclear vibrations.

$$\mathcal{H} = H_{\text{L.F.}} + Q\frac{\delta H_{\text{L.F.}}}{\delta Q_i} + \tfrac{1}{2}k_j Q_j^2 + \cdots \qquad (126)$$

k_i is the harmonic oscillator restoring force constant $= 4\pi^2 m_i c^2 \nu_i^2$ where ν_i is the frequency of the vibration (in cm^{-1}) and m_i is the reduced mass.

The effect of the third term is to convert the energy levels from Sections 2 and 3 into potential energy surfaces as in Figure 20. From the quantum mechanical treatment of vibrations, one obtains the energy solutions given by Eqn. (127) where m is the vibrational quantum number ($= 0, 1, 2, \ldots$) of the ground state (n for excited

Table 17 Fundamental Skeletal Normal Modes of Vibration

Molecular Skeleton	Point Group	Normal Modes of Vibration
ML_6	O_h	$A_{1g} + E_g + T_{2g} + 2T_{1u} + T_{2u}$
ML_6	D_3	$3A_1 + 2A_2 + 5E$
ML_5	D_{3h}	$2A_1' + 3E' + 2A_2'' + E''$
ML_5	C_{4v}	$3A_1 + 2B_1 + B_2 + 3E$
ML_4	D_{4h}	$A_{1g} + B_{1g} + B_{2g} + A_{2u} + B_{2u} + 2E_u$
ML_4	C_{4h}	$A_g + 2B_g + A_u + B_u + 2E_u$
ML_4	T_d	$A_1 + E + 2T_2$
ML_4	D_{2d}	$2A_1 + B_1 + 2B_2 + 2E$
ML_4L_2'	D_{4h}	$2A_{1g} + B_{1g} + B_{2g} + E_g + 2A_{2u} + B_{2u} + 3E_u$
ML_4L_2'	D_{2h}	$2A_g + B_{1g} + 2B_{1u} + 2B_{2u} + 3B_{3u}$
ML_3L'	C_{3v}	$3A_1 + 3E$
ML_2L_2'	C_{2v}	$4A_1 + A_2 + 2B_1 + 2B_2$

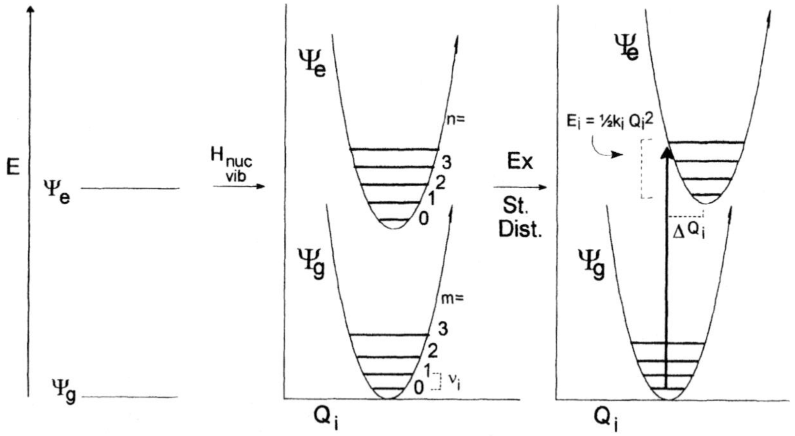

Figure 20. Potential energy surfaces for ground and excited states with the latter at the same equilibrium nuclear distance as the ground state (middle) or at a longer distance (right) in normal mode Q_i. The analysis is derived from ref. 48

states) which defines the Hermite vibrational functions, $\chi(Q_i)$.[47]

$$E_m = (m + 1/2)\upsilon_i \tag{127}$$

In the Born–Oppenheimer approximation, the total wavefunction is written as a product of electronic (ψ^{el}) and nuclear ($\chi(Q_i)$) wavefunctions:

$$\begin{aligned} \Psi_g^{\text{tot}} &= \Psi_g^{\text{el}} X_m(Q_j) \\ \Psi_e^{\text{tot}} &= \Psi_e^{\text{el}} X_n(Q_j) \end{aligned} \tag{128}$$

The second term on the right in Eqn. (126) is called the linear coupling term as it couples electronic and nuclear motions. Since H_{LF} is the ligand field potential, $\delta H_{\text{LF}}/\delta Q_i$ can be obtained by taking the first derivative of the appropriate expression in Section 2 with respect to Q_i. This operates on the electronic part of the total wavefunction, Eqn. (129). The integral in Eqn. (129) is a constant which defines the linear change in energy of Ψ^{el} with a change in geometry along normal mode Q_i.

$$E_{\text{el–nuc}} = \left\langle \Psi^{\text{el}} \left| \frac{\delta H_{\text{LF}}}{\delta Q_j} \right| \Psi^{\text{el}} \right\rangle Q_j \tag{129}$$

The total energy of a state associated with nuclear distortion is then given by:

$$E_{\text{nuc}} = \left\langle \Psi^{\text{el}} \left| \frac{\delta H_{\text{LF}}}{\delta Q_j} \right| \Psi^{\text{el}} \right\rangle Q_j + 1/2 k_j Q_j^2 \tag{130}$$

At the equilibrium geometry ΔQ_i, $\partial E_{\text{nuc}}/\partial Q_i = 0$, which gives:

$$\Delta Q_j = -\frac{\left\langle \Psi^{\text{el}} \middle| \frac{\delta H_{\text{LF}}}{\delta Q_j} \middle| \Psi^{\text{el}} \right\rangle}{k_j} \tag{131}$$

A state will distort relative to an initially assumed geometry along normal mode Q_i when $\Delta Q_i \neq 0$ and therefore

$$\left\langle \Psi^{\text{el}} \middle| \frac{\delta H_{\text{LF}}}{\delta Q_j} \middle| \Psi^{\text{el}} \right\rangle \neq 0$$

From group theory, since $Q_i(\delta H_{\text{LF}}/\delta Q_i)$ is a term in the molecular Hamiltonian (126) it must be totally symmetric, thus $\Gamma(\delta H_{\text{LF}}/\delta Q_i) = \Gamma(Q_i)$. The integral in Eqn. (131) will be nonzero and thus the electronic state will distort when $[\Gamma_{\text{el}} \times \Gamma_{\text{el}}] = \Gamma(Q_i)$ where the square brackets indicate that the symmetric direct product is taken since the integral involves an electronic state with itself.

$$\begin{aligned} &\textit{Sym. Dir. Prod.}: [\chi^2](R) = 1/2[(\chi(R))^2 + \chi(R^2)] \\ &\textit{Antisym. Dir. Prod.}: (\chi^2)(R) = 1/2[(\chi(R))^2 - \chi(R^2)] \end{aligned} \tag{132}$$

The Q_i in the symmetric direct product always include the totally symmetric irreducible representation. It is usually assumed that in the ground state

$$\left\langle \Psi_g^{\text{el}} \middle| \frac{\delta H_{\text{LF}}}{\delta Q_{a_{1g}}} \middle| \Psi_g^{\text{el}} \right\rangle = 0$$

as one already has the equilibrium structure. If the ground state is degenerate, the Q_i of the symmetric direct product will also include non-totally symmetric modes which will lower the symmetry of the molecule. This is the Jahn–Teller effect. If one initially assumes that a molecule has octahedral symmetry, and if the ground state is ${}^{2S+1}E_g$, ${}^{2S+1}T_{1g}$, or ${}^{2S+1}T_{2g}$, the Jahn–Teller theorem predicts the molecule will in fact distort since $[E \times E] = A_{1g} + E_g$; $[T_{1(2)g} \times T_{1(2)g}] = A_{1g} + E_g + T_{2g}$. Usually the E_g distortion (Figure 21) is dominant as this involves sigma interactions with the ligands, and elongated tetragonal complexes are frequently observed.

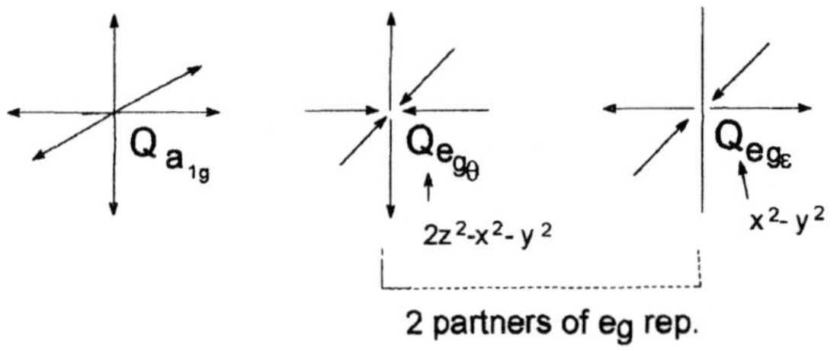

Figure 21. Normal modes of vibration which also correspond to distortion of an octahedral species along a_{1g} (left), $e_{g\theta}$(middle) and $e_{g\varepsilon}$(right) normal coordinates.

Excited states will in general distort relative to the equilibrium position of the ground state. This derives from the fact that

$$\left\langle \Psi_e^{\mathrm{el}} \middle| \left(\frac{\delta H_{\mathrm{LF}}}{\delta Q_i} \right)_0 \middle| \Psi_e^{\mathrm{el}} \right\rangle \neq 0$$

(in Eqn. (131)) and, from Group Theory, this must involve totally symmetric vibrations (as well as Jahn–Teller modes for degenerate excited states).[48] From Figure 22, if there is no excited state distortion the overlap integral of the $\chi_0(Q_i)$ Hermite func-

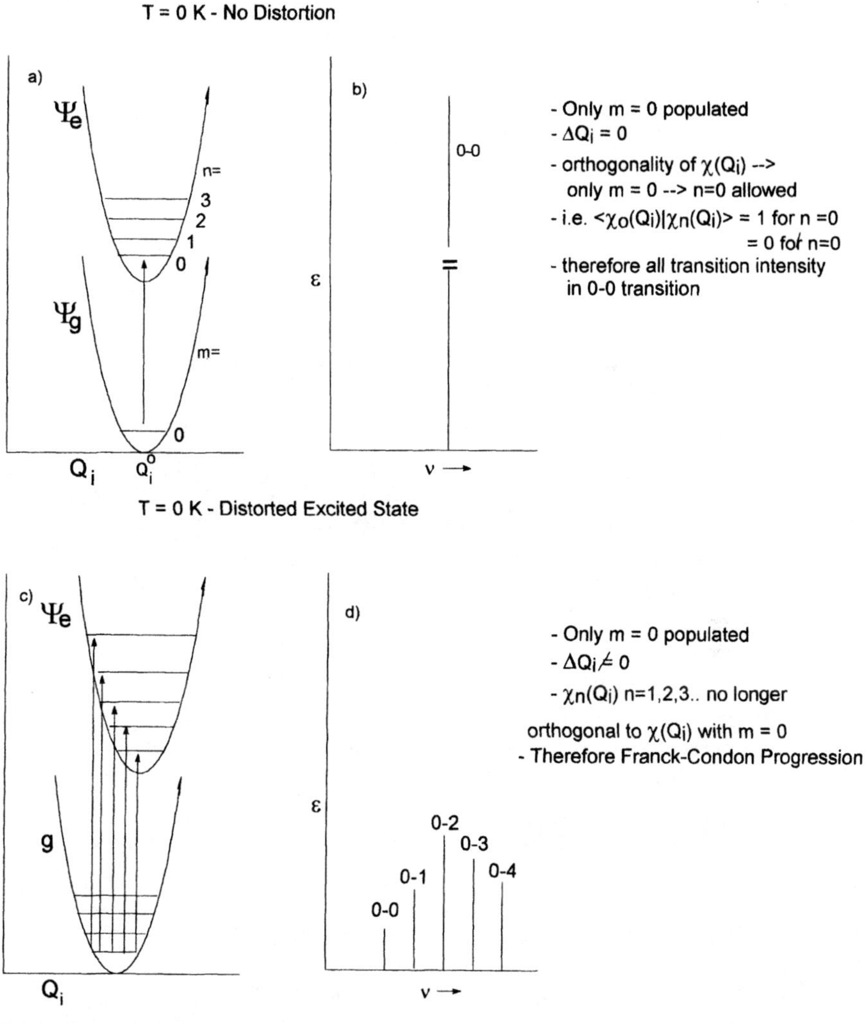

Figure 22. Bandshapes (Figure 22b) predicted for a species at 0 K with no distortion in the excited state (Figure 22a) and Franck–Condon progression (Figure 22d) when the excited state is distorted (Figure 22c). For details see insets on right of figure, see ref. 48.

tion of the ground state will only be nonzero for the $n = 0$ Hermite function in the excited state, and all electronic transition intensity (*vide infra*) will be included in the 0–0 band. However, if the excited state is distorted by an amount ΔQ_i, with a corresponding energy lowering $E_i = 1/2k_i \Delta Q_i^2$ relative to the energy of the excited state in the ground state equilibrium geometry (see Figure 20, right), the vibrational overlap term $\langle \chi_o | \chi_n \rangle \neq 0$ for $n \neq 0$ and this produces a Franck–Condon progression. This is a distribution of intensity with an energy spacing v_i in normal mode Q_i. This band shape is defined by Eqn. (133) where S_i is a Huang–Rhys parameter $= E_i/v_i$. Thus absorption (and emission) bandshapes provide insight into excited state distortion (relative to ground state geometry) which results from the change in electronic structure with excitation. Very narrow transitions are associated with little or no geometric distortion in the excited state and/or distortion only along low frequency modes, while broad transitions necessarily involve significant displacement along one or more modes. Analogous vibrational overlap integrals associated with excited state distortions also determine resonance enhancements in Raman Spectra as described in Chapter 7.

$$\frac{I_{0\to n}}{I_{0\to 0}} = \frac{S_i^n}{n!} \tag{133}$$

4.2 Magnetism

The nature of the ground state is probed experimentally by magnetic susceptibility, and EPR and magnetic Mossbauer spectroscopies. These methods require perturbation of the ground state by an external field, such as in the Zeeman effect given by Eqn. (134).

$$H_{\text{zee}} = \beta(L + 2S) \cdot H \tag{134}$$

Here $-\beta(L+2S)$ gives the magnetic moment of the ground state and β is the Bohr magneton $= e\hbar/2mc = 0.927 \times 10^{-20}$ erg/gauss $= 4.6686 \times 10^{-5}$ cm^{-1}/gauss. For a free transition metal ion, the ground state including spin–orbit coupling is $^{2S+1}L_J$ which is $2J + 1$ degenerate in M_J $(= J, J - 1, \ldots, -J)$. Note that spin–orbit coupling must be included before the Zeeman effect as the spin–orbit interaction is a much larger energy perturbation than the Zeeman effect. Application of a magnetic field to this ground state splits the M_J degeneracy, $E_{M_j} = \beta g M_J H$ where

$$g = 1 + \frac{J(J+1) - L(L+1) + S(S+1)}{2J(J+1)}$$

is the Landé factor which is the main experimental quantity of interest. Note that for $L = 0$, $J = S$ and $g = 2.0$.

For transition metal complexes only $^{2S+1}T_{1g}$ and $^{2S+1}T_{2g}$ states have orbital angular momentum ($L' = 1$). These have in-state spin–orbit coupling as given by the modified Landé formula in Eqn. (124). The g value associated with each spin–orbit state J' is given by Eqn. (135), which is a modified form of the above equation due to the change in sign of the orbital angular momentum operator (*vide supra*). Thus for T states the g values can deviate significantly from 2.00 due to orbital contributions.

$$g_{J'} = -1 + 3\left[\frac{J'(J'+1) - L'(L'+1) + S'(S'+1)}{2J'(J'+1)}\right]$$
$$g_{J'=0} = 0 \tag{135}$$

For $^{2S+1}A_{1g}$, $^{2S+1}A_{2g}$ and $^{2S+1}E_g$ states, there is no in-state orbital angular momentum, however, the g values can still deviate from 2.0 by a limited amount due to spin–orbit coupling of excited states (at $10Dq$) into the ground state which mixes some orbital angular momentum into this state. Expressions accounting for this mixing are given below:

$$A_{1g}\colon\ g = 2.00$$
$$A_{2g}\colon\ g = 2.00(1 - 4\lambda/10Dq)$$
$$E_g\colon\ g = 2.00(1 - 2\lambda/10Dq) \tag{136}$$

For lower symmetry complexes the g values will be anisotropic, that is, dependent on the orientation of the external magnetic field direction relative to the molecular axes. The combination of spin–orbit mixing with excited states and the Zeeman effect on the ground state leads to:

$$g_i = 2 - 2\lambda \sum_{n\neq 0} \frac{\langle\Psi_o|L_i|\Psi_n\rangle\langle\Psi_n|L_i|\Psi_o\rangle}{E_n - E_o}, \qquad i = x, y, z \tag{137}$$

where Ψ_o is the ground state given by ligand field theory in the absence of spin–orbit coupling and Ψ_n are the excited states at energy $E_n - E_o$. The effects of the L_i on the real d orbitals are given in Table 18.

Finally for $S > 1/2$, even orbitally non-degenerate states can split in energy in the absence of a magnetic field, if the symmetry of the complex is lower than cubic (O_h or T_d). This is described by the phenomenological spin Hamiltonian for zero field splitting (ZFS):

$$H_{\text{zfs}} = D\left[S_z^2 - \frac{1}{3}S(S+1)\right] + E\left(S_x^2 - S_y^2\right) \tag{138}$$

Table 18 Effect of L_i on Real d Orbitals

$\hat{L}_x d_{xz} = -id_{xy}$	$\hat{L}_y d_{xz} = id_{x^2-y^2} - i\sqrt{3}d_{z^2}$	$\hat{L}_z d_{xz} = id_{yz}$
$\hat{L}_x d_{yz} = i\sqrt{3}d_{z^2} + id_{x^2-y^2}$	$\hat{L}_y d_{yz} = id_{xy}$	$\hat{L}_z d_{yz} = -id_{xz}$
$\hat{L}_x d_{xy} = -id_{xz}$	$\hat{L}_y d_{xy} = -id_{yz}$	$\hat{L}_z d_{xy} = -2id_{x^2-y^2}$
$\hat{L}_x d_{x^2-y^2} = -id_{yz}$	$\hat{L}_y d_{x^2-y^2} = -id_{xz}$	$\hat{L}_z d_{x^2-y^2} = 2id_{xy}$
$\hat{L}_x d_{z^2} = -i\sqrt{3}d_{yz}$	$\hat{L}_y d_{z^2} = i\sqrt{3}d_{xz}$	$\hat{L}_z d_{z^2} = 0$

D is the axial and E the rhombic zero field splitting parameter. This zero field splitting derives from spin–orbit coupling with orbitally degenerate excited states which are split in energy by the low symmetry. D and E are thus obtained by diagonalizing the complete ligand field and spin–orbit matrices for the lower symmetry complex. Zero field splitting can be fairly large and greatly affect the ground state magnetic properties of a transition metal complex.

4.3 Spectroscopy

We experimentally probe the complete energy level diagram of a transition metal complex in Figure 19 using the range of spectroscopic methods that will be developed in detail in the following chapters of these Volumes. Ligand field and molecular orbital theories predict the energies of the states and the change in electronic structure upon excitation which leads to excited state distortions and absorption band shapes. For these states to be spectroscopically observed, the absorption intensity which is quantified by the oscillator strength, f, must be greater than 0.

$$f = (4.702 \times 10^{29})\upsilon\left[\langle\Psi_g|M|\Psi_e\rangle\right]^2 \tag{139}$$

M is the transition moment operator (see Chapter 1, Vol. II for units) which in the limit of the wavelength of light being very much greater than the radius of the electron in the complex is given by the multiple expansion in Eqn. (140).[50]

$$M = er \cdot \boldsymbol{E} + \beta(L + 2S) \cdot \boldsymbol{H} + e(r \cdot \boldsymbol{k})(r \cdot \boldsymbol{E}) + \cdots \tag{140}$$

The first term is the electric dipole operator, the second is the magnetic dipole and the third is the electric quadrupole operator. Each term is orders of magnitude smaller than the preceding one and the expression given usually suffices. $\boldsymbol{E}$ and $\boldsymbol{H}$ are the electric and magnetic vectors of the electromagnetic radiation and $\boldsymbol{k}$ is its propagation direction. These project out specific components of the associated operator which involves the electron coordinates in the transition moment integral in Eqn. (139). This leads to the selection rules and polarizations of transitions. When $\Gamma_{\psi_g} \times \Gamma_M \times \Gamma_{\psi_e}$ contains the totally symmetric irreducible representation the $\Psi_g \rightarrow \Psi_e$ transition is allowed (i.e., it can have a nonzero oscillator strength). If the complex can be oriented in a single crystal, the specific component of M_i or M_{ij}, $i, j = x, y, z$, determines the polarization of the transition (see Chapter 4).[4,41,49,51]

The quadrupole term is usually much smaller than the dipole terms and is usually neglected. However at higher photon energies (i.e., for X-rays) the wavelength of light is small (at 9000 eV at the Cu K-edge, $\lambda \sim 1.4$ Å) and higher terms in the multiple expansion can contribute. This has been shown to be the case for the $1s \rightarrow 3d$ transitions in the pre-K edge region of the X-ray absorption spectrum of transition metal complexes.[52]

The magnetic dipole term is utilized in EPR spectroscopy where there is no change in the orbital part of the electronic wave function. For transitions between the ground and excited states this term is estimated to be $\sim 10^{-4}$ of the magnitude of an electric dipole allowed transition. However, for electric dipole forbidden transitions (such as $d \rightarrow d$, *vide infra*) this term does contribute in a few cases.[41]

The electric dipole operator generally determines what is observed in absorption and emission spectroscopy. This operator, $r = x, y, z$, has u symmetry, thus only $g \rightarrow u$ and $u \rightarrow g$ transitions are parity allowed. Since d orbitals have g symmetry, ligand field transitions are all parity forbidden (also called Laporté forbidden). The ligand field can overcome this forbiddenness leading to observed spectroscopic transitions in one of two ways. In non-centrosymmetric complexes, parity allowed excited states can mix into the ligand field excited states.

$$|\Psi_e'\rangle = |\Psi_e^o\rangle + \frac{\langle\Psi_e^o|H'_{\text{odd}}|\Psi_u\rangle}{E_e - E_u}|\Psi_u\rangle$$
$$f = \frac{v_e}{v_u}\left[\frac{\langle\Psi_e^o|H'_{\text{odd}}|\Psi_u\rangle}{E_e - E_u}\right]^2 f_u \qquad (141)$$

where H'_{odd} is the term in the molecular Hamiltonian which allows mixing with Ψ_u which is an allowed excited state with oscillator strength f_u at an energy $E_u - E_e$ above the excited state. v_e and v_u are the energies of the $\Psi_g \rightarrow \Psi_e$ and $\Psi_g \rightarrow \Psi_u$ transitions. Note that as $E_e - E_u$ becomes small the mixing increases. The mixing term in brackets can be related to the amount of ligand character mixed into the antibonding d orbitals due to bonding (*vide infra*).

For centrosymmetric complexes, the same Ψ_u mixing into parity forbidden excited states occurs dynamically through vibrations of the molecule. This is called vibronic (or Herzberg–Teller) coupling. The linear coupling term from Eqn. (126) is now used to mix Ψ_e and Ψ_u (rather than evaluated within a given electronic state).

$$|\Psi_e'\rangle = |\Psi_e^o\rangle + \frac{\langle\Psi_e^o|Q_i\frac{\delta H_{\text{LF}}}{\delta Q_i}|\Psi_u\rangle}{E_e - E_u}|\Psi_u\rangle$$
$$= |\Psi_e^o\rangle + \frac{Q_i\langle\Psi_e^o|\frac{\delta H_{\text{LF}}}{\delta Q_i}|\Psi_u\rangle}{E_e - E_u}|\Psi_u\rangle \qquad (142)$$

Since the expression for the excited state ψ_e' now contains Q_i which operates on nuclear coordinates, we evaluate the electric dipole operator between the Born–Oppenheimer product ground and excited state total wavefunctions (Eqn. (128)).

$$f^{1/2} \propto \langle\Psi_g^{\text{tot}}|M(x,y,z)|\Psi_e^{\text{tot}}\rangle = \langle\Psi_g\chi_m(Q_i)|M(x,y,z)|\Psi_e'\chi_n(Q_i)\rangle$$
$$\propto \frac{\langle\Psi_e|\frac{\delta H_{\text{LF}}}{\delta Q_i}|\Psi_u\rangle}{E_e - E_u}\langle\Psi_g|M(x,y,z)|\Psi_e\rangle$$
$$\times \langle\chi_m(Q_i)|Q_i|\chi_n(Q_i)\rangle \qquad (143)$$

The first two integrals give the electronic selection rules for vibronic coupling. For these simultaneously to be nonzero (i.e., $\Gamma_e \times \Gamma_{\delta H_{\text{LF}}/\delta Q_i} \times \Gamma_u \subset A_{1g}$ and

$\Gamma_g \times \Gamma_{x,y,z} \times \Gamma_u \subset A_{1g}$ with $\Gamma_{Q_i} = \Gamma_{\delta H_{\text{LF}}/\delta Q_i}$) then $\Gamma_g \times \Gamma_{x,y,z} \times \Gamma_e \times \Gamma_{Q_i} \subset A_{1g}$ which means that the quadruple direct product must be totally symmetric. For parity forbidden transitions, $\Gamma_g \times \Gamma_e$ has g symmetry and the Q_i must match the u symmetry vibrational modes in Table 17 for the intensity to be nonzero. Thus a different set of vibrational normal modes are involved in vibronic coupling from those participating in the excited state distortions described in Section 4.1. The third integral in Eqn. (143) is the vibrational selection rule for vibronic coupling. This integral can be nonzero only if $n = m \pm 1$. Thus at low temperature only the $m = 0 \rightarrow n = 1$ transition is vibronically allowed in an odd parity vibration Q_i. This vibrational integral is proportional to the $\sqrt{m}$ or $\sqrt{n}$ whichever is larger which leads to the temperature dependence of the absorption intensity of vibronically allowed transitions given in Eqn. (144).

$$f(T) = f(0^\circ K)\coth(v/2kT) \tag{144}$$

The allowed transitions $\Psi_g \rightarrow \Psi_u$, which are either observed directly at ν_u or mix intensity into the parity forbidden transitions at v_e as described above, can either be localized on the metal ($3d \rightarrow 4p$ for the UV region, $1s \rightarrow 4p$ for the X-ray region, etc.) or involve charge transfer between the metal and ligand centers. The intensity of the former is simply obtained by evaluating the electric dipole operator between the appropriate metal centered atomic orbitals, $I = c\langle\Psi_M|r|\Psi'_M\rangle^2$ where c represents the constant in the oscillator strength expression in Eqn. (139). However in a complex, the intensity must be modified to account for the covalent reduction of the amount of metal character in the valence orbital on the metal ion, $\alpha^2 < 1$.

$$I_{(M\text{ centered})} = \alpha^2\left[c\langle\Psi_M|r|\Psi'_M\rangle^2\right] \tag{145}$$

For charge transfer transitions (which usually dominate the visible–UV spectral region of many metal complexes) the ground and excited states must both be modified to include ligand character, $\Psi = c_M\phi_M + c_L\phi_L$. The electric dipole transition moment integral is then expanded into four terms:

$$\begin{aligned}\langle\psi_g|r|\psi_e\rangle &= c_M^g c_M^e\langle\phi_M^g|r|\phi_M^e\rangle + c_M^g c_L^e\langle\phi_M^g|r|\phi_L^e\rangle \\ &\quad + c_L^g c_M^e\langle\phi_L^g|r|\phi_M^e\rangle + c_L^g c_L^e\langle\phi_L^g|r|\phi_L^e\rangle\end{aligned} \tag{146}$$

The fourth term is the ligand-ligand integral which is usually found to dominate[53,54] and can be approximated by Eqn. (147), where α are the ligands, k_α the coefficient of ligand α in ϕ_L, r_α is the position vector of ligand α taken from the metal center, and δ_{ge} is the overlap of atomic orbitals on a given ligand in the molecular orbitals associated with the ground and excited states. Within this approximation, transition intensity requires the same orbital to be present on a given ligand in both the ground and excited states.

$$\langle\phi_L^g|r|\phi_i^e\rangle = \sum_\alpha k_\alpha^g k_\alpha^e r_\alpha \delta_{ge} \tag{147}$$

Thus the metal centered electric dipole allowed transitions are found to decrease in intensity due to covalency, while charge transfer intensity (and the corresponding intensities of the ligand field transitions with which these mix) increases from zero as ligand character mixes into the metal *d* orbitals due to bonding.

4.4 Covalency

In Section 4.3. it was observed that the intensity of charge transfer and metal edge transitions reflect the covalency of the ligand–metal bond. It is useful, at this point, to list other spectroscopic parameters which directly reflect the amount of ligand character mixed into the metal *d* orbitals due to bonding. In magnetic susceptibility and EPR it is generally found that the experimental *g* values deviate significantly from those predicted by ligand field theory (Eqns. (136, 137)). This is because covalency delocalizes the unpaired electron partially onto the ligand which contributes a reduced orbital angular momentum contribution to the ground state[55,56] (also see Chapter 2). EPR is a higher resolution method which allows one to observe additional structure, in particular nuclear spin hyperfine coupling to the electron spin on the metal. If the ligand has a nonzero nuclear magnetic moment, it can undergo a ligand hyperfine interaction with the unpaired electron, called superhyperfine coupling, the magnitude of which can directly quantify the amount of ligand character in the ground state. Metal hyperfine coupling and electron spin zero field splitting are also strongly affected by covalency but in a less direct manner; and thus covalency must usually be treated as input in the analysis of these parameters.[55,57] In addition to intensity, the energies of ligand field and charge transfer states can be used through a configurational interaction (CI) model to evaluate the bonding interactions which give covalent mixing.[58] Spectroscopic transition energies can also be compared to the results of electronic structure calculations to evaluate the accuracy of the calculated description of bonding.

Photoelectron spectroscopy (PES) directly probes all the valence orbitals on the metal and ligands involved in bonding. Since metal *d* and ligand *p* atomic orbitals have qualitatively different photoionization cross sections (i.e., photon energy dependence of photoelectron intensity) use of a synchrotron to obtain variable energy PES can provide a powerful probe of bonding over all occupied valence orbitals.[59]

While the intensity of metal edge transitions will be reduced with increased covalency of the valence orbitals on the metal ion, the opposite is the case for ligand K edges. The ligand K edge can be used as a direct probe of the amount of ligand character in a metal *d* orbital, very much as with superhyperfine coupling but of more general utility as this method does not require an EPR active site and can be used for all unoccupied and half occupied metal valence orbitals.[45] The idea here is that the ligand K edge involves a transition of a $1s$ orbital centered on the ligand to the "*d*" valence orbitals centered on the metal ion. The electric dipole selection rule governs the intensity such that $1s \rightarrow np$ on the ligand is allowed. Thus intensity in the ligand K edge will reflect the amount of ligand *p* character mixed into the metal *d* orbital due to bonding, Eqn. (148), where $\alpha^2 > 0$ due to covalency and $I(L1s \rightarrow Lnp)$ is the intensity of a pure $1s \rightarrow np$ transition centered on the ligand.

$$I = \alpha^2 I(L1s \rightarrow Lnp) \qquad (148)$$

5 CONCLUDING COMMENTS

It should be obvious from the above discussion that ligand field and related bonding theories provide a firm basis for our understanding of the spectroscopic properties of transition metal complexes. These are very powerful approaches but at this point we should emphasize two limitations of the above bonding descriptions. First, they assume what is known as the frozen orbital approximation. This means that the electronic structure does not change upon excitation to the final state. This approximation, however, is not always reasonable as the electron repulsion will change in transitions between orbitals with different covalencies or upon ionization. In the ligand field and charge transfer spectroscopic region this effect can be of the order of 1 to 2 eV, and can produce significant variations in the relative energies of the ligand field and charge transfer manifolds.[53] In higher energy spectroscopies, photoelectron, and X-ray absorption, the effects of electronic relaxation can be very large. The most obvious manifestation of this is in the appearance of shake-up and shake-down satellite peaks in addition to the main spectroscopic features. These formally involve two electron transitions (in PES one electron is ionized and the second undergoes a ligand-to-metal charge transfer transition; in XAS this involves a $1s \rightarrow 4p$ excitation plus a ligand-to-metal charge transfer transition). These can have no intensity (the electric dipole operator in Eqn. (140) is a one-electron operator) unless the electronic structure significantly changes in the final state. The intensity distribution over the main and satellite peaks can, in fact, be used to quantify the change in electronic structure upon excitation.[59]

The second limitation of ligand field theory is that it utilizes a spin restricted description of bonding. This means that two electrons with different spins occupy the same electronic wavefunction. However, for metal ions with partially filled valence orbitals, electron exchange will be different for the spin-up and spin-down electrons. If the electrons of different spin are allowed to have different orbital wavefunctions, that is, a *spin unrestricted description of bonding*, this spin polarization can significantly change the energies of the spin-up relative to the spin-down states. This exchange splitting can become large as the number of unpaired electrons increases and can dominate the bonding description in the limiting case of high-spin d^5, which leads to an "inverted" bonding scheme.[60] This is experimentally observed in the Fermi contact contribution to metal hyperfine coupling[55,56] and in resonance effects in the photoelectron spectra of transition metal complexes.[61] While it is important to be aware of these limitations and their spectroscopic consequences, ligand field and related bonding theories do indeed provide a well established (by experiment) basis for understanding the spectroscopy and physical properties of transition metal complexes.

We finally conclude this introductory chapter by noting that in addition to physical properties (structure, magnetism, etc.) the electronic structure makes major contributions to chemical reactivity. Electronic structure plays a key role in determining the redox properties of molecules and their kinetics of electron transfer. Often in bioinorganic systems, one observes unique spectral features compared to small molecule inorganic complexes of the same metal ion, indicating novel electronic structures which can be responsible for activating a metal site for catalysis.

Electronic excitation changes the electronic structure of a molecule and this can lead to new reactivity (i.e., photochemistry and photophysics), and excited states can provide ideal systems for evaluating electronic structure contributions to reactivity. These themes are developed in detail in the second Volume of this series.

APPENDIX 1

Table 1A Spherical Harmonics in Cartesian Coordinates[3] Using Cordon and Shortley[39] Phases

$$Y_0^0 = \sqrt{\frac{1}{4\pi}}$$

$$Y_1^{-1} = \sqrt{\frac{3}{8\pi}}\frac{x - iy}{r}$$

$$Y_1^0 = \sqrt{\frac{3}{4\pi}}\frac{z}{r}$$

$$Y_1^1 = -\sqrt{\frac{3}{8\pi}}\frac{x + iy}{r}$$

$$Y_2^{-2} = \sqrt{\frac{5}{4\pi}}\sqrt{\frac{3}{8}}\frac{(x - iy)^2}{r^2}$$

$$Y_2^{-1} = \sqrt{\frac{5}{4\pi}}\sqrt{\frac{3}{2}}\frac{z(x - iy)}{r^2}$$

$$Y_2^0 = \sqrt{\frac{5}{4\pi}}\sqrt{\frac{1}{4}}\frac{3z^2 - r^2}{r^2}$$

$$Y_2^1 = -\sqrt{\frac{5}{4\pi}}\sqrt{\frac{3}{2}}\frac{z(x + iy)}{r^2}$$

$$Y_2^2 = \sqrt{\frac{5}{4\pi}}\sqrt{\frac{3}{8}}\frac{(x + iy)^2}{r^2}$$

$$Y_3^{-3} = \sqrt{\frac{7}{4\pi}}\sqrt{\frac{5}{16}}\frac{(x - iy)^3}{r^3}$$

$$Y_3^{-2} = \sqrt{\frac{7}{4\pi}}\sqrt{\frac{15}{8}}\frac{z(x - iy)^2}{r^3}$$

$$Y_3^{-1} = \sqrt{\frac{7}{4\pi}}\sqrt{\frac{3}{16}}\frac{(x-iy)(5z^2-r^2)}{r^3}$$

$$Y_3^{0} = \sqrt{\frac{7}{4\pi}}\sqrt{\frac{1}{4}}\frac{z(5z^2-3r^2)}{r^3}$$

$$Y_3^{1} = -\sqrt{\frac{7}{4\pi}}\sqrt{\frac{3}{16}}\frac{(x+iy)(5z^2-r^2)}{r^3}$$

$$Y_3^{2} = \sqrt{\frac{7}{4\pi}}\sqrt{\frac{15}{8}}\frac{z(x+iy)^2}{r^3}$$

$$Y_3^{3} = -\sqrt{\frac{7}{4\pi}}\sqrt{\frac{5}{16}}\frac{(x+iy)^3}{r^3}$$

$$Y_4^{-4} = \sqrt{\frac{9}{4\pi}}\sqrt{\frac{35}{128}}\frac{(x-iy)^4}{r^4}$$

$$Y_4^{-3} = \sqrt{\frac{9}{4\pi}}\sqrt{\frac{35}{16}}\frac{z(x-iy)^3}{r^4}$$

$$Y_4^{-2} = \sqrt{\frac{9}{4\pi}}\sqrt{\frac{5}{32}}\frac{(x-iy)^2}{r^4}(7z^2-r^2)$$

$$Y_4^{-1} = \sqrt{\frac{9}{4\pi}}\sqrt{\frac{5}{16}}\frac{(x-iy)}{r^4}(7z^3-3zr^2)$$

$$Y_4^{0} = \sqrt{\frac{9}{4\pi}}\sqrt{\frac{1}{64}}\frac{35z^4-30z^2r^2+3r^4}{r^4}$$

$$Y_4^{1} = -\sqrt{\frac{9}{4\pi}}\sqrt{\frac{5}{16}}\frac{(x+iy)}{r^4}(7z^3-3zr^2)$$

$$Y_4^{2} = -\sqrt{\frac{9}{4\pi}}\sqrt{\frac{5}{32}}\frac{(x+iy)^2}{r^4}(7z^2-r^2)$$

$$Y_4^{3} = -\sqrt{\frac{9}{4\pi}}\sqrt{\frac{35}{16}}\frac{z(x+iy)^3}{r^4}$$

$$Y_4^{4} = -\sqrt{\frac{9}{4\pi}}\sqrt{\frac{35}{128}}\frac{z(x+iy)^4}{r^4}$$

$$Y_5^{-5} = \sqrt{\frac{11}{4\pi}}\sqrt{\frac{63}{256}}\frac{(x-iy)^5}{r^5}$$

$$Y_5^{-4} = \sqrt{\frac{11}{4\pi}}\sqrt{\frac{315}{128}}\frac{z(x-iy)^4}{r^5}$$

$$Y_5^{-3} = \sqrt{\frac{11}{4\pi}}\sqrt{\frac{35}{256}}\frac{z(x-iy)^3}{r^5}(9z^2-r^2)$$

$$Y_5^{-2} = \sqrt{\frac{11}{4\pi}}\sqrt{\frac{105}{32}}\frac{(x-iy)^2}{r^5}(3z^3-zr^2)$$

$$Y_5^{-1} = \sqrt{\frac{11}{4\pi}}\sqrt{\frac{15}{128}}\frac{(x-iy)}{r^5}(21z^4 - 14z^2r^2 + r^4)$$

$$Y_5^0 = \sqrt{\frac{11}{4\pi}}\frac{1}{8}\frac{63z^5 - 70z^3r^2 + 15zr^4}{r^4}$$

$$Y_5^1 = -\sqrt{\frac{11}{4\pi}}\sqrt{\frac{15}{128}}\frac{x+iy}{r^5}(21z^4 - 14z^2r^2 + r^4)$$

$$Y_5^2 = \sqrt{\frac{11}{4\pi}}\sqrt{\frac{105}{32}}\frac{(x+iy)^2}{r^5}(3z^3 - zr^2)$$

$$Y_5^3 = -\sqrt{\frac{11}{4\pi}}\sqrt{\frac{35}{256}}\frac{(x+iy)^3}{r^5}(9z^2 - r^2)$$

$$Y_5^4 = \sqrt{\frac{11}{4\pi}}\sqrt{\frac{315}{128}}\frac{z(x+iy)^4}{r^5}$$

$$Y_5^5 = -\sqrt{\frac{11}{4\pi}}\sqrt{\frac{63}{256}}\frac{(x+iy)^5}{r^5}$$

Table 1B Spherical Harmonics in Polar Coordinates Using Condon and Shortley[39] Phases

$$Y_{00} = 1/(4\pi)^{1/2}$$

$$Y_{10} = (3/4\pi)^{1/2}\cos\theta$$

$$Y_{1\pm1} = \mp(3/8\pi)^{1/2}\sin\theta\, e^{\pm i\phi}$$

$$Y_{20} = (5/16\pi)^{1/2}\left(2\cos^2\theta - \sin^2\theta\right)$$

$$Y_{2\pm1} = \mp(15/8\pi)^{1/2}\sin\theta\cos\theta\, e^{\pm i\phi}$$

$$Y_{2\pm2} = (5/32\pi)^{1/2}\sin^2\theta\, e^{\pm 2i\phi}$$

$$Y_{30} = (7/16\pi)^{1/2}\left(2\cos^3\theta - 3\sin^2\theta\cos\theta\right)$$

$$Y_{3\pm1} = \mp(21/64\pi)^{1/2}\left(4\cos^2\theta\sin\theta - \sin^3\theta\right)e^{\pm i\phi}$$

$$Y_{3\pm2} = (105/32\pi)^{1/2}\sin^2\theta\cos\theta\, e^{\pm 2i\phi}$$

$$Y_{3\pm3} = \mp(35/64\pi)^{1/2}\sin^3\theta\, e^{\pm 3i\phi}$$

$$Y_{40} = 9/(256\pi)^{1/2}\left[(35/3)\cos^4\theta - 10\cos^2\theta + 1\right]$$

$$Y_{4\pm1} = 9(5/64\pi)^{1/2}\sin\theta\left[(7/3)\cos^3\theta - \cos\theta\right]e^{\pm i\phi}$$

$$Y_{4\pm2} = 3(5/128\pi)^{1/2}\sin^2\theta\left[7\cos^2\theta - 1\right]e^{\pm 2i\phi}$$

$$Y_{4\pm3} = 3(35/64\pi)^{1/2}\sin^3\theta\cos\theta\, e^{\pm 3i\phi}$$

$$Y_{4\pm4} = 3(35/512\pi)^{1/2}\sin^4\theta\, e^{\pm 4i\phi}$$

$$e^{i\alpha} = \cos\alpha + i\sin\alpha$$

$$e^{-i\alpha} = \cos\alpha - i\sin\alpha$$

APPENDIX 2

A collection of 3-j symbols listed in the form $j_1 j_2 j_3/m_1 m_2 m_3$. Recall the permutation rules described in Section 2.1.2.2 to find 3-j symbols presented here in a different order from that sought.

$j_1 j_2 j_3/m_1 m_2 m_3$

$110/000 = -\surd(1/3)$
$110/1{-}10 = \surd(1/3)$
$111/000 = 0$
$211/000 = \surd(2/15)$
$211/2{-}1{-}1 = \surd(1/5)$
$123/1{-}10 = -\surd(1/35)$
$222/20{-}2 = \surd(2/35)$
$222/000 = -\surd(2/35)$
$222/{-}121 = -\surd(3/35)$
$242/10{-}1 = \surd(8/315)$
$242/{-}220 = \surd(1/42)$
$242/24{-}2 = 1/3$

$132/12{-}3 = \surd(1/7)$
$132/{-}12{-}1 = \surd(1/105)$
$323/{-}321 = \surd(1/42)$
$330/000 = -\surd(1/7)$
$332/2{-}20 = 0$
$323/02{-}2 = -\surd(1/21)$
$321/000 = -\surd(3/35)$
$330/1{-}10 = \surd(1/7)$
$330/2{-}20 = -\surd(1/7)$
$330/3{-}30 = \surd(1/7)$
$331/1{-}21 = \surd(5/84)$
$331/2{-}31 = -\surd(1/28)$
$331/2{-}1{-}1 = \surd(1/84)$
$332/{-}202 = -\surd(1/21)$
$332/{-}101 = -\surd(1/210)$
$332/1{-}32 = \surd(1/42)$
$332/1{-}10 = -\surd(3/140)$
$332/2{-}20 = 0$
$333/{-}303 = \surd(1/42)$
$333/{-}202 = \surd(1/42)$
$333/{-}101 = -\surd(1/42)$

$j_1 j_2 j_3/m_1 m_2 m_3$

$111/{-}101 = \surd(1/6)$
$211/{-}101 = -\surd(1/10)$
$211/0{-}11 = \surd(1/30)$
$211/1{-}10 = -\surd(1/10)$
$123/10{-}1 = \surd(2/35)$
$123/000 = -\surd(3/35)$
$222/10{-}1 = \surd(1/70)$
$242/{-}13{-}2 = -\surd(1/18)$
$242/20{-}2 = \surd(1/630)$
$242/000 = \surd(2/35)$
$242/{-}121 = \surd(4/63)$

$322/{-}101 = \surd(1/35)$
$322/{-}1{-}12 = \surd(3/70)$
$321/{-}101 = \surd(2/35)$

$132/02{-}2 = -\surd(1/21)$
$323/{-}12{-}1 = \surd(2/35)$
$323/{-}220 = -\surd(1/21)$
$332/3{-}30 = \surd(5/84)$
$323/10{-}1 = -\surd(3/140)$
$323/000 = \surd(4/105)$
$321/10{-}1 = \surd(2/35)$

$331/{-}101 = \surd(1/14)$
$331/000 = 0$
$331/1{-}10 = \surd(1/84)$
$331/2{-}20 = -\surd(1/21)$
$331/3{-}30 = \surd(3/28)$
$332/{-}1{-}12 = \surd(2/35)$
$332/000 = \surd(4/105)$
$332/1{-}21 = \surd(1/28)$
$332/2{-}31 = -\surd(5/84)$
$332/3{-}30 = \surd(5/84)$
$333/{-}2{-}13 = -\surd(1/21)$
$333/{-}1{-}12 = 0$
$333/00 = 0$

$j_1 j_2 j_3/m_1 m_2 m_3$	$j_1 j_2 j_3/m_1 m_2 m_3$
$341/-34-1 = \sqrt{(1/9)}$	$343/-34-1 = \sqrt{(1/33)}$
$343/-24-2 = -\sqrt{(5/99)}$	$341/-321 = \sqrt{(1/252)}$
$341/-101 = \sqrt{(1/42)}$	$341/-121 = \sqrt{(5/84)}$
$341/000 = \sqrt{(4/63)}$	$341/-220 = \sqrt{(1/21)}$
$343/-303 = \sqrt{(1/154)}$	$343/2-20 = \sqrt{(1/462)}$
$343/-202 = \sqrt{(7/198)}$	$343/-101 = \sqrt{(1/1386)}$
$343/1-21 = -\sqrt{(20/693)}$	$343/000 = -\sqrt{(2/77)}$
$343/12-3 = \sqrt{(3/77)}$	$432/000 = 0$
$432/-202 = \sqrt{(1/21)}$	$432/-1-12 = -\sqrt{(2/63)}$
$432/-101 = -\sqrt{(1/42)}$	$432/0-22 = \sqrt{(1/63)}$
$432/0-11 = \sqrt{(5/126)}$	$432/1-32 = -\sqrt{(1/210)}$
$432/1-21 = -\sqrt{(7/180)}$	$432/1-10 = -\sqrt{(1/84)}$
$432/2-31 = \sqrt{(3/140)}$	$432/2-20 = \sqrt{(4/105)}$
$432/2-1-1 = -\sqrt{(1/252)}$	$432/3-30 = -\sqrt{(1/20)}$
$432/3-2-1 = -\sqrt{(1/140)}$	$432/3-1-2 = \sqrt{(1/18)}$
$432/4-31 = \sqrt{(1/15)}$	$432/4-2-2 = -\sqrt{(2/45)}$

APPENDIX 3

Table A3-1 Energy Matrices for Octahedral d^2 Transition Metal Complexes

$^1A_1$	t^2	$10B + 5C - 8Dq$	$\sqrt{6}(2B + C)$
	e^2	$\sqrt{6}(2B + C)$	$8B + 4C + 12Dq$
$^3A_2$	e^2	$-8B + 12Dq$	
^1E	t^2	$B + 2C - 8Dq$	$-2\sqrt{3}B$
	e^2	$-2\sqrt{3}B$	$2C + 12Dq$
$^1T_1$	et	$4B + 2C + 2Dq$	
$^3T_1$	t^2	$-5B - 8Dq$	$6B$
	et	$6B$	$4B + 2Dq$
$^1T_2$	t^2	$B + 2C - 8Dq$	$2\sqrt{3}B$
	et	$2\sqrt{3}B$	$2C + 2Dq$
$^3T_2$	et	$-8B + 2Dq$	

Table A3-2 Energy Matrices for the Configuration d^3 in a Cubic Field (Tanabe and Sugano)[43]

$^2T_2(a^2D, b^2D, ^2F, ^2G, ^2H)$					
t^3	$-12Dq + 5C$	$-3\sqrt{3}B$	$-5\sqrt{3}B$	$4B + 2C$	$2B$
$t^2(^3T_1)e$		$-2Dq - 6B + 3C$	$3B$	$-3\sqrt{3}B$	$-3\sqrt{3}B$
$t^2(^1T_2)e$			$-2Dq + 4B + 3C$	$-\sqrt{3}B$	$-\sqrt{3}B$
$te^2(^1A_1)$				$8Dq + 6B + 5C$	$10B$
$te^2(^1E)$					$8Dq - 2B + 3C$
$^2T_1(^2P, ^2F, ^2G, ^2H)$					
t^3	$-12Dq - 6B + 3C$	$-3B$	$3B$	0	$-2\sqrt{3}B$
$t^2(^3T_1)e$		$-2Dq + 3C$	$-3B$	$3B$	$3\sqrt{3}B$
$t^2(^1T_2)e$			$-2Dq - 6B + 3C$	$-3B$	$-\sqrt{3}B$
$te^2(^3A_2)$				$8Dq - 6B + 3C$	$2\sqrt{3}B$
$te^2(^1E)$					$8Dq - 2B + 3C$
$^2E(a^2D, b^2D, ^2G, ^2H)$					
t^3	$-12Dq - 6B + 3C$		$-6B\sqrt{2}B$	$-3\sqrt{2}B$	0
$t^2(^1A_1)e$			$-2Dq + 8B + 6C$	$10B$	$\sqrt{3}(2B + C)$
$t^2(^1E)e$				$-2Dq - B + 3C$	$2\sqrt{3}B$
e^3					$18Dq - 8B + 4C$
$^4T_1(^4P, ^4F)$				$^4A_2(^4F)t^3$	$-12Dq - 15B$
				$^4T_2(^4F)t^2(^3T_1)e$	$-2Dq - 15B$
$t^2(^3T_1)e$	$2Dq - 3B$	$6B$		$^2A_1(^2G)t^2(^1E)e$	$-2Dq - 11B + 3C$
$te^2(^3A_2)$		$8Dq - 12B$		$^2A_2(^2F)t^2(^1E)e$	$-2Dq + 9B + 3C$

Table A3-3 Energy Matrices for the Configuration d^4 in a Cubic Field (Tanabe and Sugano)

$^1E(a^1D, b^1D, a^1G, b^1G, ^1I)$					
t^4	$-16Dq - 9B + 7C$	$6B$	$\sqrt{2}(2B + C)$	$-2B$	$-4B$
$t^3(^2E)e$		$-6Dq - 6B + 6C$	$-3\sqrt{2}B$	$-12B$	0
$t^2(^1E)e^2(^1A_1)$			$4Dq + 5B + 8C$	$10\sqrt{2}B$	$-10\sqrt{2}B$
$t^2(^1A)e^2(^1E)$				$4Dq + 6B + 9C$	0
$t^2(^1E)e^2(^1E)$					$4Dq - 3B + 6C$

$^3T_2(^3D, a^3F, b^3F, ^3G, ^3H)$					
$t^3(^2T_1)e$	$-6Dq - 9B + 4C$	$-5\sqrt{3}B$	$\sqrt{6}B$	$\sqrt{3}B$	$-\sqrt{6}B$
$t^3(^2T_2)e$		$-6Dq - 5B + 6C$	$-3\sqrt{2}B$	$3B$	$\sqrt{2}(3B + C)$
$t^2(^3T_1)e^2(^3A_2)$			$4Dq - 13B + 4C$	$-2\sqrt{2}B$	$-6B$
$t^2(^3T_1)e^2(^1E)$				$4Dq - 9B + 4C$	$3\sqrt{2}B$
te^3					$14Dq - 8B + 5C$

$^1T_1(^1F, a^1G, b^1G, ^1I)$				
$t^3(^2T_1)e$	$-6Dq - 3B + 6C$	$5\sqrt{3}B$	$3B$	$\sqrt{6}B$
$t^3(^2T_2)e$		$-6Dq - 3B + 8C$	$-5\sqrt{3}B$	$\sqrt{2}(B + C)$
$t^2(^1T_2)e^2(^1E)$			$4Dq - 3B + 6C$	$-\sqrt{6}B$
te^3				$14Dq - 16B + 7C$

$^3E(^3D, ^3G, ^3H)$			
$t^3(^4A_2)e$	$-6Dq - 13B + 4c$	$-4B$	0
$t^3(^3E)e$		$-6Dq - 10B + 4C$	$-3\sqrt{2}B$
$t^2(^1E)e^2(^3A_2)$			$4Dq - 11B + 4C$

$^3A_2(a^3F, b^3F)$		
$t^3(^2E)e$	$-6Dq - 8B + 4C$	$-12B$
$t^2(^1A_1)e^2(^3A_2)$		$4Dq - 2B + 7C$

Table A3-3 (Continued)

$^1A_2(^1F, ^1I)$		
$t^3(^2E)e$	$-6Dq - 12B + 6C$	$6B$
$t^2(^1E)e^2(^1E)$		$4Dq - 3B + 6C$

$^5E(^5D)t^3(^4A_2)e$	$-6Dq - 21B$
$^5T_2(^5D)t^2(^3T_1)e^2(^3A_2)$	$4Dq - 21B$
$^3A_1(^3G)t^3(^2E)e$	$-6Dq - 12B + 4C$

$^3T_1(a^3P, b^3P, a^3F, b^3F, ^2G, ^3H)$							
t^4	$-16Dq - 15B + 5C$	$-\sqrt{6}B$	$-3\sqrt{2}B$	$\sqrt{2}(2B + C)$	$-2\sqrt{2}B$	0	0
$t^3(^3T_1)e$		$-6Dq - 11B + 4C$	$5\sqrt{3}B$	$\sqrt{3}B$	$-\sqrt{3}B$	$3B$	$\sqrt{6}B$
$t^3(^2T_2)e$			$-6Dq - 3B + 6C$	$-3B$	$-3B$	$5\sqrt{3}B$	$\sqrt{2}(B + C)$
$t^2(^3T_1)e^2(^1A_1)$				$-4Dq - B + 6C$	$-10B$	0	$3\sqrt{2}B$
$t^2(^3T_1)e^2(^1E)$					$4Dq - 9B + 4C$	$-2\sqrt{3}B$	$-3\sqrt{2}B$
$t^2(^1T_2)e^2(^3A_2)$						$4Dq - 11B + 4C$	$\sqrt{6}B$
te^3							$14Dq - 16B + 5C$

$^1T_2(a^1D, b^1D, a^1G, b^1G, ^1F, ^1I)$							
t^4	$-16Dq - 9B + 7C$	$3\sqrt{2}B$	$-5\sqrt{6}B$	0	$-2\sqrt{2}B$	$\sqrt{2}(2B + C)$	0
$t^3(^2T_1)e$		$-6Dq - 9B + 6C$	$-5\sqrt{3}B$	$3B$	$-3B$	$-3B$	$-\sqrt{6}B$
$t^3(^2T_2)e$			$-6Dq + 3B + 8C$	$-3\sqrt{3}B$	$5\sqrt{3}B$	$-5\sqrt{3}B$	$\sqrt{2}(3B + C)$
$t^2(^3T_1)e^2(^3A_2)$				$4Dq - 9B + 6C$	$-6B$	0	$-3\sqrt{6}B$
$t^2(^1T_2)e^2(^1E)$					$4Dq - 3B + 6C$	$-10B$	$\sqrt{6}B$
$t^2(^1T_2)e^2(^1A_1)$						$4Dq + 5B + 8C$	$\sqrt{6}B$
te^3							$14Dq + 7C$

$^1A_1(a^1S, b^1S, a^1G, b^1G, ^1I)$					
t^4	$-16Dq + 10C$	$-12\sqrt{2}B$	$\sqrt{2}(4B + 2C)$	$2\sqrt{2}B$	0
$t^3(^2E)e$		$-6Dq + 6C$	$-12B$	$-6B$	0
$t^2(^1A_1)e^2(^1A_1)$			$4Dq + 14B + 11C$	$20B$	$\sqrt{6}(2B + C)$
$t^2(^1E)e^2(^1E)$				$4Dq - 3B + 6C$	$2\sqrt{6}B$
e^4					$24Dq - 16B + 8C$

Table A3-4 Energy Matrices for the Configuration d^5 in a Cubic Field (Tanabe and Sugano)

$^2E(a^2D, b^2D, c^2D, a^2G, b^2G, {}^2H, {}^2I)$

$t^4(^1A_1)e$	$-10Dq - 4B + 12C$	$10B$	$6B$	$6\sqrt{3}B$	$6\sqrt{2}B$	$-2B$	$4B + 2C$
$t^4(^1E)e$		$-10Dq - 13B + 9C$	$-3B$	$3\sqrt{3}B$	0	$2B + C$	$2B$
$t^3(^2E)e^2(^1A_1)$			$-4B + 10C$	0	0	$-3B$	$-6B$
$t^3(^3E)e^2(^3A_2)$				$-16B + 8C$	$2\sqrt{6}B$	$-3\sqrt{3}B$	$6\sqrt{3}B$
$t^3(^2E)e^2(^1E)$					$-12B + 8C$	0	$6\sqrt{2}B$
$t^2(^1E)e^3$						$-10Dq - 13B + 9C$	$-10B$
$t^2(^1A_1)e^3$							$-10Dq - 4B + 12C$

$^2A_1(a^2F, b^2F, {}^2I)$

$t^4(^1E)e$	$-10Dq - 3B + 9C$	$-3\sqrt{2}B$	0	$6B + C$
$t^3(^2E)e^2(^1E)$		$-12B + 8C$	$-4\sqrt{3}B$	$3\sqrt{2}B$
$t^3(^4A_2)e^2(^3A_2)$			$-19B + 8C$	0
$t^2(^1E)e^3$				$10Dq - 3B + 9C$

$^2A_2(^2F, {}^2I)$

$t^4(^1E)e$	$-10Dq - 23B + 9C$	$3\sqrt{2}B$	$-2B + C$
$t^3(^2E)e^2(^1E)$		$-12B + 8C$	$-3\sqrt{2}B$
$t^2(^1E)e^3$			$10Dq - 23B + 9C$

$^4T_1(^4P, {}^4F, {}^4G)$

$t^4(^3T_1)e$	$-10Dq - 25B + 6C$	$-3\sqrt{2}B$	C
$t^3(^2T_2)e^2(^3A_1)$		$-16B + 7C$	$-3\sqrt{2}B$
$t^2(^3T_1)e^3$			$10Dq - 25B + 6C$

$^4T_2(^4E, {}^4G, {}^4D)$

$t^4(^3T_1)e$	$-10Dq - 17B + 6C$	$\sqrt{6}B$	$4B + C$
$t^3(^2T_1)e^2(^3A_2)$		$-22B + 5C$	$-\sqrt{6}B$
$t^2(^3T_1)e^3$			$-10Dq - 17B + 6C$

$^4E(^4D, {}^4G)$

$t^3(^2E)e^2(^3A_2)$	$-22B + 5C$	$-2\sqrt{3}B$
$t^3(^4A_2)e^2(^1E)$		$-21B + 5C$

$^6A_1(^6S)$	$t^3(^4A_2)e^2(^3A_2)$	$-35B$
$^4A_1(^4G)$	$t^3(^4A_2)e^2(^3A_2)$	$-25B + 5C$
$^4A_2(^4F)$	$t^3(^4A_2)e^2(^1A_1)$	$-13B + 7C$

Table A3-4 (*Continued*)

$^2T_2(a^2F, b^2F, a^2G, b^2G, {}^2H, {}^2I, a^2D, b^2D, c^2D)$										
t^5	$-20Dq-20B+10C$	$3\sqrt{6}B$	$\sqrt{6}B$	0	$-2\sqrt{3}B$	$4B+2C$	$2B$	0	0	0
$t^4({}^3T_1)e$		$-10Dq-8B+9C$	$3B$	$(\sqrt{6}/2)B$	$(-3\sqrt{2}/2)B$	$(3\sqrt{6}/2)B$	$(3\sqrt{6}/2)B$	0	$4B+C$	0
$t^4({}^1T_2)e$			$-10Dq-18B+9C$	$(3\sqrt{6}/2)B$	$(-3\sqrt{2}/2)B$	$(5\sqrt{6}/2)B$	$(-5\sqrt{6}/2)B$	C	0	0
$t^3({}^2T_1)e^2({}^3A_2)$				$-16B+8C$	$2\sqrt{3}B$	0	0	$(-3\sqrt{6}/2)B$	$(-\sqrt{6}/2)B$	0
$t^3({}^2T_1)e^2({}^1E)$					$-12B+8C$	$-10\sqrt{3}B$	0	$(3\sqrt{2}/2)B$	$(3\sqrt{2}/2)B$	$-2\sqrt{3}B$
$t^3({}^2T_2)e^2({}^1A_1)$						$2B+12C$	0	$(-5\sqrt{6}/2)B$	$(-3\sqrt{6}/2)B$	$4B+2C$
$t^3({}^2T_2)e^2({}^1E)$							$-6B+10C$	$(-5\sqrt{6}/2)B$	$(3\sqrt{6}/2)B$	$2B$
$t^2({}^1T_2)e^3({}^2E)$								$10Dq-18B+9C$	$3B$	$-\sqrt{6}B$
$t^2({}^3T_2)e^3({}^2E)$									$10Dq-8B+9C$	$-3\sqrt{6}B$
te^4										$20Dq-20B+10C$

$^2T_1({}^2P, a^2F, b^2F, a^2G, b^2G, {}^2H, {}^2I)$								
$t^4({}^3T_1)e$	$-10Dq-22B+9C$	$-3B$	$(-3\sqrt{2}/2)B$	$(3\sqrt{2}/2)B$	$(-3\sqrt{2}/2)B$	$(-3\sqrt{6}/2)B$	0	C
$t^4({}^1T_2)e$		$-10Dq-8B+9C$	$(-3\sqrt{2}/2)B$	$(3\sqrt{2}/2)B$	$(15\sqrt{2}/2)B$	$(5\sqrt{6}/2)B$	$4B+C$	0
$t^3({}^2T_1)e^2({}^1A_1)$			$-4B+10C$	0	0	$10\sqrt{3}B$	$(3\sqrt{2}/2)B$	$(-3\sqrt{2}/2)B$
$t^3({}^2T_1)e^2({}^1E)$				$-12B+8C$	0	0	$(-3\sqrt{2}/2)B$	$(-3\sqrt{2}/2)B$
$t^3({}^2T_2)e^2({}^3A_2)$					$-10B+10C$	$-2\sqrt{3}B$	$(15\sqrt{2}/2)B$	$(-3\sqrt{2}/2)B$
$t^3({}^2T_2)e^2({}^1E)$						$-6B+10C$	$(5\sqrt{6}/2)B$	$(-3\sqrt{6}/2)B$
$t^2({}^1T_2)e^3$							$10Dq-8B+9C$	$-3B$
$t^2({}^3T_1)e^3$								$10Dq-22B+9C$

REFERENCES

1. Bethe, H.A. *Ann. Physik.* **1929**, *3*, 133.
2. van Vleck, J.H. *J. Chem. Phys.* **1935**, *3*, 803, 807.
3. Ballhausen, C.J. *Introduction to Ligand Field Theory*; McGraw Hill: New York, **1962**.
4. Lever, A.B.P. *Inorganic Electronic Spectroscopy, 2nd Edition*; Elsevier Science Publishers: Amsterdam, **1984**.
5. Brink, D.M.; Satchler, G.R. *Angular Momentum*; Clarendon Press: Oxford, **1968**.
6. Flurry, R.L. *Symmetry Groups*; Prentice Hall: Englewood Cliffs, NJ, **1980**.
7. Griffith, J.S. *Theory of Transition Metal Ions*; Cambridge University Press: Cambridge, **1961**.
8. Watanabe, H. *Operator Methods in Ligand Field Theory*; Prentice Hall: Englewood Cliffs, NJ, **1980**.
9. Fano, U.; Racah, G. *Irreducible Tensorial Sets*; Academic Press: New York, **1959**.
10. Nielson, C.W.; Koster, G.F. *Spectroscopic Coefficients for the p^n, d^n and f^n Configurations*; MIT Press: Cambridge, MA, **1963**.
11. Polo, S.R. *Fundamentals of Crystal Field Theory*; RCA Laboratories: Princeton, NJ.
12. Wigner, E.P. *Group Theory and Its Application to the Quantum Mechanics of Atomic Spectra*; Academic Press: New York, **1959**.
13. Rotenberg, M.; Bivins, R.; Metropolis, N.; Woten, Jr. J.K. *The 3-j and 6-j Symbols*; The Technology Press, MIT: Cambridge, MA, **1959**.
14. Companion, A.L.; Komarynsky, M.A. *J. Chem. Educ.* **1964**, *41*, 257.
15. Schaeffer, C.E. *Struct. and Bond.* **1968**, *5*, 68.
16. Schaeffer, C.E.; Jorgensen, C.K. *Molec. Phys.* **1965**, *9*, 401.
17. Schönherr, T. *Topics Curr. Chem.* **1997**, *191*, 87.
18. McClure, D.S. *Advances in the Chemistry of the Coordination Compounds,* Ed: Kirschner, S, Macmillan: New York, **1961**.
19. Yamatera, H. *Bull. Chem. Soc. Japan,* **1958**, *31*, 95.
20. Wolfsberg, L.; Helmholz, M. *J. Chem. Phys.* **1953**, *20*, 837.
21. This analysis assumes a linearly ligating ligand. If the ligand is not linear, e.g., a water molecule, then the orientation of the water molecule should be taken into account.
22. These differences were defined as $d\sigma$ and $d\pi$ in earlier analyses of McClure and Yamatera.[18,19]
23. Ref. 4, p.52.
24. a) Smith, D.W. *Struct. and Bond.* **1978**, *35,* 87; b) Riley, M.J. *Inorg. Chim. Acta,* **1998**, 268, 55.
25. Atanasov, M.; Hitchman, M.A.; Hoppe, R.; Murray, K.S.; Moubaraki, B.; Reinen D.; Stratemeier, H. *Inorg. Chem.* **1993**, *32*, 3397.
26. Astley, T.; Gulbis, J.M.; Hitchman, M.A.; Tiekink, E.R.T. *J. Chem. Soc. Dalton* **1993**, 509.
27. Comba, P.; Hambley, T.W.; Hitchman, M.A.; Stratmeier, H. *Inorg. Chem.* **1995**, *34*, 3903.
28. Bridgeman, A.J.; Essex, S.J.; Gerloch, M. *Inorg. Chem.* **1994**, *34*, 5411.
29. Duer, M.J.; Fenton, N.D.; Gerloch, M. *Int. Rev. Phys. Chem.* **1990**, *9*, 227; Duer, M.J.; Duer, R.J.; Gerloch, M. *Inorg. Chem.* **1987**, *26*, 2573, 2578, and 2582.

30. Bridgeman, A.J.; Gerloch, M. *Inorg. Chem.* **1995**, *34*, 4370; Bridgeman, A.J.; Gerloch, M. *Progr. Inorg. Chem.* **1997**, *45*, 179; Fenton, N.D.; Gerloch, M. *Inorg. Chem.* **1990**, *29*, 3718.
31. Schaeffer, C.E.; Yamatera, Y. *Inorg. Chem.* **1991**, *30*, 2840.
32. Atanasov, M.; Schönherr, T.; Schmidtke, H.H. *Theor. Chim. Acta,* **1987**, *71*, 59.
33. Mink, H.J.; Schmidtke, H.H. *Chem. Phys. Lett.*, **1994**, *231*, 235.
34. Steffan, G.; Reinen, D.; Stratemeier, H.; Riley, M.J.; Hitchman, M.; Matthies, H.E.; Recker, K.; Wallrafen, F.; Niklas, J.R. *Inorg. Chem.* **1990**, *29*, 2123.
35. Gerloch, M.; Harding, J.H.; Woolley, R.G., *Struct. and Bond.* **1981**, *46*, 2; Gerloch, M. in *Understanding Molecular Properties*, Ed. Avery, J.S.; Dahl, J.P.; Hansen, A., Reidel, **1987**, p. 111; Gerloch, M., *Magnetism and Ligand Field Analysis*, Cambridge University Press, Cambridge, **1983**.
36. Cotton, F.A. *Chemical Applications of Group Theory*; 3rd Edition, John Wiley: New York, **1990**.
37. Mulliken, R.S. *Phys. Rev.* **1932**, *40*, 55.
38. Mulliken, R.S.; Ricke, C.A.; Orloff, D.; Orloff, H. *J. Chem. Phys.* **1949**, 1248.
39. Condon, E.U.; Shortley, G.H. *Theory of Atomic Spectra*; Cambridge University Press: Cambridge, **1935**.
40. Ballhausen, C.J. *Molecular Electronic Structures of Transition Metal Complexes*; McGraw Hill: London, **1979**.
41. Ferguson, J. *Progr. Inorg. Chem.* **1970**, *12*, 159.
42. Silver, B.L. *Irreducible Tensor Methods*; Academic Press: New York, **1976**.
43. Tanabe, Y.; Sugano, S. *J. Phys. Soc. (Japan)* **1954**, *9*, 753 and 766.
44. Jahn, H.A.; Teller, E. *Proc. Roy. Soc. London* **1937**, *A161*, 220.
45. Shadle, S.E.; Hedman, B.; Hodgson, K.O.; Solomon, E.I., *J. Am. Chem. Soc.* **1995**, 117, 2259.
46. Westre, T.E.; Kennepohl, P.; DeWitt, J.G.; Hedman, B.; Hodgson, K.O.; Solomon, E.I. *J. Am. Chem. Soc.* **1997**, *119*, 6297.
47. Wilson, E.B.; Decius, J.C.; Cross, P.C. *Molecular Vibrations*; Dover Publication: **1955**.
48. Wilson, R.B.; Solomon, E.I., *Inorg. Chem.* **1978**, 17, 1729.
49. Solomon, E.I. *Comments in Inorg. Chem.* **1985**, *3*, 225.
50. Davydov, A.S. *Quantum Mechanics*; Neo Press: **1966**.
51. Lever, A.B.P. *Inorganic Electronic Spectroscopy, 1st Edition*; Elsevier Science: Amsterdam, **1968**.
52. Hahn, J.E.; Scott, R.A.; Hodgson, K.O.; Doniach, S.; Desjardins S.R.; Solomon, E.I., *Chem. Phys. Lett.* **1982**, 88, 595.
53. Desjardins, S.R.; Penfield, K.W.; Cohen, S.L.; Musselman, R.L.; Solomon, E.I. *J. Am. Chem. Soc.* **1983**, *105*, 4590.
54. Avoird, A.; Ros, P. *Theoret. Chim. Acta* **1966**, *4*, 13.
55. Gewirth, A.A.; Cohen, S.L.; Schugar, H.J.; Solomon, E.I. *Inorg. Chem.* **1987**, *26*, 1133.
56. McGarvey, B.R. *Trans. Met. Chem.*, **1966**, 3, 90.
57. Deaton, J.C.; Gebhard, M.S.; Solomon, E.I. *Inorg. Chem.* **1989**, *28*, 877.
58. Tuczek, F.; Solomon, E.I. *J. Am. Chem. Soc.* **1994**, *116*, 6916.

59. Didziulis, S.V.; Cohen, S.L.; Gewirth, A.A.; Solomon, E.I. *J. Am. Chem. Soc.* **1988**, *110*, 250.

60. Butcher, K.D.; Gebhard, M.S.; Solomon, E.I. *Inorg.Chem.* **1990**, *29*, 2067.

61. Butcher, K.D.; Didziulis, S.V.; Briat, B.; Solomon, E.I. *J. Am. Chem. Soc.* **1990**, *112*, 2231.

62. Figgis, B.N. *Introduction to Ligand Fields,* Interscience, New York, **1966**.

63. Solomon, E.I.; Lowrey, M.D.; LaCroix, L.D.; Root, D.E. *Methods in Enzymology*, Part C ed. Riordan, J.F.; Vallee, B.L., **1993**; Vol. 226, pp. 1–33.

2 Electron Paramagnetic Resonance Spectroscopy

ALESSANDRO BENCINI
Dipartimento di Chimica
Università di Firenze
via Maragliano 75/77
50144 Firenze, Italy
E-mail: sandro@chim1.unifi.it; bencini@dada.it

DANTE GATTESCHI
Dipartimento di Chimica
Università di Firenze
via Maragliano 75/77
50144 Firenze, Italy
E-mail: gattesch@chim1.unifi.it

Inorganic Electronic Structure and Spectroscopy, Volume I: Methodology.
Edited by E. I. Solomon and A. B. P. Lever.
ISBN 0-471-15406-7.

1 INTRODUCTION

Electron Paramagnetic Resonance, EPR, is the oldest magnetic resonance technique; the first spectra were recorded by Zavoiski in 1944.[1] EPR measures the absorption of electromagnetic radiation by a paramagnetic system placed in a static magnetic field. The frequency of the absorbed radiation falls in the microwave region of the spectrum (0.3–1.5 cm^{-1}; 9.2–45 GHz) except for the more recent instruments which can work up to 3 THz. From the middle of 1950 the EPR technique, often called also ESR, Electron Spin Resonance, left the world of the physicists and started to be applied in Chemistry. Since then, in the field of Inorganic Chemistry, it has been widely used for obtaining unique information on the electronic and geometrical structure of transition metal ions and, after an initial stage in which attention was devoted to the understanding of the properties of the different metal ions, an intense exploitation followed with applications in such different fields as mineralogy and biology.

The first obvious limitation in the use of EPR is that paramagnetic systems are needed. In principle EPR is very versatile because experiments can be performed in a large variety of different environments going from single crystals, through frozen or fluid solutions to glassy matrices, and even in the gas phase or *in vivo* in biological systems. In practice, however, there are many complications which make EPR spectra much more difficult to obtain and record than, for example, NMR spectra. In fact paramagnetic systems can be very different from each other; thus making the collection of an EPR spectrum far from routine. Further, paramagnetism is a quantum phenomenon; therefore the interpretation of the EPR spectra requires the extensive use of quantum mechanical techniques. The first problem of EPR is that the time scale of the experiment (10^{-4}–10^{-8} s) is much shorter than that of NMR (10^{-1}–10^{-5} s), making the requirements of the electronics much more stringent than for NMR. This is indeed the reason why Fourier transform, FT, techniques have been developed only in the last 20 years in EPR. Finally there are many paramagnetic systems which cannot be investigated, either due to the fast relaxation of the magnetization, or because they require much higher magnetic fields than those usually available.

Despite these limitations, EPR is a very powerful technique which can provide unique structural information, not only for isolated ions but also in systems in which dipolar and exchange interactions between the magnetic centers are operative. As such, it has been largely employed in the investigation of transition metal complexes first, and, in more recent years, also in many complex environments where transition metal ions may be present. We will present some examples later, but the contribution of EPR to the determination of the structure of active sites of metallo-enzymes and metallo-proteins cannot be overemphasized.

In the following the basic instrumentation of EPR is presented (Section 2) and the principles of a few spectroscopic techniques related to EPR are given. In Section 3, the spin Hamiltonian formalism required to interpret the spectra is given, and, in Section 4, we present some selected applications of EPR.

A number of books appeared dealing with the principles of EPR applied to the spectra of transition metal ions,[2–7] after the classical text by J.S. Griffith[8] covering the theory of the electronic structure of transition metal ions. Also some review articles have been published dealing specifically with the EPR of transition metal ions[9–13] and progress in the literature is being continuously monitored by the Royal Chemical Society.[14]

2 EPR AND RELATED TECHNIQUES

An atom in the gas phase which possesses a nonzero magnetic moment μ, can interact with a static external magnetic field* characterized by the magnetic induction, **B** *via* a direct coupling term of the classical form: $-\mu \cdot \mathbf{B}$. This interaction is generally known as the *Zeeman interaction*. The atomic magnetic moment is related to the total spin, **S**, and orbital, **L**, angular momenta:

$$\mu = -\frac{e}{2m}(\mathbf{L} + g_e\mathbf{S}) \tag{2.1}$$

The quantity $\gamma_e = \frac{e}{2m_e}$, in which e and m_e are the charge and mass of the electron respectively, is the gyromagnetic ratio of the electron. The proportionality constant between μ and **S** is known as the free electron g value, $g_e = 2.002319304386(20)$.

* Magnetic fields are characterized, in the vacuum, by the magnetic induction **B** which is measured in Wb m^{-2} in the SI units of measurements. This unit is called tesla (T). In the emu and gaussian system of units of measurements, which are still widely used in the field of magnetism and EPR, **B** is measured in gauss (G): 1 T = 10^{-4} G. Magnetic fields in linear media are generally treated by separating the field generated by real electric currents from those arising from the atomic nature of the medium (atomic magnetization currents). This is formally achieved by defining a new field $\mathbf{H} = (\mathbf{B}/\mu_0) - \mathbf{M}$ which is called magnetic field (strength). **M**, the volume magnetization, is the physical quantity related to the atomic nature of the substance. In steady conditions, the rotor of **H** is just the real current density. **H** and **M** are measured in A m^{-1} in the SI units. **H** and $\mathbf{B}/\mu_0$ are equal outside the region in which **M** has a finite value and can be used indifferently to measure the magnetic field itself. This situation is evidenced in the gaussian system of units where $\mathbf{H} = \mathbf{B} - 4\pi\mathbf{M}$ and **H** and **B** have the same dimensions. It is common use in the literature to express H in gauss. The magnetic moment μ is measured in A m^2 or J T^{-1} in SI units. Definitions and relationships between physical quantities used in the rest of the chapter can be found in reference 15.

The appropriate Hamiltonian for **B** parallel to the z axis has the simple form

$$\mathcal{H} = -\mu_z B_z \tag{2.2}$$

The eigenvalues of (2.2) can easily be found[3] in the Russell–Saunders coupling scheme by using the z component of $\mathbf{J} = \mathbf{L} + \mathbf{S}$ to label the states as

$$E_{m_J} = g_J \mu_B m_J B \tag{2.3}$$

where the Landé splitting factor, g_J, is given by

$$g_J = 1 + \frac{S(S+1) + J(J+1) - L(L+1)}{2J(J+1)} \tag{2.4}$$

and

$$\mu_B = \frac{e\hbar}{2m_e} = -\gamma_e \hbar$$

is the Bohr magneton. In SI units, $\mu_B = 9.27401 \times 10^{-24}$ JT^{-1}. The Zeeman interaction removes the $2J + 1$ degeneracy of a J spin manifold.[16]† For states with $L = 0$, g_J must attain the free electron value g_e. Electric fields, like those generated by the crystal field when an atom or ion is trapped in a solid matrix or in a complex, can remove the degeneracy of the L manifold of levels and cause g_J to approach g_e thus effectively *quenching* the orbital angular momentum.

The simplest EPR experiment can be performed on a free electron spin, or on a $^2S_{1/2}$ atom in the gas phase. In this case only two Zeeman levels are possible corresponding to $m_J = m_s = \pm 1/2$. Applying an oscillating magnetic field, $\mathbf{B_1}$, perpendicular to **B** a transition can be induced between these two levels when the frequency of the oscillating field, ν_0, meets the resonance condition

$$h\nu_0 = g_e \mu_B B_0 \tag{2.5}$$

This transition is magnetic dipole in origin and obeys the selection rule: $|\Delta m_s| = \pm 1$. Equation (2.5) holds in general for any paramagnetic species with one unpaired electron when g_e is subsituted by g, a constant which is different from 2.0023 and depends on the particular orientation of **B**. EPR is useful because it records the information associated with the shifts of the resonance due to the chemical environment

† This result can be derived from the symmetry properties of $\mathcal{H}_{Ze}$. Since **B** is an axial vector it is invariant under reflection in the origin, i. On the other hand in a reflection in a plane parallel to himself it changes sign. Neglecting translations, the symmetry group of **B** is therefore generated by the inversion, and the rotations about z, and $\mathcal{H}_{Ze}$ is invariant with respect to these transformations. The rotation operator around z of the angle φ is $R_Z = e^{-i\varphi J_z}$, therefore all the transformation of the group are expressible as functions of J_z and i. Since these two observables commute ($iJ_z = J_z i$) the invariance of $\mathcal{H}_{Ze}$ under rotations and inversion requires that its eigenstates cannot be systematically degenerate, i.e. the irreducible representations of the symmetry groups are one dymensional.

seen by the unpaired electron. The relevant parameter is the g value which can be measured through the relation:

$$g = \frac{g_e B_0}{B_r} \tag{2.6}$$

where B_r is the resonant field for the unpaired electron in the chemical environment and B_0 is the resonant field of the free electron.

A closer examination of the physics of this system allows an understanding of much of the basic principles of the EPR experiment.[17] Let us label the two levels $|a\rangle$ and $|b\rangle$ and let N_+ and N_- be the number of states with $m_s = 1/2$ and $m_s = -1/2$, respectively. The probability of the transition between the $|a\rangle$ and $|b\rangle$ states induced by the oscillating field is $W \propto |\langle a|\mu \cdot \mathrm{B}_1|b\rangle|^2$ with selection rules $|\Delta m_s| = \pm 1$. This transition causes a change in the populations N_+ and N_- subject to the condition $N = N_+ + N_-$, where N is the total number of spins of the system. The rate of variation of N_+ can be written as

$$\frac{\mathrm{d}N_+}{\mathrm{d}t} = W(N_- - N_+) \tag{2.7}$$

or in a more compact form

$$\frac{\mathrm{d}n}{\mathrm{d}t} = -2Wn \tag{2.8}$$

where $n = N_+ - N_-$ is the difference in population between the two levels. Solution of (2.8) yields

$$n = n_0 e^{-2Wt} \tag{2.9}$$

where n_0 is the value of n at $t = 0$. From (2.9) one sees that the difference in population tends to vanish in time. Since the probability of a transition from $|a\rangle$ to $|b\rangle$ is equal to that from $|b\rangle$ to $|a\rangle$, the net absorption of energy in the whole process can be computed by counting the number of spins which pass from the lower energy level to the upper one and subtracting the number of spins which go in the reverse direction by emitting energy as

$$\frac{\mathrm{d}E}{\mathrm{d}t} = N_+ W\Delta E - N_- W\Delta E = Wn\Delta E \tag{2.10}$$

where $\Delta E = g_e \mu_B B_0$. $\mathrm{d}E/\mathrm{d}t$ is the power either emitted or absorbed and is measured in Watts (W). From (2.10) one sees that a net absorption of energy can occur only if $n \neq 0$. From (2.9) a fast decay of absorption is expected; but this which is contrary to common observation. The existence of non-radiative processes which restore the difference in population, n, must therefore occur. These processes can be viewed as energy exchange (e.g. heat exchange) with the environment surrounding the spin system. This exchange of energy will take place until an equilibrium population is reached. In thermodynamic terms, this will stop when the relative population N_-/N_+ reaches the value compatible with the temperature T of the lattice, which functions as a thermal reservoir, according to the Boltzmann distribution law $e^{-\Delta E/kT}$. The overall effect of this *spin-lattice relaxation mechanism* is to differentiate the probability of the $|a\rangle$ to $|b\rangle$ transition from that going in the opposite direction. The rate equation for $\mathrm{d}n/\mathrm{d}t$ taking into account both the radiative and the spin-lattice transitions takes the form

$$\frac{\mathrm{d}n}{\mathrm{d}t} = -2Wn + \frac{n_0 - n}{T_1} \tag{2.11}$$

where T_1 is a characteristic time associated with the approach to thermal equilibrium, and it is generally called *spin-lattice relaxation time*. In stationary conditions, $\mathrm{d}n/\mathrm{d}t = 0$, and

$$n = \frac{n_0}{1 + 2WT_1} \tag{2.12}$$

The alternating field does not disturb the populations from their thermal equilibrium value as long as $1 \gg 2WT_1$. Under these conditions the absorbed power is given by

$$P = \frac{\mathrm{d}E}{\mathrm{d}t} = Wn\Delta E = \Delta E \frac{n_0 W}{1 + 2WT_1} \tag{2.13}$$

Since $W \propto |\langle a|\mu \cdot \mathbf{B}_1|b\rangle|^2$, and therefore is proportional to the square of the oscillating magnetic field, B_1^2, we see that the power absorbed by the spin system is proportional to the amplitude of the alternating field. As soon as $2WT_1$ is no longer $\ll 1$, the absorbed power does not depend linearly on B_1^2. In particular when $2WT_1$ is dominant, the power absorbed decreases despite an increase of W. This effect is known as *saturation* and is related to the actual value of T_1.

The simplest EPR experiment is performed when equation (2.13) holds. This is the basis of the conventional continuous wave EPR (CW-EPR) experiment, in which the equilibrium situation is only slightly perturbed by the radiation field. According to (2.5) the ratio ν_0/B_0 for a free electron is 28.02 GHz T^{-1}. Since CW spectrometers work at a fixed frequency, most commonly $\nu_0 \cong 9.3$ GHz (X-band spectrometers), a resonance field of 0.3390 T is needed in order to observe the resonance. This field is in the range of the magnetic fields generated by common electromagnets (0–2.0 T).

Unlike NMR, the technology used in EPR is rather old[18,19] since commercial spectrometers were developed about forty years ago. Microwaves are generally propagated through waveguides, the use of coaxial components being developed much more recently, and iron-core electromagnets are used to generate the magnetic fields.[20] Only very recently have high magnetic fields obtained by superconducting magnets been applied.

A useful general classification[21] of the EPR experiment can be done following the response of the paramagnetic systems to the power of the incident microwave radiation, P_0. An EPR experiment is said to be linear if the signal is not saturated by P_0 and the lineshape is independent of P_0 and nonlinear in the alternate case. Linear EPR experiments are generally performed in the CW-EPR, also called *steady-state* EPR, while saturation effects are important in time domain EPR and multiple resonance techniques. *Time domain* or *time resolved* EPR measures the response of the paramagnetic system as a function of time to the microwave excitation. In this technique single microwave pulses, as in the *saturation recovery* or *free-induction decay* experiments, or multiple pulses, as in *electron spin echo* measurements, are used. Other excitations such as laser or thermal or acoustic pulses followed by the measurements of the transient response of some property of the system have also been used in recent years. *Double resonance* techniques use the variation of saturation of an EPR signal as a response function and can be considered nonlinear techniques. Among these, the most commonly used techniques are *Electron Nuclear DOuble Resonance* (ENDOR) and *ELectron DOuble Resonance* (ELDOR). In the following we will present the basic principles of measurements and instrumentation of CW-EPR, and the general principles of pulsed techniques and ENDOR spectra. Readers more interested in spectra and their interpretation may skip to Section 3.

2.1 CW-EPR

The basic instrumentation[21] for a commercial EPR spectrometer comprises: a) microvawe source, often a klystron or a solid state source; b) a sample resonator, commonly a resonant reflection cavity; c) an electromagnet with control of the field intensity and field sweep; d) a data acquisition apparatus, generally a current detector and a phase sensitive detector; e) a data display apparatus and a data processing system. The basic apparatus is complemented by tools[20] for measurements in the gas phase or in the solid state (polycrystalline powders and single crystals), in solution (fluid and frozen), at variable temperature and so on. A scheme of a CW-EPR is shown in Figure 1. The electromagnetic radiation is commonly transmitted through rectangular waveguides[22] which can be viewed as formed by two parallel reflecting planes in such a way that there is no electric field component in the direction of propagation, say x (Transverse Electric or TE mode of propagation), but the magnetic field has the largest value. The electric field in the directions y and z varies like sin or cos $\frac{\pi n i}{\ell}$, where n is an integer and $i = y, z$. The dimension ℓ is related to the wavelength λ by: $2\ell > \lambda > \ell$. By making $\ell < \lambda/2$ no TE mode can be propagated with the electric vector polarized in that direction.

The sample cavity is generally a resonating cavity whose volume is varied through the adjustement of a metallic iris connected to a screw. Generally rectangular or cylindrical cavities like those shown in Figure 2 are used.

A TE012 cavity used for X-band ($\nu \approx 9.5$ GHz; $\lambda \approx 3.2$ cm) measurements has typical size $2.29 \times 1.02 \times 2.54$ cm. Since the size of the waveguide and resonant cavities depend on the actual value of λ, it is necessary to vary the whole apparatus according to the microwave frequency used. Low frequency cavities are

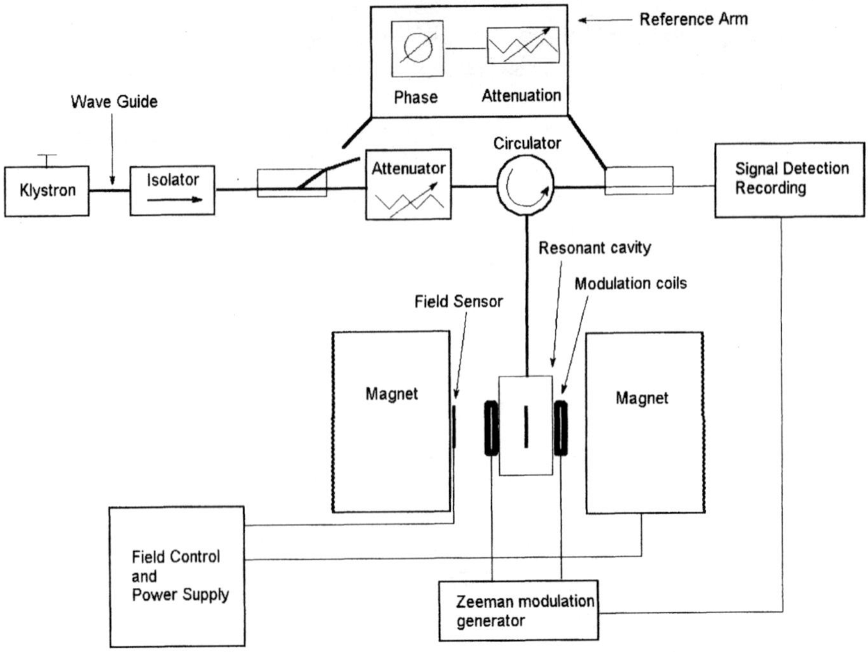

Figure 1. Principal components of a CW-EPR spectrometer.

bulky and high frequency ones are too small to be handled. Actually, for high magnetic fields no resonant cavity is used, but the micro power reflected by the sample is measured. This procedure requires the use of bulky samples and gives a general loss of sensitivity. Loop gap or split ring resonators are becoming the alternative to resonant cavities especially in the frequency range of 1–4 GHz.[23,24] The microwave radiation is transmitted through a coaxial cable. The original loop gap resonator is formed by a metallic cylinder with a narrow gap parallel to the main axis as shown in Figure 3. It can be viewed as an infinite one-turn solenoid, in which the resonant matching is achieved by moving a coupling loop circuit. The microwave magnetic field is uniformly oriented parallel to the main axis and the electrical field is mainly confined to the split. In this way dielectric losses associated with solvents are minimized.

It can be seen in Figure 1 that auxiliary coils are placed around the sample cavity. These coils modulate the intensity of the external magnetic field by coupling it to a low frequency (often 100 MHz) electromagnetic radiation. The resulting signal is encoded to the modulation frequency, ω_m, so that after phase sensitive detection the signal appears in the form of a first derivative as shown in Figure 4. By tuning the detector to $2\omega_m$ a second derivative spectrum can be recorded.

Transfer of modulation[25,26] in the resonance condition forms the basis of the saturation transfer EPR technique (ST-EPR). In this technique modulation side bands are generated through appropriate modulations frequencies. Usually signals with very

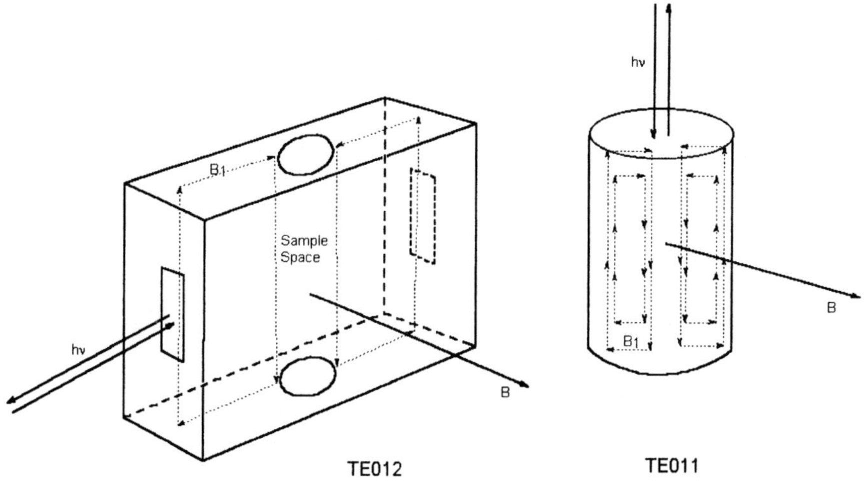

Figure 2. Schematic view of two resonant cavities: rectangular (left) and cylindrical (right). The magnetic field patterns of the incoming microwave radiation inside the cavities is also shown.

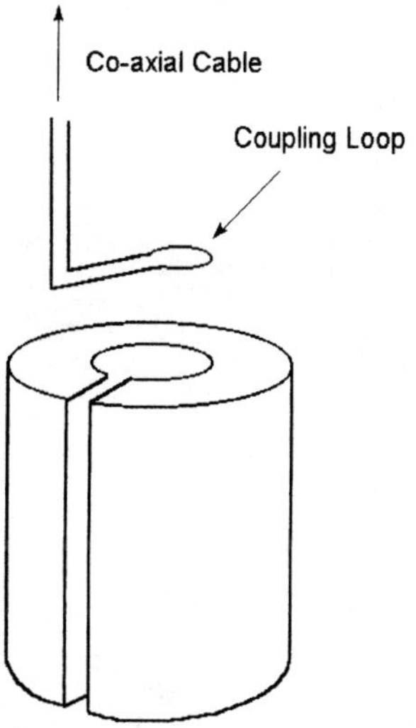

Figure 3. Schematic view of a loop-gap resonator. The resonator is shielded by a thin conducting cylinder. The relative position of the coupling loop circuit allows matching the resonance conditions.

narrow linewidths are needed in order to observe these effects (2–3 μT) and ST-EPR is not usually applied to transition metal ions.

The performance of a cavity is generally espressed by the so-called Q factor, defined as the ratio between the energy stored in the cavity and the energy loss.

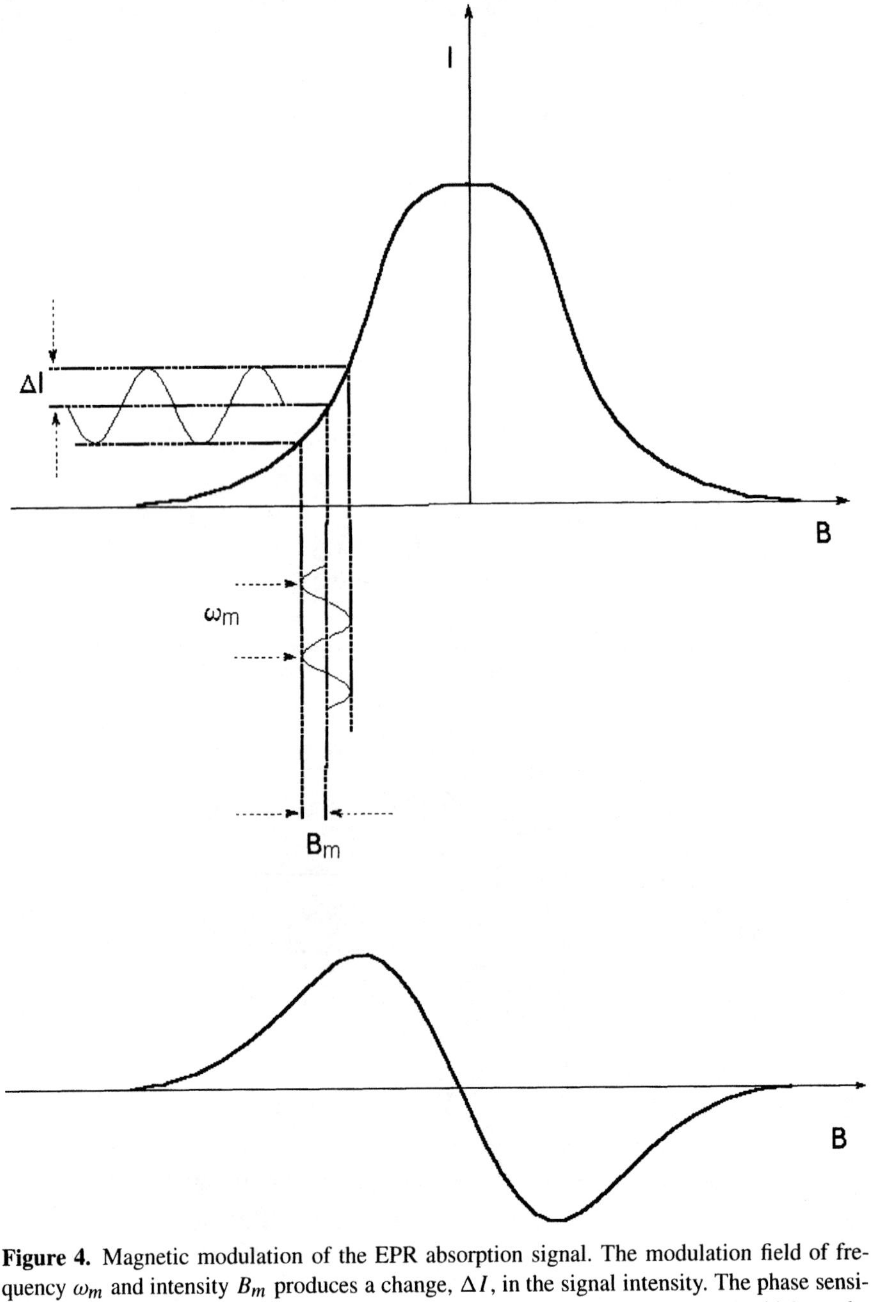

Figure 4. Magnetic modulation of the EPR absorption signal. The modulation field of frequency ω_m and intensity B_m produces a change, ΔI, in the signal intensity. The phase sensitive detector produces an output proportional to the phase and amplitude of ΔI which results in the first derivative of the absorption.

Clearly Q must be as large as possible. The other relevant factor for the quality of the EPR spectra is the filling factor, η, which takes into account the effective volume occupied by the sample ($0 < \eta < 1$). The product ηQ is the important quantity for recording an EPR signal. Experiments up to 35 GHz are characterized by high Q factor cavities (2000–5000), but with low η values ($\sim$0.1), while low frequency apparatus using loop gap resonators have η close to 1 but Q values in the range of 500–1000.

A number of EPR experiments can now be performed at frequencies higher than or within W-band (75–110 GHz) corresponding to fields of about 2.8 T and higher, for $g = 2$. This technique is generally indicated as High-Field EPR (HF-EPR). A good review of HF-EPR can be found in reference[27]. One of the main technical developments that has boosted HF-EPR is the availability of superconducting magnets, although pulsed magnetic field and resistive magnets are still used.[28,29] The main technical problem encountered in HF-EPR is the need of a high magnetic field resolution (about 10^{-5} T) combined with a large sweep of the magnetic field itself. The high field spectrometers[30–35] also use high frequencies (95–150 GHz), which can be obtained either by far infrared lasers or by Gunn diodes. The advantages of using HF-EPR are increased resolution and the possibility of observing the spectra of systems with $S > 1/2$ and large zero field splittings.

2.2 Multiple Resonance Techniques

Multiple resonance techniques are basically obtained by irradiating the sample in an EPR cavity with an extra electromagnetic radiation.[36] Using microwave radiation with frequency corresponding to a different EPR transition the electron-electron double resonance (ELDOR) is measured. When radiofrequency radiation, corresponding to transitions between nuclear spin states is used , the electron–nuclear double resonance (ENDOR) is obtained. All techniques are characterized by a pump frequency which is close to saturating an EPR transition and an observed frequency with non-saturating power. The ELDOR experiment is carried out in a bimodal cavity in which it is possible to apply the saturating frequency to one mode and the observing frequency to the other, so that transfer of saturation can pass from the pump to the observing transition.

The ENDOR apparatus couples a radiofrequency generator as used in NMR spectroscopy to a standard EPR spectrometer. The modulation coils are placed around the sample in the EPR cavity in order to generate a radiofrequency field on the sample. The pumping source uses microwave radiation to saturate an EPR transition. The purpose of the ENDOR experiment is to obtain information on the nuclear Zeeman effect and on the term representing the mutual interaction between electron and nuclear moments (hyperfine coupling). This effect will be treated in more detail in Section 3.3, but will be briefly introduced here. A simple form of the interaction Hamiltonian, useful for the present discussion, is:

$$\mathcal{H} = g\mu_B \mathbf{B} \cdot \mathbf{S} - g_N \mu_N \mathbf{B} \cdot \mathbf{I}_N + a_N \mathbf{S} \cdot \mathbf{I}_N \tag{2.14}$$

where $\mu_N = (m_e/m_p)\mu_B$ is the nuclear magneton, 5.051×10^{-27} JT^{-1}. The first two terms in (2.14) represent the isotropic electronic and the nuclear Zeeman interactions respectively; the third term represents the interaction of the spin $\mathbf{S}$ with the nuclear spin $\mathbf{I}_N$ characterized by the isotropic a_N value. The eigenvalues of (2.14) can be written as

$$E(M_S, M_I) = g\mu_B BM_S - g_N\mu_N BM_I + M_S M_I a_N \tag{2.15}$$

The nuclear Zeeman interaction is always smaller than the electronic Zeeman term. Two limiting situations can occur: either the nuclear Zeeman interaction is larger than the hyperfine coupling, $g_N\mu_N BM_I > M_S M_I a_N$ (case a), or it is smaller (case b). Both situations are pictorially shown in Figure 5 for $S = 1/2$ and $I_N = 1/2$. The energy levels look similar except for the inversion of the two higher energy states. In both cases the EPR spectrum consists of two lines given by the resonance condition

$$h\nu = g\mu_B B_0 \pm \frac{a_N}{2} \tag{2.16}$$

Irradiating the sample with the radiofrequency ν_1 changes the population of the upper energy level and changes the steady-state population created by the saturating EPR microwave field. By measuring the variation of the EPR intensity as a function of the radiofrequency the ENDOR spectrum is measured. The resonance condition for cases a and b are:

$$\begin{aligned} \text{case a: } h\nu_N &= g_N\mu_N B_0 \mp \frac{a_N}{2} \\ \text{case b: } h\nu_N &= \frac{a_N}{2} \pm g_N\mu_N B_0 \end{aligned} \tag{2.17}$$

In case a, two lines centered at the nuclear resonance condition and separated by a_N will be observed; in case b, two transitions centered at $a_N/2$ and separated by $\nu_N\mu_N B_0/h$ will be measured. Case a is most commonly encountered in organic radicals and case b in transition metal complexes.

More complicated spectral patterns are obtained in the presence of nuclear quadrupole interactions ($I > 1/2$) and when conditions a and b are not exactly matched. ENDOR spectroscopy is used to directly measure hyperfine interactions, either with the metal nucleus or with the ligand nuclei, which are often hidden under the linewidth of the inhomogeneously broadened EPR signal. The ENDOR spectrum is characterized by narrow lines whose linewidth is of the same order of magnitude of the NMR lines; therefore the resolution of the spectrum is higher than that of an EPR spectrum. The measurement of the ENDOR spectrum, however, requires a saturation, at least partial, of an EPR signal and therefore a detailed knowledge of the relaxation mechanisms is needed. Generally, saturation of the EPR signal requires the use of low temperatures which can be lower than 20–30 K if transition metals are involved, making the observation of the spectrum a difficult matter. Quadrupole interactions can also be measured directly using ENDOR.

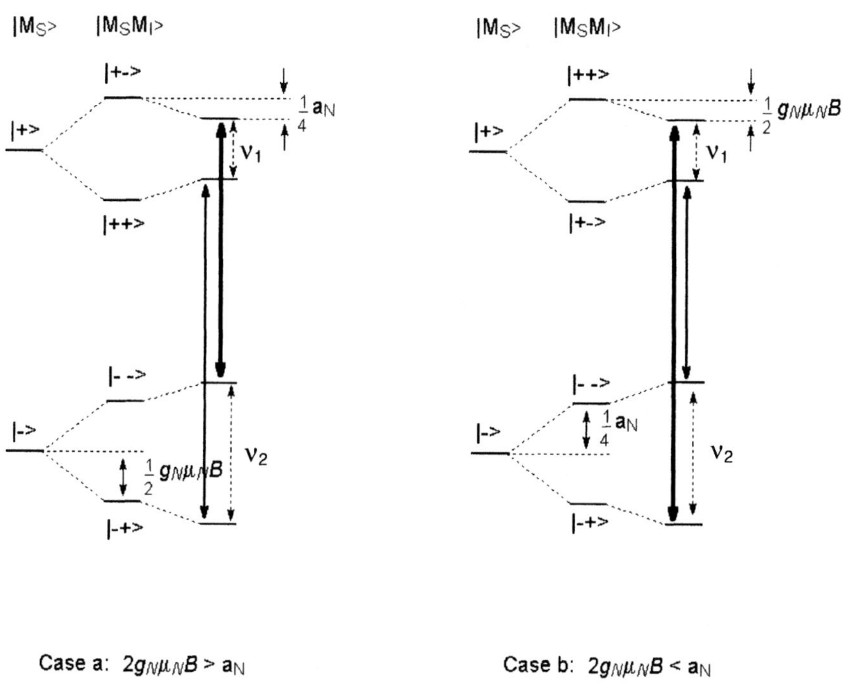

Figure 5. Energy level diagram for $S = I = 1/2$ spin system and $a_N > 0$. Case a: nuclear Zeeman interaction greater than the hyperfine coupling; Case b: hyperfine coupling greater than the nuclear Zeeman interaction. EPR transitions are indicated by solid arrows; ENDOR transitions are marked by dotted arrows. The saturating EPR frequencies are shown with heavier lines.

A number of techniques have been derived from the simple experiment described above which make use of polarized r.f. fields or two radiofrequency fields (double-ENDOR). The most important techniques[37] derived from the fundamental experiment are collected in Table 1. The response of a particular ENDOR line by sweeping the static magnetic field B can be also monitored. In this way a new EPR spectrum is observed having contributions from only one nuclear species. This technique takes the name of ENDOR induced EPR (EI-EPR).[38]

2.3 Pulsed EPR Techniques

Time domain EPR is much less common than time domain NMR although the principles are exactly the same. The essential features are the use of microwave pulses and pulse sequences in time which cause different nonlinear responses of the paramagnetic system. The way to indicate pulses and sequences of pulses is, analogous to the NMR case, easily understood in the classical picture of magnetic resonance. We look at the magnetic system as characterized by a magnetization vector **M** which is precessing around the external magnetic field at the Larmor frequency. Applying

Table 1 ENDOR Derived Techniques and Main Applications

Technique	Apparatus	Applications
Orientation selection		Gives crystal-like ENDOR spectra allowing the determination of anisotropies
EI-EPR (ENDOR Induced EPR)	Frequency correction of the ENDOR resonances	Separation of overlapped EPR spectra and of different orientations in powders
DOUBLE-ENDOR	Two r.f. fields	Separation of overlapping ENDOR spectra; determination of the relative sign of hyperfine couplings
Nuclear Spin Decoupling	Two r.f. fields	Assignment of ENDOR transitions
Double Quantum Transitions	One/two r.f. fields	Simplifies the interpretation of ENDOR spectra
CP-ENDOR (ENDOR with Circularly Polarized r.f. fields)	Circularly polarized r.f. field	Reduces the number of ENDOR lines
CP-DOUBLE-ENDOR (DOUBLE-ENDOR with Circularly Polarized r.f. fields)	One linearly polarized r.f. field and one circularly polarized r.f. field	Reduces the number of ENDOR lines
PM-ENDOR (Polarized Modulated ENDOR)	r.f. field, linearly polarized r.f. field rotating in the *xy* plane	Additional labeling of ENDOR lines; two-dimensional displaying of the spectra
ESE (Electron Spin Echo)	Pulsed microwave field	Measurement of hyperfine coupling constants
ESE-ENDOR (Electron Spin Echo ENDOR)	Pulsed microwave and r.f. fields	Measurement of hyperfine coupling constants

a magnetic field, $\mathbf{B}_1$, oscillating at the Larmor frequency, or sweeping the external field to match the resonance condition, causes **M** to precess around $\mathbf{B}_1$, in the rotating frame, and be orthogonal to **B**. By applying a field $\mathbf{B}_1$ for a time shorter than the relaxation time, in a pulse, one can cause **M** to tip into the plane perpendicular to **B**, say xy. This particular pulse is called a $\pi/2$ pulse. The free induction decay (FID) of the magnetization caused by the interaction of the paramagnetic system with the environment can then be measured as in NMR spectroscopy. This technique requires the use of pulses as short as 10 ns compared with the μs pulses used in NMR and it is still not widely applied.

Electron spin echo (ESE) methods are much more commonly used.[39] To record a spin echo signal, one applies, after a time τ from the $\pi/2$ pulse, a π pulse which causes the spin precession of **M** in the xy-plane to reverse. Since the precession of the spin is now reversed, **M** tends to return parallel to the initial position after the same time τ, therefore the decay of the echo signal is observed after the time 2τ. The measurements of echoes in the presence of unresolved hyperfine structures, particularly arising from interactions with distant nuclei, has shown that the echoes do not always decay smoothly, but a modulation of the echoes occurs.[40] This phenomenon has been called electron spin echo envelope modulation (ESEEM) and allows one to measure weak hyperfine and quadrupole interactions.[41] Additional echoes can be detected using three pulse sequences. Generally in the three pulse experiment all the pulses are $\pi/2$ pulses. The first two are spaced by the time τ, the third one is applied a time t after the second one. The interesting point is that the Fourier transform of the three-pulse ESE in the frequency domain has peaks at the ENDOR frequencies. ESEEM and ENDOR are complementary to each other since ESEEM can be applied only in the cases where the hyperfine interactions are much smaller than the nuclear Zeeman term.[42]

3 SPECTRA AND THEIR INTERPRETATION

3.1 Spin Hamiltonian and Symmetry

The EPR experiment records the response of a paramagnetic sample in a static magnetic field to irradiation with an oscillating field. Therefore in order to understand the spectrum, we need to consider all the interactions which take place between the paramagnet and the external field. For an isolated atom with zero nuclear moment or a free electron, only the Zeeman term considered in Section 2 is required. In more general cases, in addition to the electronic Zeeman interaction, the nuclear Zeeman interaction and, when more than one unpaired electron is present, interactions between electrons, and between electrons and magnetic nuclei, must be added.

First principle calculations of the electronic structure of molecules in a magnetic field are generally prohibitive since knowledge of the ground state many-electron wavefunction corrected for spin-orbit coupling is required for finding the value of the applied field. Therefore perturbation theory is used which allows the inclusion of spin-orbit coupling and magnetic field effects on the zero-field states obtained by any quantum mechanical approach to the many-electron problem. Using second order perturbation theory, it has been shown that the main magnetic interactions can be represented using effective Hamiltonians which include only spin (electronic and nuclear) operators as well as magnetic field vectors. This is the spin Hamiltonian approach, which is widely used to interpret experimental EPR spectra. A series of parameters come out from the spectra which must be correlated with the electronic structure of the molecule through quantum mechanical techniques. In this chapter we will present the most important features of the spin Hamiltonians needed to interpret the EPR spectra of transition metal ions and we will outline some of the relationships existing between the spin Hamiltonian parameters and the electronic structure of the molecules. We will always assume that the ground state of the system is orbitally non-degenerate as a consequence of crystal field effects. Under this assumption, which is satisfactorily met in many experimental systems, the spin Hamiltonian approach is particularly simple.

In the more general case a spin Hamiltonian can be considered a symmetry expansion of the interaction terms briefly outlined above, in which all the electron and nuclear coordinates are replaced by angular momentum operators, in order to exploit their well established symmetry properties. The interaction of the electronic system with the static external magnetic field, **B**, can be represented in general by a term which is linear in **B** and in **S**:

$$\mathcal{H}_{ze} = \mu_B \mathbf{B} \cdot \mathbf{g} \cdot \mathbf{S} \tag{3.1}$$

where **S** is an angular momentum operator, and **g** is a matrix.‡ In the following Section we will make clear the meaning of **g**. The Hamiltonian (3.1) operates on a basis of $2S + 1$ functions, which are eigenfunctions of $\mathbf{S}^2$ characterized by the quantum number S. These functions may correspond to the true spin of the system, as already seen in Section 2, or to an effective spin, which will become clear in the next Section. For the nuclear Zeeman term an expression similar to (3.1) is obtained:

$$\mathcal{H}_{zn} = -\mu_N \mathbf{B} \cdot \mathbf{g}_N \cdot \mathbf{I}_N \tag{3.2}$$

where $\mathbf{I}_N$ is the nuclear angular momentum operator and μ_N is the nuclear Bohr magneton. Since μ_B is 1836 μ_N, the operator (3.2) is often neglected in an EPR experiment (see Section 2.2). The appropriate eigenfunctions are characterized by an I_N nuclear quantum number. In general the terms linear in **B** and $\mathbf{S}(\mathbf{I}_N)$ are sufficient to describe the interaction of electronic and nuclear spin with external fields. Only for values of S larger than $5/2$ will higher order terms be included in the Hamiltonian (3.1); but their effects are not yet well established.

When there is more than one unpaired electron, their mutual interaction must be taken into consideration. The corresponding lowest approximation for the spin

‡ The **g** and **A** matrices (see equation 3.4) are commonly called tensors, although they are not mathematically true second rank tensors. This is due to the fact that both **g** and **A** connect two vector operators which are referred to two independent reference systems, namely **B** and **S**, and **S** and **I** for **g** and **A** respectively. Only the matrices $\mathbf{g}^\dagger \cdot \mathbf{g}$ and $\mathbf{A}^\dagger \cdot \mathbf{A}$ transform like tensors.

Hamiltonian term is:

$$\mathcal{H}_{zfs} = \mathbf{S} \cdot \mathbf{D} \cdot \mathbf{S} \quad (3.3)$$

where **D** is a symmetric traceless second rank tensor, which is called zero field splitting, zfs, tensor. This term is different from zero only when $S > 1/2$ and in symmetries lower than the cubic one. The term describing the electron–nucleus interaction is:

$$\mathcal{H}_{SI} = \mathbf{S} \cdot \mathbf{A}_N \cdot \mathbf{I}_N \quad (3.4)$$

$\mathbf{A}_N$ is now a symmetric matrix. In this case there is no restriction to the value of S and I_N. The term describing the nucleus-nucleus interaction is:

$$\mathcal{H}_{NQ} = \mathbf{I}_N \cdot \mathbf{Q}_N \cdot \mathbf{I}_N \quad (3.5)$$

Spin Hamiltonian terms such as (3.1, 3.3, 3.4) must be understood as the first term in a series expansion. They all contain simple products of angular momentum operators. A general term of the spin Hamiltonian[3] takes the form $B_i^l S_j^m I_k^n$ where $l + m + n$ is even, $m \leqslant 2S$ and $n \leqslant 2I$. These terms should be multiplied by the appropriate spin Hamiltonian parameters. For instance a term which sometimes has been used for the interpretation of the spectra of systems with $S \geqslant 2$ is the fourth order term

$$\mathcal{H}_{FO} = \mathbf{S}^2 \cdot \mathbf{O} \cdot \mathbf{S}^2 \quad (3.6)$$

where **O** is a fourth rank tensor. General expression for the spin Hamiltonian based on irreducible tensor operators have been developed.[43,44] Similar expression should hold for the nuclear terms 3.2 and 3.5, but are far less used.

3.2 Electron Zeeman Interaction

The electron Zeeman interaction is characterized by the matrix **g**, which can be made diagonal, providing the principal values and principal directions.[§] In the most general case the three principal values, g_x, g_y, and g_z are different from each other. If the system has axial symmetry $g_x = g_y = g_\perp$, and $g_z = g_\parallel$, while if the symmetry is cubic $g_x = g_y = g_z = g_{iso}$. The principal values and directions of **g** depend on the electronic structure of the paramagnet. In most cases the ground state of the system has no orbital degeneracy and is well separated from the first excited states. In zeroth order approximation the orbital contribution to the **g** tensor is completely quenched, and **g** is isotropic and identical to the free electron value $g_e = 2.0023$. At a more acceptable level we consider that spin-orbit coupling mixes excited states

§ Actually only the $\mathbf{g} \cdot \mathbf{g}$ tensor can be diagonalized and the principal values of **g**, g_i, are obtained as $g_i = \sqrt{g_{ii}{}^2} (i = x, y, z)$. The same applies for the **A** matrix defining the hyperfine coupling.

into the ground state. When the ground state is orbitally nondegenerate, second order perturbation theory provides[45,8] the following expression for the **g** tensor:

$$\mathbf{g} = g_e\mathbf{I} + 2\lambda\Lambda \tag{3.7}$$

where **I** is the identity matrix $I_{lm} = \delta_{lm}$, and Λ is given by:

$$\Lambda = \sum_n \frac{\langle g|\mathbf{L}|n\rangle\langle n|\mathbf{L}|g\rangle}{E_g - E_n} \tag{3.8}$$

In (3.8) $|g\rangle$ is the ground state of the system, whose energy is E_g, $|n\rangle$ is an excited state, whose energy is E_n. In deriving (3.8) the spin-orbit coupling operator was taken in the form $\lambda\mathbf{L}\cdot\mathbf{S}$. This is exact for $S = 1/2$ ions, while it is valid only within a given spin multiplet when $S \geqslant 1$. For transition metal ions, for example, this form of the spin-orbit coupling holds only within one Russell–Saunders term, where $\lambda = \pm\zeta/2S$, and ζ is the one-electron spin-orbit coupling constant appearing in the general spin-orbit coupling operator

$$\mathcal{H}_{SO} = \sum_{i=1}^{N} \zeta \boldsymbol{\ell}_i \cdot \mathbf{s}_i$$

The + sign applies for ions with less than five *d* electrons in the valence shell and the − sign applies for ions with more than five *d* electrons. Some typical values of ζ for selected atoms and transition metal free ions are given in Table 2. A more general formula for the spin Hamiltonian has been worked out.[46]

The physical meaning of (3.7–3.8) is that spin-orbit coupling reintroduces some orbital degeneracy into the ground states and produces deviations of g from the free electron value. The diagonal matrix elements of Λ are negative; therefore, for ions with a number of unpaired electrons smaller than those corresponding to a half-filled

Table 2 Spin-Orbit Coupling Constants ζ for Some Free Ions and Atoms (in cm^{-1})

B	C	N	O	F
11	29	76	150	270
Al	Si	P	S	Cl
61	150	299	374	587
Ti^{III}	V^{IV}	Cr^{II}	Cr^{III}	Cr^{V}
155	250	230	275	380
Mn^{II}	Mn^{III}	Mn^{IV}	Fe^{II}	Fe^{III}
300	355	415	400	460
Co^{II}	Ni^{II}	Cu^{II}	Mo^{V}	Ru^{II}
515	630	830	1030	1000

configuration the corrections to g_e are negative, and $g < g_e$, while the reverse is true for configurations more than half-filled. It is also evident from (3.7–3.8) that the deviation from the free electron value is larger for ions with large spin-orbit coupling. The other main factor determining the **g** tensor is the separation of the excited state from the ground state: the smaller the separation the larger the shift of the g value.

A simple example to appreciate the meaning of (3.7–3.8) is provided by the magic pentagon depicted in Figure 6, which shows pictorially how the components of Λ can be computed for different ground states arising from a d^1 or d^9 configuration.

Let us suppose that the ground state of a transition metal ion is described by a function which is essentially $|xy\rangle$. We see that in Figure 6 the vertex labeled $|xy\rangle$ is connected to $|x^2-y^2\rangle$ with a line labeled 8, while it is connected with $|xz\rangle$ and $|yz\rangle$ with lines labeled 2. This means that the g values expected for this ground state are:

$$g_x = g_e - \frac{2\lambda}{\Delta_{xz}}; \qquad g_y = g_e - \frac{2\lambda}{\Delta_{yz}}; \qquad g_z = g_e - \frac{8\lambda}{\Delta_{x^2-y^2}}; \qquad (3.9)$$

where Δ_{yz}, Δ_{xz} and $\Delta_{x^2-y^2}$ correspond to the energy separation between the ground state and the indicated excited state. When $|xz\rangle$ and $|yz\rangle$ have the same energy, as for instance in tetragonal ligand fields, then $g_x = g_y = g_\perp$.

The g values computed using (3.9) for various d^n configurations in geometries derived from the octahedral one are shown in Table 3. The reference point has been taken from octahedral complexes, with the addition of tetragonal or rhombic perturbations. In Table 3 we have also indicated how easy it is to observe the EPR spectra of the various metal ions. In general the conditions for easy observation depend on relaxation and zero field splitting effects. Fast relaxation causes broadening of the signal, thus making the spectra difficult to observe unless very low temperatures are reached. As a general rule, relaxation tends to be faster when several energy levels are close to the ground state. The systems which are best suited to observation *via* EPR are those with orbitally nondegenerate ground states. Systems with an E type ground state usually undergo strong Jahn–Teller distortion and often give rise to ground states which are well separated from the excited ones, so giving slow relax-

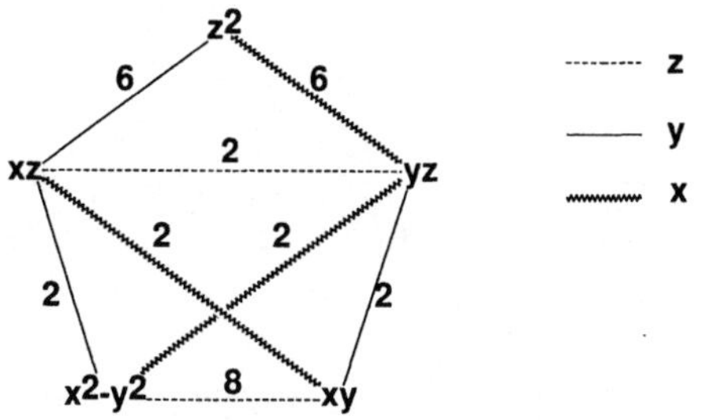

Figure 6. Schematic view of the effect of the spin orbit coupling on the d functions. The different lines represent the value of the integral $\langle i|\mathcal{H}_{SO}|j\rangle$ connecting the orbitals $|i\rangle$ and $|j\rangle$ along x, y, and z.

Table 3 Calculated *g* Values for d^n Ions in Pseudo-Octahedral Coordination

Configuration	S	Ground State	obs.[a]	g_x	g_y	g_z
d^1	1/2	$^2T_{2g}$	E	$g_e - 2\lambda/\Delta_1$[b]	$g_e - 2\lambda/\Delta_2$[b]	$g_e - 8\lambda/\Delta_3$[b]
d^2	1	$^3T_{1g}$	VD	$g_e - 9\lambda/2\Delta$[c]	$g_e - 9\lambda/2\Delta$[c]	g_e
d^3	3/2	$^4A_{2g}$	E	$g_e - 8\lambda/\Delta_1$[d]	$g_e - 8\lambda/\Delta_2$[d]	$g_e - 8\lambda/\Delta_3$[d]
d^4 HS	2	^5E_g	VD	$-6\lambda/\Delta_1$[e]	$-6\lambda/\Delta_2$[e]	g_e
				$-2\lambda/\Delta_1$[f]	$-2\lambda/\Delta_2$[f]	$-8\lambda/\Delta_3$[f]
d^5 HS	5/2	$^6A_{1g}$	E	g_e	g_e	g_e
d^5 LS	1/2	$^2T_{2g}$	D	$2[-2\alpha\gamma + \beta^2 + k\beta(\alpha^2 - \gamma^2)]$[g]	$2[-2\alpha\gamma - \beta^2 - k\beta(\alpha^2 - \gamma^2)]$[g]	$2[\alpha^2 - \beta^2 + \gamma^2 + k(\alpha^2 - \gamma^2)]$[g]
d^6 HS	2	$^5T_{2g}$	VD	$g_e + 2\lambda/\Delta_1$[h]	$g_e + 2\lambda/\Delta_1$[h]	$g_e + 8\lambda/\Delta_2$[h]
d^7 HS	3/2	$^4T_{2g}$	D	$2(5-\gamma)/3$[i]	$2(5-\gamma)/3$[i]	$2(5-\gamma)/3$[i]
				0[ℓ]	0[ℓ]	$2(3-\gamma)$[ℓ]
				4[m]	4[m]	2[m]
d^8	1	$^3A_{2g}$	VD	$g_e + 8\lambda/\Delta_1$[n]	$g_e + 8\lambda/\Delta_2$[n]	$g_e + 8\lambda/\Delta_3$[n]
d^9	1/2	^2E_g	E	$g_e + 2\lambda/\Delta_1$[o]	$g_e + 2\lambda/\Delta_2$[o]	$g_e + 8\lambda/\Delta_3$[o]
				$g_e + 6\lambda/\Delta_1$[p]	$g_e + 6\lambda/\Delta_2$[p]	g_e

[a]E = easy to observe: generally at room temperature; D=difficult to observe: usually liquid helium temperature is required; VD=very difficult to observe: zero field splitting can be large to prevent observation except in very high magnetic fields. The x and y axes of the reference system bisect the equatorial L-M-L angle. The symmetries of the d orbitals are: $|xy\rangle, |xz\rangle, |yz\rangle \in T_{2g}$ and $|x^2 - y^2\rangle, |z^2\rangle \in E_g$. λ should be taken as positive in any case.

[b]Ground state $|xy\rangle$. Δ_1 and Δ_2 are the energies of $|yz\rangle$ and $|xz\rangle$ (degenerate in tetragonal symmetry); Δ_3 is the energy of $|x^2 - y^2\rangle$.

[c] Δ is the splitting of the ground term in tetragonal symmetry.

[d]Δ_i are the states arising from the splitting of the $^4T_{2g}$ state. The **g** tensor is generally nearly isotropic even in distorted chromophores.

[e]Ground state configuration: $(t_{2g})^3(e_g)^1 \equiv (|xy\rangle)^1 (|xz\rangle)^1 (|yz\rangle)^1 (x^2 - y^2\rangle)^1 (|z^2\rangle)^0$; Δ_1 and Δ_2 are the energies of the excitations $|yz\rangle \to |z^2\rangle$ and $|xz\rangle \to |z^2\rangle$ respectively. They are degenerate in tetragonal symmetry.

[f] Ground state configuration: $(t_{2g})^3(e_g)^1 \equiv (|xy\rangle)^1 (|xz\rangle)^1 (|yz\rangle)^1 (x^2 - y^2\rangle)^0 (|z^2\rangle)^1$; Δ_1 and Δ_2 are the energies of the excitations $|yz\rangle \to |xy\rangle$ and $|xz\rangle \to |xy\rangle$ respectively. They are degenerate in tetragonal symmetry.

[g]The ground Kramers doublet arising form the $^2T_2$ state is written as $|^2T_2, \pm\rangle = \pm\alpha|\pm 1, \pm\rangle + \frac{1}{2}\beta\sqrt{2}[|2, \mp\rangle - |-2, \mp\rangle] \pm \gamma|\mp 1, \pm\rangle$. (After Griffith, J.S. *Mol. Phys.* **1971**, *21*, 135).

[h]Elongated tetragonal. Ground state $^5B_2$. Δ_1 and Δ_2 are the energies of the excitations to ^5E and $^5B_1$.

[i]O_h symmetry. $\gamma = -1$ in strong ligand fields and $-3/2$ in weak fields.

[l]Tetragonal elongated complexes.

[m]Tetragonal compressed complexes.

[n]Δ_i are the states arising from the splitting of the $^3T_{2g}$ excited state. $\Delta_1 \equiv {}^3B_{3g}$; $\Delta_2 \equiv {}^3B_{2g}$; $\Delta_3 \equiv {}^3B_{1g}$; Δ_1 and Δ_2 are degenerate in tetragonal symmetries.

[o]Complexes having $|x^2 - y^2\rangle$ have the ground state (tetragonally elongated). Δ_1 and Δ_2 are the energies of $|xz\rangle$ and $|yz\rangle$ (degenerate in tetragonal symmetry); Δ_3 is the energy of $|xy\rangle$.

[p] Complexes having $|z^2\rangle$ have the ground state (tetragonally compressed). Δ_1 and Δ_2 are the energies of $|xz\rangle$ and $|yz\rangle$ (degenerate in tetragonal symmetry).

ation. Copper(II) is typical in this respect, and as a matter of fact it is the metal ion most commonly investigated through EPR. The systems with T type ground state, in general, do not undergo strong Jahn–Teller distortions. They have *quasi*degenerate ground states, the degeneracy being removed by spin-orbit coupling, which gives non-zero first order contribution within the T manifold; and broad EPR lines due to fast relaxation. d^1 ions are often characterized by the presence of M = O groups which generate large deviations from octahedral symmetry and give nice EPR spectra. This is for instance the case for oxovanadium(IV) complexes. An additional complication for the observation of EPR spectra is associated with zfs, but we will defer this discussion until Section 3.4.

The above treatment is valid within Ligand Field theory. Implicit in this treatment is the assumption that the unpaired electrons are localized on the metal ion in a $3d$ orbital. Equations (3.7–3.8) can be applied to a larger variety of situations when LCAO wavefunctions (obtained by a restricted open shell Hartree–Fock calculation) are used. In the case of one unpaired electron, equation (3.8) then takes the form:[47]

$$\Lambda_{ij} = 2\sum_{n\neq g}\sum_{k,k'} \frac{\langle \psi_g{}^k | \xi_k(r_k) L_i{}^k | \psi_n{}^k \rangle \langle \psi_n | L_j{}^{k'} | \psi_g{}^{k'} \rangle}{E_g - E_n} \tag{3.10}$$

where $|\psi_g\rangle$ is the orbital containing the unpaired electron, k and k' label different atoms of the molecule, $\mathbf{L}^k$ is the angular momentum operator centered on nucleus k, $\xi_k(r_k)$ represents the radial factor of the spin-orbit coupling operator for atom k; $|\psi_m{}^k\rangle$ indicates that part of the m-th molecular orbital centered on k. $\langle \psi_n | L_j{}^{k'} | \psi_g{}^{k'} \rangle$ is called a orbital Zeeman matrix element and the orbital ψ_n is not confined to be on the nucleus k'. This means that two-center contributions are to be taken into account. The spin-orbit coupling matrix elements $\langle \psi_g{}^k | \xi_k(r_k) L_i{}^k | \psi_n{}^k \rangle$ contain only one-center contributions due to the r_k^{-3} dependence of $\xi(r_k)$.

The principal values of $\mathbf{g}$ can in general be easily measured from experimental spectra.[20] Since the $\mathbf{g}$ "tensors" are in general anisotropic one needs to record anisotropic spectra. This can be done on single crystals, but in general this is a complex technique, which requires notable experimental effort. In fact it is much easier to record the spectra of samples where all the possible orientations of the magnetic centers are present at the same time. This can be done either using polycrystalline powders or frozen solutions. A fluid solution in general cannot be used for this purpose because the tumbling time of small molecules in solution is typically much shorter than the characteristic time of the EPR experiment, therefore only the average g value is obtained:

$$g = \frac{1}{3}(g_x + g_y + g_z) \tag{3.11}$$

Typical spectra of a $S = 1/2$ spin system for a polycrystalline powder with isotropic, axial, and completely anisotropic $\mathbf{g}$ tensors are shown in Figure 7. Details about the procedure for computing polycrystalline powder spectra can be found in Section 3.6.

As already mentioned in Section 2.1, the usual way of recording the EPR spectra provide the first derivative of the absorption. The lineshape is generally Lorentzian for magnetically dilute systems, and the absorption maximum corresponds to the field B_r at which the derivative curve goes to zero. Using equation (2.6), the g value can be calculated from B_r. Generally the microwave frequency is not directly measured, and the g value is obtained using for g_e in equation (2.6) the g_0 value of a standard sample, independently measured. A common standard is the organic radical diphenyl picryl hydrazyl, DPPH, which has $g_0 = 2.0037$.

For axial spectra ($g_x = g_y = g_\perp$; $g_z = g_\parallel$) the appearance of polycrystalline powder or frozen solution spectra is very asymmetric, due to the fact that the probability of crystallites with the unique z axis parallel to the external filed is lower than that of crystallites with the external field perpendicular to the axis. As explained in Section 3.6, only resonances corresponding to turning points in the field dependence of the resonant fields can be observed in polycrystalline powder or frozen solution spectra. For $S = 1/2$ spin systems, these correspond to the principal values of the **g** tensor. The parallel resonance is easily recognized because it is similar to an absorption, while the perpendicular resonance is more similar to a derivative signal, as shown in Figure 7.

A similar situation is observed for completely rhombic spectra ($g_x \neq g_y \neq g_z$). In the absorption spectrum the resonance corresponding to the intermediate g value corresponds to a maximum, while the other two correspond to shoulders. This gives rise to the derivative spectrum shown in Figure 7. It is interesting to notice that, while it is possible to measure the three principal values of **g**, it is impossible to associate them to the x, y, or z axes. This can only be done, and in some cases with difficulties, by recording single crystal spectra, as outlined in Section 3.6.

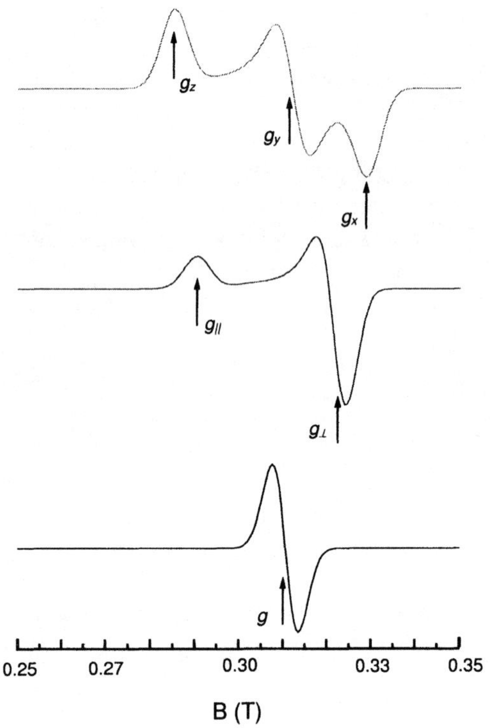

Figure 7. Typical polycrystalline powder spectra of a $S = 1/2$ spin system. The arrows indicate where the g values must be read. Bottom: isotropic spectrum with $g = 2.13$; Middle: axial spectrum with $g_\parallel = 2.28$, $g_\perp = 2.05$; Top: rhombic spectrum with $g_x = 2.01$, $g_y = 2.12$, $g_z = 2.32$.

It must be stressed that the spectra must be recorded in magnetically diluted systems, *i.e.* the paramagnetic centers must be well separated one from the other to prevent line broadening due to magnetic interactions or intermolecular averaging of the signals. These points will be discussed at some length in Section 3.5.

3.3 Electron–Nucleus Interaction

The electron–nucleus interaction is represented in equation (3.4) by the matrix **A**, which is symmetric.[48] The inclusion of the operator (3.4) in the spin Hamiltonian requires that the basis on which the spin Hamiltonian acts also includes the nuclear spin states. At the simplest level of approximation this basis can be taken as the direct product of the electronic and nuclear spin spaces, and the basis functions can be written as products of electron-spin and nuclear-spin kets as $|SM_S\Pi I_N M_{IN}\rangle$, where the product extends to all the magnetic nuclei which are present.

The simplest system to be treated is that of a $S = 1/2$ spin interacting with only one nucleus having nuclear spin $I = 1/2$. The energy levels for this case, in the assumption of isotropic interactions, are as shown in Figure 8. Often the nuclear Zeeman term of equation (3.2) is much smaller than the electron–nucleus interaction term and can be neglected in an EPR experiment. This is, of course, not possible in an ENDOR experiment (see Section 2.2.). In strong magnetic fields the transitions are magnetic dipole allowed and the selection rules are: $\Delta M_S = \pm 1$, $\Delta M_I = 0$.

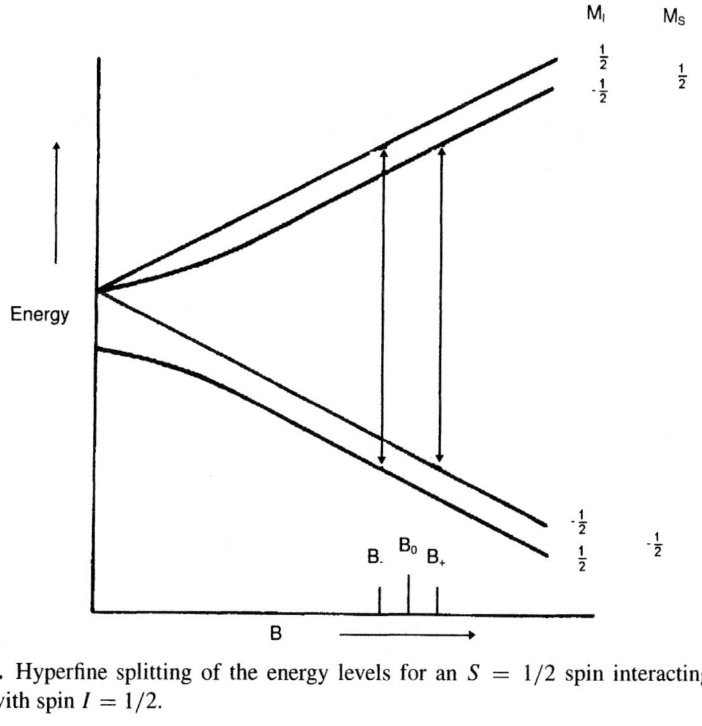

Figure 8. Hyperfine splitting of the energy levels for an $S = 1/2$ spin interacting with a nucleus with spin $I = 1/2$.

The allowed EPR transitions are those depicted in Figure 8 and in the strong field approximation which often holds, *i.e.* when the electron Zeeman interaction is much stronger than the electron–nucleus interaction, the resonances are at fields:

$$B_{\pm} = \frac{g_e}{g}\mu_B B_0 \pm \frac{1}{2}A \tag{3.12}$$

Therefore the single line at B_0 which would be observed in the absence of electron–nucleus interaction is split into two lines, separated by A. This splitting is called the *hyperfine structure* if the nucleus is where the unpaired electron is formally localized, and *superhyperfine structure* if the nucleus is different. In a transition metal complex the hyperfine structure is that associated with the metal ion, and the superhyperfine with the ligands. Note that in zero magnetic field the hyperfine interaction produces a splitting of the energy levels which can be labeled using the total spin operator $\mathbf{F} = \mathbf{S} + \mathbf{I}$. In Figure 8, the energy levels in zero field correspond to the states with $F = 1$ and 0. A general solution of the spin Hamiltonian is given by the Breit–Rabi equation[48] which can be applied to interpret EPR the spectra of systems in which the strong field approximation does not apply.

If the nucleus has $I_N > 1/2$ the hyperfine splitting yields $(2I_N + 1)$ equally spaced lines, the separation between neighboring lines being A. If there are n equivalent nuclei there are $2nI_N + 1$ lines, with intensities which can be obtained through binomial expansions. If there is also another set of m equivalent nuclei with spin I_M the total number of lines will be $(2nI_N + 1)\,(2mI_M + 1)$.

In Figure 9 we show polycrystalline powder spectra of a model system with $S = 1/2$ and $I = 3/2$. In Figure 9a only axial hyperfine interaction is observable, and the spectrum easily yields $g_{\perp} = 2.08$. The low field feature, corresponding to the $g_{\parallel}$ region ($g_{\parallel} = 2.30$), is split into four components of equal intensity, separated by $a = 100$ G (0.0100 T). This is due to the interaction of the unpaired electron with a nucleus of $I = 3/2$. The measured value of a can be converted in the more used unit (cm^{-1}) using the relationship $a(\mathrm{cm}^{-1}) = \mu_B g\, a(\mathrm{G})$, yielding $A_{\parallel} = 107 \times 10^{-4}\,\mathrm{cm}^{-1}$ ($\mu_B = 4.66864 \times 10^{-5}\,\mathrm{cm}^{-1}\,\mathrm{G}^{-1}$). Hyperfine coupling constants are often expressed in MHz. The relationships between the various units of measurement are: a(MHz) = $\mu_B/h\; g\, a(\mathrm{G}) = 1.3996\, g\, a(\mathrm{G})$; $a(\mathrm{cm}^{-1}) = a(\mathrm{MHz})/c = 0.3335641 \times 10^{-4}\, a(\mathrm{MHz})$.

In Figure 9b the four parallel features are further split into 5 lines of relative intensities 1 : 2 : 3 : 2 : 1 which correspond to the interaction of the unpaired electron with two equivalent $I = 1$ nuclei (as for example two equivalent nitrogen nuclei). The separation between the superhyperfine lines is 10 G, yielding a constant of $11 \times 10^{-4}\,\mathrm{cm}^{-1}$.

In Figure 9c and Figure 9d two sample spectra showing measurable hyperfine splitting also in the $g_{\perp}$ region are shown. The hyperfine splitting in the $g_{\perp}$ feature is $a = 20$ G, which gives $A_{\perp} = 19 \times 10^{-4}\,\mathrm{cm}^{-1}$. In Figure 9c, $g_{\parallel} = 2.18$ and the hyperfine lines overalp considerably. A better resolved spectrum is shown in Figure 9d where $g_{\parallel} = 2.30$.

There are several different physical mechanisms which contribute to the hyperfine term. In the case of one unpaired electron, using the same formalism needed

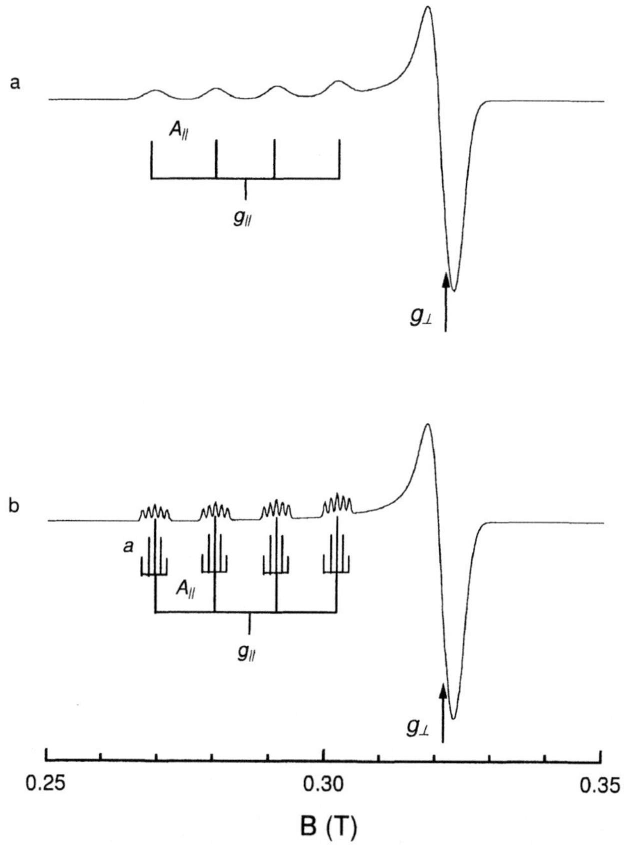

Figure 9. Typical polycrystalline powder spectra of axial copper(II) complexes ($S = 1/2$; $I = 3/2$). See text for the description.

to develop equation (3.10), the different contributions to the hyperfine tensor can be accounted for by rather simple expressions.

The first mechanism is associated with the fraction of unpaired electron directly present on the nucleus, and is called the *contact*, or *Fermi*, interaction. It originates from the symmetry mixing of s orbitals into the ground state wavefunction or from spin polarization of the core s orbitals. For transition metal ions this last mechanism is always present and often is very important in determining the spin density on the nucleus. The Fermi contact contribution can be written as:

$$A_{xx}^{Con} = A_{yy}^{Con} = A_{zz}^{Con} = \frac{8\pi}{3} P|\psi_n(0)|^2 \qquad (3.13)$$

where $P = g_e\mu_B g_N\mu_N$, and $\psi_n(0)^2$ is the unpaired spin density on the nucleus. A general way of computing $|\psi_n(0)|^2$ is to take the difference between the spin up and spin down densities obtained by a spin unrestricted calculation. This interac-

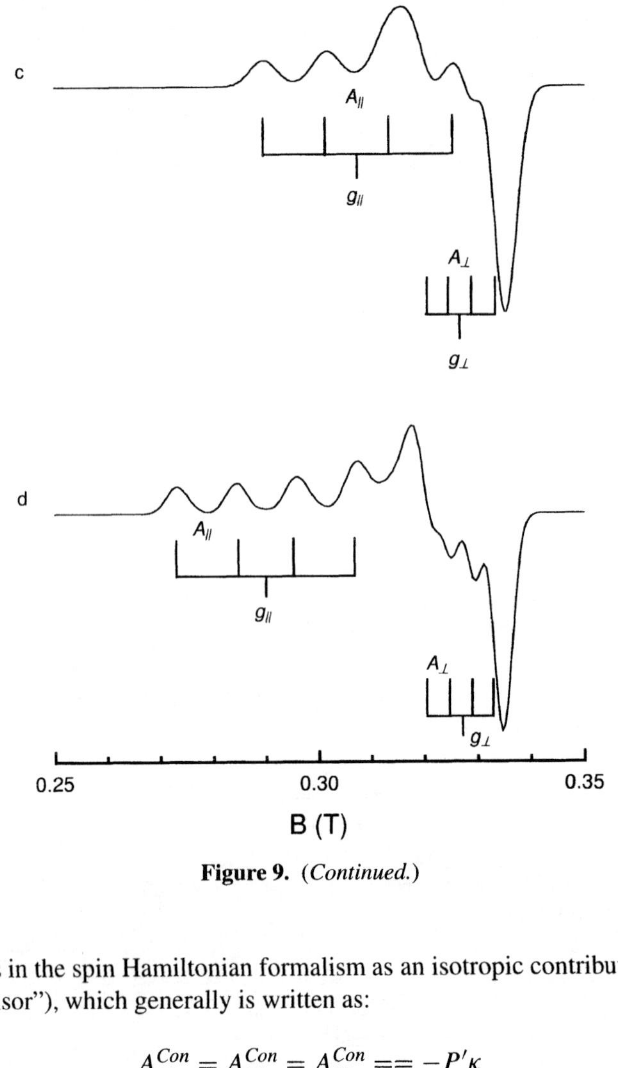

Figure 9. (*Continued.*)

tion appears in the spin Hamiltonian formalism as an isotropic contribution to the **A** matrix ("tensor"), which generally is written as:

$$A_{xx}^{Con} = A_{yy}^{Con} = A_{zz}^{Con} == -P'\kappa \tag{3.14}$$

$$P' = P\langle r^{-3}\rangle \tag{3.15}$$

where both P' and κ should be regarded as parameters to be determined from the experimental data. In Table 4 some typical values of P' and κ commonly used for various transition metal ions are shown.

The second mechanism is given by the through-space magnetic interaction between the electron and nuclear moment. It can be considered to be given by the magnetic dipolar interaction between the two magnetic moments. It is generally decomposed into two terms: one is the dipolar interaction between the nuclear moment and the spin of the electron, and the other is the interaction between the nuclear moment and the orbital angular momentum of the electron. The dipolar interaction can

Table 4 Computed Values of P′ (× 10^{-4} cm^{-1}) and κ for Selected Transition Metal Atoms[a]

Element	P'	κ
Sc	80.2	11.8
Ti	−24.6	10.6
V	146.0	9.52
Cr	−34.4	7.26
Mn	207.	8.09
Fe	32.6	7.64
Co	282.	7.03
Ni	−125.	6.65
Cu	399.	5.01
Zn	117.	5.93
Y	−20.8	20.1
Zr	−51.9	17.7
Nb	152.	14.4
Mo	−50.3	13.2
Tc	229.	13.7
Ru	−53.3	11.1
Rh	−40.4	10.1
Pd	−62.7	—
Ag	−68.3	8.94
Cd	−430.	10.6
La	79.1	25.3
Hf	41.4	35.5
Ta	149.	33.7
W	60.9	31.6
Re	386.	30.7
Os	150.	29.27
Ir	41.1	28.4
Pt	492.	23.3
Au	44.0	21.8
Hg	53.7	26.0

[a]From Weltner, W.; Jr. *Magnetic Atoms and Molecules*, Dover Publications, 1983.

be easily computed within the Ligand Field framework by taking the average value of the dipolar operator between the *d* orbitals. It depends on the nature of the orbital as shown in Table 5 and is, therefore, anisotropic. The term arising from the through-space magnetic interaction between the nuclear moment and the orbital component of the electron moment is generally expressed as a function of the **g** anisotropy. This is also anisotropic, but its trace is different from zero. Therefore it contributes to the trace of the tensor, which can be experimentally determined by measuring the spectrum in fluid solution, together with the contact term. For this reason it is sometimes called the *pseudo-contact* term.

The calculated expressions for the hyperfine tensor for d^1 and d^9 transition metal ions in pseudo-octahedral symmetry are shown in Table 6. From the analysis of these expressions the various contributions to the hyperfine are apparent. The Fermi and

Table 5 Dipolar Contributions to the Electron–Nucleus Coupling for d Orbitals in Units of P′

Orbital	A_{xx}^{dip}	A_{yy}^{dip}	A_{zz}^{dip}
z^2	$-2/7$	$-2/7$	4/7
x^2-y^2	2/7	2/7	$-4/7$
xy	2/7	2/7	$-4/7$
xz	2/7	$-4/7$	2/7
yz	$-4/7$	2/7	2/7

the pseudo-contact terms contribute to the trace of the hyperfine tensor, while the dipolar contributions (second term in all the expressions) average out to zero and do not contribute to the isotropic constant

$$a_{\text{iso}} = \frac{A_x + A_y + A_z}{3}$$

which can be measured in fluid solution spectra.

Of course, the Ligand Field approach and the formulas in Table 6 do not account for superhyperfine interactions and the more general molecular orbital treatment should be applied. In systems with only one unpaired electron, a general expression[47] representing the first order contributions is available for the dipolar interaction:

$$A_{ij}^{k,\text{Dip}} = P\left\{\left\langle \psi_n{}^k \left| \frac{F_{ij}{}^k}{r_k{}^3} \right| \psi_n{}^k \right\rangle + 2\sum_{k\neq k'} \left\langle \psi_n{}^{k'} \left| \frac{F_{ij}{}^k}{r_k{}^3} \right| \psi_n{}^k \right\rangle + \sum_{k\neq k'} \left\langle \psi_n{}^{k'} \left| \frac{F_{ij}{}^k}{r_k{}^3} \right| \psi_n{}^{k'} \right\rangle\right\} \quad (3.16)$$

where

$$F_{ij}^k = \frac{3r_ir_j - r^2\delta_{ij}}{r^2}$$

is the dipolar operator for the electron on nucleus k. The summations in (3.16) run on all the nuclei different from k. The $A_{ij}^{k,\text{Dip}}$ matrix is traceless. A non-traceless matrix is obtained when second order contributions are included, which require the mixing of excited states into the ground one *via* the spin-orbit coupling operator. These contributions include both the dipolar operator and the orbital-nuclear spin interaction represented by the operator

$$\frac{2P}{g_e}\sum_k \frac{\mathbf{I}\cdot\mathbf{L}^k}{r_k{}^3}$$

Table 6 Calculated A values for d^1 and d^9 ions in Pseudo-Octahedral Coordination[a]

Configuration	Ground state	A_x	A_y	A_z
d^1	$\|xy\rangle$	$P'\left\{-\kappa + \frac{2}{7} + \Delta g_x + \frac{3}{14}\Delta g_y\right\}$	$P'\left\{-\kappa + \frac{2}{7} + \Delta g_y + \frac{3}{14}\Delta g_x\right\}$	$P'\left\{-\kappa - \frac{4}{7} + \Delta g_z - \frac{3}{14}[\Delta g_x + \Delta g_y]\right\}$
d^9	$\|x^2 - y^2\rangle$	$P'\left\{-\kappa + \frac{2}{7} + \Delta g_x - \frac{3}{14}\Delta g_y\right\}$	$P'\left\{-\kappa + \frac{2}{7} + \Delta g_y - \frac{3}{14}\Delta g_x\right\}$	$P'\left\{-\kappa - \frac{4}{7} + \Delta g_z + \frac{3}{14}[\Delta g_x + \Delta g_y]\right\}$
	$\|z^2\rangle$	$P'\left\{-\kappa - \frac{2}{7} + \frac{15}{14}\Delta g_x\right\}$	$P'\left\{-\kappa - \frac{2}{7} + \frac{15}{14}\Delta g_y\right\}$	$P'\left\{-\kappa + \frac{4}{7} - \frac{1}{14}[\Delta g_x + \Delta g_y]\right\}$

[a]Symbols are defined in Table 3.

The matrix elements have the form:

$$A_{ij}^{k,\mathrm{Pseudo}} = P\sum_{n\neq g}\sum_{p,q}\frac{i\varepsilon_{ipq}\left[\sum_k \langle\psi_g{}^k|\xi_k(r_k)L_p{}^k|\psi_n{}^k\rangle\right]\langle\psi_n{}^k|\frac{F_{qi}^k}{r_k{}^3}|\psi_g{}^k\rangle}{\varepsilon_g-\varepsilon_n}$$
$$+P\sum_{n\neq g}\frac{2\left[\sum_k \langle\psi_g{}^k|\xi_k(r_k)L_i{}^k|\psi_n{}^k\rangle\right]\langle\psi_n{}^k|\frac{L_j^k}{r_k{}^3}|\psi_g{}^k\rangle}{\varepsilon_g-\varepsilon_n} \tag{3.17}$$

where $\varepsilon_{\mathrm{ipq}}$ is the Levi–Civita tensor defined as:

$$\varepsilon_{ipq}=\begin{cases}0 & \text{if any two } i,p,q \text{ are equal}\\ +1 & \text{if } ipq \text{ is an even permutation of } xyz\\ -1 & \text{if } ipq \text{ is an odd permutation of } xyz\end{cases}$$

The second term in 3.17 can be rewritten using the Λ_{ij} matrix element defined in 3.10 and is therefore expressible as the deviation of g_{ij} from the free electron value. This term is particularly important when dealing with the transition metal ion when it is the dominant contribution to the *pseudo*-contact termi and it is the only contribution included in the formulas of Table 6. In organic radical systems, this term is generally much smaller than the first one.

3.4 Electron–Electron Interaction

The electron–electron interaction is represented in the spin Hamiltonian formalism, see equation (3.3), by a tensor, **D**, which is symmetric and traceless. In general this property is made explicit using two parameters, D, and E, which are defined as:

$$D=\frac{3}{2}D_{zz}=-\frac{3}{2}\left(D_{xx}+D_{yy}\right) \tag{3.18}$$

$$E=\frac{1}{2}\left(D_{xx}-D_{yy}\right) \tag{3.19}$$

and the Hamiltonian (3.3) is rewritten as:

$$\mathcal{H}_{zfs}=D\left[S_z{}^2-\frac{1}{3}S(S+1)\right]+E\left(S_x{}^2-S_y{}^2\right) \tag{3.20}$$

For multiplets with $S > 1/2$ it is easy to see that (3.20) also splits the levels in the absence of an applied magnetic field. For this reason it is called the zero field splitting (zfs) Hamiltonian. The D and E parameters can be varied arbitrarily, but in practice they may be constrained within the limits $0 \leqslant |E/D| = \lambda \leqslant 1/3$. When $\lambda = 0$ the system has axial symmetry, when $\lambda = 1/3$ the system has the maximum rhombic splitting. When $D = 0$ the system has cubic symmetry. Figure 10

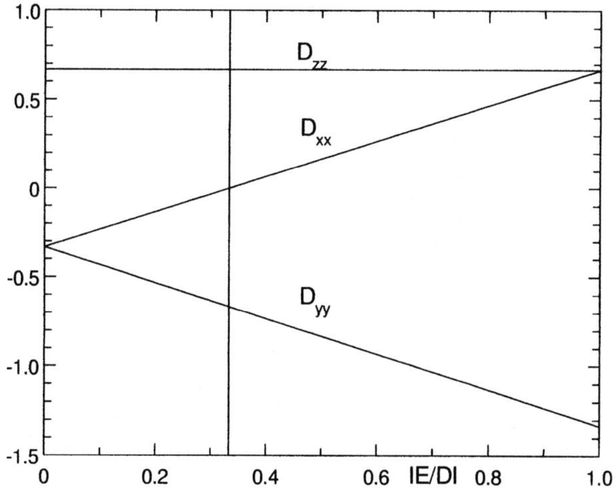

Figure 10. D_{ii} components vs. E/D. The vertical line corresponds to $E/D = 1/3$.

shows that for $E/D = 1/3$ the difference between the three D_{ii} components is at a maximum, while, on increasing it further, D_{xx} and D_{zz} become closer to each other, until at $E/D = 1$ they are identical; the system is again axial, but with unique axis y.

The effect of the zfs term on the spectrum is complex. The E term in (3.20) has matrix elements different from zero between states with $M_S = M_S \pm 1$; and in the general case a full matrix diagonalization is needed to obtain the energies of the spin levels or perturbation theory at some level of approximation must be applied. Some details on the solution of the spin Hamiltonian in the presence of zfs are given in Section 3.6.

For $S = 1$ spin systems an analytic solution of the spin Hamiltonian is possible when **g** and **D** have the same principal axes. The energies of the spin levels along the principal axes are plotted against the magnetic field in Figure 11. The transition fields can then be computed[49–51] as:

$$B_{x_1} = \frac{g_e}{g_x}\sqrt{(B_0 - D' + E')(B_0 + 2E')}; \quad B_{x_2} = \frac{g_e}{g_x}\sqrt{(B_0 + D' - E')(B_0 - 2E')}$$

$$B_{y_1} = \frac{g_e}{g_y}\sqrt{(B_0 - D' - E')(B_0 - 2E')}; \quad B_{y_2} = \frac{g_e}{g_y}\sqrt{(B_0 + D' + E')(B_0 + 2E')}$$

$$B_{z_1} = \frac{g_e}{g_z}\sqrt{(B_0 - \mathrm{D}')^2 - E'^2}; \quad B_{z_2} = \frac{g_e}{g_z}\sqrt{(B_0 + D')^2 - E'^2} \qquad (3.21)$$

where B_0 is the resonant field of the free electron, $B_0 = h\nu/\mu_B g_e$, and D' and E' are the zero field splitting parameters in gauss: $D' = D/\mu_B g_e$ and $E' = E/\mu_B g_e$. It is apparent that the actual spectrum depends on the ratio between the microwave frequency and D. Along each principal direction at most two allowed transitions

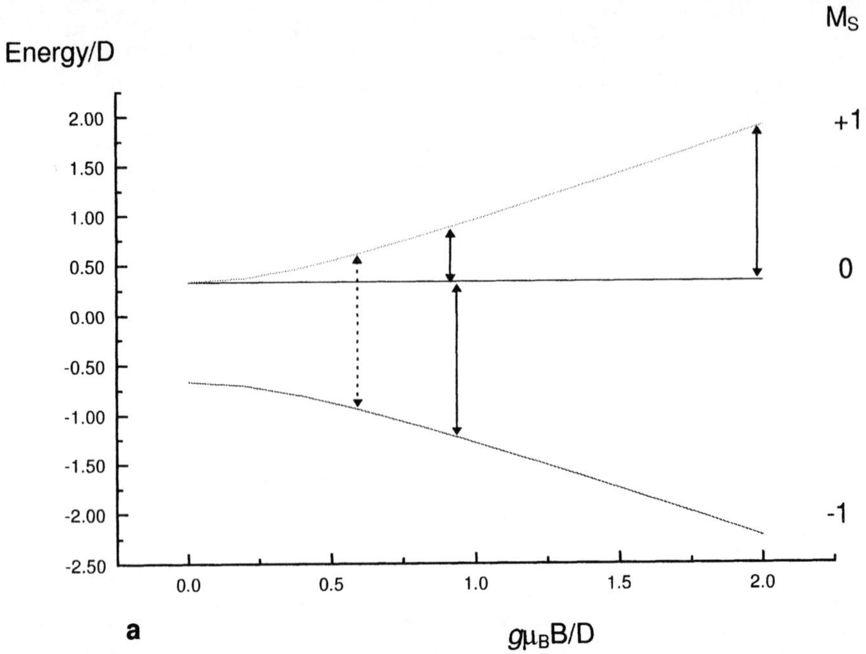

Figure 11. Variation of the energy levels of an $S = 1$ spin system on the magnetic field B. **a**: B along the x axis when $\lambda = 0$; **b**: B along the z axis when $\lambda = 0$; **c**: B along the x axis when $\lambda = 0.2$; **d**: B along the y axis when $\lambda = 0.2$; **e**: B along the z axis when $\lambda = 0.2$. The arrows indicate the allowed (full line) and forbidden (dotted line) transitions. Short arrows correspond to $h\nu = 0.5D$, long arrows correspond to $h\nu = 1.5D$.

are expected according to Figure 11. A typical example of polycrystalline powder spectrum of an $S = 1$ spin system is shown in Figure 12.

Together with the allowed transitions, some forbidden bands are often observed at low fields. In polycrystalline powder spectra, the half field transition occurring at B_{min} is a $\Delta M_S = \pm 2$ transition, that occurring at B_{dq} involves the absorption of two quanta of radiation. The observation of the transition at B_{min} is often diagnostic for the assignment of a triplet spectrum since it is related to the zero field splitting parameters through the relationship:

$$B_{min} = \frac{g_e \sqrt{\frac{B_0{}^2}{4} - \frac{D'^2 + 3E'^2}{3}}}{g_{min}} \quad (3.22)$$

where

$$g_{min} = (g_x g_y \sin^2 \alpha + g_z{}^2 \cos^2 \alpha)$$

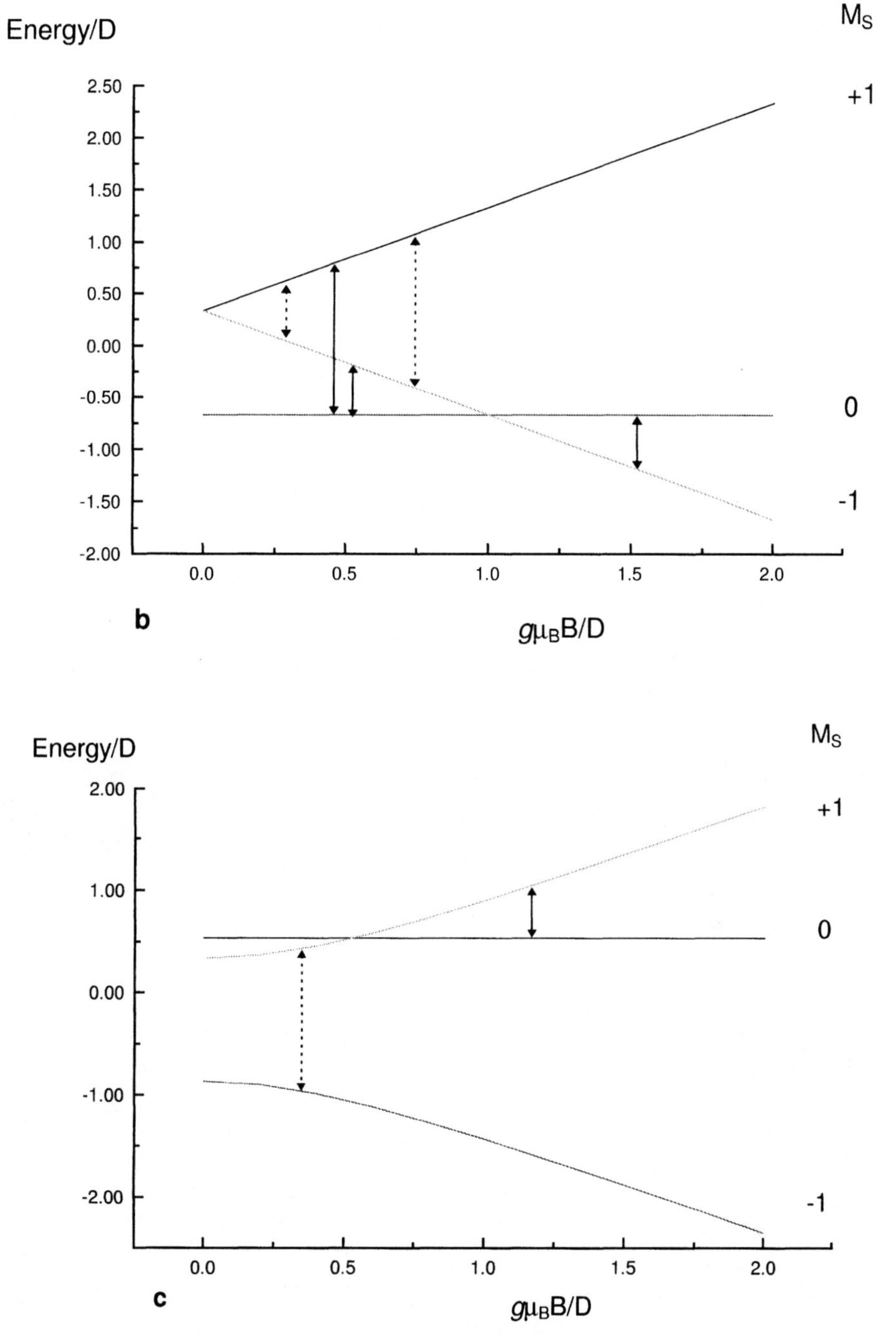

Figure 11. (*Continued.*)

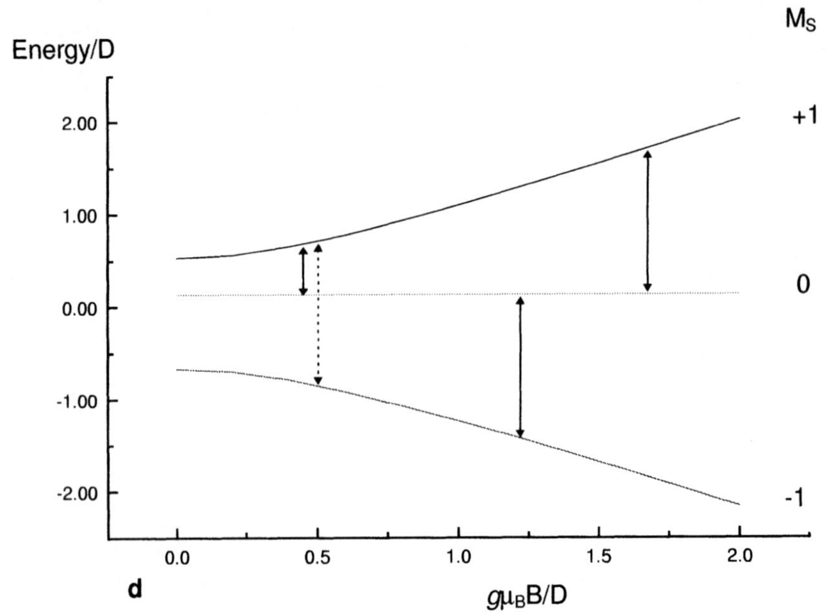

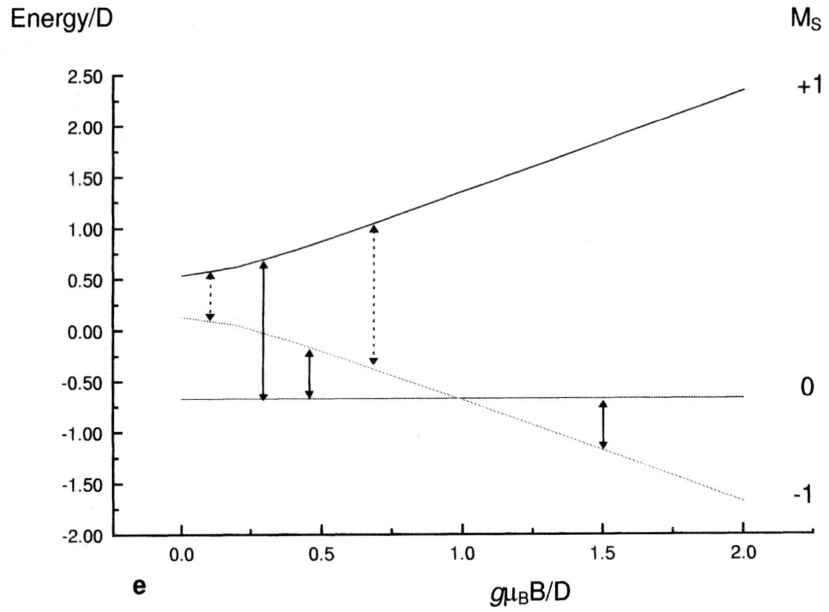

Figure 11. (*Continued.*)

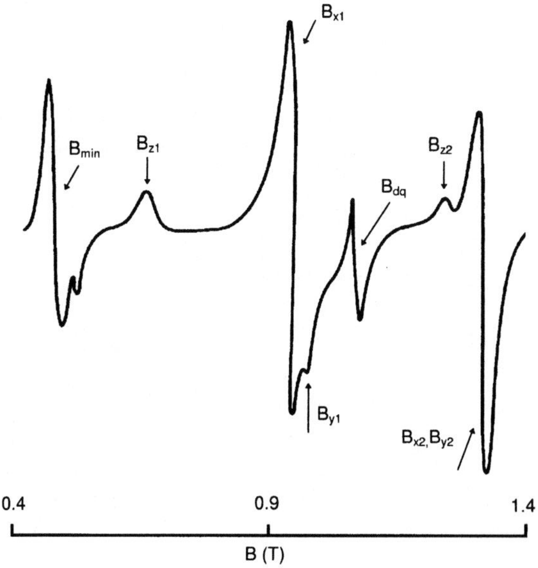

Figure 12. Polycrystalline powder spectrum of [Cu(chloroacetato)$_2\alpha$-picoline]$_2$ recorded at Q-band. (After ref. 12).

and

$$\alpha = \cos^{-1} \frac{9 - 4D^2/h^2\nu^2}{27 - 36D^2/h^2\nu^2}$$

A general solution can be obtained in closed form in axial symmetry ($E = 0$) assuming an isotropic **g** tensor. Under these conditions and provided that $D \ll h\nu$, the resonance field of the transition from the state $M_S - 1$ to M_S and its intensity at high temperature takes the simple form:

$$\begin{aligned} B(\vartheta) &= B_0 - \frac{D'}{2}(3\cos^2\vartheta - 1)(2M_S - 1) + \text{higher order terms} \\ I_{M_S-1\to M_S} &\propto (S - M_S)(S + M_S + 1) \end{aligned} \tag{3.23}$$

where ϑ is the angle between the z axis and the magnetic field. When kT becomes comparable to the Zeeman energy, $g_e\mu_B B_0$, the relative intensities can change drastically because the relative population of the M_S levels at which the transitions originate vary dramatically, as shown in Figure 13. In fact the observation of intensity polarization in the EPR transitions is a powerful tool for the determination of the sign of D'. If D' is positive the parallel transitions at high field will increase their intensities on decreasing temperature, while the reverse is true if D' is negative.[52] This technique can now be easily exploited with the high frequency-high field EPR spectrometers.

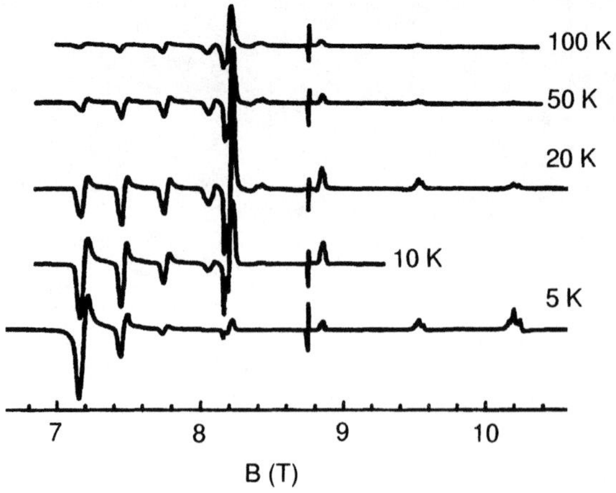

Figure 13. Polycrystalline powder spectra at variable temperature of an $S = 3$ spin system recorded at 245 GHz showing the variation of the intensities of the transition as a function of the temperature. (After reference 52).

As already seen for the $S = 1$ case, beyond the transitions which correspond to the $\Delta M_S = \pm 1$ selection rule, also transitions corresponding to $\Delta M_S = \pm 2$ ($S \geqslant 1$), ± 3 ($S \geqslant 3/2$), etc. become allowed when the field is not parallel the unique axis, or when the symmetry is lower than axial, and can be used as a diagnostic tool for assigning the spin state of the system.

In the opposite limit of zero field splitting much larger than the Zeeman term, matters differ greatly for systems depending on whether they have an odd or an even number of unpaired electrons. For the latter the levels, differing by $\Delta M_S = \pm 1$ may be separated by energy differences much larger than the microwave quantum in zero field. In order to induce a transition a very high magnetic field would be needed, which in general is not available. Therefore many systems with an even number of unpaired electrons are EPR silent. For systems with odd numbers of unpaired electrons, the minimum degeneracy of the levels in zero field is always two. The couples of these levels, which are degenerate in the absence of a magnetic field, are called Kramers doublets. The transition between the $M_S = \pm 1/2$ levels will be in principle allowed. In this case it is possible to interpret the spectra within an effective spin $S' = 1/2$, because the transitions occur within a set of two levels. However the anisotropy of the zero field splitting will still be recognized, because the effective g_{eff} values are related to the true one via simple perturbation expressions. When $\lambda = 0$, neglecting third-order terms,[53] the following relation holds:

$$g_{eff,\parallel} = g_{\parallel}$$
$$g_{eff,\perp} = \frac{2S+1}{2} g_{\perp} \tag{3.24}$$

For a system with $S = 3/2$, $g_{eff,\perp}$ will be $2g_{\perp}$, and for a system with $S = 5/2$, $g_{eff,\perp} = 3g_{\perp}$.

There is another interesting case of large zero field splitting yielding easily recognized spectra, that of a system with $S = 5/2$, and $|E/D| = 1/3$, i.e. with maximum rhombic splitting. In this case, the spectra show only one isotropic transition at an effective value $g_{eff} = 4.3$. This situation will be discussed in more detail in Section 4.1. The same conditions hold for all the multiplets with an odd number of Kramers doublets. For instance for $S = 9/2$, one expects $g_{eff} = 6.3$.

In the case when the zfs and the Zeeman terms are comparable there is not much chance for qualitative interpretation of the spectra, and extensive simulation is needed.

In principle there are two possible origins for the electron-electron interaction, namely the through space interaction between the magnetic moments (magnetic dipolar) and the spin-orbit coupling mediated crystal field effects. The former is in general negligible for transition metal ions, while it provides the main case for organic radicals. As in the justification of the deviation of the g values from the free electron value, one must remember that spin-orbit coupling admixes excited states with different S into the ground state. Therefore the energies of the M_S components will vary, and the multiplet is split in zero field. Within the ligand field approximation it is possible to express the **D** tensor using second order perturbation theory as:

$$\mathbf{D} = \lambda^2 \Lambda \tag{3.25}$$

where Λ was given in equation (3.8). When the excited states belong to the same LS manifold of the ground state, then the zero field splitting parameters are given by:

$$D = \lambda/2[g_z - 1/2(g_x + g_y)] \tag{3.26}$$

$$E = \lambda/4[g_x - g_y] \tag{3.27}$$

These equations show that sizable zero field splittings can be obtained even for small **g** anisotropies, if λ is large enough. For instance for nickel(II) $\lambda = -315\ \text{cm}^{-1}$. If $g_x = g_y = 2.25$, $g_z = 2.30$, corresponding to a moderate axial anisotropy, $D = 7.875\ \text{cm}^{-1}$. Such a large zero field splitting would certainly hamper the observation of the EPR spectra in a normal spectrometer. In fact, using equation (3.21) the calculated transition fields could be observed at fields larger than 1.4 T at X-band.

3.5 Spectra in Magnetically Non-Diluted Media

In all the above discussion we have assumed that the paramagnetic centers are magnetically dilute, i.e. they do not interact with each other. This condition can be met either by recording spectra in solution or diluting the centers in suitable diamagnetic hosts. In general if the spectra of pure solid samples are recorded, the individual centers interact by both dipolar and exchange interactions. The former tend to broaden the spectra, while the latter average out the broadening. The linewidth of the spectra depends therefore on the result of the two contrasting interactions. The dipolar interaction is due to the through space magnetic interaction between the magnetic moments $\mathbf{S}_1$ and $\mathbf{S}_2$ and can be described by the Hamiltonian:

$$\mathcal{H} = \mathbf{S}_1 \cdot \mathbf{D}_{12}^{\text{dip}} \cdot \mathbf{S}_2 \tag{3.28}$$

for the interaction of two spins. $\mathbf{D}_{12}^{\text{dip}}$ is a traceless matrix which can be calculated as:

$$\mathbf{D}_{12}^{\text{dip}} = \frac{\mu_B{}^2}{R_{12}^3}\left[\mathbf{g}_1 \cdot \mathbf{g}_2 - 3\frac{(\mathbf{g}_1 \cdot \mathbf{R}_{12})(\mathbf{R}_{12} \cdot \mathbf{g}_2)}{|R_{12}|^2}\right] \tag{3.29}$$

R_{12} being the distance between the two interacting centers having Zeeman tensors $\mathbf{g}_1$ and $\mathbf{g}_2$ respectively. A given spin, say 1, interacts with all the spins in the lattice, and, since the dipolar interaction decays slowly, it will be under the influence of many spins in the lattice. As a consequence the line is homogeneously broadened, giving rise to a gaussian line. The linewidth is proportional to the dipolar second moment:

$$\begin{aligned} M_2^{\text{dip}} &= \frac{3}{4}S(S+1)\mu_B^2 g^2 \\ &\times \sum_j R_j{}^{-6}\left\{\begin{array}{l}\left(3\cos^2\theta_j - 1)^2 + \sin^4\theta_j \exp\left[-\frac{1}{2}\left(\frac{2\omega_0}{\omega_e}\right)^2\right]\right) \\ +10\cos^2\theta_j \sin^2\theta_j \exp\left[-\frac{1}{2}\left(\frac{\omega_0}{\omega_e}\right)^2\right]\end{array}\right\} \end{aligned} \tag{3.30}$$

where the sum is over all the spins in the lattice. R_j is the distance from the spin for which the interactions are calculated and θ_j is the angle between the vector connecting the two spins and the z axis. ω_0 and ω_e are the Zeeman and exchange frequencies to be described below.

The spins can also interact with each other through an exchange mechanism. This allows the unpaired electron of one center to be delocalized onto another center, through the formation of a weak bond. In a spin Hamiltonian formalism at the simplest level of approximation the interaction between two centers can be described as:

$$\mathcal{H}_{ex} = J_{12}\mathbf{S}_1 \cdot \mathbf{S}_2 \tag{3.31}$$

where J_{12} is called the exchange coupling constant between centers 1 and 2. The effect of the exchange (3.31) is that of allowing a given electron to jump from one center to another with a frequency $\omega_e{}^2 = 2J_{12}{}^2/\hbar^2$. If the frequency of these jumps is larger than the frequency of the dipolar interaction between the two spins, the result is that of averaging the dipolar interaction to zero. It was already noticed that $\mathbf{D}_{12}^{\text{dip}}$ is a traceless matrix. If an electron is hopping fast it will see the average of the dipolar interaction, $1/3\ (D_{12,xx}^{\text{dip}} + D_{12,yy}^{\text{dip}} + D_{12,zz}^{\text{dip}})$, but since the matrix is traceless the average is zero.

If the condition that the exchange frequency is much larger than the dipolar frequency holds, the line is said to be exchange narrowed. It becomes Lorentzian with a width proportional to M_2^2/J. This is the reason why it is possible to record spectra of magnetically non-dilute systems. Although the lines can be reasonably narrow a lot of information is nevertheless lost. In fact all the hyperfine structure is wiped out. If there are centers with different orientations of the molecular axes, characterized by the tensors $\mathbf{g}_1$ and $\mathbf{g}_2$, only one average signal is recorded in the i direction when the exchange constant J_{12} is larger than $|g_{1i} - g_{2i}|\mu_B B$. The spectra in this case correspond to crystal averages and do not provide the spin Hamiltonian parameters of the individual centers. Great care must be exerted in the interpretation of the spectra of magnetically non-dilute systems.

3.6 Simulation Techniques

The accurate analysis of the EPR spectra requires the use of simulation and fitting techniques in order to fully determine the components of the tensors appearing in the spin Hamiltonian. Since a second-rank symmetric tensor is defined by 6 independent components, the full determination of the tensor requires in general the determination of transition fields on single crystals with the static magnetic field oriented in different planes. By diagonalization of the tensor the three principal values and directions with respect to a laboratory frame of reference are obtained. Polycrystalline powder or frozen solution spectra allow in general the determination of the principal values of the tensors with less experimental effort, but do not give direct information about the relative orientations of the tensors. Computer simulation of the spectra is needed for an accurate measurement of the spin Hamiltonian parameters. Calculation of polycrystalline powder or frozen solution EPR spectra is generally performed as a spatial average of single crystal spectra. In the following we will first briefly examine the appearance of single crystal spectra and later a short description of the principal methods of simulation of powder spectra will be given.

Single crystal spectra can be recorded by mounting the single crystal of the sample on a quartz or perspex rod, mechanically cut as shown in Figure 14, and attaching the rod to a goniometer for a careful measurement of the angle of rotation. By rotating the rod, spectra corresponding to different orientations of the magnetic field in the plane perpendicular to the rotation axis can be measured. By changing the crystal orientations as shown in Figure 14, this technique allows the measurements of spectra in three orthogonal planes and therefore the measurement of tensor components referred to the orthogonal system of axes which are the rotation axes (Schonland[54–56] method). In this way one determines a *crystal tensor*. Knowledge of the orientation of these axes (the laboratory frame) with respect to the crystallographic ones, allow the measurement of the *molecular tensor*, *i.e.* to link the measured tensor to molecular bonding directions (the molecular frame). The simplest experiment is performed on an effective $S = 1/2$ spin system in the absence of hyperfine interaction. In this case, if the direction of the static magnetic field in the laboratory plane, ij, is given by the angle ϑ from the i axis the resonance condition is given by

$$h\nu_0 = g_{eff}\mu_B B_0 \tag{3.32}$$

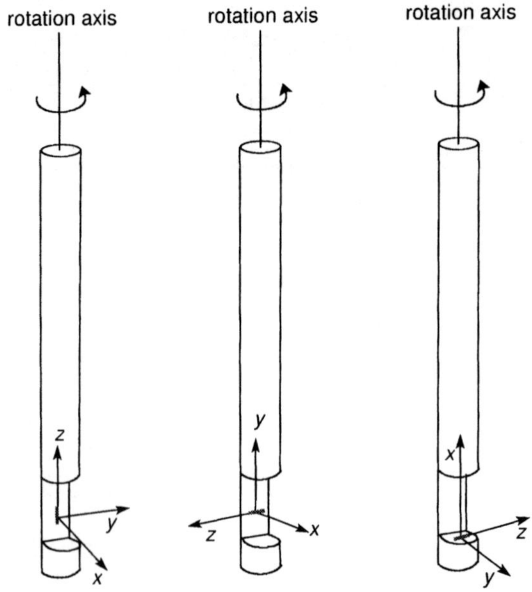

Figure 14. Simple crystal mounting to perform rotations around three orthogonal axes (the laboratory axes). From left to right: rotation around z, y, and x.

where

$$g_{eff}^2 = g_{ii}^2 \cos^2 \vartheta + g_{jj}^2 \sin^2 \vartheta + 2g_{ij}^2 \sin \vartheta \cos \vartheta \tag{3.33}$$

Measuring g_{eff} for several values of ϑ in different planes gives a series of equations of the form (3.33) which can be fitted to afford the least squares estimates of the g_{ij} elements. Equations similar to (3.33) hold for more complex situations when the Zeeman term is dominant in the spin Hamiltonian. In the case of a system with anisotropic hyperfine interaction with a single nucleus for which $g\mu_B B \gg |A_N|$ the same equation as (3.32) holds for the $A_{eff}^2 g_{eff}^2$ tensor in the place of g_{eff}^2. Perturbation solutions of the spin Hamiltonian which can be used to elaborate algebraic solution of the resonance energies have been published in several instances.[7] Also several variations of the basic Schönland method have been proposed to measure the spin Hamiltonian parameters from the single crystal spectra.[5]

Computer simulation of single crystal spectra requires knowledge of the transition fields or frequencies as a function of the spin Hamiltonian parameters and the use of a general expression for the field and frequency dependence of the EPR signal. This second problem is generally solved by using the general form[5] for a single EPR line:

$$S(B, \nu) = C\eta Q_0 \sum_{a>b} |\langle a|\mu \cdot \mathbf{B}_1|b\rangle|^2 f_\nu[\nu - \nu_0(\mathbf{B}), \sigma_\nu] \tag{3.34}$$

where $|\langle a|\mu \cdot \mathbf{B}_1|b\rangle|^2$ is the probability of the $|a\rangle$ to $|b\rangle$ transition, ν is the spectrometer frequency, $\nu_0(\mathbf{B})$ is the value of the resonance frequency at the applied external field $\mathbf{B}$, and f_ν is the lineshape function with width σ_ν. η and Q_0 are the filling factor and the unloaded Q factor, C being a proportionality constant.[5] $\nu_0(\mathbf{B})$ depends on both the strength and the orientation of the magnetic field and the lineshape is assumed to be symmetrical around $\nu_0(\mathbf{B})$. In (3.33) the summation extends over all the possible pairs of energy levels between which the transitions occur. A somewhat more complete expression has been used by Coffman[57] by adding a term to account for the relative thermal populations of the energy levels:

$$\frac{N\sum_{a>b}\exp\left(-\frac{E_a}{kT}\right)-\exp\left(-\frac{E_b}{kT}\right)}{Z} \tag{3.35}$$

where Z is the partition function. For a two level case, *e.g.* a $S = 1/2$ spin system, a simpler expression avoiding unnecessary constants is

$$S(B,\nu) \propto \frac{h\nu}{Z}|\langle a|\mu \cdot \mathbf{B}_1|b\rangle|^2 f_\nu[\nu - \nu_0(\mathbf{B}), \sigma_\nu] \tag{3.36}$$

where, in (3.36), we have used the high temperature expansion: $e^x = 1 + x + \cdots$ and the resonance condition $E_a - E_b = h\nu$. Common lineshape functions are the Lorentzian, f_L, and the Gaussian, f_G:

$$f_L = \frac{1}{\pi\sigma_\nu\left[1+\left(\frac{\nu-\nu_0}{\sigma_\nu}\right)^2\right]} \tag{3.37}$$

$$f_G = \frac{1}{\sqrt{2\pi}\sigma_\nu}e^{\left[-\frac{1}{2}\left(\frac{\nu-\nu_0}{\sigma_\nu}\right)^2\right]} \tag{3.38}$$

Equation (3.34) covers both the field and the frequency dependence of the spectrum. In the usual EPR experiment the microwave frequency is kept fixed at a constant value, $\nu = \nu_c$, while the magnetic field is swept. In the above equations ν_0 becomes the field dependent variable. A lineshape function, $f_B(B - B_0, \sigma_B)$, which explicitly includes this field dependence of the signal, is more commonly used in the place of (3.37–3.38). This field dependent lineshape is linked to f_ν by the relationship

$$f_B[B - B_0(\nu_c), \sigma_B] = \frac{\partial B}{\partial \nu_c} f_\nu[\nu_c - \nu_0, \sigma_\nu] \tag{3.39}$$

which is rigorously valid for Dirac-δ distribution and holds to a good approximation for symmetric lineshapes with finite widths. A detailed treatment of the frequency and field dependence of the EPR signal can be found in the book by Pilbrow.[5] A general expression useful for the simulation of field dependent spectra is obtained by taking the first derivative of (3.34) with respect to the magnetic field strength. The final form of this equation reads:

$$\frac{\partial S(\mathbf{B}, \nu_c)}{\partial B} = C' \sum_{a>b} |\langle a|\mu \cdot \mathbf{B}_1|b\rangle|^2 \frac{\partial f_B[B - B_0(\nu_c), \sigma_B]}{\partial B} \frac{\partial B_0}{\partial h\nu_c} \tag{3.40}$$

where C' includes all the constants not depending on B. Equation (3.40) is generally valid for computing single crystal spectra for any spin system in the absence of hyperfine or superhyperfine coupling. When these interactions are present a summation over the interacting nuclei and their isotopes should be added and each line is generally weighted by different intensity factors. Equation (3.40) assumes a simple form for a $S = 1/2$ spin state for which the last factor can easily be evaluated by the resonance condition $h\nu_c = g\mu_B B_0$ as

$$\frac{\partial B_0}{h\nu_c} = \frac{1}{g\mu_B}$$

which was introduced by Aasa and Vanngard[58] in 1975. Also the dependence of the transition probability on the orientation assumes the closed form:

$$g_1{}^2 = (g_x g_y g_z)^2 \left[\frac{(\sin\alpha\cos\gamma + \cos\alpha\cos\beta\sin\gamma)^2}{g_x{}^2} + \frac{(-\sin\alpha\cos\gamma + \cos\alpha\cos\beta\cos\gamma)^2}{g_y{}^2} + \frac{(\cos\alpha\sin\beta)^2}{g_z{}^2} \right] \sin^2\tau \tag{3.41}$$

where the angles are defined in Figure 15. From this equation it is apparent that $g_1{}^2$ is at a maximum when $\tau = \pi/2$, that is, when the static and the oscillating fields are orthogonal to each other, which is therefore the situation commonly met in the experimental apparatus.

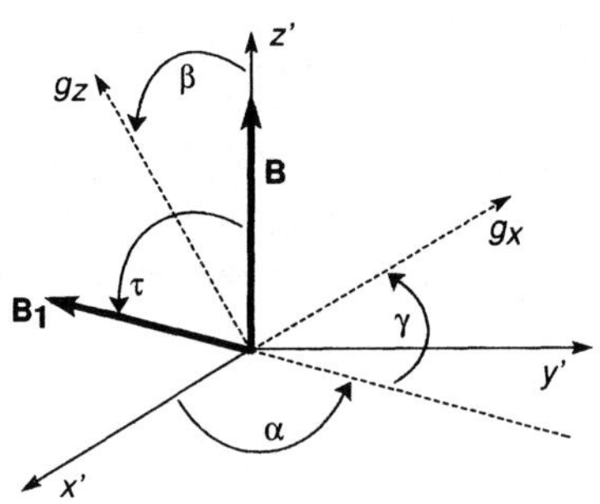

Figure 15. Euler angles (α, β, γ) defining the position of $\mathbf{B}$ and coordinates for $\mathbf{B_1}$.

Polycrystalline powder or frozen solution spectra can be conveniently computed by averaging (3.39) over space according to:

$$Q(B) = \int_0^{\pi} \mathrm{d}\varphi \int_0^{\pi} \sin\vartheta \mathrm{d}\vartheta \frac{\partial S(\mathbf{B}, \nu_c)}{\partial B} \tag{3.42}$$

Several procedures have been developed in order to properly partition the space for efficient numerical integration and to reduce the integration limits according to the symmetry of the Hamiltonian. The most common method consists in partitioning the ϑ and φ spaces into a finite number of points, $\vartheta_i = (\pi/N)i$, and $\varphi_i = (\pi/M)j$, and in approximating (3.41) with a double summation

$$Q(B) = \sum_{i=1}^{N} \sum_{j=1}^{M} \sin\vartheta_i \frac{\partial S(\mathbf{B}, \nu_c)}{\partial B} \frac{\pi^2}{NM} \tag{3.43}$$

Monte Carlo methods[59] have also been used to generate the values of ϑ and φ. For $S = 1/2$ spin systems the average transition probability can be computed in the principal axes system of **g** as:

$$\begin{aligned} \langle g_1{}^2 \rangle = \big[& g_x{}^2 g_y{}^2 \sin^2\vartheta + g_y{}^2 g_z{}^2 (sin^2\varphi + \cos^2\vartheta \cos^2\varphi) \\ & + g_z{}^2 g_x{}^2 (\cos^2\varphi + \cos^2\vartheta \sin^2\varphi) \big] / (2g^2) \end{aligned} \tag{3.44}$$

where ϑ and φ are the polar angles of **B** and g^2 is given by

$$g^2 = g_x{}^2 \sin^2\vartheta \cos^2\varphi + g_y{}^2 \sin^2\vartheta \sin^2\varphi + g_z{}^2 \cos^2\vartheta \tag{3.45}$$

Examples of powder spectra computed for a $S = 1/2$ spin system in isotropic, axial, and rhombic symmetry using finite component linewidths are shown in Figure 7. The positions in which the principal values of **g** are to be measured are also indicated. An extensive analysis of the position of the peaks in a powder EPR spectrum has been performed by van Veen[60], who showed how the peaks in the powder spectrum are found at the turning points of the angular dependence of the transition fields.

The calculation of the transition fields is the central point in any simulation program. While this is simply obtained by $h\nu_c = g\mu_B B_0$ and (3.44) for a $S = 1/2$ spin system, the calculation is more complicated when hyperfine is included or for $S > 1/2$ spins. Perturbative solutions to the spin Hamiltonian have been developed at several orders of approximations, one simple expression being given in equation (3.12). More accurate simulations require the diagonalization of the spin Hamiltonian matrix computed for any field orientation. From the eigenvalues, the difference $\nu_0(B) = [|E_a(B) - E_b(B)|]/h$ is computed for several values of the magnetic field strength, B, until the difference $\nu_c - \nu_0(B)$ becomes zero. The eigenvectors are used to compute the transition probability. This rather lengthy procedure can be overcome, at the expense of storing larger matrices, by using the eigenfield formalism developed[61] by Belford *et al.* In this formalism the transition fields are directly

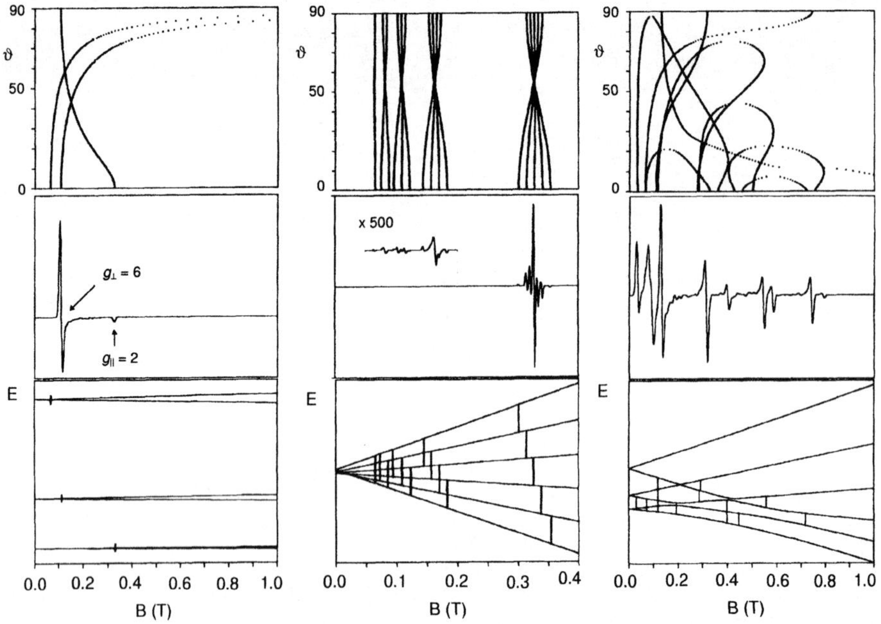

Figure 16. EPR spectra of a $S = 5/2$ spin system computed for X-band in axial symmetry using the eigenfield method. From left to right: $D \gg h\nu$; $D \ll h\nu$; $D \approx h\nu$. Top: angular dependence of the transition fields in the xz plane; Center: computed spectrum; Bottom: energy levels and transition fields at $z = 0$ (left and center) and 20 (right) respectively.

obtained by the solution of a generalized eigenvalue equation as the eigenvalues of a superoperator and the transition probabilities are easily obtained from the eigenvectors. This method requires the diagonalization of an n^2 by n^2 matrix compared to the diagonalization of an n by n spin Hamiltonian matrix.

An example of the simulation of an $S = 5/2$ spectrum is shown in Figure 16, where the connection between the peak positions and the turning points in the angular dependence of the transition fields is made clear. The spectra of $S = 5/2$ spin systems will be discussed at some length in Section 4.1.

In the methods outlined above the dependence of the transition probability and linewidths on the field strength has been neglected. More accurate algorithms treating these effects, which also include the so called g-strain, have been worked out, but will not be explicitly mentioned here. The interested reader can find detailed information in reference 7.

4 SOME EXAMPLES

In this Section we will present some examples of the application of EPR spectroscopy to various fields of interest in Inorganic Chemistry. In Section 4.1. we will consider complexes of copper(II) and iron(III). In Section 4.2. some applications of

EPR to Biological Systems are presented; and the use of EPR to characterize Inorganic Materials will be briefly covered in Section 4.3.

4.1 Simple Complexes and Clusters

Copper(II) complexes have been studied for a long time with EPR, since they offered a large variety of structural environments. Copper(II) has the valence electronic configuration $3d^9$ and possesses, therefore, one unpaired electron in any stereochemistry. The g values are always greater than the spin only value $g_e = 2.0023$ and can be explained, at least to a first approximation, by using perturbative formulas like those reported in Section 3. Two stable isotopes of copper, ^{65}Cu and ^{63}Cu, with relative abundances of 31% and 69% respectively, both have a nuclear spin $I = 3/2$ and very similar gyromagnetic ratios, so that often only one characteristic hyperfine splitting into four lines is observed. Some examples of spectra have already been reported in Figure 7; and the relevant perturbative expressions for the A values are given in Table 6. In a few cases hyperfine interaction with ligand nuclei, mainly nitrogen and oxygen, has been observed. Details on the calculation of the spin Hamiltonian parameters for copper(II) complexes can be found in references.[62–68]

Spin Hamiltonian parameters for selected copper(II) complexes in different stereochemistries are collected in Table 7. Octahedral copper(II) complexes have a ^{2}E ground state and EPR has been used for a long time to characterize the Jahn–Teller distortion of this electronic state[69] caused by the coupling of a vibrational e mode with the electronic state, the so-called $E \otimes e$ case.[70–72] Two extreme cases of Jahn–Teller effects (JT) can be easily recognized in the EPR spectra: the strong and the weak JT.

In the strong JT case a distortion from octahedral symmetry occurs leading to three equivalent minima in the potential energy surface (the 'Mexican hat' potential wells) corresponding to elongations of the octahedron along three orthogonal axes. At sufficiently low temperatures anisotropic spectra due to copper(II) centers trapped in the energy minima are observed. The spectra at higher temperatures are isotropic, fluid solution-like as a result of a dynamic reorientation of the system between static distortions. The whole experience is generally denoted as dynamic JT. Classical examples of this behavior are the copper(II) ion doped into $ZnSiF_6 \cdot 6H_2O$,[73] and into $La_2Mg_3(NO_3)_{12} \cdot 24H_2O$.[74,75]

In this latter case, a 'static' JT effect is observed at temperatures below 20 K giving rise to an anisotropic spectrum with $g_{\parallel} = 2.465$, $g_{\perp} = 2.099$, $A_{\parallel} = -122 \times 10^{-4}$ cm^{-1} and $A_{\perp} = 16 \times 10^{-4}$ cm^{-1}. At higher temperatures an almost isotropic spectrum is seen with $g_{\parallel} = 2.219$, $g_{\perp} = 2.218$, $|A_{\parallel}| = 29 \times 10^{-4}$ cm^{-1} and $|A_{\perp}| = 28 \times 10^{-4}$ cm^{-1}.

In the weak JT situation the energy minima are separated by a low energy barrier and the spectrum shows a rather complicated temperature dependence with the coexistence of isotropic and anisotropic spectra down to temperatures on the order of 10 K as observed in copper(II) doped into MgO[76,77] and into CaO.[78] These effects have been explained with moderate linear vibronic coupling by Ham[79,80] who

Table 7 Selected Examples of EPR Spectra of Copper(II)

Stereochemistry	Compound	Chromophore	g_x	g_y	g_z	$A_x \times 10^{-4}$ (cm^{-1})	$A_y \times 10^{-4}$ (cm^{-1})	$A_z \times 10^{-4}$ (cm^{-1})	Ref
Elongated Octahedral	$[Cu(pyO)_6](ClO_4)_2^a$	CuO_6	2.18 2.07	2.18 2.07	2.18 2.37				12
	$[(Cu/Zn)(Im)_6]Cl_2 \cdot 4H_2O^{a,b}$	CuN_6	2.20 2.05	2.20 2.09	2.20 2.32	≈ -7	≈ 20	156	*c*
Square Planar	$Cu(8\text{-hydroxyquinolinate})_2$	CuO_2N_2	2.05	2.05	2.21	29	29	206	*d*
	$[Cu(NH_3)_4][PtCl_4]$	CuN_4	2.05	2.05	2.22	28	28	211	*d*
Square Pyramidal	$[Cu(apy)_5](ClO_4)_2$	CuO_5	2.08	2.08	2.40	35	35	126	12
Trigonal Bipyramidal	$[(Cu/Zn)(Me_6tren)Br]Br$	CuN_4Br	2.18	2.18	1.96	100	100	80	12
	$[(Cu/Zn)(tren)NH_3](ClO_4)_2$	CuN_4N	2.18	2.18	2.03	110	110	84	*d*
Intermediate Geometries	$[Cu(bipy)_2NH_3](BF_4)_2$	CuN_4N	2.23	2.14	2.01				12
	(Cu/Zn)SALMe	CuN_3O_2	2.25	2.13	2.02	131	65	47	12
	(Cu/Zn)SALMeDPT	CuN_3O_2	2.23	2.07	2.04	150	38	66	12
Compressed Tetrahedral	$Cs_2[(Cu/Zn)Cl_4]$	$CuCl_4$	2.10	2.08	2.44	46	51	23	12
	$[Cu(bipyam)_2](ClO_4)_2$	CuN_4	2.06	2.06	2.24			145	*d*
	$(Cu/Zn)(SALtBu)_2$	CuN_2O_2	2.08	2.03	2.29	≈ 20	≈ 12	117	12

[a] The first number refers to the room temperature spectra, the second to the low temperature ones.
[b] Two inequivalent sites were detected at low temperature.
[c] Keijzers, C.P.; Jansen, T.; De Boer, E.; Van Kalkeren, G.; Wood, J.S. *J. Magn. Res.* **1983**, *52*, 211.
[d] Hathaway, B.J. in *Comprehensive Coordination Chemistry* **1987**, 5, 533, Wilkinson, G.; Gillard, R.D.; McCleverty, J.A.; Eds., Pergamon Press, Oxford.

included excitations from the anisotropic ground state to an isotropic excited level. Generalization to a three state theory was subsequently proposed.[81]

The classical example of dynamic JT effect is given by the EPR spectrum of copper(II) doped into zinc Tutton's salt enriched of ^{17}O.[82] The observed variation of the EPR parameter with temperature is reported in Table 8. The crystals are monoclinic and the geometry of the copper(II) ion is a compromise between the symmetry imposed by the lattice and vibronic coupling. The copper(II) ion shows three different bond distances to the oxygen atoms and the EPR spectrum has been interpreted with a potential surface with three inequivalent minima, the so-called sun-warped Mexican hat, shown in Figure 17.

Copper(II) complexes with geometries derived from the octahedral one, *i.e.* tetragonal octahedral and square pyramidal, always have the unpaired electron in the $d(x^2 - y^2)$ orbital and the formulas given in Section 3 can be used to compute the g and A values. The g and A value patterns are $g_{\parallel} > g_{\perp} > g_e$, and $A_{\parallel} > A_{\perp}$ as shown in Table 7. Square planar complexes, which also have the unpaired electron in the

Table 8 Spin Hamiltonian Parameters of Copper(II) Doped $K_2Zn(SO_4)_2 \cdot 6H_2O$ (from ref. 82)

				^{63}Cu ($\times 10^{-4}$ cm^{-1})			^{17}O(I) ($\times 10^{-4}$ cm^{-1})			^{17}O(II) ($\times 10^{-4}$ cm^{-1})		
T	g_1	g_2	g_3	A_1	A_2	A_3	A_1	A_2	A_3	A_1	A_2	A_3
295	2.31	2.26	2.02									
77	2.36	2.20	2.03	76	36	60	11	12	21.5	7	13.5	9
20	2.42	2.15	2.03	96	<20	60	10.9	11.9	21	9.9	16.1	8.8

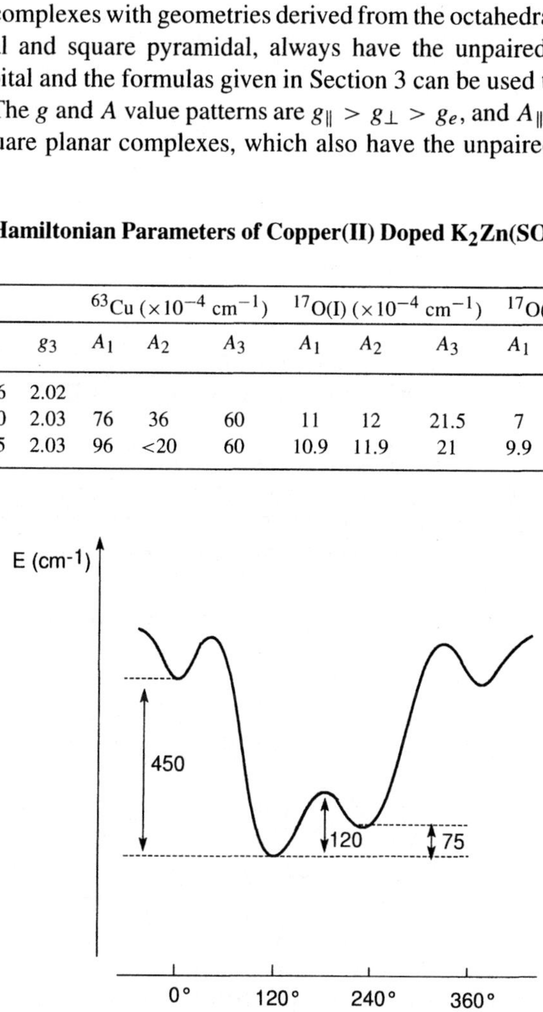

Figure 17. The potential surface for copper(II) doped into zinc Tutton's salt (After Hathaway, B.J. in *Comprehensive Coordination Chemistry* 1987, 5, 533, Wilkinson, G.; Gillard, R.D.; McCleverty J.A.; Eds., Pergamon Press, Oxford.).

$d(x^2 - y^2)$ orbital, are characterized by smaller deviations of g from g_e compared to the square pyramidal case; and also by larger A values. This is due to the stronger ligand field present in the square planar complexes state which determines a larger energy separation of the excited states from the ground state.

Five coordinate complexes with geometries derived from the trigonal bipyramid have rather different EPR spectra since the unpaired electron is in a $d(z^2)$ orbital. In this case the spin Hamiltonian parameters are given by:

$$g_{\parallel} = g_e; \qquad g_{\perp} = g_e - \frac{6\lambda}{\Delta(xz, yz)}$$
$$A_{\parallel} = P'\left[-k + \frac{4}{7} - \frac{1}{7}(g_{\perp} - g_e)\right]; \quad A_{\perp} = P'\left[-k + \frac{2}{7} + \frac{15}{14}(g_{\perp} - g_e)\right] \tag{4.1}$$

where the symbols have been defined in Section 3. The g and A value patterns $g_{\parallel} \approx g_e < g_{\perp}$, and $|A_{\parallel}| < |A_{\perp}|$ can then be used as a fingerprint to distinguish between trigonal bipyramidal stereochemistries and those derived from the octahedral ones. The actual values of $g_{\parallel}$ can be smaller than g_e because of higher order perturbation terms[83] which are not included in equation (4.1). General relationships between the spin Hamiltonian parameters for complexes with geometries intermediate between a trigonal bipyramid and a square pyramid can be found in reference 84.

Pseudo-tetrahedral copper(II) complexes, although not very numerous, are known in tetragonally distorted environments. The unpaired electron is in a $d(xy)$ orbital and an equation similar to those obtained for the tetragonal octahedral complexes can be used, as shown by the application of the pentagon of Figure 6. However, the energies of the excited states are lower for tetrahedral chromophores than for planar or octahedral chrmophores, therefore larger deviations of g from g_e are anticipated, as shown in Table 7. Another interesting feature of pseudo-tetrahedral complexes is that $A_{\parallel}$ tends to be smaller than in planar or octahedral chromophores and $A_{\perp}$ tends to be larger. The limiting case for this is given by the EPR spectrum[85–87] of copper(II) doped into Cs_2ZnCl_4, which has been interpreted with $A_{\parallel} = 25 \times 10^{-4}$ cm^{-1} and $A_{\perp} = 50 \times 10^{-4}$ cm^{-1}. These values of A are strongly dependent on the covalency of the metal–ligand bond and have been rationalized through molecular orbital calculations.[88]

Due to the large amount of experimental data on copper(II) complexes, empirical relationships between the spin Hamiltonian parameters have been found which have no direct theoretical explanation. Among these empirical correlations the plots of $g_{\parallel}$ *vs.* $A_{\parallel}$ are probably the most famous.[89] These are shown in Figure 18. The utility of these plots lies in the differentiation between different square planar chromophores which have been used in classifying copper proteins according to their EPR spectra.[90] Another interesting relationship is the observed linear dependence of $g_{\parallel}$ on the tetragonality parameter, defined as the ratio between the equatorial, R_e, and the axial, R_a, average bond lengths, found in tetragonal octahedral chromophores.[91]

As a final example of simple complexes we will briefly account for the EPR spectra of high spin iron(III) complexes. The only stable isotope of iron which possesses a nonzero nuclear spin is ^{57}Fe, I = 1/2, which has a low natural abundance (2.2%).

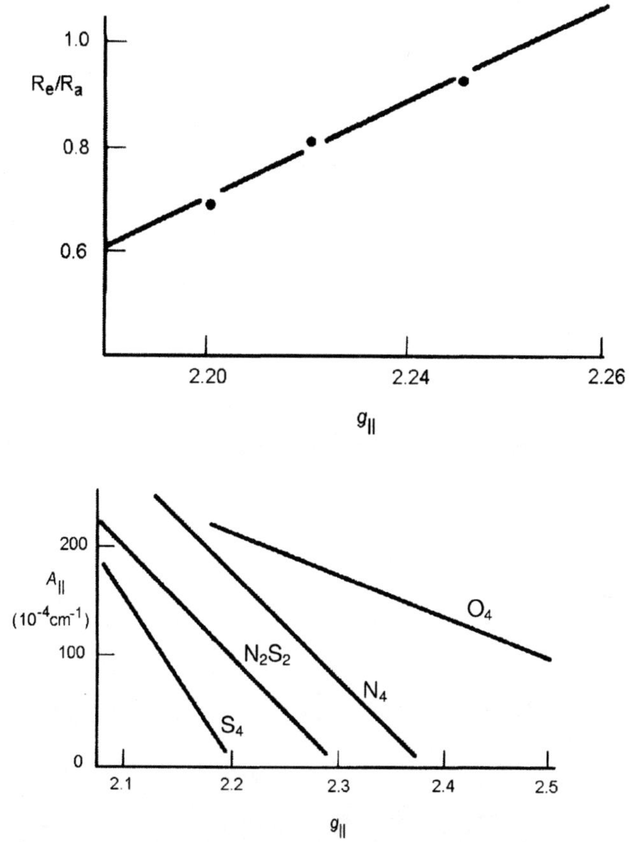

Figure 18. Empirical correlations in copper(II) complexes: (top) $g_{\parallel}$ values as a function of tetragonality in a series of $Cu(NH_3)_4X_2$ complexes; (bottom) $A_{\parallel}$ *vs.* $g_{\parallel}$ for a series of square planar chromophores (After Hathaway, B.J. in *Comprehensive Coordination Chemistry* **1987**, 5, 533, Wilkinson, G.; Gillard, R.D.; McCleverty J.A.; Eds., Pergamon Press, Oxford.).

Therefore the EPR spectra of iron complexes in general do not show hyperfine coupling. Iron(III) in the high spin state has an $^6A_1$ ground state and the appearance of the spectrum is mainly influenced by the zero field splitting term. Since the spin-orbit coupling has no first-order matrix elements within the $^6A_1$ state, the zero field splitting is caused by the mixing of the excited quartet and doublet states into the ground sextet. For geometries derived from the octahedral symmetry, simple relationships have been obtained[92,93] for the zero field splitting neglecting other excited states except the lowest $^4T_1$ state:

$$D = \frac{\zeta^2}{10}\left(\frac{2}{\Delta_1} - \frac{1}{\Delta_2} - \frac{1}{\Delta_3}\right), \qquad E = \frac{\zeta^2}{10}\left(\frac{1}{\Delta_3} - \frac{1}{\Delta_2}\right) \tag{4.2}$$

using the energies defined in Scheme 1.

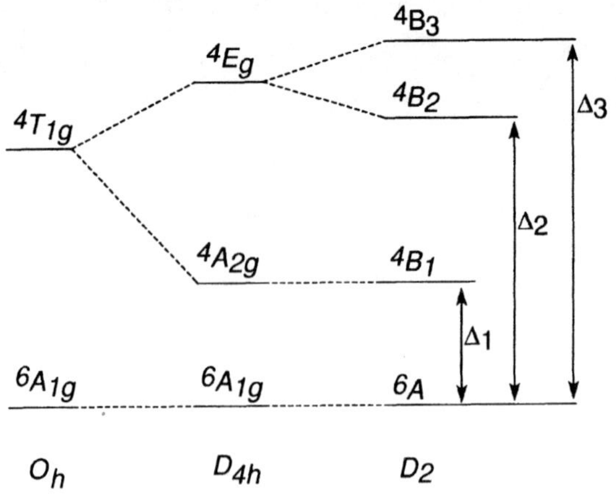

Scheme 1. *zfs* in Fe(III) pseudo-octahedral complexes.

In octahedral symmetry the zero field splitting determined by second order terms is zero and the degeneracy of the $^6A_{1g}$ state is removed only by fourth order terms of equation (3.6). In cubic symmetries only the term

$$\frac{a}{6}\left\{S_x{}^4 + S_y{}^4 + S_z{}^4 - \frac{S(S+1)(3S^2+3S-1)}{5}\right\}$$

must be included in the spin Hamiltonian, while in lower symmetries also the containing term F

$$\frac{F}{180}\{35S_z{}^4 - 30S(S+1)S_z{}^2 + 25S_z{}^2 - 6S(S+1) + 3S^2(S+1)^2\}$$

Table 9 Selected Examples of Spectra of High Spin Iron(III)[a]

Stereochemistry	Compound	Chromophore	$g_\perp$	$g_\parallel$	D (cm^{-1})	E/D
Octahedral	$[Fe(urea)_6]Cl_3$	FeO_6	2.002	2.002	0.130	0.067
	$[Fe(pyO)_6](ClO_4)_3$	FeO_6	2.002	2.002	0.36	0.233
	$[Fe(Ph_3PO)_4Br_2]FeBr_4$	*trans*-FeO_4Br_2	2.002	2.002	1.20	0.0
Square Pyramidal	$FeCl(acac)_2$	FeO_4Cl			$g_{eff} = 4.3$	
	$(NH_4)_2(Fe/Sb)F_5^b$	FeF_5	1.996	1.996	0.069	0.001
Trigonal bipyramidal	$Fe(Ph_3PO)_2(NCS)_3$	FeO_2N_3	2.002	2.002	0.37	0.10
	$Fe(Ph_3PO)_2(NO_3)_3$	FeO_2O_3	2.002	2.002	0.55	0.04
Square Planar	$[Fe(Ph_3PO)_4](ClO_4)_3$	FeO_4	2.002	2.002	0.84	0.005
	$[Fe(Ph_3AsO)_4](ClO_4)_3$	FeO_4	2.002	2.002	1.05	0.000

[a] Taken from reference 12.
[b] $a = 0.0032$ cm^{-1}; $F = 0.0006$ cm^{-1}.

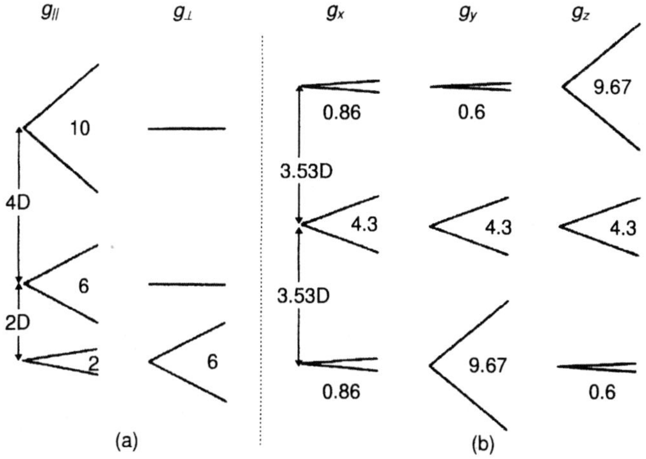

Figure 19. Splitting of the three Kramers doublets arising from an $S = 5/2$ spin state by the effect of the zero field splitting and the magnetic field. The principal values of the effective **g** tensor within the three Kramers doublets are indicated for: (a) axial symmetry, $\lambda = 0$; (b) ortorhombic symmetry, $\lambda = 1/3$.

also has to be added. The spin Hamiltonian parameters for selected high spin iron(III) complexes are shown in Table 9. Due to the small value of the spin-orbit coupling constant and the high energies of the excited states, the g values of high spin iron(III) complexes are always very close to g_e and the D parameters rarely exceed 1 cm^{-1}. Calculation of the d^5 spin Hamiltonian parameters using ligand field models has attracted interest for many years and requires the inclusion of at least all the quartet states. Recent results on the subject can be found in reference 94.

When the zero field splitting is larger than the microwave quantum, only transitions between one Kramers doublet can be observed, so that the spectra can be interpreted using an effective $S_{eff} = 1/2$ spin Hamiltonian:

$$\mathcal{H} = \mu_B \mathbf{B} \cdot \mathbf{g}_{eff} \cdot \mathbf{S}_{eff} \tag{4.3}$$

This situation is met in strong ligand fields, such as in iron porphyrin and porphyrin like systems and in high spin iron(III) proteins.[95] The effective g values depend on the actual value of the zero field splitting. Two limiting situations, the axial and the complete orthorhombic, are shown in Figure 19. In the axial situation, a g pattern corresponding to $g_{eff,\parallel} = 2$ and $g_{eff,\perp} = 6$ is found as in synthetic iron(III) porphyrins[95] or in a number of heme ferric proteins, like met Mb.[96]

In the rhombic environments usually a sharp resonance at $g_{eff} = 4.3$ with possibly a signal at $g_{eff} = 9.7$ is observed. As shown in Figure 19, the first signal arises from the transition between the middle doublet; while the second signal is expected only when **B** is parallel to y and z. Furthermore, depending on the value of D, the upper doublet can not be populated at low temperatures and the $g_{eff} = 9.7$ signal can often not be observed. Representative examples of this behavior are many non-heme proteins such as rubredoxins, which is an iron-sulfur protein containing one iron center per molecule.[97] In many other cases resonances attributable to the population of Kramers doublets other than the ground doublet have been observed[95] making the interpretation of the spectra rather difficult. Well defined situations occur which determine isotropic g_{eff} values at 4.3 and 3.3. The conditions under which these isotropic g values can be obtained[98] are collected in Table 10.

EPR spectroscopy has been applied to characterizing coupled magnetic systems since the 1950s. Coupling between copper ions in copper(II) sulfate hydrate was discovered in 1948 by Bagguley and Griffiths[99] and the binuclear nature of the copper(II) acetate dihydrate was ascertained in 1952 by Bleaney and Bowers[100] on the basis of the EPR spectra alone. The theory and application of EPR to the investigation of the properties of paramagnetic clusters can be found in references 6 and 101.

The relevant terms which must be included in the spin Hamiltonian of a pair of interacting paramagnetic centers have been discussed in Section 3.5. They are the magnetic dipolar and the isotropic exchange interactions. Other terms such as the anisotropic and anti-symmetric exchange interactions may also be necessary to fully account for the EPR spectra of clusters; and these can be found in reference 6. When magnetic dipolar interactions are the dominant terms in the Hamiltonian, the measurement of the zero field splitting parameters, D and E, can give information on the distance between the paramagnetic centers. The simplest expression relating the measured D value to the distance r between the paramagnetic centers is:

$$D = \frac{0.433g^2}{r^3} \tag{4.4}$$

when D is expressed in cm^{-1} and r in Å. Selected publications on the subject are given in reference 102–106.

Another limiting situation which is often met in practice is when the isotropic exchange interaction Equation (3.31) is the leading term in the spin Hamiltonian. In this case the electronic structure of the cluster can be represented by a series of states labeled according to the eigenvalues of the square of the total spin operator $\mathbf{S} = \sum_i \mathbf{S}_i$, $\mathbf{S}_i$ being the spin operator of the i-th paramagnetic center. In the case of

Table 10 Relationships between Spin Hamiltonian Parameters for which EPR Spectra of High Spin Iron(III) are Isotropic[a]

$g = 4.3$	$18D - 3a - F = 0$	$a + 6E = 0$
	$18D - 3a - F = 0$	$a - 6E = 0$
	$3D + a + F = 0$	$E = 0$
	$3D + 2a + F = 0$	$\forall E$
$g = 3.3$	$3D + 3a + F = 0$	$E = 0$
	$3D + F = 0$	$E = 0$

[a]From reference 98.

a binuclear system the energy interval between total spin states is given by:

$$E(S) - E(S-1) = JS \tag{4.6}$$

where J is the exchange coupling constant. When $J \gg kT$ only the ground spin state is populated at the temperature of the experiment, T. In this case the EPR spectrum can be interpreted using an effective spin Hamiltonian $\mathcal{H}_S$, acting upon the total spin space S, of the form:

$$\mathcal{H}_S = \mu_B \mathbf{B} \cdot \mathbf{g}_S \cdot \mathbf{S} + \mathbf{S} \cdot \mathbf{D}_S \cdot \mathbf{S} + \sum_k \mathbf{I}^k \cdot \mathbf{A}_S{}^k \cdot \mathbf{S} \tag{4.7}$$

The spin Hamiltonian parameters appearing in (4.7) can be related to the parameters of the ions forming the magnetically interacting couple through the equations:

$$\mathbf{g}_S = c_1 \mathbf{g}_1 + c_2 \mathbf{g}_2 \tag{4.8}$$

$$\mathbf{D}_S = d_1 \mathbf{D}_1 + d_2 \mathbf{g}_2 + d_{12} \mathbf{D}_{12} \tag{4.9}$$

$$\mathbf{A}_S^k = c_1 \mathbf{A}_1^k + c_2 \mathbf{A}_2^k \tag{4.10}$$

where 1 and 2 label the interacting centers, D_{12} accounts for anisotropic exchange and magnetic dipole interactions, k indicates the nucleus with which the unpaired electrons interact, and the c and d coefficients are defined as:

$$c_1 = \frac{1+c}{2};\ c_2 = \frac{1-c}{2};$$

$$d_1 = \frac{c_+ + c_-}{2};\ d_2 = \frac{c_+ - c_-}{2};\ d_{12} = \frac{1-c_+}{2} \tag{4.11}$$

$$c = \frac{S_1(S_1+1) - S_2(S_2+1)}{S(S+1)} \tag{4.12}$$

$$c_+ = \frac{3[S_1(S_1+1) - S_2(S_2+1)]^2 + S(S+1)[3S(S+1)}{(2S+3)(2S-1)S(S+1)} \frac{-3 - 2S_1(S_1+1) - 2S_2(S_2+1)]}{(2S+3)(2S-1)S(S+1)} \tag{4.13}$$

$$c_- = \frac{4S(S+1)[S_1(S_1+1) - S_2(S_2+1)]}{(2S+3)(2S-1)S(S+1)} \frac{-3[S_1(S_1+1) - S_2(S_2+1)]^2}{(2S+3)(2S-1)S(S+1)} \tag{4.14}$$

in (4.12–4.14) the coefficients are zero when $S = 0$ and when the denominator is zero.

The use of equations (4.8–4.14) has been found to be very useful in interpreting the EPR spectra and the magnetic properties of exchange coupled systems. Extension of these equations to larger clusters is described in some detail in reference 6. For a number of similar ions (*e.g.* copper(II) ions) equations (4.8) and (4.10) become:

$$\mathbf{g}_S = \frac{1}{2}\mathbf{g}_1 + \frac{1}{2}\mathbf{g}_2 \tag{4.15}$$

$$\mathbf{A}_S{}^1 = \frac{1}{2}\mathbf{A}_1^1 + \frac{1}{2}\mathbf{A}_2^1 \tag{4.16}$$

Equation (4.16) represent the interaction of the unpaired electrons with nucleus number 1 and an equivalent expression holds for nucleus number 2. In (4.16) $\mathbf{A}_1^1$ is the dominant term since the coupling of nucleus 1 with the electrons on nucleus 2 is generally small. The terms $\mathbf{A}_2^1$ and $\mathbf{A}_1^2$ are generally called supertransferred hyperfine coupling. Equations (4.15) and (4.16) show that, when the tensors on the right-hand side have parallel principal axes (symmetric dimer), the $\mathbf{g}_S$ tensor coincides with the single ion **g** tensors and the hyperfine coupling constant is half that of the isolated ions. Of course, the number of hyperfine lines is doubled with respect to the mononuclear case since in (4.7) we must use the total nuclear spin $\mathbf{I} = \mathbf{I}^1 + \mathbf{I}^2$. The above formalism has been widely applied to interpreting the spectra of clusters.[6] An interesting application of the above relationships was suggested to measure the spin Hamiltonian parameters of the EPR silent nickel(II) ion in different stereochemistries by measuring the spectra of Cu(II)-Ni(II) pairs. The ground state of Cu(II)-Ni(II) pairs in the case of antiferromagnetic coupling is $S = 1/2$ and EPR spectra can generally be measured. Equation (4.8) becomes in this case:

$$\mathbf{g}_{1/2} = \frac{4}{3}\mathbf{g}_{\mathrm{Ni}} - \frac{1}{3}\mathbf{g}_{\mathrm{Cu}} \tag{4.17}$$

Knowing, therefore, the g values for the copper(II) center, the values for the nickel(II) can be obtained. Selected examples are reported in Table 11.

Table 11 ***g*** **Values for Nickel(II) Complexes Obtained from EPR Spectra of Cu(II)–Ni(II) Pairs**[a]

Stereochemistry	Compound	Chromophore	g_1	g_2	g_3
Octahedral	$[CuNi(DBA)_2(py)_4]$	NiO_4N_2	2.18	2.20	2.21
	$[CuNi(dhph)_2(H_2O)_4]Cl_4{\cdot}4H_2O$	NiN_4O_2	2.16	2.17	2.21
	$[CuNi(fsa)_2en(H_2O)_2]{\cdot}H_2O$	NiO_4N_2	2.20	2.24	2.24
Square Pyramidal	$[CuNiL_2Cl_2]$	$NiO_2N_2Cl_2$	2.12	2.32	2.38
	$[Cu/Ni(pyO)_2Cl_2H_2O]_2$	NiO_2Cl_2O	2.17	2.26	2.48
Trigonal bipyramidal	$[Ni_2(dbhe)_2](ClO_4)_4$	NiO_2N_3	2.06	2.09	2.61

[a] Taken from: Gatteschi, D.; Bencini, A. in *Magneto-Structural Correlations in Exchange Coupled Systems* **1985**, Willett, R.D.; Gatteschi, D.; Kahn, O.; Eds., D. Reidel Publishing Co., Dordrecht.

4.2 Biologically Relevant Compounds

EPR has been widely used to obtain structural information on metallo-enzymes and metallo-proteins. Many reviews are available for the interested reader.[107–109] We will present here some examples in order to show the potentialities of the techniques, and the wealth of information which can come in order to deduce the structure of the active site of the proteins, or, conversely, in the presence of the structure, to understand the details of the electronic structure at the metal ion, and from this, in some cases, information on the mechanism of action of the enzyme. The metal ions which have been more widely investigated by EPR in metallo-enzymes and metallo-proteins, either in the native or in metal substituted forms, are copper(II), nickel(III), cobalt(II), manganese, iron, and oxovanadium(IV).

Copper is certainly one of the most heavily investigated metal ions, because EPR spectra of a very large number of simple model compounds are available, which allow one to fingerprint the various kinds of structure in large detail, as outlined in Section 4.1. Further, the spectra are in general readily observed at room temperature. In fact EPR and UV-Vis absorption spectra are used to classify the active sites in copper proteins. They were recently reviewed,[110,111] and this classification will be briefly introduced here.

The so-called blue copper centers, or Type 1 centers, are characterized optically by a very intense band at ca. 600 nm ($\varepsilon \approx 5,000$ mol^{-1}cm^{-1}) and by an EPR spectrum with a small copper hyperfine as shown in Figure 20a. The pattern of g values ($g_{\parallel} \rangle g_{\perp} \rangle g_e$) clearly indicates that the unpaired electron is in a $|xy\rangle(|x^2 - y^2\rangle)$ orbital. Perhaps the protein which has been investigated in more detail with EPR is copper plastocyanin. In this class of proteins the copper ions present in the active site participate in outersphere electron transfer. The copper ion is coordinated to the two nitrogen atoms of two histidine molecules and two sulfur atoms, one from a cysteine and the other from a methionine,[112] as shown in Figure 21. The Cu–N bonds are regular, while the Cu–S (cysteine) is very short (2.1 Å), and the Cu–S (methionine) is very long (2.9 Å). Detailed single crystal spectra showed $g_x = 2.042$, $g_y = 2.059$, $g_z = 2.226$. The z axis of the **g** and **A** tensors makes an angle of ca. 5° with the Cu–S (methionine) direction.[113] One of the spectroscopic fingerprints of Type 1 copper proteins is provided by the small copper hyperfine splitting, $A_z = 63 \times 10^{-4}$ cm^{-1}, similar to that previously observed in other pseudotetrahedral copper(II) complexes such as the tetrahedral $CuCl_4^{2-}$ ions, which were justified by large covalency effects[88] as already mentioned in Section 4.1. Molecular orbital calculations[113] on copper plastocyanin showed that the large delocalization towards the short bonded sulfur atom causes a reduction of the dipolar term by about 35% relative to the usual value for a localized complex. Further covalency also reduces the isotropic component of the hyperfine splitting. The beauty of the analysis is that it could exploit all the information coming from single crystal spectra of the protein (i.e. the principal values and directions of both **g** and **A**).

The Type 2, or normal copper proteins, exhibit EPR signals similar to those of Figure 20b, which are typical of tetragonal copper(II) complexes. They provide evidence that the unpaired electron is in an orbital which can be essentially described

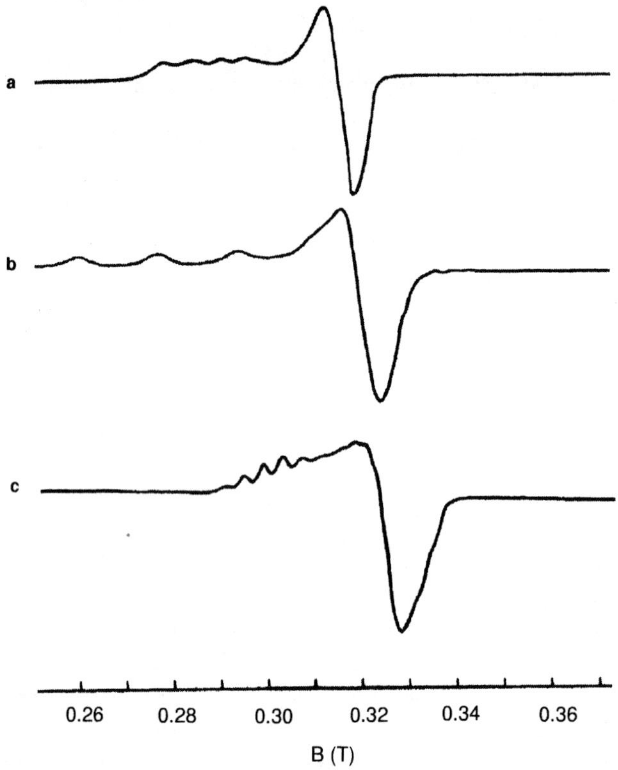

Figure 20. X-Band EPR spectra of different types of copper(II) proteins. **a**: Type 1 or blue copper protein plastocyanin; **b**: Type 2 or normal copper protein dopamine β-hydroxylase; **c**: The Cu_A site in nitrous oxide reductase. (After reference 111).

as $|xy\rangle(|x^2 - y^2\rangle)$. These spectra are observed in galactose oxidase, amine oxidase, dopamine β-hydroxylase, peptidylglycine α-amidating monooxygenase, etc. Similar spectra, but with a sizeable anisotropy in the $g_\perp$ components, are observed in Cu/Zn superoxide dismutase, in which the copper ion is distorted five coordinate, which determines some admixture of $|z^2\rangle$ orbital in $|x^2 - y^2\rangle$.

The Type 3 proteins contain two copper(II) centers, but they do not give any EPR signal due to the strong antiferromagnetic coupling determined by superexchange through bridging ligands.

Cytochrome c oxidase and nitrous oxide reductase contain a Cu_A center which has two copper ions and gives an EPR spectrum[114] as shown in Figure 20c. The parallel feature clearly shows a splitting into seven lines, a clear indication that there is one unpaired electron which interacts with two equivalent copper nuclei, suggesting a mixed valence species.

Another field where EPR and related techniques have provided important information is that of iron-sulfur proteins.[110] These are ubiquitous proteins, containing iron ions in variable numbers, ranging from 1 to 4 (neglecting the more complex

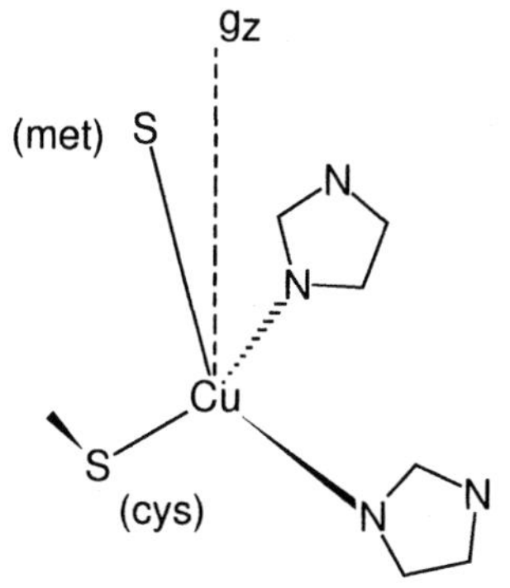

Figure 21. Coordination of the copper(II) ion in the active site of plastocyanin.

structure of nitrogenase), with general structures schematized Figure 22. The oxidation state of iron is variable. For examples proteins like rubredoxins, Rb, contain one iron ion which, in the reduced form, is high spin iron(II) and in the oxidized form is high spin iron(III). The EPR spectrum of the oxidized form of Rb from *C. Pasteurianum* is characterized by a strong signal at $g_{eff} = 4.3$, and a single peak at $g_{eff} = 9.7$, which, as shown in Section 4.1, are typical of an $S = 5/2$ spin state whit a large rhombic zero field splitting. The zfs parameters obtained by Mössbauer spectra were $D = 1.9\ cm^{-1}$ and $E/D = 0.23$, which are in good agreement with the

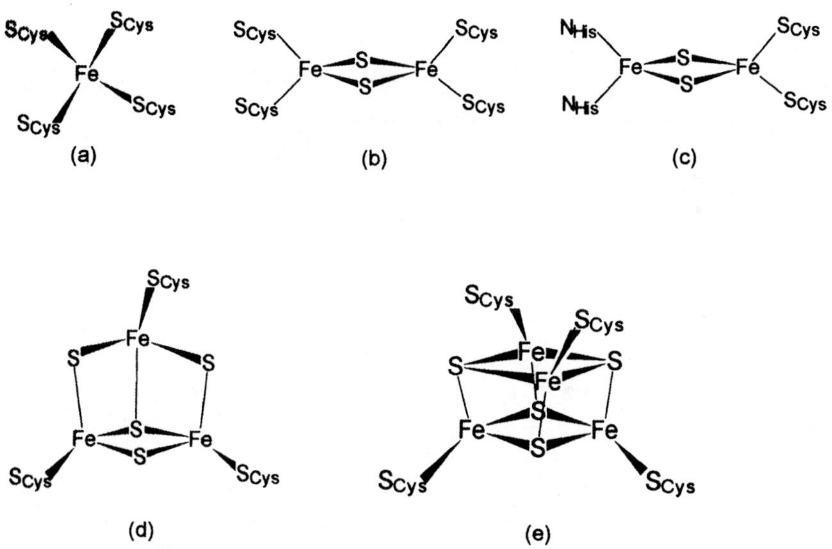

Figure 22. Representative view of the structures of the iron centers observed in iron-sulfur proteins. (a) rubredoxin (*Clostridium pausterianum*); (b) ferredoxin (*Equisetum arvense*); (c) Reiske proteins; (d) ferredoxin II (*Desulfovibrio gigas*); (e) ferredoxin I (*Azotobacter vinelandii*).

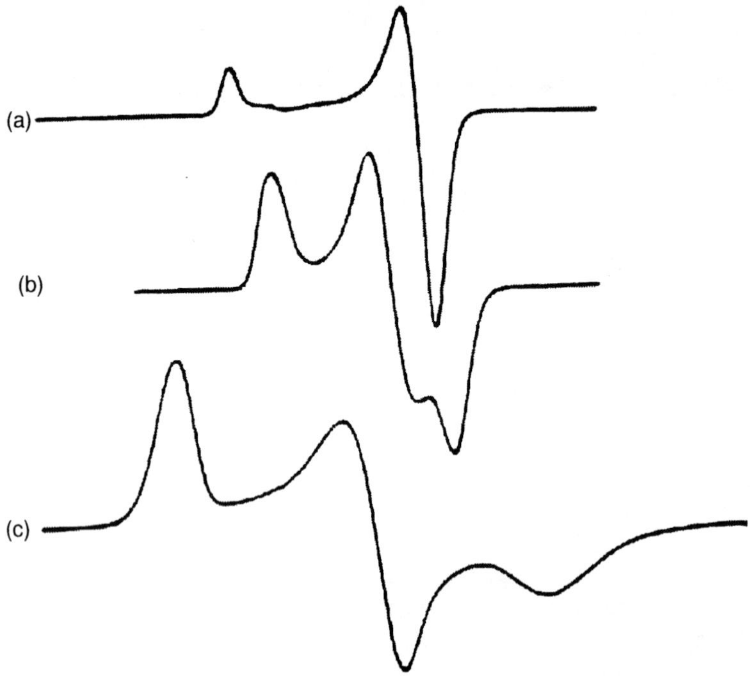

Figure 23. Some examples of EPR spectra of Fe-S proteins: (a) from bovine adrenals; (b) *Clostridium pausterianum*; (c) from spinach. (After reference 6).

EPR results.[115] The relatively large zero field splitting is typical of pseudotetrahedral iron(III) ions coordinated to four sulphur atoms.

Proteins containing two iron centers have, in the active site, dinuclear species in which the iron atoms are bridged by two μ_2-sulphide ions, as shown in Figure 22b. These proteins are generally named ferredoxins. In the reduced form of the protein one center is high spin iron(III), and the other is high spin iron(II). The valences are trapped, as shown by the Mössbauer spectra of, for example, spinach ferredoxin[116] and the ions are antiferromagnetically coupled. The ground state resulting from the exchange interaction is an $S = 1/2$ spin state, as evidenced by the EPR spectra of some ferredoxins shown in Figure 23. The spectra can vary in appearance passing from axial to rhombic, but they all have g values which are smaller than 2.0. Iron(III) has a d^5 valence configuration, and it is expected to have quasiisotropic g values very close to 2.00, while iron(II), with a d^6 configuration, has expected values of $g > 2.00$. In the pair, $g < 2.00$, which is the signature of the antiferromagnetic coupling between the two centers. In fact, if the spin of iron(III) will be oriented parallel to the external field, that of iron(II) will be oriented antiparallel. Therefore the g values of iron(II) must be subtracted from that of iron(III), thus giving $g < 2$. Quantitatively we can compute the g values expected for the $S = 1/2$ ground state

Table 12 EPR Spectra of Metal Centers in Iron Sulfur Proteins

Type of Protein[a]	Iron Center[b]	EPR Signal[95]
Fe-S	Fe(II) HS, S = 2	EPR silent
	Fe(III) HS,S = 5/2	290 K: $g_{eff} = 4.3$, $g_{eff} = 9.7$
2Fe-2S	2Fe(III), S = 0	EPR silent
	1Fe(II) + 1Fe(III), S = 1/2	77 K: $g_{av} < 2.$
3Fe-4S	3Fe(III), S = 1/2	15 K: $g_{av} > 2$
	2Fe(III) + 1Fe(II)	30 K: $g_{eff} = 12$
4Fe-4S (HiPIP)	3Fe(III) + 1Fe(II), S = 1/2	30 K; $g_{av} > 2.$
	2Fe(III) + 2Fe(II), S = 0	EPR silent
4Fe-4S (Fd)	2Fe(III) + 2Fe(II), S = 0	EPR silent
	1Fe(III)+3Fe(II), S = 1/2	15 K; $g_{av} < 2.$

[a] HiPIP = high-potential iron protein; Fd = ferredoxin.
[b] HS = high spin state; S is the total spin state of the paramagnetic moiety.

using equation (4.8) as:

$$g_{1/2} = \frac{7}{3}g_{\mathrm{Fe(III)}} - \frac{4}{3}g_{\mathrm{Fe(II)}} \tag{4.18}$$

Assuming that $g_{\mathrm{Fe(III)}} = 2.00$, the $g_{\mathrm{Fe(II)}}$ values can be calculated and associated with the electronic structure of the center.

Spectra have been measured also for proteins containing a larger number of iron-centers. General features of the spectra observed in various iron sulphur proteins are collected in Table 12. Recently very accurate EPR and ENDOR measurements have been performed on single crystals of iron sulphur complexes which can be considered as models of 4Fe-4S proteins.[117]

4.3 Materials

There are several different uses of EPR in the investigation of materials. They range from the identification of paramagnetic impurities to the use of paramagnetic probes for the investigation of phase transitions; from the investigation of the nature of charge carriers in conductors and semiconductors to the detection of microwave absorption due to Josephson junctions in ceramic superconductors; from the detection of radiation induced defects to the determination of the age of minerals. There is no way to cover all these topics, and others which can also be addressed by EPR techniques; therefore we will simply present some examples in order to show the potential of the technique.

Oxides form many important classes of materials which are used for such diverse applications as catalysis and superconductivity. For many of these applications the surfaces of oxides represent the active site. This is particularly true for catalysis, and EPR has been extensively used for characterizing the nature of the catalytic sites and of the intermediate species. A large review focusing on the use of EPR and associated techniques as tools to investigate the transition metal chemistry on oxide surfaces has been recently published.[118]

An interesting example of the use of ESEEM for obtaining information on the location of copper(II) in mesoporous materials was reported by Yamada *et al.*[119] The EPR spectra of copper(II) doped hectorite, a lattice comprising silicon, magnesium, and lithium, showed an anisotropic environment for copper(II), indicating that the metal ion is bound either to the interlayer surface or to the pillars. The ESEEM pattern, shown in Figure 24, does not show any ^{7}Li modulation, suggesting that the

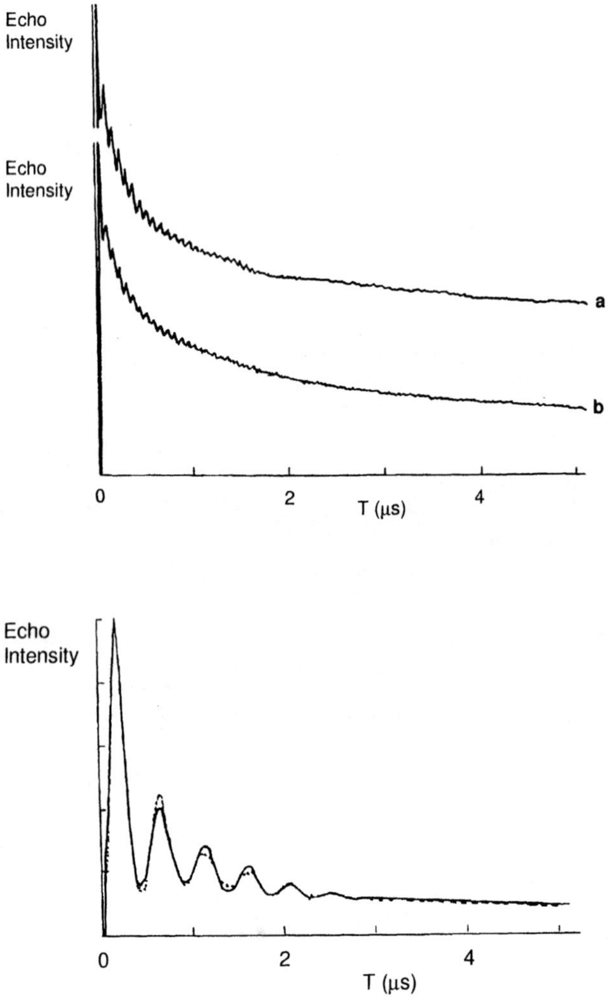

Figure 24. Three-pulse ESEEM pattern for Cu(II) doped hectorite. Upper: after air drying (**a**) and after evacuation at room temperature (**b**). Lower: with adsorbed D_2O. (After reference 119).

copper(II) ion is not located at a lithium site in the octahedral sheet of the hectorite, but at a site close to the pillar. In the ESEEM spectrum a modulation due to the interaction with H is observed at $\tau = 0.09\ \mu s$. The ESEEM spectrum of a sample treated with D_2O showed a strong signal which was simulated with a two shell model with six interacting nuclei at 2.9 Å and A = 0.16 MHz in one shell, and two nuclei at 3.7 Å in the other shell. The short distance is typical of coordinated water, thus suggesting that the three water molecules are directly bound to copper.

Zeolites are aluminosilicate materials comprised of a three dimensional network of alumina and silica tetrahedra characterized by cages or channels ranging from 2 to 20 Å and larger; this structural feature confers on them typical catalytic behavior. Transition metal ions can be easily incorporated into zeolites by ion exchange in aqueous solution. The metal ions which have been most investigated in zeolites are shown in Table 13. The asymmetry between ions with odd and even numbers of electrons is striking. In fact metal ions with an even number of unpaired electrons are often characterized by large zero field splittings, which in low symmetry completely removes the degeneracy of the S multiplets. Under these conditions the EPR transitions can be observed only by using high field and/or high frequencies, while the ions are EPR silent in conventional X- and Q-band experiments. Matters are different for metal ions with an odd number of unpaired electrons, because in this case the minimum degeneracy in the absence of an applied field is two, thus allowing in general the detection of the signal within a Kramers doublet.

One of the features of zeolites is that of stabilizing unusual oxidation states for transition metal ions. For instance in the reduction of nickel(II) by molecular hydrogen at 50–100°C, nickel(I) is stabilized in X and Y zeolites.[120] Figure 25 shows how complex the spectra recorded in this kind of material can be. The spectra refer to $CaNi_8$-Y zeolites in the presence of 1-butene. 8 gives the number of nickel(II) ions present in the unit cell. The H_2 reduced samples show species A ($g = 2.096$) and D ($g_{\parallel} = 2.151$, $g_{\perp} = 2.061$)). When the sample is exposed to butene isomers the signal of the species D disappears completely and that of A partially, and simultaneously the signals of new species B1 ($g_{\parallel} = 2.109$, $g_{\perp} = 2.032$), B2 ($g = 2.048$) and B3 ($g_{\parallel} = 1.965$, $g_{\perp} = 2.620$) are formed. The g values of the B1 and D species immediately identify them as d^9 species with a ground $|x^2 - y^2\rangle$ orbital, as can easily be checked using equation (3.9). The g values of the B3 species are more difficult to assign. They show very large deviations from the free ion value, which would require unusually large orbital contributions; however, they are of a species

Table 13 Ions Studied by EPR in Zeolites

Configuration	Spin	Ions
d^1	1/2	Ti^{III}, V^{IV}, Cr^{V}, Mo^{V}
d^3	3/2	Cr^{III}
d^5	5/2	Mn^{II}, Fe^{III}, Ru^{III}
d^7	1/2	Ni^{III}, Rh^{II}, Pd^{III}
	3/2	Co^{II}
d^9	1/2	Cu^{II}, Ni^{I}, Pd^{I}, Ag^{II}

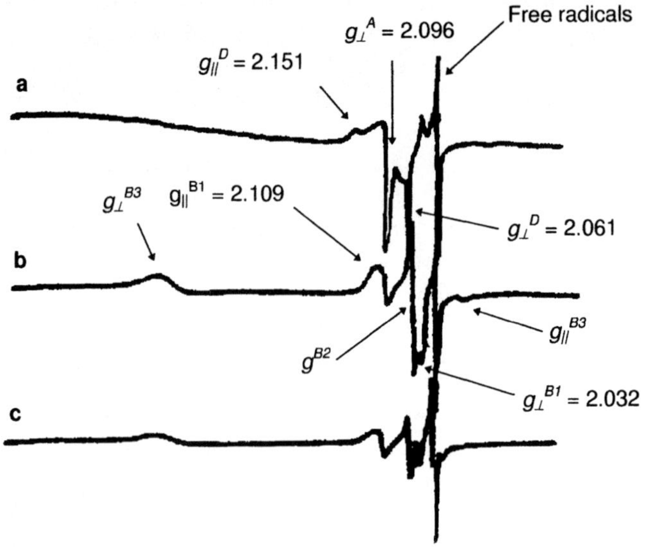

Figure 25. EPR spectra of $CaNi_8$-Y at −196° C in X-band. **a**: O_2-pretreated sample heated in H_2 at 280°C after evacuation at 100° C; **b**: after the sample is exposed to *i*-butene for 600 s; **c**: after overnight standing of **b** at 25° C. (After reference 120).

with $S = 1/2$, ruling out nickel(II). One possibility is to assume that they are due to nickel(II)-nickel(I) pairs with antiferromagnetic coupling. In fact in this case the ground state has $S = 1/2$. The g values of the pair can be expressed as a function of the g values of the individual ions with the assumption that the isotropic exchange interaction is dominant. In this hypothesis the **g** tensors of the pair are related to those of the individual ions by:

$$\mathbf{g} = \frac{4}{3}\mathbf{g}_{\mathrm{Ni(II)}}\frac{1}{3}\mathbf{g}_{\mathrm{Ni(I)}} \tag{4.19}$$

If the **g** tensor of nickel(II) is larger than the **g** tensor of nickel(I), then large deviations from the free ion value can be observed. Further deviations can originate from the fact that the single ion zero field splitting of nickel(II) is not negligible compared to the isotropic exchange interaction. In this case the g values of the pair cannot be analytically expressed, and can largely deviate from the free ion value.[121]

An important application of EPR is in the investigation of phase transitions.[122] It is possible to record both the spectra of magnetic ions which are actually involved in the transition and those of ions deliberately added to non-magnetic lattices as probes. The latter strategy has been widely used for the investigation of paraelectric to ferroelectric phase transitions. An example is provided by $SrTiO_3$ which crystallizes in the ABO_3 perovskite-like structure, which is cubic above 377 K. Below this temperature the structure changes to tetragonal, due to alternate rotations of the oxygen octahedra about [100], [010], and [001]. Details of the mechanism of the transition were obtained by doping the lattice with Fe(III), which substitutionally replaced Ti(IV) ions. As a matter of fact, it was the EPR studies which first indicated the existence of the phase transition. In fact, above $T_c = 377$ K the coordination environment of the iron(III) ion is perfectly cubic, no zero field splitting is observed, and the EPR spectra do not show any fine structure. However at the temperature of the phase transition the symmetry is reduced, and the zero field splitting parameter D becomes different from zero and a fine structure appears in the spectrum. The zero field splitting showed a temperature dependence below T_c, and EPR provided clear information about the structural dynamics in the tetragonal phase. The principal axes of the **D** tensor followed the rotation of the octahedra, therefore their experimental determination provided important structural information. The experiment were performed on single crystals. By rotating, for instance, around the [001] axis, the principal axis of **D** was determined as that direction corresponding to the maximum splitting of the fine structure lines of the spectra. Since spectra at different temperatures can be easily recorded, the dynamics of the phase transition was fully monitored. It was found that the rotation of octahedra follows a $(T_c - T)^{1/2}$ dependence, where T_c is the critical temperature.

5 SOME CONCLUDING COMMENTS

EPR is a powerful spectroscopic technique, which can provide structural information by measuring spectra in solution, in polycrystalline powders, and in single crystals. The interpretation of the spectra requires some effort, because the appearance of the spectra depends on the quantum mechanical features of the resonating centers, but the spin hamiltonian approach allows to avoid complex calculations and to obtain the parameters needed to characterize the nature of the paramagnetic center without recurring to complex calculations. In this chapter we have attempted to show how this can be achieved, in particular relating the spin Hamiltonian parameters to the results of theoretical calculations, ranging from ligand field to molecular orbital approaches. We have tried to show the potentialities of the technique as applied to simple complexes and clusters, biologically relevant compounds, and materials.

ABBREVIATIONS

apy = 2,3-dimethyl-1-phenyl-Δ^3-pyrazolin-t-one
bipy = 2, 2′-bipyridine
bipyam = di-2-pyridyl-amine
DBA = 1,5-di-phenyl-1,3,5-pentanetrionato
$H_4(fsa)_2en$ = N,N′-bis(2-hydroxy-3-carboxybenzylidene)-1,2-diaminoethane
Im = imidazole
Me_6tren = tris(2-dimethhylaminoethyl)amine
phen = *o*-phenanthroline
pyO = pyridine-N-oxide

SALMe = N-methylsalicylaldiminato
SALMeDPT = (3-salicylaldiminatopropyl)methylamine
SALtBu = N-*ter*-butylsalicylaldiminato
tren = tris(aminoethyl)amine

REFERENCES

1. Zavoiski, E. *Fiz. Zh.* **1944**, *9*, 211.
2. Van Vleck, J.H. *The Theory of Electric and Magnetic Susceptibilities*, Oxford University Press, Oxford, 1932.
3. Abragam, A.; Bleaney, B., *Electron Paramagnetic Resonance of Transition Metal Ions,* Oxford University Press, Oxford, 1970; Dover Publications, New York, 1986.
4. Wertz, J.E.; Bolton, J.R., *Electron Spin Resonance: Elementary Theory and Practical Applications*, McGraw Hill, New York, 1972.
5. Pilbrow, J.R., *Transition Ion Electron Paramagnetic Resonance*, Clarendon Press, Oxford, 1990.
6. Bencini, A.; Gatteschi, D., *Electron Paramagnetic Resonance of Exchange Coupled Systems*, Springer-Verlag, Heidelberg, 1990.
7. Mabbs, F.E.; Collison, D., *Electron Paramagnetic Resonance of d Transition Metal Compounds*, Elsevier, Amsterdam, 1992.
8. Griffith, J.S., *The Theory of Transition-Metal Ions*, Cambridge University Press, Cambridge, 1961.
9. McGarvey, B.R., in *Transition Metal Chemistry*, Carlin, R.L.; Ed., **1966**, *3*, 89, Marcel Dekker.
10. Goodman, B.A.; Raynor, B., in *Advances in Inorganic Chemistry and Radiochemistry*, Emeléus, H.J.; Sharpe, A.G.; Eds., **1970**, *13*, 135, Academic Press.
11. Kuska, H.A.; Rogers, M.T. in *Spectroscopy in Inorganic Chemistry*, Rao, C.N.R.; Ferraro, J.R.; Eds., **1971**, *2*, Academic Press.
12. Bencini, A.; Gatteschi, D. in *Transition Metal Chemistry*, Melson, G.A.; Figgis, B.N.; Eds., **1982**, *8*, 1, Marcel Dekker.
13. Trautwein, A.X.; Bill, E.; Bominaar, E.L.; Winkler, H., *Structure and Bonding (Berlin)* **1991**, *78*, 1.
14. *Electron Spin Resonance*, A Specialist Periodical Report of the Royal Chemical Society, published annually.
15. Mills, I.; Cvitaš, T.; Homann, K.; Kallay, N.; Kuchitsu, K. *Quantities, Units and Symbols in Physical Chemistry,* Blackwell Scientific Publications, Oxford, **1988**.
16. Messiah, A. *Quantum Mechanics*, North-Holland, Amsterdam, **1972**.
17. Slichter, C.P. *Principles of Magnetic Resonance*, Springer Verlach, Berlin, **1990**.
18. Gordy,W.; Smith, W.V.; Trambarulo, R.F. *Microwave Spectroscopy*, Wiley, New York, **1953**.
19. Towne, C.H.; Schawlow, A.L. *Microwave Spectroscopy*, McGraw-Hill, New York, **1955**.
20. Poole, C.P., Jr. *Electron Spin Resonance*, Wiley, New York, **1983**.
21. Sealy, R.; Hyde, J.S.; Antholine, W.E. in *Modern Physical Methods in Biochemistry*, Neuberger, A.; Van Deenen, L.L.M.; Eds., Elsevier, New York, **1985**.

22. Bleaney, B.I.; Bleaney, B. *Electricity and Magnetism*, Oxford University Press, Oxford, **1987**.

23. Froncisz, W.; Hyde, J.S. *J. Magn. Reson.* **1982**, *47*, 515.

24. Hyde, J.S.; Froncisz, W. in *Electron Spin Resonance*, Symons, M.C.R., Ed.; Specialist Periodical Reports, Royal Chemical Society, London, vol. 10A, **1986**.

25. Dalton, L.A.; Dalton, L.R. *Modulation Effects in Multiple Electron Resonance Spectroscopy* in *Multiple Electron Resonance Spectroscopy*, Dorio, M.M.; Freed, J.H., Eds., Plenum Press, New York, **1979**.

26. Freed, J.H. *Theory of Multiple Resonance and ESR Saturation in Liquids and Related Media* in *Multiple Electron Resonance Spectroscopy*, Dorio, M.M.; Freed, J.H., Eds., Plenum Press, New York, **1979**.

27. Eaton, S.S.; Eaton, G.R. *Magn. Reson. Rev.* **1993**, *16*, 157.

28. Motokawa, M. in *High Field Magnetism*, Date, M., Ed. North Holland Publishing Co., 1983.

29. Palme, W.; Ambert, G.; Boucher, J.P.; Dhalenne, G.; Revcolevschi, A. *Phys. Rev.* **1996**, *76*, 4817.

30. Lebedev, Y. *Russ. Chem. Rev.* **1983**, *52*, 850.

31. Lebedev, Y. *Appl. Magn. Reson.* **1995**, *7*, 339.

32. Prisner, T.F.; van der Est; Bittl, R.; Lubitz, W.; Stehlik, D.; Möbius, K. *Chem. Phys.* **1995**, *194*, 361.

33. Weber, R.T.; Disselhorst, A.J.M.; Prevo, L.J.; Schmidt, J.; Weckenbach, W.Th. *J. Magn. Res.* **1989**, *81*, 129.

34. Haindl, E.; Möbius, K.; Oloff, H. *Z. Naturforsh.* **1984**, *40a*, 169.

35. Lebedev, Y. in *Modern Pulsed and Continous Wave ESR*, Kevan, L.; Bowman, M.K., Wiley, New York, **1990**.

36. Kevan, L.; Kispert, L.D. *Electron Spin Double Resonance Spectroscopy*, John Wiley & Sons, New York, **1976**.

37. Schweiger, A. *Structure and Bonding* **1982**, *51*, 1.

38. Niklas, J.R.; Spaeth, J.M. *Phys. Status Sol. b* **1980**, *101*, 221.

39. Kevan, L.; Schwartz, R.N. *Time Domain Electron Spin Resonance*, John Wiley & Sons, New York, **1979**.

40. Rowan, L.G.; Hahn, E.L.; Mims, W.B. *Phys. Rev.* **1965**, *A137*, 61.

41. Reijerse, E.J.; Paulissen, M.L.H.; Keijzers, C.P. *J. Magn. Reson.* **1984**, *60*, 66.

42. Mims, W.B.; Peisach, J. in *Biological Magnetic Resonance*, Berliner, L.J.; Reuben, J., Eds., Plenum Press, New York, **1981**.

43. Buckmaster, H.A.; Shing, Y.H. *Phys. Status Sol. A* **1972**, *12*, 325.

44. Buckmaster, H.A.; Chatterjee, R.; Shing, Y.H. *Phys. Status Sol. B* **1972**, *13*, 9.

45. Pryce, M.H.L. *Proc. Phys. Soc. A* **1950**, *63*, 25.

46. Griffith, J.S. *Mol. Phys.* **1960**, *3*, 79.

47. Geurts, P.J.M.; Bouten, P.C.P.; van der Avoird, A. *J. Chem. Phys.* **1980**, *73*, 1306.

48. Breit, G.; Rabi, I. *Phys. Rev.* **1931**, *38*, 2082.

49. van der Waals, J.H.; de Groot, M.S. *Mol. Phys.* **1959**, *2*, 333.

50. van der Waals, J.H.; de Groot, M.S. *Mol. Phys.* **1960**, *3*, 130.

51. Reedijk, J.; Nieuwenhuijse, B. *Rec. Trav. Chim.* **1972**, *91*, 533.

52. Barra, A.L.; Brunel L.C.; Gatteschi, D.; Pardi, L.; Sessoli, R. *Acc. Chem. Res.* **1998**, *31*, 460.
53. Pilbrow, J.R. *J. Magn. Reson.* **1978**, *31*, 479.
54. Schonland, D.S. *Proc. Phys. Soc.* **1959**, *B73*, 788.
55. Waller, W.G.; Roger, M.T. *J. Magn. Res.* **1973**, *9*, 92.
56. Waller, W.G.; Roger, M.T. *J. Magn. Res.* **1975**, *18*, 39.
57. Coffman, R.E. *J. Phys. Chem.* **1975**, *79*, 1129.
58. Aasa, R.; Vanngard, T. *J. Magn. Res.* **1975**, *19*, 308.
59. Galindo, S.; Gonzalez-Tovany, L. *J. Magn. Res.* **1981**, *44*, 250.
60. Van Veen, G. *J. Magn. Res.* **1978**, *30*, 91.
61. Belford, G.G.; Burkhalter, J.F.; Belford, R.L. *J. Magn. Res.* **1973**, *11*, 251.
62. Maki, A.H.; McGarvey, B.R. *J. Chem. Phys.* **1958**, *29*, 31.
63. Maki, A.H.; McGarvey, B.R. *J. Chem. Phys.* **1958**, *29*, 35.
64. Hitchmann, M.A.; Olson, C.D.; Belford, R.L. *J. Chem. Phys.* **1969**, *50*, 1195.
65. Hitchmann, M.A.; Olson, C.D.; Belford, R.L. *J. Chem. Phys.* **1969**, *50*, 1195.
66. Belford, R.L.; Harrowfield, B.V.; Pilbrow, J.R. *J. Magn. Res.* **1977**, *28*, 433.
67. Keijzers, C.P.; De Boer, E. *J. Chem. Phys.* **1972**, *57*, 1277.
68. Keijzers, C.P.; Paulussen, G.F.M.; De Boer, E. *Mol. Phys.* **1975**, *29*, 1733.
69. Ham, F.S. in *Electron Paramagnetic Resonance* **1972**, Geschwind, S.; Ed., Plenum Press, New York.
70. Bersuker, I.B. *The Jahn-Teller Effect and Vibronic Interactions in Modern Chemistry* **1984**, Plenum Press, New York.
71. Bersuker, I.B. *The Jahn-Teller Effect: a Bibliographic Review* **1984**, IFI/Plenum Press, New York.
72. Bill, H. in *The Dynamical Jahn-Teller Effect in Localised Systems* **1984**, Perlin Yu. E.; Wagner M.; Eds., Elsevier, Amsterdam.
73. Abragam, A.; Price, M.H.L. *Proc. Phys. Soc.* **1950**, *A63*, 409.
74. Bleaney, B.; Bowers, K.D.; Treman, R.S. *Proc. Roy. Soc.* **1955**, *A228*, 157.
75. Brean, D.P.; Krupka, D.C.; Williams, F.I.B. *Phys. Rev.* **1969**, *179*, 241.
76. Coffman, R.E. *Phys. Lett.* **1965**, *19*, 475.
77. Coffman, R.E. *Phys. Lett.* **1968**, *21*, 381.
78. Coffman, R.E.; Lyle, D.L.; Mattison, D.R. *J. Phys. Chem.* **1968**, *72*, 1392.
79. Ham, F.S. *Phys. Rev.* **1965**, *A138*, 1727.
80. Ham, F.S. *Phys. Rev.* **1968**, *166*, 307.
81. Setser, G.G.; Barksdale, A.O.; Estle, T.L. *Phys. Rev. B* **1975**, *12*, 4720.
82. Silver, B.L.; Getz, D. *J. Chem. Phys.* **1974**, *61*, 638.
83. Barbucci, R.; Bencini, A.; Gatteschi, D. *Inorg. Chem.* **1977**, *16*, 2117.
84. Bencini, A.; Bertini, I.; Gatteschi, D.; Scozzafava, A. *Inorg. Chem.* **1978**, *17*, 3194.
85. Sharnoff, M. *J. Chem. Phys.* **1964**, *41*, 2203;
86. Sharnoff, M. *J. Chem. Phys.* **1965**, *42*, 3383;
87. Sharnoff, M.; Reimann, C.W. *J. Chem. Phys.* **1965**, *43*, 2993.
88. Bencini, A.; Gatteschi, D. *J. Am. Chem. Soc.* **1983**, *105*, 5535.
89. Peisach, J.; Blumberg, W.E. *Arch. Biochem. Biophys.* **1974**, *165*, 691.

90. Karlin, K.D.; Gultneh, Y. *J. Chem. Educ.* **1985**, *62*, 983.

91. Dudley, R.J.; Hathaway, B.J.; Hodgson, P.J.; Mulcahy, J.K.; Tomlinson, A.A.G. *Inorg. Nucl. Chem.* **1974**, *36*, 1947.

92. Griffith, J.S. *Biopolymer.* **1964**, *1*, 35;

93. Kotani, M. *Adv. Quantum. Chem.* **1968**, *2*, 227.

94. Bencini, A.; Ciofini, I.; Uytterhoeven, M.G. *Inorg. Chim. Acta*, **1998**, *274*, 90.

95. Trautwein, A.X.; Bill, E.; Bominaar, E.L.; Winkler, H. *Structure and Bonding (Berlin)* **1991**, *78*, 1.

96. Feher, G.; Isaacson, R.; Scholes, C.P.; Nagel, R. *Ann. N.Y. Acad. Sci.* **1973**, *222*, 86.

97. Blumberg, W.E.; Peisach, J. *Ann. N.Y. Acad. Sci.* **1973**, *222*, 539.

98. Golding, R.M.; Singhasuwich, T.; Tennant, W.C. *Mol. Phys.* **1977**, *34*, 1343.

99. Bagguley, D.M.S.; Griffiths, J.H.E. *Nature* **1948**, *162*, 538.

100. Bleaney, B.; Bower, K.D. *Proc. R. Soc.* **1952**, *A214*, 451.

101. Owen, J.; Harris, E. in *Electron Paramagnetic Resonance* **1972**, Geschwind, S.; Ed., Chapter 6, Plenum Press, New York.

102. Smith, T.D.; Pilbrow, J.R. *Coord. Chem. Rev.* **1974**, *13*, 173.

103. Buettner, G.R.; Coffman, R.E. *Biochim. Biophys. Acta* **1977**, *480*, 495.

104. Eaton, S.S.; More, K.M.; Sawant, B.M.; Boymel, P.M.; Eaton, G.R. *J. Magn. Reson.* **1983**, *52*, 435.

105. Eaton, S.S.; More, K.M.; Sawant, B.M.; Eaton, G.R. *J. Am. Chem. Soc.* **1983**, *105*, 6560.

106. Eaton, G.R.; Eaton, S.S. *Biol. Magn. Reson.* **1989**, *8*, 339.

107. Hoff, A.J., Ed. *Advanced EPR. Applications in Biology and Biochemistry*, Elsevier, Amsterdam, 1989.

108. Eaton, G.R.; Eaton, S.S.; Ohno, K., Eds. *EPR Imaging and in Vivo EPR*, CRC Press, Boca Raton, Fl, 1991.

109. Berliner, L.J.; Reuben, J., Eds. *EMR of Paramagnetic Molecules*, vol. 13 of *Biological Magnetic Resonance*, Plenum Press, New York, 1993.

110. Holm, R.H.; Kennepohl, P.; Solomon, E.I. *Chem. Rev.* **1996**, *96*, 2239.

111. Solomon, E.I.; Sundaram, U.M.; Machonkin, T.E. *Chem. Rev.* **1996**, *96*, 2563.

112. Guss, J.M.; Freeman, H.C.; *J. Mol. Biol.*, **1983**, *169*, 521.

113. Penfield, K.W.; Gewirth, A.A.; Solomon, E.I. *J. Am. Chem. Soc.*, **1985**, *107*, 4519.

114. Antholine, W.E.; Kastran, D.H.W.; Steffens, G.C.M.; Buse, G.; Zumft, W.G.; Kroneck, P.M.H. *Eur. J. Biochem.* **1992**, *209*, 875.

115. Debrunner, P.G.; Münck, E.; Que, L.; Schultz, C.E. in *Iron Sulphur Proteins*, Lovenberg, W. Ed.; Academic Press, New York; vol. III, p. 381, 1977.

116. Dunham, W.R.; Bearden, A.J.; Salmeen, I.T.; Palmer, G.; Sands, R.H.; Orme-Johnson, W.H.; Beinert, H. *Biochim. Biophys. Acta*, **1971**, *253*, 134.

117. Le Pape, L.; Lamotte, B.; Mouesca, J.M.; Rius, G. *J. Am. Chem. Soc.* **1997**, *119*, 9757.

118. Dyrek, K.; Che, M. *Chem. Rev.* **1997**, *97*, 305.

119. Yamada, H.; Azuna, N.; Kevan, L. *J. Phys. Chem.*, **1995**, *99*, 11190.

120. Ghosh, A.K.; Kevan, L. *J. Phys. Chem.*, **1990**, *94*, 3117.

121. Bencini, A.; Gatteschi, D. *Mol. Phys.* **1985**, *54*, 969.

122. Owens, F.J. in *Magnetic Resonance of Phase Transitions*, Owens, F.J.; Poole, C.P., Jr.; Farach, H.A. Eds.; Academic Press, New York, p. 291, 1979.

3 Mössbauer Spectroscopy

PHILIPP GÜTLICH AND JÜRGEN ENSLING
Fachbereich Chemie
Institut für Anorganische Chemie und Analytische Chemie
Johannes-Gutenberg-Universität Mainz
Staudinger Weg 9, 55099 Mainz, Germany
E-mail: p.guetlich@uni-mainz.de
ensling@iacgu7.chemie.uni-mainz.de

Inorganic Electronic Structure and Spectroscopy, Volume I: Methodology.
Edited by E. I. Solomon and A. B. P. Lever.
ISBN 0-471-15406-7.

1 INTRODUCTION

Since the discovery of the Mössbauer effect by Rudolf Mössbauer[1–3] in 1958 this nuclear spectroscopic method has found a wide variety of applications in solid state physics, chemistry, metallurgy, earth science, biophysics, and biochemistry.

Up to the present time, the Mössbauer effect has been observed for nearly 100 nuclear transitions in more than 40 different elements (cf. Figure 1). Not all of these transitions are suitable for actual studies, for reasons which will be discussed below. But ca. 15–20 elements remain for applications of which the best known Mössbauer nuclide is iron, because the 14.4 keV transition in ^{57}Fe makes it optimally suitable for the method, and also because iron is very abundant in nature.

The purpose of this article is to offer the reader unfamiliar with the field an introduction to the principles of Mössbauer spectroscopy, and also to the various kinds of chemical information that can be extracted from the electric and magnetic hyperfine interactions reflected in the Mössbauer spectrum. For pedagogical reasons we shall mainly focus on the principles of ^{57}Fe Mössbauer spectroscopy in the introductory part, and shall then present examples of typical applications from the literature as well as from our own work. The more deeply interested reader is advised to consult the various textbooks, reviews, and special compilations of original communications.[4–13]

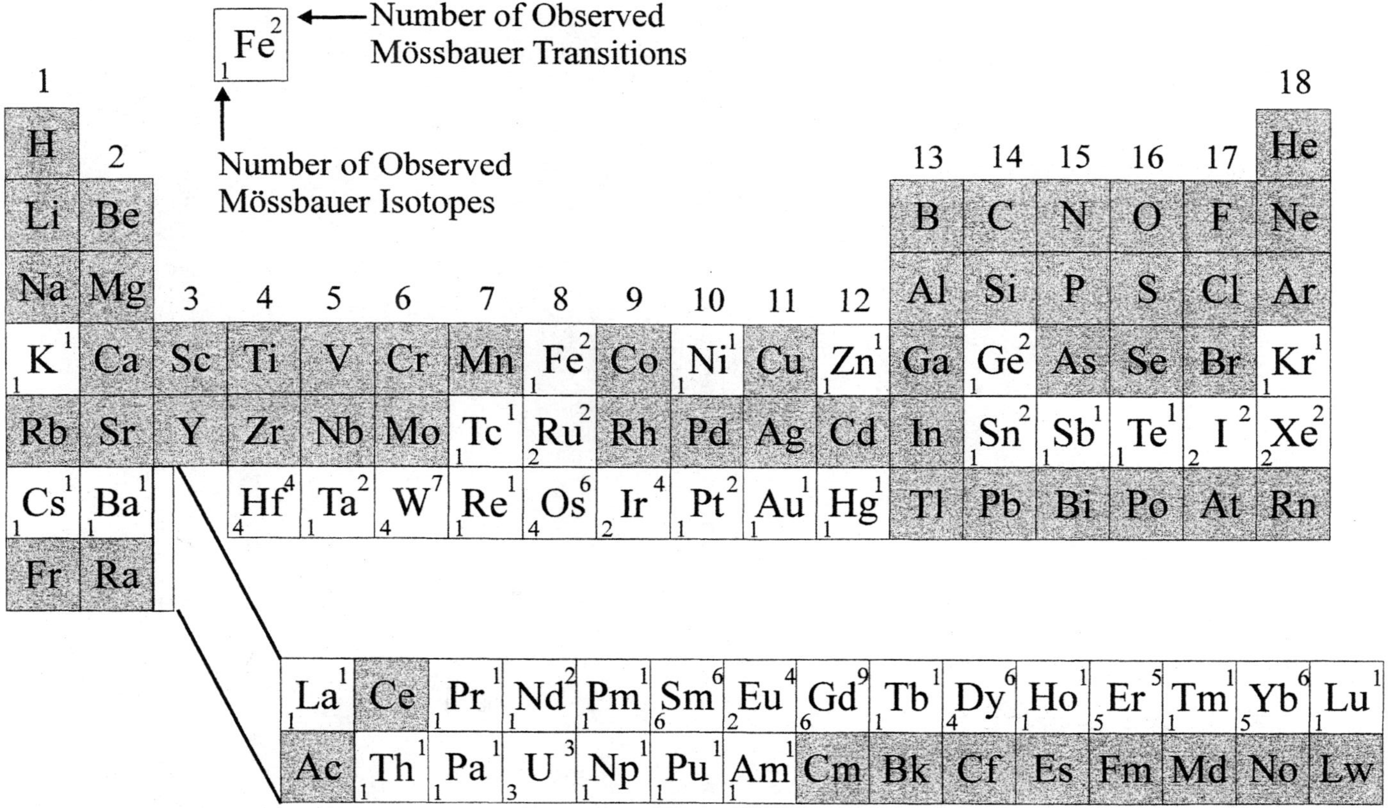

Figure 1. Periodic table of the elements; those in which the Mössbauer effect has been observed are marked appropriately.

2 PRINCIPLE AND EXPERIMENTAL CONDITIONS OF RECOIL-FREE NUCLEAR RESONANCE FLUORESCENCE

2.1 Nuclear Resonance

The resonance absorption of electromagnetic radiation is a phenomenon well known in many branches of physics. The scattering of sodium light by sodium vapor, the excitation of a tuning fork by sound, and the excitation of a dipole by r.f. radiation are some familiar examples. The possibility of observing resonanse in nuclear transitions involving γ radiation was suggested by Kuhn[14] in 1929 but for many years all experimental attempts were unsuccessful. Finally, Rudolf Mössbauer established the experimental conditions as well as the theoretical description of the nuclear resonance absorption (fluorescence) of γ rays.[1–3] In the following we shall discuss the Mössbauer effect conditions.

Consider a nucleus in an excited state of energy E_e with the mean life time τ. Transition to the ground state of energy E_g occurs by emitting a γ quantum with the energy

$$E_0 = \hbar\omega = E_e - E_g. \tag{1}$$

This γ quantum may be absorbed by a nucleus of the same kind in its ground state, thereby taking up exactly the energy difference E_0. This phenomenon is called nuclear resonant absorption of γ-rays, and is schematically sketched in Figure 2. The fact that the life time τ of an excited nuclear state is finite means that its energy has a certain broadening, Γ, which, due to the Heisenberg uncertainty relation, is connected to τ by

$$\tau\Gamma = \hbar. \tag{2}$$

Consequently, the probability for the emission of a quantum is a function of the energy, or, in other words, the energy spectrum exhibits an intensity distribution I(E) about a maximum at E_0 with a width at half maximum, Γ, which corresponds to the uncertainty in energy of the excited nuclear state. The explicit shape of this distribution is Lorentzian (cf. Figure 3) according to

$$I(E) \propto \frac{\Gamma/2\pi}{(E - E_0)^2 + (\Gamma/2\pi)^2}. \tag{3}$$

The mean life time τ of the excited states of the Mössbauer isotopes are in the range of 10^{-6} to 10^{-10} s; the corresponding uncertainties of the transition energies are between 10^{-9} and 10^{-5} eV. Nuclear resonance fluorescence is known to possess an amazingly high energy resolution given by the ratio Γ/E_0, which is on the order of 10^{-12}. As an example, the first excited state of ^{57}Fe has a mean lifetime $\tau = 141$ ns

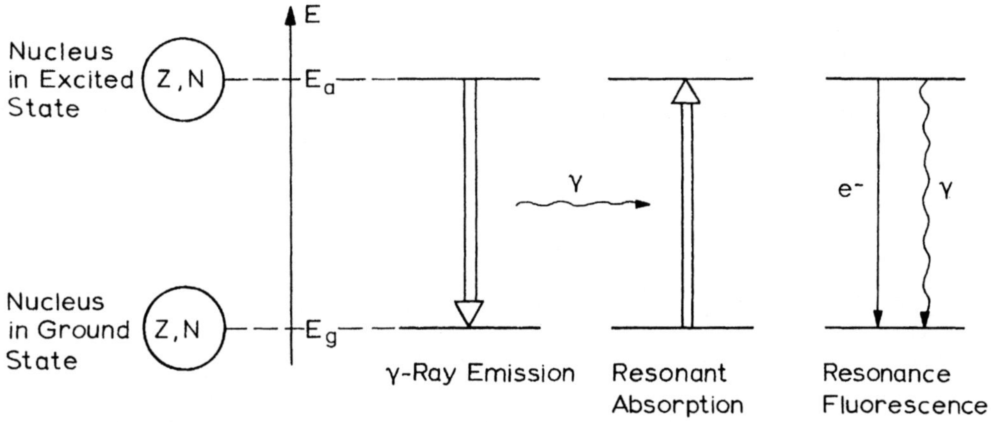

Figure 2. Schematic representation of nuclear resonance absorption of γ-rays (Mössbauer effect) and nuclear resonance fluorescence.

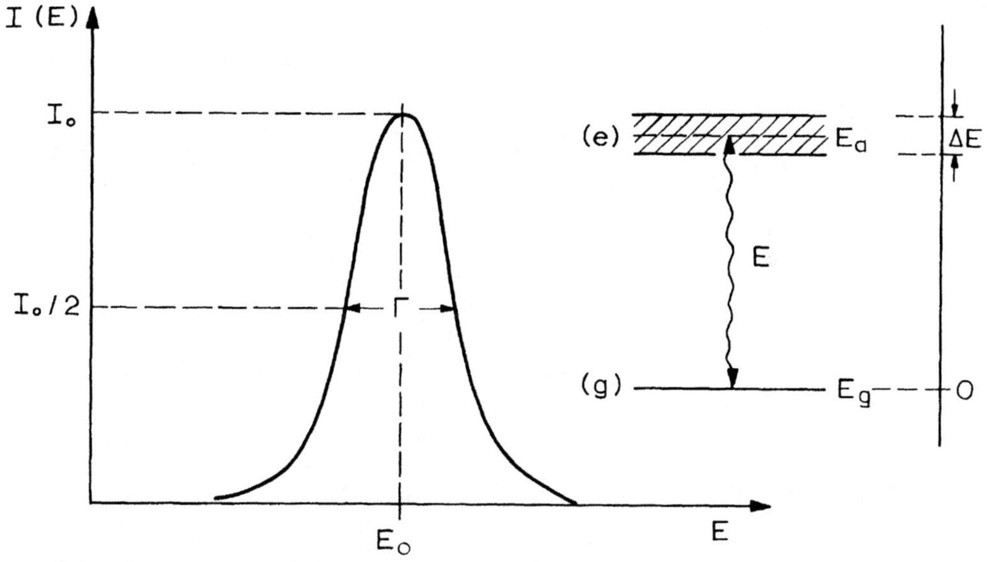

Figure 3. Intensity $I(E)$ as a function of transition energy E. $\Delta E = \Gamma = \hbar/\tau$ is the energy width of the excited state (e) with a mean lifetime τ as well as the width of the transition spectral line at half maximum (Heisenberg natural line width).

which leads to a natural linewidth $\Gamma = \hbar/\tau = 4.7 \cdot 10^{-9}$ eV and hence to a relative energy resolutionof $\Gamma/E_0 = 3.3 \cdot 10^{-13}$.

Since absorption is the time reversal of emission, the absorption line is also described by a Lorentzian curve. Nuclear resonance absorption can only be observed when the difference of the transition energies for emission and absorption are not much greater than the sum of the linewidths, that is, when the emission and absorption lines overlap sufficiently.

After resonance absorption, the excited nucleus will then decay by either emitting isotropically a γ quantum (as in the primary γ-ray emission of Figure 2) or a conversion electron e^-, preferentially from the K shell. This phenomenon is termed *nuclear resonance fluorescence* and may be used in Mössbauer scattering experiments (surface investigations).

2.2 Recoil Effect

Consider the emission of a γ quantum from an excited nucleus of mass M in a free atom or molecule (in the gas or liquid phase). Let the energy of the excited state be E_0 above that of the ground state, and the energy of the photon be E_γ. If the excited nucleus is supposed to be at rest before the decay, a recoil effect will be imparted to the nucleus upon γ emission, causing the nucleus to move with velocity v in opposite direction to that of the emitted γ-ray (see Figure 4). The resulting recoil energy is

$$E_R = \frac{1}{2}Mv^2 = \frac{p_n^2}{2M}. \tag{4}$$

Momentum conservation requires that

$$p_\gamma = E_\gamma/c = -p_n, \tag{5}$$

where p_γ and p_n are the linear momenta of the γ quantum and the recoiled nucleus, respectively, and c is the velocity of light. Thus the energy of the emitted γ quantum is

$$E_\gamma = E_0 - E_R \approx E_0. \tag{6}$$

Since E_R is very small compared to E_0, we may write for the recoil energy (in nonrelativistic approximation)

$$E_R = \frac{p_n^2}{2M} = \frac{E_\gamma^2}{2Mc^2} \approx \frac{E_0^2}{2Mc^2}. \tag{7}$$

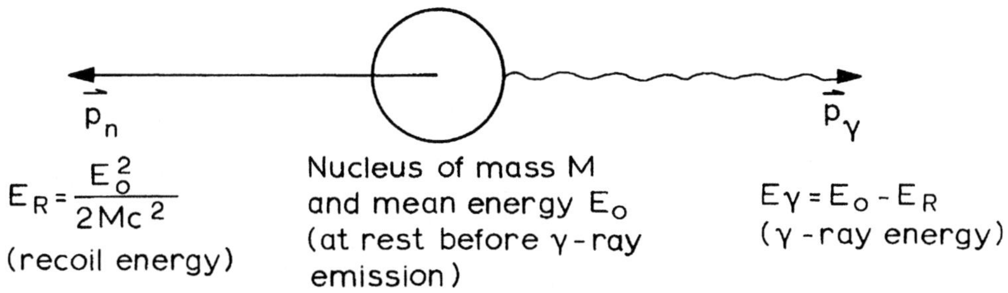

Figure 4. Recoil of momentum $\vec{p}_n$ and energy E_R imparted to an isolated nucleus upon γ-ray emission.

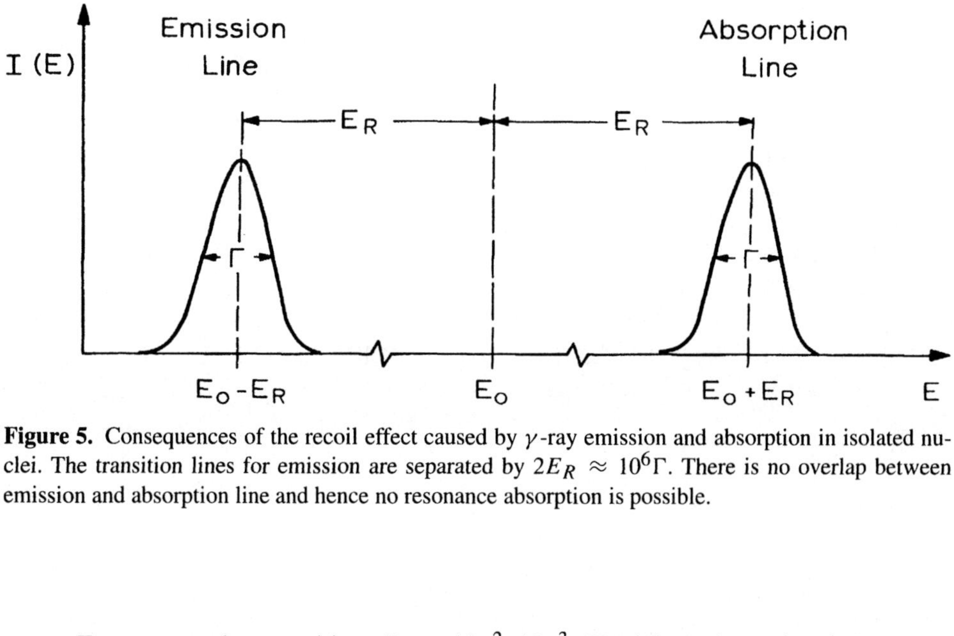

Figure 5. Consequences of the recoil effect caused by γ-ray emission and absorption in isolated nuclei. The transition lines for emission are separated by $2E_R \approx 10^6\Gamma$. There is no overlap between emission and absorption line and hence no resonance absorption is possible.

For many nuclear transitions $E_R \approx 10^{-2}$–10^{-3} eV, while E_0 has values between 10 and 100 eV. Thus, E_R is very small compared to E_0 and it is reasonable to assume that $E_\gamma \approx E_0$.

The recoil effect causes the energy of the emitted photon to be decreased by E_R and the emission line appears at an energy $E_0 - E_R$ rather than at E_0. In the absorption process, the γ-ray to be absorbed by a nucleus requires the total energy $E_0 + E_R$ to make up for both the transition from ground to excited state and the recoil effect, which causes the absorbing nucleus to move in the same direction as the incoming γ-ray photon. As shown schematically in Figure 5 the transition lines

Table 1 Comparison Between Atomic and Nuclear Transitions

Transition (eV)	Electronic	Nuclear
E_0	2.1	14.400
Γ	$4.4 \cdot 10^{-8}$	$4.7 \cdot 10^{-9}$
E_R	10^{-10}	$1.95 \cdot 10^{-3}$

for emission and absorption are separated by $2E_R$, which is about 10^6 times larger than the natural linewidth Γ. The condition for overlap of emission and absorption line, viz.

$$2E_R \leqslant \Gamma \tag{8}$$

is not fulfilled at all. Thus, the cross section of nuclear resonance absorption in isolated atoms and molecules in the gas and liquid phase is extremely small. Table 1 compares the recoil energies and linewidths of the yellow spectral line of sodium and of the 14.4 keV γ-ray of ^{57}Fe.

In the optical case the recoil energy is negligible compared to the linewidth, and the resonance condition is fulfilled for freely emitting and absorbing particles.

This conclusion is strictly valid only for atoms and molecules at rest. In a gas or liquid the motion of the atoms (or molecules) will broaden the energy of the γ-ray. Let us suppose that an atom has velocity $\vec{v}$ at an angle of θ with the direction of emission; the γ photon of energy E_γ receives a Doppler energy E_D,

$$E_D = E_\gamma \frac{v}{c} \cos\theta, \tag{9}$$

which adds to the transition energy

$$E_\gamma = E_0 - E_R + E_D. \tag{10}$$

The expression becomes simpler if we write $E_K = \frac{1}{2}Mv^2$ for the kinetic energy of the atom. Then

$$E_D = E_\gamma \frac{v}{c} \cos\theta = 2\sqrt{E_K E_R} \cos\theta \tag{11}$$

If we allow θ to take all possible values, the spectrum is broadened by an amount

$$\overline{E_D} = 2\sqrt{E_K E_R} \tag{12}$$

where $\overline{E_K}$ is the mean value of the kinetic energy of the moving atom (nucleus). $\overline{E_D}$ is in the order of E_R or larger (e.g., ^{57}Fe: E_0 = 14.4 keV, $E_R = 1.95 \cdot 10^{-3}$ eV, $\overline{E_D} \simeq 10^{-2}$ eV at 300 K). Therefore, there is some very low probability for resonance absorption even in the case of very large recoil loss; the Doppler broadened emission and absorption lines overlap to a small extent.

Before the discovery of the Mössbauer effect some sucessful experiments on nuclear resonance absorption in the gas phase had been carried out utilizing the Doppler effect. High velocities were required ($\approx 10^2$–10^3 ms^{-1}). This could be achieved either by mounting a source of emitting nuclei on a high speed rotor or by heating the source and the absorber, so that some overlap of the lines could take place.

Rudolf Mössbauer, while working on his Ph.D. thesis, carried out experiments of this kind with ^{191}Ir. But surprisingly, on lowering the temperature he found an increase in the absorption effect rather than a decrease. Mössbauer could explain this unexpected phenomenon on quantum mechanical grounds, and was awarded the Nobel Prize in 1961 for the observation and correct interpretation of *recoilless nuclear resonance absorption (fluorescence)* which is now known as the *Mössbauer Effect* and which has provided the basis for a powerful technique in solid state research.

2.3 Mössbauer Effect

We have explained in the preceeding Section that nuclear resonance absorption and fluorescence of γ rays is not possible for nuclei in freely emitting or absorbing particles. This is not so for the solid state, in which the nuclei are more or less rigidly bound to the lattice. If a γ ray photon is emitted from the excited nucleus, the concomitant recoil energy may now be assumed to consist of two parts,

$$E_R = E_{tr} + E_{vib}. \tag{13}$$

E_{tr} refers to the translational energy transferred through linear momentum to the crystallite of effective mass M_{eff}. As M_{eff} is very much larger than the mass of a single atom, the corresponding recoil energy E_{tr} imparted to the crystallite becomes negligibly small according to eq. 7; in fact it is many orders of magnitude smaller than the linewidth Γ.

Most of the recoil energy E_R will be dissipated into processes other than linear momentum, viz. to the lattice vibrational system. E_{vib} is on the order of 10^{-3} eV and thus several orders of magnitude smaller than the atom displacement energy (ca. 25 eV) and also much smaller than the characteristic lattice phonon energies (ca. 10^{-1}–10^{-2} eV). Thus E_{vib} is dissipated by heating the nearby lattice surroundings. As a consequence of this phonon creation with frequencies up to ω_i, the resulting energy of the emitted γ-ray will be

$$E_\gamma = E_0 - \sum_i n\hbar\omega_i \tag{14}$$

This would again destroy the resonance phenomenon by shifting the emission and absorption lines too far apart from each other on the energy scale. Fortunately, however, as quantum theory tells us, the lattice vibrations are quantized and E_{vib} changes the vibrational energy of the lattice oscillators by integral multiples of the phonon energy, $n\hbar\omega_i$ $(n = 0, 1, 2, \ldots)$. This means that there is a certain probability f that no lattice excitation (energy transfer of $0 \cdot \hbar\omega_i$, called zero-phonon-process) occurs during the γ-ray emission or absorption process. Mössbauer has shown that only to the extent of this probability f of unchanged lattice vibrational states during γ-ray emission and absorption does nuclear resonance absorption become possible. This has been termed the *Mössbauer effect.* f is called the recoil-free fraction and denotes the fraction of nuclear transitions which occur without recoil.

The recoil-free fraction in Mössbauer spectroscopy is equivalent to the fraction of X-ray scattering processes without lattice excitation; this fraction of elastic processes in X-ray and neutron diffraction is described by the Debye–Waller factor:

$$f = e^{-k^2\langle x^2\rangle} \tag{15}$$

where $\langle x^2\rangle$ is the mean square amplitude of vibration of the resonant nucleus in the photon propagation direction and $k = 2\pi/\lambda = E_\gamma/(\hbar c)$ is the wave number of the γ quantum. The value of f depends on the lattice properties, described by $\langle x^2\rangle$, on the one hand, and on the photon properties, represented by k^2, on the other hand. The ratio of the mean square displacement and wavelength λ of the photon must be small ($\ll 1$), in order to have a sufficiently high Debye–Waller factor f. It is also obvious that the Mössbauer effect cannot take place in liquids, where the molecular motion is characterized by unbound $\langle x^2\rangle$, causing the recoil-free fraction to vanish. The following consequences emerge:

- f decreases exponentially with k^2, and therefore also with E_γ^2. The Mössbauer effect is hardly detectible at γ-ray energies above 100–150 keV.
- The mean square displacement $\langle x^2\rangle$ should be small, in other words, the Mössbauer atom should be bound tightly to the crystal lattice. However, crystallinity of the material is not a necessary condition for the observation of the Mössbauer effect. The effect is also detectable in amorphous materials and in frozen solutions.
- The recoil-free fraction f is temperature dependent. This dependence is contained in $\langle x^2\rangle$, which increases with increasing temperature and hence forces f to decrease. Vice versa, the Debye–Waller factor f increases on lowering the temperature. However, the Mössbauer effect does not reach unity even at $T = 0$ due to the fact that $\langle x^2\rangle \neq 0$ at $T = 0$ because of the quantum-mechanical zero-point motion of the atoms (nuclei). In order to express f in terms of the usual

experimental variables, the mean square amplitude $\langle x^2 \rangle$ is calculated using lattice dynamic models (e.g. Einstein, Debye). For a more detailed instruction on the recoil-free fraction and lattice dynamics, the reader is referred to relevant textbooks.[15,6] In brief, the Debye model for solids leads to the approximations

$$f = \exp\left[-\frac{E_R}{k_B \Theta_D}\left(\frac{3}{2} + \frac{\pi T^2}{\Theta_D^2}\right)\right] \quad \text{for } T \ll \Theta_D \tag{16}$$

$$f = \exp\left(-\frac{6E_R}{k_B \Theta_D}\right) \quad \text{for } T > \Theta_D. \tag{17}$$

k_B is the Boltzmann factor and $\Theta = \hbar\omega/k_B$ the Debye temperature.

- $\langle x^2 \rangle$ can be anisotropic and hence f. As a consequence, a dependence of the intensity of absorption lines on the observation direction in single crystal experiments may be observed.

3 MÖSSBAUER EXPERIMENT

3.1 Doppler Effect

The main components of a Mössbauer spectrometer are the γ-ray source containing the Mössbauer active nuclides, the absorber (or scatterer), and a detector for low-energy γ radiation (see Figure 6) plus electronics for automatic recording of the spectrum. The source and the absorber are moved relative to each other (either by moving the source and keeping the absorber fixed or vice versa). The transmitted (scattered) γ quanta are registered as a function of the relative velocity. In this way, it is possible to stepwise trace the absorption line by the emission line utilizing the Doppler effect. A γ quantum which is emitted from the source moving at a velocity v receives a Doppler energy E_D modulation such that

$$E_\gamma = E_0 + E_D = E_0\left(1 + \frac{v}{c}\right) \tag{18}$$

where E_0 is the energy of the γ quantum emitted from the same source at rest and c is the velocity of light. This relation is valid for a moving source. The velocity is defined as positive for the source moving towards the absorber and negative for moving away from the absorber. Due to the very narrow linewidth Γ of the resonance line, it is generally sufficient to produce a small Doppler energy change E_D of the order of the linewidth Γ, by Doppler shifting the source in order to sweep over the resonance. The Doppler velocities needed in case of ^{57}Fe spectroscopy are in the range of 0 to ± 10 mm s^{-1}, and for most of the other Mössbauer nuclides velocities of less than ± 100 mm s^{-1} are sufficient. The Mössbauer spectrum, the plot of the

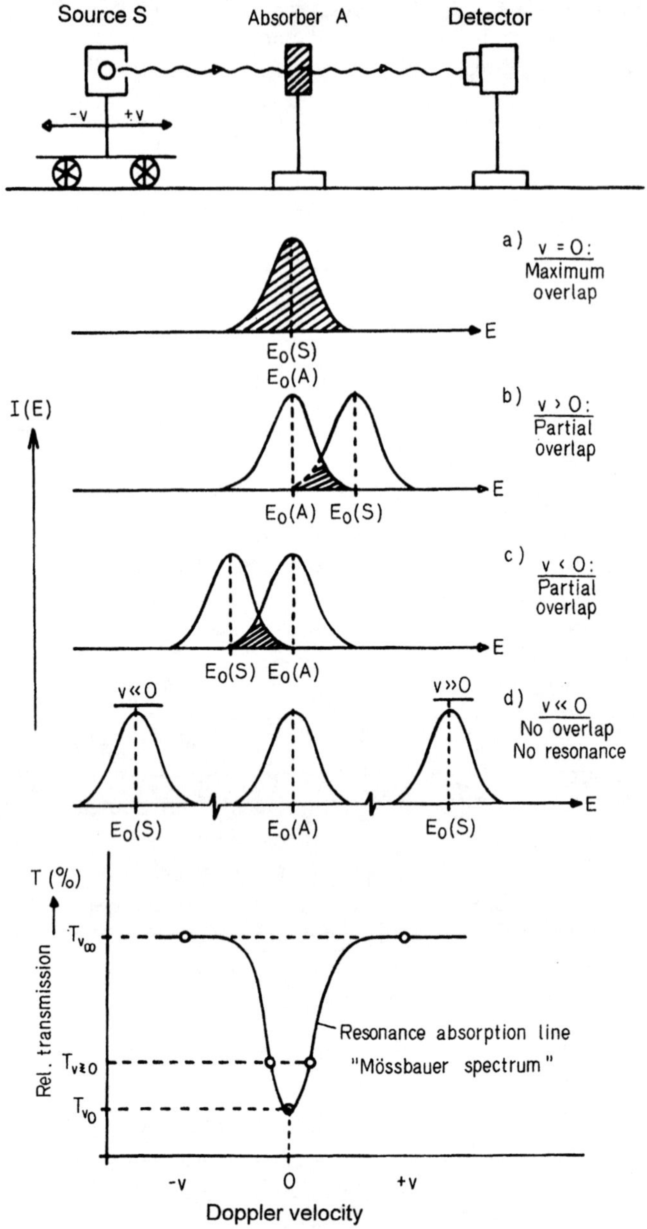

Figure 6. Schematic illustration of the experimental arrangement for recoilless nuclear resonance absorption, and relative transmission of recoilless γ quanta as a function of Doppler velocity.

relative transmission as a function of Doppler velocity, shows maximum resonance and therefore minimum relative transmission at relative velocities where, ideally, emission and absorption lines overlap (cf. Figure 6). At high positive or negative velocities the overlap of emission and absorption lines is negligible and the resonance effect practically zero. Therefore this relative transmission yields the baseline.

3.2 Mössbauer Spectrometer

The Mössbauer spectrometers which are in use nowadays generate the spectrum by the velocity sweep method. The drive system moves the source (or absorber) repeatedly through a range of velocities while simultaneously counting the γ quanta behind the absorber into synchronized channels.

The essential components of a modern Mössbauer spectrometer—as illustrated in the block diagram of Figure 7—are: the velocity transducer, the waveform generator and synchronizer, the multichannel analyzer, the γ-ray detection system, a cryostat or oven for low temperature and temperature dependent measurements, a velocity calibration device, the source, the absorber, and a readout unit.

The source (or, in fewer cases, the absorber) is mounted on the vibrating axis of an electromagnetic transducer (loudspeaker system) of the Kankeleit type[16] which is moved according to a voltage waveform applied to the driving coil of the system. The usual velocity functions are of the triangular, sawtooth, or sinusoidal form. In special cases the source may be moved with a constant velocity.

A function generator produces the desired waveform of the motion. A feedback system (feedback amplifier) operates in such a manner that the actual velocity of the source follows the reference waveform of the function generator with an accuracy better than 0.1%. Using the triangular waveform one obtains an undistorted spec-

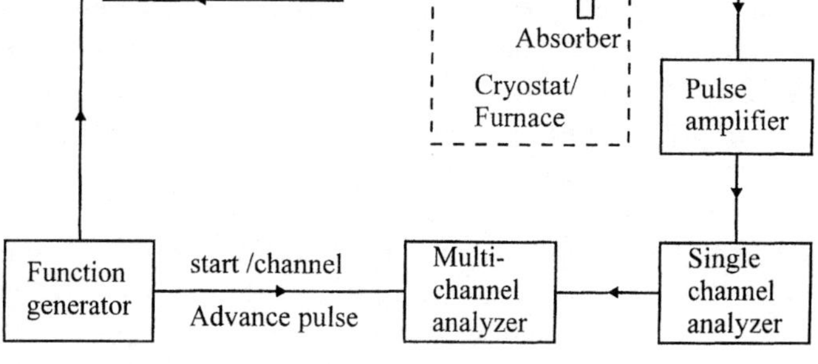

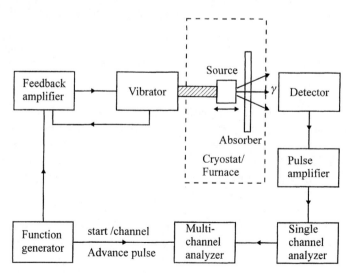

Figure 7. Block diagram illustrating the principle of a Mössbauer spectrometer.

trum, because the velocity is a linear function of time. In extreme cases, where high velocities are required or the mass to be moved is large, it is preferable to use a sinusoidal waveform. For convenience, this type of spectrum is converted after the measurement into linear form.

The counts detected behind the absorber are stored in memory (mostly in the memory of a multichannel analyzer containing typically 512 or 1024 channels). Synchronization of the channel number in the memory and the instantaneous velocity of the source (absorber) is achieved by advancing the address of the memory one by one through an external clock which subdivides the period of the waveform into the number of available channels. A start pulse, which coincides with the beginning of the waveform, triggers the multichannel analyzer to start with channel no. 1 which is subsequently advanced by the clock pulses. In this way synchronization is achieved, that is, during each period of motion a certain channel of the analyzer always corresponds to the same instantaneous velocity of the source. The whole velocity range of interest is scanned many (typically 10 to 50) times per second. As each instantaneous velocity occurs twice during one scan period, the spectrum is registered twice in the memory, where both halves of the memory appear as mirror images, symmetrical about the center of the channel series.

It is essential to have a standard reference line position which other signals can be referred to, particularly in Mössbauer spectroscopy, where often different host metals for a particular Mössbauer isotope are utilized, e.g., ^{57}Co/Cu, ^{57}Co/Pt, ^{57}Co/Rh in case of ^{57}Fe measurements with emitting γ-rays of slightly different energy. The most widely used reference material in Mössbauer spectroscopy is a natural iron foil of which the six line positions of the magnetic hyperfine split spectrum are accurately known. The distance between the two outermost lines of the iron spectrum serves as a calibration basis for the measurements of the hyperfine interactions with strengths corresponding to a Doppler velocity region of 0 to ± 10 mms^{-1}. This region is suitable for Mössbauer effect measurements in ^{57}Fe, ^{119}Sn, ^{61}Ni, ^{197}Au, and others.

The usual detectors for low-energy γ radiation are thin NaI(Tl) scintillation crystals coupled to a photomultiplier, and gas-filled proportional counters. Considerable improvements in energy resolution can be achieved with solid-state detectors such as the Li drifted Ge counter. These detectors in connection with a preamplifier, amplifier, discriminator (single channel analyzer), and the multichannel analyzer form the γ-ray detection system or nuclear channel of a Mössbauer spectrometer.

It is only with low-energy γ emitters, such as the excited state of ^{57}Fe, that the Mössbauer effect can be observed at room temperature and even above. With most Mössbauer isotopes it is necessary to cool the absorber and sometimes also the source to liquid-nitrogen or liquid-helium temperatures in order to obtain a measurable effect. Apart from these practical requirements it is often desirable to vary the temperature of the absorber in order to obtain information about phase transitions or the electronic state and the molecular symmetry of the compound of interest. Many types of cryostats are commercially available. The temperature range varies from 1.2 to 300 K in He-bath cryostats and from 6 to 300 K in He-flow cryostats.

A He-bath cryostat equipped with a superconducting magnet can be utilized for Mössbauer investigations of materials in an applied magnetic field. Magnetic fields up to 10 T are currently very common.

Above room temperature electrical furnaces are used. In order to keep the risk of oxidation of the material low, a moderate vacuum of about 10^{-2} to 10^{-3} mbar is appropriate.

4 PREPARATION OF MÖSSBAUER SOURCE AND ABSORBER

The source for a Mössbauer experiment is a radioactive isotope of reasonable half-life and appropriate nuclear transition energy and excited state lifetime. Through radioactive disintegration, the isotope populates a nuclear excited state which decays to the ground state by emitting low-energy γ radiation. Table 2 lists a selection of isotopes which are suitable for practical Mössbauer effect measurements.

As a typical example, Figure 8 shows the decay scheme of ^{57}Co which populates the 14.4 keV Mössbauer level of ^{57}Fe with a lifetime of $\tau = 140$ ns. The isotope ^{57}Co can be produced in a cyclotron by the nuclear reaction ^{56}Fe(d,n)^{57}Co. The decay of ^{57}Co occurs essentially by electron capture (99.8%) from the K shell leaving a hole in this shell which is filled from higher shells under emission of a 6.4 keV X-ray. Sources of ^{57}Co are usually prepared by electrochemically depositing the carrier-free isotope on metallic supports and then diffusing it into the metal at high temperatures.

A Mössbauer source should meet the following requirements:

- The emission line should be as narrow and intense as possible, unsplit and unbroadened.
- The recoil-free fraction should be as high as possible.
- The source material should be chemically inert during the lifetime of the source and resistant against autoradiolysis.
- The host material should not give rise to interfering X-rays, and Compton scattering and photoelectric processes should be insignificant.

Table 2 Some Important Isotopes in Mössbauer Spectroscopy

Mössbauer isotope	Abundance of stable element (%)	γ-Ray energy (keV)	Half-life of Mössbauer transition (10^{-9} s)	Parent isotope	Half-life of parent isotope
^{57}Fe	2.14	14.4	97.8	^{57}Co	270 d
^{119}Sn	8.58	23.83	17.8	^{119m}Sn	245 d
^{121}Sb	57.25	37.15	3.5	^{121m}Sn	76 y
^{129}I	0 (β^-)	27.77	16.8	^{129}Te	33 d
^{151}Eu	47.82	21.54	9.5	^{151}Sm	90 y
^{197}Au	100	77.35	1.89	^{197}Pt	18 h

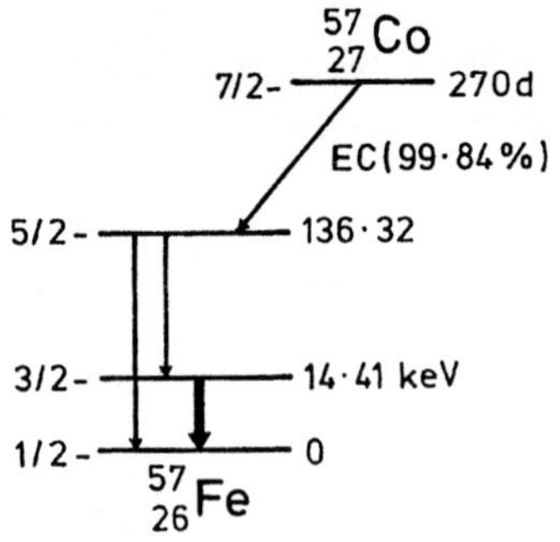

Figure 8. Decay scheme of ^{57}Co.

A uniform absorber with randomly oriented crystallites can be prepared easily by sandwiching the finely ground material between the thin windows of a specially designed perspex holder. In order to obtain undistorted lineshapes it is desirable that the absorber be thin. The natural linewidth Γ_{nat} of the Mössbauer line having Lorentzian shape (see Section 2.1, eq. 3) is determined, as already discussed, by the half-life of the excited nuclear state and the Heisenberg uncertainty principle. The experimental line (Γ_{exp}), however, is broadened by several effects, one of which is the so-called thickness broadening. This dependence can be expressed as:

$$\Gamma_{exp}/2 \simeq (1 + 0.135t)\Gamma_{nat} \quad \text{for } 0 \leqslant t \leqslant 5 \quad \text{and} \tag{19}$$

$$\Gamma_{exp}/2 \simeq (1.01 + 0.145t - 0.0025t^2)\Gamma_{nat} \quad \text{for } 4 \leqslant t \leqslant 10, \tag{20}$$

where Γ_{nat} is the natural linewidth (0.097 mms^{-1} for ^{57}Fe), Γ_{exp} the experimental linewidth, which is the sum of Γ_S and Γ_A, the linewidths of source and absorber, respectively, and t the effective thickness of the absorber. t is given by $= nf_a\sigma_0$, where n is the number of atoms of the Mössbauer isotope per cm^2, f_a the recoil-free fraction of the absorber and σ_0 the maximum resonance absorption cross section per atom. For ^{57}Fe $\sigma_0 = 2.57 \cdot 10^{-22}$ m^2. In the case of ^{57}Fe, if 10 mg cm^{-2} of iron (containing only 2.19% of ^{57}Fe) is used in an absorber, the contribution to the linewidth is $0.135\Gamma_{nat}t \leqslant \Gamma_{nat}$. The absorber is said to be thin ($t \leqslant 5$) in this case.

If heavy elements are present in the compound under investigation, the absorption decreases noticeably. These heavy elements will scatter and absorb the Mössbauer γ-rays, and/or emit photons of similar energy which are detected along with the Mössbauer photons, leading to a decrease in the signal to noise ratio. Mössbauer lines of appropriate intensity can therefore be obtained by decreasing the absorber thickness and increasing the recording time. It is therefore preferable to estimate an optimum absorber thickness for which the measuring time is as short as possible in order to achieve the maximum information from the experiment. For more details the reader is referred to the references[17,18] in which different methods of determining the absorber thickness are discussed.

5 HYPERFINE INTERACTIONS

Due to the narrow linewidth of the Mössbauer nuclear transition (on the order of 10^{-8} eV), the resonance spectrum is extremely sensitive to energy variations of γ radiation. Interactions between the nucleus and the surrounding electrons with energies comparable to the width of the transition lines ($\approx 10^{-8}$ eV) manifest themselves in the Mössbauer spectrum. It is the influence of the electronic environment on the emission and absorption lines which determines the hyperfine structure of the spectrum. This interaction of the positively charged nucleus with the electric and magnetic fields caused by the orbital electrons in the region of the nucleus is called *hyperfine interaction*. This perturbation may be such that it shifts or splits degenerate nuclear levels.

The Hamiltonian for the nucleus of a Mössbauer isotope in a solid can be written as

$$\mathcal{H} = \mathcal{H}_N + \mathcal{H}_{HFS}. \qquad (21)$$

The intranuclear forces described by $\mathcal{H}_N$ give rise to different nuclear states, which are characterized by their spin quantum numbers I. The term $\mathcal{H}_{HFS}$ describes the interaction of the nucleus with its environment and can be expressed in a multipole expansion

$$\mathcal{H}_{HFS} = \mathcal{H}_{E_0} + \mathcal{H}_{M_1} + \mathcal{H}_{E_2} + \cdots \qquad (22)$$

Only the three lowest order terms are considered here, since the higher order terms are negligibly small and cannot be detected by Mössbauer effect measurements. The first term, the so-called monopole interaction, E_0, gives rise to the *isomer shift* (δ) and shifts the energy levels of both the ground state and the excited state. It originates from the Coulombic interaction between nucleus and electrons at the nuclear site. The magnetic dipole (M_1) interaction removes the degeneracy of the nuclear levels if the nuclear spin quantum number is nonzero. It is commonly known as the magnetic hyperfine interaction and describes the influence of a magnetic field on the nuclear spin. In the Mössbauer spectrum this results in the *magnetic splitting* ΔE_M. E_2, the third term in eq. 22, represents the electric quadrupole interaction, that is, the interaction between the nuclear quadrupole moment and an electric field gradient at the nuclear site. The resulting Mössbauer parameter is called *quadrupole splitting* ΔE_Q.

As we shall see in the following, these interactions can, to a first approximation, be expressed by the product of a term containing nuclear parameters and another term with parameters of electronic origin. These electronic parameters refer to electric and magnetic effects caused by the orbital electrons in the region of the nucleus. They can be interpreted in terms of the electronic structure of the atom.

The different interactions and the kind of information one can obtain from the respective Mössbauer parameters are summarized in Table 3. In the following we shall discuss the three hyperfine interactions separately.

Table 3 The Interactions Between Nucleus and Surrounding Electrons, the Respective Mössbauer Parameters, and the Type of Information

Type of interaction	Mössbauer parameter	Information
Electric monopole interaction between nucleus and electrons at the nuclear site	Isomer shift δ	a. Oxidation state (nominal valency) of the Mössbauer atom. b. Bonding properties in coordination compounds (covalency effects between central atom/ion and ligands. Delocalization of d-electrons due to back-bonding effects, shielding of s-electrons by p- and d-electrons). c. Electronegativity of ligands.
Electric quadrupole interaction between electric quadrupole moment of the nucleus and electric field gradient at the nuclear site	Quadrupole splitting ΔE_Q	a. Molecular symmetry b. Oxidation state (nominal valency) c. Spin state d. Bonding properties
Magnetic Dipole interaction between magnetic dipole moment of the nucleus and a magnetic field at the nucleus	Magnetic splitting ΔE_M	Magnetic properties (e.g. Ferro-, antiferro-, para-, diamagnetism, absolute value and direction of local magnetic fields)

5.1 Isomer Shift

The isomer shift (also called chemical isomer shift or center shift) arises from the fact that the nucleus of an atom occupies a finite volume, and s-electrons have the ability of penetrating the nucleus and spending a fraction of time inside the nuclear region (p-, d-, and f-electrons do not have this ability; if relativistic effects are considered, $p_{1/2}$ electrons are also able to spend a small fraction of time inside the nucleus.).

The electric monopole energy of a uniform spherical nucleus with radius R and charge Ze in a constant electron charge density $-e|\psi(0)|^2$ is given by

$$\mathcal{H}_{E_0} = \frac{2}{5}\pi Ze^2|\psi(0)|^2 R^2. \tag{23}$$

The electronic state of the atom, and consequently $|\psi(0)|^2$, does not change during a standard Mössbauer experiment. The nuclear radii of the ground and excited states, in general, differ by a small amount δR. This results in a gain or loss of Coulombic energy δE of the nucleus, when it emits a γ quantum. This energy is transferred to the γ-ray and therefore produces a shift. The same argument holds for the absorption process, so that the total shift, δ, is expressed by

$$\delta = \frac{4}{5}\pi Ze^2 R^2(\delta R/R)\left\{|\psi_a(0)|^2 - |\psi_s(0)|^2\right\} \tag{24}$$

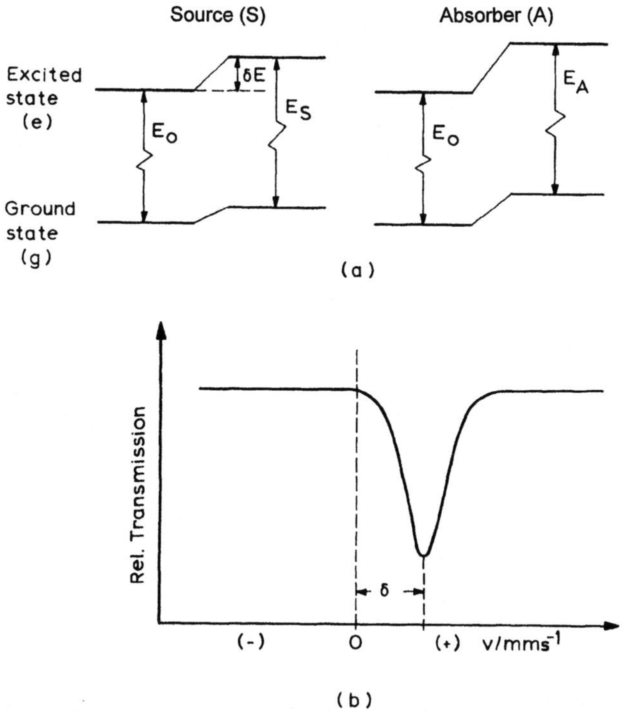

Figure 9. Origin of the isomer shift. (a) Electric monopole interaction in source and absorber shifts the nuclear energy levels without affecting the degeneracy; (b) resultant Mössbauer spectrum (schematic).

where $\delta R = R_e - R_g$ is the change in nuclear radius on going from the excited (e) to the ground state (g). The subscripts a and s refer to absorber and source, respectively. Figure 9 shows schematically the origin of the isomer shift. From eq. 24 it is obvious that the isomer shift is always a relative shift, between absorber and source.

Therefore, the measured isomer shift is always given with respect to a standard material; this can be the Mössbauer source used in the particular experiment or any conventional absorber material (in ^{57}Fe spectroscopy metallic iron and sodium nitroprusside dihydrate, $Na_2[Fe(CN)_5NO]\cdot 2H_2O$ (SNP), are commonly used as standards).

Thus, the isomer shift is expressed by a nuclear term $\delta R/R$ and an electronic term $\{|\psi_a(0)|^2 - |\psi_s(0)|^2\}$, which measures the electron density at the nucleus of an absorber *relative to a given source.* For a given Mössbauer atom the nuclear factor is constant; the isomer shift is thus exclusively dependent on the difference of the electron densities between absorber and source. For ^{57}Fe $\delta R/R$ is negative and the nuclear radius in the excited state is smaller than in the ground state; in the case of ^{119}Sn the reverse is true.

With the isomer shift we have a spectroscopic quantity which can detect minute changes in the electron density at the nucleus. In a nonrelativistic approximation, the electron density at the nucleus is large only for electrons with zero angular momentum (s-electrons) and can be approximated by $|\psi(0)|^2$. Thus, the isomer shift of the Mössbauer spectra is a relative measure of the total s-electron density at the nuclear probe in a compound.

The total s-electron density of an atom in a chemical compound may be expressed as the sum of contributions from the inner and outer shells:

$$|\psi(0)|^2 = |\psi(0)_{inner\ shells}|^2 + |\psi(0)_{outer\ shells}|^2. \tag{25}$$

The contribution from inner electron shells comes from the filled s orbitals of the atom. The contribution from the outer electron shells (valence shells) arises from those external s orbitals of the element which are partially occupied by electrons from the surrounding ligands. This last contribution is sensitive to changes in the chemical environment of the element (e.g. change of charge state, change of bonding properties by electron delocalization etc.) and will exert a *direct* influence on the isomer shift. The valence p-, d-, and f-electrons have only an *indirect* influence on the isomer shift by their shielding effect on the outer s-electron shells against the positive nuclear charge.

The isomer shift δ reflects very small changes in the electron density at the nucleus due to changes in the valence orbital population of the Mössbauer atom. The measurement of the isomer shift gives information on

- the oxidation state (e.g., Fe(II), Fe(III), Fe(IV), etc.),
- the spin state (high spin, low spin),
- the bond properties (covalency, ionicity),
- the electronegativity of the ligands

of a Mössbauer atom in solid material (coordination compounds, alloys, intermetallic phases, amorphous materials, etc.). Simple arguments from molecular orbital theory and atomic structure calculations are often helpful in understanding the bond properties and electronic configurations derived from isomer shift correlations.

As an example, the isomer shift values of high spin iron(II) compounds ($S = 2$) appear in a region of the Doppler velocity scale well separated from that of the high spin iron(III) ($S = 5/2$) and low spin iron(II) compounds ($S = 0$). Figure 10 shows approximate ranges of the isomer shifts determined in iron compounds with different oxidation and spin states of the central metal ion.

Results from Hartree–Fock atomic structure calculation by Watson[19] have shown that the electronic density difference on going from the configuration $3d^6$ (Fe^{2+}) to $3d^5$ (Fe^{3+}) essentially originates from changes in the $3s$ shell. The filled $1s$ and $2s$ orbitals remain practically constant for both configurations. The removal of the $3d$ electron leads to an increase in the electron density at the nucleus. This is due to the less effective shielding of the $3s$ electrons from the nuclear charge by the $3d$ electrons in Fe^{3+} as compared to Fe^{2+}, and—since $\delta R/R$ is negative for ^{57}Fe—causes the isomer shift to become more negative for Fe^{3+} than for Fe^{2+} (with respect to the same standard). The wide spread of the ranges of δ values for each oxidation state

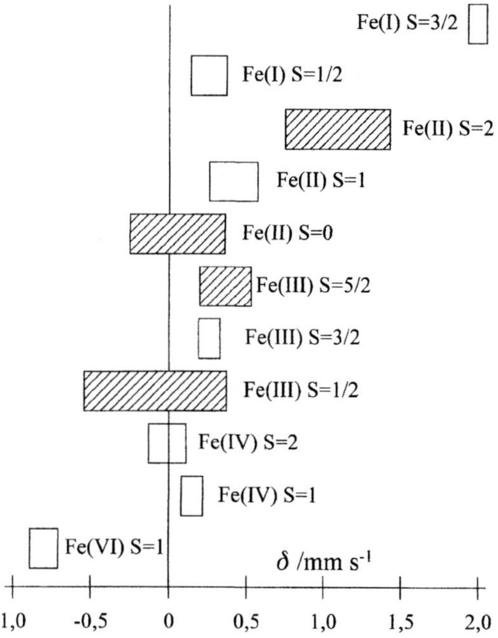

Figure 10. Approximate ranges of isomer shifts observed in iron compounds (relative to metallic iron). S refers to the electronic spin quantum number. Shadowed ranges belong to more frequently met configurations. Note that $(\delta R/R) < 0$ for ^{57}Fe.

is a direct consequence of the nature of the chemical bond. The electron distribution in the molecular orbitals is the result of the variable abilities of the ligands to donate electrons to the metal ion via σ-bonding and to accept electrons from the metal ion via π-bonding.

Similar correlation diagrams to those shown in Figure 10 are available for numerous Mössbauer atoms. Further details can be found in the monograph by Shenoy and Wagner.[20]

5.2 Magnetic Hyperfine Interaction

An atomic nucleus, either in the ground state or excited state, with nuclear spin quantum number $I > 0$, possesses a magnetic dipole moment $\vec{\mu}$ and interacts with a magnetic field $\vec{B}$ at the nuclear site. The magnetic field $\vec{B}$ may arise from the electronic environment or from an external magnet. This interaction is called *magnetic dipole interaction* and is described by the Hamiltonian

$$\mathcal{H}_{M_1} = -\vec{\mu}\vec{B} = -g_N \mu_N \hat{\vec{I}}\vec{B}, \tag{26}$$

where g_N is the nuclear Landé factor and $\mu_N = e\hbar/2Mc$ (M: mass of the nucleus) is the nuclear magneton. The magnetic dipole interaction splits a nuclear state $|I\rangle$ into

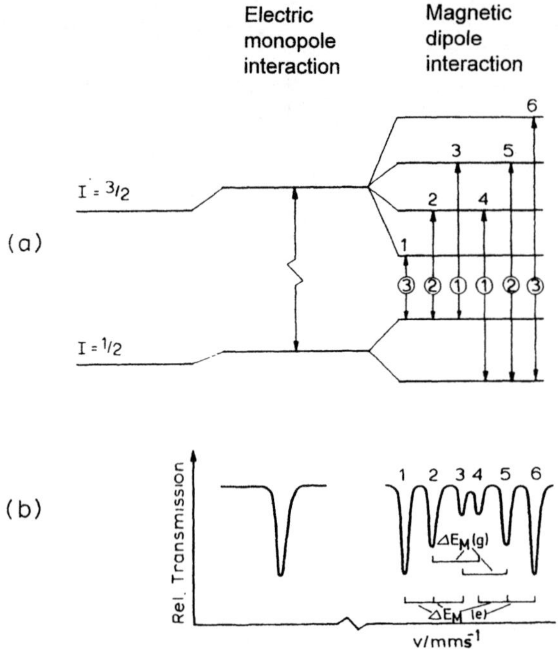

Figure 11. Origin of the magnetic hyperfine splitting. (a) Magnetic dipole interaction in ^{57}Fe in source and absorber. (The center of gravity of the nuclear levels are shifted by electric momopole interaction, which is always present.) (b) Resultant Mössbauer spectrum (schematic); $\Delta E_M(g) = g_g \beta_N H$ refers to the splitting of the ground state, $\Delta E_M(e) = g_e \beta_N H$ refers to the splitting of the excited state.

$2I + 1$ equally spaced substates $|I, m_I\rangle$, each of these being characterized by the nuclear magnetic spin quantum number $m_I = -I, -I + 1, \ldots, I$ (nuclear Zeeman effect). The energies of the substates $|I, m_I\rangle$ are the eigenvalues of the first-order perturbation matrix given by

$$E_M(m_I) = -\mu B M_I / I = -g_N \mu_n B m_I. \tag{27}$$

Since the Mössbauer transition in ^{57}Fe is of the magnetic dipole type (M_1), only those nuclear sublevels with $\Delta m_I = 0, \pm 1$ and $\Delta I = \pm 1$ are coupled by the γ-ray. This selection rule yields only six transitions between the two ground state sublevels ($I = \frac{1}{2}$) and the four excited state sublevels ($I = \frac{3}{2}$). Figure 11 illustrates the splitting of the nuclear levels, the allowed transitions, and the resulting Mössbauer spectrum, a sextet due to the magnetic hyperfine interaction. The effective magnetic hyperfine field $\vec{B}_{eff}$ that the nucleus experiences can originate from the ion electron shell ($\vec{B}_{ion}$) or can be generated by paramagnetic ions in the nearby lattice ($\vec{B}_{lat}$), or can be due to an applied magnetic field ($\vec{B}_{app}$):

$$\vec{B}_{eff} = \vec{B}_{ion} + \vec{B}_{lat} + \vec{B}_{app}. \tag{28}$$

Dipolar interactions of the moments of the surrounding paramagnetic ions are responsible for the field $\vec{B}_{lat}$. This field contains both the demagnetization and the Lorentz field and can be calculated at site i by summing over the contributions of the individual ionic spins S_j on sites j according to

$$\vec{B}_{lat} = g\mu_B \sum_i \left\{ \frac{\vec{S}_j}{|\vec{r}_{ij}|^3} - \frac{3\vec{r}_{ij}(\vec{S}_j \cdot \vec{r}_{ij})}{|\vec{r}_{ij}|^5} \right\}, \tag{29}$$

where g and μ_B are the electronic g-factor and the Bohr magneton, respectively, and $\vec{r}_{ij}$ is the difference vector between lattice sites i and j.

The first term in eq. 28 may contain three contributions:[21]

$$\vec{B}_{ion} = \vec{B}_{FC} + \vec{B}_{dip} + \vec{B}_{orb}. \tag{30}$$

$\vec{B}_{FC}$ is the so-called Fermi contact field and represents the most important contribution to the $\vec{B}_{ion}$ field felt by the individual Mössbauer ion. This field is the direct coupling between the nucleus and the unpaired s-electron density at the nucleus,[22]

$$\vec{B}_{FC} = -\frac{16\pi}{3}\mu_B \left\langle \sum \left(\uparrow\psi_s^2(0) - \downarrow\psi_s^2(0) \right) \right\rangle \tag{31}$$

where $\uparrow\psi_s^2(0)$ is the spin density at the nucleus of s-electrons with spin parallel to the total spin of the valence d-electrons and $\downarrow\psi_s^2(0)$ that with spin antiparallel to the total spin of the d-electrons.

The exchange interaction between the spin-up polarized d-shell and the spin-up s-electron is attractive, while that between the spin-up d-shell and the spin-down s-electron is repulsive; this causes the radial parts of the two s-electrons to distort, one being closer to the nucleus and the other being more distant. As a consequence, the spin density with spin-down is enhanced at the nucleus. The direction of the Fermi contact field is antiparallel to this spin, and consequently the Fermi contact field is negative by defintion, which means anti-parallel to the magnetic moments of the d-electrons.

The other contributions arising from the spins and the electron angular momenta of the Mössbauer atom itself are usually smaller. $\vec{B}_{dip}$ is the dipolar field which for electrons with spin $\vec{S}_j$ at position $\vec{r}_{ij}$ relative to the nucleus at site i is given by eq. 29. $\vec{B}_{orb}$, represents the orbital contribution to the total field in the classical description as

$$\vec{B}_{orb} = -2\mu_B \frac{\vec{L}}{|\vec{r}|^3}. \tag{32}$$

Since the spin-orbit coupling aligns the electronic spin parallel to the orbital momentum, the field is positive, and therefore parallel to the magnetic moment of the d-electrons.

For high spin ferric compounds with electron configuration $3d^5$, ^{6}S, the dipolar field due to its own electrons and the contribution of the orbital momentum to the total field are zero. The Fermi contact term estimated to be -11 T per unpaired $3d$ electron dominates the contributions of the neighboring dipoles and the applied field. In high spin ferrous compounds, however, the negative Fermi contact term is opposed by a positive orbital contribution of the same order of magnitude. The positive dipolar contribution is in general quite small. In this case it is impossible to predict the sign of the internal hyperfine field, B_{ion}. This problem can be solved by applying an external magnetic field that aligns the $3d$ moments. Sign and absolute value of the internal field can then be obtained from the vector sum of the applied and the internal field.

The hyperfine structure of the Mössbauer spectra depends essentially on the magnetic state of the sample: paramagnetic, ferromagnetic, ferrimagnetic, or antiferromagnetic. With ferro-, ferri-, and antiferromagnetic materials, the electronic spin-spin coupling is much stronger than the nuclear Zeeman coupling. As a consequence, the nuclear spin interacts with the average value of the internal field and the Mössbauer spectrum shows magnetic hyperfine structure. For paramagnetic samples, it is important to consider the time t which elapses between two successive flips of the electronic spin. These flips can be due to electronic relaxation processes in paramagnetic compounds. If the frequency $1/t$ of the flips is much larger than the Larmor frequency of the nuclear spin in the internal field, the hyperfine splitting is absent, since the mean value of the internal field "seen" by the nucleus is zero. In contrast, if the frequency of the flips is comparable to or smaller than the Larmor frequency of the nuclear spin, the spectrum will show a magnetic hyperfine pattern. Examples of the latter situation are known, for example, for Fe^{3+} ions diluted in corundum[23] and in a number of trivalent iron compounds.

There are two types of electronic relaxation processes in paramagnetic substances, namely spin-lattice relaxation processes and spin-spin relaxation processes. Spin-lattice interaction originates from the interaction between paramagnetic ions and phonons. The spin-lattice relaxation time depends on temperature, because the population of the phonon states is temperature dependent. Spin-spin relaxation arises from the magnetic interaction between the paramagnetic ions. The spin-spin relaxation time depends on the concentration of the paramagnetic ions, but not on temperature. Mössbauer spectroscopy enables one to differentiate between the two types of paramagnetic relaxation processes by varying the temperature and the concentration.

5.3 Quadrupole Splitting

The third term in the multipole expansion (cf. eq. 22) describes the *electric quadrupole interaction*. This type of nuclear-electronic interaction is also electrostatic in nature and occurs if the following conditions are fulfilled:

(i) The Mössbauer nucleus must posses a measurable electric quadrupole moment eQ, which arises from nonspherical nuclear charge distribution leading to spin quantum numbers $I > 1/2$. A measure of the deviation from spherical symmetry is given by the electric quadrupole moment, eQ, which can be calculated by the spacial integral

$$eQ = \int dv\rho r^2(3\cos^2\theta - 1), \tag{33}$$

where e is the proton charge and ρ is the charge density at the spherical coordinates r and θ. The sign of the nuclear quadrupole moment refers to the shape of the distorted nucleus: Q is negative for a flattened and positive for an elongated nucleus.

(ii) The electric field at the Mössbauer nucleus must be inhomogeneous, measured by the electric field gradient EFG $\neq 0$. This arises from a noncubic charge distribution of the electrons and/or of the neighboring ions around the nucleus.

A point charge q at a distance $r = (x^2 + y^2 + z^2)^{1/2}$ from the nucleus gives rise to a potential $V(r) = q/r$ at the nucleus. The electric field $\vec{E}$ at the nucleus is the negative gradient of the potential, $-\nabla V$, and the gradient of the electric field, finally, is given by

$$EFG = \vec{\nabla}\vec{E} = -\vec{\nabla}\vec{\nabla}V = -\begin{pmatrix} V_{xx} & V_{xy} & V_{xz} \\ V_{yx} & V_{yy} & V_{yz} \\ V_{zx} & V_{zy} & V_{zz} \end{pmatrix}, \tag{34}$$

where

$$V_{ij} = \partial^2 V/\partial i \partial j \quad (i, j = x, y, z). \tag{35}$$

The Laplace equation requires the tensor to be traceless,

$$V_{xx} + V_{yy} + V_{zz} = 0. \tag{36}$$

Additionally, it is possible to define a principal coordinate system along the main axes of the EFG, transforming the tensor into a diagonal one. With respect to the principal axes, the EFG tensor is described by only two independent parameters, usually chosen as $V_{zz} = eq$ and $\eta = (V_{xx} - V_{yy})/V_{zz}$.

It is common practice to choose $|V_{zz}| \geqslant |V_{xx}| \geqslant |V_{yy}|$, which confines the asymmetry parameter η to $0 \leqslant \eta \leqslant 1$. η becomes zero for axial symmetry. This is the case for a fourfold or threefold axis passing through the Mössbauer nucleus and coinciding with the V_{zz} direction. The EFG vanishes completely if two three- or higher-fold rotation axes are mutually perpendicular (e.g., in cubic symmetry).

Due to the electrostatic attraction between the negative electron charges and the positive nuclear charge, different orientations of the nucleus described by different values of the nuclear magnetic spin quantum number m_I result in different energies. The quadrupole splitting arising from this interaction between the nuclear quadrupole moment and the nonzero EFG is described by the Hamiltonian

$$\mathcal{H}_{E_2} = \frac{eQV_{zz}}{4I(2I-1)}\left\{3\hat{I}_z^2 - I(I+1) + \frac{\eta}{2}(\hat{I}_+^2 + \hat{I}_-^2)\right\}, \tag{37}$$

where $\hat{I}_z$ and I are the nuclear spin operator and the nuclear spin quantum number, respectively. $\hat{I}_+$ and $\hat{I}_-$ are the raising and lowering operators. Using first-order perturbation theory, the eigenvalues of $\mathcal{H}_{E_2}$ are found as

$$E_Q(I, m_I) = \frac{eQV_{zz}}{4I(2I-1)}\{3m_I^2 - I(I+1)\}\left(1 + \frac{\eta^2}{3}\right)^{1/2}. \tag{38}$$

As an example, the effect of the electric quadrupole interaction in a Mössbauer nucleus with $I = 3/2$ in the excited state and $I = 1/2$ in the ground state, as is the case in ^{57}Fe and ^{119}Sn, is pictured in Figure 12. In the absence of the magnetic interaction the quadrupole interaction gives rise to a doublet with the splitting

$$\Delta E_Q = \pm\frac{1}{2}eQV_{zz}\left(1 + \frac{\eta^2}{3}\right)^{1/2}. \tag{39}$$

The interaction is positive if the $\pm\frac{3}{2}$ state of the excited level is higher, and negative when it is lower than the $\pm\frac{1}{2}$ state of this level.

For a powder sample the quadrupole splitting yields only the absolute value of the product of the nuclear moment and the electric field gradient, but not its sign. This is due to the fact that the two allowed transitions (1,2) have equal probabilities (intensities) in randomly oriented particles. An exception is the so-called Goldanskii–Karyagin effect,[24] which interprets the occurrence of different line intensities for the $|\pm 3/2\rangle \longleftrightarrow |\pm 1/2\rangle$ and $|\pm 1/2\rangle \longleftrightarrow |\pm 1/2\rangle$ transitions as arising from anisotropic recoilless radiation.

The quadrupole splitting in a Mössbauer spectrum is a highly sensitive probe for the chemical environment that determines the EFG at the nucleus of the Mössbauer atom. In general, there are two fundamental sources which can contribute to the total EFG:

1. Charges on ions surrounding the Mössbauer probe atom in noncubic symmetry give rise to the so-called *lattice contribution*.
2. Noncubic distribution of the electrons in the partially filled valence orbitals of the Mössbauer atom generate the *valence contribution*.

The filled inner shells have spherical symmetry and make no contribution to the field gradient; however, both the charges external to the atom and the unpaired valence electrons can polarize these inner shells which will then contribute to the EFG at the nucleus. This magnification of the electric field gradient is called anti-shielding, and

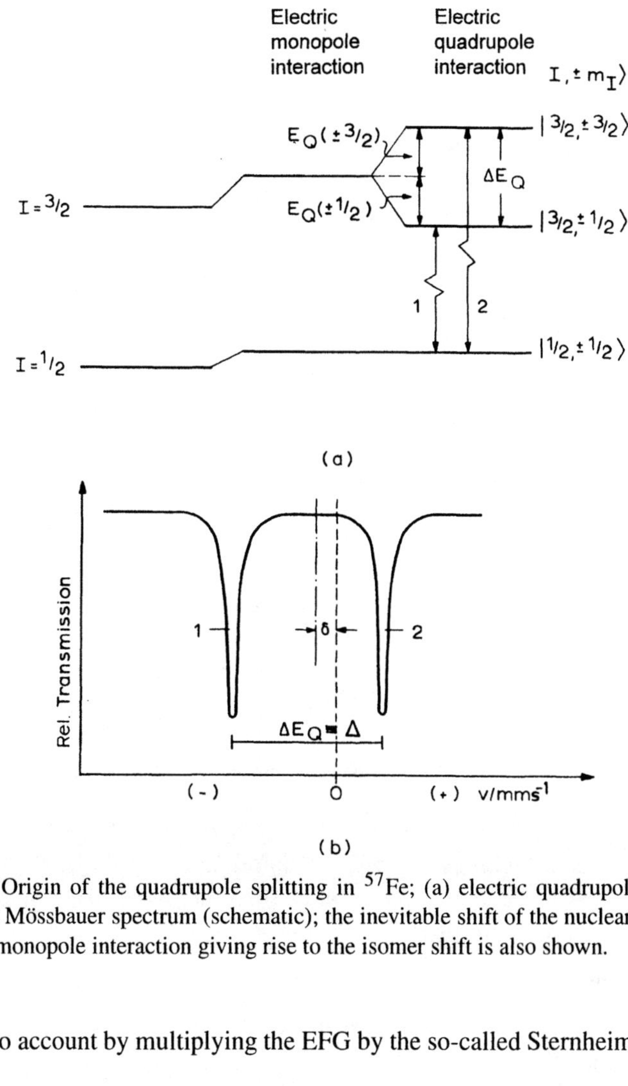

Figure 12. Origin of the quadrupole splitting in ^{57}Fe; (a) electric quadrupole interaction, (b) resultant Mössbauer spectrum (schematic); the inevitable shift of the nuclear levels due to the electric monopole interaction giving rise to the isomer shift is also shown.

is taken into account by multiplying the EFG by the so-called Sternheimer shielding factors.[25]

$$V_{zz} = (1 - \gamma_\infty)V_{zz}^{lat} + (1 - R)V_{zz}^{val}, \tag{40}$$

where *lat* and *val* refer to the lattice and valence contributions, respectively. γ_∞ is called the Sternheimer anti-shielding factor and R the Sternheimer shielding factor. The factor $(1 - \gamma_\infty)$ has been estimated to be on the order of 10 for iron compounds, and R to be 0.2–0.3 for iron and tin.

The lattice contribution can be estimated by a point charge lattice sum

$$V_{zz}^{lat} = \sum_i q_j \frac{3\cos^2\theta_j - 1}{r_j^3}, \tag{41}$$

$$\eta^{lat} V_{zz}^{lat} = \sum_{i} q_j \frac{3 \sin^2 \theta_j \cos 2\phi_j}{r_j^3}, \tag{42}$$

where q_j is the charge of an ion with spherical coordinates r_j, θ_j and ϕ_j. The sum runs in principle over the whole lattice, but normally very good convergence is obtained within a sphere of several nm. Reasonable agreement between lattice sum and experimentally determined EFG is observed in ionic salts.

The valence electron contribution to V_{zz} is evaluated simply by taking the expectation value of the quantity $-e(3\cos^2\theta_j - 1)r^{-3}$ for each electron in the valence orbital state $|l_i m_i\rangle$ and summing over all valence electrons i,

$$V_{zz}^{val} = -e \sum_{i} \left\langle l_i m_i \right| (3\cos^2 - 1) r^{-3} \left| l_i m_i \right\rangle. \tag{43}$$

The expectation value for $1/r^3$, $\langle r^{-3}\rangle$, is normally factored out from the above integral, since it is generally not calculable with satisfactory precision; it may be obtained from experiment. Thus,

$$V_{zz}^{val} = -e \textstyle\sum_i \langle l_i m_i | 3\cos^2\theta - 1 | l_i m_i \rangle \langle r_i^{-3} \rangle. \tag{44}$$

$$\eta^{val} V_{zz}^{val} = -e \textstyle\sum_i \langle l_i m_i | 3\sin^2\theta \cos 2\phi | l_i m_i \rangle \langle r_i^{-3} \rangle. \tag{45}$$

Similar expressions exist for the other two diagonal EFG elements. The principal component and asymmetry parameter due to an unpaired electron with wavefunction $|l_i m_i\rangle$ can thus be calculated and are given in Table 4 for p- and d-electrons.

According to the different origins (valence/lattice contribution) it is expected that observed quadrupole splittings may reflect information about the electronic structure (oxidation state, spin state), bond properties, and molecular symmetry. As an example the spectra of three iron coordination compounds are shown in Figure 13:

a) The observed large quadrupole splitting of $\Delta E_Q = 3.4$ mms^{-1} for $FeSO_4 \cdot 7H_2O$ is essentially determined by the valence contribution V_{zz}^{val}. This compound is a typical high spin compound where the Fe(II) ion is octahedrally surrounded by

Table 4 Values of the z-Component of the Total EFG and η for p- and d-electrons

Orbital	$\frac{V_{zz}^{val}}{e\langle r^{-3}\rangle}$	η	Orbital	$\frac{V_{zz}^{val}}{e\langle r^{-3}\rangle}$	η
p_x	+2/5	−3	d_{z^2}	−4/7	0
p_y	+2/5	+3	$d_{x^2-y^2}$	+4/7	0
p_z	−4/5	0	d_{xy}	+4/7	0
			d_{yz}	−2/7	−3
			d_{zz}	−2/7	+3

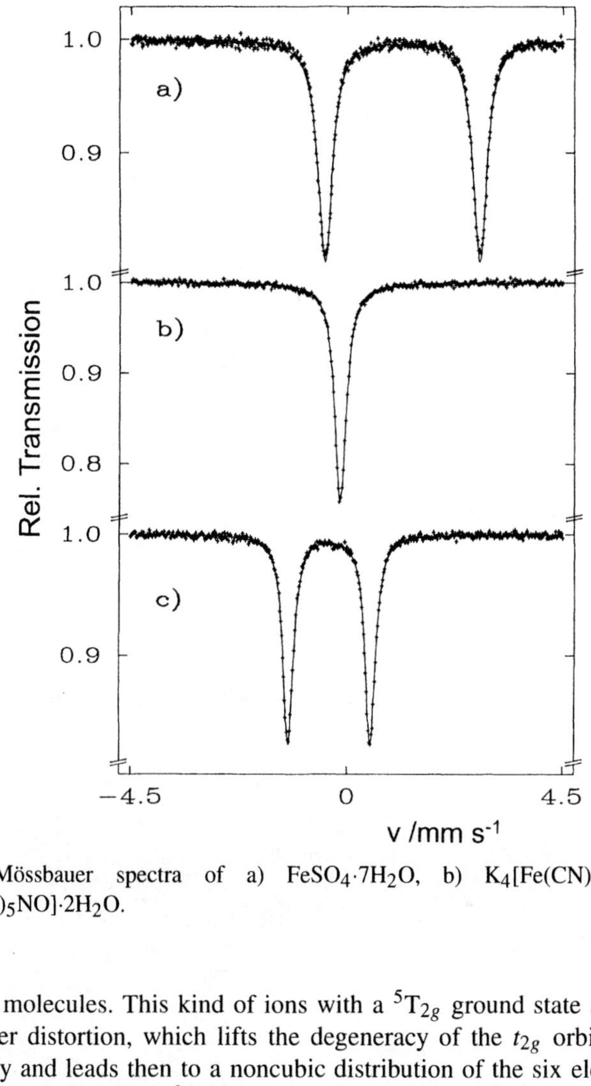

Figure 13. Mössbauer spectra of a) $FeSO_4{\cdot}7H_2O$, b) $K_4[Fe(CN)_6]{\cdot}3H_2O$, and c) $Na_2[Fe(CN)_5NO]{\cdot}2H_2O$.

six water molecules. This kind of ions with a $^5T_{2g}$ ground state are subject to Jahn–Teller distortion, which lifts the degeneracy of the t_{2g} orbitals partly or completely and leads then to a noncubic distribution of the six electrons in the $3d$ shell of the $[Fe(H_2O)_6]^{2+}$ ions. The doubly occupied lowest d_{xy} orbital gives rise to the relatively large quadrupole splitting in the tetragonally compressed $[Fe(H_2O)_6]^{2+}$ octahedron. Further details can be found in reference.[12]

b) $K_4[Fe(CN)_6]{\cdot}3H_2O$ is a typical low spin Fe(II) compound. The single line Mössbauer spectrum results from the absence of a quadrupole interaction due to the cubic charge distribution of the electron configuration t_{2g}^6 which consequently leads to zero valence contribution V_{zz}^{val}.[12] The lattice contribution V_{zz}^{lat} also vanishes since the $[Fe(CN)_6]^{4-}$ complex ion forms a regular octahedron.

c) The relatively large quadrupole splitting in the case of $Na_2[Fe(CN)_5NO]{\cdot}2H_2O$ is solely due to the fact that the presence of different ligand molecules can no longer form a regular octahedron as in the case of $[Fe(CN)_6]^{4-}$. The $[Fe(CN)_5NO]^{2-}$

complex has C_{4v} symmetry and the magnitude of the quadrupole splitting is mainly determined by the lattice contribution; a contribution from the valence electrons can be excluded since the distorted t_{2g} orbitals are still completely filled for this Fe(II) low spin compound.

5.4 Combined Quadrupole and Magnetic Hyperfine Interactions

Quite frequently electric quadrupole interaction and magnetic dipole interaction may be present in addition to the electric monopole interaction which is always active. The Hamiltonian for the combined electric quadrupole and magnetic dipole interaction can be written as

$$\mathcal{H} = \frac{eQV_{zz}}{4I(2I-1)}\left\{3\hat{I}_z^2 - I(I+1) + \frac{\eta}{2}(\hat{I}_+^2 + \hat{I}_-^2)\right\} - g_N\mu_N \vec{B}_{eff}\hat{\vec{I}}. \qquad (46)$$

For comparable interaction strengths, $E_Q \approx E_M$, the interpretation of the Mössbauer spectrum may become complicated. In most cases, however, one can apply a perturbation treatment assuming either $E_M \ll E_Q$ or $E_Q \ll E_M$.

Suppose the latter case applies, where the electric quadrupole interaction can be treated as a perturbation of the stronger magnetic dipole interaction. If the EFG is axially symmetric and the principal component of the EFG, V_{zz}, makes an angle ϑ with the direction of the magnetic field, the eigenvalues of the nuclear sublevels are given by

$$E_{M,Q}(I, M_I) = -g_N\mu_N B m_I + (-1)^{|m_I|+1/2}(eQV_{zz}/8)(3\cos^2\vartheta - 1). \qquad (47)$$

We find in the case of ^{57}Fe, which is depicted in Figure 14, that the sublevels $|3/2, \pm 3/2\rangle$ are shifted by an amount $E_Q(\pm m_I)$ to higher energy and the sublevels $|3/2, \pm 1/2\rangle$ are shifted by E_Q to lower energy, if V_{zz} is positive. These energy shifts by E_Q are reversed if V_{zz} is negative. As a result, the sublevels of the excited state $I = 3/2$ are no longer equally spaced, unless $\cos\vartheta = \sqrt{1/3}$. Combined electric quadrupole and magnetic dipole interaction generally manifests itself in the Mössbauer spectrum as an asymmetrically split sextet (for $I = 3/2 \longleftrightarrow I = 1/2$ transitions as in ^{57}Fe) as pictured in Figure 14. As the sublevel spacings and thus the asymmetry of the spectrum is directly correlated with the sign of V_{zz}, one can determine the sign of the EFG of a polycrystalline materials from the magnetically split hyperfine spectrum. The effect of a magnetic dipole interaction (applied magnetic field) as a perturbation of the electric quadrupole splitting is described in detail in reference.[27]

These three types of hyperfine interactions with the relevant Mössbauer parameters are most important in solid state research (chemistry, physics, metallurgy, materials science, biology, etc.). In addition, one often extracts further helpful information from the temperature and pressure dependence of the Mössbauer parameters, the shape and width of the resonance lines (relaxation phenomena) and the second-order Doppler shift (lattice dynamics).

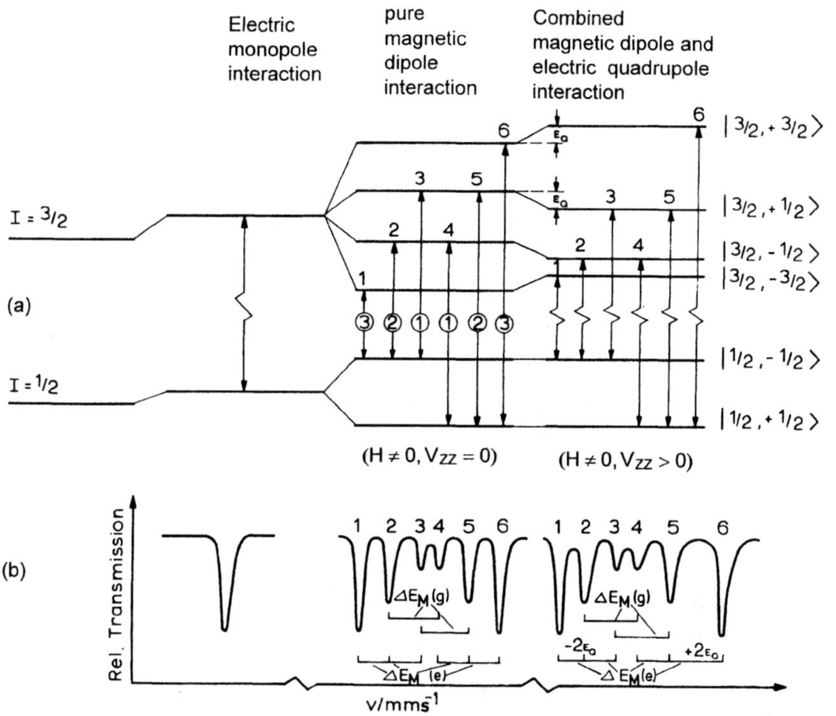

Figure 14. Combined quadrupole and magnetic hyperfine interaction; (a) splitting and shifts, (b) resultant Mössbauer spectrum (schematic).

5.5 Relative Intensities of Resonance Lines

The relative intensities of the lines in a Mössbauer spectrum provide additional information about the orientation of the quantization axes of the hyperfine operators with respect to the source-absorber axis. The relative intensities of the several Mössbauer transitions between the sublevels $|I_e m_e\rangle$ of the excited state and the sublevels $|I_g m_g\rangle$ of the ground state are determined by an angular independent term and an angular dependent term.

$$I_{g \to e} \propto (CG)^2 \times F_L^{\Delta m}(\theta, \phi) \tag{48}$$

where CG are the Clebsch–Gordon coefficients $CG = \langle I_e m_e L \Delta m | I_g m_g \rangle$, L is the angular momentum of the photon, and $\Delta m = m_g - m_e$. θ and ϕ are the polar and azimuthal angles of the z-direction (defined by the direction of the magnetic field or the field gradient tensor) and the direction of the γ-ray emission.

The Mössbauer transitions in ^{57}Fe and ^{151}Eu are both magnetic dipole transitions with $L = 1$. In these cases, the angular factor, F, is only θ-dependent. The square Clebsch–Gordon coefficients of the $1/2 \to 3/2$ transition and the angular factors with $L = 1$ are given in Table 5.

Table 5 Normalized Squared Clebsch–Gordon Coefficients $(I_e m_e L \Delta m \mid I_g m_g)^2$, and Normalized Angular Factors $F_L^{\Delta m}(\theta)$ for the 1/2 → 3/2 Transition in ^{57}Fe ($I_e = 3/2$, $I_g = 1/2$, $\Delta m = m_g - m_e$)

Magnetic hyperfine spectra						
m_g	m_e	Δm	$(CG)^2$	$F_{L=1}^{\Delta m}(\theta)$	$\theta = 90°$	$\theta = 0°$
+1/2	+3/2	−1	3	$1 + \cos^2\theta$	3	6
+1/2	+1/2	0	2	$2\sin^2\theta$	4	0
+1/2	−1/2	+1	1	$1 + \cos^2\theta$	1	2
+1/2	−3/2	+2	0	0	0	0
−1/2	+3/2	−2	0	0	0	0
−1/2	+1/2	−1	1	$1 + \cos^2\theta$	1	2
−1/2	−1/2	0	2	$2\sin^2\theta$	4	0
−1/2	−3/2	+1	3	$1 + \cos^2\theta$	3	6

Quadrupole spectra				
Transitions	$(CG)^2$	$F_{L=1}^{\Delta m}(\theta)$	$\theta = 90°$	$\theta = 0°$
±1/2, ±1/2	1	$2 + 3\cos^2\theta$	5	2
±1/2, ±3/2	1	$3(1 + \cos^2\theta)$	3	6

For the case of a magnetically split spectrum we derive from this table the relative intensity ratios 3:2:1:1:2:3 for the hyperfine components of the Zeeman pattern of a powder sample. We take into account that for a polycrystalline absorber the average $\overline{F_L^{\Delta m}(\theta, \phi)}$, given by the integration of $F_L^{\Delta m}(\theta, \phi)$ over all orientations, is 4/3 for all allowed transitions (with $\overline{\sin^2\theta} = 2/3$ and $\overline{\cos^2\theta} = 1/3$). On the other hand a single-crystal absorber does have angular properties and the intensities of the hyperfine lines will be affected by the orientation. In order to obtain the relative intensities the square of the Clebsch–Gordon coeffiecients have to be multiplied by the angular dependent factors for the $\Delta m = 0$ and $\Delta m = \pm 1$ transitions giving rise to the numbers in the sixth and seventh column of Table 5. It is obvious that the relative intensity of the $|\pm 1/2\rangle \rightarrow |\pm 3/2\rangle$ transition to the $|\pm 1/2\rangle \rightarrow |\mp 1/2\rangle$ transition is always 3:1, whereas the relative intensity of the $|\pm 1/2\rangle \rightarrow |\pm 1/2\rangle$ transition is dependent on θ and ranges from 0 to 4.

In the lower part of Table 5 the results for a pure quadrupole spectrum are depicted showing that a dipole $1/2 \rightarrow 3/2$ transition leads to 1:1 intensities for a powder sample and 5:3 and 1:3 intensity ratios for a single-crystal with the x-axis perpendicular and parallel to the principal axis of a symmetric EFG tensor, respectively. Due to this angular dependence of the intensity of the quadrupolar split lines we can determine which of the two lines must be assigned to the $|\pm 1/2\rangle \rightarrow |\pm 3/2\rangle$ transition. From this information the sign of the principal component of the EFG tensor can be determined, a quantity which cannot be obtained from a powder spectrum in the absence of an applied field.

Thus far we have only dealt with transitions between "pure" states. The situation becomes more complicated if both magnetic dipole and electric quadrupole interactions are present and of comparable strength. In this case the states are no longer pure $|Im_I\rangle$ states but rather linear combinations of these. It is not possible to use the earlier arguments, and a more fundamental treatment must be applied. For more details the reader is referred to the references.[28,29]

6 EVALUATION OF MÖSSBAUER SPECTRA

The evaluation of a Mössbauer spectrum implies the determination of the physical parameters, such as isomer shift, quadrupole splitting, magnetic hyperfine splitting, line intensities and widths, parameters which help to characterize the material under investigation. In order to obtain precise values of the peak parameters (position and intensity), it is essential to compute the spectra. This procedure is only applicable when the lineshapes are known precisely, the variance or standard deviation of the counts are known, the off-resonance region (baseline) is linear or a well defined function of velocity, and the spectral output is available in digital form. These conditions are ideally fulfilled in Mössbauer spectroscopy, and the area, width, and position of a spectral line along with their standard deviation can readily be obtained. The lineshape can in most cases be considered Lorentzian, the deviation of a count N is equal to $\sqrt{N}$, the baseline is linear in case of a triangular drive waveform, and the spectrum is available from the multichannel analyzer in digital form.

Especially in the case of poorly resolved or complex hyperfine spectra due to the overlap of a number of different line patterns caused by the presence of different chemical species or several lattice sites the use of a computer program is definitely necessary. The general method used to fit Mössbauer spectra will be briefly described.

The fitting is generally performed by utilizing the least squares method; in other words, the weighted mean square deviation χ^2 between the experimental data points y_i^{exp} and the corresponding theoretical values y_i^{theo} has to be minimized,

$$\chi^2 = \sum_{i=1}^{N} \frac{1}{\sigma_i^2}\left[y_i^{theo} - y_i^{exp}\right]^2 \stackrel{!}{=} Minimum \tag{49}$$

σ_i^2 is the square of the statistical variation of the count rate in the ith channel of the multichannel analyzer with N channels. In the most simple case y_i^{theo} is the sum of n uncorrelated Lorentzian lines with different intensities and halfwidths

$$y_i^{theo} = B - \sum_{j=1}^{n} \frac{\Gamma_j^2}{4} \frac{A_j}{(v_i - \delta_j)^2 + \Gamma_j^2/4} \tag{50}$$

B is the off resonance count rate, A_j the amplitude (height) and Γ_j the full width at half maximum, and δ_j the position of resonance line j. The baseline B may be due to geometry effects like the periodic change of the solid angle during the movement of the transducer as a function of the velocity (channel i).

For complex spectra, it is often recommended to keep some parameters constant or constrain a number of parameters to be equal during the fitting procedure. Without these constraints, it is in many cases not possible to reach convergence of the fit of a complicated spectrum. There are cases (thick absorber, relaxation effects, etc.) where the simple model function of equation 50 no longer holds. There are a number of computer programs which handle such problems employing appropriate model functions. Highly recommended are the very versatile programs developed by Müller[30] or Brand.[31]

7 SELECTED APPLICATIONS

Since the discovery some 40 years ago, Mössbauer spectroscopy has become an extremely powerful analytical tool for the investigation of various kinds of materials. In most cases, only two parameters are needed, viz. the isomer shift and the quadrupole splitting, to identify a specific sample. In case of magnetically ordered materials, the magnetic dipole interaction is a further helpful parameter for characterisation. In the following we shall discuss some applications selected from different research areas. In Appendix A a compilation of Mössbauer parameters of selected types of iron compounds is given.

7.1 Mixed-Valence Compounds

Mössbauer spectroscopy provides an effective method to elucidate the electronic and structural properties of so-called mixed-valence compounds. Examples are the color pigments "Prussian Blue and Turnbull's Blue". The deep blue compound Prussian Blue was first mentioned in 1704. For a long time, there were controversial discussions whether Prussian Blue, which is formed in the reaction of Fe(III) salts with $[Fe(CN)_6]^{4-}$ in aqueous solution, and Turnbull's Blue, which conversely precipitates by mixing Fe(II) salts and an aqueous solution of $[Fe(CN)_6]^{3-}$, are identical substances. Already in 1963 Fluck[32] could prove by means of Mössbauer spectroscopy that in both reactions the product is ferric-hexacyanoferrate(II); the empirical formula is $Fe_4[Fe(CN)_6]_3 \cdot 15\ H_2O$. Maer[33] could show using ^{57}Fe doped samples that a fast electron exchange occurs during the reaction in solution. This was ascertained by doping the Fe(II) and Fe(III) salts which were used in both reactions. In both cases only the quadrupole doublet of high spin Fe(III) ions could be detected in the Mössbauer spectrum. Bonnette et al.[34] refined the study by using enriched ^{56}Fe for the ferrous salts; in this way only the singlet spectrum of $[Fe(CN)_6]^{4-}$ in which the ferrous ions are in the $^1A_1$ low spin state appeared in the spectrum.

The results of the spectroscopic and crystallographic studies show, that Prussian Blue is to be assigned to class II of the mixed-valence compounds (classification of Robin & Day[35]). Typically for these compounds, in which the metal centers only slightly differ crystallographically, is the observation of an Intervalence-CT transition, responsible for the deep color of the material. However, the rate of the inter-

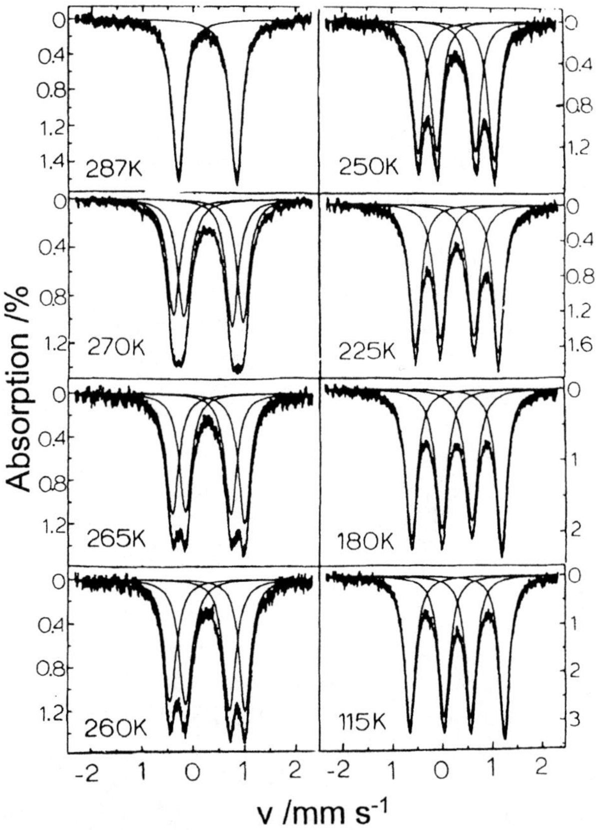

Figure 15. Temperature dependence of the Mössbauer spectra of a microcrystalline sample of 1′, 1″-Diethylbiferrocenium-triiodide. Adapted from Dong et al.[36]

valence charge transfer is slow compared to the Mössbauer time window giving rise to the observation of the resonance lines for both metal centers (Fe(II)/Fe(III)) separately.

Among the mixed-valence compounds there are examples in which the rate of the intervalence charge transfer is fast compared to the Mössbauer time window. Figure 15 shows the temperature dependence of the Mössbauer spectra of 1′, 1″-Diethylbiferrocenium-triiodide.[36] At low temperatures the two quadrupole doublets characteristic of Fe(II) (outer two lines) and Fe(III) (inner two lines) can be identified. Increasing the temperature a gradual collapse to only one doublet (above about 270 K) is observed, of which the Mössbauer parameters (δ, ΔE_Q) are approximately the mean of those of the discrete doublets seen at low temperatures.

On account of its behavior this mixed-valence compound is a member of class III of the Robin and Day classification. Both metal centers are crystallographically identical and the physical properties of the material are determined by the time average of the relevant quantities of the centers.

7.2 Biological Systems (Ferredoxin II)

The importance of Mössbauer spectroscopy in the study of biological systems mainly results from the role of iron in a large number of biomolecules. These large and complex molecules have this Mössbauer atom in their active centers at which the biologically significant reactions occur. It is this situation that Mössbauer spectroscopy in combination with other relevent techniques has been successfully used to study bioinorganic systems. The information studied relates to the oxidation and spin state of the iron atom, to the nature and coordination of ligands, to a possible spin coupling and magnetic ordering, and to the equivalence or nonequivalence of all the iron atoms in the system. The application of Mössbauer spectroscopy to biosystems ranges from simple isolated biomolecules to complete organisms under *in vivo* conditions. Out of this variety of biological Mössbauer studies we have selected an extensively investigated example of isolated biomolecules.

7.2.1 ***Ferredoxin II*** Iron-sulfur proteins are important in many biological processes, such as photosynthesis, nitrogen fixation, and respiration. Their role in these processes is in electron transfer leading to changes in the oxidation state of the system. They are known to contain either one, two, or four iron atoms, but now it has become evident that some of these proteins exist with three-iron centers. Among these, the bacterium *Desulfovibrio gigas*, also known as ferredoxin II, has been investigated by Huynh et al.[37] using Mössbauer spectroscopy. In Figure 16 spectra of two redox states of ferredoxin II are shown. The single quadrupole doublet of the oxidized state at 77 K indicates that all the iron atoms are equivalent. The Mössbauer parameters are typical for ferric ions in the tetrahedral sulfur coordination found in iron-sulfur proteins (e.g. oxidized rubredoxin or plant-type ferredoxin). In the reduced form the spectrum consists of two quadrupole doublets with a spectral area ratio of 2 : 1. Approximately one third of the ferric ions in the protein remain unchanged upon

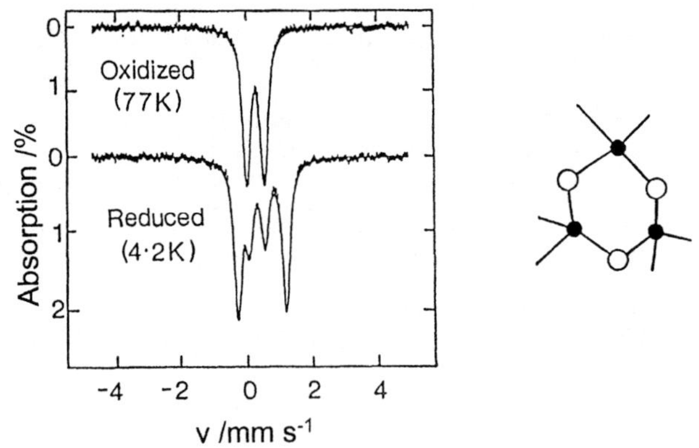

Figure 16. Left: Mössbauer spectra of *Desulfovibrio gigas* ferredoxin II. Adapted from Huynh et al.;[37] right: three-iron center (solid circles represent iron atoms, open circles sulfur atoms).

reduction. The parameters of this doublet correspond essentially to those of the oxidized protein. The Mössbauer characteristics of the second doublet, with two thirds of the area fraction, the larger splitting and the isomer shift of 0.45 mms^{-1} which is in between the values for ferric and ferrous ions in rubredoxin, for example, indicate that the extra electron due the reduction process is equally shared by two of the iron atoms in the three-iron center of ferredoxin II leading to a mixed-valence state intermediate between ferric and ferrous. Applied magnetic field experiments show that both redox states of the protein give rise to a magnetic hyperfine splitting in their Mössbauer spectra. This behavior rules out a spin coupling to a nonmagnetic state in both redox states.

7.3 Thermal and Light-Induced Spin Transition

It is well established that coordination compounds of $3d$ transition metal ions with $3d^4$ up to $3d^7$ electron configuration may undergo thermal spin transition (spin crossover), if the Gibbs free enthalpy difference between the high spin (HS) and low spin (LS) states involved in the crossover is comparable to thermal energy k_BT. Extensive studies have been performed with iron(II) spin crossover compounds,[38] for which the transitions between the HS state, $^5T_2(O_h)$, with $t_{2g}^4e_g^2$ electron configuration and the LS state, $^1A_1(O_h)$, with t_{2g}^6 configuration can be followed elegantly using Mössbauer spectroscopy. An example is demonstrated by the temperature dependent Mössbauer spectra of $[Fe(2\text{-pic})_3]Cl_2{\cdot}EtOH$ (2-pic = 2-picolylamine) in Figure 17.

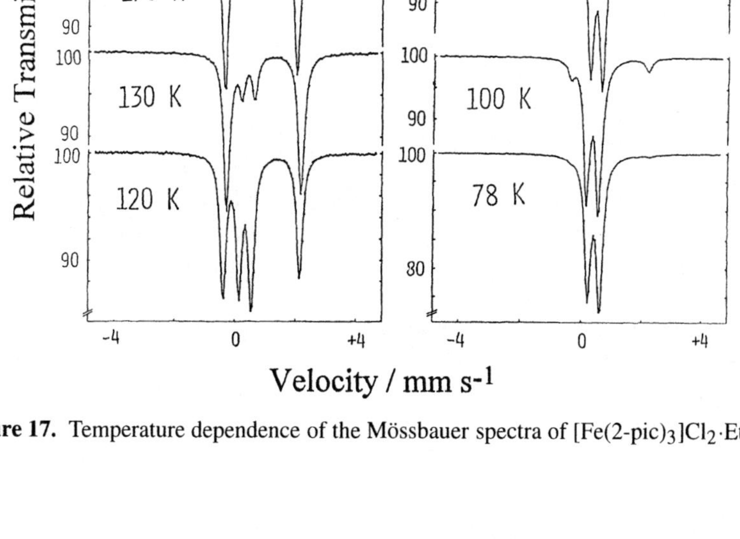

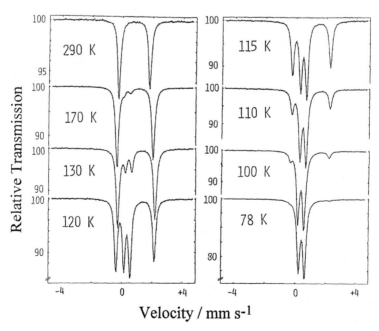

Figure 17. Temperature dependence of the Mössbauer spectra of $[Fe(2\text{-pic})_3]Cl_2{\cdot}EtOH$.

Near room temperature the spectra reflect the pure Fe(II)-HS state, with the typically large quadrupole splitting arising from the dominating valence contribution to the electric field gradient (EFG), and an isomer shift on the order of 1 mm s^{-1} (relative to α-iron). The transition to the LS state becomes readily evident in the spectra by the appearance of a new signal. The quadrupole doublet of Fe(II)-LS typically has smaller splitting energy, because the valence contribution to the EFG vanishes, and only a lattice contribution arising from symmetry lowering due to bidentate ligand coordination remains.

Switching between LS and HS states in such spin crossover complexes can also be effected by irradiation with light,[39] and followed by Mössbauer spectroscopy. The five spectra in Figure 18 were recorded for $[Fe(2\text{-pic})_3]Cl_2{\cdot}EtOH$ at 4.2 K before irradiation (A), after irradiation with green light (B), after heating the polycrystalline sample to ca. 30 K, where HS $\rightarrow$ LS relaxation sets in, which is clearly seen by the gradual disappearance of the doublet of the metastable HS state in favor of the growing intensity of the LS doublet (C, D, E).

More on thermal and light-induced spin transition with applications of other physical techniques will be discussed in a special chapter in Volume II of this series.

7.4 Magnetic Materials and Magnetic Phase Change

7.4.1 ***Magnetic Ordering in Magnetite*** Magnetite, Fe_3O_4, is the only pure iron oxide of mixed valence. The structure of magnetite at room temperature is that of an inverse spinel, that is, the iron atoms occupy interstitial sites in the close-packed arrangement of oxygen atoms. There exist two types of sites; the tetrahedral sites (A) are occupied by ferric cations, and the octahedral sites (B) half by ferric and half by ferrous cations. In the unit cell there are twice as many B-sites as A-sites. The structural formula can therefore be written: $(Fe(III))_{tet}[Fe(II)Fe(III)]_{oct}O_4$. There is little distortion in the cubic close packing of the oxygens, so that the structure is close to ideal.

Magnetite has a Curie temperature of 840 K. In the magnetically ordered phase the iron atoms on the A-sites are antiferromagnetically coupled to those on the B-sites resulting in a complete canceling of the moments of the Fe(III) ions. Thus, a ferromagnetic moment, originating in the Fe(II) ions on the A-sites, does not cancel out. On the whole, magnetite shows ferrimagnetic behavior.

The statistical distribution of the di- and trivalent ions of iron on equivalent lattice sites (B) explains many unusual properties of magnetite, especially its extremely high electric conductivity comparable to that of other metals. Investigations show that the conductivity is caused by electron transfer and not by ion transport. A fast electron-transfer process (electron hopping) between ferrous and ferric ions on the octahedral B-sites takes place.[40] The Mössbauer spectrum of magnetite at room temperature (cf. Figure 19) confirms this interpretation.

The spectrum consists of two more or less overlapping six-line spectra with internal magnetic fields of 49 and 46 Tesla and a relative area ratio of approximately 1 : 2, which originate from the iron ions on sites A (Fe(III)) and B (average field of Fe(II) + Fe(III)).

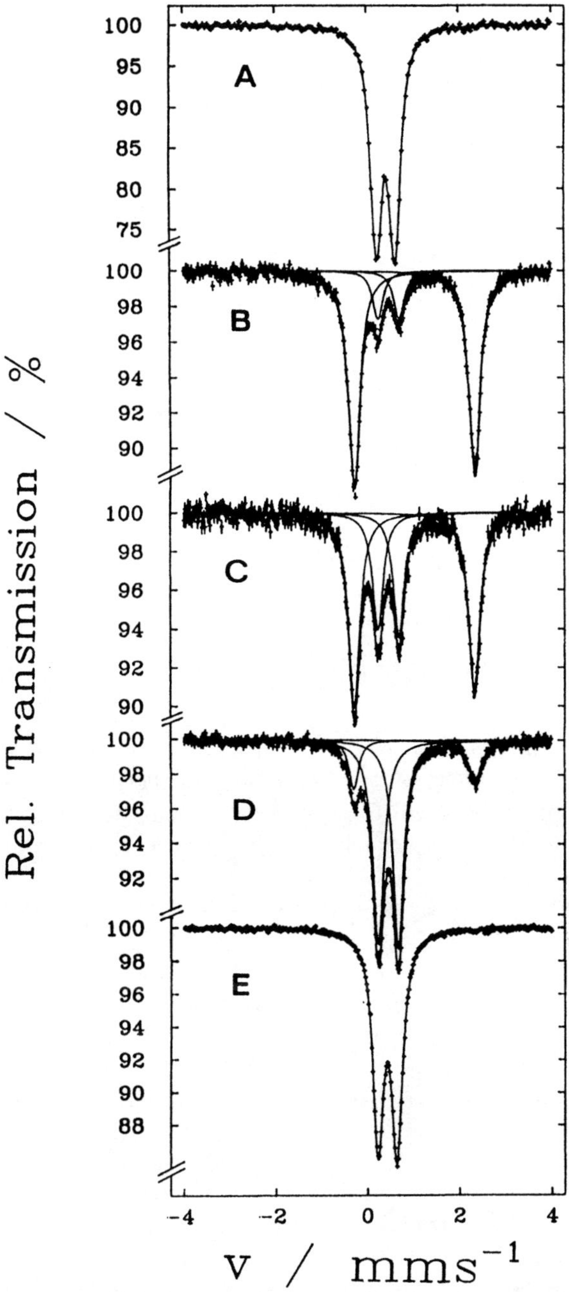

Figure 18. ^{57}Fe Mössbauer spectra of [Fe(2-pic)$_3$]Cl$_2$·EtOH, at 4.2 K before irradiation (A), at 4.2 K after irradiation with green light (B), after heating to 30 K and following the HS → LS relaxation within hours (C, D, E).

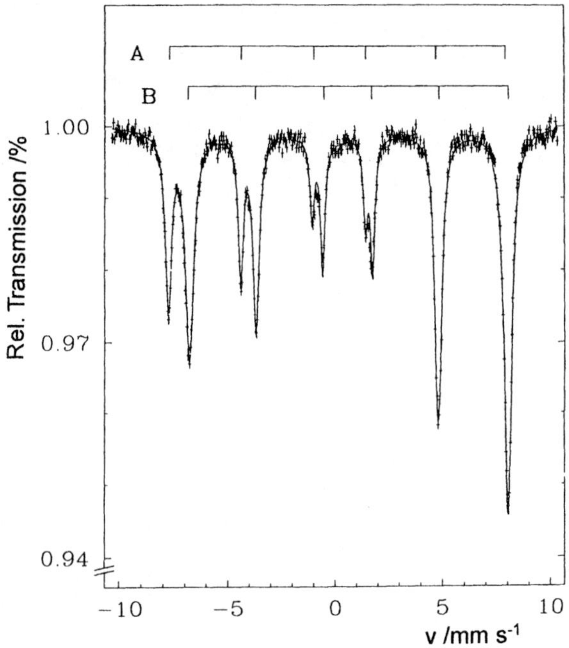

Figure 19. Room temperature Mössbauer spectrum of magnetite; subspectra of lattice site A and B are indicated by the stick diagram.

The Mössbauer parameters (δ, ΔE_Q and H_{int}) for the sextet stemming from the site B ions are intermediate between those of oxidic Fe(II) and Fe(III) compounds; concerning the iron ions on lattice site B, we are dealing with a mixed-valence system of Class III in the Robin–Day classification (cf. Section 7.1). The resonance lines of this spectrum are about 50% broader than those of site A. This was first interpreted in terms of a relaxation process due to the fast electron hopping between Fe(II) and Fe(III) ions (with relaxation times similar to the Mössbauer time window) on the octahedral sites. But later it was shown[41,42] that this line broadening is not due to a relaxation effect, but rather produced by the existence of several slightly different magnetic fields at B-sites (inhomogeneous broadening).

At 120 K, magnetite shows the so-called Verwey transition. The rate of the electron hopping between the Fe(II) and Fe(III) ions on the lattice B sites slows down such that it becomes slower than the reciprocal of the Mössbauer time window. This causes a significant change in the magnetically split subspectra.[43,44]

7.4.2 ***Magnetic Phase Transition in $RbFeF_3$*** A nice example for a magnetic phase transition as a function of temperature in an inorganic compound is shown in Figure 20. $RbFeF_3$ has an ideal cubic perovskite structure at ambient temperature with Fe(II) ions octahedrally coordinated by fluorine. The perfect cubic symmetry ascertains the absence of an electric field gradient, and a single resonance line is seen in the paramagnetic state. Below 103 K antiferromagnetic ordering of the ferrous

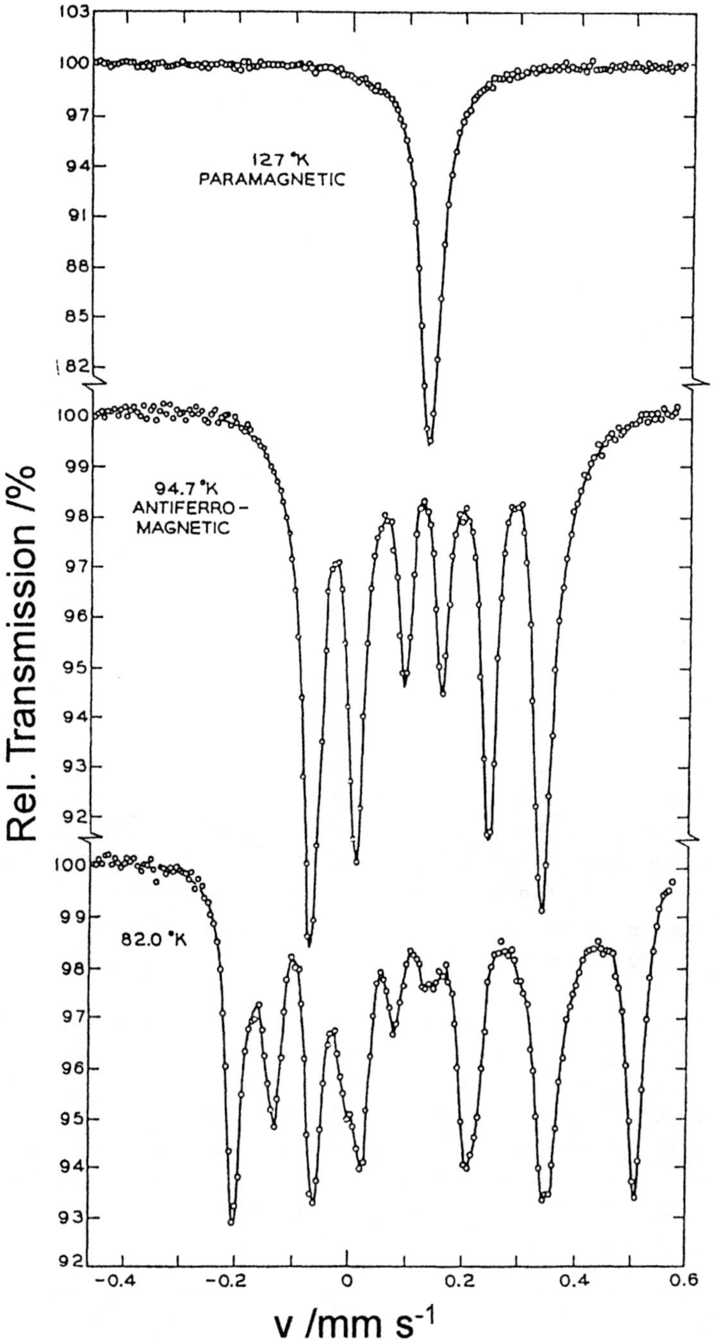

Figure 20. Mössbauer spectra of $RbFeF_3$ in the paramagnetic, antiferromagnetic, and ferrimagnetic region. Adapted from Wertheim et al.[45]

ion spins occurs and a six-line magnetic spectrum, slightly perturbed by quadrupolar interaction, is observed. Below 87 K a further transition to a ferrimagnetic state with two inequivalent ferrous ion sites takes place. The two sites possess different local magnetic fields; the resulting two six-line spectra superimpose and generate a complex spectrum of the ferrimagnetic state.[45]

7.4.3 ***Magnetic Ordering in Nickel Spinel Compounds*** Making use of the $I = 5/2 \leftrightarrow I = 3/2$ nuclear transition in ^{61}Ni, which is suitable for Mössbauer measurements, one has found that the local magnetic field in nickel-containing spinel compounds splits the $I = 5/2$ excited state into six, and the $I = 3/2$ ground state into four sublevels. Altogether 12 transitions between the sublevels of the $I = 5/2$ excited state and the $I = 3/2$ ground state are allowed, which appear in four groups in the Mössbauer spectrum with three lines in each. It has been found that the local magnetic field at ^{61}Ni in tetrahedral sites as in $NiCr_2O_4$ is five times larger (ca. 45 T) than that in octahedral sites as in $NiFe_2O_4$ (ca. 9 T).[46] It should be pointed out that performing ^{61}Ni Mössbauer measurements is considerably more difficult than measurements with ^{57}Fe, ^{119}Sn, ^{151}Eu, etc. The reasons are (i) the short lifetime of the parent nuclide ^{61}Co ($t_{1/2} = 99$ min), (ii) the necessary access to a nearby electron accelerator for the production of the short-lived ^{61}Co via the nuclear reaction ^{62}Ni(γ,p)^{61}Co, and (iii) the necessity of performing low temperature measurements because of the transition energy of 67 keV.

7.5 Corrosion Products

^{57}Fe Mössbauer spectroscopy has proven to be an extremely powerful tool for the analysis of iron-containing corrosion products on steel and technical alloys. Depending on the corrosion conditions a variety of iron oxides and oxihydroxides may be formed. Mössbauer spectroscopy enables the differentiation between the various phases on the basis of their different magnetic behavior. Figure 21 illustrates this with a series of six corrosion products. Fe_3O_4 (magnetite), shows the superposition of two six-line spectra as described in Section 7.4. α-Fe_2O_3 (hematite), γ-Fe_2O_3 (maghemite), and α-FeOOH (goethite) differ in the size of the internal magnetic field yielding different spacings between corresponding resonance lines. A problem seems to arise with β- and γ-FeOOH, because they have nearly identical Mössbauer parameters at room temperature and cannot be distinguished by their Mössbauer spectra. However, lowering the temperature solves the problem: β-FeOOH appears to be magnetically ordered at 80 K, whereas γ-FeOOH remains in the paramagnetic state.

Such Mössbauer studies may be carried out in a nondestructive manner and are even possible with highly dispersed particles which are no longer amenable to X-ray diffraction measurements.

7.6 Catalysis

The successful use of Mössbauer spectroscopy for catalyst studies is demonstrated in an important technical example (catalyst system). In Figure 22 a series of spectra

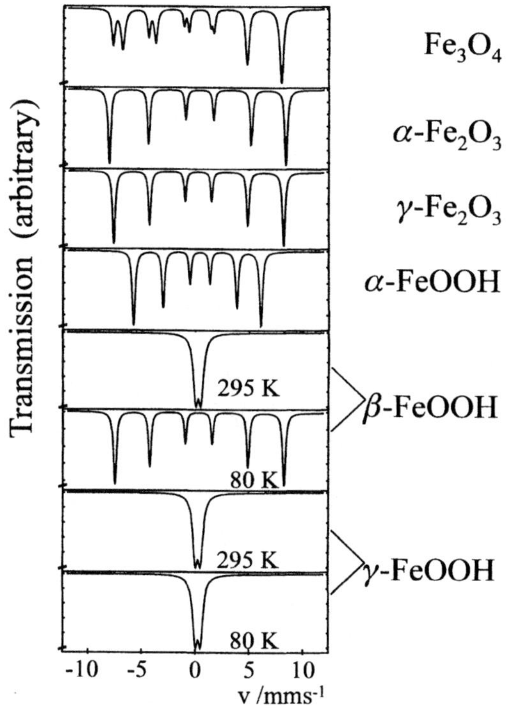

Figure 21. Mössbauer spectra of the most important corrosion products.

of the Fe/TiO_2 catalyst system[47] is shown. The top spectrum represents the freshly prepared catalyst; it is obvious that iron is present in the oxidation state 3+, possibly in the form of finely dispersed oxide or oxyhydroxide on TiO_2. In reducing atmosphere (H_2 at 675 K) this iron phase is essentially converted into metallic iron (six line pattern). Some residue of Fe(II) and Fe(III) contribute to the spectrum as doublets. Under Fischer–Tropsch synthesis (FTS) conditions at 575 K in CO and H_2, metallic iron is transformed into the Hägg carbide χ-Fe_5C_2. The unreduced iron is now present as Fe(II) which is oxidized to ferric iron when the catalyst is exposed to air at ambient temperature. The carbide phase is left unchanged.

7.7 Surface and Thin Layer Studies

Thin films of oxidic and metallic iron can be prepared by the Langmuir–Blodgett (LB) technique.[48,49] Using Conversion Electron Mössbauer spectroscopy (CEMS) it is possible to investigate physical and chemical properties of such ultrathin films. A surface sensitivity of less than a single monolayer up to about 20 nm is attainable. In Figure 23 the CEM spectra of two different multilayers of metallic iron produced by the LB method are compared with the CEM spectrum of a vapor-deposited thin iron film. It is obvious that with increasing film thickness the ratio of paramagnetic to ferromagnetic metallic iron decreases.[50]

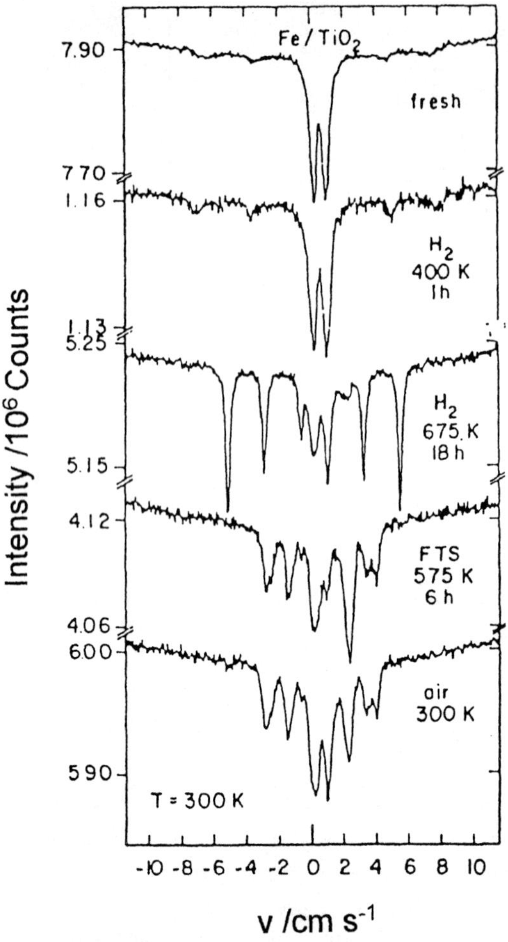

Figure 22. ^{57}Fe Mössbauer spectra at room temperature (from top to bottom) of the catalyst system Fe/TiO_2 after different treatments (cf. text). Reproduced from Applied Catalysis.[47]

In another example[51] ultrathin Fe/Ni films formed by this technique were investigated with regard to the magnetic behavior of metallic iron as a function of incorporated Ni. Figure 24 shows some CEM spectra of samples with different Ni concentrations. In the pure Fe film, besides some oxidic phases which originate from re-oxidation of the surface in contact with air, only one magnetically ordered metallic phase with parameters close to α-Fe appears. In the mixed films, however, an additional single line is detected, which can be attributed to the γ-phase of Fe. The relative amount of this γ-phase increases with increasing Ni concentration. Obviously, a martensitic phase transition of the Fe phase from bcc (α-phase) to fcc (γ-phase) takes place in these thin films, a phenomenon which is known from Fe/Ni bulk alloys.

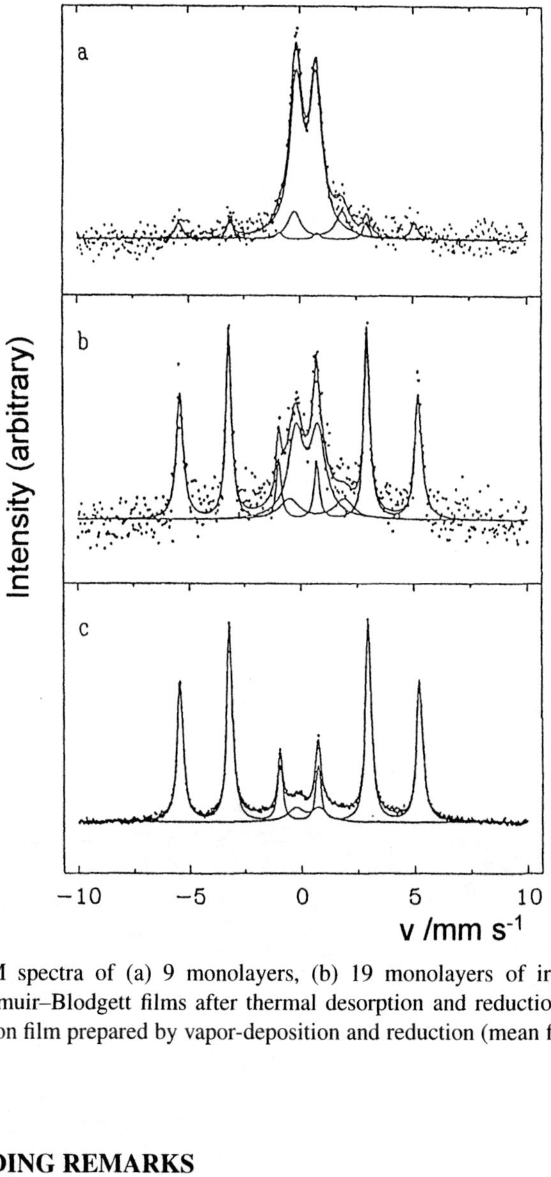

Figure 23. CEM spectra of (a) 9 monolayers, (b) 19 monolayers of iron formed from Fe-stearate Langmuir–Blodgett films after thermal desorption and reduction in H_2, and (c) a sample of an iron film prepared by vapor-deposition and reduction (mean film thickness ca. 3 nm).

8 CONCLUDING REMARKS

Among the large variety of analytical techniques that are in use nowadays to characterize different kinds of materials (inorganic and metal organic compounds, alloys, metals, etc.) Mössbauer spectroscopy has manifested itsef as a powerful tool for studying in a nondestructive manner the electronic and molecular structural properties, magnetism, phase changes, solid state reactions, etc. The basis for study are three kinds of hyperfine interactions (electric monopole, electric quadrupole, and magnetic dipole interactions) between the nuclei and electric and magnetic fields

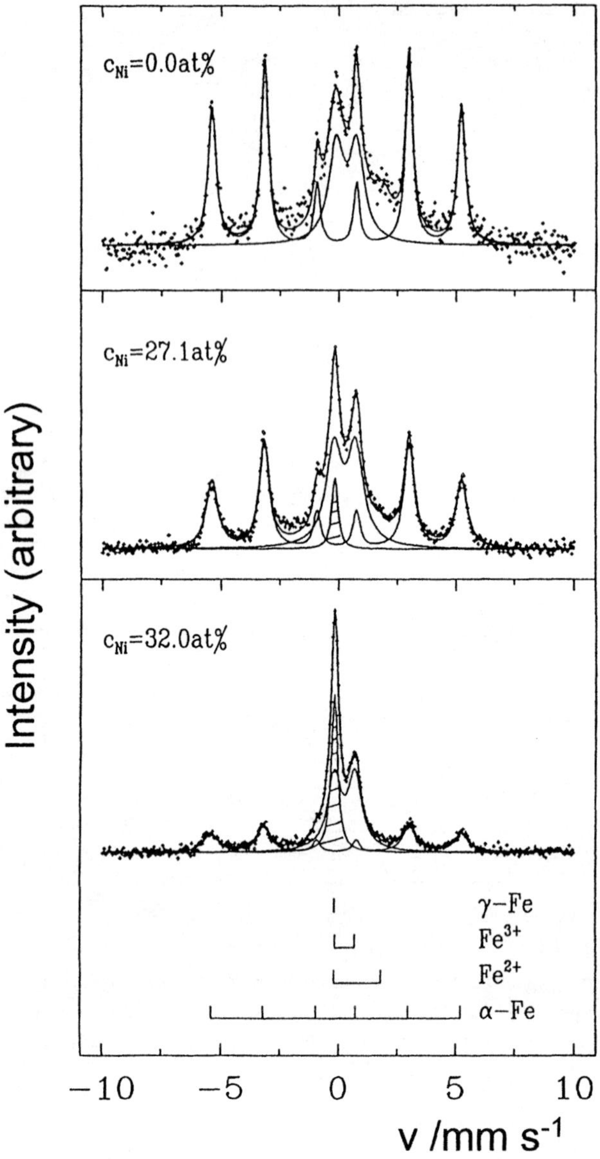

Figure 24. CEM spectra of thin films of Fe/Ni formed by the Langmuir–Blodgett technique.

generated by the electron structure of the material. The technique is highly sensitive to energy changes on the order of 10^{-8} eV (10^{-4} cm^{-1}). Applications of this technique range from purely fundamental research, such as studies of valence state and bond properties in connection with sophisticated MO calculations; to applications such as the fingerprint technique for phase analysis in industrial research. We wish

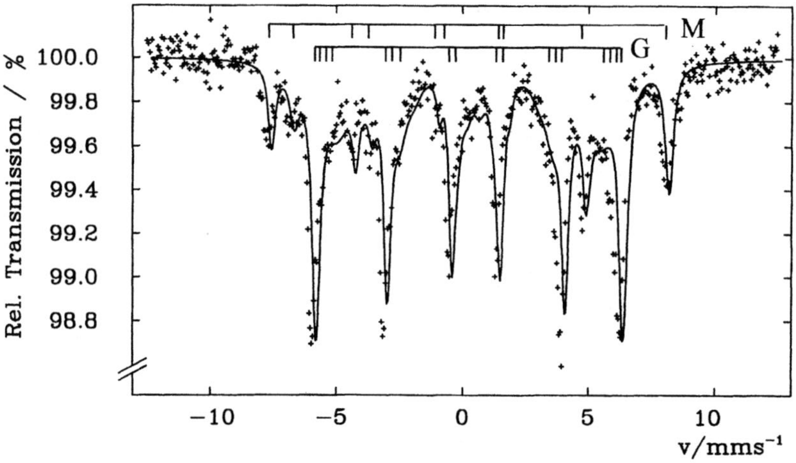

Figure 25. ^{57}Fe Mössbauer spectrum of a one dollar bill; the stick bars indicate the spectral components: M = magnetite (Fe_3O_4), G = goethite (α-FeOOH).

to conclude with an amusing application, i.e. the ^{57}Fe Mössbauer spectrum of a one dollar bill (cf. Figure 25), which yields the composition of 20 wt-% of magnetite and 80 wt-% of goethite as iron containing color pigments.

APPENDIX

Table A.1 Compilation of Room Temperature Mössbauer Parameters of Iron Compounds Selected from Listings Given in Ref. 8 (Isomer Shifts Are Given Relative to α−Fe)

Compound	δ /mm s^{-1}	ΔE_Q /mm s^{-1}	H /kG
High spin iron(II) halides			
FeF_2	1.37	2.79	—
$KFeF_3$	1.33	0.0	—
$CsFeF_3$	1.36	1.55	—
	1.36	0.47	—
$FeCl_2 \cdot H_2O$	1.13	2.03	—
$FeCl_2 \cdot 2H_2O$	1.03	2.50	—
$FeCl_2 \cdot 4H_2O$	1.22	2.98	—
$FeBr_2 \cdot 2H_2O$	1.14	2.49	—
$FeBr_2 \cdot 4H_2O$	1.22	2.83	—
High spin iron(II) cations bonding to oxygen			
$FeSO_4$	1.26	3.22	—
$[Fe(H_2O)_6]SO_4 \cdot H_2O$	1.24	1.72	—
$FeCO_3$	1.24	1.79	—

Table A.1 (*Continued*)

Compound	δ /mm s^{-1}	ΔE_Q /mm s^{-1}	H /kG
$Fe_3(PO_4)_2 \cdot 8H_2O$	1.21	2.50	—
	1.23	2.98	—
High spin iron(II) complexes with pyridine			
$Fe(py)_4Cl_2$	1.04	3.14	—
$Fe(py)_4Br_2$	1.07	2.52	—
$Fe(py)_4I_2$	1.16	0.65	—
High spin iron(II) complexes with 1,10-phenanthroline			
$Fe(phen)_2Cl_2$	1.06	3.28	—
$Fe(phen)_2Br_2$	1.07	3.26	—
$Fe(phen)_2I_2$	1.05	2.80	—
$Fe(phen)_2(NCS)_2$	1.04	3.09	—
$Fe(phen)_2(NCSe)_2$	1.03	2.52	—
Octahedral halogen complexes of high spin iron(III)			
FeF_3	0.48	0.0	—
$FeF_3 \cdot 3H_2O$	0.45	0.48	—
$FeCl_3$	0.44	0.0	—
$FeCl_3 \cdot 6H_2O$	0.45	0.97	—
$FeBr_3$ (78K)	0.55	0.0	—
$[Co(NH_3)_6]FeCl_6$ (78K)	0.54	0.0	—
Tetrahedral halogen complexes of high spin iron(III)			
Me_4NFeCl_4 (77K)	0.30	0.0	479
Et_4NFeBr_4 (77K)	0.36	0.0	420
$Et_4NFe(NCO)_4$ (77K)	0.34	0.86	394
Iron (II) and iron(III) low spin complexes			
$K_4[Fe(CN)_6] \cdot 3H_2O$	−0.05	0.0	—
$Ag_4[Fe(CN)_6]$	−0.12	0.0	—
$K_3[Fe(CN)_6]$	−0.124	0.280	—
$Ag_3[Fe(CN)_6]$	−0.147	0.767	—
$Na_2[Fe(CN)_5NO] \cdot 2H_2O$	−0.258	1.705	—
$Na_2[Fe(CN)_5NO_2]$	+0.005	0.855	—
$Na_3[Fe(CN)_5NO_2]$	−0.09	1.78	—
cis − $[Fe(CN)_2(CNMe)_4]$	−0.09	0.24	—
trans − $[Fe(CN)_2(CNMe)_4]$	−0.09	0.44	—
cis − $[Fe(CN)_2(CNEt)_4]$	−0.04	0.29	—
trans − $[Fe(CN)_2(CNEt)_4]$	−0.04	0.59	—
$[Fe(phen)_3](ClO_4)_2$	0.31	0.15	—
$[Fe(5-Me-phen)_3](ClO_4)_2$ (77K)	0.334	0.19	—
$[Fe(5-Cl-phen)_3](ClO_4)_2$ (77K)	0.336	0.24	—
$[Fe(bipy)_3](ClO_4)_2$ (77K)	0.325	0.39	—
$[Fe(phen)_3](ClO_4)_3 \cdot H_2O$	0.05	1.09	—
$[Fe(bipy)_3](ClO_4)_3 \cdot 3H_2O$	0.03	1.69	—
Iron-organic complexes			
$Fe(CO)_5$	−0.09	2.57	—
$Fe_2(CO)_9$	0.16	0.42	—
$Fe_3(CO)_{12}$	0.11	1.13	—
	0.05	0.13	—
$Na_2[Fe(CO)_4]$	−0.18	0.00	—

Table A.1 (*Continued*)

Compound	δ /mm s^{-1}	ΔE_Q /mm s^{-1}	H /kG
$[Et_4N][Fe(CO)_4H]$	−0.17	1.36	—
$(Cp)_2Fe$	0.53	2.37	—
$(Cp)_2FeBr$	0.43	≈ 0.2	—
$[CpFe(C_5H_4)]_2$	0.52	2.36	—
$CpFe(CO)_2I$	0.23	1.83	—
$CpFe(CO)_2Cl$	0.24	1.88	—
$CpFe(CO)_2Br$	0.2	1.87	—
Iron oxides and hydroxides			
$\alpha - Fe_2O_3$ (Hematite)	0.38	0.12	515
$\gamma - Fe_2O_3$ (Maghemite)	0.27	0.0	488
	0.41	0.0	499
$Fe_{0.93}O$ (Wüstite)	0.91	0.46	—
	0.86	0.78	—
Fe_3O_4 (Magnetite)	0.27	0.0	491
	0.63	0.0	454
$\alpha - FeOOH$ (Goethite)	0.35	−0.15	384
$\beta - FeOOH$ (Akaganeite)	0.35	0.55	—
$\gamma - FeOOH$ (Lepidocrocite)	0.35	0.55	—
Iron(IV) coordination complexes			
$[Fe(Me_2dtc)_3]BF_4$	0.19	−2.06	—
$[Fe(Et_2dtc)_3]BF_4$	0.22	−2.23	—
$[Fe(pyrdtc)_3]BF_4$	0.22	−2.30	—
Iron(IV) porphyrine and phthalocyanine complexes			
$[(TTP)Fe]_2N$ (131K)	0.18	1.08	—
$[(TTP)Fe]_2CCl_2$ (131K)	0.10	2.28	—
$[(Pc)Fe]_2N$ (77K)	0.06	1.76	—
$[(Pc)Fe]_2C$ (77K)	−0.03	2.69	—
Iron(VI) compounds			
K_2FeO_4	−0.89	0.0	140
Rb_2FeO_4 (77K)	−0.81	0.0	149
$SrFeO_4$	−0.86	0.0	142
Hemoglobin derivatives			
HbCO (194K)	0.18	0.36	—
Hb reduced (194K)	0.90	2.40	—
HbO_2 (195K)	0.20	1.89	—
HiH_2O (195K)	0.20	2.00	—
HiCN (195K)	0.17	1.39	—

Abbreviations	
phen	1,10-phenanthroline
bipy	2,2′-bipyridine
dtc	dithiocarbamate
pyrdtc	pyrrolidyldithiocarbamate
TTP	$\alpha, \beta, \gamma, \delta$-tetraphenylporphyrine
Pc	phthalocyanine
Hb	Fe(II) hemoglobin
Hi	Fe(III) hemoglobin

REFERENCES

1. Mössbauer, R.; Z. Phys., **1958**, *151*, 124
2. Mössbauer, R.; Naturwissenschaften, **1959**, *45*, 538
3. Mössbauer, R.; Z. Naturforsch., **1959**, *14 a*, 211
4. Frauenfelder, H.; "The Mössbauer Effect". **1962**, Benjamin, New York
5. Wertheim, G.K.; "Mössbauer Effect: Principles and Applications". **1964**, Academic Press, New York
6. Wegener, H.; "Der Mössbauer-Effekt und seine Anwendung in Physik und Chemie." **1965**, Bibliographisches Institut, Mannheim
7. Goldanskii, V.I.; Herber, R. (eds.); "Chemical Applications of Mössbauer Spectroscopy." **1968**, Academic Press, New York
8. Greenwood, N.N.; Gibb, T.C.; "Mössbauer Spectroscopy". **1971**, Chapman and Hall Ltd., London
9. Bancroft, G.M.; "Mössbauer Spectroscopy, An Introduction for Inorganic Chemists and Geochemists". **1973**, McGraw-Hill, New York
10. Gonser, U. (ed.); "Mössbauer Spectroscopy", in Topics in Applied Physics, Vol. 5. **1975**, Springer, Berlin, Heidelberg, New York
11. Gibb, T.C.; "Principles of Mössbauer Spectroscopy". **1976**, Chapman and Hall, London
12. Gütlich, P.; Link, R.; Trautwein, A.X.; "Mössbauer Spectroscopy and Transition Metal Chemistry", in Inorganic Chemistry Concepts, Vol. 3. **1978**, Springer, Berlin, Heidelberg, New York
13. Stevens, J.G.; Mössbauer Effect Reference and Data Journal, Mössbauer Data Center, University of North Carolina, Asheville, USA
14. Kuhn, W.; Phil. Mag., **1929**, *8*, 625
15. Barb, D.; Grundlagen und Anwendungen der Mössbauerspektroskopie. **1980**, Akademie-Verlag, Berlin
16. Kankeleit, E.; Rev. Sci. Instr., **1964**, *35*, 194
17. Shimony, U.; Nucl. Instr. Methods, **1965**, *37*, 348
18. Nagy, S.; Levay, B.; Vertes, A.; Acta. Chim. Acad. Sci. Hung., **1975**, *85*, 273
19. Watson, R.E.; Phys. Rev. **1960**, *118*, 1036
20. Shenoy, G.K.; Wagner, F.E.; "Mössbauer Isomer Shifts." **1978**, North Holland Publishing Company, Amsterdam, New York, Oxford
21. Abragam, A.; The Principles of Nuclear Magnetism. **1961**, Oxford University Press, Oxford
22. Marshal; Phys. Rev., **1958**, *110*, 1280
23. Wertheim, G.K.; Remeika, J.P.; Phys. Letters, **1964**, *10*, 14
24. Goldanskii, V.I.; Gorodinskii, G.M.; Karyagin, S.V.; Korytko, L.A.; Krizhanskii, L.M.; Makarov, E.F.; Suzdalev, I.P.; Khrapov, V.V.; Doklady Akad. Nauk S.S.S.R., **1962**, *147*, 127
25. Sternheimer, R.M.; Foley, H.M.; Phys. Rev., **1953**, *92*, 1460; Sternheimer, R.M.; Phys. Rev., **1954**, *96*, 951; **1957**, *105*, 158
26. Ono, K.; Ito, A.; J. Phys. Soc. Japan, **1964**, *19*, 899
27. Collins, R.L.; J. Chem. Phys. **1965**, *42*, 1072

28. Viegers, T.; Thesis **1976**, University of Nijmegen, Netherlands
29. Gedikli, A.; Winkler, H.; Gerdau, E.; Z. Phys. **1974**, *267*, 61
30. Müller, W.; "MOSFUN—Mössbauer Spectra Fitting Program for Universal Theories, Description and User's Guide", **1980**, Institut für Anorganische Chemie und Analytische Chemie, Johannes Gutenberg-Universität, Mainz, Germany
31. Brand, R.; "NORMOS", Laboratorium für Angewandte Physik, Gerhard-Mercator-Universität GH, Duisburg, Germany
32. Fluck, E.; Kerler, W.; Neuwirth, W.; Angew. Chem. **1963**, *75*, 461
33. Maer Jr., K.; Beasley, M.L.; Collins, R.L.; Milligan, W.O.; J. Am. Chem. Soc. **1968**, *90*, 3201
34. Bonnette, A.K.; Allen, J.F.; Inorg. Chem. **1971**, *10*, 1613
35. Robin, M.B.; Day, P.; "Mixed Valency Chemistry—A Survey and Classification." Adv. Inorg. Chem. Radiochem. **1967**, *10*, 247, Academic Press, New York and London
36. Dong, T.-Y.; Hendrickson, D.N.; Iwai, K.; Cohn, M.J.; Geib, S.J.; Rheingold, A.L.; Sano, H.; Motoyama, I.; Nakashima, S.; J. Am. Chem. Soc. **1985**, *107*, 7996
37. Huynh, B.H.; Moura, J.J.G.; Moura, I.; Kent, T.A.; LeGall, J.; Xavier, A.V.; Münck, E.; J. Biol. Chem. **1980**, *225*, 3242
38. Gütlich, P.; Hauser, A.; Spiering, H.; Angew. Chem. Int. Ed. Engl. **1994**, *33*, 2024
39. Decurtins, S.; Gütlich, P.; Köhler, C.P.; Spiering, H.; Hauser, A.; Chem. Phys. Lett., **1984**, *105*, 1
40. Vervey, E.J.W.; Nature **1939**, *144*, 327
41. van Diepen, A.M.; Phys. Lett. **1976**, *57A*, 354
42. van Diepen, A.M.; Physica **1977**, *86-88B*, 955
43. Hargrove, R.S.; Kundig, W.; Solid State Comm. **1970**, *8*, 330
44. Rubinstein, M.; Forester, D.W.; Solid State Comm. **9**, *9*, 1675
45. Wertheim, G.K.; Guggenheim, H.J.; Williams, H.J.; Buchanan, D.N.E.; Phys. Rev., **1967**, *158*, 446
46. Gütlich, P.; Hasselbach, K.M.; Rummel, H.; Spiering, H.; J. Chem. Phys., **1984**, *81*, 1396
47. van der Kraan, A.M.; Nonnekens, R.C.H.; Stoop, F.; Niemantsverdriet, J.W., Appl. Catal. **1986**, *27*, 285
48. Langmuir, I.; Trans. Faraday Soc., **1920**, *15*, 62
49. Blodgett, K.B.; J. Am. Chem. Soc., **1935**, *57*, 1007
50. Faldum, T.; Meisel, W.; Gütlich, P.; Appl. Phys. **1996**, *A 62*, 317
51. Faldum, T.; Meisel, W.; Gütlich, P.; Fresenius J. Anal. Chem., **1995**, *353*, 723

4 Polarized Absorption Spectroscopy

MICHAEL A. HITCHMAN
Chemistry Department
University of Tasmania
Hobart, Tasmania 7001, Australia
E-mail: Michael.Hitchman@utas.edu.au

MARK J. RILEY
Chemistry Department
University of Queensland
St Lucia, Queensland 4072, Australia
E-mail: riley@chemistry.uq.edu.au

Inorganic Electronic Structure and Spectroscopy, Volume I: Methodology.
Edited by E. I. Solomon and A. B. P. Lever.
ISBN 0-471-15406-7.

1 THEORY UNDERLYING POLARIZED ABSORPTION SPECTROSCOPY

Measurement of the electronic absorption spectrum provides probably the most convenient way of investigating the electronic energy levels and bonding characteristics of a compound. This is particularly true for transition metal complexes, both because these generally absorb in the readily accessible near infrared/visible/near ultraviolet part of the spectrum, and because a well-developed theoretical framework exists to interpret the spectra. The general features of the electronic spectra of complexes are covered in several texts.[1–3] The purpose of the present chapter, which builds on several previous review articles,[4–6] is to highlight the particular advantages gained by measuring the absorption spectra of single crystals using polarized light. Aspects of importance are illustrated using a range of examples. Polarized electronic absorption spectra often provide information of interest outside the field of spectroscopy—for instance, the assignment of the bands in low symmetry complexes may yield detailed metal–ligand bonding parameters, while vibrational fine structure can be used to show how the geometry of a complex is related to its electronic structure.

In what follows we will loosely associate an "intensity" with a spectral transition, though strictly speaking this intensity refers to the absorption band associated with the electronic transition. The intensity of light absorption **I** associated with an electronic transition from state ψ_1 to ψ_2 is proportional to the square of the transition moment integral, **P**:

$$\mathbf{I} \propto \left|\langle\psi_2|\mathbf{R}|\psi_1\rangle\right|^2 = \mathbf{P}^2 \tag{1}$$

where **R** is an operator representing the interaction of the electromagnetic radiation with the molecule. The probability that a photon will be absorbed is given by the oscillator strength **f**, which is proportional to the band area observed in the electronic

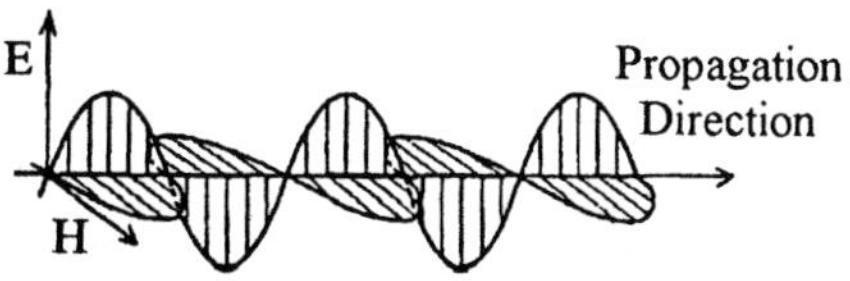

Figure 1. Waveform of polarized light; **E** and **H** represent the electric and magnetic vectors, respectively.

absorption spectrum corresponding to the integral:

$$\mathbf{f} = 4.32 \times 10^{-9} \int_{\overline{\nu_1}}^{\overline{\nu_2}} \varepsilon(\overline{\nu})\, d\overline{\nu}. \quad (2)$$

Here, $\varepsilon(\overline{\nu})$ is the molar extinction coefficient given in $\text{mol}^{-1}\text{Lcm}^{-1}$ at wavenumber $\overline{\nu}$ (in cm^{-1}), assuming that the band stretches from $\overline{\nu}_1$ to $\overline{\nu}_2$. Intensities are usually quantified in terms of ε measured at the band maximum.

Electromagnetic radiation consists of oscillating electric (**E**) and magnetic (**H**) field vectors, perpendicular both to each other and to the direction of propagation of the light (Figure 1). A transition will involve contributions from the electric and magnetic dipole components of the radiation. In principle, electric quadrupole and even higher-order terms should be considered, but these usually have a negligible effect. For a fully allowed electronic transition, the electric dipole intensity is $\sim 10^4$ times the magnetic dipole intensity, while that from the electric quadrupole interaction is even lower. Except for very weak transitions, the electric dipole mechanism therefore dominates. As an approximate guide, magnetic dipole intensity is unlikely to contribute significantly to bands with $\varepsilon > \sim 2\ \text{mol}^{-1}\text{L cm}^{-1}$. In certain circumstances, the two contributions to absorption may be distinguished experimentally, and this procedure will be discussed in Section 2.3. The electronic spectra involving d-electrons largely involve the electric dipole mechanism, and the present discussion will be limited to this situation except where mentioned specifically. The electric dipole operator may be represented by a unit vector **r** and the transition moment integral then becomes $\mathbf{P} = \langle \psi_2 | \mathbf{r} | \psi_1 \rangle$ with **P** varying in magnitude according to the direction of **r** within the molecule. This aspect forms the basis of polarized absorption spectroscopy which involves measurement of the anisotropy of the light absorption with respect to the orientation of the electric vector of light. The use of polarized light often greatly enhances the resolution of the bands in a spectrum and provides a powerful way of assigning excited electronic states.

1.1 Band intensities

1.1.1 ***Parity-Allowed Transitions*** The intensity of light absorption by an electric dipole mechanism is given by selection rules based upon two approximations. The first is that the spin and orbital parts of the electronic wavefunctions may be considered independently. The transition moment integral may then be written:

$$\mathbf{P} = \langle \Phi_2 | \mathbf{r} | \Phi_1 \rangle \langle \theta_{2,s} | \theta_{1,s} \rangle \tag{3}$$

where Φ_1 and $\theta_{1,s}$ represent the orbital and spin components of the ground state wavefunction, respectively. This approximation, the basis of the Russell–Saunders coupling scheme, works well for metal ions of the first transition series, but becomes progressively poorer for second and third row transition metal ions.[1] For a spin-allowed transition, the spin quantum numbers of the ground and excited states $\theta_{1,s}$ and $\theta_{2,s}$ must be equal. The second rule, often called the Laporte rule, concerns the parity of the ground and excited state wavefunctions Φ_1 and Φ_2. The transition moment will be nonzero only if the direct product of the irreducible representations of Φ_1 and Φ_2 with the irreducible representation of the electric dipole operator, **r**, gives or contains the totally symmetric representation:

$$\Gamma(\Phi_1) \times \Gamma(\mathbf{r}) \times \Gamma(\Phi_2) \subset A_1(\mathrm{g}). \tag{4}$$

Here the $\times$ and $\subset$ symbols represent the direct product and "contains" respectively. A transition will be much more intense in z than in x or y polarization if this condition is satisfied for $r = z$, but not for $r = x$ and y. The application of the Laporte rule to interpret the absorption of light polarized along the different axes of a molecule provides probably the best method of assigning the peaks in its absorption spectrum.

For a centrosymmetric complex, the representations are all either gerade (g) or ungerade (u) depending upon whether they are symmetric or antisymmetric to inversion across the symmetry center. The electric dipole vector is a u-function, so the direct product can only yield the totally symmetric representation if the ground and excited state wavefunctions are of different parity. For first row transition metal ions, spin-forbidden transitions are typically ~100 times less intense than spin-allowed transitions, unless they are close enough to 'borrow' intensity from a nearby spin-allowed transition. Parity-allowed transitions exhibit a wide range of intensities, depending on the precise nature of the ground and excited state wavefunctions. However, the simple group theoretical rule given by equation (4) may generally be used to interpret the band polarizations satisfactorily.

To interpret the intensities of spin-forbidden transitions, and for those metal ions for which Russell–Saunders coupling is a poor approximation, it is no longer appropriate to separate the wavefunctions ψ_1 and ψ_2 into their orbital and spin components. Instead, the coupling between the spin and orbital momenta of the electrons to give states characterized by a set of quantum numbers **J** must be considered, and the wavefunctions described in terms of the spin-orbit irreducible representations of the double point group. The symmetry properties of the resulting wavefunctions $\psi_{\mathbf{J}1}$, $\psi_{\mathbf{J}2}$ may then be used to determine whether a transition is allowed in a particular polarization via the relationship:

$$\Gamma(\psi_{\mathbf{J}1}) \times \Gamma(\mathbf{r}) \times \Gamma(\psi_{\mathbf{J}2}) \subset A_1(\mathrm{g}) \tag{5}$$

with the analysis now being carried out using the appropriate double group of the molecule.

1.1.2 ***Vibronically-Allowed Transitions*** Parity-forbidden transitions are ~100 times or more weak than fully allowed transitions and typically have absorption coefficients of $\varepsilon \approx 2$–$50\ \mathrm{mol^{-1}Lcm^{-1}}$ when they are spin-allowed. The band intensities can only be interpreted by including the effects of vibrations, and it is necessary to consider not the time-averaged geometry of the complex, but rather the way in which the geometry changes with time. Since, as stated by the Franck–Condon principle,[7] electronic transitions occur much more rapidly than the timescale of vibrational motion, from a semiclassical viewpoint at any instant of time the electronic spectrum will be the superposition of the spectra of complexes having the complete range of geometries defined by the normal vibrational modes. The selection rules which describe the polarization properties of each transition for a centrosymmetric complex, for example, may be derived simply by applying the group-theoretical criteria considered in the preceding section to the particular geometry produced by each ungerade vibration.[8] This semiclassical approach provides a simple understanding of the reason why parity-forbidden transitions have nonzero intensity and can also be useful in correlating the intensities with those of asymmetrically-distorted complexes.[9] However, historically the treatment of this phenomenon has proceeded along rather different lines,[10] as from a quantum mechanical viewpoint the transitions occur between the vibronic wavefunctions of stationary states.

The vibronic intensity mechanism considers that the j ungerade normal modes of a complex, Q_{uj}, act to mix excited ungerade electronic states, Φ_{ui}, into each g-wavefunction. For the formally forbidden transition $\Phi_{g1} \rightarrow \Phi_{g2}$, if i u-states are available, the g-wavefunctions are transformed into two new wavefunctions wavefunctions Φ_1 and Φ_2:

$$\Phi_1 = \Phi_{g1} + \sum_i \sum_j c_{1ij}\Phi_{ui}; \qquad \Phi_2 = \Phi_{g2} + \sum_i \sum_j c_{2ij}\Phi_{ui}. \tag{6}$$

The mixing coefficients are obtained by first expanding the electronic Hamiltonian, H_e, as a Taylor series about the equilibrium geometry of the ground state, Q_o, of the molecule. The coefficients are then obtained by first order perturbation theory in what is often called Herzberg–Teller coupling.

$$c_{1ij} = Q_{uj}\left\langle \Phi_{g1} \left| \left(\frac{\partial H_e}{\partial Q_{uj}}\right)_{Q_o} \right| \Phi_{ui} \right\rangle \Big/ (E_{ui} - E_{g1}), \tag{7a}$$

$$c_{2ij} = Q_{uj}\left\langle \Phi_{g2} \left| \left(\frac{\partial H_e}{\partial Q_{uj}}\right)_{Q_o} \right| \Phi_{ui} \right\rangle \Big/ (E_{ui} - E_{g2}). \tag{7b}$$

Substitution of the expanded wavefunctions (6) into the transition moment integral

(3) yields the nonzero terms:

$$\mathbf{P} \propto \sum_i \sum_j c_{1ij} \langle \Phi_{g2} | \mathbf{r} | \Phi_{ui} \rangle + c_{2ij} \langle \Phi_{ui} | \mathbf{r} | \Phi_{g1} \rangle. \tag{8}$$

Considering the first term in equation (8), the integral $\langle \Phi_{g2} | \mathbf{r} | \Phi_{ui} \rangle$ is nonzero if $\Gamma(\Phi_{g2}) \times \Gamma(\mathbf{r}) \times \Gamma(\Phi_{ui}) \subset A_{1g}$. As the operator $\partial H_e / \partial Q$ transforms in the same way as Q, equation (7a) shows that the coefficient c_{1ij} is nonzero if $\Gamma(\Phi_{g1}) \times \Gamma(Q_{uj}) \times \Gamma(\Phi_{ui}) \subset A_{1g}$. For the transition to be vibronically allowed, *both* conditions must be satisfied, and combining the expressions yields the condition:

$$\Gamma(\Phi_{g2}) \times \Gamma(\mathbf{r}) \times \Gamma(\Phi_{g1}) \times \Gamma(Q_{uj}) \subset A_{1g}. \tag{9}$$

The same requirement is obtained for the second term in equation (8). The representation of the normal mode (or modes) that induce intensity into the electronic transition is then given by the expression:

$$\Gamma(Q_{uj}) = \Gamma(\Phi_{g2}) \times \Gamma(\mathbf{r}) \times \Gamma(\Phi_{g1}) \tag{10}$$

If the molecule has at least one vibration of this symmetry type, the transition is vibronically allowed in **r** polarization, if not, it is vibronically forbidden. As the character tables and direct products of the normal modes are tabulated for all molecular point groups,[11] it is a relatively straightforward procedure to derive simple vibronic selection rules for any molecule. The measurement of spectra using polarized light thus provides a powerful means of assigning electronic transitions (see Section 3.1.).

While the above selection rule usually has the strongest influence on the polarizations of vibronically-allowed transitions, other factors are also important. The coefficients c_{2ij} in equation (8) will generally be significantly larger than c_{1ij} because excited states of u symmetry are closer in energy to the Φ_{g2} excited state than to the Φ_{g1} ground state; that is, $(E_{ui} - E_{g1}) > (E_{ui} - E_{g2})$ in equations (7a) and (7b). This means that the vibronic intensity mechanism essentially operates by the borrowing of intensity from the parity-allowed transitions $\langle \Phi_{ui} | \mathbf{r} | \Phi_{g1} \rangle$ via the vibrational coupling of the excited d-states with states of u symmetry: $\langle \Phi_{g2} | \partial H_e / \partial Q_u | \Phi_{ui} \rangle$. This can influence band polarizations in two ways. If one parity-allowed transition happens to be very close in energy to the excited d-states, or is particularly intense, then this will dominate the vibronic intensity. Every vibronically-allowed *d–d* transition will then be particularly intense in the polarization of this parity-allowed transition. Secondly, if one u vibration is especially active (or inactive) in mixing intensity into the *d–d* transitions, then those transitions allowed by this vibration will be particularly intense (or weak). The spectrum discussed in Section 2.5 illustrates these aspects.

In the previous discussion, the Q_{uj} in equations (7) are odd-parity ungerade normal coordinates which causes vibronic coupling *between* electronic states. We now consider the actual form of the potential energy surface of an excited state. If the Taylor series expansion of H_e about the equilibrium geometry of the ground state, Q_o is evaluated *within* a single nondegenerate electronic state and the Born–Oppenheimer approximation is assumed, this then defines the potential energy surface of that state:

$$V(Q) = V(Q_o) + \sum_i \left\langle \Phi \left| \left(\frac{\partial H_e}{\partial Q_i} \right)_{Q_o} \right| \Phi \right\rangle Q_i + \cdots . \tag{11}$$

The potential energy surface, $V(Q)$, transforms as the totally symmetric irreducible representation in the point group of the molecule. The coefficients of the linear terms in equation (11) are integrals which will vanish unless the direct product of the irreducible representations, $\Gamma(\Phi) \times \Gamma(\partial H_e/\partial Q_i) \times \Gamma(\Phi)$, is totally symmetric. Since we are assuming that Φ is a non-degenerate electronic state, $\Gamma(\Phi) \times \Gamma(\Phi)$ is totally symmetric. The cofficients $\langle \Phi|(\partial H_e/\partial Q_i)|\Phi\rangle$ will therefore be zero unless $(\partial H_e/\partial Q_i)$ transforms as totally symmetric. The term $\partial H_e/\partial Q_i$ transforms as the same irreducible representation as Q_i. This is the reason why the displacement of the potential energy surface of a non-degenerate excited state with respect to the ground state equilibrium geometry can *only* occur when Q_i is totally symmetric.

1.2 Electronic and Vibronic Origins

For a parity-allowed transition at low temperature there will be one or more peaks due to the pure electronic transition between the lowest vibrational levels of the ground and excited state, possibly split by spin-orbit coupling or low-symmetry components of the ligand field. For parity-forbidden transitions, assuming that the vibrations are harmonic, the selection rules discussed in Section 1.1.2 require that the vibrational quantum number v of each intensity-inducing mode change by ± 1 in conjunction with the electronic transition. At low temperature, when the molecule is in the $v = 0$ state, the electronic transition will be to the $v = 1$ level of the excited electronic state. At higher temperatures, some molecules will be in the $v = 1$ level of the ground state, and for these, vibronically allowed transitions will occur to the $v = 0$ and $v = 2$ levels of the excited electronic state. The peaks due to these transitions are called origins because they often form the basis of extended progressions due to the absorption of several vibrational quanta. If the electronic origin of the excited state is at $\mathbf{E_0}$, the vibronic origin will be at $\mathbf{E_0} + \overline{\nu}'$ at low temperature, with a new so-called hot vibronic origin at $\mathbf{E_0} - \overline{\nu}$ appearing at higher temperatures. Here, $\overline{\nu}$ and $\overline{\nu}'$ are the energies of the intensity-inducing mode in the ground and excited electronic state, respectively. As these are usually similar, the hot bands are approximate mirror-images of the cold bands observed at low temperature, reflected to lower energy of the $\mathbf{E_0}$ electronic origin. The electronic origin itself is not observed by the electric dipole mechanism for a parity-forbidden transition. The energies of the origins for parity- and vibronically-allowed transitions are summarized in equations (12a, b, c).

parity-allowed origin:	$\mathbf{E_0}$	(12a)
vibronically-allowed origin:	$\mathbf{E_0} + \overline{\nu}'$	(12b)
vibronically-allowed 'hot' origin:	$\mathbf{E_0} - \overline{\nu}$	(12c)

Each origin is usually followed by at least one well-developed progression in a totally symmetric mode (see following section) so that the structure in *d–d* bands is often complicated and not resolved. These problems are much less pronounced in the *f–f* spectra of lanthanides and actinides (these are parity-forbidden for centrosymmetric complexes for the same reason that *d–d* transitions are forbidden).

1.3 Band Progressions

In addition to parity and vibronically allowed origins, progressions involving transitions to higher vibrational levels of the excited electronic state may occur. The intensity of these vibronic transitions can be calculated within the Born–Oppenheimer approximation that wavefunctions may be separated into electronic and vibrational components[12] (this aspect is further covered in Chapter 7). Transitions between vibronic states may then be written as:

$$\Psi_{i,v} = \Phi_i(q, Q)\chi_{i,v}(Q) \tag{13}$$

where Φ and χ are the electronic and vibrational parts of the vibronic function Ψ. The coordinates q and Q correspond to electronic and nuclear coordinates respectively and Φ depends only parametrically on Q. For simplicity we will use the Herzberg–Teller electronic states given in equation (6) and consider the case of a single vibrational coordinate, Q, and a single odd-parity excited state, Φ_u, involved in vibronic coupling with the excited state:

$$\begin{aligned} \Psi_{1,v} &= \Phi_1\chi_{1,v}, \\ \Psi_{2,v} &= (\Phi_2 + c_2\Phi_u)\chi_{2,v}. \end{aligned} \tag{14}$$

Further, we will consider transitions from only the lowest vibrational level of the ground electronic state, corresponding to the low temperature limit. Substituting the vibronic states into equation (1), with $\mathbf{R} = \mathbf{r}$ for an electric dipole mechanism, results in the following expression for the overall band intensity:

$$\begin{aligned} I(\Psi_{1,0} \rightarrow \Psi_{2,v}) &\propto \left|\langle\Psi_{2,v}|\mathbf{r}|\Psi_{1,0}\rangle\right|^2 \\ &= \left|\mathbf{P}_{1,2}\langle\chi_{2,v}|\chi_{1,0}\rangle + \mathbf{P}_{1,u}\langle\chi_{2,v}|c_2|\chi_{1,0}\rangle\right|^2. \end{aligned} \tag{15}$$

Here, the electronic transition moments are given by $\mathbf{P}_{1,2} = \langle\Phi_2|\mathbf{r}|\Phi_1\rangle$ and $\mathbf{P}_{1,u} = \langle\Phi_u|\mathbf{r}|\Phi_1\rangle$. The mixing coefficient c_2 in equation (14) is assumed small so in *allowed* transitions the first term of equation (15) dominates. In this case the intensity of each member of a vibrational progression, $v = 0, 1, 2, \ldots$, is determined by the overlap integrals $\langle\chi_{2,v}|\chi_{1,0}\rangle$ between the vibrational wavefunctions of a totally symmetric vibration.

In *forbidden* transitions, $\mathbf{P}_{1,2} \approx 0$ and the second term of equation (15) dominates. The vibronic intensity is provided by:

$$\langle\chi_{2,v}|c_2|\chi_{1,0}\rangle = \langle\Phi_u|\partial H_e/\partial Q_u|\Phi_1\rangle/(E_u - E_g) \times \langle\chi_{2,v}|Q_u|\chi_{1,0}\rangle. \tag{16}$$

Here the χ are now the vibrational wavefunctions of an odd-parity mode. In this case, at low temperature, the progression is built upon the vibronic origins (equation (12b)), while at higher temperatures progressions built upon hot origins (equation (12c)) will appear. We now consider the vibronic fine structure due to the $\langle \chi_{2,v} | \chi_{1,0} \rangle$ and $\langle \chi_{2,v} | Q_u | \chi_{1,0} \rangle$ type overlap integrals in more detail.

If the ground and excited state potential energy surfaces are harmonic and undisplaced, and the vibrational frequency is equal in the ground and excited states, then the functions $\chi_{1,v}$ and $\chi_{2,v}$ are orthogonal. This results in all intensity being solely in the electronic origin: $\Psi_{1,0} \rightarrow \Psi_{2,0}$. However, the potential energy surfaces are usually displaced in the totally symmetric metal–ligand stretching coordinate and this is what gives rise to the vibrational progressions observed in electronic spectra. The intensities of the individual members of the vibrational progression are represented by the Franck–Condon factors:

$$|\langle \chi_v | \chi_0 \rangle|^2 = D^{2v}/(2^v \times v!) \exp(-D^2/2) \tag{17}$$

which are given in terms of the displacement of the normal coordinate, D, expressed in dimensionless units. These dimensionless units are related to those given in length by the conversion:

$$Q = \sqrt{h/4\pi^2 c\overline{\nu}\mu} \times D = 580.648/\sqrt{\overline{\nu}\mu}\ \text{pm} \tag{18}$$

where $\overline{\nu}$ and μ are in cm^{-1} and amu respectively. Strictly speaking, the mass, μ, is the eigenvector of the normal coordinate, obtained from a normal coordinate analysis, multiplied by the inverse of the G matrix. However for molecules such as $ML_6(O_h)$, $ML_4(T_d)$, $ML_4(D_{4h})$, $ML_3(D_{3h})$, $ML_2(D_{\infty h})$ where L is a simple ligand like a halide or water, there is only one totally symmetric vibration and the metal lies at the center of mass of the molecule. Then μ is just the mass of one ligand. Often a Franck–Condon analysis of a more complex molecule will assume an "effective" skeletal vibration with an "effective" μ.

Equation (17) has been derived (and periodically rediscovered) by many people[13] and assumes that the ground and excited state potential energy surfaces are both harmonic and have the same vibrational frequency. The generalization to unequal frequencies results in a more complicated expression and is usually given as a recursion relation.[14] The case of undisplaced potential energy surfaces with unequal frequencies is given by:

$$|\langle \chi_v | \chi_0 \rangle|^2 = 2\frac{\sqrt{\overline{\nu}\,\overline{\nu}'}}{\overline{\nu} + \overline{\nu}'} \frac{v!}{2^v\left(\frac{v}{2}!\right)^2} \left(\frac{\overline{\nu} - \overline{\nu}'}{\overline{\nu} + \overline{\nu}'}\right)^v \cdots \tag{19}$$

where $\overline{\nu}$ and $\overline{\nu}'$ are the harmonic vibrational frequencies in the ground and excited electronic state respectively and v is restricted to even values. When the potential energy surfaces are anharmonic the Franck–Condon overlaps must be calculated numerically by variational methods.[14]

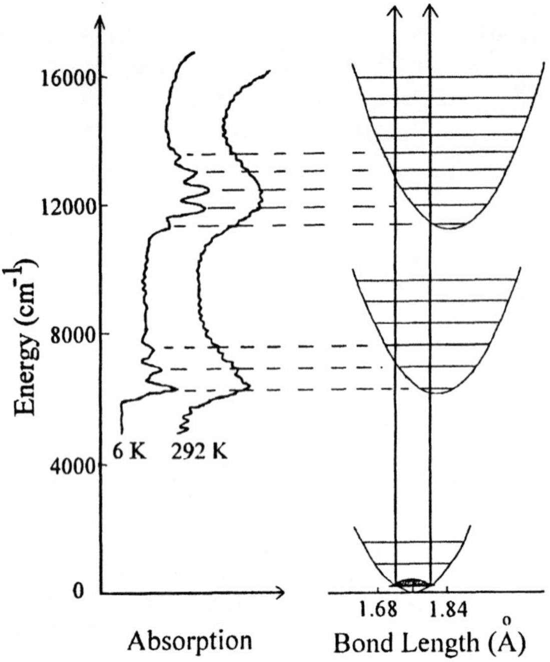

Figure 2. Correlation of the *d–d* bands observed for K_3NiO_2 with the energy levels of the totally symmetric stretching mode in the ground state and excited state potential surfaces. The spectrum is the sum of the *xy* and *z* spectra of the linear complex and the spectra are rather noisy because of the extremely small size of the crystal (adapted from reference 17 with permission).

The Franck–Condon factors are rather sensitive to the displacement in the ground and excited state potential surfaces when this is small. Analysis of the band shape, or better still the relative intensities of the members of the progression, if these are resolved, therefore provides a powerful way of estimating the change in geometry accompanying an electronic transition.[15,16] Because electronic transitions take place much more rapidly than the motion of the nuclei (the Franck–Condon principle), the vibrational overlaps are related to the vertical projection of the ground state vibrational wavefunction onto the wavefunctions of the various vibrational levels of the excited state. This means that energy levels observed in an absorption spectrum correspond to the geometry of the electronic ground state, rather than the equilibrium geometry of the excited state. The dependence of the intensities of band members upon the displacement of the excited state potential surface is illustrated in Figure 2. On the left side of the figure, the absorption spectrum of a simple complex, the linear NiO_2^{3-} ion in K_3NiO_2,[17] is plotted vertically, and the members of each band are correlated with the vibrational levels of the excited state. The potential surface of the first excited state is displaced by only a modest amount with respect to the ground state, and here the vibronic origin is the most intense member of the progression, which dies away fairly rapidly. The second excited state involves a much larger displacement, and here the intensities of the members of the progression rise and then fall as the vibrational quantum number in the excited state increases. At room temperature, the vibrational structure is not resolved, but the band shapes of the two transitions still reflect the different displacements in the excited state potential surfaces. Whereas the lower energy band is asymmetric, rising steeply and falling gradually with increasing energy, the higher energy band is broader but quite symmetric, approximating a Gaussian lineshape.

As discussed in Section 1.1, to the first order a displacement of a non-degenerate excited state potential energy surface with respect to the ground state equilibrium geometry *can only occur along the totally symmetric coordinates.* Most bands therefore involve progressions in α_1 modes (or Jahn–Teller active modes for orbitally degenerate excited states). However, occasionally the equilibrium geometry of the excited state may differ from that of the ground state by a displacement in some other mode. The absorption spectra of several planar Cu^{2+}, Ni^{2+}, and Pt^{2+} complexes provide evidence for a distortion towards a tetrahedral geometry in some excited states.[18] The basic group theoretical arguments based upon an electric dipole mechanism, expressed in equations (4) and (9), still apply. This means that to a first order approximation, light will only be absorbed when the vibrational quantum number changes by an even number of units, since the product of a representation with itself gives the totally symmetric representation. From the lowest vibrational level of the ground electronic state, a progression involving every second excited state vibrational level is therefore expected. For excited state potential energy surfaces with double minima such as that shown in Figure 21, the excited state energy levels below the barrier height occur as closely spaced pairs, so the progression will resemble that of a normal displaced single well potential. However, anharmonicity effects will cause the pattern to be more complicated near the barrier between the wells, which is the region vertically above the ground state geometry. The vibrational structure observed at low temperature on the highest energy transition of the planar $CuCl_4^{2-}$ complex in $(Naem)_2CuCl_4$ (Section 3.1.) has been interpreted in this fashion.[19] A basic progression in the α_{1g} mode is observed, plus an additional structure due to a progression in the out-of-plane mode of β_{2u} symmetry which carries the complex towards a distorted tetrahedral geometry.

An alternative interpretation of band progressions in terms of an electric octapole mechanism has been proposed by Hollebone, and it has been used to interpret the vibrational fine structure observed in the polarized spectra of several complexes.[20]

1.4 Influence of Temperature on Absorption Spectra

1.4.1 ***Intensities*** Though the extinction coefficients of parity-allowed transitions usually increase substantially on cooling from 300 K to 10 K, this is accompanied by a concomitant decrease in halfwidth, so that the oscillator strength remains essentially unaltered. The intensity of the parity-forbidden bands of centrosymmetric molecules, on the other hand, is derived via coupling with u-vibrations. This will be greater when upper levels $v = 1, 2, \ldots$ of the modes inducing intensity are thermally populated as the intensity is proportional to v (i.e. the harmonic oscillator matrix element $\langle \chi_v | Q | \chi_{v+1} \rangle = \sqrt{(v+1)/2}$). Semiclassically, the intensity increase can be thought of as due to the greater amplitude of the coupling vibration.

In addition the intensity increases due to the participation of the "hot" bands $v = 1 \rightarrow v = 0, 2$; $v = 2 \rightarrow v = 1, 3$; etc. When the vibronic origins can be resolved over a temperature range, these two effects may be studied independently, but this is normally only the case for the very sharp f–f transitions of lanthanides and actinides.[21] For the spin-allowed transitions in d–d spectra, the progressions built on vibronic origins almost always overlap to give featureless bands at higher temperatures. If a Boltzmann distribution over the vibrational levels of the ground state is assumed, and these are considered to be harmonic oscillators, then the total band intensity $f(T)$ at temperature T due to coupling with mode of energy $\overline{\nu}_u$ is given by[6]:

$$f(T) = f(0)\left[1 + \exp\left(-\overline{\nu}_u/\mathrm{k}T\right)\right]/\left[1 - \exp\left(-\overline{\nu}_u/\mathrm{k}T\right)\right] \qquad (20)$$

Here, $f(0)$ is the intensity at 0 K, and k is the Boltzmann constant. It has been shown[22] that this relationship should be obeyed irrespective of the geometry of the molecule in the excited state. Equation (20) may be rearranged to give the so-called $\coth^{-1}$ rule:

$$\coth^{-1}\left(f(T)/f(0)\right) = \overline{\nu}_u/(2\mathrm{k}T) \qquad (21)$$

which implies that a plot of $\coth^{-1}(f(T)/f(0))$ against the inverse of temperature should give a straight line passing through the origin, with a slope proportional to the energy of the intensity-inducing vibration.

The temperature dependence of the intensity is largest for low energy active vibrations, and the variation of the intensity with temperature for modes of several energies is shown in Figure 3. When, as is often the case, the intensity is derived by coupling with more than one vibration, the band area is the sum of the intensities produced by each mode. The observed temperature variation will then correspond to an effective vibration with an energy intermediate between those of the active modes, though, as may be seen from Figure 3, lower energy modes have more influence than those of higher energy. If the intensity of a band decreases on cooling, this usually indicates that it is due to a vibronically-allowed transition, and hence that the molecule has an inversion center (though occasionally other effects may influence intensity, as discussed in Sections 3.6 and 3.7). Unless all the intensity-inducing vibrations are of very high energy, the converse is also true; if the intensity of a transition is temperature independent this usually indicates that it is a parity-allowed transition.

1.4.2 ***Bandwidths and Energies*** The bands of parity-allowed transitions normally consist simply of progressions in modes of α_1 symmetry built upon the electronic origin. Because higher levels of these vibrations have larger amplitudes, their thermal population leads to the development of a broader band, as illustrated in Figure 4. That is, the Franck–Condon factors are spread over a larger range of excited

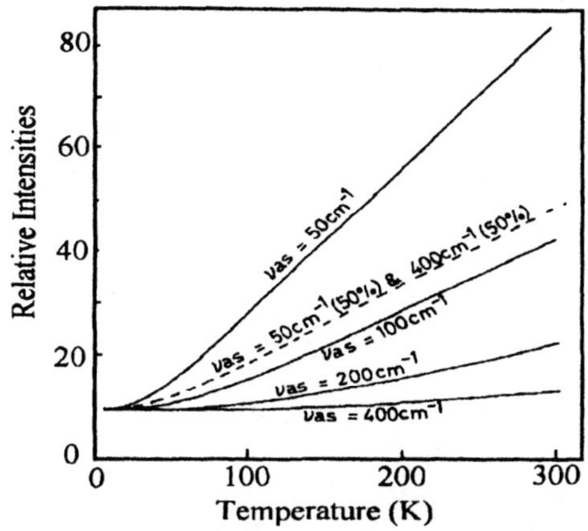

Figure 3. Effect of the energy of the intensity-inducing vibration upon the temperature dependence of a vibronically-allowed band.

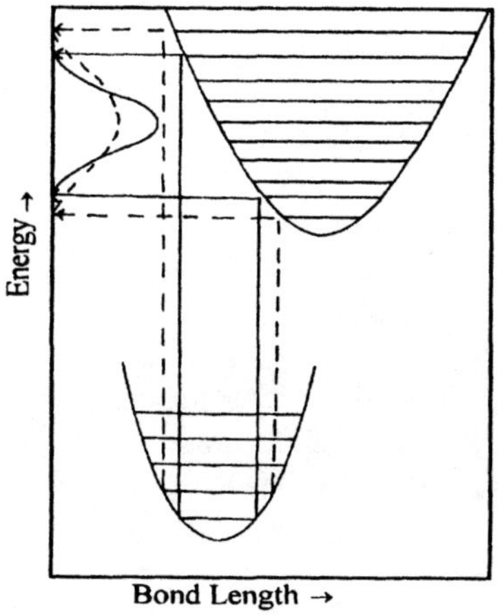

Figure 4. Mechanism by which the thermal population of higher vibrational levels of the α_1 metal–ligand stretching vibration influences the bandwidth of an electronic transition.

state vibrational levels. The halfwidth thus increases as a function of temperature, and if the progression is in just one totally symmetric mode of energy $\overline{\nu_{\alpha 1}}$, the halfwidth at temperature T, $\delta(T)$, is related to that at 0 K, $\delta(0)$ by the expression:

$$\delta(t)^2 = \delta(0)^2 \coth(\overline{\nu_{\alpha 1}}/2\mathrm{k}T) \tag{22}$$

The bandwidths of vibronically-allowed transitions are also expected to broaden upon thermal population of upper levels of the α_{1g} vibrations causing the progressions. In this case, additional broadening will occur due to the development of hot bands built upon origins shifted to lower energy by $\sim 2\overline{\nu}_u$ with respect to the parent vibronic origin, where $\overline{\nu}_u$ is here the average energy of the intensity-inducing mode in the ground and excited state. Because of these hot bands, parity-forbidden transitions commonly give rise to peaks which are asymmetric at high temperature, with a larger halfwidth on the low-energy side.

The development of hot bands also causes the energy of the band maximum of vibronically-allowed transitions to shift to low energy as the temperature is increased. If the shift from the electronic origin at temperature T is denoted by $E(T)$, this will change from $E(T = 0) \approx \overline{\nu}_u$ at absolute zero, to $E(T \rightarrow \infty) \approx 0$ in the high temperature limit. The shift can also be expressed as a hyperbolic function[14]:

$$E(T) \approx \overline{\nu}_u \tanh(\overline{\nu}_u/2kT) \tag{23}$$

As thermal energy at room temperature corresponds to ~200 cm^{-1}, this shift will only occur for modes of energy ~200–300 cm^{-1} or less, and energy changes of this magnitude are indeed often observed for vibronically-allowed *d–d* transitions. A further factor which may cause bands to move to lower energy on warming, for both parity-allowed and parity-forbidden transitions, is a lengthening of the metal–ligand bonds. For discrete complexes, such a lengthening is only expected if the metal–ligand totally symmetric stretching vibration is anharmonic,[23] when the mean metal–ligand distance increases slightly as a function of the vibrational quantum number. For compounds in which the metal ions are linked by ligands, the metal–ligand distance will increase as the lattice expands on warming, and this will cause a modest red shift of the band maxima. Exceptions are the unusual transitions in which the excited state geometry has shorter bond lengths than the ground state, in which case a blue shift on warming will be observed.

Occasionally much larger red shifts may occur in *d–d* spectra due to other mechanisms. An example is given in Section 3.5, where the planar $CuCl_4^{2-}$ complex has a double minimum excited state potential along a non-totally symmetric coordinate which leads to a very large red shift of the band maxima.

2 EXPERIMENTAL ASPECTS

2.1 Advantages of Using Single Crystals and Polarized Light

The advantage of using a single crystal, rather than a solution or powder, is that the molecule is held in a fixed position in space. The anisotropy of the absorption spectrum may therefore be investigated by positioning the electric vector of linearly polarized light in different directions within the molecule. This not only improves the resolution of overlapping peaks, but often greatly assists in their assignment. Moreover, if the crystal thickness and density are known, the intensities of the absorption bands may be determined quantitatively, which is difficult for powdered samples. The magnitude of the intensity is of value in deciding the mechanism by which the transition gains intensity. Spectral resolution can be significantly improved by cooling the single crystal, something which is difficult with a solution. At low temperature, the number of transitions between the vibrational levels of the ground and excited state is reduced, so that vibrational structure superimposed upon the electronic transition can sometimes be resolved.

2.2 Measurement of Spectra

To obtain high quality spectra, it is important that crystals are of good optical quality, and this is best tested by seeing that they extinguish sharply under crossed polarizers.[24] The characteristics of the spectrophotometer place significant constraints upon the physical dimensions of crystals which can be studied. It is desirable that the crystal face which is perpendicular to the light beam be as large as possible, though modern commercial instruments allow quite small crystals to be studied if the region of interest coincides with the maximum sensitivity of the spectrophotometer, generally ~500 nm. Sensitivity is lower in the near infrared (NIR) region, 800–2500 nm, so that here larger crystals may be required. However, many Fourier-transform infrared spectrometers have a range extending to 12,000 cm^{-1} or even higher, and provide an alternative option for the measurement of electronic spectra in the NIR region. The optical density, or light absorbance A of a crystal will normally obey the Beer–Lambert law:

$$A = \varepsilon c d \tag{24}$$

where c is the molar concentration, d is the crystal thickness, and ε the molar extinction coefficient. For typical crystals, commercial spectrophotometers can normally handle samples with ε values between ~1 and ~100 $mol^{-1}L\ cm^{-1}$. The more intense bands observed for highly asymmetric complexes and charge transfer transitions can be studied by cleaving or grinding the crystals, or doping the complex of interest into a transparent host lattice. A major advantage of custom built equipment, incorporating very sensitive detectors and/or a microscope using reflecting optics, is that far smaller crystals may be studied, giving greater flexibility for selecting a sample with a suitable thickness and quality. A detailed description of a system specifically designed for high-resolution single crystal studies over a wide spectral range has been reported,[25] and the fine structure shown in Figures 7 and 18 exemplifies the results which can be obtained using equipment of this kind.

Absorption normally involves the interaction of the electric vector with a molecule. The direction of the electric vector may be controlled by placing a polarizer in the light beam, with the sample being placed on a suitable mask so that only light passing through the crystal reaches the detector. It is convenient to hold the crystal fixed in the center of the light beam and to rotate the polarizer. By convention the polarization direction is taken as the direction of the electric vector of polarized light.

The light absorption of the crystal must conform to the symmetry of its unit cell.[24] For all orientations of cubic crystals, and for uniaxial crystals when the light beam is parallel to the unique crystal axis, the absorption is isotropic with respect to the direction of the electric vector, so nothing is gained by measuring a spectrum using polarized light. For other orientations of uniaxial crystals, and all orientations of biaxial crystals, light becomes depolarized on passing through the crystal except

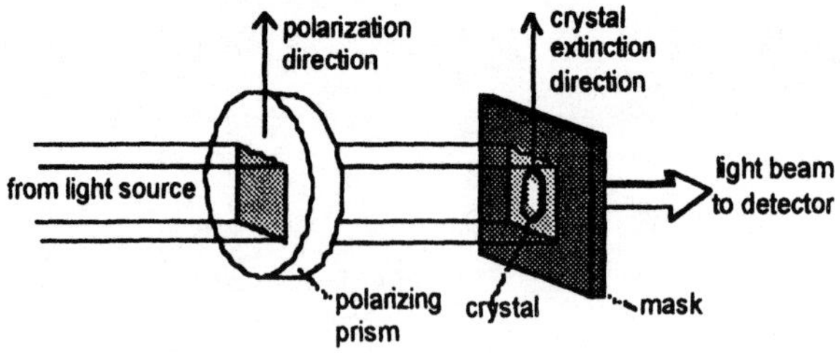

Figure 5. Experimental setup used to measure a polarized absorption spectrum of a crystal.

when the electric vector is parallel to a so-called "extinction direction."[24] The polarizer must therefore be aligned as illustrated schematically in Figure 5. Each crystal face has two orthogonal extinction directions, so that only two polarized absorption spectra may be measured for any crystal face. If the crystal face contains a symmetry axis, the extinction directions are constrained to lie along and perpendicular to this direction.[26] Otherwise, their orientation may vary with wavelength and, in principle, with temperature, and this should be borne in mind when working with the spectra of crystals belonging to low symmetry space groups. For most crystals, only a small fraction of the light beam is utilised. Because the spectrophotometer slit-width generally alters as a function of wavelength, this fraction changes as the spectrum is measured, producing a sloping "baseline," which must be compensated for by measuring the "spectrum" of the mask with the crystal removed.

Resolution can be significantly improved by cooling the crystal. Except for continuous lattices, where very low energy "phonons" may couple with the electronic transitions, metal–ligand vibrations usually have energies in the range 100–400 cm^{-1}, and a significant improvement can be achieved by cooling to ~100 K, with maximum resolution being reached by ~15 K. Flow-tubes using boiled-off liquid nitrogen or helium provide a relatively simple way of cooling the sample. Alternatively, commercial closed-cycle cryostats allow samples to be cooled to any specified temperature down to ~12 K. While very easy to use, the latter have the disadvantage that the sample is in a vacuum. This means that cooling is achieved largely by conduction, and it is therefore most important to check that the crystal is in good thermal contact with the sample block.

2.3 Discrimination Between Electric Dipole and Magnetic Dipole Mechanisms for Uniaxial Crystals

Generally, the electronic transitions of transition metal ions occur by an electric dipole mechanism, though the polarized spectra of some octahedral Ni, Co, and Cu complexes with marked ionic character do indicate the presence of significant magnetic dipole intensity for certain transitions.[3] For uniaxial crystals the dominant mechanism may be deduced by comparing the band intensity for three different orientations of the crystal with respect to the polarized light beam, as illustrated in Figure 6. When the propagation direction of the light is parallel to the unique crys-

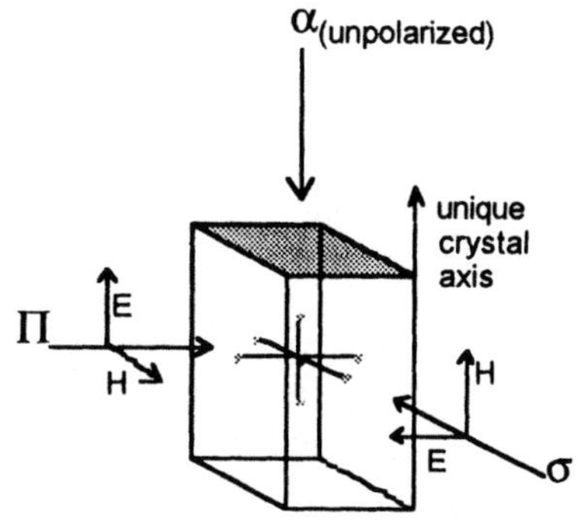

Figure 6. Orientation of the electric (**E**) and magnetic (**H**) vectors of polarized light for the α, π, and σ spectra of the CuF_6^{4-} complex formed when Cu^{2+} is doped into the tetragonal crystal K_2ZnF_4.

tal axis, c, both the electric (E) and magnetic (H) vectors are perpendicular to this axis. For this spectrum, designated α, it is unnecessary to use polarized light. When the propagation direction of the light is perpendicular to the unique axis, the electric vector can either be parallel or perpendicular to the c axis. When $E||c$ ($H \perp c$) this is termed π polarization, and conversely when $E \perp c$ ($H||c$) it is termed σ polarization. Transitions allowed by an electric dipole mechanism will therefore exhibit bands for which $\alpha = \sigma \neq \pi$, while magnetic dipole transitions will have $\alpha = \pi \neq \sigma$. The polarized spectrum[27] of the CuF_6^{4-} complex formed when a small percentage of Cu^{2+} substitutes for the Zn^{2+} ions in K_2ZnF_4 (Figure 7) illustrates the application of these rules. The sharp peaks between 7500 cm^{-1} and 9000 cm^{-1} are magnetic dipole transitions, while the underlying broad bands are predominantly electric dipole in character. As expected, the magnetic dipole peaks have a smaller area, and hence oscillator strength, than the electric dipole bands. Their relatively high extinction coefficients, and consequent prominence in the spectra, is due solely to the fact that they are relatively sharp, having half-widths of ~14 cm^{-1}.[27] The nature of the fine structure, and use of the polarizations to assign the spectrum of this example, is discussed in Section 3.2.

2.4 "Oriented Gas" Model of Crystal Spectra

For many compounds, the crystal spectrum is simply the sum of the spectra of the molecules comprising the lattice *ie* the system behaves like an "oriented gas". However, when molecules are relatively closely packed in the lattice, the change in dipole moment associated with the excitation of one molecule to form an "exciton" will affect the energy states of the neighbouring molecules, causing a so-called "Davidov" splitting of the excited states.[28,29] This splitting depends both on the closeness of the chromophores and on the intensity of the transitions. Behaviour of this kind has been reported for a number of compounds of low-spin d^8 metal ions which crystallize with the planar complexes forming stacks with quite short (~3.3 Å) metal–metal

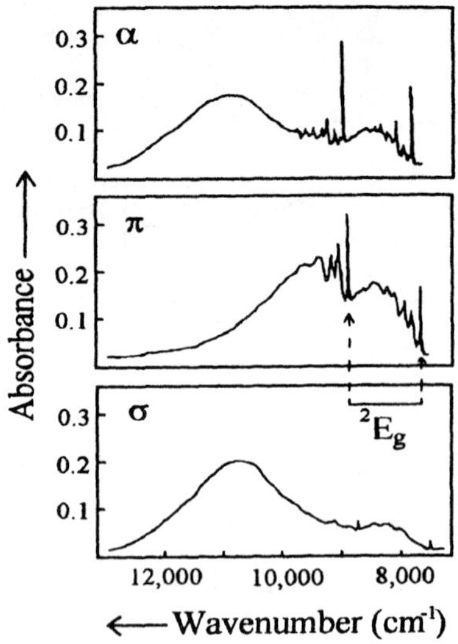

Figure 7. Crystal spectrum of ~2% Cu^{2+} doped into the tetragonal crystal K_2ZnF_4 (adapted from reference 27 with permission). See Figure 6 for the orientation of the electric and magnetic vectors of polarized light. The two sharp magnetic-dipole origins of the components of the ^2E_g state split by spin-orbit coupling are indicated by arrows.

contacts. The crystal spectra must then be interpreted not in terms of the spectra of individual molecules, but rather of the unit cell as a whole. Such interactions cause the crystal spectrum to differ significantly from that of a solution. A striking example is provided by salts of the $Pt(CN)_4^{2-}$ ion,[29] which are intensely coloured in the solid state. Here, Davidov splittings cause a very intense parity allowed transition to shift from the ultraviolet into the visible or near ultraviolet region of the spectrum. The magnitude of the shift correlates with the metal–metal separation, which depends on the nature of the counter cation. In contrast, solutions of the complex are colourless.

Coupling between the vibrational and electronic wavefunctions ("vibronic" coupling) often influences electronic spectra, and while this is usually dominated by the metal–ligand vibrations, coupling with lattice modes may also be important. When this is the case, analysis of the vibrations must be undertaken in terms of the unit cell, rather than an isolated molecule.[30] Magnetic interactions between the metal ions in a lattice can also significantly influence electronic spectra. For instance, relatively weak magnetic interactions have a significant effect on the spin-forbidden transitions in continuous lattices such as $KNiF_3$,[3] and on low energy charge transfer transitions in dimeric copper(II) complexes.[31] Polarized absorption spectroscopy provides a powerful method of investigating the metal–metal interactions in cluster complexes.[1] Note that complexes of this kind still fall within the framework of the "oriented gas" model, but with the unit of interest being the cluster rather than an isolated metal centre.

2.5 Derivation of Molecular Spectra from Crystal Spectra

Assuming that the oriented gas model is valid, and any lattice effects may be treated as perturbations to the basic molecular spectrum, the light absorption of a crystal is simply the sum of the absorbance of the molecules comprising the unit cell. The light absorption of a molecule is a tensor defined by three principal values and an orientation in space. For a molecule of very low symmetry, both the orientation and the principal values of the absorption ellipsoid need to be determined by a process analogous to that used for the g-tensor or magnetic susceptibility tensor.[32] However, this procedure is often impossible because of the limitation that measurements can only be made when light is polarized parallel to an extinction direction. While a method has been described for analysing the spectra of molecules in which just one direction of the absorption ellipsoid is defined by symmetry,[33] it is complicated, and polarized absorption spectroscopy is normally limited to relatively high symmetry molecules where the orientation of the absorption ellipsoid may be assumed.

For a particular orientation of the electric vector, the crystal absorbance A is related to the molecular absorption values, A_x, A_y, A_z by the expression:

$$A = x^2 A_x + y^2 A_y + z^2 A_z \qquad (25)$$

Here x^2, etc., are the squares of the normalised projection made on the molecular axis, averaged over the different orientations in the unit cell (i.e., for two molecular orientations, $x^2 = (x_1^2 + x_2^2)/2$). To obtain all three molecular absorbance values, the crystal spectrum must be measured for at least three orientations of the electric vector. As only two independent spectra can be measured using one crystal face, with the electric vector along each extinction direction, at least two different crystal faces must be studied to completely define the molecular absorption spectrum. It is sometimes sufficient simply to correlate the observed spectra with the molecular projections. However, except when the extinction directions happen to lie very close to the molecular axes, quantitative molecular absorbances must be obtained by substituting the values of A obtained from two crystal faces into equation (25), and solving the resulting four equations. The molecular spectra are generated by carrying out this procedure over a range of wavelengths. The thicknesses of the crystal faces must be taken into account (equation (24)), though this problem may be avoided by utilizing the *ratios* of the absorbances, rather than their absolute values.[34]

The spectrum of the planar complex Cu(benzac)$_2$, where benzac is the 1-phenylacetylacetonate anion,[34] illustrates the advantage of deriving the molecular spectrum quantitatively. Formally, this complex belongs to the point group C_{2h}, which would require the orientation of the absorption ellipsoid in the plane of the complex to be determined experimentally. However, if the asymmetric substitution of the ligand is neglected, the symmetry rises to D_{2h}, and the molecular axes may then be defined as illustrated in Figure 8. The compound crystallizes in a monoclinic unit cell, space group $P2_1/c$, containing two molecules related by a mirror plane. The polarized spectra measured at 8 K for the (001) and (011) crystal faces are shown in Figure 9, the molecular projections averaged over the two molecules in the unit cell being given in the figure caption. Note that for the (001) face the extinction directions are

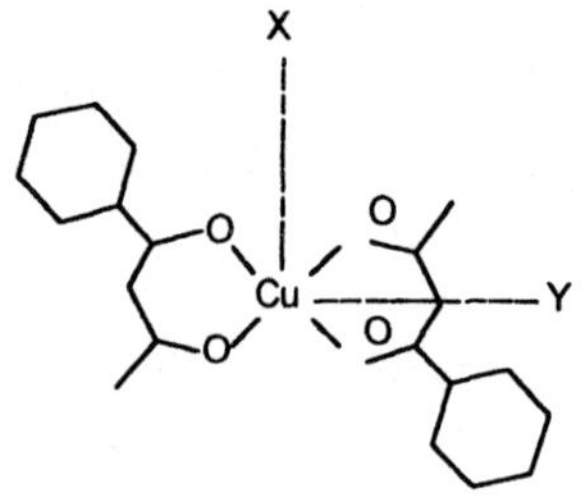

Figure 8. In-plane molecular axes of Cu(benzac)$_2$.

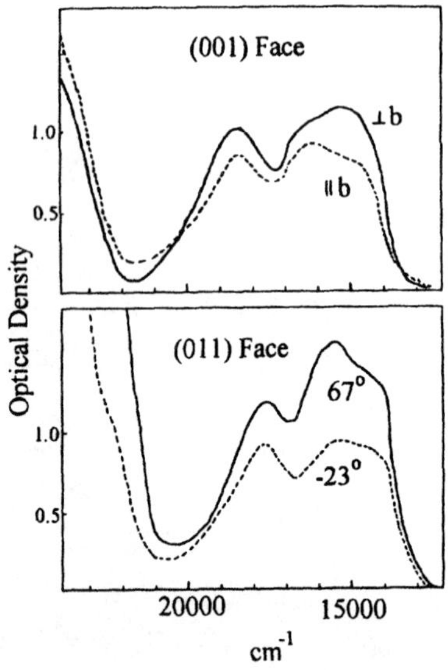

Figure 9. Crystal spectra of Cu(benzac)$_2$ at ~10 K (adapted from reference 34 with permission). The electric vector makes angles of $0.634x^2 + 0.192y^2 + 0.174z^2$ and $0.000x^2 + 0.488y^2 + 0.512z^2$ when $\|b$ and $\|a$, respectively, for the (001) face. For the (011) face the projections are $0.476x^2 + 0.454y^2 + 0.070z^2$ and $0.090x^2 + 0.258y^2 + 0.652z^2$, respectively, when the electric vector makes angles of 67° and −23° with the *a* axis.

required by symmetry to lie parallel and perpendicular to the unique *b* crystal axis, while for the (011) face no such symmetry restriction applies. It is apparent that while the crystal spectra show polarization effects, these cannot be readily correlated with particular molecular axes. Transforming the crystal spectra in the manner outlined above yields the molecular spectra shown in Figure 10. Here, the thickness of one crystal face was used to convert the intensities from optical density units to molar extinction coefficients.

The overall spectrum is strongly *y* polarized, and this illustrates the way in which *d–d* spectra sometimes reflect the polarization properties of the parity-allowed transitions from which intensity is borrowed (Section 1.1.2). The vibronic selection rules

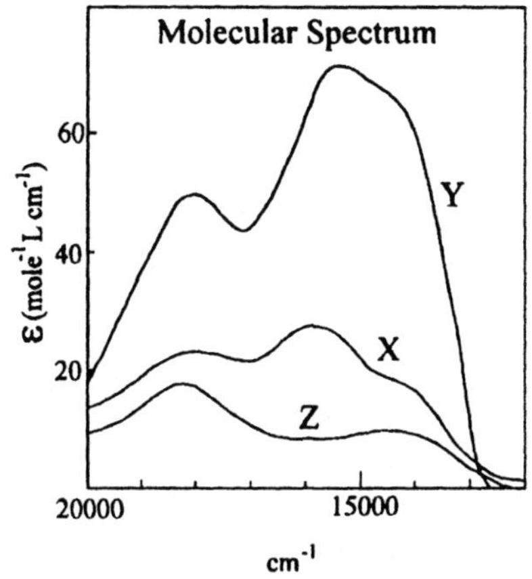

Figure 10. Molecular spectrum of Cu(benzac)$_2$ (adapted from reference 34 with permission).

show that for the D_{2h} point group of this complex a vibration is available to induce intensity into every d–d transition in each polarization. The high intensity of the y spectrum in this and other similar complexes occurs because of the close proximity of a y-polarized ligand→metal parity-allowed charge transfer transition to the excited d-states, with most of the active vibrations being approximately equally effective in the intensity mechanism.[34,35] The fact that the middle band of Cu(benzac)$_2$ is absent in z polarization suggests that here the particular vibrations inducing intensity for this transition are ineffective.

The accuracy with which molecular spectra may be derived from crystal spectra depends upon the extent to which the set of equations (25) are linearly independent. In the ideal situation, each crystal spectrum is dominated by a different molecular spectrum, which is to say that one polarization direction yields the x molecular spectrum, another y, and a third z. Of course, it is then unnecessary to carry out a transformation because the crystal spectra yield the molecular spectra directly. In the opposite extreme, the molecular packing may be such that it is *impossible* to fully derive the molecular spectra from the crystal spectra. This is the case when an axis of one molecule in a unit cell is approximately parallel to a different axis of a second molecule. Such a packing problem occurs for the complex Cu(acac)$_2$, where acac is the acetylacetonate anion. This complex has a geometry identical to that shown in Figure 8, except that the phenyl groups are replaced by methyl groups. Cu(acac)$_2$ also crystallizes with two molecules in the unit cell, but in this case the y axis of one molecule is almost exactly parallel to the z axis of the second molecule, so that these two components of the molecular spectrum cannot be separated.[36] To avoid such problems, the molecular orientations should be checked before measuring the polarized absorption spectra of a crystal. If a complex is of particular interest, it may be advantageous to crystallize it in several lattices, for instance with different counterions, to see which gives the best resolved molecular spectra. A quick guide to promising candidates can be obtained by viewing the behavior of crystals upon rotation under a polarizing microscope.[24] The spectra of highly dichroic crystals will be strongly polarized in the visible region.

3 TYPICAL APPLICATIONS OF POLARIZED ABSORPTION SPECTROSCOPY

3.1 Band Polarizations

As a simple example, we consider the VIS/NIR polarized absorption spectrum (Figure 11) of the $(10\bar{1})$ crystal face of the compound [N-(2-ammonioethyl)morpholinium]$CuCl_4$,[9] (Naem)$CuCl_4$. This contains $CuCl_4^{2-}$ ions of two kinds, a centrosymmetric planar complex, and a non-centrosymmetric complex with the highly distorted tetrahedral geometry illustrated in Figure 12. The absorptions in this region are due to *d–d* transitions. As the *d*-orbitals are symmetric functions, for the planar complex the ground and excited *d*-states all transform as gerade representations of the D_{4h} group, so that every transition is forbidden by the parity selection rule (equation (4)). In agreement with this, the absorption peaks of this complex, which occur in the region 14,000–18,000 cm^{-1}, are rather weak and their interpretation requires the consideration of vibronic coupling.

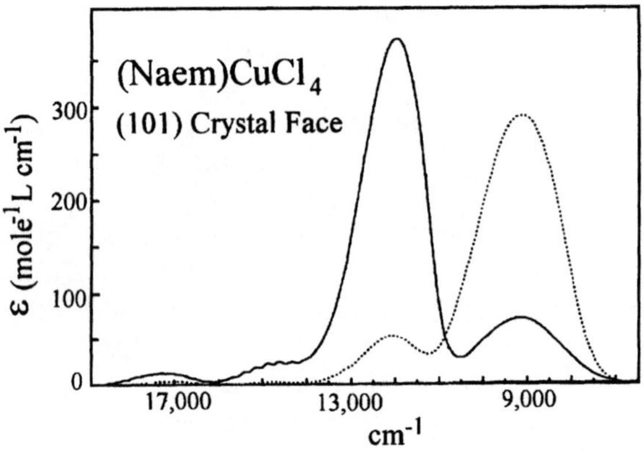

Figure 11. Crystal spectrum of (Naem)$CuCl_4$ at ~12 K with the electric vector parallel to the *b* axis (dashed line) and along [101] (full line) (adapted from reference 9 with permission).

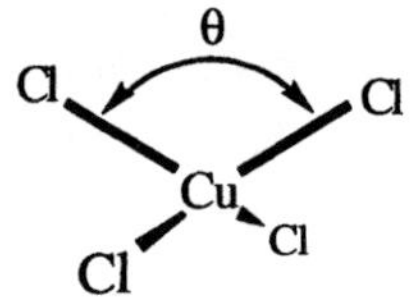

Figure 12. Geometry of a $CuCl_4^{2-}$ ion of D_{2d} symmetry.

Table 1 Polarization Properties of the Electronic Transitions of the $CuCl_4^{2-}$ Ion (D_{2d} point group)

Transition:	$^2B_2 \rightarrow {}^2E$	$^2B_1$	$^2A_1$
Polarization:	xy	—	z

The distorted tetrahedral complex, on the other hand, belongs to the point group D_{2d}. Here, the *d–d* transitions are parity-allowed, and the peaks in the region 7000–13,000 cm^{-1} due to this species are indeed some 20 times more intense than those due to the planar complex (Figure 11). The intensity of each band is strongly dependent on the orientation of the electric vector in the complex. Note that as it has axial symmetry, x and y are equivalent, and the molecular spectrum may be completely characterized by measuring just the two polarized spectra of one crystal face. The ground state is of $^2B_2$ symmetry, and the excited states are of ^2E, $^2B_1$, and $^2A_1$ symmetry. The band polarizations are interpreted by considering whether the direct product of the representations of the ground and excited state wavefunctions yields the totally symmetric representation (equation (4)). This leads to the selection rules in Table 1.

When the electric vector is in the xy plane, the only allowed transition is $^2B_2 \rightarrow {}^2E$, while when it is along z only $^2B_2 \rightarrow {}^2A_1$ is allowed. The transition $^2B_2 \rightarrow {}^2B_1$ is parity-forbidden for both orientations. The arrangement of the distorted $CuCl_4^{2-}$ ion in the crystal lattice is highly favorable for spectroscopy as all the complexes have a similar orientation, with z almost exactly parallel to the [101] extinction direction. The band at 12,000 cm^{-1} has a high intensity in this polarization, so it may be assigned to the transition to the $^2A_1$ excited state. When the electric vector is parallel to the [010] direction it lies almost exactly in the xy molecular plane. Here, the band at ~9000 cm^{-1} is intense, so this may be assigned to the $^2B_2 \rightarrow {}^2E$ transition. The parity-forbidden band due to the $^2B_2 \rightarrow {}^2 B_1$ transition is not resolved. This assignment conforms with the expected energy levels for a copper(II) complex with this geometry.[9] The residual intensity observed when each transition is formally parity forbidden is probably mainly due to spin-orbit coupling between the $^2A_1$ and ^2E states, which means that the $^2B_2 \rightarrow {}^2E$ transition may borrow intensity from the $^2B_2 \rightarrow {}^2A_1$ transition, and vice versa.

The distortion of a four coordinate complex away from a planar geometry is conveniently described by the angle θ between *trans* metal–ligand bonds (Figure 12). A value of $\theta = 109.5°$ corresponds to the highly asymmetric tetrahedral geometry, and $\theta = 180°$ to the centrosymmetric planar geometry. The present example, where $\theta = 146.2°$, lies almost exactly between these two extremes. The band intensities are expected to be a function of the deviation from planarity, and the polarized absorption spectra of complexes with a wide range of distortions have been reported,[9] indicating the way the intensities and energy levels change as a function of the coordination geometry.

While the *d–d* transitions of centrosymmetric complexes are parity-forbidden, the charge-transfer and intra-ligand transitions may be parity-allowed. In practice, spin-

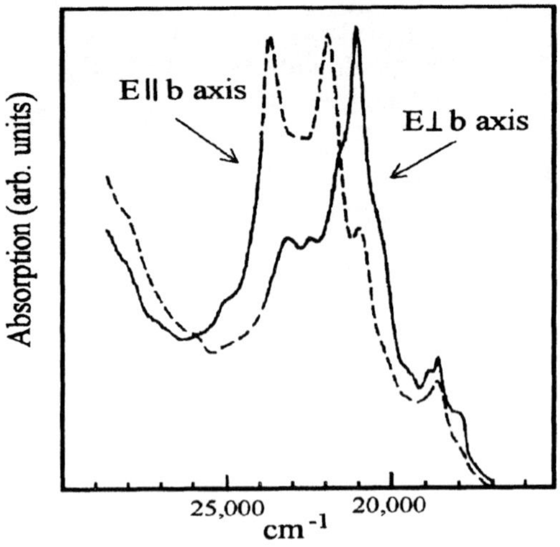

Figure 13. Polarized absorption spectra at 6 K of ~0.01% Ru in a crystal of racemic $[Zn(bpy)_3](ClO_4)_2$ (adapted from reference 38 with permission).

allowed transitions of this kind are generally too intense to be readily measured for a single crystal unless the chromophore is diluted into a transparent host lattice. A series of complexes which have been studied in great detail is that formed by bipyridyl (bpy) with Fe^{2+}, Ru^{2+}, and Os^{2+}. These have metal → ligand charge transfer transitions in the visible region, and are of particular interest because they undergo a range of photochemical reactions in the excited electronic state.[37] The transitions are very intense, and part of the spectrum of ~0.01% Ru^{2+} doped into a crystal of racemic $[Zn(bpy)_3](ClO_4)_2$ is shown in Figure 13 with the light propagated down the *c* crystal axis.[38] The complexes involve two ligands which are crystallographically equivalent, while the third ligand has a slightly different lattice environment. When the polarization direction is parallel (||) to the *b* axis it lies exactly along the vector connecting the metal and the centre of the unique ligand, while when it is perpendicular (⊥) to *b* it makes an angle of ~30° with the metal–ligand vectors of the two crystallographically equivalent ligands. Charge-transfer transitions are polarized along the direction of electron transfer, so the spectra suggest that the double peak at ~22,500 cm^{-1} is associated with the excited state in which an electron is promoted to the unique ligand (the cause of the band splitting is not yet fully understood). The polarization of the lower energy band at ~21,000 cm^{-1} implies that this largely involves electron transfer to the other pair of ligands. Detailed analysis suggests that for the parity-allowed bands there is some transfer of intensity between the transitions involving the different ligands, though for the much weaker triplet excited states occurring at ~17,000 cm^{-1} (almost unobservable on the scale of Figure 13) the transitions are to completely *localized* electronic states in which an electron has been transferred to a particular ligand.[39]

Another series of compounds in which the polarized absorption spectra may be used to distinguish charge-transfer from *d–d* transitions are those containing [Co–

Table 2 Symmetries of the Vibrations which will Induce Intensity into the *d–d* Transitions of a Cr^{3+} Complex of D_{4h} Symmetry

Transition	Polarization	
	xy	z
$^4B_{1g} \rightarrow {}^4E_g$	α_{2u}, β_{2u}	ε_u
$^4B_{2g}$	ε_u	$(\alpha_{1u})^*$
$^4A_{2g}$	ε_u	$(\beta_{1u})^*$

* The complex has no vibrations of this symmetry.

O–O–Co]$^{4+}$ groups.[40] Here, the band polarizations suggest that the lowest energy bands, in the region 15,000–20,000 cm^{-1}, are due to ligand → metal charge transfer transitions from peroxide π^* orbitals to empty *d*-orbitals on the Co(III) ion. In this case, the information is especially valuable, because the bands are of comparable intensity ($\varepsilon \approx 100$ $mol^{-1}Lcm^{-1}$) to the *d–d* transitions in a typical asymmetric complex. This illustrates the fact that not all parity-allowed transitions are very intense. Sometimes, as on this occasion, the orbital overlap is such that the transition moment integral (equation (1)) is small.

As an example of a spectrum involving parity-forbidden *d–d* transitions, we consider the compound [*trans* $Cr(en)_2F_2]ClO_4$, en = 1,2 diaminoethane. Neglecting minor effects due to the chelate ring, the point group of the $Cr(en)_2F_2^+$ complex is D_{4h}. The ground state is of $^4B_{1g}$ symmetry, and the $^4T_{2g}$ and $^4T_{1g}(F)$ excited states of the parent octahedral complex split into the components $^4E_g(T_{2g})$ and $^4B_{2g}$, and $^4E_g(T_{1g})$ and $^4A_{2g}$, respectively. The symmetries of the vibrations inducing intensity into each transition for xy and z polarized light are shown in Table 2. A molecule of D_{4h} symmetry has u-vibrations of α_{2u}, β_{2u}, and ε_u symmetry, but no vibrations of α_{1u} or β_{1u} symmetry. The transitions $^4B_{1g} \rightarrow {}^4B_{2g}$ and $^4B_{1g} \rightarrow {}^4A_{2g}$ are therefore expected to have very low intensity in z polarization. The visible spectra of a crystal face with the electric vector of the incident light along the two extinction directions is shown in Figure 14. Four bands are observed, with two of these being absent in one polarization. It was therefore inferred[41] that for polarization **I** the electric vector is close to the z molecular axis, and for **II** it lies in the xy plane. The first and third bands are then due to transitions to the ^4E_g excited states, while the second and fourth are due to the $^4B_{1g} \rightarrow {}^4B_{2g}$ and $^4B_{1g} \rightarrow {}^4A_{2g}$ transitions. Note that each band decreases significantly in intensity on cooling to 5 K. This is expected if the intensity is derived by vibronic coupling, as the vibrational amplitudes will decrease on cooling. While the band positions had been characterized earlier from the solution spectrum, their assignment had been the subject of some controversy.[42,43] Analysis of the band energies indicates that fluoride is a very strong π-donor in this complex, and is actually a stronger σ-donor than the amine.[41,43] The study of band polarizations is a powerful tool in the assignment of electronic spectra, and hence in the derivation of metal–ligand bonding parameters.[44]

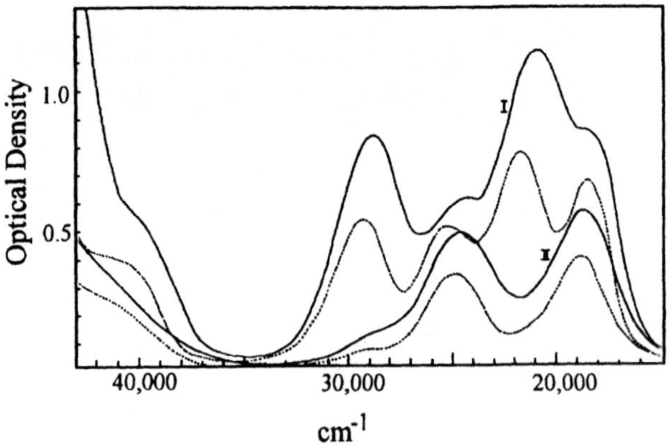

Figure 14. Polarized absorption spectra of the prominent crystal face of [*trans* $Cr(en)_2F_2]ClO_4$ with the electric vector along the two extinction directions (I and II) at 295 K (full lines) and 4 K (dashed lines) (adapted from reference 41 with permission).

3.2 Electronic and Vibronic Origins

The electronic origins of parity-forbidden transitions are usually too weak to be observed, and the two pronounced sharp magnetic-dipole allowed peaks in the α and π polarized spectra of Cu^{2+} doped K_2ZnF_4 (Figure 7) represent unusual examples in this respect.[27] In fact, these peaks confirm the proposal,[45] based upon the EPR spectrum, that the guest CuF_6^{4-} species in the lattice has the unusual tetragonally compressed octahedral geometry, rather than the tetragonally elongated stereochemistry more commonly observed for copper(II) complexes. The electronic spectrum involves transitions to the split components of the $^2T_{2g}$ level of the parent octahedral complex. An elongated tetragonal geometry would place the ^2E_g state higher in energy than $^2B_{2g}$, while the reverse is expected for the compressed tetragonal geometry. The two sharp peaks clearly represent the electronic origins of the components of the ^2E_g state split by spin-orbit coupling, showing that the latter situation applies. The weaker sharp peaks to higher energy of the electronic origins consist mainly of progressions in the totally symmetric α_{1g} vibration, as discussed in the following section. The magnitude of the observed splitting, 1207 cm^{-1}, may be interpreted satisfactorily when proper account is taken of both the spin-orbit interaction and the Jahn–Teller vibronic coupling causing the distortion.[27] The polarization behavior of the highest energy band is also consistent with this assignment. Irrespective of the nature of the distortion, the electric-dipole selection rules, derived as discussed in Section 1.1.2, suggest that the only transition which is vibronically forbidden is that to the $^2B_{2g}$ state in z polarization.[27] As the z molecular axis is parallel to the c crystal axis (Figure 6), the fact that the highest energy peak is absent when the electric vector is along this direction confirms its assignment to the transition to the $^2B_{2g}$ level.

When hot bands are observed, this allows the position of the electronic origin to be deduced even when it cannot be resolved directly. This may be seen in the crystal

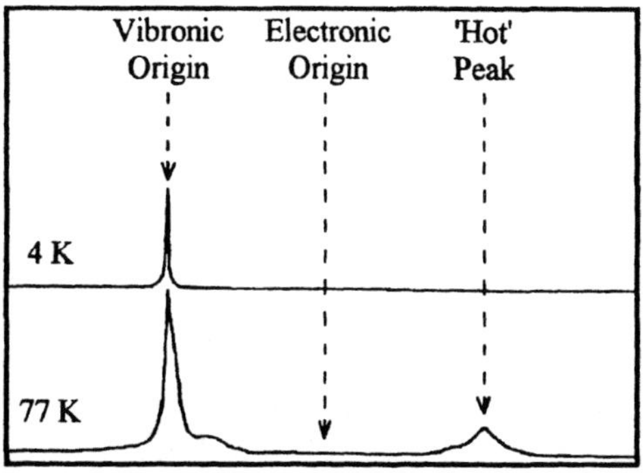

Figure 15. Transition to one of the $^3H_6$ levels of the UCl_6^{2-} ion illustrating the appearance of a hot peak at higher temperature (adapted from reference 21 with permission).

spectrum of the centrosymmetric UCl_6^{2-} ion at 4 K and 77 K, part of which is shown in Figure 15.[21] The pronounced peak present at both temperatures is a vibronic origin due to electronic excitation plus a quantum of the *u* metal–ligand vibration of energy 93 cm^{-1}. The weak peak appearing 93 cm^{-1} below the electronic origin at 77 K is the hot transition from the $v = 1$ vibrational level of the ground state to the $v = 0$ level of the excited state. The parity-forbidden transition of the electronic origin itself is too weak to be observed.

While the bulk of the intensity of parity-forbidden transitions is derived from metal–ligand modes, sometimes coupling with modes which are formally internal ligand vibrations also makes a contribution. This is most readily apparent for O–H and N–H stretching modes, since these are high enough in energy to produce peaks well separated from the vibronic origins produced by metal–ligand vibrations. Weak features of this kind have been observed on the high energy side of the *d–d* transitions of a range of transition metal hydrate[46] and ammine[47] complexes. An unusually striking example occurs in the polarized crystal spectrum of the *trans*-$V(H_2O)_4Cl_2^+$ complex in $Cs_3VCl_6.4H_2O$.[48] Here, a band centered at 19,320 cm^{-1} was observed to shift to lower energy by 520 cm^{-1} on deuteration, implying that it involves an O–H stretching vibration of the water molecules coupled to the $^3A_{2g} \rightarrow {}^3T_{2g}$ electronic transition which gives rise to a band centered at 16,430 cm^{-1}. The two bands are comparable in intensity and it is not known why coupling with an internal water vibration is so effective in this complex—usually bands of this kind are much less intense than those derived from coupling with metal–ligand vibrations.

3.3 Band Progressions

The relative intensities of the members of a band progression depend upon the displacement between the potential surfaces of the upper and lower electronic states in the normal coordinate involved in the progression, usually the totally symmetric vibration. Spin-allowed *d–d* transitions and charge transfer transitions involve rather

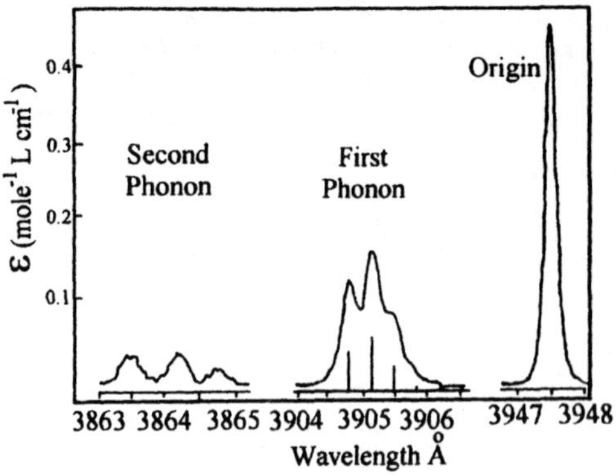

Figure 16. Band structure observed for the transition to the $^4T_2$ state of a crystal of Cs_3MnCl_5 at 4 K (adapted from reference 49 with permission).

large displacements, and typically exhibit bands with an approximately Gaussian bandshape. Some spin-forbidden transitions, on the other hand, involve excited states with potential surfaces lying almost exactly above that of the ground state. Here, the intensity is concentrated largely in the origin, and the peaks can be very narrow. A good example occurs for a transition to one of the spin-orbit components of the $^4T_2$ state of the $MnCl_4^{2-}$ ion in Cs_3MnCl_5 (Figure 16).[49] Here, the peaks are so sharp that the vibrations associated with different isotopes of the ligand are resolved! The transition is electric dipole allowed, and the transition to the electronic origin occurs as a single symmetrical peak. The progression is in the α_1 Mn-Cl stretching mode, and even in the first member, structure due to the different vibrational frequencies associated with the ^{35}Cl and ^{37}Cl isotopes is observed. As expected, the splitting is even more pronounced in the second member of the progression. Because the isotopes occur with natural abundances of 75.4% and 24.6%, the compound contains complexes ranging from $Mn^{35}Cl_4^{2-}$, through $Mn^{35}Cl_3^{37}Cl^{2-}$ etc., to $Mn^{37}Cl_4^{2-}$. Each of these will have a slightly different reduced mass, and hence α_1 vibrational frequency. The observed intensities correlate with the expected statistical distribution of the isotopes.

For parity-allowed transitions, the progressions are built directly upon the electronic origin, while for parity-forbidden transitions, they are built upon the vibronic origins of the intensity inducing vibrations. The polarized absorption spectrum of the transition $^1A_{1g} \rightarrow {}^1T_{1g}$ of $[Co(NH_3)_6](ClO_4)_2Cl.KCl$ at low temperature shows both types of progression (Figure 17).[50] The transition to the electronic origin is observed as a very weak peak, with the intensity being due to a small non-centrosymmetric component to the ligand field. The progression labeled A_1, A_2, etc. is built upon this origin. A second progression, labeled A′, is built upon an origin 89 cm^{-1} above this, and this is assigned to coupling with a lattice vibration. Two further progressions, labeled B and C, are built upon vibronic origins 218 cm^{-1} and 304 cm^{-1} above the electronic origin. The modes active in inducing the intensity of the latter two progressions are the metal–ligand rocking vibrations of τ_{2u} and τ_{1u}

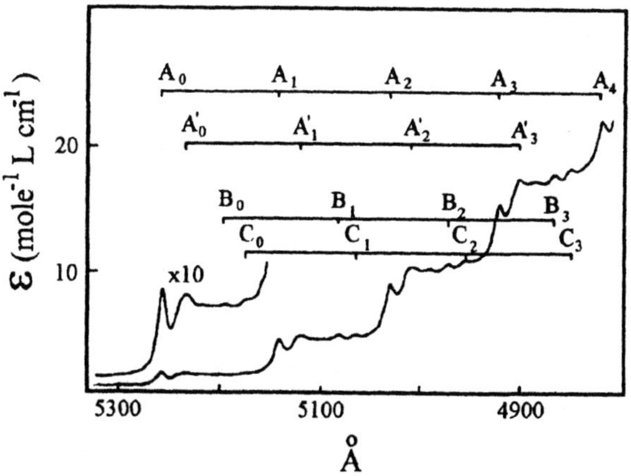

Figure 17. Onset of the band due to the $^1A_{1g} \rightarrow {}^1T_{1g}$ transition of $[Co(NH_3)_6](ClO_4)_2$ Cl.KCl at 8 K with the electric vector parallel to the unique axis of the crystal (adapted from reference 50 with permission).

symmetry. The bands are thought to be a superposition of progressions in the α_{1g} and Jahn–Teller active ε_g vibrations, which are expected to be close in energy for this complex (note that the $^1T_{1g}$ excited state is orbitally triply degenerate, and hence subject to a Jahn–Teller distortion). The interval of the progressions, 429 cm^{-1}, is similar to the energy of the metal–ligand α_1 breathing vibration in the ground state, 490 cm^{-1}.

Because the relative intensities of the members of a band depend upon the displacement of the excited state potential surface in the mode giving rise to the progression, their measurement provides a powerful method of studying the change in geometry which accompanies an electronic transition.[15,16,51] By this means it was deduced[17] from the electronic spectrum of the linear NiO_2^{3-} ion (Figure 2) that each Ni–O bond length changes by $\Delta r = 0.048$ Å and 0.085 Å in the first and second excited states, respectively. The analysis does not indicate the direction of the displacements, but it may safely be assumed that the bonds are longer in the excited states because these have greater anti-bonding character. This will normally always be the case for spin-allowed transitions, though for some spin-forbidden transitions where the energy is dominated by inter-electron repulsion terms the bonds may be shorter in the excited state. Progressions in the α_1 metal–ligand stretching mode have also been observed on charge transfer transitions. A particularly well resolved structure has been reported for the $PtCl_4^{2-}$ ion doped into $ZrCl_6^{2-}$,[52] and the MnO_4^- ion doped into KIO_4.[53]

For *d–d* transitions, the bond length change may be estimated by balancing the effect due to the electron rearrangement against the shift in total energy as parameterized by the force constant f of the α_1 mode.[15,54] If the energy difference E between the orbitals involved in the transition depends inversely upon some power n of the metal–ligand bond length r, then the change in total energy U due to the displacement in normal coordinate Q is given by:

$$U = fQ^2/2 + mE\left[(r_0 + CQ)/r_0\right]^{-n} \tag{26}$$

where r_o is the initial metal–ligand bond length, m is the number of electrons involved in the transition and C is a normalization constant. For small changes in the bond length:

$$U \approx fQ^2/2 + mE[1 - nCQ/r_0] \tag{27}$$

and the displacement corresponding to the optimum geometry of the excited state is obtained by noting that here $\mathrm{d}U/\mathrm{d}Q = 0$. Thus, to a first approximation:

$$Q \approx mnCE/fr_0 \tag{28}$$

For simple ligands such as halide or ammonia, the change in each bond length is given by $\Delta r = CQ$, and for the totally symmetric mode $C = 1/N^{1/2}$, so that:

$$\Delta r \approx mnE/Nfr_0 \tag{29}$$

Here, N is the number of ligands bound to the metal. Both experiment and theory suggest that the d-orbital splitting varies inversely as about the fifth power of the bond distance.[55] Substitution of $n = 5$ into equation (29) produces Δr values which agree well with those derived by analyzing the progressions observed in polarized absorption spectra as may be seen from the typical values in Table 3.

It is apparent that the energy separation between the orbitals occupied in the ground and excited states has a strong influence on the change in the metal–ligand

Table 3 Comparison of Bond Length Changes Δr (Å) Estimated from Progressions in the α_1 Modes Observed in Crystal Electronic Spectra with those Calculated Using Equation (29)

Complex	Shape	E(cm^{-1})	m	f(gr.)[a]	f(exc.)[a]	Δr(exp.)	Δr(calc.)
NiO_2^{3-}	linear	6500	1	4.21	3.40	0.048	0.045
NiO_2^{3-}	linear	12500	1	4.21	2.70	0.085	0.086
$NiCl_4^{2-}$	tet.	3045	~2	1.78	1.70	0.03	0.04
$CuCl_4^{2-}$	planar	12150	1	1.59	~1.48	0.079	0.085
$CuCl_4^{2-}$	planar	16500	1	1.59	~1.48	0.106	0.114
$PdBr_4^{2-}$	planar	20051	1	1.70	1.30	0.10	0.12
$Co(NH_3)_6^{3+}$	oct.	24000	1	2.49	2.31	0.07	0.08
$Co(NH_3)_6^{3+}$	oct.	24000	~2	2.49	1.90	0.12	0.16

[a] Force constants of the α_1 mode (mdyne Å^{-1}) in the ground (gr.) and excited (exc.) state; 1 mdyne Å^{-1} = 0.5035×10^5 cm^{-1} Å^{-2}.

bond lengths. Comparison of the two excited states of NiO_2^{3-} and $CuCl_4^{2-}$ shows an essentially linear relationship, as predicted by equation (29). Moreover, a small bond length change accompanies the d-excitation of the tetrahedral $NiCl_4^{2-}$ complex, which is in line with the low value of the d-orbital splitting associated with this stereochemistry. The number of electrons involved in the electronic transition has a similar effect, as may be seen by comparing the bond length changes for two excited states of $Co(NH_3)_6^{3+}$. The transition which approximates closely to a two-electron jump causes a much larger bond length change. (Note that a two-electron transition is normally forbidden, but mixing with other electronic states causes this restriction to be relaxed.) Although Δr depends inversely upon the number of ligands in the complex, in practice this factor has little effect as a low coordination number is associated with short, strong bonds with high force constants.

The progressions also indicate the vibrational energies of molecules in their excited states and the force constants are compared with their ground state values in Table 3. As expected, lower values occur in the excited states, with the decrease paralleling the increase in the bond lengths. It is not uncommon to observe progressions involving eight or more members in optical spectra, providing a good opportunity to study the anharmonicity of the α_1 potential surface. Perhaps surprisingly, the vibrations are generally observed to be highly harmonic, though recently a detailed analysis of the resonance Raman lineshape and temperature dependence of the bandshape observed for the mixed-valence transition of a dimeric iron complex has revealed significant anharmonicity in two of the α_1 modes.[56]

For orbitally degenerate excited states, distortions also occur in Jahn–Teller active vibrations in addition to modes of α_1 symmetry. An expression analogous to equation (29) may be used to interpret the Jahn–Teller distortions derived from the intensity distribution of the members of a band progression.[15,57] While much less data are available than for the totally symmetric mode, agreement with the results derived from electronic spectra appears satisfactory. For trivalent metal ions such as $Cr^{3+}$and Co^{3+} the Jahn–Teller active vibration is expected to have a similar energy to the α_1 metal–ligand stretching vibration, and a single progression is observed which is thought to encompass both modes.[50,58] For the $^4T_{2g}$ excited state of the $Cr(NH_3)_6^{3+}$ complex, the splitting of the electronic origin was used in conjunction with the relative intensities of the members of the band progression to estimate the displacement in both the ε_g Jahn–Teller mode and the α_{1g} stretch.[58] If the Jahn–Teller and α_1 vibrations have significantly different energies, separate progressions may be observed in the two modes.[59] However, the patterns associated with two vibrations sometimes produce a single progression which has an interval corresponding to the energy of neither vibration. This situation may apply if a displacement occurs in two or more modes of α_1 symmetry, and has been termed the missing mode effect.[60] It has been investigated for a number of complexes by using the time-dependent theory of electronic spectroscopy.[61] Progressions of this kind may also be treated by more conventional bandshape analysis using the overlap integrals of wavefunctions obtained from a variational calculation. For instance, the absorption spectrum of the complex *trans* $CuCl_2$(2,6-dimethylpyridine)$_2$ exhibits progressions with an interval of $\sim$110 cm^{-1}, and these may be simulated satisfactorily in terms of displacements in two modes of α_{1g} symmetry with energies of 140 cm^{-1} and 245 cm^{-1}.[62] The former vibration largely involves motion of the bulky amine ligands, and the latter the lighter chloride ions. When the geometry in the ground and excited state differs by a displacement in two or more modes of the same symmetry, the normal coordinates in the excited electronic state are a mixture of the ground electronic state normal coordinates. This rotation of the normal coordinates, or Dushinsky effect, has been discussed in detail elsewhere.[63]

On occasion, the energy interval of a vibrational progression can aid in the assignment of an electronic transition. Thus, band progressions observed in the electronic spectra of complexes such as $Re_2Cl_8^{2-}$ provide useful information on whether an electronic transition occurs between levels associated with the metal–metal bond, or the metal–halide bond.[64] For $Re_2Cl_8^{2-}$, a band built upon an electronic origin at 14,183 cm^{-1} exhibits a well developed progression with an interval of 248 cm^{-1}. This clearly corresponds to the Re-Re α_{1g} stretching mode, which occurs at 274 cm^{-1} in the ground state, implying that the dominant structural change in the excited state involves an increase in the metal–metal separation. The band is therefore assigned to the spin-allowed $\delta \rightarrow \delta^*$ transition. In agreement with this, the band is polarized along the metal–metal direction. Higher energy bands exhibit progressions in the Re-Cl stretching mode, suggesting that these involve excitation to an orbital which points along the metal–halogen bond directions.

Polarized absorption spectroscopy has also been of value in studying the metal–metal bonding in the series $M_2X_9^{3-}$, M = Cr, Mo, W; X = Cl, Br, I. These dimers have D_{3d} symmetry, with three bridging and six terminal halide ions. The spectra of the Cr^{3+} complexes are basically similar to those of isolated octahedral complexes, with additional features due to double excitations in which electrons are excited simultaneously on both metal ions in the weakly coupled pairs.[65] The W^{3+} complexes, on the other hand, involve strong metal–metal bonds.[66] The Mo^{3+} dimers represent an interesting and complicated situation in which the metal–metal, metal–ligand, and interelectron repulsion interactions are comparable in magnitude, with spin-orbit coupling also playing a significant role. The polarized spectra exhibit a series of richly structured bands, as may be seen from the spectrum of $Cs_3Mo_2Cl_9$ shown in Figure 18. This illustrates the large amount of detailed information yielded by some spectra, and it says much for the current state of theory that most aspects of the spectra can be explained satisfactorily.[67]

Vibrational structure may sometimes reveal bands in the visible region that are due not to *d–d* transitions, but rather to charge transfer or ligand-centered transitions. A number of complexes of aromatic amines exhibit bands at ~22,000 cm^{-1} which have been identified as ligand-centered $\pi \rightarrow \pi^*$ transitions,[68] while nickel nitrite complexes provide examples of both kinds of phenomena. The single crystal spectra of a range of complexes containing nitrite ions coordinated via either one or both oxygen atoms exhibit a weak band at ~22,000 cm^{-1}, and a stronger band at ~28,000 cm^{-1}.[69] Both bands exhibit progressions in two vibrations, with energies of ~600 cm^{-1} and ~1050 cm^{-1}. These vibrations are assigned to the nitrite bend and symmetric stretch, respectively, which have energies of 829 cm^{-1} and 1325 cm^{-1} in the ground state of sodium nitrite. The band at ~28,000 cm^{-1} is assigned to a spin-

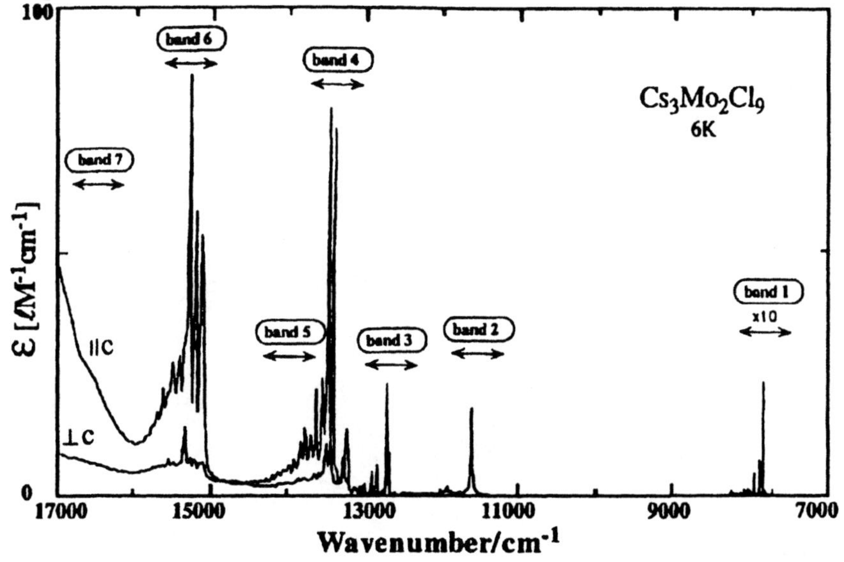

Figure 18. Polarized absorption spectrum of $Cs_3Mo_2Cl_9$ at 6 K with the electric vector parallel and perpendicular to the *c* crystal axis (adapted from reference 67 with permission).

allowed $n \rightarrow \pi^*$ transition in which an electron is excited from an oxygen-based nonbonding orbital into the nitrite π^* orbital. The band at ~22,000 cm^{-1} is the spin-forbidden analog of this. In $NaNO_2$ this band is extremely weak, gaining intensity in the complexes by virtue of the large spin-orbit coupling constant of the metal ion. When the nitrite binds to the metal via nitrogen, either as a monodentate or a bridging ligand, the band at 22,000 cm^{-1} is not observed, presumably because the intensification of the spin-forbidden $n \rightarrow \pi^*$ nitrite-centered transition requires the oxygen to be coordinated to the metal. However, a peak does occur at ~20,000 cm^{-1} in these nickel(II) nitro-complexes. This peak has an intensity similar to spin-allowed, parity-forbidden *d–d* transitions, and early work assigned it as such. However, single crystal studies of several complexes revealed a well developed progression on this band, with an energy interval of ~600 cm^{-1}.[70,71] Metal–ligand vibrations have a considerably lower energy than this, and it was concluded that the progression probably involves the internal nitrite bending vibration. Because this is substantially reduced from the value in $NaNO_2$, it was inferred that the band at ~20,000 cm^{-1} is probably due to a weak metal → ligand π^* charge transfer transition.[70] This is only observed when the ligand bonds via nitrogen because this provides optimum overlap between the metal and nitrite π^* orbitals.

3.4 Temperature Dependence of Band Intensities

The variation of the spectrum of the centrosymmetric $CuCl_4^{2-}$ complex as a function of temperature in *z* polarization is shown in Figure 19.[72] A least-squares fit of the

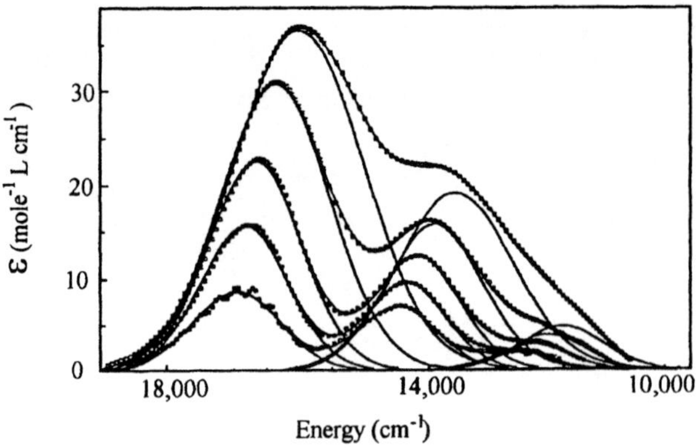

Figure 19. Temperature dependence of the spectrum of the planar $CuCl_4^{2-}$ ion in z polarization (adapted from reference 72 with permision). In order of increasing intensity, the spectra were measured at 10, 60, 100, 200, and 290 K.

observed band intensities to equation (20) is shown in Figure 20a. The fit suggests energies of 62±6, and 110±20 cm^{-1} for the effective energies of the modes inducing intensity into the bands centered at ~16,500 and ~14,000 cm^{-1}, respectively. The vibronic selection rules show that the former band, due to the transition to the $^2A_{1g}$ excited state, is allowed by coupling with modes of β_{2u} symmetry (Section 1.1.2.). The complex has only one such vibration, the out-of-plane bending vibration which carries the complex towards a tetrahedral geometry. As this is both infrared and Raman inactive, the energy cannot be measured directly, so that the only experimental estimate is that derived from the temperature dependence of the polarized spectrum. The intensity of the band at ~14,000 cm^{-1}, due to the transition to the ^2E_g state, is derived by coupling with modes of ε_u symmetry. The complex has two such modes, an in-plane stretch and an in-plane bend, having energies of ~300 and 178 cm^{-1}, respectively. The energy derived from the intensity variation shown in Figure 20a is even lower than that of the bending vibration, suggesting that this vibration dominates the intensity mechanism, supplemented by one or more low energy lattice modes. The weak band centered at ~12,000 cm^{-1} in Figure 19 is due to the transition to the $^2B_{2g}$ state which is vibronically forbidden in z polarization, the residual intensity probably being borrowed from the other bands via spin-orbit coupling.

3.5 Bandwidths and Energies

The temperature dependence of the bandwidth may be used to estimate the energy of the α_1 mode producing the progression and several studies of the temperature dependence of ligand → metal charge transfer transitions have found that the half-widths obey equation (22) for values of $\overline{\nu_{\alpha_1}}$ which correlate well with those observed by Raman spectroscopy.[5,73,74] Further examples are provided by the mixed-valence transitions in compounds such as Cs_2SbCl_6.[75] This contains $SbCl_6^-$ and $SbCl_6^{3-}$ ions, and exhibits a broad band at ~17,000 cm^{-1} due to excitation of an electron from

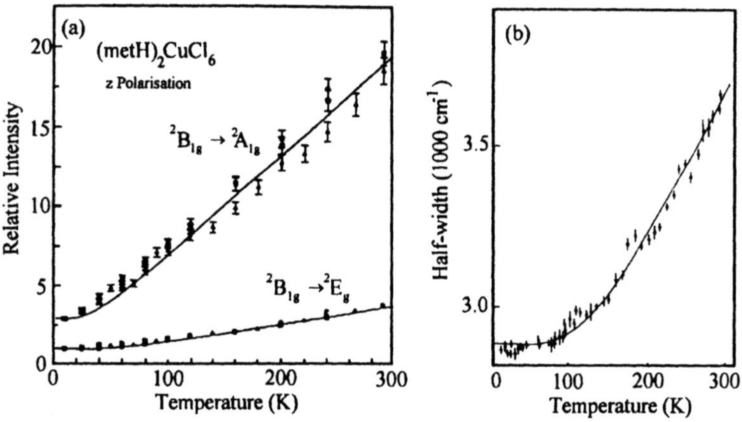

Figure 20. (a) Calculated and observed temperature dependence of the band intensities observed for the planar $CuCl_4^{2-}$ ion in z polarization (adapted from reference 72 with permission). (b) Calculated and observed temperature dependence of the halfwidth of the band due to the mixed-valence transition between $SbCl_6^{3-}$ and $SbCl_6^-$ ions (adapted from reference 73 with permission).

the Sb(III) to the Sb(V) complex. A plot of the band halfwidth as a function of temperature observed for these species doped into a crystal of $(CH_3NH_3)_2SnCl_6$ (Figure 20b) obeys the relationship given in equation (22) for $\overline{\nu_{\alpha_1}} = 290\ cm^{-1}$. The excited state involves a $SbCl_6^{2-}$ complex which is expected to have a bond length intermediate between those of the $SbCl_6^-$ and $SbCl_6^{3-}$ ions of the ground state, and the estimated vibrational energy, 290 cm^{-1}, is indeed between the values observed for the α_{1g} mode of these two species, 327 cm^{-1} and 267 cm^{-1}, respectively.[75] In mixed-valence compounds involving bridging ligands, the geometry of the excited state will often involve displacements in more than one mode. Use of equation (22) to analyze the variation of the bandwidth as a function of temperature then yields an effective energy which reflects the relative contributions these modes make to the change in geometry. In the case of a mixed-valence iron dimer, a geometry change involving displacement in two normal modes was deduced by normal coordinate analysis.[56]

For some compounds which contain the planar $CuCl_4^{2-}$ ion, large shifts in band maxima are observed. Here, the maximum of the highest energy band shifts to lower energy by ~900 cm^{-1} in z polarization on warming from 10 K to 300 K (Figure 19). This is thought to be due mainly to the fact that the complex has a distorted tetrahedral geometry in the ^2A_g excited state.[14] In z polarization the intensity of the band derives from coupling with vibrations of β_{2u} symmetry. The complex has only one such vibration, and this is also the normal coordinate which carries the complex into the equilibrium geometry of the excited state. At low temperature, only the $v = 0$ level of the β_{2u} mode is occupied, but as the temperature is raised, the $v = 1, 2$ etc. levels become populated, leading to nonzero overlaps with a broader range of vibra-

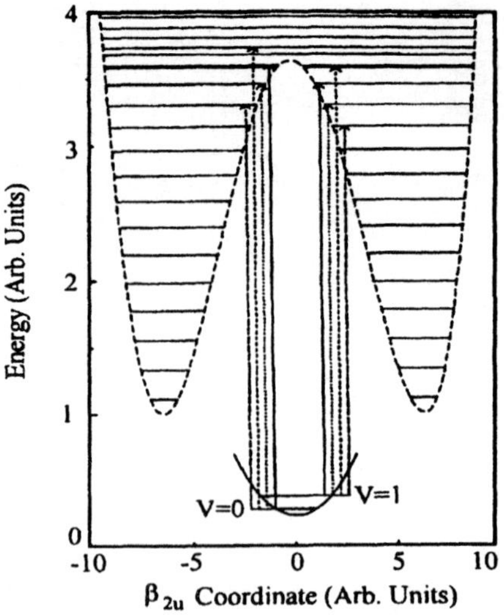

Figure 21. Ground and $^2A_{1g}$ excited state potential surfaces in the out-of-plane β bending vibration for the planar $CuCl_4^{2-}$ ion in (methadonium)$CuCl_4^{2-}$ (adapted from reference 93 with permission). The four most intense vibronic transitions originating from the $v = 0$ and $v = 1$ levels of the ground state are indicated.

tional levels of the ^2A_g state. This is illustrated in Figure 21, which shows the ground and excited state potential surfaces in the β_{2u} coordinate for the planar $CuCl_4^{2-}$ ion in (methodonium)$_2CuCl_4$. The transitions giving rise to the four most intense vibronic origins are illustrated for the $v = 0$ and $v = 1$ levels of the ground state. It may be seen that thermal population of higher vibrational level leads to vibronic origins at lower energies, and hence a red shift of the band maximum. The extent of the distortion is lattice dependent.[14,19] For (methadonium)$_2CuCl_4$ good agreement with experiment (Figure 22) is obtained for an excited state of the form shown in Figure 21, where each minimum corresponds to the geometry shown in Figure 12, with *trans* Cl-Cu-Cl angles of ~163°.[14] Excited state distortions of this kind are only expected when, as in the present case, the vibration involved is of very low energy.

3.6 Structural Changes

The structure of some complexes is altered as a function of temperature. This will affect the electronic spectrum, which therefore provides a way of investigating the nature of the change. An important class of compounds studied in this way are those in which a thermal equilibrium occurs between two spin states of a complex. The spin change is usually accompanied by a significant alteration of the geometry and the electronic spectrum, and the temperature dependence of the single crystal absorption spectrum has provided a useful method of studying the change in concentration of the spin-isomers.[76]

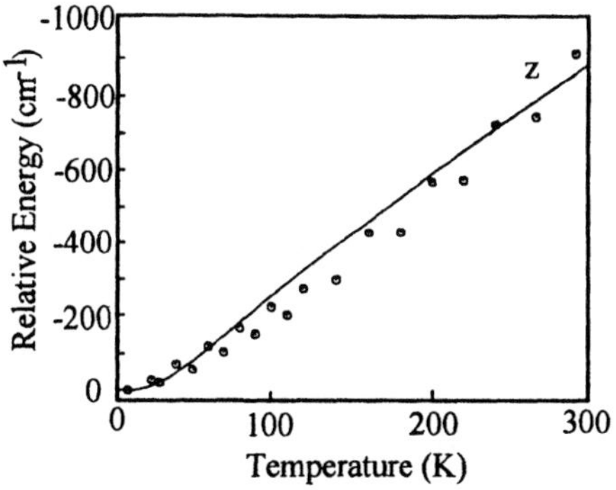

Figure 22. Temperature dependence of the band maximum of the $^2B_{1g} \rightarrow ^2A_{1g}$ transition of the planar $CuCl_4^{2-}$ ion in (methadonium)$_2$$CuCl_4$ (adapted from reference 94 with permission).

A particularly striking example of a temperature dependent structural change is provided by the complex *trans* Ni(cyanurate)$_2$(NH_3)$_4$ formed by the cyanurate anion[77]:

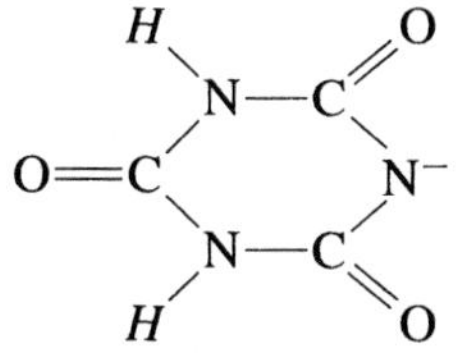

In the room temperature crystal structure the metal and two cyanurate ions all lie in a plane bisecting the NH_3-Ni-NH_3 bond angles, and the complex belongs to the point group D_{2h}. On cooling, the cyanurate ligands progressively bend out of this plane as shown in Figure 23, until by 139 K each makes an angle of 16° with the plane. At low temperature the complex belongs to the point group C_{2v}. In the centrosymmetric D_{2h} point group, the electronic transitions are parity-forbidden, but certain transitions become parity-allowed in the non-centrosymmetric point group C_{2v}. The temperature dependence of the polarized absorption spectrum conforms to these expectations. The complexes have identical orientations in the crystal, and the molecular axes of the C_{2v} point group (Figure 23) are parallel to the extinction directions. The bands in the electronic spectrum are strongly polarized, as may be seen from the low temperature spectra in x and z polarization shown in Figure 24(a). The ground state of the distorted complex is $^3B_2$ and the electric dipole selection rules show that transitions to states of $^3A_2$ symmetry are allowed in x polarization. Three transitions increase in intensity on cooling, and ligand field calculations suggest that these are indeed to the states of $^3A_2$ symmetry (Figure 24(b)). It is noteworthy that the energy of the transitions do not change dramatically on cooling. In fact, the shifts are not larger than may be explained by the mechanisms discussed in the preceding section. This is because the structural change does not involve the primary coordination sphere

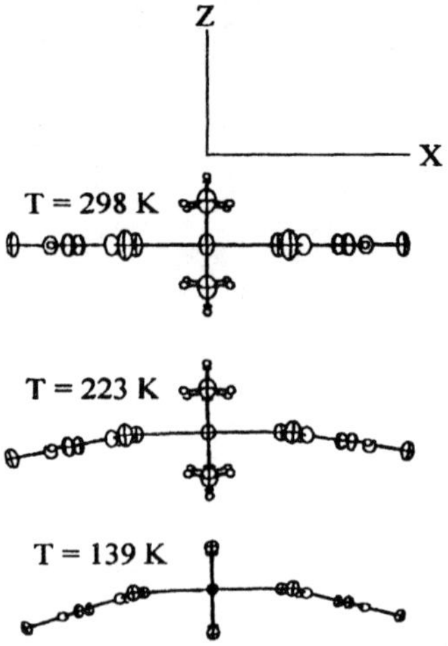

Figure 23. Variation of the geometry of the complex Ni(cyanurate)$_2$(NH$_3$)$_4$ on cooling (adapted from reference 77 with permission).

of the complex. Some of the infrared stretching and bending combination peaks due to N-H vibrations in the near infrared region also alter on cooling. For instance, the single peak at ~4900 cm^{-1} at 300 K progressively splits into two below 260 K (see the insert to Figure 24(b)). This suggests that the movement of the cyanurate groups is associated with a change in the hydrogen bonding network in the lattice.

3.7 Ground State Splittings

When the ground state of a complex is split by an amount comparable to thermal energies, the intensities of the electronic transitions originating from these levels will vary as a function of temperature. Spin-allowed transitions are normally too broad to resolve such splittings, but sharp spin-forbidden transitions have provided evidence of the zero-field splitting of the ground state of several complexes. The $^3A_1 \rightarrow {}^1A_1$ transition of RuO_4^{2-} doped into a crystal of $BaSO_4$ shows just such an effect.[78] Here, lattice forces distort the tetrahedral RuO_4^{2-} complex, and second-order spin-orbit coupling with the excited states splits the $^3A_1$ ground state into three spinor levels, Γ_1, Γ_2, and Γ_4. The polarized absorption spectra with the electric vector along the extinction directions parallel to the three crystal axes are shown in Figure 25(a) measured at 6 K, 15 K, and 30 K. At the lowest temperature, just one peak, largely polarized $\|c$, is observed at 7329 cm^{-1}. This represents the transition from the lowest spin-orbit component of the ground state, Γ_1, to the $^1A_1$ state. On raising the temperature, hot bands appear 13 cm^{-1} and 18 cm^{-1} to lower energy in the other two polarizations, due to the thermal population of the Γ_2 and Γ_4 levels of the ground state, respectively. The energies of these levels are confirmed by plots of the areas of the two hot peaks as a function of temperature (Figure 25(b)), calculated assuming a Boltzmann population distribution. Ground state splittings of this kind are usually

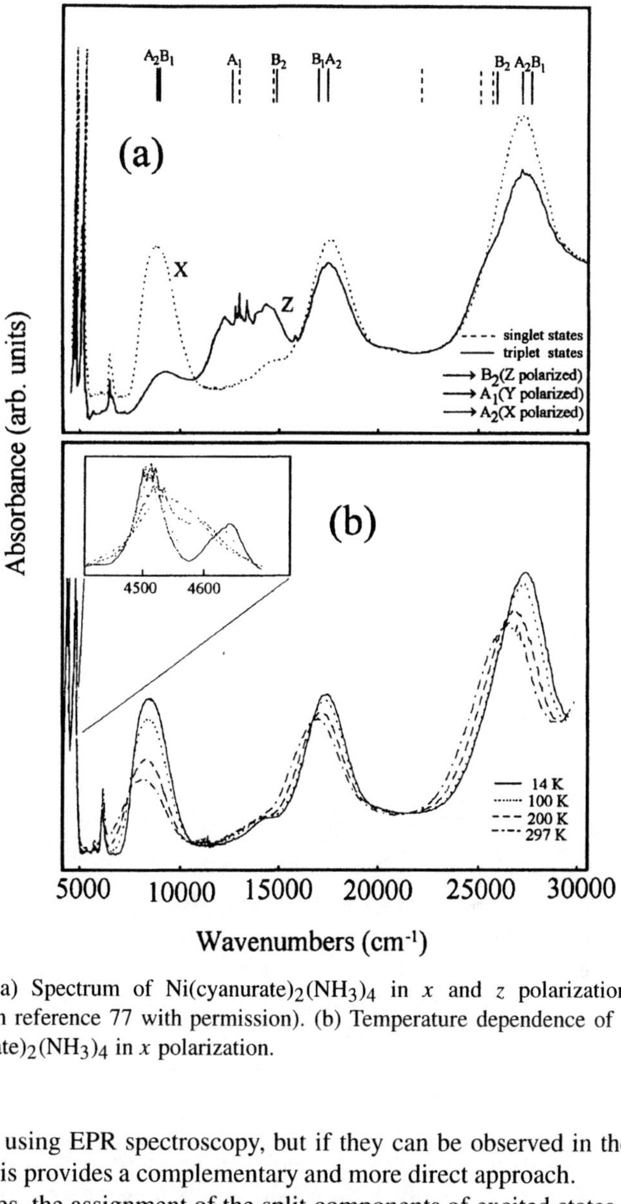

Figure 24. (a) Spectrum of Ni(cyanurate)$_2$(NH$_3$)$_4$ in x and z polarization at $\sim$12 K (adapted from reference 77 with permission). (b) Temperature dependence of the spectrum of Ni(cyanurate)$_2$(NH$_3$)$_4$ in x polarization.

investigated using EPR spectroscopy, but if they can be observed in the electronic spectrum, this provides a complementary and more direct approach.

Sometimes, the assignment of the split components of excited states revealed by band polarizations may be used to infer ground state splittings which are too small to be studied directly by optical methods alone. This is the case for the distorted tetrahedral $FeCl_4^-$ complex in (C_6H_5)$FeCl_4$.[79,80] Here, a conventional ligand field analysis based upon the band energies implies a ground state splitting corresponding to a zero-field parameter $D_{zfs} \approx 0.22$ cm^{-1}.[79] Moreover the band assignments were confirmed by observing the effect of a high magnetic field on the optical spectrum.[80] The zero field splitting obtained in this way differs in sign from the estimate derived from

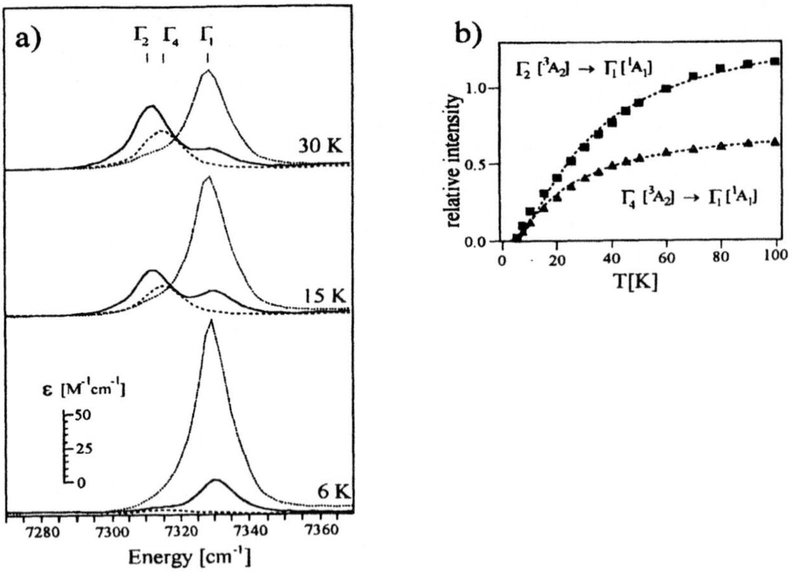

Figure 25. (a) Temperature dependence of the $^3A_2 \rightarrow {}^1A_1$ band of $RuO_4^{2-}/BaSO_4$ (adapted from reference 78 with permission). The spectra with the electric vector parallel to the *a*, *b*, and *c* crystal axes are shown as solid, broken, and dotted lines, respectively, and the spinor representations indicate the ground state level giving rise to each component of the band. (b) Temperature dependence of the peak intensities of the transitions to Γ_2 (squares) and Γ_4 (triangles) relative to Γ_1 for the $^1A_1$ absorption of $RuO_4^{2-}/BaSO_4$ (adapted from reference 78 with permission). The broken lines show the calculated Boltzmann fit for these two levels at 13 and 18 cm^{-1} above the lowest level, respectively.

EPR measurements, $D_{zfs} \approx -0.04\ cm^{-1}$ at 4.2 K. However, a ground state splitting in agreement with the EPR experiment is obtained if the calculations are extended to include the unpaired spin density delocalized into ligand orbitals.[80] Interestingly, the EPR suggests that the zero field splitting decreases in magnitude significantly at higher temperatures, and a corresponding decrease occurs in the band splittings observed in the polarized absorption spectrum. This implies that the distortion from tetrahedral symmetry becomes smaller as the temperature increases.

Magnetic interactions between metal ions linked by bridging ligands also cause small ground state splittings, so that the population of these levels is quite sensitive to temperature in the range 2–300 K. This can have a strong influence on the absorption spectrum, which may therefore be used to study the magnetic interactions.[81] The coupling in chromium(II)[82] and copper(II)[31,83] cluster compounds has been the subject of considerable interest, and the polarized spectra provide a useful way of probing these interactions. As well as the band envelope due to the *d–d* transitions in the near infrared part of the spectrum, which are quite similar in energy and intensity to monomers with a similar coordination geometry, copper(II) dimers usually have an additional characteristic band in the visible region. A typical example is $Cu_2Cl_6^{2-}$, which shows a band centered at $\sim$20,000 cm^{-1} in a number of compounds.[31] This is markedly temperature dependent, and for anti-ferromagnetic $KCuCl_3$ the dramatic increase in intensity on cooling may be correlated directly with the depopulation of the spin-triplet component of the ground state of the dimer. One possible cause of the 20,000 cm^{-1} band is a double excitation, in which electrons are simultaneously excited on both metal ions of the dimer. Such transitions are sometimes observed for compounds containing coupled metal ions,[81] but they exhibit bandshapes which differ from those of the corresponding single electron transitions.[84] As this is not the case for the present compounds, such an assignment is unlikely. The observed temperature dependence and polarization of the band are, however, consistent with it being due to the ${}^1A_g \rightarrow {}^1B_{3u}$ charge transfer transition in which an electron is excited from a nonbonding π-molecular orbital on the terminal chloride ligands to the half-filled d-orbital on a Cu^{2+} ion. Interestingly, the corresponding spin-triplet transition occurs at considerably higher energy. Theoretical calculations support this interpretation,[31] and the characteristic "green" band observed for copper(II) carboxylate dimers apparently originates from a similar type of transition.[83] The model has subsequently been extended to treat the band splittings observed for peroxide-bridged copper(II) complexes,[85] and satisfactorily explains the important features observed in the charge-transfer spectrum of oxyhemocyanin.

4 GENERAL CONCLUSIONS

Measurement of the polarized absorption spectra of single crystals is a powerful way of characterizing the excited states of low symmetry molecules and provides a range of chemical insights not readily available from other techniques. The assignments obtained from band polarizations have proved invaluable in deriving σ- and π-bonding parameters for a wide range of ligands,[44] providing quantitative information on the bonding characteristics conventionally used to describe chemical behavior. Similar studies[72,86] have been instrumental in showing that interactions between the d-orbitals and the higher energy metal s-orbital influence the properties of complexes which depart drastically from cubic symmetry.[87] Investigations of the polarized spectra of simple model complexes have also played a vital role in elucidating the electronic structure of a number of biologically important molecules.[79,85,88]

Because relatively simple absorption spectra may be obtained using polarized light, vibrational fine structure is often observed, and this may be analyzed to provide information not only on the vibrational properties of molecules in their ground and excited states, but also on the changes in geometry which accompany excitation, as discussed in Section 1.3. Similarly, measurement of the influence of temperature on absorption bands, which is usually only practical in single crystal studies, is able to yield much information on the vibrational and electronic energy levels of molecules in their ground states. When taken in conjunction, the polarization behavior of the bands observed in the absorption spectra of a range of copper(II) dimers, and the way in which the band intensities change with temperature, have formed the basis for a detailed theoretical study of the magnetic exchange mechanisms operating in both the ground and excited states of the complexes.[31,85]

When the major interest is assignment of the optical spectrum, measuring the band polarizations alone may be sufficient. However, as testified by the other chapters in this series, a wide range of techniques is now available to probe the energy levels in complexes, and it is often the case that several of these are required to solve the increasingly complicated problems of current interest. For very intense transitions, it is often impossible to obtain thin enough crystals to allow measurement of the absorption spectrum. However, the band intensities and polarizations may be derived from the specular reflectance of the crystal, and the charge transfer spectra of a range of $Cu_2Cl_6^{2-}$ dimers have been characterized in this way.[31] Where it can be measured, the luminescence spectrum also provides a powerful complement to the absorption spectrum, since here the initial energy levels correspond to the equilibrium geometry of the excited rather than the ground electronic state (see Chapter 7). Thus, the luminescence spectrum of the $Ru(bpy)_3^{2+}$ complex discussed in Section 3.1. has given important evidence concerning the nature of the metal → ligand charge transfer excited state in this complex.[89] Similarly, the luminescence spectra of complexes similar to $Cr(en)_2F_2^+$ have yielded valuable information on the nature of the excited states,[90] complementing that from the absorption spectra outlined in Section 3.1. Resonance Raman spectroscopy provides a powerful way of probing the structural changes associated with excitation to particular excited states, and this technique has been applied to the charge-transfer bands in $[Co\text{-}O\text{-}O\text{-}Co]^{4+}$ groups, as discussed briefly in Section 3.1.[40] The influence of a magnetic field can also greatly extend the range of information deduced from the absorption spectrum. When peaks are sharp enough, the Zeeman splitting will allow the spin and orbital angular momenta of an excited state to be determined, as for the $[(pyrazole)Cu(acetate)_2]$ dimer.[83] In addition, the effect of the magnetic field on the absorption of circularly polarized light, as measured by magnetic circular dichroism (MCD), provides a powerful method of characterizing excited states when taken in conjunction with the absorption of linearly polarized light. This allowed the complete assignment of the *d–d* and charge-transfer transitions in the visible spectrum of the complicated copper(II) acetate dimer system,[83] and has provided confirmation of the band assignments for the $FeCl_4^-$ ion.[80] In general, the range of techniques needed to interpret a spectrum usually depends upon its complexity, and it is not surprising that the rich polarized absorption spectrum of the $Mo_2Cl_9^{3-}$ ion shown in Figure 18 required supplementation by measurement of the luminescence, Zeeman, and MCD spectra before yielding its secrets![67]

The band intensities observed in polarized absorption spectra are generally interpreted using the group theoretical selection rules described in Section 1.1. However, theoretical models have been formulated which provide a more quantitative interpretation. For electric dipole allowed transitions, a ligand polarization model has been developed by Mason and co-workers.[91] This utilizes the fact that the intensity derives largely from the admixture of ligand-based charge transfer excited states into the ground and excited *d*-states. Other approaches have utilized the angular overlap model (AOM) of the bonding in complexes. A specific application considered the parity-allowed $^2B_2 \rightarrow {}^2E$ transition of the distorted tetrahedral $CuCl_4^{2-}$ ion, where the intensity was estimated as a function of the distortion angle θ (Figure 12).[9] The calculation included the planar limit approached as θ tends to 180°, when the intensity is derived via vibronic coupling. Recently, Gerloch and co-workers[92] have developed a quite general treatment of band intensities in terms of AOM parameters analogous to those widely used to describe the energies of metal–ligand bonding interactions. This has been incorporated in a computer program which can handle both parity- and vibronically-allowed transitions for complexes of any geometry. The intensity-parameters obtained in this way should provide a framework for the analysis of band intensities which will greatly enhance the value of the information provided by the polarized absorption spectra of transition metal complexes.

ACKNOWLEDGMENTS

The assistance of Dr. Horst Stratemeier and Dr. Elmars Krausz in the preparation of this article is gratefully acknowledged.

REFERENCES

1. Lever, A.B.P. *Inorganic Electronic Spectroscopy (second edition)*; Elsevier: Amsterdam, **1984**.
2. Jørgensen, C.K. *Absorption Spectra and Chemical Bonding in Complexes*, Pergamon: Oxford, **1962**.
3. Ferguson, J. *Prog. Inorg. Chem.* **1970**, *12*, 159.
4. Ferguson, J. *Electronic States of Inorganic Compounds: New Experimental Techniques*, P. Day Ed., Reidel.: Dordrecht, The Netherlands, **1974**.
5. Day, P. *Angew. Chem. Int. Ed. Engl.* **1980**, *19*, 290.
6. Hitchman, M.A. *Transition Metal Chemistry*; Marcel Dekker: New York, **1985**; vol. 9, chapter 1.
7. Atkins, P.W. *Quanta, A Handbook of Concepts*, Oxford University Press, London, **1974**, p. 78.
8. Note that the orientation of the cartesian axes may change when molecules belonging to different point groups are compared.
9. McDonald, R.G, Riley, M.J., Hitchman, M.A. *Inorg. Chem.* **1988**, *27*, 894.
10. Flint, C.D. *Coord. Chem. Rev.* **1974**, *14*, 47; Cieslak-Golonka, M., Bartechi, A., Sinka, S. *Coord. Chem. Rev.* **1980**, *31*, 251.
11. See, for example, Wilson, E.B., Decius, J.C., Cross, P.C. *Molecular Vibrations*, McGraw-Hill, New York, **1955**.
12. Ballhausen, C.J. and Hansen, A.E. *Ann. Rev. Phys. Chem.* **1972**, *13*, 15.
13. Waldenstrom, S. and Naqvi, K.R. *Chem. Phys. Lett.* **1982**, *85*, 581.
14. Riley, M.J., Hitchman, M.A. *Inorg. Chem.* **1987**, *26*, 3205.
15. Schmidtke, H.-H., Degen, J. *Struct. Bonding* **1989**, *71*, 100.

16. Yersin, H., Otto, H., Zink, J.I., Glieman, G. *J. Am. Chem. Soc.* **1980**, *102*, 951; note that an error occurs in the equations given in this reference. See ref. 6 p. 59.
17. Möller, A., Hitchman, M.A., Krausz, E., Hoppe, R. *Inorg. Chem.* **1995**, *34*, 268.
18. Martin, D.S., Jr., Tucker, M.A., Kassman, A.J. *Inorg. Chem.* **1965**, *4*, 1682; Ballhausen, C.J., Bjerrum, N., Dingle, R., Eriks, K., Hare, C.R. *Inorg. Chem.* **1965**, *4*, 514; McDonald, R.G., Riley, M.J., Hitchman, M.A. *Chem. Phys. Letters* **1987**, *142*, 529.
19. McDonald, R.G., Riley, M.J., Hitchman, M.A. *Inorg. Chem.* **1989**, *28*, 752.
20. Hollebone, B.R., Langford, C.H., Serpone, N. *Coord. Chem. Rev.* **1981**, *39*, 181; Hollebone, B.R. *Theor. Chim. Acta* **1980**, *56*, 45; Hoggard, P.E., Albin, M. *Inorg. Chem.* **1981**, *20*, 4413.
21. Johnston, D., Satten, R.A., Wong, E. *Optical Properties of Ions in Crystals* H.M. Crosswhite, H.W. Moos, Eds., Wiley Interscience, New York, **1967**, p. 429.
22. Lohr, L.L., Jr. *J. Chem. Phys.* **1969**, *50*, 4596.
23. Lee, J.L., Lever, A.B.P. *J. Mol. Spect.* **1968**, *26*, 189.
24. Wahlstrom, E.E. *Optical Crystallography*, Wiley, New York, **1969**.
25. Krausz, E. *Aust. J. Chem.* **1993**, *14*, 1041.
26. See reference 6, p. 12 for a more detailed discussion of this aspect.
27. Riley, M.J., Dubicki, L., Moran, G., Krausz, E., Yamada, I. *Chem. Phys.* **1990**, *145*, 363.
28. Davidov, A.S. *Theory of Molecular Excitons* translated by M. Kasha and M. Oppenheimer, McGraw-Hill, New York, **1962**.
29. Day, P. *J. Mol. Struct.* **1980**, *59*, 109 and references therein.
30. Flint, C.D. *Coord. Chem. Rev.* **1974**, *14*, 47.
31. Desjardin, S.R., Wilcox, D.E., Musselman, R.L., Solomon, E.I. *Inorg. Chem.* **1987**, *26*, 288.
32. Dawson, K., Hitchman, M.A., Prout, C.K., Rossotti, F.J.C. *J. Chem. Soc. Dalton Trans.* **1972**, 1509; Marathe, V.R., Mitra, S. *Inorg. Chem.* **1975**, *5*, 970.
33. Hitchman, M.A. *J. Chem. Soc. Trans. Faraday Trans. II* **1976**, *72*, 54.
34. Hitchman, M.A., Belford, R.L. *Inorg. Chem.* **1971**, *10*, 984.
35. Belford, R.L., Carmichael, J.W., Jr. *J. Chem. Phys.* **1967**, *46*, 4515.
36. Ferguson, J., Belford, R.L., Piper, T.S. *J. Chem. Phys.* **1962**, *37*, 1569.
37. Anderson, S., Seddon, K.R. *J. Chem. Res. (S)* **1979**, 74.
38. Krausz, E., Riesen, H., Rae, A.D. *Aust. J. Chem.* **1995**, *48*, 929.
39. Krausz, E., Ferguson, J. *Prog. Inorg. Chem.* **1989**, *37*, 293.
40. Tuczek, F., Solomon, E.I. *Inorg. Chem.* **1992**, *31*, 944.
41. Dubicki, L., Hitchman, M.A., Day, P. *Inorg. Chem.* **1970**, *9*, 188.
42. Glerup, J., Schäffer, C.E. *Progress in Coordination Chemistry*, M. Cais, Ed., Elsevier, Amsterdam, **1968**, p. 500.
43. Dubicki, L., Day, P. *Inorg. Chem.* **1971**, *10*, 2043; Lever, A.B.P. *Can. J. Chem.* **1971**, *49*, 192; Lever, A.B.P. *Can. J. Chem.* **1973**, *51*, 3690.
44. See for example, ref. 1 chapter 9.
45. Riley, M.J., Hitchman, M.A., Reinen, D. *Chem. Phys.* **1986**, *102*, 11.
46. Harding, M.J., Briat, B. *Mol. Phys.* **1973**, *25*, 745; Solomon, E.I., Ballhausen, C.J. *Mol. Phys.* **1975**, *29*, 279; McDonald, R.G., Hitchman, M.A. *Inorg. Chem.* **1990**, *29*, 3081.
47. Hamm, D.J., Schreiner, A.F. *Inorg. Chem.* **1975**, *14*, 519.

48. McCarthy, P.J., Lauffenburger, J.C., Schreiner, M.M., Rohrer, D.C. *Inorg. Chem.* **1981**, *20*, 1571.

49. Tacon, J., Day, P., Denning, R.G. *J. Chem. Phys.* **1974**, *61*, 251.

50. Wilson, R.B., Solomon, E.I. *J. Am. Chem. Soc.* 1980, *102*, 4085.

51. Solomon, E.I. *Comments on Inorg. Chem.* 1984, *3*, 225.

52. Harrison, T.G., Patterson, H.H., Godfrey, J.J. *Inorg. Chem.* **1976**, *15*, 1291.

53. Cox, P.A., Robbins, D.J., Day, P. *Mol. Phys.* **1975**, *30*, 405.

54. Hitchman, M.A. *Inorg. Chem.* **1982**, *21*, 821.

55. Drickamer, H.G., Frank, C.W. *Electronic transitions and the high pressure chemistry and physics of solids* Chapman and Hall: London, **1973**; Bajermo, M., Pueyo, L. *J. Chem. Phys.* **1983**, *78*, 854.

56. Gamelin, D.R., Bominar, E.L., Mathonière, C., Kirk, M.L., Wieghardt, K., Girerd, J.-J., Solomon, E.I. *Inorg. Chem.* **1996**, *35* , 4323.

57. Deeth, R.J., Hitchman, M.A. *Inorg. Chem.* **1986**, *25*, 1225.

58. Wilson, R.B., Solomon, E.I. *Inorg. Chem.* **1978**, *17*, 1729.

59. Pfeil, A. *Theor. Chim. Acta (Berlin)* **1971**, *20*, 159.

60. Tutt, L., Tannor, D., Heller, E.J., Zink, J.I. *Inorg. Chem.* **1982**, *21*, 3858.

61. Reber, C., Zink, J.I. *Comments on Inorg. Chem.* **1992**, *13*, 177; Wexler, D., Zink, J.I., Reber, C. *J. Phys. Chem.* **1992**, *96*, 8757.

62. McDonald, R.G., Hitchman, M.A. *Inorg. Chem.* **1990**, *29*, 3074.

63. Roche, M., Jaffé, H.H. *Chem. Soc. Rev.* **1976**, *5*, 165.

64. Trogler, W.C., Cowman, C.D., Gray, H.B., Cotton, F.A. *J. Am. Chem. Soc.* **1977**, *99*, 2993; Bursten, B.E., Cotton, F.A., Fanwick, P.E., Stanley, G.G. *J. Am. Chem. Soc.* **1983**, *105*, 3082.

65. Briat, B., Russel, M.F., Rivoal, J.C., Chapelle, J.P., Kahn, O. *Mol. Phys.* **1977**, *34*, 1357; Dean, N.J., Maxwell, K.J. *Mol. Phys.* **1982**, *42*, 551.

66. Cotton, F.A., Walton, R.A. *Multiple Bonds Between Metal Atoms*, Oxford University Press, **1993**; Stranger, R., Macgregor, S.A., Lovell. T., McGrady, J.E., Heath, G.A. *J. Chem. Soc. Dalton Trans.* **1996**, 4485.

67. Stranger, R., Dubicki, L., Krausz, E. *Inorg. Chem.* **1996**, *35*, 4218.

68. Wallace, L., Woods, C., Rillema, D.P. *Inorg. Chem.* **1995**, *34*, 2875.

69. Walker, I.M., Lever, A.B.P., McCarthy, P.J. *Can. J. Chem.* **1980**, *58*, 823.

70. Hitchman, M.A., Rowbottom, G.L. *Inorg. Chem.* **1982**, *21*, 823.

71. Hitchman, M.A., Rowbottom, G.L. *Coord. Chem. Rev.* **1982**, *42*, 55.

72. McDonald, R.G., Hitchman, M.A. *Inorg. Chem.* **1986**, *25*, 3273.

73. Day, P., Grant, E.A. *J. Chem. Soc. Chem. Commun.* **1969**, 123.

74. Day, P., Grant, E.A. *J. Chem. Soc. A* 1970, 100; Day, P., Diggle, P.J., Griffiths, G.A. *J. Chem. Soc. Dalton Trans.* **1974**, 1446.

75. Prassides, K., Day, P. *J. Chem. Soc. Faraday Trans.* **1984**, *80*, 85.

76. See, for example, Gütlich, P., Hauser, A., Spiering, H. *Angew. Chem. Int. Ed. Engl.* **1994**, *33*, 2024.

77. Falvello, L.R., Palacio, F., Pascual, I., Schultz, A.J., Young, D.M., Hitchman, M.A., Stratemeier, H. *J. Am. Chem. Soc.* (in press).

78. Brunold, T., Güdel, H.U. *Inorg. Chem.* **1997**, *36*, 1946.

79. Deaton, J.C., Gebhard, M.S., Koch, S.A., Millar, M., Solomon, E.I. *J. Am. Chem. Soc.* **1988**, *110*, 6241.

80. Deaton, J.C., Gebhard, M.S., Solomon, E.I. *Inorg. Chem.* **1989**, *28*, 877.

81. McCarthy, P.J., Güdel, H.U. *Coord. Chem. Rev.* **1988**, *88*, 69.

82. See, for example, Janke, E., Wood, T.E., Day, P. *J. Phys. C: Solid State Phys.* **1982**, *15*, 3809.

83. Ross, P.K., Allendorf, M.D., Solomon, E.I. *J. Am. Chem. Soc.* **1989**, *111*, 4009.

84. Shugar, H., Solomon, E.I., Cleveland, W.L., Goodman, I. *J. Am. Chem. Soc.* **1975**, *97*, 6442.

85. Tuczek, F., Solomon, E.I. *J. Am. Chem. Soc.* **1994**, *116*, 6916.

86. Hitchman, M.A., Bremner, J.B. *Inorg. Chim. Acta* **1978**, *27*, L61–L63.

87. Vanquickenborne, L.G., Ceulemans, A. *Inorg. Chem.* **1981**, *20*, 796; Riley, M.J. *Inorg. Chim. Acta* **1998**, *268*, 55.

88. Desjardins, S.R., Penfield, K.W., Cohen, S.L., Musselman, R.L., Solomon, E.I. *J. Am. Chem. Soc.* **1983**, *105*, 4590.

89. Riesen, H., Gao, Y., Krausz, E. *Chem. Phys. Letters* **1994**, *228*, 610; Riesen, H., Krausz, E. *J. Chem. Phys.* **1993**, *99*, 7614.

90. Flint, C.D., Mathews, A.P. *Inorg. Chem.* **1975**, *14*, 1008.

91. Mason, S.F. *Struct. Bonding (Berlin)* **1980**, *39*, 43.

92. Bridgeman, A.J., Gerloch, M. *Inorg. Chem.* **1994**, *33*, 5411; Bridgeman, A.J., Gerloch, M. *J. Mol. Phys.* **1993**, *79*, 1195 and references therein.

93. McDonald, R.G., Riley, M.J., Hitchman, M.A. *Chem. Phys. Letters* **1987**, *142*, 529.

94. R.G. McDonald, Ph.D. thesis, University of Tasmania, **1988**, chapter 4.

5 Luminescence Spectroscopy

THOMAS C. BRUNOLD
Department of Chemistry
Stanford University
Stanford, CA 94305-5080, USA
E-mail: tbrunold@chem.stanford.edu

HANS U. GÜDEL
Departement für Chemie und Biochemie
Universität Bern, Freiestrasse 3
CH-3000 Bern 9, Switzerland
E-mail: Guedel@iac.unibe.ch

Inorganic Electronic Structure and Spectroscopy, Volume I: Methodology.
Edited by E. I. Solomon and A. B. P. Lever.
ISBN 0-471-15406-7.

1 INTRODUCTION

Luminescence, the spontaneous emission of light upon electronic excitation, is a rare phenomenon among inorganic compounds. This is due to the predominance of nonradiative relaxation processes. An electronic excitation of a complex or a metal center in a crystal usually ends up as vibrational energy and eventually as heat. In those cases where spontaneous light emission does occur, its spectral and temporal characteristics carry a lot of important information about the metastable emitting state and its relation to the ground state. Luminescence spectroscopy is thus a valuable tool to explore these properties. By studying the luminescence properties we can gain insight not only into the light emission process itself, but also into the competing nonradiative photophysical and photochemical processes.

Luminescence spectroscopy can be done on various levels of sophistication. In this Chapter we address both the generalist, who is using a luminescence spectrum as one of several techniques of characterization; and the specialist, who is interested in extracting the maximum amount of information from a highly resolved low temperature luminescence spectrum. In Section 2 we introduce the most important theoretical expressions governing the spontaneous emission of light in a complex or crystal environment. This introduction is based on Chapters 1 and 4, in which the formalism for the description of stationary states in transition metal complexes and their excitation by the absorption of light is presented. Section 3 is devoted to the measuring principle and the most important experimental setups and techniques for obtaining luminescence spectra. In Section 4 we present a number of selected examples. They were chosen to illustrate the wide scope of topics that can be addressed by luminescence spectroscopy. The spectra are analyzed by making extensive use of the formulas in Section 2. Effects such as luminescence line narrowing and spectral hole burning, which are of great importance in modern luminescence spectroscopy, have been purposely left out here because they are treated in Chapter 8.

2 THEORY

2.1 Emission of Light

In this Section we consider transitions between electronic states of luminescent ions. Absorption and emission processes involve similar formalisms, and we refer the reader to Chapter 4 for a more detailed treatment of the interaction between an ion and a radiation field.

In order to avoid confusion in the terminology of luminescence processes we first want to define the different terms we use in this Chapter. *Luminescence* is the phenomenon of spontaneous emission of light. In organic systems it is usual to distinguish between *fluorescence* and *phosphorescence*, corresponding to spin allowed and spin forbidden spontaneous emission processes, respectively. In inorganic systems the spin quantum number is not as well defined due to spin-orbit coupling, and the terms fluorescence and phosphorescence are less meaningful. The light emission process stimulated by a radiation field is called *induced* or *stimulated emission*. Although the emphasis of this Chapter will be on luminescence, we will also briefly discuss stimulated emission in order to understand the principle of laser action.

Let us consider a two-level system consisting of a ground state $|a\rangle$ and an excited state $|b\rangle$. After excitation the system is found in state $|b\rangle$ at an energy $\hbar\omega_{ba} = E_b - E_a$ above the ground state. Suppose that this system is exposed to an oscillating electromagnetic field with a frequency ω. The resulting perturbation can be expressed as

$$\mathcal{H}'(q,t) = 2\mathcal{H}'(q)\cos\omega t = \mathcal{H}'(q)\big(e^{i\omega t} + e^{-i\omega t}\big). \tag{1}$$

In Eq. 1 q stands for all the spatial variables. Time-dependent perturbation theory leads to the following expression for the induced transition rate $W_{b\to a}$:[1]

$$W_{b\to a} = \frac{2\pi}{\hbar^2}|H_{ba}|^2\rho_N(\omega_{ba}) = \frac{1}{\hbar^2}|H_{ba}|^2\rho_N(\nu_{ba}), \tag{2}$$

where

$$H_{ba} = \langle a|\mathcal{H}'(q)|b\rangle \tag{3}$$

and $\rho_N(\nu_{ba})$ is the density of frequencies in the radiation field at ν_{ba}. Eq. 2 is known as *Fermi's Golden Rule* for induced emission.

The perturbation $\mathcal{H}'(q)$ can be specified depending on the physical nature of the interaction, electric dipole (μ^{ED}) or magnetic dipole (μ^{MD}), respectively:

$$\mu^{\text{ED}} = \sum_i e\mathbf{r}_i, \tag{4a}$$

$$\mu^{\text{MD}} = \frac{e}{2m}\sum_i (\mathbf{l}_i + 2\mathbf{s}_i). \tag{4b}$$

$\mathbf{r}_i$ is the position of the i-th electron of the system. With the definition for the *transition dipole moment*:

$$\mu_{ba} = \langle a|\mu|b\rangle, \tag{5}$$

where μ is either μ^{ED} or μ^{MD}, the emission rate stimulated by an isotropic radiation field becomes:

$$W_{b\to a} = \frac{1}{6\varepsilon_0\hbar^2}\left|\mu_{ba}\right|^2 \rho(\nu_{ba}). \tag{6}$$

$\rho(\nu_{ba})$ is the energy density of the radiation field at the frequency ν_{ba} and ε_0 is a constant ($\varepsilon_0 = 8.85 \times 10^{-12}\ \text{J}^{-1}\text{C}^2\text{m}^{-1}$).

This is usually rewritten and we get for a stimulated emission process:

$$\textit{Stimulated emission:} \quad W_{b\to a} = B_{ba}\rho(\nu_{ba}). \tag{7}$$

B_{ba} is called the Einstein coefficient for stimulated or induced emission. From a comparison with Eq. 6 we see that B_{ba} is proportional to the square of the transition dipole moment as follows:

$$B_{ba} = \frac{1}{6\varepsilon_0\hbar^2}\left|\mu_{ba}\right|^2. \tag{8}$$

Denoting the degeneracies of the ground and excited states by g_a and g_b, respectively, the coefficients for the corresponding induced absorption and emission processes are related by:

$$g_a B_{ab} = g_b B_{ba}. \tag{9}$$

Einstein suggested another emission process to occur which is not stimulated by the radiation field:

$$\textit{Spontaneous emission:} \quad W_{b\to a} = A_{ba}, \tag{10}$$

where A_{ba} is the Einstein coefficient for spontaneous emission. From the relation

$$\frac{A_{ba}}{B_{ba}} = \frac{8\pi h\nu^3}{c^3} \tag{11}$$

we see that the relative importance of spontaneous emission *vs* induced emission increases with the third power of the transition frequency. This is a very important result with regard to the realization of high frequency lasers and explains why X-ray lasers are rather elusive.

The rate at which the population of the $|b\rangle$ excited state decays by spontaneous emission is given by

$$\frac{\mathrm{d}N_b}{\mathrm{d}t} = -N_b A_{ba}. \tag{12}$$

Integration yields

$$N_b(t) = N_b(0)\exp\left[-t/\tau_{\mathrm{rad}}\right], \tag{13}$$

where

$$\tau_{\mathrm{rad}} = \frac{1}{A_{ba}} \tag{14}$$

is the *radiative decay time* for $|b\rangle \rightarrow |a\rangle$ spontaneous emission.

The intensity of an absorption process $|a\rangle \rightarrow |b\rangle$ is frequently expressed in terms of the oscillator strength f, defined by:

$$f = \frac{4\pi m_e \nu_{ab}}{3e^2\hbar}\,|\mu_{ab}|^2, \tag{15}$$

$$= 4.319 \times 10^{-9} \int \varepsilon(\tilde{\nu})\mathrm{d}\tilde{\nu}, \tag{16}$$

where $\tilde{\nu}$ is in [cm^{-1}] and $\varepsilon(\tilde{\nu})$ is the molar extinction coefficient in [l/mole cm]. Since the A_{ba} and B_{ba} coefficients are related by Eq. 11, f is related to the radiative lifetime τ_{rad} of the corresponding spontaneous emission process by:[2]

$$\tau_{\mathrm{rad}} = \alpha \frac{\lambda_{ba}^2}{n\left[(n^2+2)/3\right]^2}\frac{g_b}{g_a}\frac{1}{f}. \tag{17}$$

α is a constant (1.5×10^4 s·m^{-2}), λ_{ba} is the average emission wavelength, g_b and g_a are the degeneracies of the $|b\rangle$ and $|a\rangle$ states, respectively, and n is the refractive index averaged over all polarizations. The intensity of the spontaneously emitted light is proportional to $N_b(t)$ and therefore—in the absence of nonradiative decay—τ_{rad} is an experimentally accessible quantity. It ranges from $\approx 10^{-9}$ s for fully allowed electric dipole (ED) transitions to seconds for spin forbidden ligand centered transitions.

On the basis of Eq. 17 we can in practice determine whether a measured luminescence decay time τ_{obs} is purely due to radiative processes or whether there are nonradiative contributions. And using the obvious Eq. 18 we can then calculate the $|b\rangle \rightsquigarrow |a\rangle$ nonradiative transition rate:

$$W_{b \rightsquigarrow a} = \frac{1}{\tau_{\text{obs}}} - \frac{1}{\tau_{\text{rad}}}. \tag{18}$$

Another important experimental quantity is the natural linewidth of a transition. It is one of the key parameters in Chapter 6, which deals with laser spectroscopy. According to Heisenberg's uncertainty principle the finite lifetime of the excited state $|b\rangle$ gives rise to some uncertainty in energy. This leads to a Lorentzian bandshape for the $|a\rangle \leftrightarrow |b\rangle$ transition with a full width at half maximum δE given by:

$$\delta E = \frac{\hbar}{\tau_{\text{rad}}} = \hbar A_{ba}. \tag{19}$$

Typically for an allowed ED transition $\tau_{rad} \approx 10^{-8}$ s, and therefore its natural linewidth is $\delta E \approx 0.5 \times 10^{-3}\ \text{cm}^{-1}$. Observed linewidths in the luminescence spectra of complexes or crystals are usually broader than δE by orders of magnitude. This is due to both *homogeneous* and *inhomogeneous* broadening effects in condensed systems.

2.2 Born–Oppenheimer Approximation

Luminescence spectra of ions in solids consist, characteristically, of bands with widths ranging from a wavenumber to several thousands of wavenumbers. At low temperatures these bands often show vibrational structure which is of considerable interest, since it reveals in a direct way information on the interaction between the optically active center and the crystalline environment.

The difficulty with the Hamiltonian describing an optically active ion in a complex or a crystal is that it contains a term depending on both electronic and nuclear coordinates and thus couples the electronic motion with nuclear displacements. Therefore the Schrödinger equation is not solvable analytically, and it is useful to follow the approach of Born and Oppenheimer, which is known as *simple-adiabatic approximation* or *Born–Oppenheimer* (BO) *approximation*. We do not provide a formal treatment of this approximation here, but rather refer to ref 3 and give a summary of the results.

The essence of the BO approximation is that electronic and nuclear variables can be separated, and the total wavefunction can be factorized into an electronic part and a vibrational part. The BO wavefunction for an electronic state $|k\rangle$ in its vibrational state v is then written as:

$$\Psi_{k,v}^{\text{BO}}(q, Q) = \Phi_k^Q(q)\chi_{k,v}(Q). \tag{20}$$

q and Q represent all the electronic and nuclear variables, respectively. The important property of Eq. 20 is that each of the two factors on the right-hand side is only a function of one coordinate. $\Phi_k^Q(q)$ does depend on the nuclear coordinates, but only in a parametric way, such that for a given nuclear configuration it is only a function of the electronic coordinates q. The physical rational for this separation lies in the fact that electronic motions are three orders of magnitude faster than nuclear motions.

On the basis of the BO approximation it is possible to plot the energy of a given electronic state as a function of the nuclear displacement along one of the normal coordinates of the complex. In order to discuss spectroscopic transitions we plot the two electronic states $|a\rangle$ and $|b\rangle$ in this so-called *single-configurational coordinate* (SCC) *diagram* in Figure 1. This kind of representation is very important and widely used in chemistry and physics. It provides the key for an understanding of both radiative and nonradiative processes. In Figure 1 we use the harmonic approximation and assume identical force constants in the two states. The energy minima of the ground and

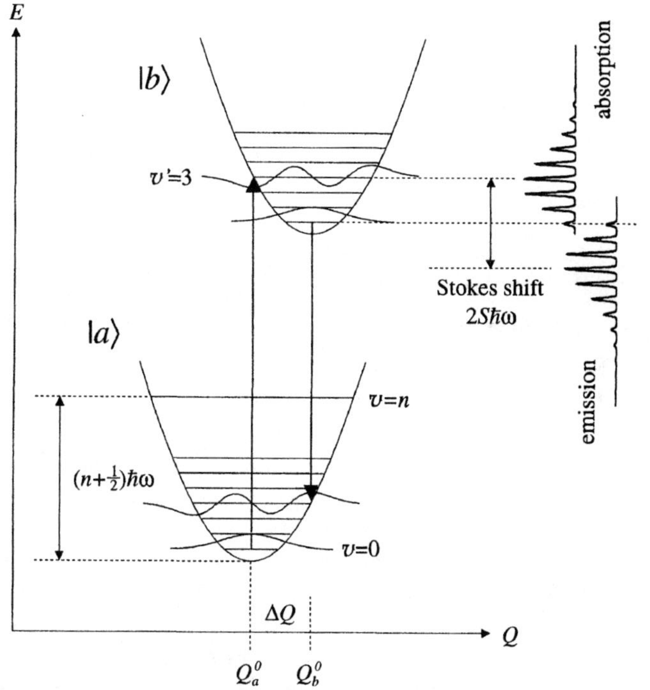

Figure 1. Single configurational coordinate (SCC) diagram in the harmonic approximation for two electronic states $|a\rangle$ and $|b\rangle$ with identical force constants. The corresponding low temperature absorption and emission spectra are shown on the right-hand side. The difference between the absorption and emission band maxima is called Stokes shift and corresponds to $2S\hbar\omega$, where S is the Huang–Rhys parameter.

excited states are at the positions Q_a^0 and Q_b^0, respectively, displaced along the relevant coordinate Q. These approximations are often made to simplify the following formalism. The vibrational functions within each potential are harmonic oscillator functions with equidistant energy spacings.

The Hellmann–Feynman theorem states that the shift of the excited state potential with respect to the ground state potential is caused by the force:[4]

$$F = -\left\langle \Psi_b \left| \left(\frac{\delta W_b}{\delta Q} \right) \right| \Psi_b \right\rangle. \tag{21}$$

Here W_b is the potential energy of the system in the excited state $|b\rangle$. If F is nonzero for the equilibrium nuclear configuration of the ground state $|a\rangle$, the configuration is unstable and the excited state will spontaneously distort along Q until F is zero. Assuming that Ψ_b is electronically nondegenerate, that is, a non-Jahn–Teller situation, we see that Q must transform totally symmetric to cause a displacement. The displacement $\Delta Q = Q_b^0 - Q_a^0$ is usually characterized by the so-called *Huang–Rhys parameter* S as follows:

$$S = \frac{k(\Delta Q)^2}{2\hbar\omega} \tag{22}$$

where k is the force constant.

The displacement of the two potentials determines the width and shape of the absorption and emission bands. Within the BO approximation the intensity distribution among the various vibrational components $\Psi_{a,v} \leftrightarrow \Psi_{b,v'}$ can be easily calculated.

2.3 Intensity Distribution

The intensity of a luminescence transition $\Psi_{b,v'} \rightarrow \Psi_{a,v}$ is proportional to

$$\left|\langle \Psi_{a,v} | \mu | \Psi_{b,v'} \rangle\right|^2 = |\mu_{ba}|^2 \left|\langle \chi_{a,v} | \chi_{b,v'} \rangle\right|^2. \tag{23}$$

In Eq. 23 $|\mu_{ba}|^2$ is the square of the electronic transition dipole moment, which determines the *total intensity* of the $|b\rangle \rightarrow |a\rangle$ emission, whereas $|\langle \chi_{a,v} | \chi_{b,v'} \rangle|^2$, the so-called *Franck–Condon* (FC) *factors*, govern the *intensity distribution*. Each FC factor is the square of an overlap integral of two vibrational functions, one in the ground state and one in the excited state. The most intense components will be the ones indicated by vertical arrows in Figure 1, $v = 0 \rightarrow v' = 3$ in absorption and $v' = 0 \rightarrow v = 3$ in emission. This is the well-known *Franck–Condon principle*.

At low temperatures only the $v' = 0$ vibrational level is thermally populated in the excited state, and the relevant FC factors for emission can be expressed in terms of the Huang–Rhys parameter as follows:

$$\left|\langle \chi_{a,v} | \chi_{b,0} \rangle\right|^2 = \frac{e^{-S} S^v}{v!}. \tag{24}$$

With the normalization

$$\sum_v |\langle \chi_{a,v} | \chi_{b,0} \rangle|^2 = 1, \tag{25}$$

the intensity distribution is then governed by the following relations:

$$\frac{I_v}{I_{v-1}} = \frac{S}{v}, \tag{26a}$$

$$\frac{I_v}{I_0} = \frac{S^v}{v!}, \tag{26b}$$

$$\frac{I_0}{I_{\text{total}}} = e^{-S}. \tag{26c}$$

Note that according to Eq. 25 the intensity of the entire band is independent of S.

In Figure 1 the computed intensity distributions for absorption and emission are shown on the upper right. The common $v = 0 \leftrightarrow v' = 0$ line serves as electronic origin for both progressions. The energy difference between the absorption and emission maxima is called *Stokes shift*. It is easy to see from Figure 1 that

$$\textit{Stokes shift} = 2S\hbar\omega. \tag{27}$$

Very often the vibrational structure is not resolved in the spectra of transition metal ions in complexes or crystals. But with a reasonable assumption about the energy of the totally symmetric metal–ligand breathing mode an estimate of the Huang–Rhys parameter and thus the excited state distortion (Eq. 22) can still be made from the Stokes shift by using Eq. 27.

The predicted bandshapes for different values of S obtained with Eq. 24 are plotted in Figure 2. For $S = 0$ all the intensity is contained in the electronic origin with $v = 0$. As the Huang–Rhys parameter increases the band maximum shifts to larger values of v, that is, to $v \approx S$. Simultaneously the bandshape changes from a Poissonian to a Gaussian curve and the bandwidth increases.

In centrosymmetric complexes the situation is slightly more complicated. Ligand field ($d \rightarrow d$) transitions are strictly Laporte forbidden, and this selection rule can be relaxed by coupling of the electronic system with odd parity vibrations. These distort the complex in such a way that the center of inversion is destroyed. In the framework of the *Herzberg–Teller coupling scheme* the Hamiltonian is expanded into a Taylor series at the equilibrium position.[5] The odd parity vibrations $Q_{\tilde{\Gamma}}$ mix some *dynamic* odd parity character into the d wavefunctions. The intensity distribution of an emission band for a Laporte forbidden $|b\rangle \rightarrow |a\rangle$ transition is then determined by the square of the *Herzberg–Teller matrix elements*:

$$|\langle \chi_{a,v} | Q_{\tilde{\Gamma}} | \chi_{b,v'} \rangle|^2 = |\langle \chi_{a,n_{\tilde{\Gamma}}} | Q_{\tilde{\Gamma}} | \chi_{b,n'_{\tilde{\Gamma}}} \rangle|^2 \prod_{\Gamma = A_{1g}} |\langle \chi_{a,n_\Gamma} | \chi_{b,n'_\Gamma} \rangle|^2, \tag{28}$$

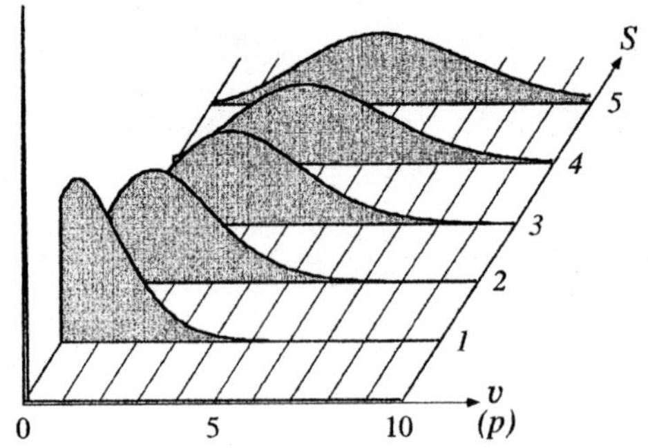

Figure 2. Franck–Condon (FC) factors at low temperatures for different values of the Huang–Rhys parameter S obtained with Eq. 24. For radiative transitions the abscissa denotes the number of vibrational quanta v in the final state and the FC factors govern the intensity distribution in the optical spectra. In the case of nonradiative transitions, the FC factors are plotted as a function of the reduced energy gap p (see Eq. 38) and relate to the multiphonon relaxation rate.

where $n_{\tilde{\Gamma}}, n'_{\tilde{\Gamma}} = 0, 1, 2, \ldots$ denote the number of quanta in mode $Q_{\tilde{\Gamma}}$ in the ground and excited state, respectively. The $Q_{\tilde{\Gamma}}$ dependent matrix element is zero unless $n_{\tilde{\Gamma}} = n'_{\tilde{\Gamma}} \pm 1$. Hence, at low temperatures the $|b\rangle \rightarrow |a\rangle$ emission band is expected to be made up of lines separated by one, and only one, quantum $\hbar\omega_{\tilde{\Gamma}}$ of the *enabling modes* from the electronic origin. On these so-called *false* or *vibronic origins* the progressions in the totally symmetric vibrations are built. Compared to allowed transitions the Stokes shift will thus be enlarged by twice the energy of the enabling mode $\hbar\omega_{\tilde{\Gamma}}$.

In our discussion of intensity distribution so far we have assumed electronically nondegenerate states $|a\rangle$ and $|b\rangle$. In high symmetry complexes degeneracies may occur and lead to some additional effects. The *Jahn–Teller* (JT) *theorem* states that if Ψ_b is a degenerate state the matrix element in Eq. 21 can exist for nontotally symmetric modes.[6] This is the case when the direct product $\Gamma(\Psi_b) \otimes \Gamma(Q) \otimes \Gamma(\Psi_b)$ contains the totally symmetric representation Γ_1, that is, when $\Gamma(Q)$ is contained in the symmetric product $[\Gamma(\Psi_b)^2]$. Exactly the same is true, of course, if the ground state Ψ_a is Jahn–Teller active. In cubic symmetry the relevant direct products decompose as follows:[7]

$$[E^2] = A_1 \oplus E, \tag{29a}$$

$$[T_{1(2)}^2] = A_1 \oplus E \oplus T_2. \tag{29b}$$

Besides a_1 coupling the following JT couplings are thus possible:

$$E \otimes e, \quad T_{1(2)} \otimes e, \quad T_{1(2)} \otimes t_2. \tag{30}$$

Linear $E \otimes e$ coupling leads to the well-known "Mexican hat" potential in the Q_e variable space. This is a completely different situation from the one depicted in Figure 1, and the resulting intensity distribution for a transition from or to such a state will significantly deviate from a regular progression. If a $T_{1(2)}$ state couples to an e mode the problem becomes remarkably simple. The potential surface in Q_e space consists of three separate parabolas, one for each orbital electronic state. The spectrum for a transition from or to a $T_{1(2)}$ state coupling to an e mode thus consists of a Poissonian band with a harmonic progression in the e mode. This is completely analogous to the coupling of a totally symmetric mode in a non-JT system. $T_{1(2)} \otimes t_2$ coupling, on the other hand, leads to a rather complicated picture. In this case the system distorts along the trigonal axes, and the bands for transitions from or to a $T_{1(2)} \otimes t_2$ JT active state are found to be very different from those in Figure 1.

Let us briefly mention an alternative principle and technique to derive information on excited state displacements from the intensity distribution. This approach is based on the work of Heller.[8] Instead of carrying out the analysis in frequency space, the time evolution of the system after excitation is calculated. Fourier transformation into frequency space then yields an intensity distribution among the various vibrational components which can be fitted to the experimental spectrum. This method gives results that are identical with those of a Franck–Condon analysis. However, thinking in the time domain makes it easier to connect electronic and resonance Raman spectroscopy. In addition, a situation with multiple distortions along various normal modes is more easily treated with this formalism. Illustrating multimode coupling in terms of a configurational diagram is difficult, so we are restricted to a single mode (see Figure 1).

According to this time-dependent theory the vibrational wavepacket ϕ of the initial electronic state is projected onto the potential energy surface of the final state, which can be displaced along *multiple* normal coordinates Q_Γ. Consequently, the wavepacket $\phi(t)$ is not a stationary state of the final surface and evolves according to the time-dependent Schrödinger equation. The optical spectrum in the frequency domain is the Fourier transform of the overlap of the initial and evolving wavepackets $\langle\phi|\phi(t)\rangle$ in the time domain:

$$I(\omega) = C \cdot \omega^x \int_{-\infty}^{\infty} \mathrm{e}^{i\omega t} \langle\phi|\phi(t)\rangle \mathrm{d}t. \tag{31}$$

C is a constant and $I(\omega)$ is the number of photons absorbed ($x = 1$) or emitted ($x = 3$) per unit time at the frequency ω. The overlap $\langle\phi|\phi(t)\rangle$ has a simple form if it is assumed that (i) the potential surfaces are harmonic with identical force constants, (ii) the transition dipole moment μ_{ab} is constant, and (iii) the multiple modes Q_Γ, along which distortions occur, are not mixed in the final state. In the low-temperature limit the overlap then takes the form

$$\langle\phi|\phi(t)\rangle = \exp\left\{\sum_\Gamma \left[-\frac{\Delta_\Gamma^2}{2}(1-\exp(-i\omega_\Gamma t)) - \frac{i\omega_\Gamma t}{2}\right] - iE_0 t - \Lambda^2 t^2\right\}. \tag{32}$$

ω_Γ and Δ_Γ are the vibrational frequency and the dimensionless displacement of normal mode Q_Γ, respectively. E_0 is the energy of the electronic origin, i.e.

$E_0 = E_{b,0} - E_{a,0}$, and Λ is a damping factor which governs the resolution of the vibrational structure. Δ_Γ is a fit parameter that is related to the Huang–Rhys parameter S_Γ and the relative displacement ΔQ_Γ by:

$$\Delta_\Gamma = \sqrt{2S_\Gamma} = \sqrt{\frac{k_\Gamma}{\hbar\omega_\Gamma}}\Delta Q_\Gamma. \tag{33}$$

Eq. 33 thus relates time-dependent theory to the semiclassical Franck–Condon approach.

2.4 Nonradiative Processes

Luminescence is a very rare phenomenon among inorganic compounds. It is exceptionally widespread in lanthanide and actinide systems. Nonradiative relaxation processes usually dominate the radiative processes to the point that no luminescence is observable even with the most sophisticated instrumentation at cryogenic temperatures. There is a multitude of possible quenching processes, and luminescence is only observed in those instances in which all these nonradiative channels are efficiently blocked.

For simplicity, as we did before, we consider a system with only two stationary electronic states $|a\rangle$ and $|b\rangle$. In analogy to the radiative transitions treated in Sections 2.1–2.3 the rate for the nonradiative transition $|b\rangle \rightsquigarrow |a\rangle$ can be expressed in terms of *Fermi's Golden Rule* for nonradiative transitions:[1]

$$W_{b\rightsquigarrow a} = \frac{2\pi}{\hbar}|H_{ba}|^2\rho(E_{a,v})\delta(E_{a,v} - E_{b,v'}), \tag{34}$$

where

$$|H_{ba}|^2 = \left|\langle a|\mathcal{H}'(q)|b\rangle\right|^2 \tag{35}$$

is the so-called *electronic factor*. $\mathcal{H}'(q)$ represents the time-independent perturbation which couples the states $|a\rangle$ and $|b\rangle$. v and v' denote the vibrational levels in the $|a\rangle$ and $|b\rangle$ states, respectively, and $\rho(E_{a,v})$ is the density of final states. The δ factor ensures energy conservation.

Even though they are formally equivalent and can both be represented by Eq. 34, one usually makes a distinction between relaxation processes within a given complex, so-called *multiphonon relaxation*, and processes involving more than one optically active center, so-called *energy transfer processes*.

2.5 Multiphonon Relaxation

The processes taking place immediately after an electronic excitation of a single complex or an isolated metal center in a crystal or glass are called multiphonon relaxation processes. The electronic excitation energy or part of it is transformed into vibrational energy and eventually into heat. The time scale of these processes corresponds to the highest energy vibrations of the system, typically picoseconds. In luminescent systems, with the exception of lanthanides and actinides, this cascade may stop at the lowest energy excited state for a radiative transition to take over. This is known as *Kasha's rule*, which was originally formulated for organic compounds.[9] In lanthanide and actinide compounds multiphonon relaxation among f levels is generally less competitive. As a consequence luminescence is also possible from higher excited states.

The quantity which is a measure of the competition between radiative and nonradiative processes in photoluminescence is the so-called *quantum yield* η. It is defined as the ratio of the number of emitted photons to the number of absorbed photons. In practice, the determination of η is a very difficult task. If the final state of the transition under consideration is the ground state $|a\rangle$, we can use Eq. 17 to calculate the radiative lifetime τ_{rad} on the basis of the $|a\rangle \rightarrow |b\rangle$ absorption spectrum. From the measured decay time τ_{obs} for $|b\rangle \rightarrow |a\rangle$ luminescence we can then calculate the quantum yield using the following expression:

$$\eta = \frac{\tau_{\text{obs}}}{\tau_{\text{rad}}}. \tag{36}$$

Let us consider the multiphonon relaxation from an excited state $|b\rangle$ to some lower state $|a\rangle$ in more detail. We use the BO approximation and the single-configurational coordinate (SCC) model shown in Figure 1. This means that the effect of the various vibrational modes in the relaxation is collected in one *effective accepting mode* Q, the abscissa in Figure 1. $|b\rangle$ and $|a\rangle$ are described by BO functions of the type Eq. 20. Multiphonon relaxation processes are then calculated in terms of a non-adiabatic perturbation $\mathcal{H}'(q)$, arising from interactions that are not diagonal in the adiabatic basis, for example, spin-orbit coupling. In the *Condon approximation*, if the integral over the electronic coordinates is regarded as independent of Q, the transition rate Eq. 34 can be expressed as follows:[10]

$$W_{b\rightsquigarrow a} = \frac{2\pi}{\hbar^2\omega} \left|\langle \Phi_a \left|\mathcal{H}'\right| \Phi_b\rangle\right|^2 g_a F_p(T)\delta(v, v' + p), \tag{37}$$

where

$$p = (E_{b,0} - E_{a,0})/\hbar\omega \tag{38}$$

is the so-called *reduced energy gap*. $\hbar\omega$ is the vibrational frequency of the effective accepting mode Q, and g_a is the electronic degeneracy of the ground state. The δ function ensures energy conservation. $F_p(T)$ is the thermally averaged Franck–Condon factor that is given by the following expression:

$$F_p(T) = \frac{\sum\limits_{v'} |\langle \chi_{a,v} | \chi_{b,v'}\rangle|^2 \exp\left[-v'\hbar\omega/k_B T\right]}{\sum\limits_{v'} \exp\left[-v'\hbar\omega/k_B T\right]}, \tag{39}$$

where summation is over all vibrational levels v' of the excited state. At low temperatures only the $v' = 0$ vibrational level is thermally populated, and Eq. 39 simplifies to:

$$F_p(T = 0) = |\langle \chi_{a,p} | \chi_{b,0} \rangle|^2 = \frac{e^{-S} S^p}{p!}, \tag{40}$$

where S is the Huang–Rhys parameter defined by Eq. 22.

For a small displacement of the ground and excited state potentials along Q, S becomes much smaller than p. This is the so-called *weak coupling case* which is characteristic for lanthanide and actinide ions. Eq. 37 then simplifies to the following expression:[11]

$$W_{b \rightsquigarrow a}(0) \propto \exp(-\beta p), \tag{41}$$

where $\beta = \ln(p/S) - 1$. Eq. 41 is the so-called *energy gap law*. The low-temperature $W_{b \rightsquigarrow a}$ rate thus exponentially *decreases* with increasing energy gap p. The relaxation between f levels in lanthanides and actinides is usually very well described by the energy gap law. As a rule of thumb for $f \rightarrow f$ relaxation of lanthanides in crystals multiphonon relaxation is competitive for $p \leqslant 6$. For the calculation of p the highest energy vibrations are relevant.

For main group and transition metal ion systems the Huang–Rhys factor S can vary by more than an order of magnitude, depending on the nature of the states $|a\rangle$ and $|b\rangle$. Spin-flip transitions, such as the $^2E_g \rightarrow {}^4A_{2g}$ transition of Cr^{3+} in ruby, are similar to the $f \rightarrow f$ situation just discussed. The two potentials have similar equilibrium distances, S is close to zero, and we are in the weak coupling limit.

In contrast, if $|b\rangle \rightarrow |a\rangle$ corresponds to a spin allowed $d \rightarrow d$ transition, a charge transfer transition, a $p \rightarrow s$ or $d \rightarrow f$ transition, we usually have $S \geqslant 1$ or $S \gg 1$ situations, corresponding to *intermediate* and *strong coupling*, respectively. The dependence of the nonradiative decay rate at low temperature on the parameters S and p is then given by Eq. 40. As shown in Figure 2, for small values of S the FC factor Eq. 40 is significant only when p is small, that is, for a small reduced energy gap. As S increases the maximum shifts to larger values of p. This is completely analogous to the situation discussed in Section 2.3 where we considered the variation of the absorption and emission bandshapes with v. Note that in the so-called *strong coupling limit* with $S \gg p$ the nonradiative relaxation rate increases as p increases. This is known as the inverse energy gap law.

Usually, in the intermediate and strong coupling cases, $W_{b \rightsquigarrow a}$ shows a strong increase at higher temperatures. Various models have been proposed to account for this T dependence, for example, the activation energy law[12] and the Struck and Fonger model.[13] This topic is treated in detail in refs 14,15.

In principle, *selection rules* can be derived from the electronic matrix element in Eqs. 35 and 37 in a similar way to the selection rules for radiative transitions. The spin selection rule $\Delta S = 0$ has some validity in systems with very small spin-orbit coupling. Thus the terms *internal conversion* and *intersystem crossing* have been introduced to distinguish between $\Delta S = 0$ and $\Delta S = 1$ nonradiative transitions in organic systems. In complexes and crystals of transition and rare earth metal ions this distinction is less meaningful because of the importance of spin-orbit coupling. For V^{3+} in a variety of host lattices the electronic factors for the nonradiative transitions ${}^3T_{2g} \rightsquigarrow {}^3T_{1g}$ and ${}^1T_{2g} \rightsquigarrow {}^3T_{1g}$ were estimated to differ by no more than a factor of two.[16]

By analogy to the enabling modes for vibronically induced radiative transitions (Eq. 28) so-called *promoting modes* have been introduced to induce nonradiative transitions.[17] They are to be distinguished from the *accepting modes* represented by the effective mode Q in Figure 1. The vibronic selection rule requires that the representations of the promoting modes occur in the direct product $\Gamma(\Psi_b) \otimes \Gamma(\Psi_a)$. The nonradiative transition ${}^3T_{2g} \rightsquigarrow {}^3T_{1g}$ for V^{3+} in an octahedral environment is thus vibronically allowed with e_g and t_{2g} as promoting modes, while ${}^4T_{2g} \rightsquigarrow {}^4A_{2g}$ in Cr^{3+} is forbidden because there is no t_{1g} normal mode in an octahedral complex. Yet, experimentally the luminescence quenching behavior was found to be very similar for analogous V^{3+} and Cr^{3+} doped chloride and bromide elpasolite lattices.[16] There is no thorough understanding of the underlying mechanisms, and it is generally true that the level of understanding of nonradiative relaxation processes is far more rudimentary than for the corresponding radiative processes.

It must be mentioned here that an excited state can also be deactivated nonradiatively by chemical reactions. This *chemical quenching* is due to the fact that the chemical reactivity of a molecule or complex may be quite different in an electronically excited state. And thus the excited state may decay (e.g. by dissociation) in a unimolecular reaction, or it may react with a partner and thus be removed nonradiatively. This belongs to the field of photochemistry which is not treated here.

2.6 Excitation Energy Transfer

In Section 2.5 we considered a single spectroscopic center diluted in a solution, glass, or host crystal. In undiluted solids there are additional relaxation mechanisms that are potentially competitive. These are energy transfer processes involving more than one spectroscopic center. All or part of the energy of an initially excited center is transferred away from that center and either emitted somewhere else or converted into heat by multiphonon relaxation.

We can again, as we did above, simplify the usually rather complex situation by considering only one step in a possible sequence of energy transfer steps:

$$D^* + A \rightarrow D + A^*, \tag{42}$$

where D and A stand for donor and acceptor, respectively, and * means electronically excited. If $D \rightarrow D^*$ can be photoexcited and $A^* \rightarrow A$ is a luminescent process, D and A are called *sensitizer* and *activator*, respectively. On the other hand, if $D^* \rightarrow D$ is a luminescent process and A^* decays nonradiatively, A is a so-called *luminescence quencher*.

The rate for the transfer process Eq. 42, based on Fermi's Golden Rule, can be expressed as:[1]

$$W_{\mathrm{DA}} = \frac{2\pi}{\hbar} \left|\langle \mathrm{D}, \mathrm{A}^* | \mathcal{H}' | \mathrm{D}^*, \mathrm{A}\rangle\right|^2 \int g_{\mathrm{D}}(E) g_{\mathrm{A}}(E)\, \mathrm{d}E, \tag{43}$$

where $\mathcal{H}'$ represents the physical interaction leading to the transfer. $g_{\mathrm{D}}(E)$ and $g_{\mathrm{A}}(E)$ are the normalized lineshape functions for the $\mathrm{D}^* \rightarrow \mathrm{D}$ emission and $\mathrm{A} \rightarrow \mathrm{A}^*$ absorption transitions, respectively. The last factor in Eq. 43 is the so-called *spectral overlap integral*. It represents both the energy conservation and the density of states factor of the general equation 34. With Eq. 43 we have implicitly assumed a *resonant energy transfer* process. Such processes can also be *phonon assisted*, and the expression for the transfer becomes more complicated.

It is important to emphasize that we are treating a process in which no radiation field is involved. This nonradiative process, in which the deactivation of D* and the excitation of A occur simultaneously, has to be clearly distinguished from the radiative energy transfer process, in which a photon is emitted by a donor and subsequently absorbed by an acceptor. The latter process, which is often called *re-absorption*, depends on the geometry of the experiment, whereas the nonradiative process does not.

Various physical interactions $\mathcal{H}'$ in Eq. 43 can lead to energy transfer. Förster treated the case of electric multipole↔multipole interactions, of which the electric dipole↔dipole (EDD) term is the most efficient in promoting transfer:[1]

$$W_{\mathrm{EDD}} = \left(\frac{1}{4\pi\varepsilon_0}\right)^2 \frac{3\pi\hbar e^4}{n^4 m_e^2 \omega^2} \frac{1}{R^6} f_{\mathrm{D}}(\mathrm{ED}) f_{\mathrm{A}}(\mathrm{ED}) \int g_{\mathrm{D}}(E) g_{\mathrm{A}}(E)\, \mathrm{d}E. \tag{44}$$

n is the refractive index, ω is the central frequency of the transition and R is the distance between D* and A. $f_{\mathrm{D}}(\mathrm{ED})$ and $f_{\mathrm{A}}(\mathrm{ED})$ are the oscillator strenghts of the $\mathrm{D}^* \rightarrow \mathrm{D}$ emission and $\mathrm{A} \rightarrow \mathrm{A}^*$ absorption transitions, respectively. We recognize two important dependences. W_{EDD} is proportional to both oscillator strengths, and it depends on the separation of D and A with the factor R^{-6}. For the analogous electric dipole↔quadrupole and quadrupole↔quadrupole mechanisms the dependences are R^{-8} and R^{-10}, respectively.

If the donor D* and acceptor A ions are close enough for direct orbital overlap, energy transfer can also be induced by exchange interactions. This was introduced by Dexter, and the transfer rate is given by:[1]

$$W_{\mathrm{EX}} = \frac{2\pi}{\hbar} \left|\langle \mathrm{D}, \mathrm{A}^* | \mathcal{H}'_{\mathrm{EX}} | \mathrm{D}^*, \mathrm{A}\rangle\right|^2 \int g_{\mathrm{D}}(E) g_{\mathrm{A}}(E)\, \mathrm{d}E. \tag{45}$$

Exchange interactions are very short range, and the transfer matrix element $\langle \mathrm{D}, \mathrm{A}^* | \mathcal{H}'_{\mathrm{EX}} | \mathrm{D}^*, \mathrm{A}\rangle$ in Eq. 45 may reasonably be written as $J_0 \exp[-R/R_0]$, where J_0 is usually approximated by the diagonal exchange term between D* and A ions in nearest-neighbor positions separated by $R = R_0$. The important difference from

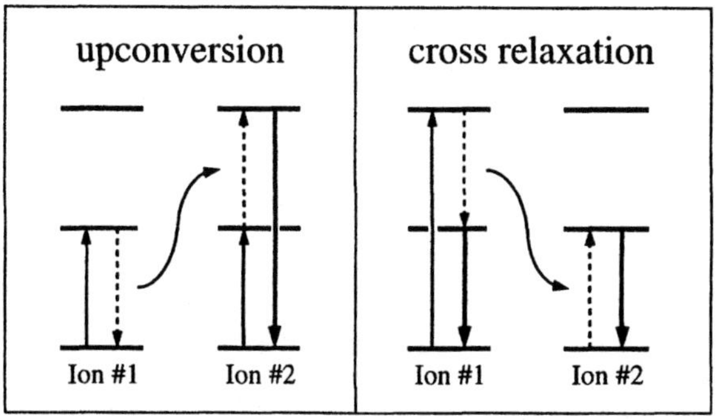

Figure 3. Schematic illustration of energy transfer upconversion and cross relaxation processes in lanthanide systems.

Eq. 44 is the independence of W_{EX} from the dipole strength of the $D^* \rightarrow D$ and $A \rightarrow A^*$ transitions.

In the real chemical and physical world there are usually not just one donor and one acceptor interacting with each other. Very often a sequence of single $D^* \rightarrow A$ transfer steps leads to energy migration. Donors and acceptors can be identical or they can be different. These cases are often referred to as energy migration and energy transfer, respectively. But the treatments of all these complicated situations are all based on the elementary steps briefly discussed above. Clever synthetic chemists have also been able to build isolated donor–acceptor complexes that correspond directly to the one-step formulas given above.

In lanthanide systems, in which there is often a multitude of metastable excited states, so-called *cross-relaxation* and *upconversion* processes are very common. They are schematically shown in Figure 3 for a three-level situation with equidistant energies. In the cross-relaxation mechanism part of the high energy excitation of one ion is transferred on the partner and both emit a low energy photon or relax by a multiphonon process. The upconversion process is exactly the reverse. Two low energy excitations on neighboring ions combine to create a single high energy excitation which may then be emitted.

3 INSTRUMENTATION AND EXPERIMENTAL TECHNIQUES

In Section 3.1 we briefly describe a typical photoluminescence setup and point to the importance of correcting luminescence spectra for system response. Various techniques of luminescence spectroscopy are then presented in Section 3.2.

3.1 Photoluminescence Setup

In a standard photoluminescence experiment the system under study is excited using a high intensity light source in the near-infrared (NIR), VIS, or UV spectral regions. For efficient excitation it is important that the output of the light source closely matches the absorption spectrum of the sample. The emitted light is dispersed by either a prism or a diffraction grating monochromator and detected with an appropriate detector. Luminescence spectra are recorded by controlling the dispersive element with a computer that also serves for data acquisition.

We do not provide a detailed description of the various components in a luminescence setup, but rather refer to ref 18 and present a rough overview. Typical broadband light sources are mercury or xenon arc lamps in the blue and UV regions and tungsten lamps in the VIS and NIR regions. An appropriate color filter must be put in front of the sample to block the excitation light in the spectral region of the luminescence. A $CuSO_4$ solution is often used for the excitation of red and NIR luminescence. Sharp-line excitation at a fixed wavelength is achieved by using lasers. Common laser light sources in the near UV, VIS, and NIR regions are various excimer lasers (around 300 nm, pulsed operation), N_2 (337 nm, pulsed operation), Ar^+ (several lines in the 455–529 nm range) and Kr^+ (lines between 407 and 676 nm) gas lasers, the third (355 nm) or second (532 nm) harmonic of a Nd: YAG laser (pulsed or continuous wave), or semiconductor diode lasers (mainly in the red and NIR) such as GaAs. Tunable lasers such as dye lasers (VIS, NIR) or a Ti: sapphire laser (700–1050 nm) allow sharp-line excitation at wavelengths which closely match the absorption spectrum of a spectroscopic center.

In luminescence spectroscopy cooling of the sample is very important, both because low temperature spectra are better resolved and may exhibit fine structure which is unresolved at room temperature, and because the luminescence quantum efficiency very often decreases with increasing temperature. The temperature dependence of the luminescence intensity and lifetime is important for an understanding of the radiative and nonradiative decay processes of the system under study. Different ways of cooling the sample are described in Chapter 4, Section 2.2. Luminescent samples can be solids, liquids, or gases. Solids can be either single crystals, polycrystalline samples, or glasses. Many luminescent metal complexes can be dissolved in solvent mixtures which form homogeneous isotropic glasses upon cooling.

We distinguish between instruments based on a dispersion element and interferometers. The former include prism and grating monochromators (spectrometers). Prism monochromators have the advantage of low light scattering, do not require order sorting, and do not produce ghosts. Their maximum resolution is about 0.01 nm in the VIS and considerably poorer in the NIR. Higher resolution ($\leqslant$0.001 nm) is achievable with grating monochromators. These can be used from the vacuum UV to the far IR, their dispersion being independent of the wavelength. Their light throughput depends on the blaze angle of the grating, but it is generally much higher than for prism monochromators. The disadvantages of grating monochromators are their high cost, the occurence of ghosts, and the necessity of removing higher orders. Nevertheless, they are the most important dispersion elements in luminescence spectroscopy. This is best demonstrated by the fact that all the spectra presented in Section 4 were measured using a grating monochromator. Interferometers combine a very high light throughput with a very high resolution of some tens of MHz at optical frequencies. They are, however, rather difficult to use, transmit many orders simultaneously, and their transmission maximum can be scanned only over a narrow wavelength range.

Detectors are mounted on the exit of the monochromator. Area sensitive multidetectors such as diode arrays or CCD cameras cover the full wavelength range defined by the dispersion of the monochromator. Scanning instruments use entrance and exit slits for wavelength selection and a grating or prism drive to scan the wavelength. The single detector is mounted behind the exit slit. The choice of an appropriate detector will be governed by factors such as the wavelength region of the radiation to be detected and the time response required to resolve dynamic events. For the wavelength region between 200 and 900 nm it is most convenient to use gallium arsenide photomultipliers that combine a high quantum efficiency with a low dark current. For measurements in the NIR a lead sulfide cell, a cooled germanium photodetector, or a special III-V semiconductor photodiode such as InSb can be used.

In general, detectors do not respond with equal efficiency to photons of different wavelengths. Also, in an experimental setup, photons of different wavelengths are not directed to the detector with equal efficiency. Thus, the intrinsic emission spectrum of the sample is only one factor in the intensity distribution measured by the detector. Hence, to obtain the intrinsic emission of the sample it is imperative to correct the experimental spectrum for system response.

Luminescence spectra can be expressed either in terms of the number of photons emitted per unit time, $\mathcal{J}$, or the total energy of photons emitted per unit time, $\mathcal{I}$. In addition, regardless of whether the spectrum is expressed in terms of $\mathcal{J}$ or $\mathcal{I}$, the corrections depend on whether the abscissa is linear in wavelength or energy. Many users of commercial luminescence instruments are not aware of the distinction between $\mathcal{J}$ and $\mathcal{I}$. A rational representation is a plot of $\mathcal{J}$ versus energy E expressed in [cm^{-1}]. The integrated intensity of an emission band is then proportional to the number of photons emitted within this band per unit time.

In a first step this correction procedure requires the determination of a so-called response curve $\mathcal{R}$ [cm^{-1}] of the system. This is achieved by (i) measuring the spectrum $\mathcal{B}'$ [nm] of a black body source (note that monochromators usually operate in wavelength mode), (ii) conversion of $\mathcal{B}'$ [nm] to $\mathcal{B}$ [cm^{-1}], and (iii) dividing the theoretical spectrum of a black body $\mathcal{J}_{\mathrm{bb}}$ [cm^{-1}], given by:[19]

$$\mathcal{J}_{\mathrm{bb}} \propto E^2 \frac{1}{\exp\left[E/k_{\mathrm{B}}T\right] - 1}, \tag{46}$$

by the experimental curve $\mathcal{B}$ [cm^{-1}]:

$$\mathcal{R} = \frac{\mathcal{J}_{\mathrm{bb}}}{\mathcal{B}}. \tag{47}$$

Once the response curve $\mathcal{R}$ is determined, it can be used for the correction of any luminescence spectrum measured under similar conditions as $\mathcal{B}'$ [nm]: The experimental curve $\mathcal{E}'$ [nm] is converted first to $\mathcal{E}$ [cm^{-1}] and then to the corrected spectrum $\mathcal{J}_{\mathrm{exp}}$ [cm^{-1}] using the following straightforward expression:

$$\mathcal{J}_{\mathrm{exp}} = \mathcal{E} \cdot \mathcal{R}. \tag{48}$$

3.2 Techniques

In the following we present various techniques that are used in luminescence spectroscopy. More details are given in refs 1,20, and additional techniques, such as those used in laser spectroscopy, are described in Chapter 8 of this book.

3.2.1 ***Continuous Wave Luminescence Spectroscopy*** This is the most common technique used to study luminescence from a spectroscopic center. The sample is optically excited by a continuous light source (see Section 3.1), and the steady state luminescence intensity is recorded as a function of wavelength or energy by scanning the dispersive element. For a physical interpretation the spectra have to be corrected for system response according to the procedure described in Section 3.1.

3.2.2 ***Time-Resolved Luminescence Spectroscopy*** In order to obtain information about dynamic processes, such as radiative and nonradiative decay or energy transfer processes, it is necessary to measure the time dependence of the luminescence spectrum after an excitation pulse. This is usually done in one of two ways. The time dependence is recorded for a given fixed luminescence wavelength. Rise and decay curves can thus be obtained. A typical simple situation is the exponential decay of an excitation created by a short laser pulse; this leads to the luminescence lifetime τ_{obs}. Alternatively, the full luminescence spectrum is recorded at various constant time delays after the excitation pulse. This so-called time-resolved spectroscopy is particularly useful to explore energy transfer processes. Recent achievements in laser physics have led to the development of sub-20 fs Ti: sapphire laser pulses.[21] Dynamic processes in these very short time scales are usually studied by applying a so-called pump-probe technique: The beam of a pulsed laser is split into a pump beam to excite the sample and a probe beam which is temporally delayed with respect to the pump beam by means of a translation stage.

3.2.3 ***Excitation Spectroscopy*** Excitation spectra are obtained by monitoring the luminescence intensity, either totally integrated or in a specific wavelength range, as a function of the wavelength of excitation. If the absorbing and luminescent species of a sample are identical and if the absorbance is low, an excitation spectrum has a profile which is very similar to that of the absorption spectrum. For luminescent samples that are difficult to crystallize or only very weakly absorbing, a study of the excitation spectrum may be the only means of measuring the absorption spectrum in any detail and with high resolution. Furthermore, excitation spectroscopy allows resolution of the absorption spectrum of a weakly absorbing luminescent species in the presence of strongly absorbing nonluminescent species. It also assists in the identification of luminescent impurities, multiple sites, and energy transfer processes in general.

3.2.4 ***Perturbation Luminescence Spectroscopy*** Perturbation spectroscopy comprises all applications of external constraints on the system under study. These include the application of hydrostatic pressure, which preserves the symmetry of the spectroscopic center but changes the strength of interaction with its surroundings.

This can affect the energies of electronic states, the spectroscopic transition probabilities, and the importance of nonradiative relaxation processes. Other applications are uniaxial stress as well as static magnetic or electric fields, all of which may result in shifts and splittings of the participating energy levels. In these latter cases the external perturbation is anisotropic, and the analyses of these perturbations have obvious similarities. Splitting of luminescence lines by a static magnetic field is referred to as the *Zeeman effect* and is related to the removal of the spin degeneracy of the levels involved in the transition. The splitting of spectral lines by an applied electric field is known as the *Stark effect*.

3.2.5 ***Optically Detected Magnetic Resonance (ODMR)*** The investigation of the magnetic properties of an isolated spectroscopic center in a solid is often done by means of electron paramagnetic resonance (EPR) spectroscopy. EPR investigations are essentially restricted to the study of ground state properties. EPR measurements in excited states are very difficult to achieve and require a long lifetime of the excited state of interest, $\tau \gg 1$ ms. For shorter-lived excited states, however, it is possible to use the much more sensitive technique of optically detected magnetic resonance (ODMR). In an ODMR measurement, a sufficiently high population in a luminescent magnetic excited state is created by optical pumping. Microwave transitions are then induced between the Zeeman or zero-field split levels of this excited state and detected by the change in intensity distribution or polarization of the luminescence. A very nice example of an ODMR experiment is given in ref 22, where this technique is used to study in great detail the luminescent triplet state of VO_4^{3-}.

4 EXAMPLES

In this Section we present and discuss some examples. The material is organized in a topical and not systematic way. The examples are chosen to illustrate some specific applications of luminescence spectroscopy mainly in the area of transition metal and rare earth metal complexes and crystals.

4.1 Determination of Excited State Geometries from Emission Spectra

According to the Franck–Condon principle, the emission process occurs without any changes in the positions of the nuclei. This is indicated by the vertical arrow in the SCC diagram in Figure 1, see Section 2.3. As a result of different bonding properties the ground and excited states may have different geometries. This leads to a relative displacement of their potential energy surfaces along some normal coordinates of the complex. If these differences are substantial the two potentials will be considerably displaced relative to each other, and the emission spectrum will consist of a broad band, see Figure 2. In case of a sufficiently resolved fine structure, the distorting modes are readily identified; they correspond to the progression forming modes. The intensity distribution can be analyzed in terms of a Franck–Condon analysis using Eq. 24 to determine the Huang–Rhys parameter S. On the basis of Eq. 22 S is then correlated to excited state distortions of the complex. Note that the accuracy of such an analysis critically depends on the resolution of the vibrational structure in the emission spectrum.

4.1.1 ***Cs_2SO_4: Mn^{6+} and Sr_2VO_4Cl: Mn^{5+}*** In Figure 4 we compare the 10 K luminescence spectra of Mn^{6+} in Cs_2SO_4[23] and Mn^{5+} in Sr_2VO_4Cl.[24] In these hosts both the Mn^{6+} and Mn^{5+} ions are in a tetrahedral oxo coordination. Despite the fact that there is only one electron more in MnO_4^{3-} than in MnO_4^{2-} the two luminescence spectra have completely different bandshapes. On the basis of the broad intensity distribution with long progressions we recognize an intermediate coupling case for MnO_4^{2-}, whereas the MnO_4^{3-} spectrum is an example of weak electron-phonon coupling. Both spectra are well resolved and thus allow detailed analyses.

Mn^{6+} has the $3d^1$ electron configuration, and luminescence is due to the $^2T_2 \rightarrow {}^2E$ ($d \rightarrow d$) transition, which involves an electronic transition from the ($\sigma + \pi$) antibonding set of t_2 molecular orbitals to the π antibonding set of e molecular orbitals. In Cs_2SO_4 the orbital degeneracy of the ^2E ground state is lifted, and the spectrum in Figure 4 is actually composed of two overlapping band systems built on the electronic origins I and II. For transition II the sidebands in O–Mn–O bending

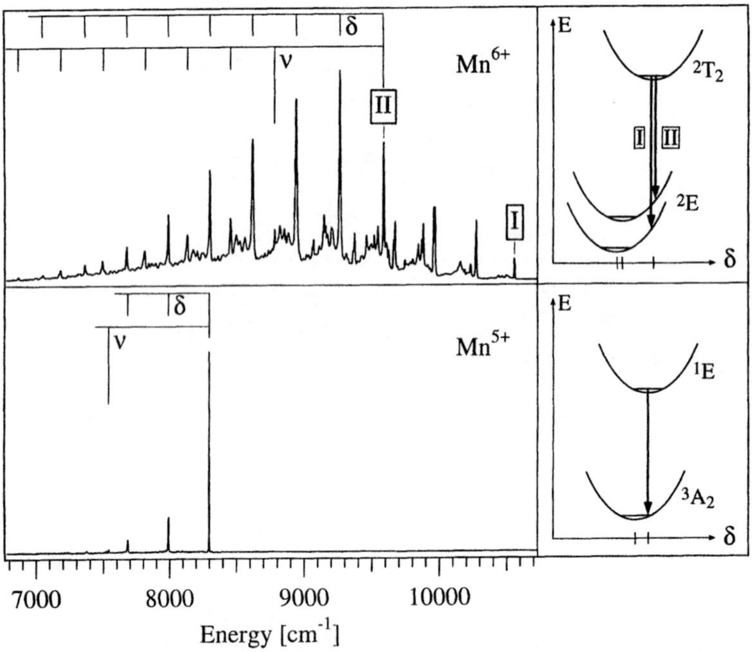

Figure 4. 10 K luminescence spectra of Cs_2SO_4: Mn^{6+} (top) and Sr_2VO_4Cl: Mn^{5+} (bottom), adapted from refs 23,24. Two electronic origins I and II, respectively, are assigned in the upper spectrum. Progressions in O–Mn–O bending (δ) and Mn–O stretching (ν) modes are indicated. The corresponding SCC diagrams along the δ coordinate are shown on the right-hand side.

(δ) and Mn–O stretching (ν) modes are indicated on the top of the spectrum. The dominant progression forming mode is the bending mode, whereas coupling to the stretching mode is less pronounced. Although the situation is slightly complicated by a weak $E \otimes e$ Jahn–Teller effect in the ground state (see Eq. 30 for the possible Jahn–Teller couplings), the corresponding Huang–Rhys parameters can be reasonably well determined in the framework of a simple Franck–Condon analysis. These are $S(\delta) = 2.4$ and $S(\nu) = 0.18$ for transition II and $S(\delta) \approx 3.5$ and $S(\nu) = 0.2$ for transition I. A schematic SCC diagram along the δ coordinate for Mn^{6+} in Cs_2SO_4 is shown in the upper graph on the right-hand side of Figure 4. As a result of a strong $T_2 \otimes e$ Jahn–Teller effect (Eq. 30) the $^2T_2$ state is considerably displaced relative to the ground state along the δ coordinate. Hence the MnO_4^{2-} ion is substantially distorted in its $^2T_2$ excited state; bond angles in the ground and excited states differ by more than 10°.

The Mn^{5+} ion has the $3d^2$ electron configuration. Both its $^3A_2$ ground and ^1E first excited states arise from the e^2 electron configuration. Luminescence is thus due to the intraconfigurational $^1E \rightarrow {}^3A_2$ spin-flip transition, and the bonding properties are not expected to change significantly. This is in accordance with the dominance of the electronic origin line in the luminescence spectrum (see Figure 4). The Huang–Rhys parameters for the O–Mn–O bending (δ) and Mn–O stretching (ν) modes are small, $S(\delta) = 0.5$ and $S(\nu) = 0.06$. Accordingly, the $^3A_2$ and ^1E potentials are only slightly displaced relative to each other (see the lower graph on the right-hand side of Figure 4).

4.1.2 *$Cs_2NaInCl_6$: Cr^{3+}* Cr^{3+} complexes and Cr^{3+} centers in crystals have played an eminent part in the development of inorganic luminescence spectroscopy and its application to questions of photophysics and photochemistry. This is because the majority of Cr^{3+} complexes exhibit luminescence around 700 nm, which is an experimentally accessible wavelength region. This luminescence is due to the $^2E_g \rightarrow {}^4A_{2g}$ transition. It is a long-lived sharp-line emission as expected for a spin forbidden spin-flip transition. In weak ligand fields there is a crossover to a $^4T_{2g}$ first excited state, and there are a few examples in the literature of $^4T_{2g} \rightarrow {}^4A_{2g}$ luminescence. This is of interest because it allows a direct probing of the properties of the $^4T_{2g}$ state. The involvement of this excited state in the photochemistry of Cr^{3+} complexes is not yet fully understood.

A number of detailed optical spectroscopic studies of the $^4T_{2g}$ state in Cr^{3+} systems of high symmetry have been reported.[25,26] In Figure 5 we show the 10 K $^4T_{2g} \rightarrow {}^4A_{2g}$ luminescence spectrum of Cr^{3+} in the cubic $Cs_2NaInCl_6$ elpasolite host.[27,28] It exhibits an unusually well resolved fine structure that can be quantitatively analyzed to obtain a detailed picture of the $^4T_{2g}$ state. The Cr^{3+} ion occupies a site of exact octahedral symmetry and the luminescence, which corresponds to a $d \rightarrow d$ transition, is thus Laporte forbidden. Hence, the intensity distribution of the emission is expected to be determined by Eq. 28, consisting of a weak magnetic dipole (MD) electronic origin and several more intense electric dipole (ED) vibronic origins. This is exactly what is observed (see Figure 5). The dominant intensity giving vibrations are the t_{2u} and t_{1u} enabling modes of the $CrCl_6^{3-}$ octahedron. Har-

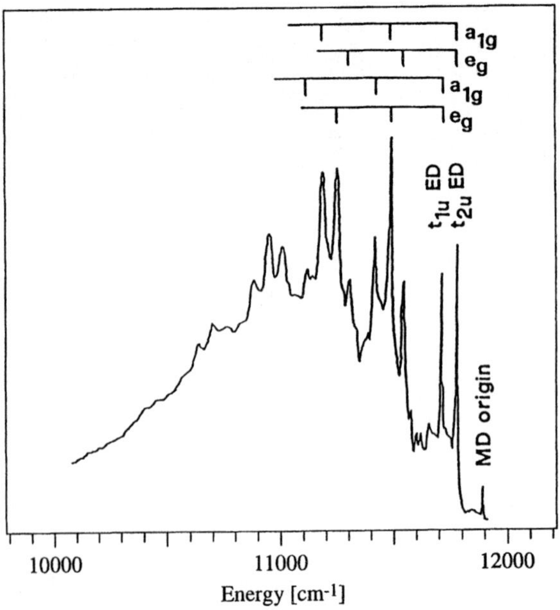

Figure 5. 10 K luminescence spectrum of $Cs_2NaInCl_6$: Cr^{3+}, adapted from ref 28. The positions of the weak magnetic dipole (MD) electronic origin and the t_{1u} and t_{2u} electric dipole (ED) vibronic origins are indicated, as are the first members of a_{1g} and e_g progressions built on the ED origins.

monic progressions built on the false origins appear in the a_{1g} as well as in the e_g mode, indicating simultaneous distortions in the $^4T_{2g}$ excited state along totally symmetric and Jahn–Teller coordinates. The corresponding Huang–Rhys parameters are $S(a_{1g}) = 1.6$ and $S(e_g) = 1.1$. This can be related to structural changes of the $CrCl_6^{3-}$ unit occuring upon $^4T_{2g}$ excitation: an axial compression by 0.02 Å and an equatorial elongation by 0.09 Å. Similar values were found in other host lattices, and it was suggested that the magnitude of these distortions might increase in a solution environment and thus provide a qualitative explanation for the photoreactivity of some Cr^{3+} complexes.[25]

Recently the excited state properties of some Cr^{3+} doped elpasolites were studied using density functional theory (DFT).[29] The total energy of the $^4T_{2g}$ state was found to decrease significantly when the CrX_6^{3-} complex (X=Cl or Br) was allowed to relax along the e_g coordinate. The calculated relaxed geometries were found in very close agreement with the spectroscopically derived ones. This suggests that the DFT technique may have some predictive value for excited state properties of coordination compounds and thus potentially for their photochemical properties.

4.1.3 ***Me₂Au(hfac)*** Let us finally consider an example of a structured emission which was analyzed in terms of the time-dependent theory of electronic spectroscopy. Metal β-diketonate complexes are of considerable interest because of their

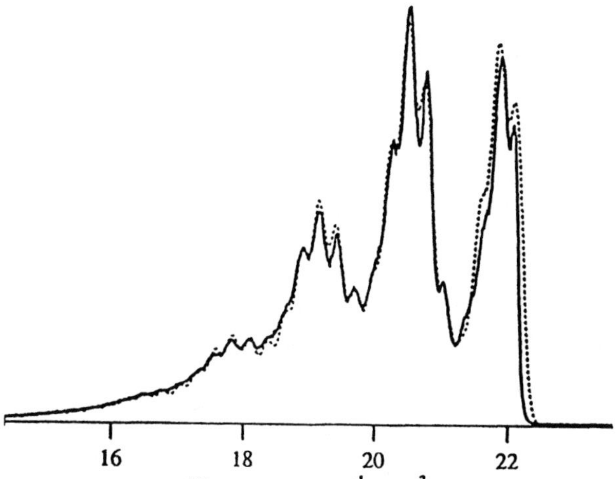

Figure 6. Comparison of the experimental (solid line) and simulated (dotted line) luminescence spectrum of frozen hexafluoroacetylacetonato dimethylgold(III), $Me_2Au(hfac)$, at 15 K. Taken from ref 31.

potential as precursors for the thermal and photochemical production of metal films via metal organic chemical vapor deposition. Irradiation of hexafluoroacetylacetonato dimethylgold(III), $Me_2Au(hfac)$, in the gas phase leads to a fragmentation of the molecule to produce single Au atoms and dimers.[30] In condensed media, on the other hand, the molecule exhibits luminescence which, at low temperatures, is fairly resolved.[31] The luminescence spectrum of $Me_2Au(hfac)$ at 15 K is shown in Figure 6 (solid line). The dominant progression forming modes are a low frequency vibration of 262 cm^{-1} and a high frequency vibration of 1352 cm^{-1}. These were assigned to totally symmetric deformation and stretching modes of the hfac ligand, respectively. This is consistent with the assignment of the emission to a ligand-centered $\pi^* \rightarrow \pi$ transition.

For an accurate determination of the excited state geometry of $Me_2Au(hfac)$, a simulated bandshape obtained in the framework of the time–dependent theory of electronic spectroscopy (see Section 2.3) was fitted to the experimental spectrum in Figure 6. Using Eq. 32 to calculate the overlap $\langle\phi|\phi(t)\rangle$ resulted in only poor correspondence between the simulated and observed spectra. The reason for the poor match is the increasing relative sideband intensity for the progression in the deformation mode with increasing quantum number of the stretching mode (see Figure 6). A much better fit was obtained when these two modes were quadratically coupled, corresponding to a distortion of the otherwise harmonic potential of the excited state. The best fit is shown by the dotted line in Figure 6. The fitted Huang–Rhys parameters for the progressions in the deformation and stretching modes are $S = 1.04$ and 1.16, respectively, indicating moderate distortions of the hfac ligand upon $\pi \rightarrow \pi^*$ excitation. The coupling between the two modes was explained in terms of a dependence of the force constant of the stretching mode on the ring bond angles that are modulated by the deformation mode. This example demonstrates that a thorough analysis of an emission spectrum may bring out details—here mode coupling—which otherwise are overlooked.

4.2 Ground and Excited State Splittings

The well-known Tanabe–Sugano diagrams show the energy levels arising from the action of a cubic ligand field on the metal *d* orbitals. Very often the symmetry of a complex or a crystal site is lower than cubic, and orbitally degenerate *E* and *T* states may be subject to splittings into two or three orbital components. Measurable splittings may also occur in spin and orbitally degenerate electronic states due to spin-orbit coupling and in polynuclear systems due to exchange interactions between the metal ions. Under favorable circumstances, energy splittings of the ground state and, if the split components are thermally accessible, of luminescent excited states can be directly determined by luminescence spectroscopy.

A first example of a mononuclear system with a split ground state is provided by MnO_4^{2-} doped into Cs_2SO_4 which was discussed in Section 4.1. In this host the site symmetry is reduced from T_d to C_s, and the orbital degeneracy of the ^2E ground state of MnO_4^{2-} is thus lifted. From the luminescence spectrum shown in Figure 4 we find a splitting of 969 cm^{-1}, corresponding to the separation between the two electronic origins I and II. In this case the luminescence spectrum provides the only way of a direct determination of this splitting.

4.2.1 ***$MgCl_2$: Ti^{2+}*** A nice example illustrating the power of luminescence spectroscopy to determine ground state splittings is provided by Ti^{2+} doped into $MgCl_2$.[32] In octahedral symmetry the Ti^{2+} ion with the $3d^2$ electron configuration has a $^3T_{1g}$ ground state. The nature of the first excited state depends on the ligand field strength. There is a crossover from a spin triplet ($^3T_{2g}$) at weak fields to a spin singlet ($^1T_{2g}$) at stronger fields. In $MgCl_2$ we are to the right of this crossing point, and at low temperatures the luminescence is due to the $^1T_{2g} \rightarrow {}^3T_{1g}$ spin-flip transition. The 10 K luminescence spectrum of $MgCl_2$: 0.1% Ti^{2+} in the NIR is shown in Figure 7. It is dominated by a series of sharp lines between 6800 and 7700 cm^{-1}, reflecting the intraconfigurational nature of these transitions. In the $MgCl_2$ host the site symmetry is reduced from O_h to D_{3d}, and the $^3T_{1g}$ ground state of Ti^{2+} is split into an upper (^3E_g) and a lower ($^3A_{2g}$) orbital component. These components are further split by the action of spin-orbit coupling; the full splitting pattern is indicated on the top of the spectrum in Figure 7. Accordingly, the six exceedingly sharp lines in the emission spectrum correspond to pure magnetic dipole transitions to the six spinor components of the ground state. These origin lines are accompanied by broader vibronic sidebands of electric dipole character, labeled vib.$^3A_{2g}$ and vib.^3E_g in Figure 7. Note that the center of inversion is retained in D_{3d} symmetry.

On warming up, the shape of the $MgCl_2$: Ti^{2+} luminescence changes drastically. The sharp-line spectrum (Figure 7) decreases rapidly, and there is a concomitant rise of a broad emission band centered around 6400 cm^{-1}. The decrease of the sharp-line to broadband intensity ratio between 40 and 100 K parallels a sharp drop of the

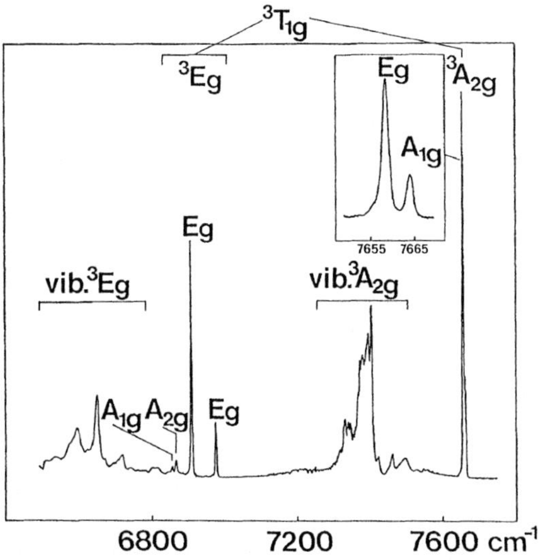

Figure 7. Unpolarized luminescence spectrum of $MgCl_2$: 0.1% Ti^{2+} at 10 K. The $O_h \rightarrow D_{3d}$ splitting of the $^3T_{1g}$ ground state is indicated. Spinor designations are given in the D_{3d} point group. The inset shows the high resolution spectrum of the $^3A_{2g}$ (D_{3d} notation) origins. Taken from ref 32.

luminescence lifetime which is shown in Figure 8. In contrast, the total intensity of the luminescence is roughly constant in the 10–200 K range. From this example we see that a drop of the luminescence lifetime is not necesserily accompanied by a drop of the quantum efficiency. In the case of $MgCl_2$: Ti^{2+} this can be explained in terms of a simple model assuming two excited states with an energy separation of ΔE and different radiative transition rates contributing to the emission: the ^1E_g component of the orbitally split $^1T_{2g}$ first excited state and the $^3T_{2g}$ state that can be considered as unsplit in a good approximation. This situation is shown in the inset of Figure 8. In the absence of nonradiative relaxation processes the decay rate can then be expressed as:

$$\tau_{\text{obs}}^{-1} = P_S \tau_S^{-1} + P_T \tau_T^{-1} \coth(\hbar\omega_{\text{av}}/2k_B T), \tag{49}$$

where

$$P_S = 2/Z, \tag{49a}$$

$$P_T = 9\exp(-\Delta E/k_B T)/Z, \tag{49b}$$

$$Z = 2 + 9\exp(-\Delta E/k_B T). \tag{49c}$$

P_S and P_T are the Boltzmann populations of the twofold degenerate ^1E_g and ninefold degenerate $^3T_{2g}$ states, respectively, and τ_S and τ_T are their respective radiative

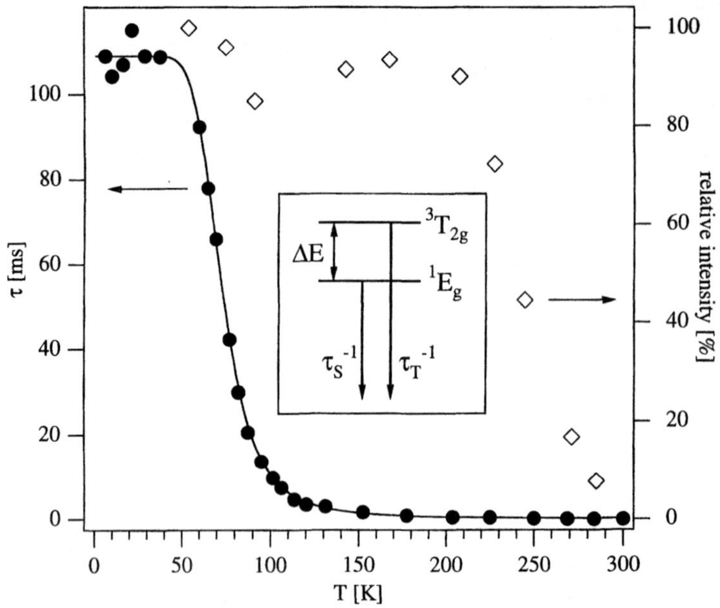

Figure 8. Temperature dependence of the luminescence lifetime (•, left-hand axis) and total intensity (◊, right-hand axis) for $MgCl_2$: 0.1% Ti^{2+}. The solid line shows a fit of Eq. 49 to the experimental data, the corresponding two-level emitting model is sketched in the inset. Adapted from ref 32.

lifetimes. Also included is a coth factor to account for the temperature dependence of the vibronic transition probability, where $\hbar\omega_{av}$ represents a properly weighted average of enabling vibrational energies (see Chapter 4). The result of a least-squares fit of Eq. 49 to the experimental lifetimes is shown by the solid line in Figure 8. The fitted parameters are $\tau_S = 109$ ms, $\tau_T = 189$ μs, $\Delta E = 401$ cm^{-1}, and $\hbar\omega_{av} = 180$ cm^{-1}. The unusually long lifetime $\tau_S = 109$ ms of ^1E_g reflects the extreme forbiddenness of the spin-flip transition to the ground state. In absorption, this transition is too weak to be observed even in crystals containing 5% Ti^{2+}. The luminescence spectrum in Figure 7 thus represents an invaluable source of information about energy splittings and intensity mechanisms. On the basis of Eq. 17, the oscillator strength f for the MD transition from the $^3A_{2g}$ component of the ground state to the ^1E_g first excited state can be calculated from τ_S. For MD transitions, in contrast to ED transitions, there is no need for a local field correction which accounts for the fact that the radiation polarizes the neighboring atomic environment,[1] and the denominator $n[(n^2+2)/3]^2$ in Eq. 17 simplifies to n^3. Using $\lambda_{ba} = 1300$ nm, $n = 1.6$, $g_b = 2$, and $g_a = 3$ and estimating that in Figure 7 the $^1E_g \rightarrow {}^3A_{2g}$ origins account for 6.5% of the total luminescence intensity, we obtain $f_{MD} = 2.5 \times 10^{-9}$ for this transition. This very low value of f is a consequence of the small spin-orbit coupling parameter for Ti^{2+}, $\zeta = 93$ cm^{-1} in this chloride environment.

4.2.2 ***$[(NH_3)_5Cr(OH)Cr(NH_3)_5]Cl_5 \cdot 3H_2O$*** Let us now turn to exchange-coupled polynuclear complexes. Such systems have attracted much interest by both physicists and chemists. The principal experimental methods used to study their ground state energy level schemes are magnetic susceptibility measurements, inelastic neutron scattering, and EPR spectroscopy. However, in some cases luminescence spectroscopy provides a means to determine such splittings with very high accuracy.

We use a hydroxo-bridged dinuclear Cr^{3+} complex, $[(NH_3)_5Cr(OH)Cr(NH_3)_5]Cl_5 \cdot 3\,H_2O$ (acid rhodo), to illustrate the principle and the obtained results. In this complex the Cr^{3+} ion has an orbitally nondegenerate (spin-only) $^4A_2$ ground state, and the exchange splitting should be describable by the well-known Heisenberg–Dirac–van Vleck operator:[33]

$$\mathcal{H}_{gs} = -2J\,(\mathbf{S}_a \cdot \mathbf{S}_b)\,. \tag{50}$$

The corresponding eigenvalues are

$$E(S) = -J\,[S(S+1) - S_a(S_a+1) - S_b(S_b+1)]\,, \tag{51}$$

giving rise to a simple *Landé pattern*. $\mathbf{S}_a$, $\mathbf{S}_b$ are the spin angular momenta of the two Cr^{3+} ions, $S_a = S_b = 3/2$. The coupled pair can thus have total spin $S = 3, 2, 1, 0$. The exchange parameter J is negative for antiferromagnetic coupling and positive for ferromagnetic coupling.

The 7 K luminescence spectrum of the acid rhodo complex is shown in Figure 9.[34] It consists of four sharp lines corresponding to transitions from the $^5B_1$ lowest energy component of the singly excited $^2E\,^4A_2$ configuration to the exchange-split levels of the $^4A_2\,^4A_2$ ground state. The energy splitting pattern is shown in the inset of Figure 9. It matches the Landé pattern Eq. 51 with $J \approx -13\ \text{cm}^{-1}$ reasonably well, indicating a moderate antiferromagnetic coupling. The deviations from the regular pattern are indicative of biquadratic exchange and can thus be analyzed. From the relative intensities of the four lines in Figure 9 we can draw some conclusions about intensity mechanisms. The most intense line is due to the only spin and symmetry allowed transition $^5B_1 \rightarrow {}^5A_1$. This transition is predominantly polarized along the Cr–Cr axis, demonstrating its exchange-induced nature. The $^5B_1 \rightarrow {}^7B_2$, $^3B_2$, and $^1A_1$ transitions gain their intensity by spin-orbit coupling.

4.3 Charge-Transfer and Ligand-Centered Luminescence

In Sections 4.1 and 4.2 we studied the luminescence spectra of some transition metal ion complexes that all exhibit luminescence due to metal centered $d \rightarrow d$ transitions. In this Section we discuss examples of systems for which the ligand orbitals directly participate in the luminescence transition.

4.3.1 ***VO_4^{3-}*** A great number of transition metal ion complexes with a formally empty d shell are known to exhibit broadband luminescence. One such system is VO_4^{3-}, in which the V^{5+} ion is tetrahedrally coordinated by four oxygens. Its highest

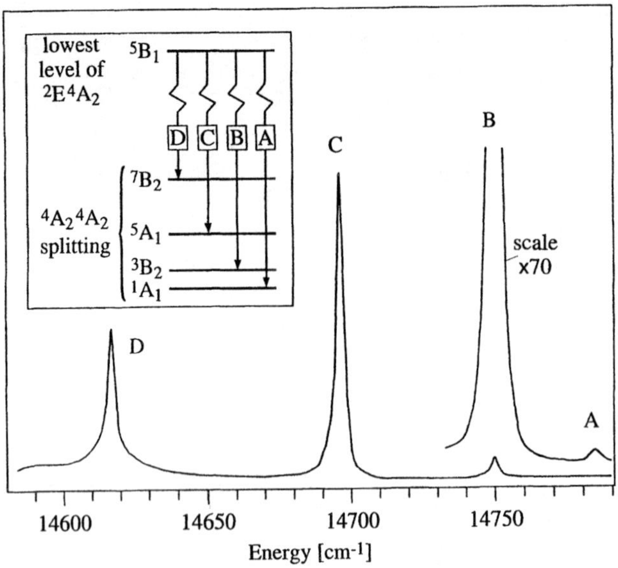

Figure 9. 7 K luminescence spectrum of $[(NH_3)_5Cr(OH)Cr(NH_3)_5]Cl_5 \cdot 3\,H_2O$ (acid rhodo). The inset shows the exchange splitting pattern and observed luminescence transitions. Adapted from ref 34.

energy fully occupied set of molecular orbitals (MO's) transforms as t_1 and is purely oxygen centered. The vanadium d orbitals are split into a set of antibonding e and t_2 orbitals, separated by an energy difference that equals the ligand field strength $10Dq$. In this one-electron picture the lowest energy excitation is $t_1 \rightarrow e$, that is, a *ligand-to-metal charge transfer* (LMCT) *transition*. However, we will show below that the effective amount of charge transfer is usually very small and the terminology thus rather confusing.

Figure 10 shows the 12 K luminescence spectrum of VO_4^{3-} adsorbed on a silica surface.[35] It consists of a broad band with a fairly resolved vibrational progression in the totally symmetric V–O stretching (a_1) mode of 970 cm^{-1}. On the basis of Eq. 22, we can relate the vibrational structure and intensity distribution in the spectrum to the V–O bond length difference Δr between the ground and first excited state of VO_4^{3-}:

$$\Delta r = \frac{1}{2}\Delta Q(a_1) = \frac{1}{2}\sqrt{\frac{S(a_1)\cdot \hbar}{\pi \cdot c \cdot \tilde{\nu}(a_1)\cdot \mu}}. \tag{52}$$

$S(a_1)$ and $\tilde{\nu}(a_1)$ are the Huang–Rhys parameter and vibrational frequency (in $[cm^{-1}]$) for the progression in a_1, respectively, and μ is the reduced mass which, in this VO_4^{3-} complex, simply corresponds to the oxygen mass for the a_1 mode.[36] From a simple Franck–Condon (FC) analysis (Eqs. 26a–26c) of the emission in Figure 10 we obtain $S(a_1) = 5.4$ and thus $\Delta r = 0.08$ Å. Since S depends on the square of ΔQ (Eq. 22), it is not possible to determine the sign of ΔQ from a FC analy-

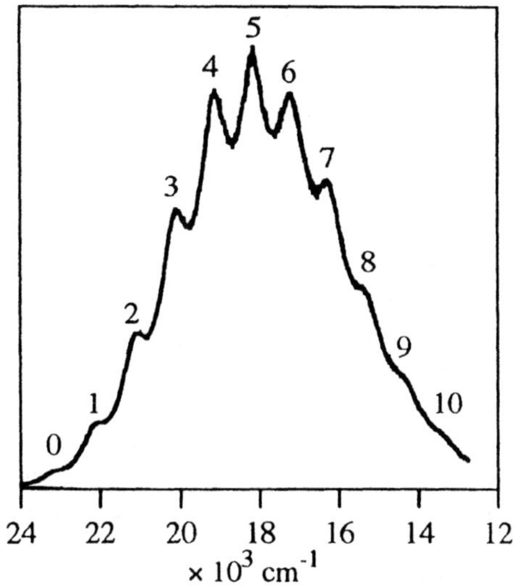

Figure 10. 12 K luminescence spectrum of SiO_2: 0.1% V^{5+}. Vibrational sidebands for the progression in the totally symmetric V–O stretching mode of VO_4^{3-} are designated. Adapted from ref 35.

sis. However, since this LMCT excitation involves an electronic transition from a nonbonding (t_1) to an antibonding (e) set of MO's, we can safely assume that it is accompanied by a V–O bond length increase.

It is remarkable that the increase of the V–O bond length in VO_4^{3-} for this nominal charge transfer excitation is only $\Delta r = 0.08$ Å and thus smaller than $\Delta r = 0.09$ Å for the equatorial Cr–Cl bonds in $CrCl_6^{3-}$ upon $^4T_{2g}$ ($d \rightarrow d$) excitation (see Section 4.1). This is a clear indication that the one-electron picture, which suggests a transfer of a full charge from oxygen to vanadium upon LMCT excitation of VO_4^{3-}, is rather too simple. Calculations based on both Hartree–Fock and local density functional principles on the isoelectronic MnO_4^- ion showed a substantial reorganization of electron density upon $t_1 \rightarrow e$ and $t_1 \rightarrow t_2$ excitations. As a result only a small fraction of a full charge is actually transferred in these nominal LMCT transitions.[37,38]

4.3.2 ***[Ir(thpy)₂bpy]⁺*** In octahedral chelate complexes with π-accepting ligands we have a completely different situation. The $t_2 \leftrightarrow e$ ligand field splitting can be so large that a π^* ligand orbital or even both the π and π^* ligand orbitals lie between the t_2 and e metal orbitals. The nature of the first excited state in such complexes is either *metal-to-ligand charge transfer* (MLCT) or *ligand centered* (LC) $^3\pi$–π^*. The majority of complexes in this family have the $4d^6$ or $5d^6$ electron configuration, and most of them luminesce both in the solid state and in fluid solution. The metastable first excited state of many of these complexes is of interest because of its photocatalytic and photosensitizer properties.[39,40] Luminescence spectroscopy is used to probe the properties of this state. The best studied complex in this family is $[Ru(bpy)_3]^{2+}$ (bpy=2,2′-bipyridine) which shows a bright orange luminescence in liquid solution at room temperature. It has been assigned to a ^{3}MLCT emission. For a distinction between MLCT and LC first excited states the luminescence bandshape is often used as a simple criterion. $^3\pi$–π^* emissions are generally structured and relatively narrow, while MLCT transitions are broad and characterized by large Stokes shifts.

An interesting situation arises when the first $^3\pi$–π^* and MLCT excited states approach each other. Such a situation occurs in the mixed ligand complex $[Ir(thpy)_2$-$bpy]^+$, where thpyH stands for the cyclometalating ligand 2-(2-thienyl)pyridine.[41] Figure 11 shows the luminescence spectra of this complex in different media and at different temperatures. The 300 K luminescence spectrum in CH_2Cl_2 is characterized by a broad structureless band with a maximum at 16670 cm^{-1}. Imbedding the complex into a rigid host drastically changes the luminescence bandwidth and bandshape. At 300 K, the luminescence spectrum in a PMMA glass (not shown) and in the crystalline host $[Rh(ppy)_2bpy]PF_6$ consists of a fairly structured band with its first maximum at 18500 and 18700 cm^{-1}, respectively. On cooling the sample down to cryogenic temperatures, the PMMA spectrum shows only a minor reduction of bandwidths. In contrast, the $[Ir(thpy)_2bpy]^+$ luminescence shows a very strong temperature dependence in the crystalline host $[Rh(ppy)_2bpy]PF_6$; at 5 K it exhibits a great deal of well resolved fine structure with individual linewidths of less than 6 cm^{-1}.

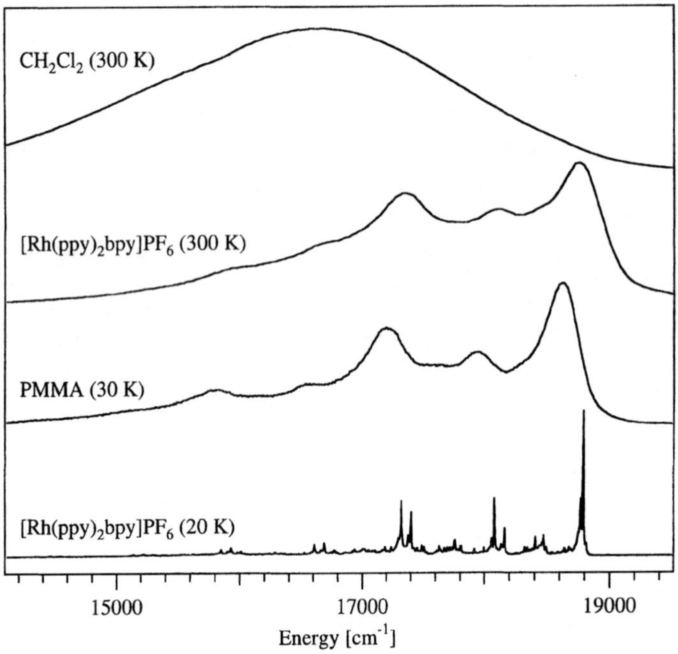

Figure 11. Luminescence spectra of $[Ir(thpy)_2bpy]PF_6$ (thpyH=2-(2-thienyl)pyridine, bpy=2,2′-bipyridine) in different media at different temperatures. The concentrations are 10^{-5} M in CH_2Cl_2, 1 mol % in $[Rh(ppy)_2bpy]PF_6$, and 10^{-3} M in poly(methylmetacrylate), PMMA. Adapted from ref 41.

The broad band in the CH_2Cl_2 solution spectrum was ascribed to a MLCT transition, whereas the structured band in the solid media was interpreted as a $^3\pi$–π^* emission. The sharp fine structure that develops in the crystal spectrum on cooling is typical of a LC transition, and the bottom trace in Figure 11 strongly resembles the luminescence spectrum of the free thpyH ligand at cryogenic temperatures. The environment of the complex obviously determines whether the emitting state is a $^3\pi$–π^* or a ^{3}MLCT in this case. MLCT transitions are expected to be more susceptible to changes of the environment than LC transitions. In the fluid CH_2Cl_2 solution the lowest ^{3}MLCT state is therefore thought to shift below the first $^3\pi$–π^* state to become the emitting state.

The low temperature luminescence spectra of $[Ir(thpy)_2bpy]^+$ in PMMA and $[Rh(ppy)_2bpy]PF_6$ shown in Figure 11 nicely illustrate the phenomenon of *inhomogeneous broadening* in glasses. In the crystalline host each complex occupies a well defined site, and the energies of the complexes in a given state thus fall within a very narrow range. Cooling of the sample to cryogenic temperatures lowers the thermal occupancy of lattice modes, the homogeneous linewidths decrease and the vibrational structure in the luminescence spectrum becomes exceedingly sharp. This structure allows an unambiguous identification of the ligand on which the lowest LC excitation is localized. In the present case $thpy^-$ was identified as the active ligand from the vibrational sideband pattern. In the PMMA glass, on the other hand, there is a distribution of complex environments, leading to a spread of energies of the complexes in a given state. Therefore, the emission spectrum corresponds to the superposition of a large number of slightly different luminescence spectra, one for each complex, and cooling of the sample does not lead to a pronounced narrowing of the bands. There is, however, an experimental technique called *luminescence line narrowing* (LLN) that can be applied to resolve the fine structure in such instances. By using a laser wavelength in the electronic origin region of the luminescence at very low temperatures a small subset of the inhomogeneous distribution can be selectively excited. This subset is then characterized by a narrow site distribution which leads to a narrowed luminescence spectrum. Examples of LLN experiments will be presented in Chapter 8.

The experimental lifetime of the LC excited state of $[Ir(thpy)_2bpy]^+$ in $[Rh(ppy)_2bpy]PF_6$ at 10 K is 17 μs and essentially radiative. This is about 3 orders of magnitude lower than for the free $thpy^-$ ligand, suggesting that the LC transition acquires considerable charge transfer character in the complex. This is fully supported by the occurrence of two vibrational sidebands separated by 297 and 315 cm^{-1} from the electronic origin in the low temperature crystal spectrum (Figure 11). These lines fall into a region where metal–ligand bending and stretching modes are expected to occur and thus clearly manifest some charge transfer character of the nominally ligand-centered emission.

A phenomenon called *dual emission* has been the subject of much debate in the inorganic literature. In the early 1970's it was postulated that in transition metal complexes with unfilled *d* shells luminescence should always originate from the lowest excited state or thermally populated higher excited states.[42] Later it was suggested that nonradiative relaxation between excited states of different orbital parentage may be hindered,[43] thus leading to dual emission from these two thermally nonequilibrated excited states. The main characteristics of dual emission would be biexponential decay curves and bandshapes that depend on the excitation wavelength. This has been observed for several complexes, in particular in glassy matrices. Nevertheless, there is a physically more appealing explanation for this phenomenon than nonequilibrated excited states. First we note that nonequilibrated excited states can only exist if both radiative and nonradiative transitions from one into the other are slower than the radiative and nonradiative transitions to the ground state. Such situations are well documented in organic systems in which spin-orbit coupling is very weak and thus the $S_1 \rightarrow T_1$ relaxation very slow compared to the $S_1 \rightarrow S_0$ emission process (where S_1 and T_1 refer to singlet and triplet excited states, respectively, and S_0 is the ground state). It is also well known that in glassy environments the thermal equilibration within the first excited triplet state T_1 in organic systems can be extremely slow at very low temperatures due to a slowing down of spin-lattice relaxation processes. The physical situation in metal complexes is very different, however, and dual emission is most likely the result of the coexistence of two sets of complexes in the glass, one exhibiting one type of emission and the other one another type. The $[Ir(thpy)_2bpy]^+$ complex discussed above is a complex whose emitting state depends on the environment. In a glass we have a distribution of environments, ranging from fluid-like to solid-like. As a consequence some complexes may have a MLCT and others a LC emitting state. There are several examples of dual emitters in the literature for which this is a likely explanation.[44]

4.4 Competition Between Radiative and Nonradiative Processes

In this Section we discuss the use of luminescence spectroscopy for studying nonradiative processes, first those taking place within an isolated complex and then in systems where energy transfer processes occur.

4.4.1 *$RbGd_2Cl_7$: Er^{3+} and $RbGd_2Br_7$: Er^{3+}* Figure 12 shows the 300 K luminescence spectra of Er^{3+} doped $RbGd_2Cl_7$ (top) and $RbGd_2Br_7$ (bottom).[45] Er^{3+} is a lanthanide ion with the $4f^{11}$ electron configuration, and luminescence is due to $f \rightarrow f$ transitions. These are the first lanthanide spectra presented in this Chapter and we open the discussion with some general considerations. Lanthanide ions are characterized by an incompletely filled $4f$ shell which is well shielded from its surroundings by the filled $5s$ and $5p$ orbitals. Their optically active $4f$ electrons are thus only weakly affected by neighboring ligands. As a result, the gross features of the $f \rightarrow f$ absorption spectra do not depend on the chemical composition or site symmetry of the host lattice, and the well-known diagrams of *Dieke*[46] can be used for an assignment of the multiplets. In a SCC diagram (Figure 1) the potentials of the ground and $f \rightarrow f$ excited states are essentially undisplaced for any normal coordinate of the complex. S is thus very small, typically $S \leqslant 0.1$. This weak electron-phonon coupling results in a low efficiency of nonradiative relaxation processes (see Eq. 40 and Figure 2), and luminescence is a very common phenom-

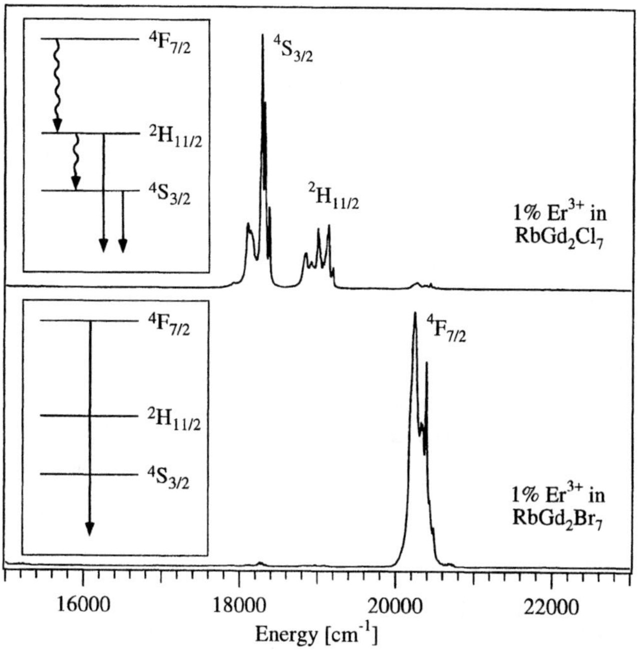

Figure 12. Room temperature upconversion luminescence spectra of $RbGd_2Cl_7$: 1% Er^{3+} (top) and $RbGd_2Br_7$: 1% Er^{3+} (bottom) for ${}^4I_{15/2} \rightarrow {}^4I_{11/2}$ excitation. The two insets show a schematic energy level diagram and the dominant luminescence transitions to the ${}^4I_{15/2}$ ground state. Adapted from ref 45.

enon in complexes and crystals containing lanthanide ions. In contrast to transition metal ions, luminescence usually occurs from more than one excited state.

The spectra in Figure 12 have the typical appearance of $f \rightarrow f$ luminescence spectra. They consist of several sharp multiplets corresponding to different electronic transitions. Their fine structure is due to crystal field splittings of some hundreds of wavenumbers. Interestingly, the two luminescence spectra of Er^{3+} doped $RbGd_2Cl_7$ and $RbGd_2Br_7$ in Figure 12 are strikingly different, although $4f$ electrons are not expected to be strongly affected by the environment; the absorption spectra are very similar in this region. In the chloride the luminescence is essentially due to transitions from the ${}^4S_{3/2}$ and ${}^2H_{11/2}$ excited states to the ${}^4I_{15/2}$ ground state, whereas in the bromide ${}^4F_{7/2} \rightarrow {}^4I_{15/2}$ luminescence accounts for more than 90% of the total intensity. The explanation is given in the two insets of Figure 12.

Upon ${}^4F_{7/2}$ excitation efficient multiphonon relaxation processes followed by luminescence from the thermally equilibrated ${}^2H_{11/2}/{}^4S_{3/2}$ states occur in the chloride. In contrast, the multiphonon relaxation is not efficient in the bromide and ${}^4F_{7/2}$ is the emitting state. This is also reflected in the observed ${}^4F_{7/2}$ luminescence decay times at 295 K: $\tau_{obs} = 6\ \mu s$ and $110\ \mu s$ for the chloride and bromide, respectively.

From the spectra in Figure 12 we recognize a weak coupling case and can thus use Eq. 41 for a discussion of multiphonon relaxation processes in the two hosts. The important quantity that determines the multiphonon relaxation rate $W_{b \rightsquigarrow a}(0)$ is the reduced energy gap p. The energy separation $\Delta E(^4F_{7/2} \leftrightarrow {}^2H_{11/2})$ is about 1250 cm^{-1} in both compounds. The highest energy vibration determined by Raman spectroscopy, on the other hand, decreases from 264 cm^{-1} in $RbGd_2Cl_7$ to 180 cm^{-1} in $RbGd_2Br_7$. This will be the most effective accepting mode, and according to Eq. 41 we obtain $p = 4.7$ and 6.9 for the chloride and bromide, respectively. With $S \approx 0.1$ it follows from Eq. 41 that $W_{b \rightsquigarrow a}(0)$ decreases by more than 3 orders of magnitude. The radiative rate constant for $^4F_{7/2} \rightarrow {}^4I_{15/2}$ lies in between $W_{b \rightsquigarrow a}(0)$ of the chloride and bromide, which makes these systems behave so differently.

As a result of the small radiative and nonradiative transition rates, there is a multitude of metastable excited states in lanthanide systems, and several processes are known which lead to the population of highly excited states after excitation in the NIR. Most of these so-called *upconversion* processes rely either on an energy transfer upconversion (ETU) or an excited state absorption (ESA). They are schematically shown in Figure 13 for a three level situation with equidistant energies. In the ETU process an excited ion nonradiatively transfers its energy to another excited neigh-

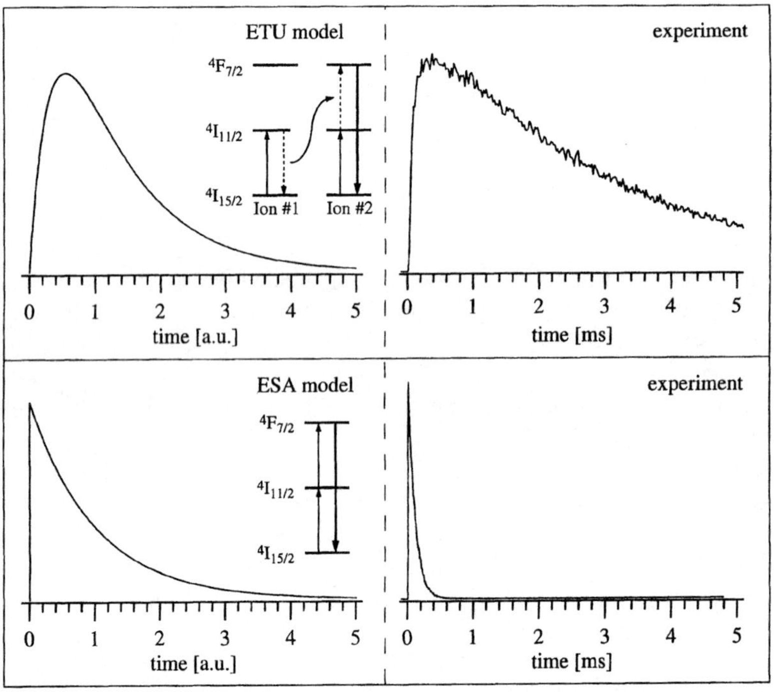

Figure 13. Left-hand side: Schematic illustration of energy transfer upconversion (ETU, top) and excited state absorption (ESA, bottom) processes for a three-level situation with equidistant energies. The corresponding upconversion transients for excitation by a short pulse are also shown. Right-hand side: $^4F_{7/2} \rightarrow {}^4I_{15/2}$ upconversion transients at 10 K of Er^{3+} doped $RbGd_2Br_7$ for two slightly different excitation wavelengths in the NIR. Adapted from ref 47.

boring ion. In contrast, the ESA upconversion mechanism is a single ion process, in which an excited ion absorbs a second photon and eventually emits a single high energy photon. In practice, perfect resonance between the involved transitions is very rare, and in the ETU process energy mismatches are compensated by the absorption or emission of low frequency phonons.

By using excitation pulses in the ns range or shorter it is possible to distinguish between ETU and ESA processes. The nonradiative ETU process can proceed after the excitation pulse, whereas the radiative ESA process must occur within the laser pulse. The expected rise and decay behavior of the luminescence after an excitation pulse is shown for both ETU and ESA processes on the left-hand side of Figure 13.

On the right-hand side of Figure 13 we compare the observed $^4F_{7/2} \rightarrow {}^4I_{15/2}$ upconversion transients at 10 K of Er^{3+} doped $RbGd_2Br_7$ (see luminescence spectrum in Figure 12) for two slightly different excitation energies in the NIR.[47] From a comparison with the model curves in Figure 13 we can readily identify the involved upconversion mechanisms, an ETU process in the upper case and ESA in the lower case. The single exponential decay of the ESA transient corresponds to the intrinsic lifetime $\tau_{obs} = 110\ \mu s$ of $^4F_{7/2}$. In comparison, the ETU transient is characterized by a much slower decay. This is because both the decay rate of the intermediate level and the energy transfer rate are significantly smaller than the intrinsic $^4F_{7/2}$ decay rate. This leads to the interesting situation that the rise part of the ETU transient (upper trace) actually corresponds to the decay of the $^4F_{7/2}$ state (see lower trace).[48]

4.4.2 ***$OsBr_6^{2-}$*** In transition metal ion systems interexcited state luminescence is very rare. The optically active *d* electrons are more strongly coupled to the phonons than *f* electrons. There are, however, a few examples of transition metal ion doped crystals in the literature which exhibit interexcited state luminescence. From the above discussion it is clear that this can only occur in those instances in which multiphonon relaxation channels are efficiently blocked, namely if the frequency of the highest energy vibration is low, the energy gap to the next lower levels large and the corresponding Huang–Rhys parameters S close to zero. A nice example of interexcited state luminescence is provided by $OsBr_6^{2-}$ in Cs_2SnBr_6, where no less than ten luminescence transitions can be definitely assigned.[49] The combination of a large spin-orbit coupling constant, large crystal field splitting, and a small interelectron repulsion parameter generates several states below 20000 cm^{-1} which derive from to the same t_{2g}^4 strong field configuration as the ground state. Transitions between these states do not result in any change in the number of nonbonding or antibonding electrons and therefore are characterized by very small Huang–Rhys parameters S. In the Cs_2SnBr_6: Os^{4+} system many transitions that were not easily accessible by absorption spectroscopy could be examined using luminescence spectroscopy.

4.4.3 ***$(bpy)_2Ru(BAB)Os(bpy)_2$*** Our last example in this Section is devoted to a photoinduced energy transfer process. In the natural photosynthetic process the solar energy is absorbed by supramolecular arrays of molecules that are strongly light absorbing and able to channel the excitation to the active site. Numerous attempts have been made to model this situation in artificial supramolecular devices based on transition metal complexes. Polynuclear complexes containing both Ru^{2+} and Os^{2+} with polypyridine-type ligands are ideally suited for this purpose, and a considerable effort has been made to understand energy- and electron-transfer processes in such systems.[39,50,51] In both Ru^{2+} and Os^{2+} bipyridyl complexes oxidation processes are metal centered, reduction processes are ligand centered, and the lowest energy excited state is a ^{3}MLCT state. It serves as the initial state for broadband luminescence to the $^1A_1$ ground state. The Ru and Os systems nicely complement each other in that the Os^{2+} complexes can be oxidized at less positive potentials than the Ru^{2+} analogues. The luminescent ^{3}MLCT state of an Os^{2+} complex lies lower in energy than that of the analogous Ru^{2+} complex. In a polynuclear mixed Ru^{2+}/Os^{2+} complex the spectral overlap of the Ru^{2+} MLCT emission and Os^{2+} MLCT absorption (Eq. 43) is thus nonvanishing. The interaction between the bridged metal centers in this type of complex critically depends on the size, shape, and electronic nature of the bridging ligands.

The room temperature absorption spectra in an acetonitrile solution of $(bpy)_2$-$M(BAB)M'(bpy)_2$ (abbreviated by M⋈M′ in the following), where M, M′ are Ru^{2+} and/or Os^{2+} and BAB is a polypyridine-type bridging ligand separating M and M′ by 14–15 Å, are shown in Figure 14.[52] They consist of a series of very intense bands in the UV below 350 nm corresponding to ligand-centered (LC) $\pi \rightarrow \pi^*$ transitions

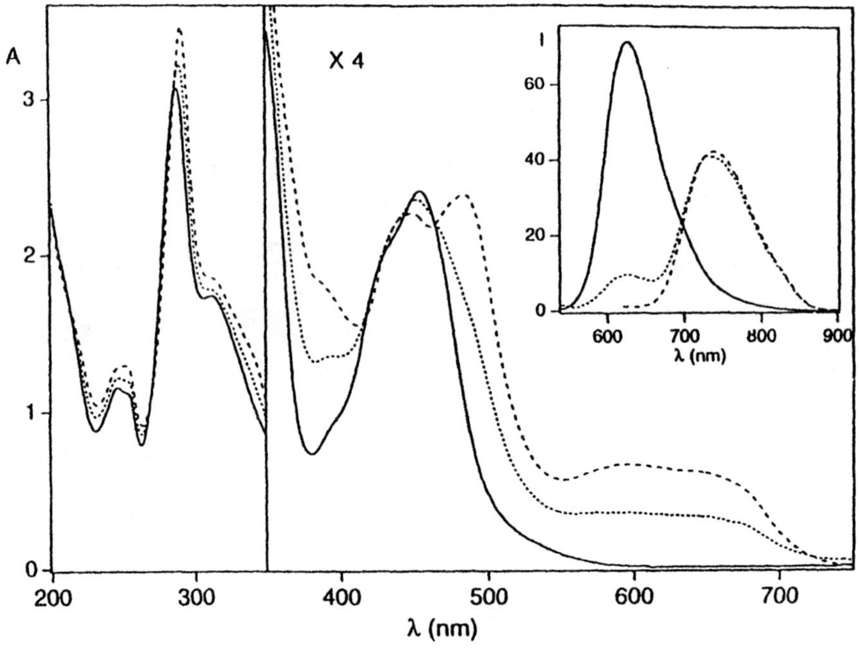

Figure 14. Room temperature absorption and (inset) emission spectra in acetonitrile solution of $(bpy)_2M(BAB)M'(bpy)_2$, M⋈M′: Ru⋈Ru (solid line), Ru⋈Os (dotted line), and Os⋈Os (broken line). BAB is a bridging ligand which separates M and M′ by 14–15 Å. Taken from ref 52.

and the characteristic $^1A_1 \rightarrow {}^1$MLCT band centered at about 450 nm. In the heterometallic Ru⋈Os and homometallic Os⋈Os complexes an additional weak band is observed in the 550–700 nm region that was assigned to the lowest energy spin forbidden $^1A_1 \rightarrow {}^3$MLCT transition of the Os unit. Its analog on the Ru unit is not observable in the absorption spectrum of Ru⋈Ru due to the lower spin-orbit coupling constant of Ru^{2+}.

Using absorption spectroscopy it was demonstrated that the metal-metal interaction in M⋈M′ complexes is very weak: (i) the absorption spectra of the Ru⋈Os complex and the 1:1 mixture of the analogous homometallic complexes are identical and (ii) there are no observable intervalence transfer bands in the mixed-valence compounds. However, even an interaction of a few wavenumbers, which cannot be noticed in optical absorption experiments, is sufficient to cause important energy- and electron-transfer processes between the metal centers which can be probed by luminescence spectroscopy.

Luminescence spectra of isoabsorptive acetonitrile solutions of Ru⋈Ru, Os⋈Os, and the heterometallic Ru⋈Os complex are shown in the inset of Figure 14. Excitation of Ru⋈Os leads to ^{3}MLCT luminescence of both the Ru and Os units. However, the intensity of the Ru luminescence in Ru⋈Os is much weaker than in Ru⋈Ru, while the intensity of the Os luminescence is comparable with that in Os⋈Os. This is a clear indication of Ru* → Os excitation transfer processes in the Ru⋈Os system since otherwise the luminescence spectrum would simply look like the average of the Ru⋈Ru and Os⋈Os spectra.

In principle, quenching of the Ru luminescence in Ru⋈Os can be due to Ru* ↔ Os electron transfer or Ru* → Os energy transfer. Electron transfer either leads to $Ru^+ ⋈ Os^{3+}$ or $Ru^{3+} ⋈ Os^+$, whereas energy transfer results in Ru⋈Os*. The relative increase of the Os luminescence runs parallel to the relative decrease of the Ru luminescence in Ru⋈Os, and an electron transfer process can thus be ruled out. Further evidence for energy transfer is provided by time-resolved luminescence experiments. The quenching of the Ru luminescence intensity is accompanied by a corresponding reduction of the decay time. At the same time the Os luminescence shows a rise time which is the same as the Ru decay time. On the basis of the relative intensities as well as rise and decay times it was possible to determine the rate for Ru* → Os energy transfer $W_{Ru\rightarrow Os} = 6 \times 10^8\ s^{-1}$ at room temperature. This is two orders of magnitude larger than the decay rate for the Ru emission, as determined in the Ru⋈Ru complex: $(\tau_{obs})^{-1} = 5 \times 10^6\ s^{-1}$. The nonradiative Ru* → Os energy transfer process thus competes very favorably with the radiative Ru emission process in Ru⋈Os.

4.5 Application of Hydrostatic Pressure

The behavior of luminescence spectra under an external physical perturbation such as hydrostatic pressure, uniaxial stress, a magnetic or an electric field can be of great help in characterizing emitting states or energy transfer processes.[53,54] We illustrate this with two examples in which hydrostatic pressure was used as a perturbation.

4.5.1 ***KZnF$_3$: Cr^{3+}*** Luminescence from Cr^{3+} in an O_h ligand field is typically very sharp with $S \ll 1$ since it corresponds to the spin forbidden $^2E_g \rightarrow {}^4A_{2g}$ spin-flip transition. In a weak ligand field, however, a crossover from ^2E_g to a $^4T_{2g}$ first excited state occurs, and the resulting interconfigurational $^4T_{2g} \rightarrow {}^4A_{2g}$ luminescence is broad; an example of this was discussed in Section 4.1 (see Figure 5). There are numerous chromium complexes, such as $Cr(H_2O)_6^{3+}$ and $Cr(urea)_6^{3+}$, which exhibit emission from both excited states.[55] With increasing temperature the relative intensity of the broadband component in these dual emitting systems increases drastically while the sharp-line component decreases.

In the $KZnF_3$ host Cr^{3+} is octahedrally coordinated by six fluorides. The 90 K luminescence spectrum of $KZnF_3$: Cr^{3+} at ambient pressure is shown in Figure 15, lower curve.[56] It consists of a broad band centered at about 13000 cm^{-1}, exhibiting some fine structure in the origin region. Fluoride is a weak-field ligand, and under ambient pressure conditions luminescence of $KZnF_3$: Cr^{3+} is due to $^4T_{2g} \rightarrow {}^4A_{2g}$. This is sketched in the inset of Figure 15, solid lines. Although the Cr^{3+} ions predominantly occupy a cubic site in $KZnF_3$, trigonally and tetragonally distorted centers are also present. The $^4T_{2g} \rightarrow {}^4A_{2g}$ band origin (labeled C in Figure 15) belongs to cubic sites, and we focus on these centers in the following.

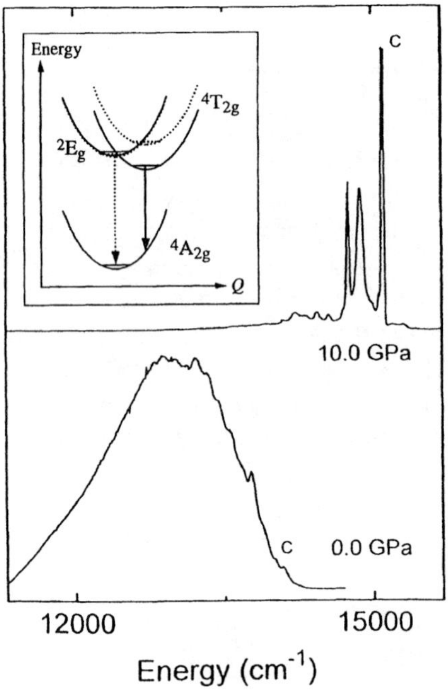

Figure 15. 90 K luminescence spectra of $KZnF_3$: Cr^{3+} under different hydrostatic pressure conditions: ambient pressure (bottom) and 10 GPa (top), adapted from ref 56. The inset shows a schematic SCC diagram for zero external pressure (solid lines) and 10 GPa (dotted lines). Note the pressure induced crossover from a $^4T_{2g}$ to a ^2E_g emitting state.

Under hydrostatic pressure, the broad luminescence band of $KZnF_3$: Cr^{3+} shows a marked blue-shift. Up to about 5 GPa it shifts linearly with increasing pressure without changing its shape. Above 5 GPa the shape changes significantly with the broad band disappearing and a sharp-line spectrum appearing. The 90 K luminescence spectrum of $KZnF_3$: Cr^{3+} taken at 10 GPa is shown in Figure 15, upper curve. The highest energy peak (labeled C) is assigned to the origin of the $^2E_g \rightarrow {}^4A_{2g}$ transition of Cr^{3+} ions occupying the cubic site. Hence, in the 0–10 GPa pressure range, a pressure-induced crossover from a $^4T_{2g}$ to a ^2E_g first excited state occurs in the $KZnF_3$: Cr^{3+} system.

Pressure increases the ligand field splitting energy Δ and thus blue shifts the $^4T_{2g}$ emission whose energy is proportional to Δ. In contrast, the ^2E_g emission shows a very slight red-shift with increasing pressure, as this spin-flip transition depends on other factors than Δ. The inset of Figure 15 shows the corresponding potentials in a single-configurational coordinate diagram. The dotted excited state potentials represent the 10 GPa situation. $^4T_{2g}$ is no longer the lowest energy excited state, it relaxes nonradiatively to ^2E_g that has become the emitting state. Note that pressure mainly affects the *vertical position* of those excited state potentials which have a different electron configuration than the ground state, whereas the horizontal positions remain essentially unchanged.

4.5.2 ***$Eu_2[Pt(CN)_4]_3 \cdot 18H_2O$*** Very favorable conditions for the study of pressure effects on the luminescence properties are provided by tetracyanoplatinate(II) compounds, $M_x[Pt(CN)_4] \cdot n\,H_2O$ (M = Na, K, Ba, Mg, ...).[57] In dilute aqueous solution the isolated $Pt(CN)_4^{2-}$ complex has very strong absorption bands above 35000 cm^{-1}. It does not appear to have any further absorption bands, even weak ones, at lower energies. Nevertheless, many $Pt(CN)_4^{2-}$ salts are intensely colored; depending on the cation their color varies from light yellow to deep red. In the solid state the square planar $Pt(CN)_4^{2-}$ complexes stack to form quasi one-dimensional structures with short intrachain Pt–Pt distances in the 3.1–3.8 Å range, which is close to the Pt–Pt distance of 2.75 Å in Pt metal. All these $Pt(CN)_4^{2-}$ compounds exhibit broadband luminescence. The position of the absorption and emission band maxima is correlated with the Pt–Pt distance. The emission maximum varies from the near UV (e.g. 26750 cm^{-1} in $Na_2[Pt(CN)_4] \cdot 3\,H_2O$ with a Pt–Pt distance of 3.67 Å) over the entire visible range into the NIR (e.g. 15600 cm^{-1} in $Sr[Pt(CN)_4] \cdot 3\,H_2O$ with a Pt–Pt distance of 3.09 Å).

The strong absorptions of the isolated $Pt(CN)_4^{2-}$ complex in the UV were assigned to transitions with considerable $Pt^{2+} \rightarrow CN^-$ charge transfer character. The pronounced red shift upon stacking was interpreted in terms of an excitonic interaction.[58] Polarized reflection spectra of tetracyanoplatinate compounds demonstrate that the visible absorption consists of a single broad band that is polarized almost entirely parallel to the stacking (**c**) axis with an oscillator strength approaching unity. The allowed character of this singlet-singlet transition is reflected by the very short lifetime of the corresponding $\mathbf{E}\|\mathbf{c}$ polarized emission: $\tau_{obs} < 300$ ns. The situation is complicated by the presence of a second, $\mathbf{E} \perp \mathbf{c}$ polarized broadband emission at slightly lower energy than the corresponding $\mathbf{E}\|\mathbf{c}$ emission. In this case the emitting state has a strong triplet character and a long lifetime: e.g. $\tau_{obs}(1.7\text{ K}) \approx 5$ ms for $Ba[Pt(CN)_4]\cdot 4H_2O$. Hence, the tetracyanoplatinates are dual emitters with thermally nonequilibrated emitting states. The radiative transition from the singlet excited state to the ground state is fast enough to compete favorably with the multiphonon relaxation to the triplet first excited state.

The Pt–Pt distance and thus the luminescence maximum of the tetracyanoplatinates are not only susceptible to chemical variations but also to external pressure. Red shifts of the emission maximum up to 4000 cm^{-1}/GPa have been observed, the largest values found for solid state compounds so far.

When lanthanide ions are used as counterions the tetracyanoplatinate compounds additionally exhibit very interesting energy transfer properties. The excitation of the $Pt(CN)_4^{2-}$ chains can be trapped by the lanthanides, and in the following we illustrate how this can be tuned by external pressure. Figure 16 shows the 100 K luminescence spectra of $Eu_2[Pt(CN)_4]_3 \cdot 18H_2O$ for $\mathbf{E} \perp \mathbf{c}$ at three different pressures (left-hand side).[59] The $\mathbf{E}\|\mathbf{c}$ polarized emission is not shown for clarity; the lifetime of the corresponding emitting state is too short to allow for appreciable Pt→Eu energy transfer. At ambient pressure there is a nonvanishing spectral overlap between the $\mathbf{E} \perp \mathbf{c}$ polarized $Pt(CN)_4^{2-}$ emission and the $^7F_0 \rightarrow {}^5D_0, {}^5D_1$ absorptions of Eu^{3+}. This leads to Pt→Eu energy transfer (see Section 2.6) which results in the presence of $^5D_1$ and $^5D_0$ sharp-line emissions of Eu^{3+} in the top trace of Figure 16. At about 20 kbar (2 GPa) the $Pt(CN)_4^{2-}$ emission is red-shifted to an extent that the spectral overlap with

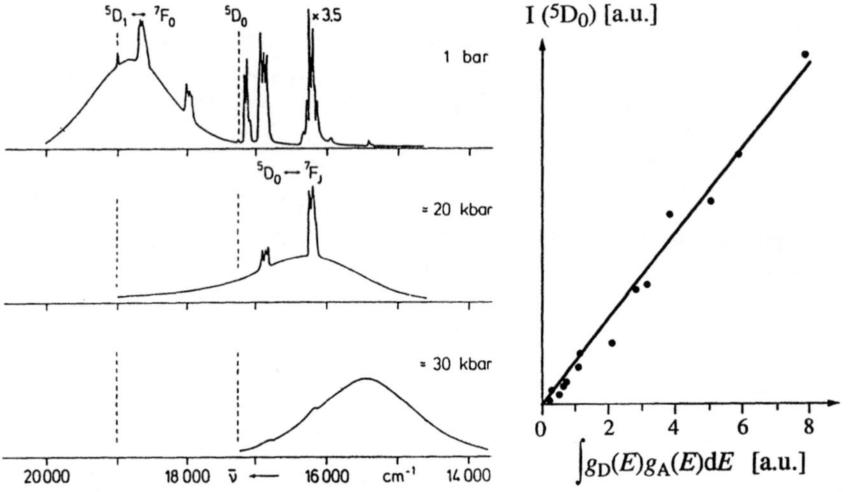

Figure 16. Left-hand side: $\mathbf{E} \perp \mathbf{c}$ polarized luminescence spectra at (100 ± 10) K of $Eu_2[Pt(CN)_4]_3 \cdot 18H_2O$ for three different pressures. The dashed vertical lines indicate the positions of the $^7F_0 \rightarrow {}^5D_J$ absorptions of Eu^{3+}. The peak intensities have to be multiplied by the given factors. Right-hand side: Integrated $^5D_0$ emission intensity I($^5D_0$) of Eu^{3+} versus the spectral overlap integral of the $\mathbf{E} \perp \mathbf{c}$ emission of $Pt(CN)_4^{2-}$ and the $^7F_0 \rightarrow {}^5D_0$ absorption of Eu^{3+}. Adapted from ref 59.

$^7F_0 \rightarrow {}^5D_1$ is zero but nonzero with $^7F_0 \rightarrow {}^5D_0$, and all the remaining Eu^{3+} emission originates from $^5D_0$. At 30 kbar the spectral overlap of the $Pt(CN)_4^{2-}$ emission with all possible acceptor absorptions is drastically reduced, and the Eu^{3+} emission has almost completely disappeared.

The right-hand side of Figure 16 shows the integrated $^5D_0$ emission intensity of Eu^{3+} versus the spectral overlap of the $\mathbf{E} \perp \mathbf{c}$ polarized $Pt(CN)_4^{2-}$ emission and the $^7F_0 \rightarrow {}^5D_0$ absorption of Eu^{3+}. In order to exclude the complication of Pt→Eu energy transfer via the $^5D_1$ level of Eu^{3+}, the data relate to pressures $\geqslant$20 kbar. From the general expression for energy transfer (Eq. 43) we expect a linear relation between the Eu^{3+} emission intensity and the spectral overlap. This is in excellent agreement with the experimental results in Figure 16.

4.6 Transition Metal and Lanthanide Ion Laser Materials

In Sections 4.1–4.5 we have discussed examples in which luminescence spectroscopy was used as a tool to study ground and excited state properties of inorganic luminophors. It is important to note that there are numerous light-emitting inorganic materials in which the light emission itself is of primary interest. Luminescent materials have important applications as phosphors in lighting and imaging or as scintillators in X-ray detection systems. We refer to the recent book by Blasse and Grabmaier[60] which shows the scientific background of the most important devices. Even though, strictly speaking, it is not within the scope of the present Chapter, some examples of transition metal and lanthanide ion doped laser materials will be presented in the following. For an extensive overview of the physics and properties of laser crystals we refer to the book by Kaminskii.[61] The relevance to "luminescence" is given by the interplay and competition of spontaneous and stimulated emission processes. Among other factors this is an important determinant whether a luminescent material is a potential laser material.

Let us briefly return to the two-level system introduced at the beginning of Section 2.1. $|a\rangle$ and $|b\rangle$ are the two nondegenerate stationary states of our active metal center. The intensity of a light beam with frequency ν_{ba} passing through the material changes according to the Beer–Lambert law (see Chapter 4):

$$I(l) = I_0 \exp(-\alpha(\nu_{ba})l). \tag{53}$$

l is the penetration depth and $\alpha(\nu_{ba})$ is the absorption coefficient which is proportional to the population difference $N_a - N_b$. To obtain amplification of the light beam requires $\alpha(\nu_{ba})$ to be negative and consequently $N_b > N_a$, that is, the excited state population must exceed the ground state population. This condition is referred to as *population inversion* and is never observed at thermal equilibrium.

In order to obtain the necessary population inversion for visible wavelengths, optical pumping is necessary, and various schemes have been devised for this purpose. They all involve more than two levels, and the most important three- and four-level laser schemes are shown in Figures 17 (a) and (b). In the scheme (a) the population inversion $N_b > N_a$ is achieved by an efficient pumping step to the intermediate level

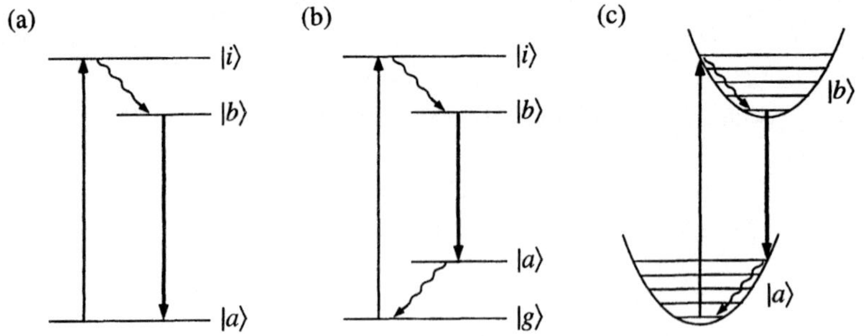

Figure 17. Schematic three-level (a) and four-level (b) laser schemes. (c) shows a single configurational coordinate diagram indicating the four-level nature of a vibronic laser.

$|i\rangle$ followed by a fast nonradiative relaxation $|i\rangle \rightsquigarrow |b\rangle$. In the very favorable four-level situation (b) an additional fast nonradiative relaxation $|a\rangle \rightsquigarrow |g\rangle$ from the final laser level to the ground state leads to an additional depletion of the level $|a\rangle$. In the scheme (c) which refers to so-called vibronic lasers, only two electronic states are involved, but as a result of the displacement of the corresponding potentials along some relevant vibrational coordinates, it effectively becomes a four-level system because of the very fast internal conversion processes.

4.6.1 ***Ruby*** The ruby laser, Al_2O_3: Cr^{3+}, was not only the first laser to operate,[62] it is also a prototype three-level laser. Pumping is achieved by the spin allowed $^4A_{2g} \rightarrow {}^4T_{2g}$ and $^4T_{1g}$ excitations in the green and blue, respectively. Multiphonon relaxation processes down to the metastable ^2E_g state (τ_{obs}(300 K) = 3.0 ms[63]) occur on a picosecond timescale, leading to the required high ^2E_g population. Typically, this is achieved in a pulsed pumping mode using the very high output from a xenon flashlamp. Since the $^2E_g \rightarrow {}^4A_{2g}$ transition corresponds to a spin flip within the t_{2g}^3 electron configuration, it is a sharp-line emission, and the tuning range of the ruby laser is thus limited to a few wavenumbers. On the other hand, it can easily be operated at room temperature, because the $^2E_g \rightsquigarrow {}^4A_{2g}$ multiphonon relaxation is least efficient in the weak coupling case. One of the most serious loss processes in transition metal systems is thus practically eliminated. Other potential loss processes are ground state absorption (GSA) and excited state absorption (ESA) at the laser wavelength or ESA at the pump wavelength. GSA is not efficient in ruby, because of the small oscillator strength of the spin forbidden $^4A_{2g} \rightarrow {}^2E_g$ transition, $f = 2 \times 10^{-6}$.[55] The good agreement between the measured and calculated emission cross sections indicates that no appreciable ESA occurs at the laser wavelength.

4.6.2 ***Nd^{3+}: YAG*** The Nd^{3+}: YAG laser (YAG = $Y_3Al_5O_{12}$) is a prototype four-level laser. The laser transition is $^4F_{3/2} \rightarrow {}^4I_{11/2}$ at 1064 nm, and the terminal level lies some 2000 cm^{-1} above the $^4I_{9/2}$ ground state. As usual for $f \rightarrow f$ transitions, the lines are sharp and thus the tuning range very limited. Another problem is the relatively low oscillator strength of $f \rightarrow f$ transitions which makes an efficient pumping rather difficult. This is usually overcome by co-doping the laser material with a sensitizer such as Cr^{3+}, which efficiently transfers the excitation to the Nd^{3+} activators by a nonradiative process. These disadvantages are more than compensated by the advantages of Nd^{3+}: YAG. The YAG host is hard, of good optical quality, and has a high thermal conductivity. The cubic structure of YAG favors a narrow linewidth of the emission which results in a high gain and low threshold for laser operation. There is no GSA at the laser wavelength, as usual in a four-level system, and the $^4F_{3/2}$ state has a quantum efficiency for emission approaching 100%. The fundamental 1064 nm wavelength of the Nd^{3+}: YAG laser can very efficiently be doubled, tripled, or even quadrupled in nonlinear optical materials, thus producing high energy laser lines that are ideally suited for the pumping of tunable lasers throughout the VIS and NIR regions.

4.6.3 ***Ti^{3+}: sapphire*** The Ti^{3+}: sapphire laser (sapphire=Al_2O_3) is an example of a vibronic laser. Ti^{3+} has the $3d^1$ electron configuration, and the two levels $|a\rangle$ and $|b\rangle$ in Figure 17 (c) correspond to the two ligand field states $^2T_{2g}$ and ^2E_g, respectively. The $d \rightarrow d$ absorption and emission peak at about 20500 and 12700 cm^{-1}, respectively.[64] This enormous Stokes shift is the result of a displacement of the two states not only along the totally symmetric TiO_6 breathing coordinate, but simultaneously along vibrational modes with e_g symmetry. This is the result of Jahn–Teller couplings of the $^2T_{2g}$ and ^2E_g ligand-field states with the e_g mode.[65] We thus have nonvanishing Huang–Rhys parameters $S(a_{1g})$ and $S(e_g)$, and the effective total S corresponds to the sum of the two. S thus becomes very large, while $S(a_{1g})$ and $S(e_g)$ both remain small enough to keep the multiphonon relaxation rates small compared to the radiative rates. The luminescence quantum efficiency thus exceeds 80% at room temperature, an extremely high value for a vibronic NIR emitter. Due to the extremely large Stokes shift there is almost no overlap between the absorption and emission profiles, and thus GSA is very small over the entire luminescence range. ESA is also very small, because the first O→Ti charge transfer transition from the ^2E_g excited state lies higher in energy than the $^2T_{2g} \leftrightarrow {}^2E_g$ origin. This leads to an exceptionally broad tuning range of the Ti^{3+}: sapphire laser extending from 670 to about 1070 nm.

5 CONCLUDING REMARKS

In this chapter we have shown that luminescence spectroscopy provides a powerful tool to probe the electronic structure of inorganic complexes and metal ions in crystals and glasses. Insight may be obtained into excited state processes, such as energy transfer, multiphonon relaxation as well as photochemical pathways, that are difficult to address by other spectroscopic methods. Excitation spectroscopy can be employed to obtain the absorption spectrum of luminescent species that are weakly absorbing, difficult to crystallize, or in heterogeneous samples in the presence of strongly absorbing nonluminescent species.

The luminescence phenomenon in itself is of great interest with respect to the application of luminescent materials as phosphors in lighting and imaging, as scintillators in X-ray detection systems, or as solid state laser materials. Research efforts in this area have led to the discovery of the Ti^{3+}: sapphire laser, an almost ideal four-level system, and there is a continuing search for new materials that exhibit broadband emission in the near-IR region.

REFERENCES

1. Henderson, B.; Imbusch, G.F. *Optical Spectroscopy of Inorganic Solids*; Oxford Science Publications: Oxford, 1989.
2. Imbusch, G.F.; Kopelman, R. in *Laser Spectroscopy of Solids*; Yen, W.M.; Selzer, P.M. (eds.); Springer-Verlag: Berlin, 1981; p. 3.
3. Ballhausen, C.J. *Introduction to Ligand Field Theory*; McGraw Hill: New York, 1962.
4. (a) Hellmann, H. *Quantenchemie*; Denticke: Leipzig, 1937; p. 285. (b) Feynman, R.P. *Phys. Rev.* **1939**, *56*, 340.
5. Ballhausen, C.J.; Hansen, A.E. *Ann. Rev. Phys. Chem.* **1972**, *13*, 15.
6. Jahn, H.A.; Teller, E. *Proc. Roy. Soc.* **1937**, *A161*, 220.
7. Bersuker, I.B. *The Jahn–Teller Effect and Vibronic Interactions in Modern Chemistry*; Plenum Press: New York, 1984.
8. Heller, E.J. *J. Chem. Phys.* **1975**, *62*, 1544.
9. Kasha, M. *Discuss. Faraday Soc.* **1950**, *9*, 14.
10. Donnelly, C.J.; Imbusch, G.F. in *Advances in Nonradiative Processes in Solids*; Di Bartolo, B. (ed.); Plenum Press: New York, 1991; p. 175.
11. Riesberg, L.A.; Moos, H.W. *Phys. Rev.* **1968**, *174*, 429.
12. Mott, N.F. *Proc. Roy. Soc. (London)* **1938**, *A 167*, 384.
13. Struck, C.W.; Fonger, W.H. *J. Lumin.* **1975**, *10*, 1.
14. *Radiationless processes*; Di Bartolo, B. (ed.); Plenum Press: New York, 1980.
15. *Advances in Nonradiative Processes in Solids*; Di Bartolo, B. (ed.); Plenum Press: New York, 1991.
16. Reber, C.; Güdel, H.U. *J. Lumin.* **1990**, *47*, 7.
17. Robbins, D.J.; Thomson, A.J. *Mol. Phys.* **1973**, *25*, 1103.
18. Moore, J.H.; Davis, C.C.; Coplan, M.A. *Building Scientific Apparatus*, 2nd ed.; Addison-Wesley Publishing Company: New York, 1989.
19. Ejder, E. *J. Opt. Soc. Am.* **1969**, *59*, 223.
20. Hamilton, T.D.S.; Munro, I.H.; Walker, G. in *Luminescence Spectroscopy*; Lumb, M.D. (ed.); Academic Press: New York, 1978; p. 149.
21. *OSA Trends in Optics and Photonics on Advanced Solid-State Lasers*; Payne, S.A.; Pollock, C.R. (eds.); Optical Society of America: Washington, 1996.
22. van Tol, J.; van Hulst, J.A.; van der Waals, J.H. *Molec. Phys.* **1992**, *76*, 547.
23. Brunold, T.C.; Güdel, H.U.; Riley, M.J. *J. Chem. Phys.* **1996**, *105*, 7931.
24. Oetliker, U.; Herren, M.; Güdel, H.U.; Kesper, U.; Albrecht, C.; Reinen, D. *J. Chem. Phys.* **1994**, *100*, 8656.
25. Wilson, R.B.; Solomon, E.I. *Inorg. Chem.* **1978**, *17*, 1729.
26. Andrews, L.J.; Hitelman, S.M.; Kokta, M.; Gabbe, D. *J. Chem. Phys.* **1986**, *84*, 5229.

27. Güdel, H.U.; Snellgrove, T.R. *Inorg. Chem.* **1978**, *17*, 1617.
28. Knochenmuss, R.; Reber, C.; Rajasekharan, M.V.; Güdel, H.U. *J. Chem. Phys.* **1986**, *85*, 4280.
29. Gilardoni, F.; Weber, J.; Bellafrouh, K.; Daul, C.; Güdel, H.U. *J. Chem. Phys.* **1996**, *104*, 7624.
30. Wexler, D.; Zink, J.I.; Tutt, L.W.; Lunt, S.R. *J. Phys. Chem.* **1993**, *97*, 13563.
31. Wexler, D.; Zink, J.I. *J. Phys. Chem.* **1993**, *97*, 4903.
32. Jacobsen, S.M.; Güdel, H.U.; Daul, C.A. *J. Am. Chem. Soc.* **1988**, *110*, 7610.
33. McCarthy, P.J.; Güdel, H.U. *Coord. Chem. Rev.* **1988**, *88*, 69.
34. Güdel, H.U. in *Magneto-Structural Correlations in Exchange Coupled Systems*; Willett, R.D.; Gatteschi, D.; Kahn, O. (eds.); Reidel: Dordrecht, 1983; p. 297.
35. Herren, M.; Yamanaka, K.; Morita, M. *Technical Reports of Seikei University* **1995**, *32*, 61.
36. Cyvin, S.J. *Molecular Vibrations and Mean-Square Amplitudes*; Elsevier: Amsterdam, 1968.
37. Ziegler, T.; Rauk, A.; Baerends, E.J. *Chem. Phys.* **1976**, *16*, 209.
38. Stückl, A.C.; Daul, C.A.; Güdel, H.U. submitted to *J. Chem. Phys.*
39. Juris, A.; Balzani, V.; Barigelleti, F.; Campagna, S.; Belser, P.; von Zelewsky, A. *Coord. Chem. Rev.* **1988**, *84*, 85.
40. Krausz, E.; Ferguson, J. *Progr. Inorg. Chem.* **1989**, *37*, 293.
41. Colombo, M.G.; Güdel, H.U. *Inorg. Chem.* **1993**, *32*, 3081.
42. a) Demas, J.N.; Crosby, G.A. *J. Am. Chem. Soc.* **1970**, *92*, 7262. b) Demas, J.N.; Crosby, G.A. *J. Am. Chem. Soc.* **1971**, *93*, 2841.
43. Watts, R.J.; White, T.P.; Griffith, G.P. *J. Am. Chem. Soc.* **1975**, *97*, 6914.
44. Colombo, M.G.; Hauser, A.; Güdel, H.U. *Top. Curr. Chem.* **1994**, *171*, 143.
45. Riedener, T.; Krämer, K.; Güdel, H.U. *Inorg. Chem.* **1995**, *34*, 2745.
46. Dieke, G.H. *Spectra and Energy Levels of Rare Earth Ions in Crystals*; John Wiley: New York, 1968.
47. Riedener, T.; Güdel, H.U. manuscript in preparation.
48. Buisson, R.; Vial, J.C. *J. Phys. (Paris) Lett.* **1981**, *42*, L115.
49. Flint, C.D.; Paulusz, A.G. *Mol. Phys.* **1980**, *41*, 907.
50. a) Kober, E.M.; Caspar, J.V.; Sullivan, B.P.; Meyer, T.J. *Inorg. Chem.* **1988**, *27*, 4587. b) Belser, P.; Dux, R.; Baak, M.; De Cola, L.; Balzani, V. *Angew. Chem. Int. Ed.* **1995**, *34*, 595. c) Riesen, H.; Krausz, E. *Comments Inorg. Chem.* **1995**, *18*, 27.
51. Kalyanasundaram, K. *Photochemistry of Polypyridine and Porphyrin Complexes*; Academic Press: London, 1991.
52. De Cola, L. *Chimia* **1996**, *50*, 214.
53. *Physics of Solids Under High Pressure*; Schilling, J.S.; Shelton, R.N. (eds.); North-Holland: Amsterdam, 1981.
54. Drickamer, H.G.; Frank, C.W. *Electronic Transitions and the High Pressure Chemistry and Physics of Solids*; Chapmann and Hall: London, 1973.
55. Lever, A.B.P. *Inorganic Electronic Spectroscopy*, 2nd ed.; Elsevier: Amsterdam, 1984; pp. 417–429.
56. Freire, P.T.C.; Pilla, O.; Lemos, V. *Phys. Rev.* **1994**, *B49*, 9232.

57. Gliemann, G.; Yersin, H. *Struct. Bonding* **1985**, *62*, 87.
58. Day, P. *J. Am. Chem. Soc.* **1975**, *97*, 1588.
59. Yersin, H.; v. Ammon, W.; Stock, M.; Gliemann, G. *J. Lumin.* **1979**, *18/19*, 774.
60. Blasse, G.; Grabmaier, B.C. *Luminescent Materials*; Springer-Verlag: Berlin, 1994.
61. Kaminskii, A.A. *Laser Crystals, Their Physics and Properties*; Springer-Verlag: Berlin, 1990.
62. Maiman, T.H. *Nature* **1960**, *187*, 493.
63. Koechner, W. *Solid-State Laser Engineering*, 3rd ed.; Springer-Verlag: Berlin, 1992.
64. Moulton, P.F. *J. Opt. Soc. Am.* **1986**, *B3*, 125.
65. Macfarlane, R.M.; Wong, J.Y.; Sturge, M.D. *Phys. Rev.* **1968**, *166*, 250.

6 Laser Spectroscopy

ELMARS KRAUSZ AND HANS RIESEN*
Research School of Chemistry
The Australian National University
Canberra, ACT 0200, Australia
E-mail: krausz@rsc.anu.edu.au
h-riesen@adfa.edu.au

* Present address: University College, University of New South Wales, Australian Defence Force Academy, Canberra, ACT 2605, Australia.

Inorganic Electronic Structure and Spectroscopy, Volume I: Methodology.
Edited by E. I. Solomon and A. B. P. Lever.
ISBN 0-471-15406-7.

1 INTRODUCTION

Optical spectroscopy is evolving into a rapidly improving and expanding range of specialized techniques and methods, some of which are represented in this volume (Chapters 4, 5, and 7). Improvements in technologies, particularly with laser sources and light detectors, have allowed measurements to be made that approach fundamental limits of sensitivity, resolution, and timescale.

Recently it has become possible to detect and measure spectral properties of ions, atoms, and single molecules.[1,2] The direct observation of very fast inter- and intramolecular electronic processes also becomes possible with short pulse lasers.

1.1 Interaction of Light with Matter

The interaction of the electromagnetic radiation field with matter is well understood. Descriptions of the interaction process become progressively more subtle.[3–9] In the usual consideration of molecular systems at conventional light intensities, an incident photon is either absorbed or transmitted. An excited species either emits spontaneously or by radiative stimulation. Alternatively, the excited state is deactivated by nonradiative processes.

In a more general approach to the light-matter interaction, the process is considered as various aspects of a scattering process. Incident photons may be scattered elastically (for example the Rayleigh process) or inelastically (absorption, Raman). However, at the power levels and spectral densities available with laser sources, the radiation field itself may modify the optical properties of the system and a range of nonlinear techniques become enabled.

Ultimately it becomes necessary to consider the coherent radiation field and the excitation of the system to be closely coupled. A quantum electrodyamical formalism is required to fully describe and analyze coherent high order optical processes that can be induced.

1.2 Raison d'Etre

Optical spectroscopy is of fundamental importance in the study of both chemical and physical bonding interactions. The information yielded is at a level beyond that of ground state bond distances and angles as obtained by diffraction methods. In general:

- The entire manifold of electronic excited states may, in principle, be identified and characterized.
- A study of the characteristics of vibrational sidelines built upon electronic excitations provides detailed information about potential surfaces. Sideline structure provides invaluable information regarding the coupling of electronic and nuclear motions—the breakdown of the central Born–Oppenheimer approximation (see Chapter 5 of Volume II). In many inorganic systems, vibronic coupling between electronic and nuclear motions is large, and this leads to special spectral properties of side structure (see Chapter 4). Vibronic coupling can have an intimate connection to electron transfer processes and more generally to photochemical and chemical reactivities.

Inorganic materials have a wide range of optical properties. They are generally more photostable and have higher damage thresholds than other materials. Inorganics may occur as crystalline, amorphous or bio-inorganic systems, as supramolecular systems or nanostructures.

The environment in which measurements of a particular chromophore are made can be of great significance. A chromophoric unit in a crystal lattice usually has a well defined geometry, environment and orientation. Spectra of chromophores in crystals (Chapter 4) suffer less from inhomogeneous broadening and features with linewidths of 1–10 cm^{-1} are often observed.

Inhomogeneous broadening is more severe in amorphous environments. This may not lead to a straightforward spread of a fixed spectral pattern. A fixed spectral pattern is assumed in the rigid shift approximation. The rigid shift approximation is, for example, often used in the analysis of MCD spectra (Chapter 5).

A chemically well defined species may have apparently anomalous properties, particularly in an amorphous environment. Inhomogeneity has alternatively been called microheterogeneity—each chromophore experiences a slightly different environment at the microscopic level. There are variations occurring at the intramolecular level that we have termed nanoheterogeneity. When a chromophore contains a number of nominally equivalent subunits (for example ligands), nanoheterogeneity addresses the independent variation of subunit environments. This effect can contribute significantly to the lowering of the symmetry of the species. Laser spectroscopies allow this interesting and sometimes critical phenomenon in inorganic spectroscopy to be unraveled and studied in detail.

Electronic excitations of measurable intensity in inorganic systems may be either metal centered, ligand centered, charge transfer, or excitonic in character. These excitations may be either parity and/or spin allowed/forbidden and furthermore may have either electric dipole, magnetic dipole, or even electric quadrupole transition mechanisms. It is essential to have the ability to make "model free" assignments. For example, a charge transfer state is best characterized by its response to an applied electric field. Correspondingly, a spin triplet state is naturally best identified by its response to a magnetic field. Inhomogeneous spectral linewidths often do not allow the direct observation of excited state splittings or shifts.

Methods such as MCD or electrochromism identify small changes in the spectra (relative to linewidths) associated with an applied field. Analyses of spectra usually depend on a model and assumptions such as the rigid shift approximation (mentioned above) which may well not be valid. Laser spectroscopies can enhance spectral resolutions to the extent that shifts and splittings may be observed directly, providing unambiguous assignments with very modest applied fields.

Laser spectroscopy can probe the various components of an inhomogeneously broadened electronic excitation and any variation in electronic properties of chromophores within the distribution may be independently and directly identified.

1.3 Outline

In this chapter, an overview of some of the important principles and some recent developments in lasers and laser spectroscopies is provided. Applications of laser techniques in the spectroscopy of inorganic systems are given, concentrating on the techniques of fluorescence/excitation line narrowing and spectral hole-burning.

These techniques are relatively simple and often enable very narrow spectral features to be obtained from inhomogeneously broadened spectra. Narrow features can be measured in a wide range of inorganic chromophores in an equally wide range of environments. The ability of these techniques to provide sharp line features reflects but a fraction of the information available from laser spectroscopies. Variations of linewidths and lineshapes with time, temperature and excitation, and the presence of narrow and/or broad side features provide a great deal of invaluable information.

2 LASER SOURCES AND METHODS

Advances in optical spectroscopy over the last decade have been enabled by significant developments in the technology of lasers and laserlike sources. There have been continuing refinements and enhancements of the well established gas discharge based laser systems and some improvements of organic dye based tunable lasers. New laser sources, both pulsed and continuous, from the UV to the infrared have become available. Developments have been mostly in the area of inorganic solid state, tunable sources. High resolution or "single frequency" (Sections 2.2 and 2.3) scanning laser systems have become commercially available and more easily and routinely operable. This is particularly true for tunable semiconductor laser diodes.

Optical Parametric Oscillators (OPOs) and Optical Parametric Amplifiers (OPAs) represent a radically different way in which coherent radiation is generated. These devices are becoming increasingly important as pulsed sources tunable over an extremely wide range. The interested reader may need to consult fuller descriptions of laser and OPO operation, and is referred to the new edition of the classic tome of Demtröder[10] and other texts.[11–13]

2.1 Some Basics

The term laser is an acronym for the process of *l*ight *a*mplification by *s*timulated *e*mission of *r*adiation between an excited state E_e and a state E_g. The probability of absorption and stimulated emission between these levels is the same. The rate of stimulated absorption or emission is given by the Einstein coefficient B, multiplied by the number of species in the initial state N and the light intensity $\rho(\nu)$

$$B_{eg} N_e \rho(\nu),\ B_{ge} N_g \rho(\nu) \quad \text{with } B_{eg} = B_{ge} \tag{1}$$

The rate of spontaneous emission from the excited to ground state is given by

$$A_{eg} N_e \tag{2}$$

independent of light intensity.

The competition between spontaneous and induced processes is governed by the ratio of the Einstein coefficients for a single radiation mode[10] (see also the corresponding ratio for isotropic radiation in Chapter 7).

$$A_{eg}/B_{eg} = 4h\left(\frac{\nu}{c}\right)^3 \tag{3}$$

In most operating lasers (unless "seeded" with light from an external source) light emitted spontaneously (Equation 2) provides resonant radiation which may be amplified by the stimulated process (Equation 1). When the amplified light is contained within an optical resonator, and passed through the active medium again and again, the light intensity $\rho(\nu)$ may continue to increase until limited by the pump process.

The prime condition whereby light amplification and laser action between E_e and E_g occurs is that $N_e > N_g$ (there is a population inversion). This is achieved by an external energy input into the system, provided by an electrical discharge, flashlamp or a (pump) laser system.

The quality of the resonant laser cavity (Q) is given by the following ratio in an oscillation period

$$Q = \frac{\text{light energy stored}}{\text{light energy lost}} \tag{4}$$

and can be as high as 10^4. Materials with transitions of relatively low oscillator strength can be used in a laser cavity of high Q. If the Q of the cavity is reduced by increasing optical losses, laser action will become inhibited and the pumping threshold is increased.

Controlling intracavity losses provides the basis of Q switching pulsed lasers. The Q of the cavity is artificially held low by a spoiling electro-optical element or by other methods. This allows the excited state population in the laser *medium* to increase to its saturation level. The Q of the cavity is subsequently returned to its normal value. A very intense laser pulse ensues.

The titanium sapphire (Ti:S) laser is a good example of a very effective laser with a relatively low dipole intensity of the very broad, parity forbidden *d–d* luminescence of Ti^{3+} in the host lattice. As a consequence of the low B coefficient, the gain of the laser medium for a given length is relatively low. Laser action from this system was obtained over two decades ago. The production of Ti^{3+} doped sapphire crystals with exceptionally high purity and optical quality was required before this system formed the basis of a practical laser.

When losses are reduced sufficiently, the Q is high and the contained radiation within the cavity is greatly increased. Eventually, an efficient extraction of excited state energy can occur. The light intensity within a laser cavity itself is invariably higher than that in the output beam. This is particularly true for oscillators based on low gain media, such as the Ti:S laser.

Nonlinear elements such as doubling crystals become more effective within a laser cavity, unless they lead to unwanted and excessive optical losses. Intracavity doubling has become a feature in allowing krypton and argon ion lasers to generate UV radiation at fixed frequencies (244 nm, 257 nm, ...)

At higher energies, in the deep blue and UV regions, the *spontaneous* emission process competes more and more effectively with the stimulated process (Equation 3). The ratio changes very rapidly varying as the ν^3 factor. The pumping threshold for laser action eventually becomes prohibitively high and eventually exceeds damage thresholds of the laser gain media or the associated optical components.

Short wavelength radiation, as in the particular case mentioned above of argon ion lasers, can be obtained by harmonic generation. Other nonlinear optical conversion processes such as frequency tripling, combined with related processes of OPO and OPA action, can be utilized. These are discussed in Section 2.3. The guiding principle is that intense laser light from relatively reliable and efficient long wavelength lasers is converted, by nonlinear processes, to useful laser radiation at shorter wavelengths.

2.2 Laser Oscillator Mode Structure

Laser oscillators are either operated in standing wave Fabry–Pérot geometry, with mirrors at each end of the optical cavity, or in ring mode with light being reflected by three or more mirrors and with radiation circling in a continuous loop (Figure 1).

2.2.1 ***Longitudinal Mode Structure*** For light in laser cavities to be sustained, amplified, and give rise to laser action, the *phase* of the lightwave must remain constant. This can occur, for a standing wave cavity, when the length of the cavity L_{sw} is an integral number of half wavelengths

$$L_{sw} = n\frac{\lambda}{2}. \tag{5}$$

For a ring cavity, the corresponding condition is that the cumulative length of the ring L_{ring} is an integral number of full wavelengths

$$L_{\text{ring}} = n\lambda \tag{6}$$

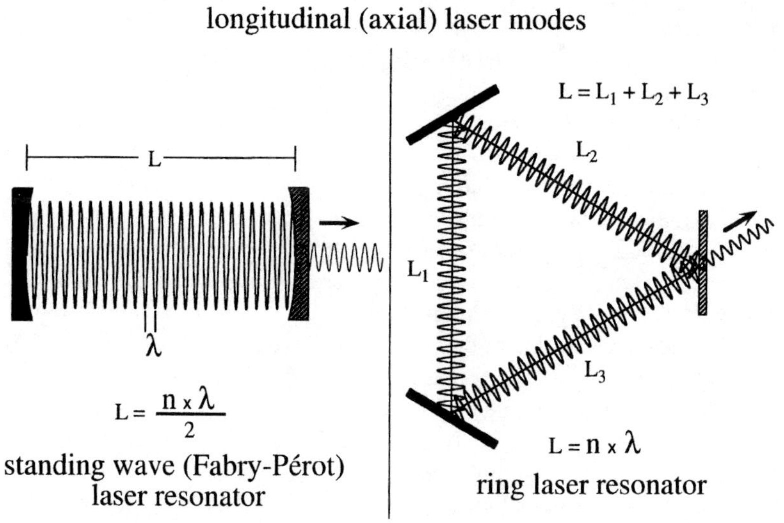

Figure 1. A comparison of the longitudinal mode structures in standing wave and ring oscillators.

The mode spacing is determined by the wavelength difference between adjacent longitudinal modes, that is between $n = k$ and $n = k \pm 1$ (Figure 1). It is c/L_{sw} or c/L_{ring} with c being the velocity of light. The longer a resonator is, the smaller the frequency difference between adjacent modes becomes.

Noting the conversion factor of 30 GHz $\leftrightarrow$ 1 cm^{-1}, we quickly see that a 1 m cavity length gives rise to a mode spacing of $(3 \times 10^8\ ms^{-1})/(2 \times 1\ m) = 150$ MHz for a standing wave laser. This figure is typical for an argon ion laser. A value of twice that (300 MHz) is obtained for a Ti:S ring laser of the same overall cavity length. A semiconductor laser diode with a typical cavity length of 0.5 mm has a much larger mode spacing[10] of about 50 GHz (1.7 cm^{-1}).

2.2.2 ***Transverse Mode Structure*** As well as longitudinal (axial) mode structure (see 2.2.1), lasers also develop transverse electromagnetic mode (TEM) structure. The light intensity transverse to the propagation direction develops, in a general case, complex nodal patterns, details of which depend on the many aspects of the operation of the laser. By limiting the diameter of the beam and choosing the appropriate cavity mirror curvatures, it is possible to operate lasers in their fundamental or TEM_{00} mode. The beam profile then becomes radially symmetric with a Gaussian cross section. A beam with such characteristics is known as a Gaussian beam. Careful control of the transverse mode structure can be *particularly* important if a laser beam has to be tightly focussed on a sample or is further used to pump a second laser or to drive an OPO/OPA system.

2.3 Tunable and Single Frequency Laser Operation

Without wavelength selective elements in a laser, the optical cavity may oscillate in many longitudinal modes. Any wavelength that has gain in the laser medium and satisfies the longitudinal cavity criteria (Equations 5 and 6) may be present. Mode spacings are very small compared to the tuning ranges of dye lasers, which may be as large as 3000 cm^{-1} $\leftrightarrow$ 90 000 GHz. Even a typical gas phase laser, such as the argon ion laser, which operates on atomic transitions that are Doppler broadened (due to their relative motions at high temperature) in the active plasma, will have $\approx$70 modes (spaced by about 150 MHz) operating over its $\approx$10 GHz gain bandwidth.

In a standing wave laser, as the name implies, there are (static) nodes and antinodes in the electric field intensity of each mode. A severe limit to the stability of the mode structure in standing wave condensed phase lasers is spatial hole-burning. This phenomenon occurs in regions of high light intensity within the active medium where the excited state concentration of the active medium becomes strongly depleted. The efficiency of a strong laser mode is thereby reduced. Other modes with antinodes in regions with higher gain become favored.

These latter modes will themselves eventually suffer from spatial hole-burning. Mode structure in such a standing wave laser is not inherently stable. A ring laser has a natural advantage in that lightwaves travel around the ring with a moving phase front. Spatial hole-burning cannot occur. In general only two modes coexist, one propagating clockwise and a second counterclockwise. It is possible to suppress one of these directions by use of an optical diode (Faraday rotator) within the cavity. Single frequency operation is readily achieved.

2.3.1 ***Wavelength Selection*** The wavelength selection and linewidth of tunable lasers (as well as other characteristics) can be progressively enhanced and tailored to match the needs of a particular scientific question via the relevant spectroscopic experiments. It may not be appropriate to use a laser of the narrowest bandwidth if one is interested in, for example, broad side structure. For reasons of sensitivity, a laser whose linewidth is compatible with the phenomenon of interest may be better.

Coarse wavelength selection is provided by the choice of dielectric multilayer coatings on the cavity mirrors and a wavelength selective device such as a prism, diffraction grating, or birefringent plate within the cavity. It is critical that the introduction of any element into the laser cavity does not seriously compromise its quality factor Q and introduce disastrous light losses. Barely visible defects of optical elements or contamination of surfaces within the cavity can lead to drastic reductions in laser output and/or stability. Lasers are best operated in clean, stable and dust-free environments. Dust accumulation is a particular problem with OPOs and OPAs.

Windows and other optical elements within an optical oscillator cavity, particularly in a condensed phase gain medium such as a dye jet or Ti:S crystal, are operated with their surfaces at the Brewster angle $\theta = \tan^{-1}(\eta)$, η being the refractive index of the material. At the Brewster angle, the polarization perpendicular to the axis of rotation of the window is transmitted without reflection losses.

A simple, low loss and very effective wavelength selective element is a birefringent (crystal quartz) plate assembly, itself inserted into the cavity at the Brewster angle. When the plate assembly is rotated (but held at the Brewster angle), at each angle of the plate there is a narrow range of wavelengths at which the crystal birefringence does not significantly change the state of polarization of the light. The birefringent assembly inhibits all other wavelengths from having gain. These wavelengths will have their state of polarization rotated on each pass through the birefringent filter and will consequently suffer reflection losses at all Brewster angle elements within the cavity. These losses effectively inhibit laser action at all but the selected wavelength(s).

2.3.2 ***Single Frequency Selection*** Coarse wavelength selective elements as described above can reduce the effective linewidth of a tunable laser to around 1 cm^{-1} (=30 GHz) or less. This corresponds to ~100 longitudinal modes in a 50 cm standing wave cavity. In order to further reduce the number of active longitudinal modes, one can introduce Fabry–Pérot étalons into the cavity. Étalons consist of two parallel, partly reflecting surfaces. A good quality quartz window can serve as an étalon.

Light reflecting between the surfaces of an étalon combines constructively or destructively depending on the spacing. The transmission of the étalon undulates from a high value with little loss, to a lower value. The distance between the reflective surfaces determines the free spectral range (FSR) of the étalon which is the interval with which the transmission pattern repeats with wavelength. A thin étalon varies more slowly, just as a short laser cavity has more widely spaced longitudinal modes. The reflectivity of the surfaces determines the finesse of the étalon, via the transmission characteristics of the device. The finesse is defined as the ratio of the FSR to the transmission bandwidth.

It is possible to operate a ring laser in single frequency using only one étalon, but it is preferable to use a combination of two interlocked étalons to ensure that mode hops do not occur. If two adjacent modes have similar gain, the laser may alternate between operating in one of the two modes upon minute variations in operating conditions of the system.

2.3.3 ***Single Frequency Scanning*** Once single frequency operation of the laser is assured, two tasks remain. First, one needs to be able to scan the laser over the range of spectroscopic interest, while maintaining single frequency operation. Second, one may need to stabilize and calibrate the laser frequency.

Fine adjustments of the operating wavelength of a laser may be achieved by minute changes of the cavity length L as indicated in Equations 5 and 6. A cavity mirror may be translated by having it on a piezoelectric mount, which can provide the submicron translations required. The insertion of a glass wedge into the beam can also provide a change in the optical path length. Coordinated adjustments of the cavity étalon(s) and the wedge (or related assembly) are needed to ensure that the laser remains in the same mode. The adjustments can be achieved by feedback electronics or active computer control.

2.3.4 ***Stability*** Unwanted changes in the optical pathlength of a cavity lead to frequency drifts and/or mode hops. They can arise from a number of sources. First, unless the laser assembly is constructed on a low temperature coefficient base such as super-invar, the dimensions of the cavity will change significantly with small temperature changes of the environment. In the argon ion laser example, the mode spacing of 150 MHz corresponds to just a $\sim 0.25\ \mu$ change in the cavity length. The cavity length must be constant to well within this dimension. Many lasers are operated as open (to air) Fabry–Pérot resonators (see Figure 1) and pressure changes lead to a change in the refractive index of air and thus the optical pathlength. The change corresponds to a shift of 188 MHz per Torr, independent of cavity length. An excellent discussion of Fabry–Pérot resonators is given by Demtröder.[10]

Temperature changes in étalons themselves also lead to changes in their optical pathlength. These arise from both a dimensional change in the element and a change in the refractive index of the material from which the étalon is constructed. A temperature change of 1 K in a quartz étalon corresponds to a 3 GHz shift in its transmission maximum. Thus, within a 0.1 K temperature range, the laser frequency abruptly changes when an adjacent longitudinal mode develops higher gain. This is called mode hopping.

Mechanical vibrations can obviously lead to transient dimensional changes and corresponding changes in the laser output frequency. Stabilized optical tables are available but adequate passive stability at the 10 MHz level can also be achieved using less expensive granite slabs.

A laser may be actively stabilized by locking its frequency to that of an external reference cavity and/or source. It is possible to lock a laser to a reference to within a frequency difference of mHz.[14] The stability and absolute calibration of the reference source to which the laser is locked is then critical. The resolution of a laser system may eventually approach or even exceed 1 in 10^{15}.

The output wavelength of a tunable laser can be roughly estimated on a monochromator or determined to higher accuracy using a wavemeter. The latter may have a resolution of 0.1 to 0.0001 nm. Absolute accuracy is often not required for selective spectroscopy as the exact energy at which a line narrowing technique is performed within a broad inhomogeneous distribution may not be critical.

Spectral shifts, due to the influence of external fields, temperature, pressure, etc. are far more important. These shifts are often conveniently calibrated using the mode spacing of the laser itself. Mode spacings are often evident in spectra for other reasons or can at least be easily obtained. If spectral resolution comparable to the mode spacing is adequate, the laser may be tuned discontinuously by tilting the étalon, which induces the laser to hop from one mode to the next. The spectrum obtained can be calibrated by the known mode spacing.

2.3.5 ***Laser Diodes*** Difficulties associated with the use of scanning single frequency dye lasers and associated pump lasers are to some extent being eclipsed by technological progress in the area of solid state semiconductor laser diodes. They are small, require no pump laser or large power supply, and are relatively inexpensive. They can be tuned quite simply over 10 nm or so by varying the current in the diode, the temperature, or externally with a grating. At present they are only readily available in the red and near IR regions. As time progresses one can expect diode lasers to be available covering a large fraction of the optical spectrum, greatly enhancing the ease of laser operation and significantly reducing the expense. The recent development of blue laser diodes based on GaN supports this view.

The small integrated laser diode structure does not suffer from spatial hole-burning and has wide mode spacing, characteristic of its short cavity length. It can operate in a single frequency without the complex arrangements needed in dye lasers. Spectral diffusion (Section 3.4) has been measured[15] using the rapid $\approx\mu$s) tuning rate of a laser diode.

2.4 Laser Pulses: Time and Intensity Dependent Phenomena

In this section we consider both the creation of laser pulses and their relationship to spectroscopic and photophysical or photochemical questions at hand. A light pulse of duration comparable to the phenomenon being studied and with power compatible with the nature and temperature of the sample should be used. The rise time and fall of the pulse is best kept relatively short. Data obtained using high power laser pulses such as Q switched YAG lasers need to be interpreted with care.

Light pulses of relatively long duration minimize the peak laser power needed to make a measurement, minimizing unwanted heating and nonlinear effects. The use of lower energy excitation, where possible, also photochemical damage, though questions of excited state absorption may be relevant in particular systems. The use of an appropriate detector and good data collection efficiency as well as good quality samples are helpful in keeping excitation powers low.

2.4.1 ***Seconds to Microseconds*** The creation of laser pulses of relatively long duration are important in many inorganic systems as lowest excited state lifetimes can be long. Such pulses are best created by interrupting a CW laser beam such as an argon or dye laser beam, either with a mechanical chopper, an acousto-optical (AO), or electro-optical (EO) modulator.

Mechanical choppers are generally useful to ms time regimes although they can be extended to μs rates. Besides their simplicity, choppers have the great advantage that they can provide a very high rejection ratio. This is particularly useful in resonant FLN experiments. AO modulators create a transient diffraction grating in the (transparent) material of the modulator by means of a driving radio-frequency signal. The grating deflects a fraction of the input beam. As the rise time of the grating can be as short as 0.1 ns, pulses of short risetime and arbitrary duration can be created. The on/off (rejection) ratio is somewhat limited and in critical cases EO modulators (Pockels cells) or a series of AO modulators in tandem may be used.

2.4.2 ***Nanoseconds*** Nanosecond laser pulses can be routinely obtained from fast gas discharge lasers such as N_2 or XeF and related eximer systems. Another very common route starts with an efficient Q switched YAG lasers at 1.06 μm which can be doubled or tripled. Energies per pulse of these systems may be higher than 1 J and peak powers well over 100 MW with repetition rates higher than 100 Hz. Measurements require fast electronics and corrections may need to be made for risetimes and propagation delays. However, measurements of ns kinetics can be routinely and very conveniently made with digital oscilloscopes.

2.4.3 ***Picoseconds and Femtoseconds*** Electronic and EO devices do not function well beyond nanosecond response times. The shortest laser pulses need to be created in a quite different fashion. Also, time dependent phenomena in this domain need to be measured using different methods. In 1 ps light travels 0.3 mm. Measurements of kinetics can be made by introducing variable time delays of the light beams via the pump-probe technique. High resolution spectroscopy in the frequency domain is no longer possible with very short pulses. A 1 ps pulse has a Fourier transform limited spectral width of $\approx$5 cm^{-1}.

A continuous train of very short light pulses can be created by mode locking a CW laser.[10] Active mode locking is achieved by placing an AO modulator in a standing wave cavity and matching its driving frequency with the round trip time of a light pulse in the laser cavity

$$\left(\nu = \frac{2L_{sw}}{c}\right).$$

For a 1 m cavity this corresponds to 150 MHz. The Q of the cavity is spoilt by the AO modulator except for a short period in each cycle. This forces all the longitudinal modes of the laser oscillator to be locked together in phase, rather than being independent and thus having random phases. The modes can only build up in the short periods when the Q of the laser is not spoiled.

The phase locking of the modes defines the laser output to be a sequence of pulses spaced by the cavity round trip time. This is easily confirmed by looking at the Fourier transform of such a series of short pulses. The pulse width is inversely proportional to the number of modes that are locked in the process. Picosecond experiments are best performed by averaging the effect of many pulses of very modest energy. A 1 nJ pulse ($\approx 10^9$ photons) in 1 ps still has a peak power of 1 kW.

Single picosecond pulses can be extracted from a mode locked pulse train, amplified or further time compressed to shorter pulses using specialized techniques. The extremely short duration of femtosecond pulses lead to high powers even for very small pulse energies and nonlinear effects are easily observed. A corollary is that experiments on condensed phases are often performed near or above the damage threshold of a material. Processes such as self focusing and defect accelerated damage as well as limited spectral resolution may limit the application of very short pulse laser techniques to inorganic systems at low temperatures to studies on semiconductors. The techniques may find applications in the study of fast processes in liquid solution, where the sample can, in effect, be replaced between pulses.

2.4.4 ***Time Resolution versus Spectral Resolution*** Very narrow spectral features are best observed in time domain measurements such as in photon echo or quantum beat experiments (Section 4.3). Slow, coherent transients correspond to the highest resolution when transformed to the frequency domain. When spectra have $\approx 1\ \text{cm}^{-1}$ features, transients occur on a timescale of ≈ 30 ps. Observation of such fast transients are technically difficult. Working in the frequency domain then becomes more effective.

2.5 Light Mixing

A theme of this chapter is that the interaction of light with matter takes on entirely new dimensions with laser sources. When the irradiance of light becomes sufficiently high (typically 10^4–10^6 W/cm^2 although effects in special materials can occur at far lower power), laser light itself can significantly alter the normal optical characteristics of the material through which it propagates.

In a classical description, the light-matter response is described through the generalized nth rank tensorial susceptibility $\chi^{(n)}$. The dielectric polarization P of the material associated with the electric field of the light $\widetilde{E}$ can be expanded in a power series[6,8]

$$\widetilde{P} = \varepsilon_0 \chi^{(1)} \widetilde{E} + \varepsilon_0 \chi^{(2)} \widetilde{E} \cdot \widetilde{E} + \varepsilon_0 \chi^{(3)} \widetilde{E} \cdot \widetilde{E} \cdot \widetilde{E} + \cdots . \tag{7}$$

The first term corresponds to the normal anisotropic (α) susceptibility term whose real and imaginary parts determine the refractive index and linear absorption of the medium, respectively. The second (hyper-polarizability or β) term is responsible for harmonic generation for example. The cubic (second hyperpolarizability or γ) term describes self focusing of light and other phenomena. In a centro symmetric medium the even (β) terms are necessarily zero.

The full quantum mechanical description of nonlinear interaction of radiation with matter can be treated through the density matrix formulation.[7,9] This describes, in full, power broadening, dynamic Stark (external electric field) effects, and many other coherent processes seen in high resolution optical transitions. A simpler approach is developed in Section 3.1.

A useful heuristic semiclassical description of some of the nonlinear processes is provided through the concept of four wave mixing (FWM). At low intensities, two coherent light beams propagating in a medium remain purely harmonic. The polarization intensities will combine additively to produce an interference pattern. If the intensity of the (two pump) beams is increased sufficiently, the interfering beams, through nonlinear terms in Equation 7, can modulate the optical properties of the medium, creating a grid like structure in the medium (Figure 2). The structure is a type of hologram and will persist as long as the nonlinear polarization. A third beam (probe beam) can be introduced and be modulated by the grating, giving rise to a (fourth) scattered beam.

In the overall light mixing process, both the total (scalar) energy E and (vector) momentum $\tilde{k}$ must be conserved. The momentum of light in a medium is determined by its velocity and frequency and thus connected to the refractive index of light (of a

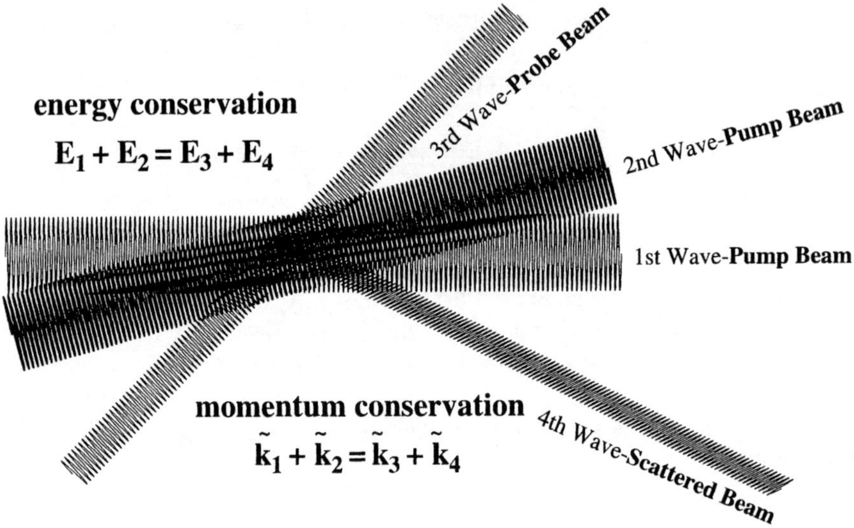

Figure 2. A Four Wave Mixing (FWM) schematic. Two pump beams provide a grating from which a third beam is scatterred to create a fourth beam.

particular wavelength) in the medium. The momentum conservation condition provides a strong constraint. In order that it be conserved, it is necessary to find a direction in a nonlinear material in which the refractive indices at different wavelengths (thus beam momenta) are the same. This process is called index matching.

At sufficiently high light intensities, all materials suffer irreversible physical damage. A criterion for any useful nonlinear medium is that its β or γ value be high enough that damage does not occur before the nonlinear process of interest becomes significant. These conditions are best met by very high quality crystals of simple, highly transparent noncentrosymmetric inorganic salts such as potassium dihydrogen phosphate (KDP), lithium niobate or β-barium borate (BBO). Many materials have higher nonlinear coefficients but their use is limited by damage thresholds. Increasing the fluence of a laser to compensate for lower nonlinear coefficients is relatively easily achieved by beam collimation or amplification.

2.5.1 ***Harmonic Generation*** The general case of FWM involves four distinct beams. However, two or more beams may be of the same wavelength and may even be components of the same beam. Harmonic generation can be considered to be a special case of FWM. The (two) pump beam(s) of frequency ω co-propagate identically creating a grating with spacing λ. The probe beam is also the pump beam. Passing through the grating, the beam's polarization becomes modulated at ω creating an output component at frequency 2ω. If the 2ω component propagates in phase with the pump beam, the 2ω component will build in intensity to a significant level.

Harmonic generation is a very useful way in which to generate short wavelength radiation from an efficient infrared or red laser. Harmonic generation is easily achieved with pulsed lasers and is also achievable to a significant extent with CW lasers by tight focusing or via resonant cavity operation, in which power levels are far higher.

2.5.2 ***Sum-Difference Mixing*** When the emergent beams from a harmonic generation process at ω and 2ω are combined in a nonlinear medium, an analogous FWM process can occur. It becomes possible to create an intense beam at frequency 3ω. Again, momentum (index) matching is required for the ω, 2ω, and 3ω beams for effective conversion. This particular example of a sum mixing process, often called tripling, is commonly used to create intense 355 nm radiation from 1.06 μ YAG light combined with doubled YAG light at 532 nm.

In a more general FWM process it is possible to combine light at ω_1 and ω_2 to create beams at $(\omega_1 + \omega_2)$ and $(\omega_1 - \omega_2)$. The conversion process must reach a limiting value, as eventually the increasingly intense $(\omega_1 + \omega_2)$ and $(\omega_1 - \omega_2)$ beams will combine to regenerate light at ω_1 and ω_2 in the inverse mixing process.

2.5.3 ***OPO and OPA Action*** The inverse of the mixing process outlined in the previous section may be used to split intense light at ω_{pump} into beams at ω_1 and ω_2. In order to conserve energy the $\omega_1 + \omega_2 = \omega_{pump}$ condition must be met. The process becomes more useful when the beams (called signal and idler respectively with

$\omega_1 > \omega_2$) are folded within an optical cavity also containing the nonlinear medium. This leads to a (nonlinear) gain in the intensity of the conversion processes from ω_{pump} to ω_1 and ω_2. The stimulated nature of the process maintains the phase and coherent, well defined beams at frequencies ω_1 and ω_2 are generated in this device which is called an OPO.

Strong wavelength constraints on the signal (and thus idler) outputs are provided by momentum conservation (phase matching). Specific orientations of the beams with respect to the nonlinear OPO crystal amplify the particular signal output wavelength which has a phase matched condition. The output wavelength of an OPO can be tuned over a very wide range by rotating a nonlinear crystal. The crystal must however have the appropriate dispersion of refractive indices and a useful value of its hyperpolarizability.

Finer wavelength control of the output can be provided by placing other wavelength selective components within in the OPO cavity and it is possible to achieve close to transform limited tunable single frequency performance from these systems.

An OPO functions in a way analogous to a laser except that there are in general two output wavelengths. An optical parametric amplifier (OPA) functions in a way analogous to a laser amplifier. In an OPA the conversion of an intense pump beam into signal and idler beams is seeded by an input beam of well defined frequency. The nonlinear conversion process is effected in a single light pass through the crystal. The output wavelength is determined entirely by the seed wavelength but again phase matching, achieved by careful alignment of the OPA crystal(s) is essential to allow an efficient conversion process. More that 20% of the pump light can be converted to the output light.

Useful OPO and OPA action is very dependent on the quality of the pump beam. It is essential that the pump beam be as monochromatic, uniform, and stable as possible. Small fluctuations in intensity of the pump are amplified greatly by the nonlinear action of the OPO and the device can quickly become ineffectual to a spectroscopist. At best, shot to shot fluctuations of the output of an OPO/OPA system are around 10%. This modest level of stability requires a rather sophisticated YAG pump laser having exceptional beam quality.

The advantage of the OPO technique is its ease of operation over a very wide frequency range while maintaining relatively high and uniform output pulse energies. Tunability from 420 nm to beyond 2000 nm, with single frequency scanning and close to transform limited linewidth (0.01 cm^{-1}, 300 MHz) is available. With the high output pulse powers available in these systems, harmonic conversion of light to cover the 210 nm to 420 nm region is easily achievable.

2.5.4 ***Raman Shifting and Raman Lasers*** The Raman scattering process is weak, yet with focused pulsed lasers, the Stokes and anti-Stokes scattering can seed a stimulated Raman scattering process. This not only shifts a significant fraction of the input energy to $\omega_l - \omega_r$ and $\omega_l + \omega_r$ frequencies, but the process is effective enough to convert a significant fraction of the shifted energy to $\omega_l - 2\omega_r$ and $\omega_l + 2\omega_r$ light and beyond.

The laser irradiance conditions needed for effective Raman shifting to occur can be sustained in liquids and gases at high pressure. Simple substances such as H_2 and N_2 are used as they can survive the focused laser beam and give rise to large Raman shifts, ω_r, as their mode frequency is high. Momentum is conserved in the emergent beam by different frequencies propagating in different directions. Index matching in isotropic media is not possible.

It is possible to construct lasers based on Raman gain.[10] A wealth of non-linear spectroscopies are describable as FWM processes. Coherent antistokes Raman (CARS) and hyper-Raleigh and hyper-Raman spectroscopies afford examples. Hyper-Raleigh scattering occurs at $2\omega_l$ and hyper-Raman at $2\omega_l \pm \omega_r$.

3 LINEWIDTHS AND LINESHAPES

A fully quantum mechanical treatment of this subject is beyond the scope of this chapter. The interested reader is referred to a number of relevant texts.[7,9]

3.1 Homogeneous Lineshape

Consider the simplest case of an electronic excitation—namely the excitation of a single electron between two well isolated, purely electronic levels. The system is then impulsively disturbed at $t = 0$ with an arbitrarily short pulse. The response of the system at $t > 0$ can be modeled classically as a simple one dimensional damped harmonic oscillator.[10]

$$\frac{\partial^2 x}{\partial t^2} + \gamma \frac{\partial x}{\partial t} + \omega_0^2 x = 0. \tag{8}$$

In Equation 8, $\omega_0^2 = D/m_e$, where D is the restoring force constant and m_e is the electron mass. The damping term γ is associated with both radiative and nonradiative energy dissipation from the excited state Ψ_e to the ground state Ψ_g whose energy difference is $E_e - E_g$.

The solution of the differential Equation 8 is given by

$$x(t) = x_0 \exp\left(-\frac{\gamma}{2}t\right)\left\{\cos\omega t + \frac{\gamma}{2\omega}\sin\omega t\right\}. \tag{9}$$

The frequency

$$\omega = \sqrt{\omega_0^2 - \frac{\gamma^2}{4}}$$

is lower than the resonant frequency ω_0 of the undamped case. The angular frequency $\omega = 2\pi\nu$ of optical transitions is a few thousand THz ($\sim 10^{15}$ Hz). For example, a transition at 500 nm has ω of $\sim$3800 THz. The lifetime T_1 for the lowest excited state Ψ_e is usually > 1 ns thus $\gamma < 1$ GHz. The second term in Equation 9 can then be neglected leaving

$$x(t) = x_0 \exp\left(-\frac{\gamma}{2}t\right) \cos \omega t. \tag{10}$$

The damping causes a distribution of frequencies ω as can be deduced from the Fourier transform of $x(t)$

$$A(\omega) = \int_0^{\infty} x(t) \exp(-i\omega t)\, dt. \tag{11}$$

The lower integration limit in the Fourier integral (Equation 11) is zero since $x(t) = 0$ for times $t < 0$. Integration of Equation 11 leads to

$$A(\omega) = \frac{x_0}{\sqrt{8\pi}} \left\{ \frac{1}{i(\omega - \omega_0) + \frac{\gamma}{2}} + \frac{1}{i(\omega + \omega_0) + \frac{\gamma}{2}} \right\}. \tag{12}$$

The intensity $I(\omega)$ of the transition is proportional to $(A\omega)A^*(\omega)$, $A^*(\omega)$ being the complex conjugate. Since $(\omega - \omega_0)^2 \ll \omega_0^2$ one can neglect terms containing $\omega + \omega_0$ and thus

$$I\big(\omega - \omega_0\big) = I_0 \frac{1}{(\omega - \omega_0)^2 + (\gamma/2)^2}. \tag{13}$$

Upon normalization one obtains

$$g\big(\omega - \omega_0\big) = \frac{1}{2\pi} \frac{\gamma}{(\omega - \omega_0)^2 + (\gamma/2)^2} \tag{14}$$

or in terms of frequency ν

$$g\big(\nu - \nu_0\big) = \frac{1}{2\pi} \frac{\Gamma}{\big(\nu - \nu_0\big)^2 + (\Gamma/2)^2}. \tag{15}$$

This is the normalized Lorentzian lineshape with full width at half maximum Γ (in Hz) of

$$\Gamma = \frac{\gamma}{2\pi} = \frac{1}{2\pi T_1}. \tag{16}$$

The function $x(t)$ defined in Equation 10 and its Fourier transform are illustrated in Figure 3. Equation 16 is a direct manifestation of the Heisenberg uncertainty principle.

In condensed phases, both ground and excited states interact with the environment. As well as the (T_1) lifetime broadening outlined above, further linewidth con-

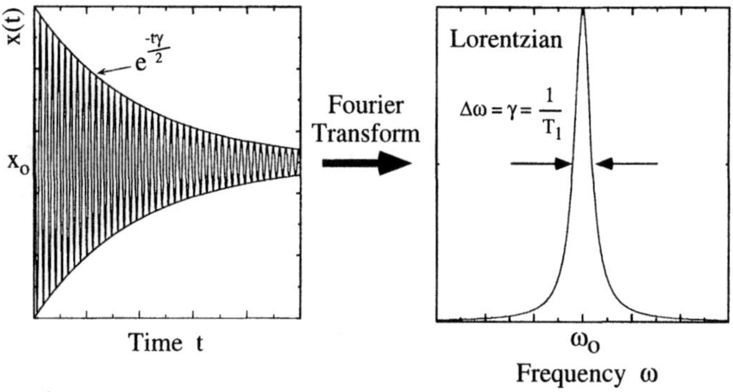

Figure 3. Fourier transform of the damped harmonic oscillator. The damping constant is denoted as $\gamma = 1/T_1$. The full width at half maximum of the frequency distribution is indicated.

tributions due to the dephasing of the time dependent wavefunctions of both states by electron-phonon interactions and also by fluctuations of electronic and nuclear spins need to be considered. Additionally, the electromagnetic field used to probe the transition ultimately has an effect on the linewidth of an excitation.

A more complete treatment leads to the following expression for the homogeneous linewidth Γ_{hom} of a two level system in condensed phases.[16,17]

$$\Gamma_{\text{hom}} = \frac{1}{\pi T_2}\sqrt{(1 + \omega_1^2 T_1 T_2)} \tag{17}$$

where

$$\frac{1}{T_2} = \frac{1}{2T_1} + \frac{1}{T_2^*}. \tag{18}$$

The effective dephasing time, T_2, is determined by the lifetime T_1 of the excited state and the pure dephasing time T_2^*. T_1 is determined from both radiative, k_r, and nonradiative, k_{nr}, rates:

$$\frac{1}{T_1} = k_r + k_{nr}. \tag{19}$$

Pure dephasing (T_2^*) is in general strongly temperature dependent. T_1 is also often temperature dependent. T_2^* processes occur in both ground and excited states.

The Rabi frequency ω_1 is given by

$$\omega_1 = \frac{\tilde{\mu} \cdot \tilde{E}}{\hbar} \tag{20}$$

where $\tilde{\mu}$ is the transition dipole moment and $\tilde{E}$ is the electric field vector of the radiation field. The Rabi frequency is related to the time needed to drive an ensemble originally in the ground state to equal ground and excited state populations. The power broadening due to the Rabi term can be a significant contribution to the linewidth if the laser power used is high. It is essential to estimate the laser power at which this term becomes significant.

Consider an electric dipole transition with an oscillator strength f. The electric field vector $\widetilde{E}$ of an electromagnetic wave traveling in direction $\widetilde{k}$ is directly related to the irradiance $\widetilde{J}$ which has units of power per unit area:

$$\widetilde{J} = \frac{E^2\widetilde{k}}{Z}. \tag{21}$$

The impedance Z is given by

$$\sqrt{\frac{\mu_r\mu_0}{\varepsilon_r\varepsilon_0}} \tag{22}$$

and has the value of 377 Ω in vacuum ($\varepsilon_r = \mu_r = 1$). Light sources are characterized by their root-mean-squared irradiance, $J_{\rm rms}$. Equation 23 relates $J_{\rm rms}$ of linearly polarized light to the electric field strength and the photon flux.

$$J_{\rm rms} = \frac{\sqrt{3}E_0^2}{2\sqrt{2}Z} \quad \text{or} \quad J_{\rm rms} = n_p h\nu \quad \text{and} \quad E_0^2 = \frac{2\sqrt{2}}{\sqrt{3}} Z n_p h\nu \tag{23}$$

where n_p is the number of photons per unit area per unit time.

The electric field is ~1 V/cm for a 1 mW laser with a 1 mm beam diameter. This irradiance is easily achieved in most CW lasers. A pulsed laser providing 100 mJ in 10 ns in the same beam diameter corresponds to ~300 MW/cm^2 which translates to a field strength of ~5×10^5 V/cm. The electric field acting on an electron in an atom is of the order of 5×10^{10} V/cm. The field intensity in a focused pulsed laser can easily exceed this value, leading to ionization and destruction of most materials.

For very low temperatures the pure dephasing time T_2^* may become long and T_2 converges to T_1. Using the expression relating oscillator strength f to dipole strength[18], the Rabi term can then be expressed by

$$\omega_1 T_1^2 \approx 5 \times 10^{-16} J_{\rm rms} \frac{\lambda^5}{f} \tag{24}$$

in the absence of radiationless losses, where $J_{\rm rms}$ and λ are in units of [mW/cm^2] and [nm], respectively. Power broadening becomes more significant in long wavelength transitions and is relatively more severe for weaker transitions. A transition at 500 nm with an oscillator strength of 10^{-2} is broadened by a factor of $\sqrt{2}$ when the laser power is 0.6 mW/cm^2. A clear consequence is that power broadening may be difficult to avoid in transient hole-burning and fluorescence line narrowing (FLN) experiments.

When the dephasing time is very short, or for low laser powers, $\omega_1^2 \ll 1/(T_1 T_2)$ and $\Gamma_{hom} = 1/(\pi T_2)$. For temperatures below 2 K, T_2^* characteristically becomes long relative to T_1. Γ_{hom} is then determined solely by T_1. By contrast, pure dephasing (T_2*) may well be the dominant contribution (~10–100 cm^{-1}) to the homogeneous linewidth at room temperature.

3.2 Inhomogeneous Broadening

In a perfect crystal lattice, symmetry related chromophores would be *identical*. They would all absorb at the same energy and would have the same (homogeneous) linewidth. However, all condensed phases suffer from inhomogeneous broadening. Each individual chromophore has a somewhat different environment. This may be due to mechanical strain, variations in electrostatic or van der Waals interactions, or other influences from chromophore to chromophore.

Both ground and excited state energies are a function of these local fields. This leads to a distribution of the transition energies of optical excitations. The distribution often (although not always) takes a Gaussian lineshape. Typical inhomogeneous widths of electronic origins in crystals are of the order of magnitude of 10 cm^{-1}.

In amorphous hosts, the lack of long range order contributes to a wider range of variations of the chromophoric unit. Inhomogeneous widths are larger and usually in the range of 100 cm^{-1} to 1000 cm^{-1}.

As mentioned in the introduction, a particular aspect of inhomogeneity, termed nanoheterogeneity, has important consequences in the spectroscopy of nominally symmetrical inorganic chromophores such as metal tris(diimine) complexes. At a finer scale than placing each chromophore in a different environment, each of the three ligands, and thus their π^* orbitals, become independently varied.[19–22] Importantly, this effect can make a significant contribution to breaking the symmetry of the chromophore.

The variation of metal–ligand distances provides a specific source of inhomogeneous broadening in coordination compounds by varying the crystal field. The crystal field transition of an octahedral complex is characterized by $10Dq$. Within the crystal field approximation for $3d$ ions[23]

$$Dq = Ze^2 \frac{\langle r^4 \rangle_{3d}}{6R^5} \tag{25}$$

where Z is the effective charge of the ligands, $\langle r^4 \rangle$ is the expectation value of r^4, r being the distance of a $3d$ electron from the metal nucleus. R is the metal–ligand distance.

The R^5 factor determines that a change of just 0.1% of the metal–ligand separation leads to a change in the overall ligand field transition energy of 0.5%. If $Dq = 2000$ cm^{-1} with $R = 2$ Å, a variation in R, $\partial R = 0.002$ Å changes $10Dq$ by -100 cm^{-1}. Minute changes in the metal–ligand distances lead to considerable variations of electronic energies.

If variations of the six metal–ligand distances are independent to an appreciable extent, the symmetry of the crystal field will be lowered. This in turn may lead to extra broadening. Spin-allowed ligand-field transitions are often structureless due to inhomogeneous broadening.

Transitions which occur within the same electronic configuration are less vulnerable to inhomogeneous broadening because they are independent on the ligand field strength in first order. For example, the $^4A_2 \leftrightarrow {}^2E$ spin-flip transition in chromium(III) systems is usually found to be (relatively) sharp. Inohomogeneous broadening is also minimal in f–f transitions of rare earth materials in which the excitation energies are not strongly influenced by crystal fields.

3.3 Temperature Dependence of the Homogeneous Linewidth

3.3.1 ***Crystalline Environments*** Four mechanisms which lead to a temperature dependence of the homogeneous linewidth are presented schematically in Figure 4. The direct process, which is also discussed in Chapter 7, is often the main source of broadening of electronic origins of higher lying excited states. The direct process can be very fast when the energy gap is near the maximum of the density of phonon states.[18] When an energy gap is within the range of phonon frequencies, the rate for the absorption, k_A, or emission, k_E, of a phonon of frequency ω is given by

$$k_E = k_0(\overline{n} + 1) \quad \text{and} \quad k_A = k_0\overline{n} \tag{26}$$

where

$$\overline{n} = \frac{1}{\exp\left(\frac{\hbar\omega}{kT}\right) - 1}$$

Usually, only the lowest-excited state of a transition metal complex is luminescent. Most materials have a sufficient number of high frequency vibrational modes (localized phonons) to act as energy acceptors in bridging the gap between higher-lying excited states and the lowest excited state. The probability of a multiphonon relaxation is strongly dependent on the number of vibrational quanta which have to be created.[24,25] Systems containing high frequency vibrational groups such as N–H, C–H and O–H will not usually show luminescence from higher lying excited states as relatively few vibrational quanta are needed to bridge the energy gap. A corollary is that when the lowest excited state lies below 10 000 cm^{-1}, the quantum efficiency for luminescence may become very low.

Exceptions are provided by materials that have large energy gaps between excited states but no high frequency vibrations. One such example is Ni^{2+} in $KZnF_3$ which shows luminescence from higher excited states.[26] By contrast, rare earth systems often luminesce from higher lying excited states. The electron-phonon interaction in rare earths is minimal but electronic energy spacings can be substantial.

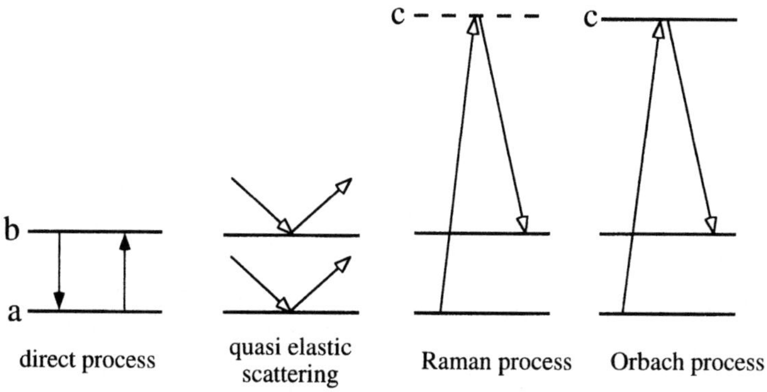

Figure 4. Electron-phonon interactions which lead to the broadening of electronic origins. Level c is a real state in the Orbach process but a virtual state in the Raman process.

Fast multiphonon relaxation to the lowest excited state substantially shortens the lifetime T_1 of higher-lying excited states and thus can be a major contribution to their homogeneous linewidths.

The energy gap between electronic levels often lies within the range of vibrational frequencies. Direct one-phonon relaxation is then possible. Within the Debye approximation, which is a simple practical description for the density of phonon states (lattice modes) in the solid state, the rate k_0 of a direct process at $T = 0$ K is proportional to the third power of the gap.[18] This reflects the vanishing density of phonon states at low energies. The relaxation between very close-lying states ($<0.5\ \text{cm}^{-1}$) may become slow compared to luminescence lifetimes.

Within the Debye approximation,[4] the temperature dependence of the Raman process (Figure 4) between close lying electronic states, when the gap is smaller than the Debye cut off frequency ω_D, is given by[27]

$$k_{\text{Raman}} = \overline{\alpha}\left(\frac{T}{T_D}\right)^7 \int_0^{T_D/T} \frac{x^6 \exp(x)}{(\exp(x) - 1)^2}\,\mathrm{d}x \tag{27}$$

where $T_D = \hbar\omega_{\text{db}}/k$ and $\overline{\alpha}$ is a constant describing the electron–phonon coupling. Levels a and b will be broadened by the same amount by the Raman process. At low temperatures the rate varies as T^7. Thus the Raman process may rapidly broaden an electronic transition in the temperature range from 1.5 K to 50 K.

The density of phonon states $\rho(\omega)$ may be modeled from sideband structure in luminescence spectra.[28] The density of phonon states and the intensity $I(\omega)$ in the sidebands of the luminescence spectrum are related by

$$I(\omega) = |C|^2 \frac{\rho(\omega)}{(\hbar\omega)^2} \tag{28}$$

The (angular) frequency shift ω is measured as a displacement from the zero phonon line. The constant C is discussed in Reference 28. Using this empirical approximation for the density of phonon states, the contribution to the temperature dependence of the homogeneous linewidth can be approximated by

$$\Gamma(T) = \overline{\alpha} \int_0^\infty \rho(\omega)^2 \overline{n}(\omega) \frac{1 + \overline{n}(\omega)}{2\pi} \, d\omega \tag{29}$$

The product of $\overline{\alpha}$ and $|C|^2$ is best used as a scaling parameter in an analysis of experimentally determined linewidths.

When $\Delta E_{bc} = E_c - E_b \gg kT$ the temperature dependence of the Orbach process is given by

$$k_{\text{Orbach}} = k^0 \exp\left(-\frac{\Delta E_{bc}}{kT}\right) \tag{30}$$

where k^0 is determined by the electron–phonon coupling of the levels a, b, and c. Both levels a and b are broadened by the same extent as the relaxation rate $a \to b$ is equal to the rate $b \to a$. Figure 4 identifies the difference between Orbach and Raman process as scattering involving a real or imaginary excited state (level c), respectively.

The quasi-elastic scattering of phonons, also called intrinsic Raman scattering, provides a further mechanism for broadening of electronic origins. The scattering of phonons induces small variations in the phase of otherwise well defined electronic wavefunctions. Consequently, a broadening of the excitation between such states results, as any modulation in the phase of the transition oscillator leads to a spread of output frequencies. The temperature dependence of this process is the same as that of the Raman relaxation process.

3.3.2 ***Amorphous Environments*** The four broadening processes outlined above also take place in amorphous hosts. However, the concept of phonons and consequently the density of phonon states becomes more difficult in systems without long range order. The theoretical treatment of the temperature dependence of homogeneous linewidths in amorphous media remains an active area of research.

The translational symmetry properties of a crystalline system require the potential energy along a coordinate q to be periodic. This periodicity is absent in amorphous systems. The potential energy along an effective coordinate q is irregular, as shown schematically in Figure 5. Unlike in a crystal, components (atoms, ions, or molecules) may not have well defined positions or orientations. Two (or more) configurations may exist differing only slightly in displacement or rotation and lead to a range of conformers with small energy differences. Energy barriers may be small enough to allow quantum mechanical tunneling from one configuration to the other.

Each pair of adjacent potential minima (for example the shaded section in Figure 5) can be considered to constitute an effective double well potential (two level) system or TLS. There will be, in general, a wide range of TLSs with different barrier heights, potential differences, etc. The variation in TLS parameters is often describable by Gaussian distributions.

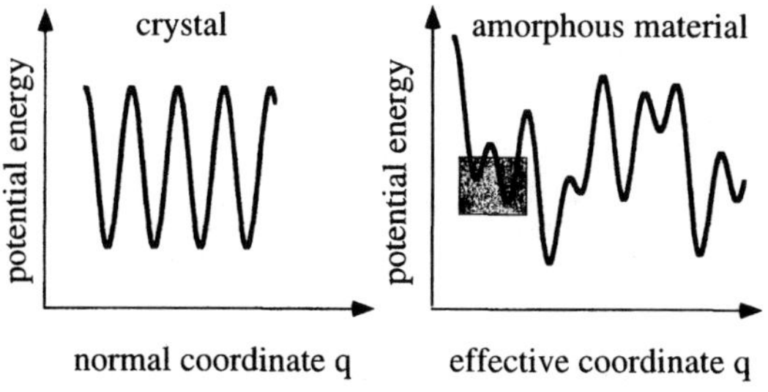

Figure 5. A comparison of potential energies in crystals and amorphous materials.

A wide range of chromophores embedded in amorphous hosts have homogeneous linewidths following the functional dependence[29]

$$\Gamma_{\text{hom}}(T) \propto T^{1.3\pm0.1} \tag{31}$$

The TLS description, initially developed[30,31] to model specific heat data of amorphous systems at low temperatures, can account for this dependence.

3.4 Spectral Diffusion

Homogeneous linewidths of chromophores embedded in amorphous hosts as derived from photon-echo (see Section 4.3) experiments were often seen to be substantially smaller than linewidths observed in spectral hole-burning (see Section 4.2) experiments.[32] Photon-echo processes are optical analogues to the very informative spin-echo techniques in NMR. A major difference between spectral hole-burning and photon-echo experiments is the timescale on which they are performed. Linewidths of order of magnitude of MHz, observed in photon-echo experiments, are determined from measurements made on the ns timescale. In contrast, persistent spectral holes are usually created and measured on the 100 s or longer timescale.

Amorphous hosts at the lowest temperatures are far from equilibrium. Consequently, there are subtle changes in the environment of an embedded chromophore occurring on a very wide range of timescales, from ns to days or months. These relaxation processes are a major source of spectral diffusion, as the transition energy of a each particular chromophore shifts with these changes. Linewidths observed in hole-burning or FLN experiments will suffer from spectral diffusion, dependent on the timescale of the experiment.

Radiative and nonradiative excitation energy transfer between chromophoric units can also be a major source of spectral diffusion. Furthermore, nuclear and electronic spin fluctuations around a chromophore may lead to significant spectral diffusion. These latter two mechanisms occur both in crystalline and amorphous systems. Spectral diffusion is, in general, more severe in amorphous hosts, due to their nonequilibrated nature. At high temperatures, particularly in amorphous media, it becomes progressively more difficult to perform hole-burning experiments, as spectral diffusion becomes more rapid.

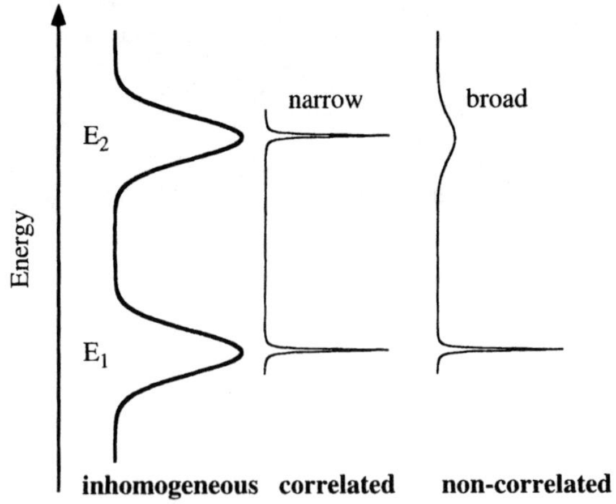

Figure 6. Correlation and noncorrelation of two energy levels having an inhomogeneous distribution.

3.5 Energy Level Correlation

The correlation of energy levels plays an important part in the laser spectroscopy of chromophores in the solid state.[19,33–35] Figure 6 portrays the cases of both full correlation and no correlation between energies E_1 and E_2 in an inhomogeneously broadened system. In a correlated system, the E_1 to E_2 spacing is constant and thus selection of a particular E_1 leads to a precisely defined E_2 value. In a poorly correlated system, E_2 can have a value anywhere within its inhomogeneous spread. In general, levels tend to be relatively well correlated in crystals (often to within 1%) whereas in glasses correlation is often absent.

The precise energy of a state is, in general, dependent on a number of parameters. In amorphous hosts the parameters vary independently to some extent. Chromophores with quite different environments may have the same transition energy. This is but an accidental degeneracy between two inequivalent chromophores. Consequently, when a particular transition energy is selected by a laser, a wide distribution of chromophores is probed if the correlation of energy levels is poor. Laser spectroscopies in these systems are energy selective and not site selective.

4 LASER TECHNIQUES

4.1 Fluorescence and Excitation Line Narrowing

Fluorescence line narrowing (FLN) was the first technique to be applied in an effort to observe the homogeneous linewidth of an optical transition in the solid state. FLN experiments are portrayed schematically in Figure 7. In an FLN experiment, the chromophores which have their transition in resonance with a narrow band laser are selectively excited. Their subsequent luminescence may be observed either res-

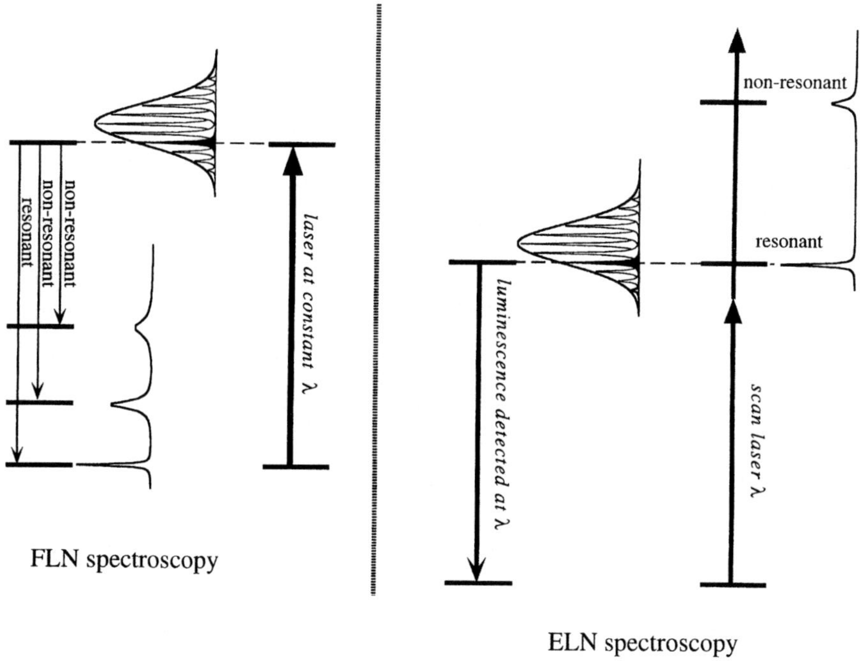

Figure 7. Schematics of the FLN and ELN techniques.

onantly (at the laser wavelength) or nonresonantly. Narrow transitions are observed due to the selective excitation. Homogeneous linewidths of the electronic origin can only be observed in the resonant FLN experiment. The limiting linewidth observable in an ideal FLN experiment is $2 \times \Gamma_{hom}$ as it involves two photons, one photon to excite the chromophore and a second emitted photon. Nonresonantly detected features are broadened both by the lack of correlation (Figure 6) between the resonant and nonresonant levels and relaxation broadening (broadening due to the lifetimes of the levels involved, see Section 3.1) of the nonresonant energy levels.

In excitation line narrowing experiments (ELN), which may be particularly informative, luminescence is detected at relatively high resolution within an inhomogeneously broadened electronic origin or vibrational sideband. The excitation laser is then scanned in order to afford a narrowed excitation spectrum. Spectral features are obtained of all chromophores that emit strongly at the detected wavelength. ELN is also portrayed in Figure 7.

Denisov and Kizel[36] were the first to report FLN experiments. They used mercury lines as the excitation source to narrow the luminescence spectrum of Eu^{3+} in a Na_2O:B_2O_3 (1:2) glass. The first laser-induced FLN experiments in the solid state were performed by Szabo.[37] He observed a narrowing of the ruby R_1 line to a width of 0.002 cm^{-1} upon resonant ruby-laser excitation. Figure 8 provides a schematic of a typical FLN apparatus.

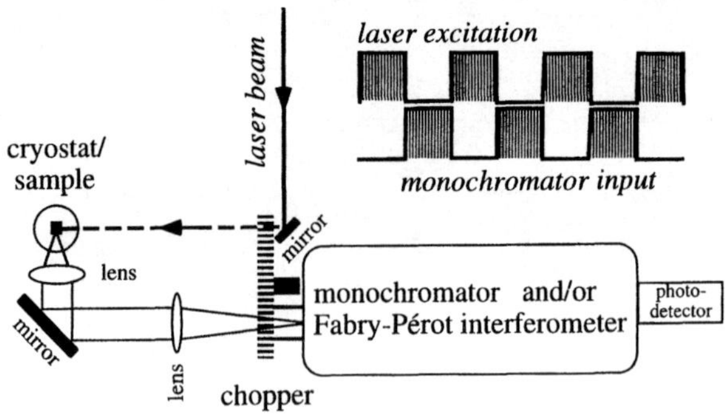

Figure 8. A typical FLN/ELN apparatus. The chopper creates pulses of exciting laser light but synchronously blocks the input of the monochromator/interferometer during these periods (inset).

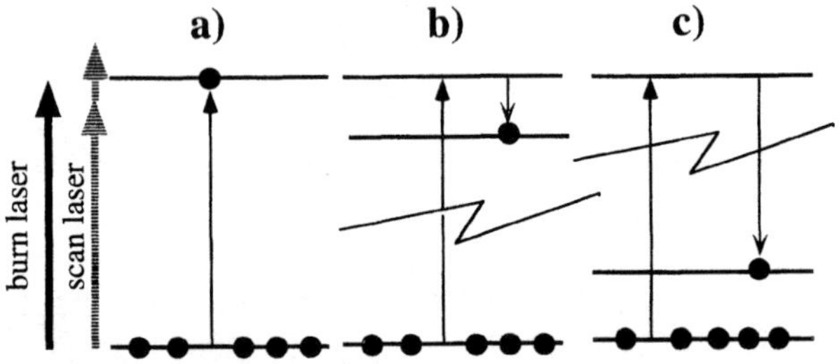

Figure 9. Transient spectral hole-burning by, a) direct depletion of the ground state, b) relaxation of the excited state to a metastable excited state, or c) relaxation to a metastable sublevel of the ground state.

4.2 Spectral Hole-Burning

Laser excitation can deplete the number of species absorbing at a particular wavelength within the inhomogeneous distribution. In absorption and/or excitation spectra, a dip or hole appears. This process is called hole-burning and can arise from quite distinct mechanisms. The hole-burning method is complementary to the FLN experiments in allowing spectral details to be observed in systems which are otherwise obscured by inhomogeneous broadening.

4.2.1 ***Transient Spectral Hole-Burning*** Upon excitation of a subset of chromophores with a particular transition energy, the excited state lifetime may be long enough in order to enable transient spectral hole-burning experiments. Alternatively, the initially excited state may relax to a metastable excited state or a metastable sublevel of the ground state multiplet. Transient hole-burning experiments are easily performed in many rare earth systems. The electronic fine structure in the ground state may provide a way to deplete the population of the ground state (from which the excitation occurs) since relaxation rates between closely spaced levels are slow.

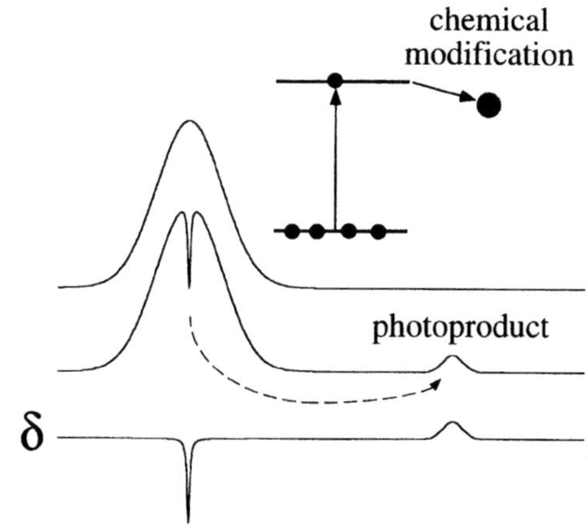

Figure 10. Photochemical hole-burning and its photoproduct. δ denotes the difference between spectra before and after hole-burning.

Figure 9 depicts mechanisms for transient hole-burning. Sufficient laser irradiance needs to be used so that a significant bleaching of the ground state occurs. However, care needs to be taken to avoid power broadening (Section 3.1).

Spaeth and Soy[38] were the first to report transient spectral hole-burning in a condensed phase. They investigated the bleaching of cryptocyanine in methanol at room temperature. They observed spectral holes of less than 2 nm width. The first transient spectral hole-burning experiment in a solid was reported in 1975.[39] Experiments were performed on the R_1 line in ruby at 4.2 K in zero field and in an applied field of 400 G.

4.2.2 ***Persistent Spectral Hole-Burning*** There are two main mechanisms which lead to persistent spectral hole-burning; photochemical and nonphotochemical (photophysical) hole-burning. In the first case the molecule undergoes a chemical transformation upon photoexcitation. The chromophores which are selectively excited by the laser are converted to new chemical species with different spectra. This is illustrated in Figure 10. Photoionization is often a mechanism in photochemical hole-burning. More complex photochemical reactions such as conformer interconversion, photodecomposition and tautomerism also occur.[40–42]

In contrast, nonphotochemical hole-burning involves subtle changes of host-guest interactions. Transition frequencies of the affected chromophores are close to the burn frequency. This is depicted in Figure 11. Nonphotochemical hole-burning is a ubiquitous phenomenon in amorphous hosts.[43,44] The phenomenon has been interpreted[45,46] in terms of the TLSs (Figure 12) (see Section 3.3.2). After photoexcitation, the system tunnels to the second well and then relaxes to the corresponding well in the ground state. Subsequently, tunneling can occur in the double well of the ground state leading to spontaneous hole-filling. Nonphotochemical hole-burning may also occur in crystalline systems.[47]

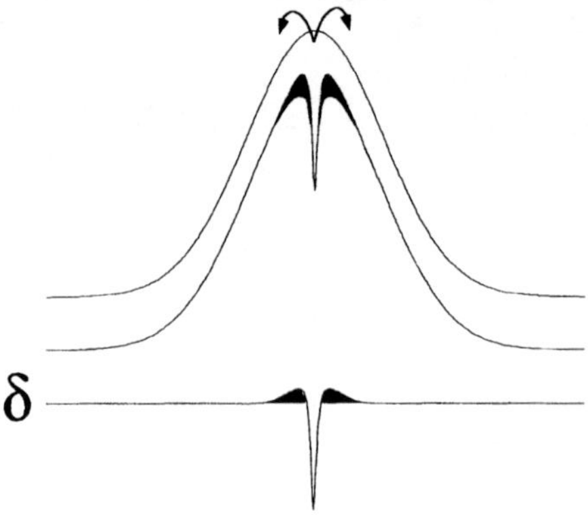

Figure 11. Nonphotochemical hole-burning and the (typical) distribution of its photoproduct (shaded area).

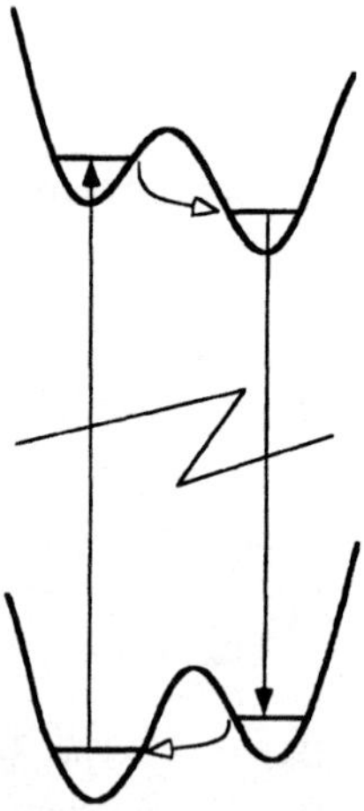

Figure 12. Nonphotochemical hole-burning via a two level system (TLS).

The first observations of persistent spectral hole-burning were independently reported in 1974 by Gorokhovskii et al.[48] and Kharlamov et al.[49] The first group observed a spectral hole in absorption and luminescence spectra of the electronic origin of H_2-phthalocyanine in *n*-octane at 5 K after exposure to a free-running ruby laser of 0.1 J pulse energy and 0.5 ms pulse duration. The second group reported holes in the electronic origin of perylene and 9-aminoacridine in ethanol at 4.2 K.

4.2.3 ***Readout Techniques for Spectral Holes*** Persistent spectral holes may be measured in either excitation or transmission. Excitation readout is very sensitive, allowing samples with low optical densities to be examined but is naturally restricted to luminescent systems. A typical experimental setup for both methods is given in Figure 13.

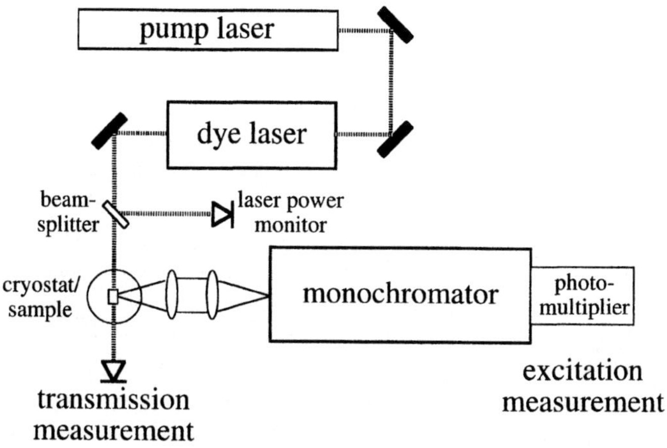

Figure 13. Schematic of a hole-burning apparatus. After a burn period at constant wavelength, the single frequency laser is scanned across the spectral hole with reduced laser power to avoid further burning. The excitation or transmission signal is normalized against the laser power.

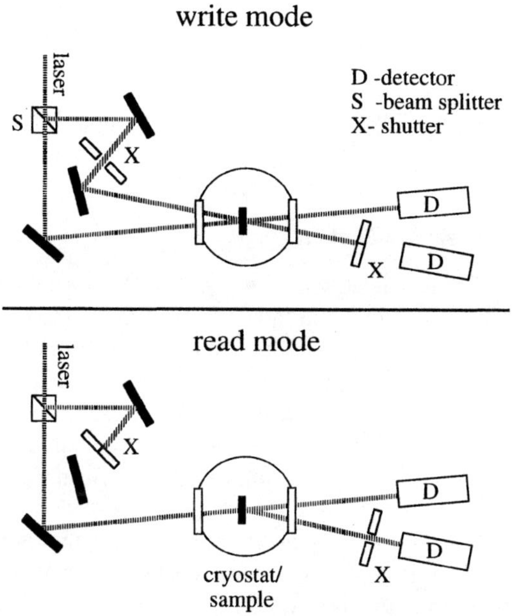

Figure 14. Schematic of the holographic technique for writing and reading spectral holes (see text).

Holography provides an elegant method with which to both write and read persistent spectral holes.[50] In the write process, two coherent beams are superimposed upon the sample (Figure 14). Spectral hole-burning will occur in the constructive interference regions. Thus a spatial grating is created at the burn wavelength. For the readout, one of the write beams is blocked and the other attenuated. The diffracted beam intensity as a function of the laser wavelength provides the spectral hole.

As the grating is only present at the burn wavelength, this method is free of background and is particularly useful for shallow holes. Naturally there are stringent requirements regarding the optical quality of the sample and the coherence of the laser. The holographic detection scheme of spectral holes can be considered to be a special case of FWM (Section 2.5). After writing a grating with two waves, a third wave (which in this case is collinear with the first wave) is used to read out the spectral hole by its diffraction to a fourth wave.

Transient holes can be read out in transmission, excitation, or by time resolved holography (FWM). The most direct method involves two single frequency lasers. The first laser is used to excite chromophores at a particular frequency and the second laser is then scanned across the inhomogeneously broadened transition to reveal a dip at the frequency of the first laser. Alternatively, the two laser frequencies can be provided by the use of a single frequency laser and two AO modulators. In a third method the laser is first held at the burn wavelength but then rapidly (within the lifetime of the excited state) scanned across the hole.

Varying of the transition frequency of the chromophore via the application of external electric or magnetic fields provides an alternative scheme to the methods outlined above.[51,52] Sample sweeping can be applied to both persistent and transient holes. Electric fields in particular can be rapidly scanned, facilitating transient hole-burning. The laser needs only be held at a constant wavelength rather than being swept.

4.3 Photon Echo Measurements

Whereas spectral hole-burning and fluorescence line narrowing experiments rely on the frequency stability of the exciting light source, photon echoes depend on the coherence of the exciting light. The laser linewidth must be broader than the homogeneous width in order to excite chromophores of different resonance frequencies from within the inhomogeneously broadened distribution. Thus the technical requirements regarding the frequency stability of the laser do not exist in photon echo experiments. Consequently, a photon echo experiment is readily performed when the homogeneous linewidth becomes very narrow. This is just the situation when corresponding spectral hole-burning and the FLN experiments become difficult to perform.

The two pulse photon echo in a two level system is schematically portrayed in Figure 15 using the gyroscopic model originally developed for magnetic resonance phenomena.[53–55] The optical excitation of each chromophore is described as a rotation of its pseudo-dipole moment. Initially, all chromophores are in the ground state. This is denoted as the $S_z = -1/2$ state in Figure 15. Upon application of a laser pulse, the pseudo-dipole moment is rotated into the x–y plane

$$\Theta = \int_0^{\tau} \widetilde{\mu} \cdot \widetilde{E}(t)\,\mathrm{d}t. \tag{32}$$

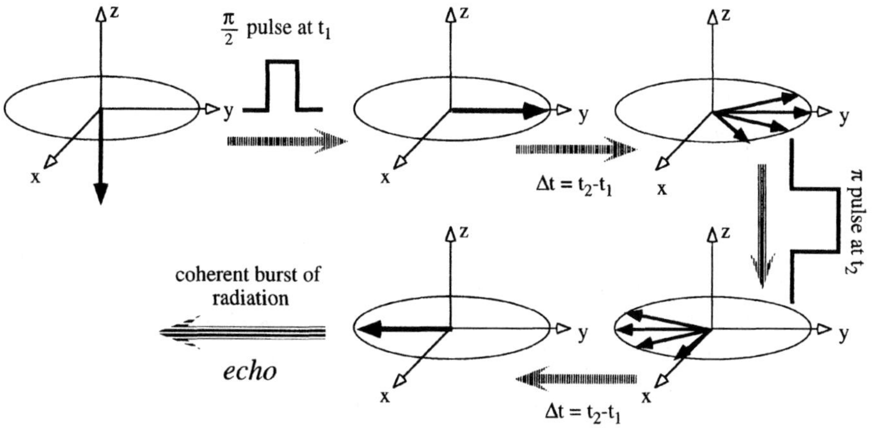

Figure 15. Schematic of the two pulse photon echo process (see text).

If the electric vector $\widetilde{E}$ of the radiation field is parallel the x axis and the pulse area Θ defined in Equation 32 is equal to $\pi/2$, all the dipoles are aligned along the y axis at $t = \tau$. This is denoted as $S_y = 1/2$ and corresponds to a coherent superposition of the ground state and the excited state of the ensemble.

The macroscopic oscillating dipole created in this way will rapidly decay into microscopic dipoles on a timescale given by the inverse of the energy distribution of excited chromophores. For example, if an inhomogeneously broadened line (width >1 cm^{-1}) is excited with a laser pulse of 1 cm^{-1} (=30 GHz) width, the macroscopic dipole will decay in $\sim 1/30$ GHz $= 33$ ps.

The chromophores which have their resonance frequency at the center of the laser pulse frequency will remain parallel to y, but others will either spread clockwise or counterclockwise, depending on their frequencies relative to the center frequency.

If a pulse of area $\Theta = \pi$ (Equation 32) is then applied, the microscopic dipoles are rotated around the x axis by 180 degrees. This rotation corresponds to a time reversal and the microscopic dipoles will rephase at the time given by the pulse separation, leading to a coherent burst of radiation, the photon echo. The amplitude of the echo will be reduced by dephasing processes of the microscopic dipoles. The echo intensity is expected to decay according to

$$I_{\text{echo}} = I_0 \exp\left(-4\frac{\Delta t}{T_2}\right) \tag{33}$$

where T_2 is the dephasing time defined in Equation 18. Hence, a plot of the echo intensity versus the delay time Δt affords the dephasing time and thus the homogeneous linewidth of an optical transition.

The $\pi/2$–π pulse sequence provides the maximal echo amplitude. However, echoes can also be observed with non-optimal pulse areas. Often, both pulses are of the same area. The first photon-echo experiments were performed on a ruby sample.[56]

4.4 Two Photon Spectroscopy

Early in the development of quantum mechanics it was recognized[57] that it would be possible for the usual resonance condition, $E_e - E_g = \hbar\omega$ to be extended to include cases involving the simultaneous absorption of two or more photons. The simplest case involves the simultaneous absorption of two photons each of frequency $\hbar\omega/2$ although this is not a general requirement. Two photon absorption is distinct from processes involving the sequential absorption of two photons, with the first leading to the creation of an excited state. The former process has a far smaller cross section than one photon absorption. The absorption rate is quadratic in the intensity of the radiation field. The first photon can be thought to briefly create a virtual state, which then absorbs the second photon.

For a two photon transition involving a light beam with frequency $\hbar\omega/2$, the transition strength from Ψ_g to Ψ_e is proportional to the term

$$\sum_i \frac{\langle\Psi_e|\widetilde{\mu}\cdot\widetilde{E}|\Psi_i\rangle\cdot\langle\Psi_i|\widetilde{\mu}\cdot\widetilde{E}|\Psi_g\rangle}{\omega_g - \omega_i} \tag{34}$$

which sums the product of two dipole matrix elements, over all states i of the system. Equation 34 is easily extended to the case of two radiation fields of different frequencies, but with $\omega_1 + \omega_2 = \omega$.

The symmetry properties, and thus selection rules and polarizations, in two photon processes are distinctly different to single photon processes and can provide quite complementary information. For an electric dipole transition as in Equation 34, a $g \leftrightarrow g$ transition is allowed via a two photon process while the normally parity allowed $g \leftrightarrow u$ transition is two photon forbidden.

The first observation[58] of two photon absorption (Eu^{2+} in CaF_2) needed to wait for the very high intensities available in a pulsed laser. The system was irradiated with a pulsed ruby laser at 694 nm with an irradiance of 7×10^{23} photons/cm^2 sec (2×10^5 W/cm^2). Even then, only 1 in 10^7 of the photons were absorbed. This weak absorption was monitored by measuring Eu^{2+} luminescence subsequent to two photon absorption. A difficulty with the two photon absorption method is that the effect often becomes measurable only at levels close to the damage thresholds of a material. The system should also be strongly luminescent to allow excitation detection of the weak absorption. Crystal hosts should be centro-symmetric as otherwise harmonic generation (see Section 2.4.1) may interfere.

4.5 Single Molecule Spectroscopy

Single molecule spectroscopy (SMS) is a dramatic recent development,[2] arising from the initial observations of Moerner and Kador.[59] It has become possible to perform a range of spectroscopic measurements on single molecules in condensed phases. This has provided direct insights into the nature of inhomogeneous broadening, spectral diffusion etc. So far, only a very limited number of organic molecules have been studied using this technique. We present the basis of this extraordinary technique as we feel that it is a significant challenge yet only a matter of time before the method is applied to inorganic systems.

For linearly polarized laser light and a linearly polarized transition dipole, the peak absorption cross section σ of the electronic origin of a single chromophoric unit is given by[60]

$$\sigma(T) = F(T)\frac{\Gamma_{rad}}{\Gamma_{hom}}(T)\frac{\lambda_0^2}{2\pi}\,3\cos^2\theta \tag{35}$$

where $\Gamma_{rad} = \frac{1}{2\pi T_1}$ and θ is the angle between the electric vector of the radiation field and the transition dipole moment direction. σ is *independent* of the dipole strength of the transition (Figure 16).

The factor $F(T)$ describes the fraction of the total intensity remaining in the purely electronic origin (zero phonon line). In condensed phases, a significant fraction of the total intensity may be distributed among excitations which simultaneously excite a lattice phonon. The degree of coupling is characterized by the Debye–Waller factor in an analogous way that the Huang–Rhys factor (see Chapter 7) measures the coupling to localized high frequency phonons (i.e., vibrations). Coupling to lattice phonons and vibrational sidelines both reduce the purely electronic cross section.

In organic molecules $F(T)$ is typically 0.1 whereas in inorganic complexes it may approach 1. At lowest temperatures Γ_{hom} may approach Γ_{rad}. Taking θ to be

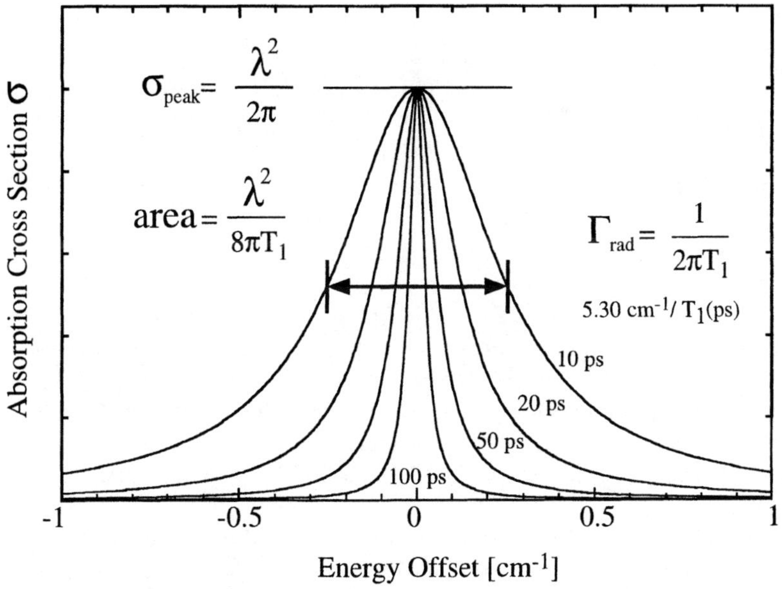

Figure 16. Idealized absorption cross section σ for an isotropic transition between purely electronic levels whose linewidth is determined by Γ_{rad}.

zero (a linearly polarized transition) provides a maximal value of $\sigma(T) \approx 0.1\ \mu m^2$ for transitions near 500 nm when $F(T)$ is 1. This can be compared to the spot size ϕ of a focused laser. Using a Gaussian beam the result is

$$\phi = \frac{f\lambda}{\pi w} = f\delta\theta \tag{36}$$

where w is the radius at the beam waist, $\delta\theta$ is the beam divergence and f is the focal length of the focusing lens. A beam of 1 mm diameter and 500 nm wavelength can be focused to a spot size $\phi \approx 3\ \mu m$ with a $f = 10$ mm lens.

SMS is feasible when the area of the focused laser beam, $\pi(\phi/2)^2$, approaches the cross section $\sigma(T)$. Experiments are best performed in the near field to minimize ϕ. Relatively few molecules may then be probed, with each maintaining a substantial probability (on resonance) of absorbing a photon from the laser beam. A cubic micron of 10^{-5} M sample contains about 6000 molecules. Rather than measuring an inhomogeneously broadened continuum, small volumes of dilute samples show individual lines associated with a statistical distribution of excitation energies of single molecules. Each probed volume will have distinct structure.

There are strong constraints on the type of system that can be studied via SMS. Short excited state lifetimes, high quantum efficiencies, low hole-burning efficiencies and minimal bottlenecks preventing the system returning to the ground state are important. Inorganic systems have a number of advantages and disadvantages here, but the prospect of being able to directly observe the spectroscopy and photophysics of single inorganic species is most attractive.

5 EXAMPLES

5.1 FLN

FLN and ELN spectroscopies can be particularly valuable in the study of vibrational sidelines in the emission and excitation spectra of inorganic complexes in amorphous hosts. Sidelines in nonselective spectra are most often not resolved due to the inhomogeneous broadening. An FLN spectrum of a ligand-centered $^3\pi$–π^* transition in $[IrCl_2(5,6\text{-}Me_2\text{-}phen)_2]^+$ is presented in Figure 17. The narrowed spectrum shows remarkable detail in comparison with the nonselectively excited luminescence.[61] The chromophore is considered to be a case in which the lowest ^{3}MLCT and $^3\pi$–π^* states are indeed quite close in energy. Detailed information regarding the nature of vibrational sidelines coupled to the luminescent state and the zero field splittings (ZFS) of the luminescent state are available from FLN and ELN and greatly assist assignments.

Figure 18 shows the Zeeman pattern of the narrowed, ligand-centered $^3\pi$–π^* luminescence of $[Ru(3,3'\text{-biisoquinoline})_3]^{2+}$. Here, any MLCT states are at significantly higher energy. A five line pattern characteristic of a pure spin triplet is observed. Energy spacings establish the g value to be isotropic and with a value very

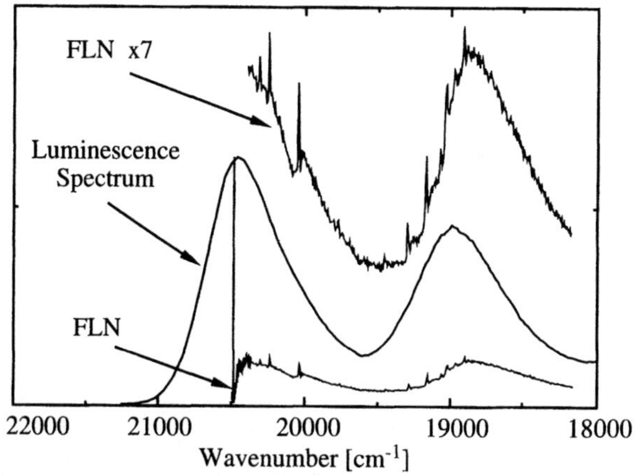

Figure 17. Nonresonantly excited luminescence and a FLN spectrum, excited at 488 nm, of $[IrCl_2(5,6\text{-}Me_2\text{-}phen)_2]^+$ in glycerol at 1.8 K. Vibrational sidelines become apparent in the narrowed spectra. Adapted from Reference 61.

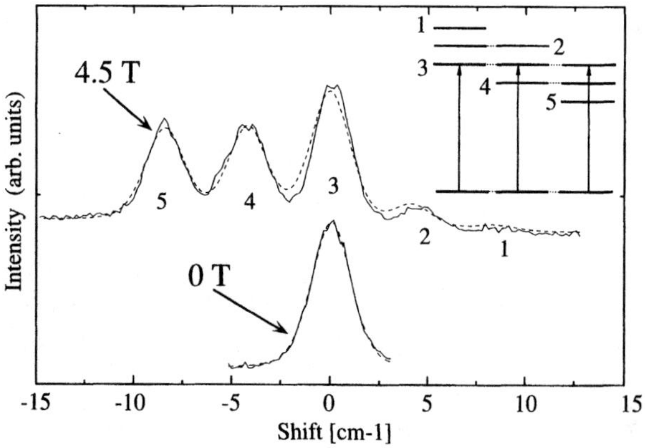

Figure 18. Resonant FLN spectra, excited at 538.5 nm, of $[Ru(3,3'\text{-biisoquinoline})_3]^{2+}$ in 4:1 ethanol/methanol glass at 4.2 K. The characteristic 5 line pattern of a $g = 2$, $S = 1$ state is apparent in an applied magnetic field. The dashed line is a modeled fit. The resolution is instrumentally limited. Adapted from Reference 62.

close to 2. Additionally, the ZFS is small compared to the Zeeman energy[62] consistent with the luminescent state being a relatively unperturbed $^3\pi^*$ state.

FLN spectroscopy can also provide the enhanced resolution needed to investigate ZFSs in the ground state. In Figure 19 contrasting ZFS behavior for two Cr^{3+} systems is displayed. For $Cr(acac)_3$ doped in $Al(acac)_3$, the $^4A_2$ ground state splitting is very well correlated[63] (see Figure 6); that is to say it has a single value for all the chromophores within the inhomogeneous distribution. The corresponding FLN sidelines of $[Cr(bpy)_3]^{3+}$ in a nafion film are significantly broadened compared to the

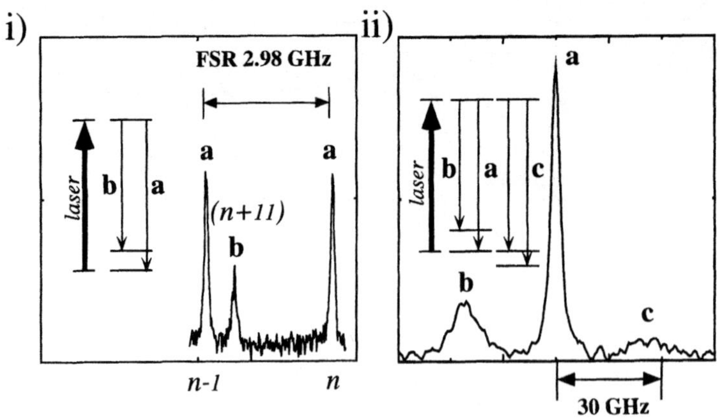

Figure 19. Resonant FLN spectra, obtained with a Fabry–Pérot interferometer, exciting into the R_1 line region of i) $Cr(acac)_3$ in $Al(acac)_3$ (Reference 63) and ii) $[Cr(bpy)_3]^{3+}$ in nafion film at 1.5 K (adapted from Reference 35). In i), transition **b** is seen in a higher order $(n + 11)$ of the interferometer. The *a*–*b* spacing is 35.1 GHz.

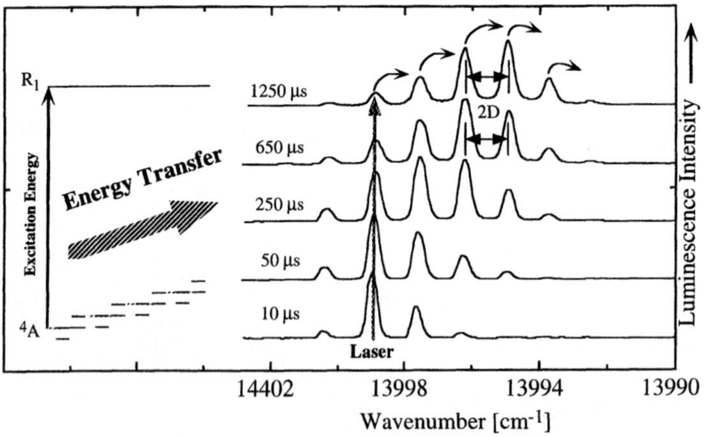

Figure 20. Time dependent FLN spectra, excited in the region of the R_1 line of $[Rh(bpy)_3][NaCr(ox)_3]ClO_4$ at 1.8 K. $2D$ is the ZFS of the $^4A_2$ ground state. Adapted from Reference 64.

resonant feature.[35] This is a consequence of noncorrelation of ZFSs. Species with a single absorption energy have a wide range of ZFSs. Measurements of ZFS values and spreads can be compared with values taken from EPR and other techniques. Such measurements provide important information regarding the local symmetry of chromophores, particularly in amorphous environments.

The FLN technique is a uniquely powerful tool with which to probe excitation energy transfer processes. An example of a beautiful, stepwise resonant excitation energy transfer within an inhomogeneously broadened range of chromophores is given

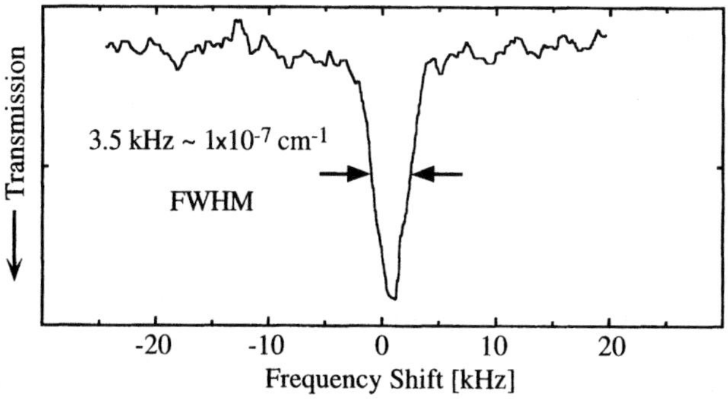

Figure 21. A narrow spectral hole in the $^7F_0 \leftrightarrow {}^5D_0$ transition of Eu^{3+} in Y_2O_3. Adapted from Reference 70.

in Figure 20. Time-resolved FLN spectra, following excitation into the R_1 line of chromium(III) in $[Rh(bpy)_3][NaCr(ox)_3]ClO_4$ show this phenomenon very clearly.[64]

Other illustrative examples of the application of FLN/ELN spectroscopy to inorganic systems include rare earth ions[65] in glasses and crystals, binuclear and trinuclear chromium(III) complexes[66–68] and the ^{3}MLCT transition of $[Ru(bpy)_3]^{2+}$ in crystals.[19,69]

5.2 Hole-Burning

Spectral hole-burning has been applied to a range of inorganic systems. We briefly present a few illustrative examples. The narrowest spectral hole reported,[70] with a hole width ~3 kHz ($=10^{-7}$ cm^{-1}), has been obtained for an f–f transition of Eu^{3+} in Y_2O_3 (Figure 21). Note that the spectral resolution here approaches 1 part in 10^{12}!

Photochemical hole-burning by photoionization[71] is well illustrated for the example in Figure 22. A spectral hole was burnt into the origin of the $^4A_2 \leftrightarrow {}^4T_1$ d–d transition of tetrahedrally coordinated Co^{2+} in $LiGa_5O_8$. The transition has an inhomogeneous width of about 20 cm^{-1}. Spectral holes can be burnt with a 660.4 nm laser in ~1 s with a fluence of 1 W/cm^2. The burning efficiency is greatly enhanced (×20) by also irradiating the sample with 673.4 nm light. This gating implies that a two step photoionization process is the burning mechanism. The first photon is used to excite the $^4T_1$ state and the second photon is then used to photoionize Co^{2+} from this excited state. Photon gated hole-burning is of interest as unwanted hole-burning can be avoided in the readout process. This is of particular advantage in any application of hole-burning techniques to the optical storage of data.[44]

A beautiful example of photochemical hole-burning based on photo-induced donor-acceptor electron transfer is shown in Figure 23. When meso-tetra(p-tolyl)-Zn-tetrabenzoporphyrin (TZT) is embedded in thin films of PMMA also containing an electron acceptor such as $CHCl_3$, photon-gated hole burning occurs in the Q band.[72] The proposed mechanism for this process is illustrated.

Time-resolved spectral hole-burning can provide important information regarding spectral diffusion. This is illustrated in Figure 24 for a spectral hole burnt into the

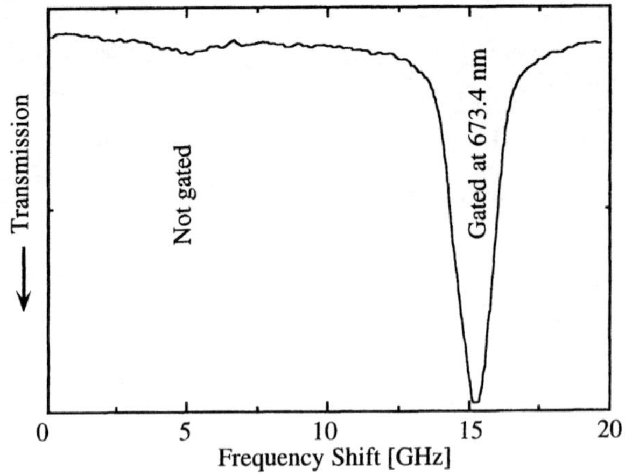

Figure 22. Photon gated hole-burning of Co^{2+} in $LiGa_5O_8$ (see text). Adapted from Reference 71.

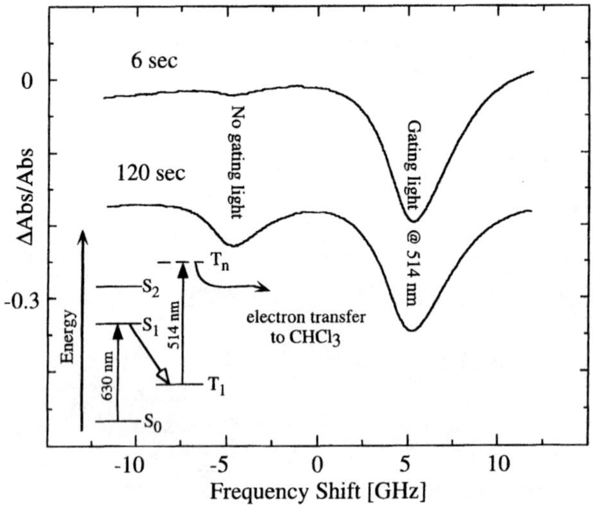

Figure 23. Photo-induced donor-acceptor electron transfer hole-burning of TZT (see text). Adapted from Reference 72.

$^4I_{15/2} \leftrightarrow {}^4F_{9/2}$ transition of Er^{3+} in $YLiF_4$. Spectral diffusion in this system is caused by spin fluctuations of fluoride ions. The hole width increases with the delay time between the burn and read processes.[73]

An example of nonphotochemical hole-burning in an inorganic system is shown in Figure 25. Deep spectral holes can be burnt into the *R* line region of $[Cr(en)_3]^{3+}$ in ethylene glycol/water. The distribution of the photoproduct indicates that the hole-burning mechanism is nonphotochemical.[74]

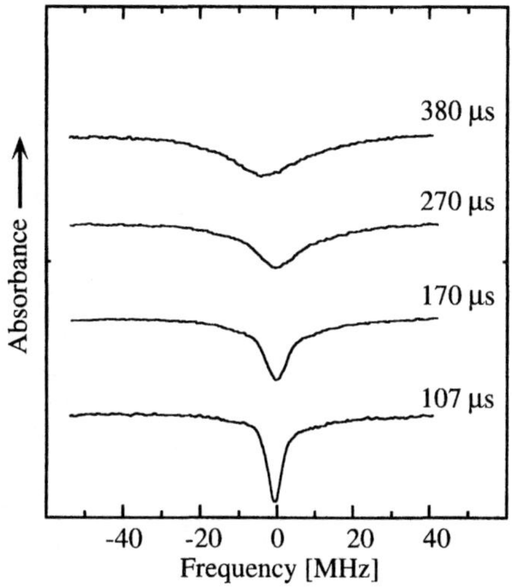

Figure 24. Spectral diffusion of Er^{3+} in $YLiF_4$. The hole-width increases with delay (see text). Adapted from Reference 73.

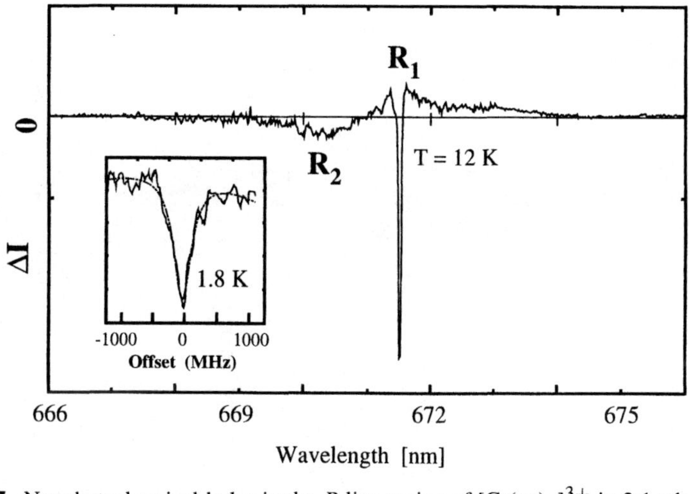

Figure 25. Nonphotochemical holes in the *R* line region of $[Cr(en)_3]^{3+}$ in 2:1 ethylene glycol/water glass. Adapted from Reference 74.

The large pseudo-Stark splitting of the electronic ^{3}MLCT origins in $[Ru(bpy)_3]^{2+}$ in $[Zn(bpy)_3](ClO_4)_2$ facilitated elegant transient hole-burning experiments.[69] These experiments utilized a fixed laser frequency with the applied electric field providing a sample sweep (see Section 4.2.3).

5.3 Photon Echo

As discussed in Section 4.3, photon-echo measurements are particularly useful for the study of the very narrowest spectral features.[55] The homogeneous linewidth of the $^7F_0 \leftrightarrow {}^5D_0$ transition of Eu^{3+} in Y_2SiO_5 was found[75] to be just ~120 Hz ($=4 \times 10^{-9}$ cm^{-1}) at 1.4 K in an applied magnetic field of 0.01 T. The applied field is needed to reduce the broadening influences associated with local fluctuations of nuclear spins.

In an exciting recent development, the interaction of neighboring ions have been studied[76] in a sample of Y_2SiO_5 codoped with Eu^{3+}, Pr^{3+} and Nd^{3+}. A photon echo associated with the $^3H_4 \leftrightarrow {}^1D_2$ transition of the Pr^{3+} ion was monitored. The echo intensity was found to be reduced (thus the term echo demolition spectroscopy) upon the excitation of the codoped ions (Eu^{3+} and Nd^{3+}) with a second laser beam. A study of this process allows both the detection of specific codoped species and the interaction process between the ions. There is an analogy in these studies with polarization transfer processes in multinuclear NMR or EPR.

5.4 Two Photon

A elegant example of two photon absorption spectroscopy is provided by work on the $[UO_2]^{2+}$ ion.[77] There has been a long lasting debate concerning the electronic structure of this chromophore, which is best described as having axial symmetry. Figure 26 shows absorption spectra of a nitrate salt of this ion at a trigonal site. There is very rich electronic and vibronic structure extending beyond the region depicted in Figure 26, both in this and other salts. A comparison of conventional and two

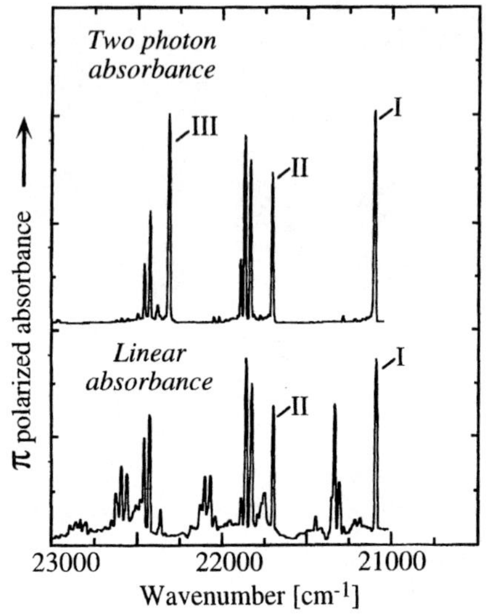

Figure 26. Normal (upper) and two photon (lower) absorption of a single $CsUO_2(NO_3)_2$ crystal at 5 K. I, II and III identify electronic origins. Adapted from Reference 77.

photon spectra clearly identifies the lowest origins I, II, and III. Very different relative intensities are observed in the two spectra. A detailed analysis confirms the proposed electronic model.

In a very recent development,[78] SMS has been extended to two photon absorption. In the usual study of two photon absorption, such as on the $[UO_2]^{2+}$ ion (above) a high power pulsed dye laser was used. In a very tightly focused beam, as is required to create the small probe volume needed in SMS, two photon absorption of single diphenyloctatetraene molecules in *n*-tetradecane was observed using a CW laser beam. This remarkable achievement provides hope that the two photon technique can be applied to more systems and that SMS can perhaps be extended to inorganic chromophores.

6 OUTLOOK

We feel that the application of laser spectroscopy in inorganic chemistry will flourish. Laser based techniques will continue to enable a far more detailed and precise understanding of the electronic structure of ground and excited states of inorganic systems as well as providing answers to important questions regarding photophysical and photochemical processes. New phenomena will doubtless arise and find important scientific and technical applications. A spectacular recent example[79] is the observation of surface enhanced Raman spectra of single dye molecules, adsorbed on gold colloid particles.

Applications will evolve from some of the more advanced techniques, outlined only briefly in this chapter, and now being developed in a number of laboratories. As a corollary to the input laser spectroscopy has to inorganic chemistry and photophysics, it seems likely that most optical, sensor, and laser materials will continue to be fabricated from robust inorganic insulators and semiconductors.

REFERENCES

1. Dehmelt, H.; Paul, W.; Ramsey, N.F.; Rev. Mod. Phys., **1990**, *62*, 525.
2. Basché, T.; Moerner, W.E.; Orrit, M.; Wild, U.P.; eds.; Single Molecule Optical Detection, Imaging and Spectroscopy; VCH Publishers: Weinheim, 1996.
3. Henderson, B.; Imbusch, G.F.; Optical Spectroscopy of Inorganic Solids; Oxford Press: Oxford, 1989.
4. Di Bartolo, B.; Optical Interactions in Solids; John Wiley: New York, 1968.
5. Louden, R.; The Quantum Theory of Light; Oxford Press: Oxford, 1973.
6. N. Bloembergen; Nonlinear Optics, 4th ed; World Scientific: 1996.
7. Cohen-Tannoudji, C.; J. Dupont-Roc, J.; Grynberg, G.; Atom Photon Interactions; John Wiley: New York, 1992.
8. Mukamel, S.; Principles of Nonlinear Optical Spectroscopy; Oxford Press: Oxford, 1995.
9. Di Bartolo, B., ed.; Non-linear Spectroscopy of Solids. Advances and Applications, NATO ASI series, Series B: Physics Vol. 339; Plenum Press: New York, 1994.

10. Demtröder, W.; Laser Spectroscopy. Basic Concepts and Instrumentation; Springer-Verlag: Berlin, 1996.
11. Schäfer, F.P., ed.; Dye Lasers, Topics in Applied Physics Vol. 1 Third Edition; Springer-Verlag: Berlin, 1990.
12. Stuke, M., ed.; Dye Lasers: 25 years. Topics in Applied Physics, Vol. 70; Springer-Verlag: Berlin, 1992.
13. Budgor, A.B.; Esterowitz, L.; Deshazer, L.G., eds.; Tunable Solid State Lasers II, Springer Series in Optical Sciences Vol. 52; Springer-Verlag: Berlin, 1986.
14. Hall, J.L.; Frequency stabilized lasers, a driving force for new spectroscopies, in Proc. International School of Physics "Enrico Fermi," Course CXX, Frontiers in Laser Spectroscopy; T.W. Hänsch and M. Inguscio, eds.; North Holland: Amsterdam, 1994; pp. 217–239.
15. Wannemacher, R.; Smorenburg, H.E.; Schmidt, Th.; Völker, S.; J. Lumin., **1992**, *53*, 266.
16. Schenzle, A.; Brewer, R.G.; Phys. Rev. A, **1976**, 14, 1756.
17. Shimoda, K. in High-Resolution Laser Spectroscopy, Topics in Applied Physics, Vol. 13; K. Shimoda, ed.; Springer-Verlag: Berlin, 1976; p. 11.
18. Imbusch, G.F. in Laser Spectroscopy of Solids, Topics in Applied Physics, Vol. 49; W.M. Yen and P.M. Selzer, eds.; Springer-Verlag: Berlin, 1981; p. 1.
19. a) Riesen, H.; Wallace, L.; Krausz, E.; Int. Rev. Phys. Chem., **1997**, 16, 291; b) Riesen, H.; Krausz, E.; Comments Inorg. Chem., **1995**, 18, 27.
20. Riesen, H.; Krausz, E.; J. Chem. Phys., **1993**, *99*, 7614.
21. Riesen, H.; Rae, A.D.; Krausz, E.; J. Lumin., **1994**, *62*, 123.
22. Krausz, E.; Higgins, J.; Riesen, H.; Inorg. Chem., **1993**, *32*, 4053.
23. Sugano, S.; Tanabe, Y.; Kamimura, H.; Multiplets of Transition-Metal Ions; Academic Press: New York, 1970.
24. (a) Kupka, H.; Mol. Phys., **1980**, *39*, 849.
(b) Jortner, J.; Mol. Phys. **1976**, *32*, 379.
25. Fong, F.K; Theory of Molecular Relaxation. Applications in Chemistry and Biology; John Wiley: New York, 1975.
26. Ferguson, J.; Masui, H.; J. Phys. Soc. Japan, **1977**, *42*, 1640.
27. McCumber, D.E.; Sturge, M.D.; J. Appl. Phys., **1963**, *34*, 1682.
28. Imbusch, G.F.; Yen, W.M.; Schawlow, A.L.; McCumber, D.E.; Sturge, M.D.; Phys. Rev., **1964**, *133*, A1029.
29. Völker, S.; Ann. Rev. Phys. Chem., **1989**, *40*, 499.
30. Anderson, P.W.; Halperin, B.I.; Varma, C.M.; Phil. Mag., **1972**, *25*, 1.
31. Phillips, W.A.; J. Low Temp. Phys., **1972**, *7*, 351.
32. Berg, M.; Walsh, C.A.; Narasimhan, L.R.; Littau, K.A.; Fayer, M.D.; J. Chem. Phys., **1988**, *88*, 1564.
33. Griesser, H.; Wild, U.P.; J. Chem. Phys., **1980**, *73*, 4715.
34. Lee, H.W.H.; Walsh, C.A.; Fayer, M.D.; J. Chem. Phys. **1985**, *82*, 3948.
35. Riesen, H.; Krausz, E.; J. Chem. Phys., **1992**, *97*, 7902.
36. Denisov, Yu.V.; Kizel, V.A.; Opt. Spectr., **1967**, *23*, 251.
37. Szabo, A.; Phys. Rev. Lett., **1970**, *25*, 924.
38. Spaeth, M.L.; Soy, W.R.; J. Chem. Phys., **1968**, *48*, 2315.

39. Szabo, A.; Phys. Rev. B., **1975**, *11*, 4512.
40. Felder, P.; Günthard, H.H.; Chem. Phys., **1984**, *85*, 1.
41. De Vries, H.; Wiersma, D.A.; Phys. Rev. Lett., **1976**, *36*, 91.
42. Völker, S.; van der Waals, J.H.; Mol. Phys., **1976**, *32*, 1703.
43. Friedrich, J.; Haarer, D.; Angew. Chem. Int. Ed. Engl., **1984**, *23*, 113.
44. Moerner, W.E., ed.; Persistent Spectral Hole-Burning: Science and Applications, Topics in Current Physics, Vol. 44; Springer-Verlag: Berlin, 1988.
45. Hayes, J.M.; Small, G.J.; Chem. Phys., **1978**, *27*, 151.
46. Hayes, J.M.; Stout, R.P.; Small, G.J.; J. Chem. Phys., **1980**, *73*, 4129.
47. a) Patterson, F.G.; Lee, H.W.H.; Olson, R.W.; Fayer, M.D.; Chem. Phys., **1981**, *84*, 59. b) Holliday, K; Manson, N.B.; J. Phys. Condens Matter, **1989**, *1*, 1339.
48. Gorokhovskii, A.A.; Kaarli, R.K.; Rebane, L.A.; JETP Lett., **1974**, *20*, 216.
49. Kharlamov, B.M.; Personov, R.I.; Bykovskaya, L.A.; Opt. Commun., **1974**, *12*, 191.
50. Meixner, A.J.; Renn, A.; Bucher, S.E.; Wild, U.P.; J. Phys. Chem., **1986**, *90*, 6777.
51. Shoemaker, R.L.; Ann. Rev. Phys. Chem., **1979**, *30*, 239.
52. Muramato, T.; Nakanishi, S.; Hashi, T.; Opt. Commun., **1977**, *21*, 139.
53. Lippert, E.; Macomber, J.D., eds.; Dynamics During Spectroscopic Transitions; Springer-Verlag: Berlin, 1995.
54. Hegarty, J.; J. Lumin., **1987**, *36*, 273.
55. Macfarlane, R.M.; Shelby, R.M.; in Spectroscopy of Solids Containing Rare Earth Ions, Kaplyanskii, A.A.; Macfarlane, R.M., eds.; Elsevier Science Publishers B.V.: Lausanne, 1987.
56. Kurnitt, N.A.; Abella, I.D.; Hartmann, S.R.; Phys. Rev. Lett., **1964**, *13*, 567.
57. Göpert-Mayer, M.; Ann. Physik, **1931**, *9*, 273.
58. Kaiser, W.; Garrett, C.G.B.; Phys. Rev. Lett., **1991**, *7*, 229.
59. Moerner, W.E.; Kador, L.; Phys. Rev. Lett., **1989**, *62*, 2535.
60. Rebane, K.K.; Rebane, I.; J. Lumin., **1993**, *56*, 39.
61. Riesen, H.; Krausz, E.; Wallace, L.; J. Phys. Chem., **1992**, *96*, 3621.
62. Riesen, H.; Krausz, E.; Chem. Phys. Lett., **1990**, *172*, 5.
63. Riesen, H.; unpublished work.
64. von Arx, M.E.; Hauser, A.; Riesen, H.; Pellaux, R.; Decurtins, S.; Phys. Rev. B, **1996**, *54*, 15800.
65. Yen, W.M.; Selzer, P.M., eds.; Laser Spectroscopy of Solids, Topics in Applied Physics, Vol. 49; Springer-Verlag: Berlin, 1981.
66. Riesen, H.; Güdel. H.U.; Chem. Phys. Lett., **1987**, *133*, 429.
67. Riesen, H.; Güdel. H.U.; Mol. Phys., **1986**, *58*, 509.
68. Dubicki, L.; Ferguson, J.; Williamson, B.; Inorg. Chem., **1983**, *22*, 3220.
69. Riesen, H.; Krausz, E.; Chem. Phys. Lett., **1993**, *212*, 347.
70. Manson, N.B.; Sellars, M.J.; Fisk, P.T.H.; Meltzer, R.S.; J. Lumin., **1995**, *64*, 19.
71. Macfarlane, R.M.; Vial, J.-C.; Phys. Rev. B, **1995**, *34*, 1.
72. Carter, T.P.; Bräuchle, C.; Lee, V.Y.; Manavi, M.; Moerner. W.E.; Opt. Lett., **1987**, *12*, 370.

73. Wang, Y.P.; Landau, D.P.; Meltzer, R.S.; Macfarlane, R.M.; J. Opt. Soc. Am. B, **1992**, *9*, 46.
74. a) Riesen, H.; Manson, N.B.; Chem. Phys. Lett., **1989**, *161*, 131; b) Riesen, H.; Manson, N.B.; Krausz, E.; J. Lumin., **1990**, *46*, 345.
75. Equall, R.W.; Sun, Y.; Cone, R.L.; Macfarlane, R.M.; **1994**, 72, 2179.
76. Altner, S.; Wild, U.P.; Mitsunaga, M.; Chem. Phys. Lett., **1995**, *237*, 406.
77. Barker, T.J.; Denning, R.G.; Thorne, J.R.G.; Inorg. Chem., **1992**, *31*, 1345.
78. Plakhotnik, T.; Walser, D.; Renn, A.; Wild, U.P.; Phys. Rev. Lett., **1996**, *77*, 5365.
79. Kneipp, K.; Wang, Y.; Kneipp, H.; Perelman, L.T.; Itzkan, I.; Ramachandra, R.; Dasari, R.R.; Feld, M.S.; Phys. Rev. Lett., **1997**, *78*, 1667.

7 IR, Raman, and Resonance Raman Spectroscopy

ROMAN S. CZERNUSZEWICZ
Department of Chemistry
University of Houston
Houston, TX 77204, USA
E-mail: roman@uh.edu

THOMAS G. SPIRO
Department of Chemistry
Princeton University
Princeton, NJ 08544, USA
E-mail: spiro@chemvax.princeton.edu

Inorganic Electronic Structure and Spectroscopy, Volume I: Methodology.
Edited by E. I. Solomon and A. B. P. Lever.
ISBN 0-471-15406-7.

1 INTRODUCTION

Infrared and Raman spectroscopy are important probes of molecular structure and dynamics. Both spectroscopies provide a record of the vibrations of molecules. Because the vibrational frequencies depend on bond strengths and bond angles, as well as on the masses of the atoms, the vibrational spectra can provide sensitive structural information. Chemical groups have characteristic frequencies, which can be used for identification and quantitation, or can serve as monitors of environmental effects on structure. The analysis of vibrational spectra is a highly developed discipline[1–11] and provides an interpretive framework for new applications.[12–15]

These applications have historically been hampered by technical limitations, especially low sensitivity and spectral crowding in complex molecules. However, these limitations have been greatly eased by important instrumental advances. The power of IR spectroscopy has been markedly enhanced by the development of Fourier transform interferometry, which speeds spectral acquisition, and which permits highly accurate difference spectroscopy, so that sensitivities at the level of single residues in proteins are attainable.[16] Raman spectroscopy has been revolutionized by the advent of the laser, which produces scattered photons from a focused point, and allows for flexible sampling geometries and acquisition of spectra from colored samples. The latter attribute permits exploitation of the resonance Raman effect, which amplifies scattering from chromophoric parts of the molecule. This effect provides a sort of zoom lens into complex molecules, focusing only on the chromophore vibrations. Both spectroscopies can now be adapted to microscope optics, providing spectra of microscopic samples, or spectrally-filtered microscope images.

Technical advances continue to proliferate, and applications will certainly multiply. The purpose of this chapter is to lay out the principles behind IR and Raman spectroscopy, and to illustrate them with examples from recent research, especially in the context of inorganic chemistry.

The laser also provides access to nonlinear optical effects, which can be applied in a variety of configurations to the detection of vibrational transitions. These methods, while fascinating from theoretical perspectives, have yet to gain wide application because they are technically demanding, and because interpretation of the data is often not straightforward. They will not be covered in this chapter, but the interested reader is referred to the excellent book by Long.[17]

2 INFRARED AND RAMAN SPECTROSCOPY

2.1 Complementary Nature of Infrared Absorption and Raman Scattering

Molecular vibrational frequencies (10^{-13}–10^{-14} Hz) fall in the infrared region of the electromagnetic spectrum. These are commonly measured using infrared or Raman spectroscopy. *Infrared spectroscopy* is a direct method for monitoring transitions between quantized vibrational levels induced by absorption of light in the infrared region. *Raman spectroscopy* is a light scattering method in which the molecular vibrations are excited by inelastic collisions of the sample molecules with the quanta of light in the ultraviolet, visible, or near-infrared regions. Figure 1*a* is a diagram of the IR and Raman processes for a diatomic molecule with one vibrational coordinate like HF. Infrared absorption and Raman scattering both promote the molecule to the same vibrationally excited state (υ'), although the incident radiation is of very different frequency in the two excitation methods.

In infrared spectroscopy the vibrational frequency is therefore observed as a peak in the absorption spectrum at the absolute frequency of the absorbed infrared radiation (Figure 1*b*). The spectrum is generally recorded by passing a beam of polychromatic infrared light through a sample and monitoring the radiant power (or intensity) of the transmitted light at each frequency of the source. One measures alternately the radiant power transmitted through a sample cell (Φ_s) and that transmitted by a reference cell (Φ_r) (Figure 1*c*) and the ratio, called the transmittance (T), is used to calculate the absorbance ($A = -\log T$). The absorbance is directly proportional to the concentration of the absorbing molecules through the Lambert–Beer law,

$$A(\nu) = -\log T(\nu) = -\log\left[\Phi_s(\nu)/\Phi_r(\nu)\right] = \varepsilon(\nu)dc \qquad (1)$$

where c is the concentration of the sample (in mol L^{-1}), d is the cell thickness (in cm), and $\varepsilon(\nu)$ is the molar absorptivity (in L mol^{-1} cm^{-1}).

In Raman spectroscopy the exciting photon has much higher energy than the molecular vibration (Figure 1*a*). As a result of the inelastic collision, part of the incident energy, equal to the vibrational quantum, $\upsilon \rightarrow \upsilon'$, is retained by the vibrating molecule, while the scattered photon emerges with lower frequency, $\nu_0 - \nu_{\upsilon,\upsilon'}$, and energy $h(\nu_0 - \nu_{\upsilon,\upsilon'})$. The molecular vibration is therefore encoded in the Raman spectrum as the frequency separation between the incoming photon (or the Rayleigh peak) and the scattered photon (Figure 1*d*). Hence, a Raman experiment requires illuminating the sample with *monochromatic* radiation and analyzing the polychromatic scattered radiation. In practice, monochromatic radiation of any convenient frequency from the laser is shone on to the sample and the absolute radiant power of the scattering can be measured at any convenient angle with respect to the excitation beam; 90° collection is illustrated in Figure 1*e*. The Raman intensity is proportional to the sample concentration:

$$I_{sc} \propto \nu_{sc}^4 E_0 c \qquad (2)$$

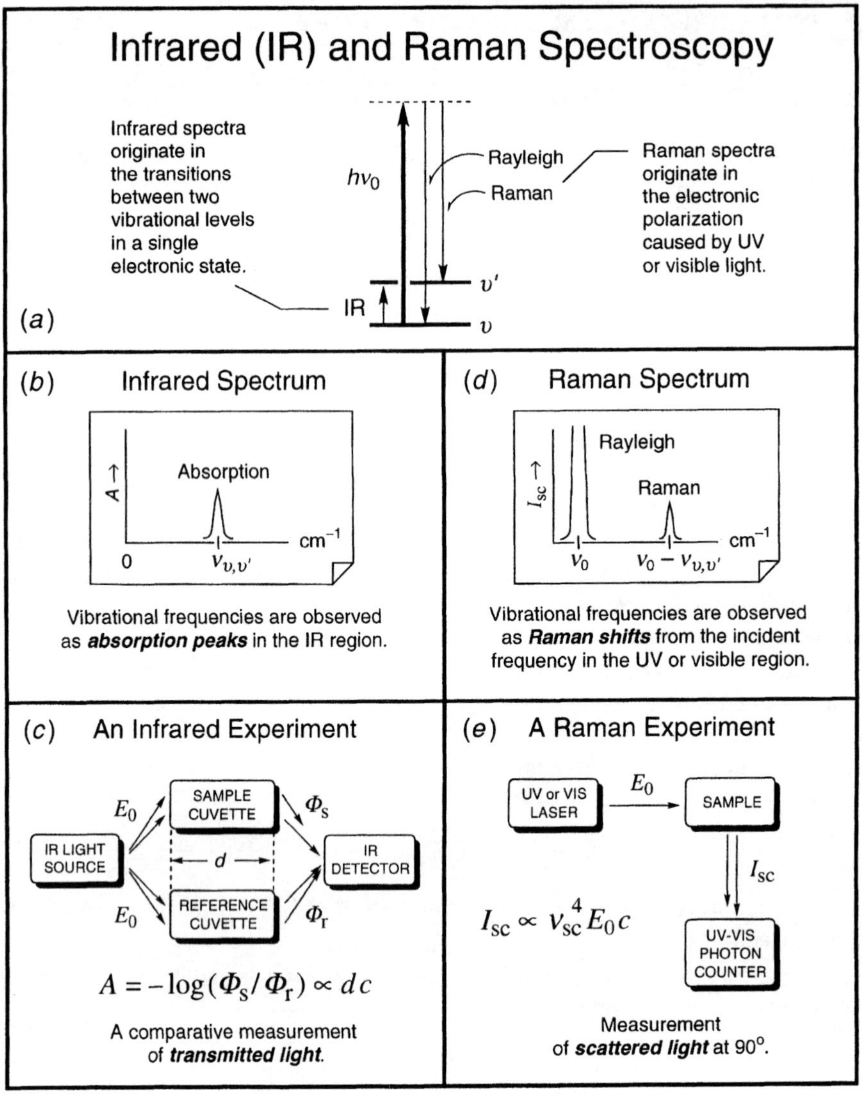

Figure 1. Comparison of IR absorption and Raman scattering: Absorption of an IR photon raises the molecule from the ground (υ) to an excited (υ') vibrational state (a), and produces a peak at $\nu_{\upsilon,\upsilon'}$ in the IR spectrum (b). In IR experiment attenuation of polychromatic IR radiation in the measuring and reference beams is alternatingly monitored as a function of radiation wavenumber (cm^{-1}) (c). The readout signal is proportional to the transmittance ($T = \Phi_s/\Phi_r$) or absorbance ($A = -\log T$). Scattering of UV or visible photons produces an intense Rayleigh peak and weaker Raman peaks, displaced from the Rayleigh by $\nu_{\upsilon,\upsilon'}$ (d). These arise because of energy transfer from the incident photon $h\nu_0$ to the molecule, which is raised to an excited vibrational level (a). The excitation for the Raman experiment (e) must be monochromatic. The readout signal is proportional to the radiant power of the scattered light, I_{sc}.

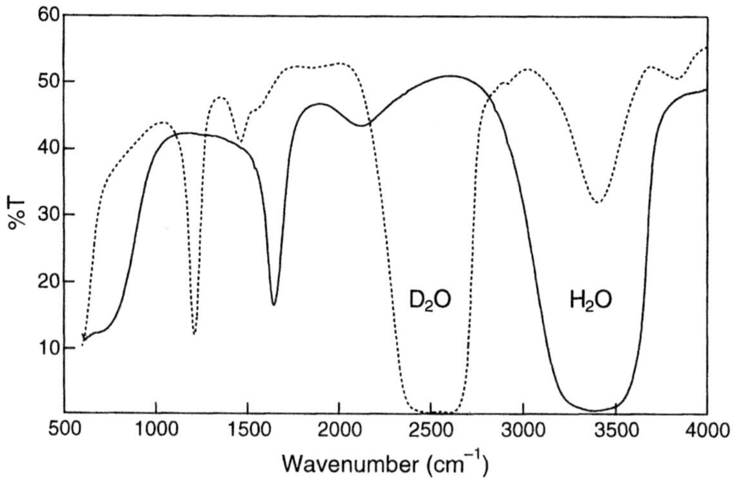

Figure 2. Infrared spectra of liquid H_2O (solid line) and D_2O (dashed line).

where ν_{sc} is the frequency of the scattered radiation, E_0 is the laser irradiance, and c is the volume concentration of scattering molecules.

All molecules except homonuclear diatomic molecules (e.g., H_2, O_2, N_2, and the halogens) absorb infrared light. All molecules including the homonuclear diatomics are Raman scatterers. The infrared spectrum of a polyatomic molecule can be quite complex due to the large number of possible *fundamental* vibrational modes (up to $3N - 6$ for an N-atomic nonlinear molecule) and the existence of nonfundamental transitions such as *overtones*, sum and difference *combination bands*, and *Fermi resonance* bands. Raman signals are ordinarily devoid of nonfundamental transitions; however, overtone and combination bands involving fundamental vibrations of a chromophoric group within a molecule can be strongly enhanced in the *resonance Raman* spectrum (see Section 3). Whereas an associated liquid like H_2O is an excellent infrared light absorber except in certain window regions,[18] it is a poor Raman scatterer, giving Raman spectroscopy an advantage for aqueous solution studies. The infrared spectra of liquid H_2O and D_2O are shown in Figure 2. The Raman spectra of liquid and frozen water (H_2O, D_2O, $H_2{}^{18}O$, $D_2{}^{18}O$) are shown in Figure 3.

Because infrared and Raman peaks are excited by different physical mechanisms (direct absorption of light versus inelastic scattering of light) their relative intensities can differ greatly; a weak infrared band may have an intense Raman counterpart, and vice versa. For instance, the vibrations of polar bonds such as O–H, N–H, S–H, and C=O are more readily observed in the infrared spectrum, while vibrations of less polar bonds are better seen in the Raman spectrum (C≡C, C=C, P=S, S–S, and C–S). Moreover, not all transitions between molecular vibrational levels are allowed. Some transitions can appear only in the infrared spectrum, some only in the Raman spectrum, and some in both infrared and Raman spectra at coincidental frequencies; still others cannot be observed in either of the spectra. To see whether the transition is IR- or Raman-*allowed* or *forbidden*, the relevant *selection rules* must be exam-

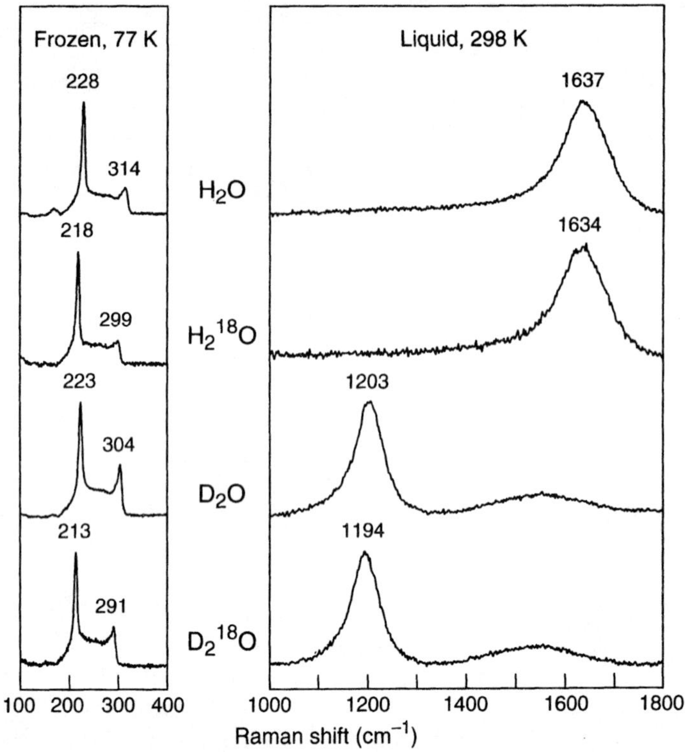

Figure 3. Raman spectra (488.0 nm excitation) of H_2O, $H_2{}^{18}O$, D_2O, and $D_2{}^{18}O$: (*left*) as ice at 77 K with a liquid N_2 cell,[22] 135° backscattering directly of the ice surface, laser power 250 mW, spectral slit width 6 cm^{-1}; and (*right*) as neat liquid at room temperature in a capillary tube, 90° scattering, laser power 250 mW, spectral slit width 10 cm^{-1}. The scattered light was dispersed by a SPEX 1403 double monochromator equipped with 1800 holographic gratings and detected by a cooled Hammamatsu 928 photomultiplier tube under the control of a SPEX DM3000 data station as described elsewhere.[56]

ined. These are determined by the *symmetry* of the molecule. For instance, totally symmetric bond stretching vibrations preserve the symmetry of the molecule and are most intense in the Raman spectrum, whereas antisymmetric stretching and deformation modes distort the molecule and are intense in the infrared spectrum. For highly symmetrical molecules and ions such as $TiCl_4$ (T_d), C_6H_6 (D_{6h}), and $PtCl_4^{2-}$ (D_{4h}) both infrared and Raman spectroscopies are required to obtain the full set of vibrational frequencies. If a molecule or an ion possesses a center of inversion, there is a *rule of mutual exclusion*; no fundamental vibration which is active in the infrared absorption can be active in the Raman scattering, and no fundamental vibration which is active in the Raman scattering can be active in the infrared absorption. Thus infrared and Raman spectroscopies are complementary and overlapping techniques for investigating the vibrations of molecules. The basic aspects of infrared and Raman spectroscopy are summarized in Table 1.

Table 1 Basic Characteristics of Infrared (IR) and Raman Spectroscopy[a]

Parameter	Infrared Spectroscopy	Raman Spectroscopy
Spectroscopic phenomenon	Absorption of light: $h\nu_{IR} = \Delta E_{vibr}$	Inelastic scattering of light: $h\nu_0 - h\nu_{sc} = \Delta E_{vibr}$
Allowed transition	$\Delta\upsilon = +1, +2, +3, \ldots$	$\Delta\upsilon = \pm1, \pm2, \pm3, \ldots$ (transitions for $\Delta\upsilon = +2, +3, \ldots$ i.e., overtones are considerably less conspicuous than in IR)
Excitation	Polychromatic IR radiation	Monochromatic radiation (ν_0) in the UV, visible, or near IR
Molecular origin	Dipole moment: $\mu = q\boldsymbol{r}$	Induced dipole moment: $\boldsymbol{P} = \alpha \boldsymbol{E}$
Requirement for vibrational activity	Change in dipole moment during vibration: $(\partial\mu/\partial Q_k)_0 \neq 0$	Change in polarizability during vibration: $(\partial\alpha/\partial Q_k)_0 \neq 0$
Band intensity	$I_{IR}^{1/2} \propto (\partial\mu/\partial Q_k)_0$	$I_{R}^{1/2} \propto (\partial\alpha/\partial Q_k)_0$
Frequency measurement	Absolute: $\nu_{vibr} = \nu_{IR}$	Relative to the excitation frequency: $\nu_{vibr} = \nu_0 - \nu_{sc}$
Readout signal	Comparative: transmittance ($T = \Phi_s/\Phi_r$) or absorbance ($A = -\log T$)	Absolute: radiant power or intensity of scattered radiation
Spectral plot	Linear in %T or logarithmic in A vs. wavenumber (cm^{-1})	Linear: Raman intensity vs. wavenumber shift (cm^{-1})
Dominant spectral feature	Vibrations destroying molecular symmetry: antisymmetric stretching and deformation modes	Vibrations preserving molecular symmetry: symmetric stretching modes
Inactive molecule	Homonuclear diatomics	None
Centrosymmetric molecule	Only "*u*"-symmetry modes active	Only "*g*"-symmetry modes active
Medium	Water is a strong absorber and is a poor solvent for IR studies	Water is a weak scatterer and is a good solvent for Raman studies

[a]h, Planck's constant; ΔE_{vibr}, energy difference of vibrational levels; ν, photon frequency; $\Delta\upsilon$, change in vibrational quantum number; q, charge; $\boldsymbol{r}$, charge spacing; α, molecular polarizability; $\boldsymbol{E}$, electric field; Φ_s and Φ_r, radiant powers transmitted by the sample and reference cells, respectively; Q_k, vibrational normal coordinate ($k \leqslant 3N - 6$); "*g*" and "*u*", normal modes of vibration symmetric (*gerade*) and antisymmetric (*ungerade*) with respect to the molecular center of inversion.

2.2 Requirements for Infrared Absorption

For a molecular vibration to absorb infrared radiation, it must (1) have a frequency equal to that of the electromagnetic radiation, and (2) change the magnitude and/or direction of the molecular dipole moment. The first criterion, that of matching frequencies, applies to any induced (or stimulated) absorption and emission phenomena. A particular molecule can exist in a variety of electronic, vibrational, rotational, etc., energy states and can move from one state to another only by a sudden jump involving a finite amount of energy (*quantization of the energy*). Consequently, if a molecule is in the path of infrared light, transfer of radiant energy to the molecule will occur only when:

$$\Delta E_{\text{vibr}} = E_{\upsilon'} - E_{\upsilon} = h\nu = hc(1/\lambda) = hc\tilde{\nu} \tag{3}$$

Here ΔE_{vibr} is the difference in energy between quantized vibrational states υ (lower) and υ' (upper) (see Figure 1*a*), h is the Planck's constant (6.6262×10^{-27} erg s), c is the speed of light ($2.99792458 \times 10^{10}$ cm s^{-1}), and ν, λ, and $\tilde{\nu}$ are the frequency (in Hz), wavelength (in cm), and wavenumber (in cm^{-1}), respectively, of the absorbed infrared radiation.

The second requirement, that of a dipole moment change, reflects the fact that a vibration must interact with the electromagnetic field in order to absorb the radiation into the molecule. This interaction is provided by an oscillating electric dipole moment μ (Figure 4). The dipole moment (μ) is a vector that is oriented from the *center of gravity* of the positive charges to that of the negative charges, and is defined as the product of the size q and the spacing $\boldsymbol{r}$ between these charges:

$$\mu = q \times \boldsymbol{r} \tag{4}$$

In a heteronuclear diatomic molecule μ is nonzero and varies periodically in phase with the stretching vibration, reaching a maximum at the greatest internuclear separation and a minimum at the closest approach (Figure 4*b*). Thus there is a net oscillatory displacement of charges or a *net change in dipole moment* when HF vibrates along the stretching coordinate. The energy for the vibration is provided by the electric field of the infrared radiation during the absorption process. As a result, the H–F stretching frequency will appear in the infrared spectrum.

However, the stretching frequency of a homonuclear molecule such as F_2 will not appear in the infrared because symmetry demands that $\mu = 0$ for all internuclear separations. Since there is no displacement of charge, no net work that would require a transfer of energy from the incident infrared radiation will be done and F_2 has no vibrational absorption activity. Indeed, all homonuclear diatomic molecules are transparent in the infrared region.

On the other hand, the presence of a permanent dipole moment is not required for absorption of infrared radiation by polyatomic molecules. A polyatomic molecule has more than one degree of vibrational freedom and some or all of its vibrations will change the magnitude or direction of the dipole moment.

The intensity of a particular absorption band is proportional to $|[\mu]_k|^2$, the square of the vibrational transition moment $[\mu]_k$ (*transition probability*) where

$$[\mu]_k = \langle \upsilon'_k | \mu | \upsilon_k \rangle \tag{5}$$

are the matrix elements of the electric dipole moment operator μ. For vibrational motion the electric dipole moment μ can be expressed as a rapidly converging power

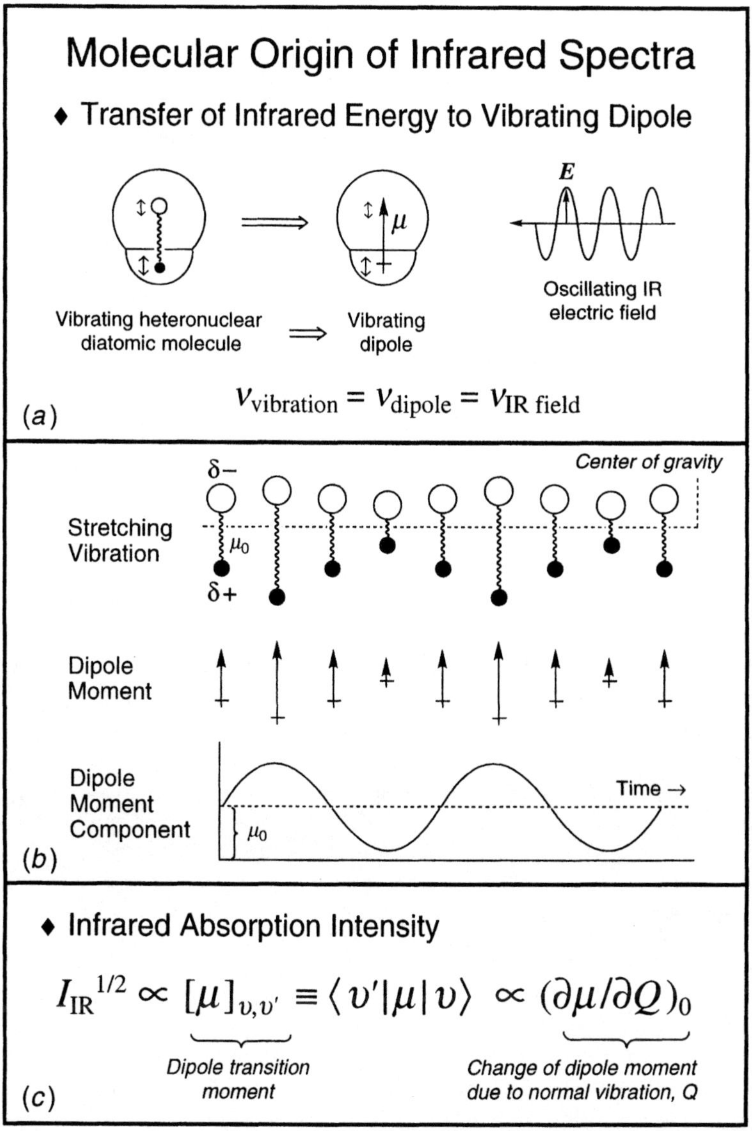

Figure 4. Schematic illustration of molecular origin and requirements for infrared absorption: When a heteronuclear molecule such as HF vibrates at a particular frequency, the molecular dipole moment μ also oscillates about its equilibrium value (*a*) as the two charged ($\delta\pm$) atoms move back and forth (*b*). The oscillating dipole can absorb energy from an oscillating electric field $\boldsymbol{E}$ only if (1) the field also oscillates at the same frequency (*a*) and (2) the motion of the atoms produces a periodic alteration in the dipole spacing (*b*). The intensity of absorption depends on the square of the change in the dipole moment $(\partial\mu/\partial Q)_0^2$ caused by the vibration, Q (*c*).

series in the atomic displacements

$$\mu = \mu_0 + \left(\frac{\partial \mu}{\partial Q_k}\right)_0 Q_k + \frac{1}{2!}\left(\frac{\partial^2 \mu}{\partial Q_k^2}\right)_0 Q_k^2 + \cdots \tag{6}$$

where μ_0 is the permanent dipole moment, Q_k is the normal coordinate of the vibration k (which is R (internuclear distance) $-R_0$ (equilibrium bond distance) in the diatomic case), and the derivatives are evaluated at $Q_k = 0$, the equilibrium position. If we combine (5) and (6) and look for off-diagonal matrix elements of μ, we find that the only nonvanishing ones are

$$\left(\frac{\partial \mu}{\partial Q_k}\right)_0 \langle \upsilon_k' | Q_k | \upsilon_k \rangle \sim \left(\frac{\partial \mu}{\partial Q_k}\right)_0 \delta(\upsilon_k', \upsilon_k \pm 1) \tag{7}$$

which gives the fundamental absorption with $\Delta\upsilon_k = +1$ if $(\partial\mu/\partial Q_k)_0 \neq 0$,

$$\left(\frac{\partial^2 \mu}{\partial Q_k^2}\right)_0 \langle \upsilon_k' | Q_k^2 | \upsilon_k \rangle \sim \left(\frac{\partial^2 \mu}{\partial Q_k^2}\right)_0 \delta(\upsilon_k', \upsilon_k \pm 2) \tag{8}$$

which gives much weaker overtone absorption with $\Delta\upsilon_k = +2$ if $(\partial^2\mu/\partial Q_k^2)_0 \neq 0$, and so on. The diagonal matrix elements of μ vanish since μ_0 is a constant and $\langle \upsilon_k' | \upsilon_k \rangle = \delta(\upsilon_k', \upsilon_k) = 0$ because of orthogonality. Thus the transition probability of fundamentals in infrared absorption is proportional to the square of the dipole moment gradient along the normal coordinate of the vibration. Accordingly, if there is no change in dipole moment when the vibration occurs, i.e., $(\partial\mu/\partial Q_k)_0 = 0$, the intensity of a vibrational band in the infrared spectrum equals zero and the vibration is said to be infrared inactive.

2.3 The Raman Effect

The Raman effect—first predicted from theoretical considerations by A. Smekal (1923) and H.A. Kramers and W. Heisenberg (1924)—is a *light scattering* phenomenon with a change of frequency. If the incident light frequency, ν_0, is in a transparent region of the molecular spectrum, much of the light passes through the sample, but a small fraction ($\sim 10^{-3}$) is scattered in all directions. Most of the scattered light emerges at the same frequency as the incident light; this is *elastic* or *Rayleigh scattering*; the photons neither lose nor gain energy in their collisions with the molecules. Rayleigh scattering carries no information about molecular vibrations. A few collisions, however, are *inelastic*. A net transfer of quantized energy ($\Delta E = h\nu_k$) occurs and some scattered light ($\sim 10^{-6}$) obtains new modified frequencies ($\nu_0 \pm \nu_k$). The frequency changes are equal to the frequencies associated with transitions between vibrational (υ_k) or, less frequently, rotational or electronic levels of the system. These two forms of light scattering with change of frequency were first observed by C.V. Raman and K.S. Krishnan in liquids[19,20] and independently by G. Landsberg and L. Mandelstam in quartz[21] in 1928, and are now called *Raman scattering*. Radiation scattered with a frequency lower than that of the incident light (i.e., $\nu_0 - \nu_k$) is referred to as *Stokes scattering*, while that at the higher frequency (i.e., $\nu_0 + \nu_k$) is called *anti-Stokes scattering*. The magnitude of the frequency shift from the incident radiation provides a measure of the vibrational energy level spacing. In practice, the lower spectral region is used for this purpose because the Stokes Raman scattering is *always* more intense than the Raman anti-Stokes at normal temperatures.

The transfer of energy results from the perturbation of the electronic wavefunction of the molecule by the rapidly changing electric field of the photon. The wavefunction of the perturbed system can be expressed as a linear combination of all possible wavefunctions of the unperturbed molecule, with time-dependent coefficients. For a very short time interval ($<10^{-14}$ s) the photon loses identity and becomes indistinguishable from the kinetic and potential energy of the perturbed electrons. Formally, the molecule is regarded as having attained a higher non-stationary energy level (*virtual state*); it returns to a stationary state by re-emitting a photon (Figure 5). If the molecule returns to its original stationary state, the re-emitted photon has the same frequency as the incident photon (Rayleigh scattering, $\nu_{sc} = \nu_0$). But if the final state differs from the initial state by a vibrational quantum ($h\nu_k$), the photon is re-emitted with shifted frequencies. At temperatures normally used a decrease in frequency (Stokes emission) is more probable because there are few thermally excited molecules. Although we can speak of a sequence of events in which the emission of a photon from the virtual state follows an absorption process, it is important to appreciate that these processes are actually simultaneous and cannot be separated in time. This means that the Rayleigh and Raman scattering are two-photon processes in which one photon is annihilated and one created at the same time (*induced secondary emission processes*).

The scattered radiation has a spectrum characteristic of the molecules in the sample, with an intense band at the incident frequency, ν_0, resulting from Rayleigh scattering, and fainter Raman bands on both sides of ν_0 at distances corresponding to the molecular vibrational frequencies, ν_k. Typical examples of such bands are seen in Figure 5, which displays the Raman spectra of pure carbon tetrachloride, CCl_4, at 298 (liquid) and 77 K (frozen liquid) upon excitation with an Ar^+ ion laser line at 488.0 nm (20,492 cm^{-1}).[22,23] A number of general observations can be made about the Raman effect and its display from these two spectra:

1. The measured spectral frequencies are expressed in wavenumber shifts $\Delta\widetilde{\nu}$, or *Raman shifts*, which are defined as the difference in wavenumbers (cm^{-1}) between the incident and scattered rays, $\Delta\widetilde{\nu} = (\widetilde{\nu}_0 - \widetilde{\nu}_{sc})$. Hence, $\Delta\widetilde{\nu} = \widetilde{\nu}_0 - \widetilde{\nu}_0 = 0\ cm^{-1}$ for the Rayleigh peak and $\Delta\widetilde{\nu} = \widetilde{\nu}_0 - (\widetilde{\nu}_0 \pm \widetilde{\nu}_k) = \pm\widetilde{\nu}_k\ cm^{-1}$ for the Raman peaks. Since Stokes transitions occur at frequencies $\nu_0 - \nu_k$, they exhibit *positive* Raman shifts (an *increase* of energy by the scattering system); anti-Stokes transitions occur at $\nu_0 + \nu_k$, and exhibit *negative* Raman shifts (a *decrease* of energy by the scattering system). That is, Stokes bands of CCl_4 are found at 218, 314, and 459 cm^{-1} while their anti-Stokes counterparts occur at −218, −314, and −459 cm^{-1}, *regardless of the wavelength of excitation.*

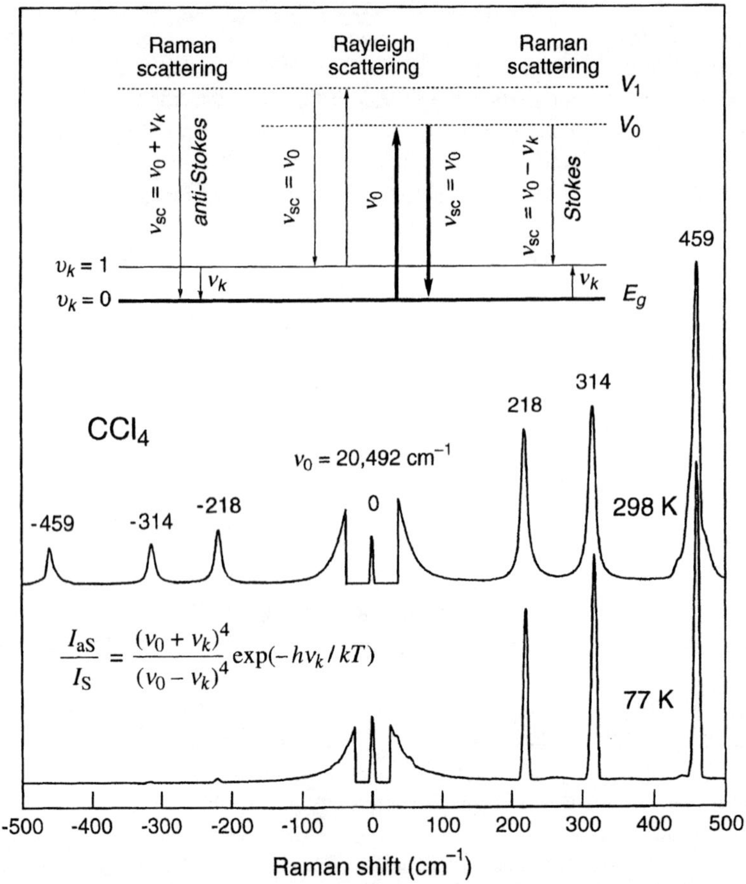

Figure 5. Energy transfer diagram illustrating Rayleigh and Raman scattering (*top*), and Raman spectra for CCl_4 excited at room (298 K) and liquid-N_2 (77 K) temperatures by Ar^+ ion laser radiation of $\lambda_0 = 488.0$ nm or $\tilde{\nu}_0 = 20{,}492\ cm^{-1}$ (*bottom*). The number above the peaks is the Raman shift, $\Delta\nu = (\nu_0 - \nu_{sc}) = \nu_k\ cm^{-1}$. Since the fraction of molecules occupying excited vibrational states depends on the Boltzmann factor ($kT = 207\ cm^{-1}$ at 298 K), the intensity of anti-Stokes lines falls off rapidly with decreasing temperature ($kT = 54\ cm^{-1}$ at 77 K) and increasing vibrational frequency ν_k.

2. Rayleigh scattering is stronger than Raman scattering, typically by a ratio of ~1000:1 for strong Raman bands such as those of CCl_4. The most probable event is the energy transfer to molecules in the ground state ($\upsilon_k = 0$) and re-emission by the return of these molecules to the ground state, as indicated by the heavy dark arrows in the energy-level diagram of Figure 5.

3. The Stokes Raman bands are more intense than the anti-Stokes Raman bands, because more molecules are in the ground than in the excited vibrational states ($\upsilon_k = 1$). For colorless substances such as CCl_4, the ratio of the intensities of

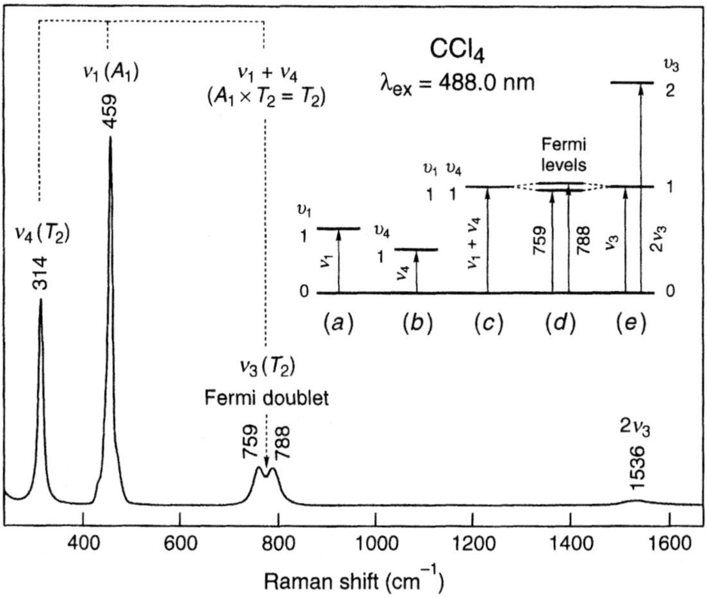

Figure 6. Vibrational energy level diagram illustrating fundamental (a, b, e), overtone (e), combination (c), and Fermi resonance (d) transitions, and the Raman spectrum of liquid CCl_4 (488.0 nm excitation, 150 mW laser power, and 5 cm^{-1} slit widths) showing splitting of the $\nu_3(T_2)$ stretching band in CCl_4 due to Fermi resonance.

the anti-Stokes signal (I_{aSt}) to the Stokes (I_{St}) is directly related to the thermal population of the excited molecular vibrational levels (υ_k'):

$$\frac{I_{aSt}}{I_{St}} = \frac{(\widetilde{\nu}_0 + \widetilde{\nu}_k)^4}{(\widetilde{\nu}_0 - \widetilde{\nu}_k)^4} \exp(-\widetilde{\nu}_k/kT) \tag{9}$$

where k is Boltzmann's constant (0.695039 cm^{-1} K^{-1}), T is the sample absolute temperature (in K), and $\widetilde{\nu}_0$ and $\widetilde{\nu}_k$ are the wavenumbers (in cm^{-1}) of the exciting radiation and the molecular vibration, respectively.

4. The intensity of anti-Stokes relative to Stokes Raman scattering decreases rapidly with increase in the molecular vibrational frequency ν_k (e.g., compare the 218 and 459 cm^{-1} bands in both spectral regions), due to the $\exp(-\nu_k/kT)$ Boltzmann population factor in (9). Since kT (thermal energy) is approximately 207 cm^{-1} at room temperature (298 K), the anti-Stokes bands become vanishingly weak ($<0.02 I_{Stokes}$) for $\nu_k > 800$ cm^{-1}.
5. Raman scattering is also strongly dependent on temperature, and the anti-Stokes bands lose most of their intensity on cooling the sample (e.g., note that the 459 cm^{-1} anti-Stokes band has zero intensity at 77 K). This is again due to the Boltzmann population factor; $kT \approx 54$ cm^{-1} at liquid N_2 temperature (77 K), and the anti-Stokes transitions become vanishingly weak for $\nu_k > 200$ cm^{-1}.

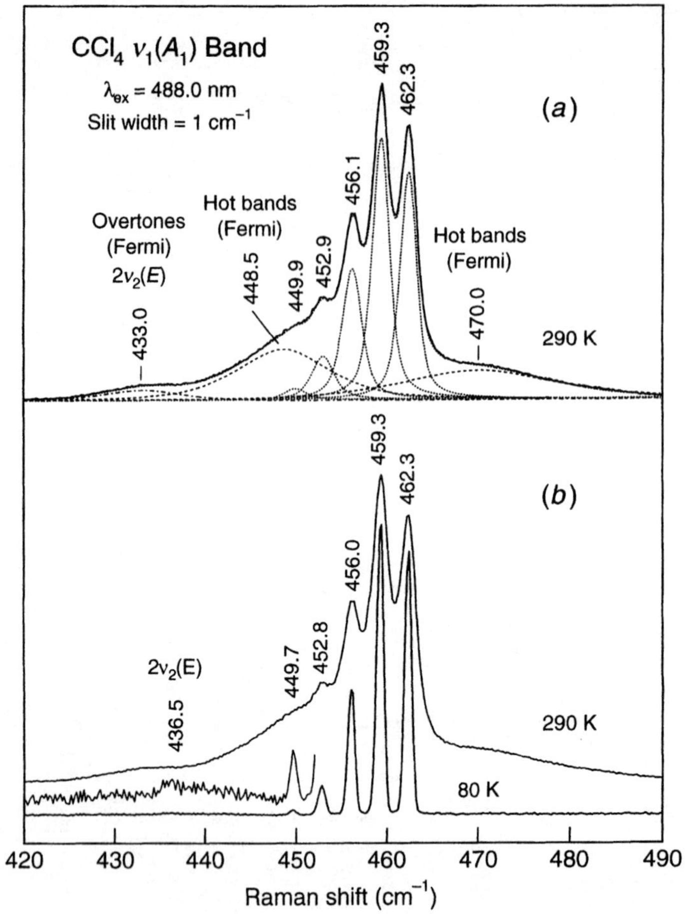

Figure 7. Isotopic structure of the $\nu_1(A_1)$ Raman band of CCl_4 excited with 488.0 nm Ar^+ ion laser radiation at 1 cm^{-1} slit widths: (*a*) in the liquid phase at room temperature (290 K), and (*b*) in the frozen state at 80 K. Spectra were recorded from a capillary (neat liquid) and a liquid N_2 cell (frozen liquid) as described in Figure 3. Temperatures are calculated values using Equation 10. The room temperature spectrum in (*a*) was analyzed with an IBM PC version of LabCalc (Galactic Industries, Inc.). The CURVEFIT routine was used to deconvolute overlapped peaks into a combination of Gaussian curves.

Thus only the lowest frequency CCl_4 vibrational mode at 218 cm^{-1} is clearly seen in the anti-Stokes region of frozen CCl_4 at 77 K.

6. Since the Raman spectrum is temperature dependent, one can use the Stokes/anti-Stokes intensity ratio for a given Raman vibrational mode (ν_k) to calculate the sample temperature (T) using:

$$T(\mathrm{K}) = \frac{1.438768\tilde{\nu}_k}{\ln(I_{\mathrm{St}}/I_{\mathrm{aSt}}) - 4\ln[(\tilde{\nu}_0 - \tilde{\nu}_k)/(\tilde{\nu}_0 + \tilde{\nu}_k)]} \tag{10}$$

7. In general, only the *fundamental* vibrational modes show up in the Raman spectrum because of the much greater probability of transitions involving the two lowest energy vibrational states ($\upsilon_k = 0,\ \upsilon'_k = 1$) of a molecule. *Non-fundamental* vibrational modes such as *overtones* (which occur when a mode is excited beyond the $\upsilon'_k = 1$ level by one photon) and *combination modes* (which occur when more than one vibration are excited by one photon) are usually weak or absent (Figure 6), but they can gain Raman intensity through vibrational mixing when an overtone ($2\nu_k$) or a combination mode ($\nu_k + \nu_l$) has the same symmetry and the same or nearly the same frequency as another fundamental (ν_m). In such a case the frequencies move apart and two separate bands with more or less equal intensities are observed. This phenomenon is known as the *Fermi resonance* and is a common occurrence in both the Raman and IR spectra. An example is two similarly intense Raman bands (*Fermi doublet*) of CCl_4 at 759 and 788 cm^{-1} in Figure 6, instead of one strong CCl_4 antisymmetric stretching band at ~768 cm^{-1}, $\nu_3(T_2)$, and one weak nonfundamental band at 773 cm^{-1} associated with the combination between modes $\nu_1(A_1)$ at 459 cm^{-1} (symmetric stretching) and $\nu_4(T_2)$ at 314 cm^{-1} (angle deformation), $\nu_1 + \nu_4$ ($A_1 \times T_2 = T_2$). The first overtone of ν_3 appears weakly at ~1536 cm^{-1}; thus the frequency of the ν_3 fundamental ($1536 \div 2 = 768$ cm^{-1}) is in close resonance with the $\nu_1 + \nu_4$ combination mode ($459 + 314 = 773$ cm^{-1}).
8. Finally, Raman scattering from CCl_4 exhibits one more fascinating spectral feature. When the slits of the spectrometer are decreased to a spectral bandpass of 1 cm^{-1} (i.e., the resolution is improved), the ν_1 band of CCl_4 at 459 cm^{-1} splits into a few narrow components between 450 and 462 cm^{-1}, as shown in Figure 7. This splitting originates from *isotopic effects*. Since the natural abundance of the ^{35}Cl and ^{37}Cl isotopes is 3:1, five Raman lines arise from various isotopomers: $C^{35}Cl_4$ (31.6%), $C^{35}Cl_3\,^{37}Cl$ (42.2%), $C^{35}Cl_2\,^{37}Cl_2$ (21.1%), $C^{35}Cl^{37}Cl_3$ (4.7%), and $C^{37}Cl_4$ (0.4%). Intensity and frequency patterns of a multiplet closely follow the natural abundance and mass changes in the $C^{35}Cl_n\,^{37}Cl_{4-n}$ isotopomers; the frequency decreases incrementally (~3 cm^{-1} per one ^{37}Cl) on going from $C^{35}Cl_4$ ($n = 4$) to $C^{37}Cl_4$ ($n = 0$) because the mass increases in the same order, while the most intense peak arises from $C^{37}Cl^{35}Cl_3$ (459.3 m^{-1}) because $C^{37}Cl^{35}Cl_3$ is the most abundant isotopomer in the mixture (42.2%). $C^{37}Cl_4$ is the least abundant (0.4%) and the heaviest isotopomer and thus produces the weakest peak at the lowest frequency (449.9 cm^{-1}). Detection of $C^{37}Cl_4$ and $C^{35}Cl^{37}Cl_3$ is difficult at room temperature, however, because they are partly obscured by a broad scattering contribution due to *hot bands* in Fermi resonance with ν_1 (Figure 7*a*). (A hot band occurs when an already excited vibration is further excited ($\upsilon'_k = 1 \rightarrow \upsilon'_k = 2$), normally at a frequency that is slightly less than its fundamental transition, ($\upsilon_k = 0 \rightarrow \upsilon'_k = 1$).) This contribution is absent in the liquid-N_2 temperature spectrum, where all five chlorine isotope lines are completely resolved (Figure 7*b*) because cooling of CCl_4 depopulated the hot bands. Remarkably, an additional Fermi resonance occurs between the overtone of the CCl_4 $\nu_2(E)$ bending mode (which appears at 218 cm^{-1}, Figure 5) and the Fermi resonance enhanced hot bands at ~450 cm^{-1}, producing a broad feature

at 433 cm^{-1} (Figure 7*a*); again, this assignment is confirmed by the low temperature spectrum, where the 433 cm^{-1} feature vanishes and an unperturbed $2\nu_2$ appears weakly near 437 cm^{-1} at high magnification (Figure 7*b*).

2.4 Requirements for Raman Scattering

While IR absorption requires a change in the dipole moment, Raman scattering requires a change in molecular polarizability, as a result of the vibrations. When a fluctuating electric field $\boldsymbol{E}$ of frequency ν_0 (laser beam) $\gg$ ν_k (vibrational frequency) strikes a molecule, it sets the electron cloud into oscillatory motion, as shown schematically for a homonuclear diatomic molecule like F_2 in Figure 8*a*. In its distorted form, the molecule is *polarized*, that is, it undergoes a momentary charge displacement, forming an *induced electric dipole moment* $\boldsymbol{P}$ which oscillates at the frequency, ν_0, of the incoming laser beam. The vibrations of the molecule induce additional frequency components of the polarization, at $\nu_0 - \nu_k$ or $\nu_0 + \nu_k$. The polarized molecule re-emits photons of these frequencies, as well as ν_0. The scattered intensity varies directly with the square of the induced dipole moment $\boldsymbol{P}$.

The magnitude of $\boldsymbol{P}$ depends both on the strength of the perturbing electric field $\boldsymbol{E}$ and the ease with which $\boldsymbol{E}$ can distort the electrons, i.e., the degree of the electronic polarization. We may write

$$\boldsymbol{P} = \alpha \boldsymbol{E} \tag{11}$$

where α is the *polarizability* of the molecule, a volumetric measure of its electron cloud distortion in the path of an incident light, relative to the nuclear framework. The polarizability α is closely related to the structure and bonding properties of the molecule and is *always* nonzero; it increases with decreasing electron density, decreasing bond strength, and increasing bond length.

The polarizability of F_2 and any other linear molecule is *anisotropic*, i.e., their electrons are more polarizable (larger α) along the chemical bond axis than in the direction perpendicular to it. We can represent the polarizability graphically by drawing a *polarizability ellipsoid*, as in Figure 8*b*. The ellipsoid is a three-dimensional surface whose distance from the center of gravity of the molecule is proportional to $1/\sqrt{\alpha_i}$ where α_i is the polarizability in the i-th direction. The polarizability ellipsoid shrinks as the F_2 bond stretches (larger α), but expands as the F_2 bond contracts (smaller α). The size of the polarizability ellipsoid varies periodically at the F_2 stretching frequency, reaching a minimum at the largest internuclear separation and a maximum at the closest approach. The energy for this work can come from the oscillating dipole moment induced by the incoming electric field of the radiation, resulting in absorption of the vibrational energy and re-radiation of the reminder of the energy. Molecules already excited can give up excess vibrational energy to the oscillating induced dipole moment, so that an increased amount of energy is radiated as light of higher frequency. As a result, the F–F stretching frequency will appear in the Raman spectrum on either side of the Rayleigh (excitation) frequency.

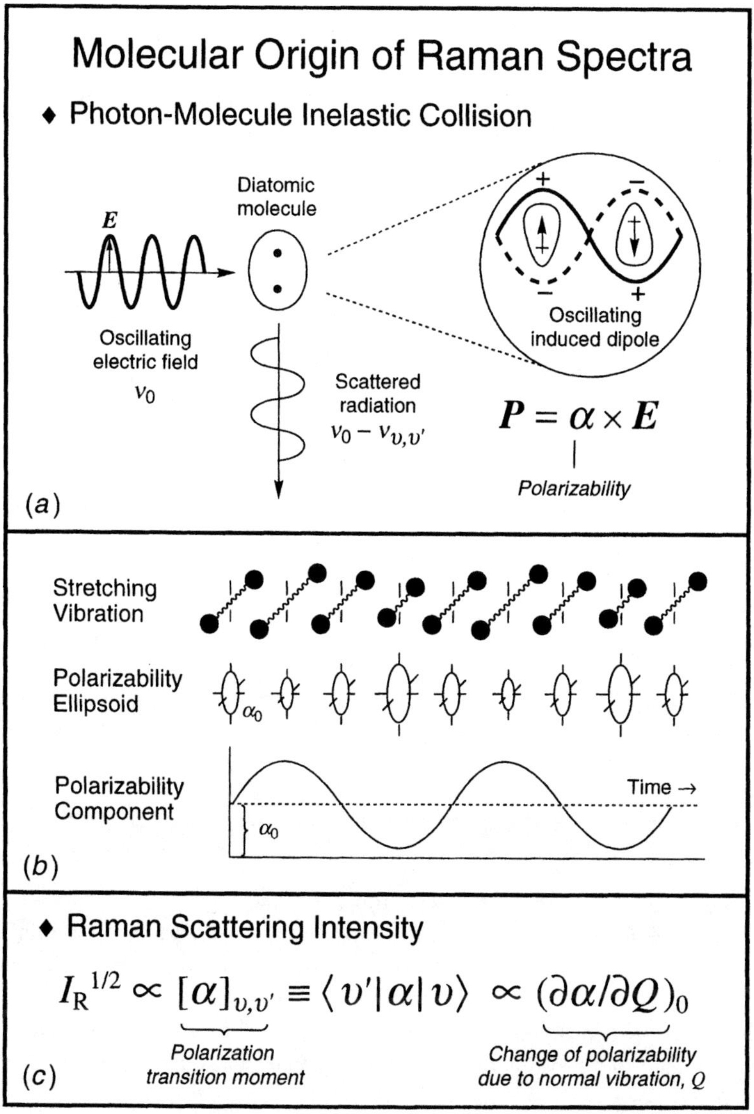

Figure 8. Schematic illustration of molecular origin and requirements for Raman scattering: Electric field $\boldsymbol{E}$ of the monochromatic UV or visible radiation distorts the electron cloud when it collides with a molecule and induces an oscillating dipole with a moment $\boldsymbol{P}$ directly related to $\boldsymbol{E}$ by the polarizability, α, of the bond (*a*). If the motion of the atoms produces a periodic alteration in α, as in (*b*), this dipole will absorb a quantum of vibrational energy $(h\nu_{\upsilon,\upsilon'})$ from $\boldsymbol{E}$ and re-radiate the excess energy as light of frequency lower than the incident light, $\nu_0 - \nu_{\upsilon,\upsilon'}$, (Raman scattered radiation). The intensity of scatter depends on the square of the change in the polarizability $(\partial\alpha/\partial Q)_0^2$ caused by the vibration, Q (*c*).

To determine the scattered radiation spectrum of an oscillating molecule, we again must look at the matrix elements of the vibrational transition moment; in particular this time we must consider how the induced electric dipole moment or polarizability ($\boldsymbol{P} = \alpha \boldsymbol{E}$)—rather than the intrinsic electric dipole moment in infrared absorption—varies with normal mode of vibration. If we write the matrix elements of the polarizability operator α

$$[\alpha]_k = \langle \upsilon_k' | \alpha | \upsilon_k \rangle \tag{12}$$

and expand α in a series similar to that of (6)

$$\alpha = \alpha_0 + \left(\frac{\partial \alpha}{\partial Q_k} \right)_0 Q_k + \frac{1}{2!} \left(\frac{\partial^2 \alpha}{\partial Q_k^2} \right)_0 Q_k^2 + \cdots \tag{13}$$

we find that the only nonvanishing elements are

$$\alpha_0 \langle \upsilon_k' | Q_k | \upsilon_k \rangle \sim \alpha_0 \delta(\upsilon_k', \upsilon_k) \tag{14}$$

which gives the dominant Rayleigh scattering with $\Delta \upsilon_k = 0$ (α_0 is always $\neq 0$),

$$\left(\frac{\partial \alpha}{\partial Q_k} \right)_0 \langle \upsilon_k' | Q_k | \upsilon_k \rangle \sim \left(\frac{\partial \alpha}{\partial Q_k} \right)_0 \delta(\upsilon_k', \upsilon_k \pm 1) \tag{15}$$

which gives the weaker fundamental Raman scattering with $\Delta \upsilon_k = \pm 1$ if $(\partial \alpha / \partial Q_k)_0 \neq 0$,

$$\left(\frac{\partial^2 \alpha}{\partial Q_k^2} \right)_0 \langle \upsilon_k' | Q_k^2 | \upsilon_k \rangle \sim \left(\frac{\partial^2 \alpha}{\partial Q_k^2} \right)_0 \delta(\upsilon_k', \upsilon_k \pm 2) \tag{16}$$

which gives much weaker overtone Raman scattering with $\Delta \upsilon_k = \pm 2$ if $(\partial^2 \alpha / \partial Q_k^2)_0 \neq 0$, and so on. Thus the probability of transition for fundamentals in the Raman spectrum is proportional to the square of the polarizability gradient along the normal coordinate of the vibration. Accordingly, if there is no net change in polarizability when the vibration occurs, i.e., $(\partial \alpha / \partial Q_k)_0 = 0$, the intensity of a vibrational band in the Raman spectrum equals zero and the vibration is said to be Raman inactive.

2.5 Molecular Symmetry and Vibrational Activity

In Figures 9 and 10 we show the three fundamental modes ν_1 (in-phase or symmetric stretching), ν_2 (bending), and ν_3 (out-of-phase or antisymmetric stretching) of the triatomic molecules CS_2 (linear, $D_{\infty h}$ symmetry) and SO_2 (bent, C_{2v} symmetry). Whereas for SO_2 all three frequencies are obtainable from either infrared absorption or Raman scattering, this is not so for a linear CS_2; its ν_1 mode frequency is from the Raman spectrum but the ν_2 and ν_3 frequencies are from the infrared spectrum.

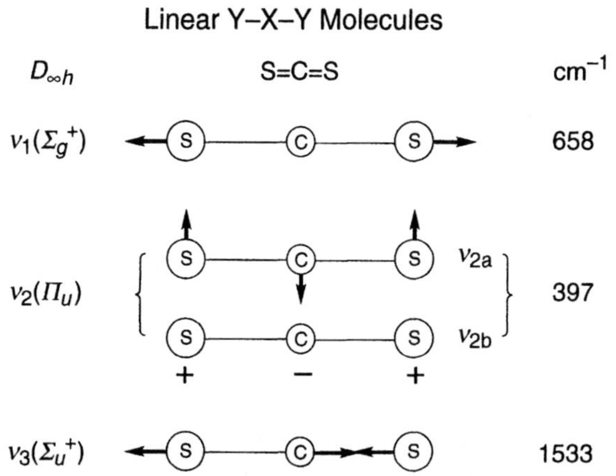

Figure 9. Normal modes of vibration and frequencies (cm^{-1}) in a linear triatomic molecule CS_2: $3 \times 3 - 5 = 4$ normal vibrations but three distinct frequencies; ν_1, in-phase or symmetric stretching (both C–S bonds are stretched simultaneously); ν_2, doubly degenerate bending (ν_{2a} occurs in the plane of the paper, ν_{2b} occurs in the plane perpendicular to the paper); ν_3, out-of-phase or antisymmetric stretching (one C–S bond is compressed while the other is stretched).

Figure 11 illustrates changes in the electric dipole moment (μ) and molecular polarizability (α) produced by the ν_1, ν_2, and ν_3 vibrations of CS_2. In the equilibrium position ($Q = 0$), $\mu = 0$ for CS_2 (required by the symmetry of the molecule) whereas $\alpha \neq 0$ (the molecule suffers some distortion in the applied field $\boldsymbol{E}$, the positively charged nuclei being attracted towards the negative pole of the field, the electrons to the positive pole). The polarizability ellipsoid in a linear polyatomic molecule has a circular cross section along the molecular axis and an elliptical one in the direction perpendicular to it. The symmetrical, in-phase stretching vibration ν_1 does not change the symmetry in the distorted molecules (both C–S bonds stretch ($+Q$) or contract ($-Q$) simultaneously), and consequently $(\partial\mu/\partial Q)_0 = 0$ for this vibration. The ν_1 is thus infrared *inactive* (the $\boldsymbol{E}$ field of the IR photon cannot pull the two negatively charged S atoms in opposite directions as required during this vibration). However, since the CS_2 molecule changes size during the ν_1 motion, there is a corresponding fluctuation in the size of the polarizability ellipsoids; the $(\partial\alpha/\partial Q)_0 \neq 0$ and the motion is thus Raman *active.*

Both the antisymmetric, out-of-phase stretching vibration ν_3 (one C–S bond stretches while the other contracts) and the bending vibration ν_2 (which is doubly degenerate) change the symmetry in the distorted molecules (by destroying the center of inversion), and cause the dipole moment to oscillate along (ν_3) or across (ν_2) the molecular axis. $(\partial\mu/\partial Q)_0 \neq 0$ for ν_3 and ν_2, and both are thus infrared *active* (the $\boldsymbol{E}$ field of the IR photon pulls the positively charged C atom in one direction and the negatively charged S atoms in the opposite direction along or across the molecular axis as required to excite ν_3 and ν_2). In contrast, although the ν_3 and ν_2 motions also change the polarizability ellipsoids (the shape changes most), the respective change

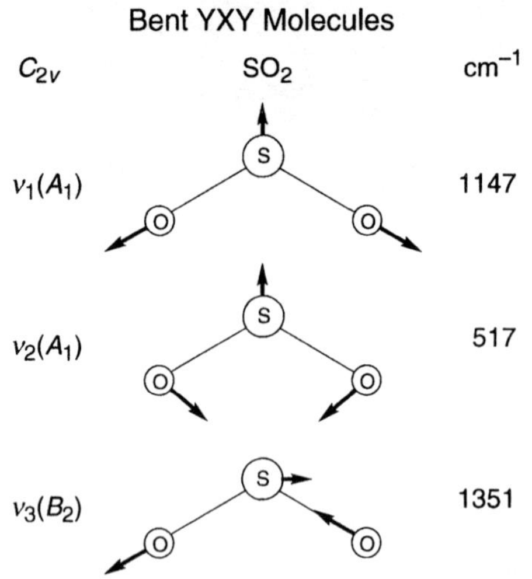

Figure 10. Normal modes of vibration and frequencies (cm^{-1}) in a nonlinear triatomic molecule SO_2: $3 \times 3 - 6 = 3$ normal vibrations and three distinct frequencies; ν_1, symmetric stretching (both S–O bonds are stretched simultaneously); ν_2, in-plane bending (any attempt to construct an out-of-plane bending of SO_2, like ν_{2b} of CS_2, would result in a rotation); ν_3, antisymmetric stretching (one S–O bond is compressed while the other is stretched).

is exactly the same for both positive and negative vibrational displacements. As a result $(\partial\alpha/\partial Q)_0 = 0$ for ν_3 and ν_2, and both vibrations are Raman *inactive.*

Now consider the bent triatomic molecule SO_2, whose ν_1, ν_2, and ν_3 fundamentals have been shown in Figure 10; in Figure 12 we sketch the extreme and equilibrium configurations of the molecule and their approximate dipole moments and polarizability ellipsoids. In this molecule both $\mu \neq 0$ and $\alpha \neq 0$ in the equilibrium position, the permanent dipole moment lying on the C_2 rotational axis and the polarizability ellipsoid having *all* cross sections elliptical (in SO_2 α is different along all three major molecular axis, which lie along the C_2 rotational axis and the normals of the two σ_v planes of the molecule). The symmetrical stretching motion ν_1 does not change the molecular symmetry ($C_{2\upsilon}$) but the molecule increases in size while the S–O bonds stretch ($+Q$) and decreases while they compress ($-Q$), and so there is a corresponding fluctuation in the magnitude of the dipole moment (which oscillates along the C_2 axis) and the size of the polarizability ellipsoid (the shape remains approximately constant during compression and expansion of the S–O bonds). On the other hand, although the bending vibration ν_2 also does not change the symmetry in the distorted molecules, it causes a periodical change in shape of the molecule that, in turn, causes the dipole spacing to oscillate along the C_2 axis and the polarizability ellipsoid to change its shape at different rates along the ellipsoid principal axes (at $+Q_2$ the molecule approaches a linear configuration, while at $-Q_2$ it approximates a diatomic one, with their axes rotated by 90°). Finally, while undergoing the antisymmetrical motion ν_3 (which destroys the molecular $C_{2\upsilon}$ symmetry), it is the dipole direction and the direction of the major axis of the polarizability ellipsoid which

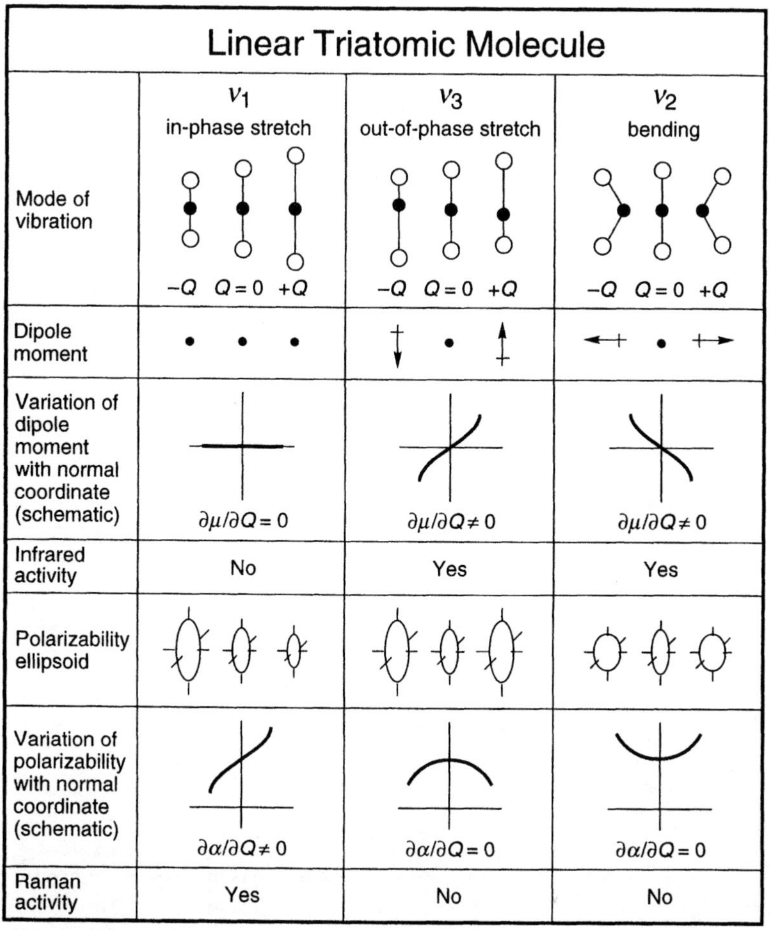

Figure 11. Dipole moment (μ) and polarizability (α) variations during normal vibrations for linear centrosymmetric triatomic molecules (e.g. CS_2) and the IR and Raman activities. The center column in each panel (ν_i) shows the equilibrium position ($Q = 0$) of the molecule while the right ($+Q$) and left ($-Q$) are the extremes of each vibration.

change most (middle panel). Consequently $(\partial\mu/\partial Q)_0 \neq 0$ and $(\partial\alpha/\partial Q)_0 \neq 0$ for all three normal vibrational modes in SO_2 (at least one aspect of μ and α changes), and all are thus infrared as well as Raman *active*, unlike CS_2 whose ν_1 mode is only Raman active and the ν_2 and ν_3 modes are only infrared active.

These visual arguments become harder to produce for more complex molecules, and we rely instead on the formal rules of group theory. The matrix elements of the dipole moment vector are

$$[\mu]_k = \langle \upsilon_k' | \mu_x | \upsilon_k \rangle + \langle \upsilon_k' | \mu_y | \upsilon_k \rangle + \langle \upsilon_k' | \mu_z | \upsilon_k \rangle \quad (17)$$

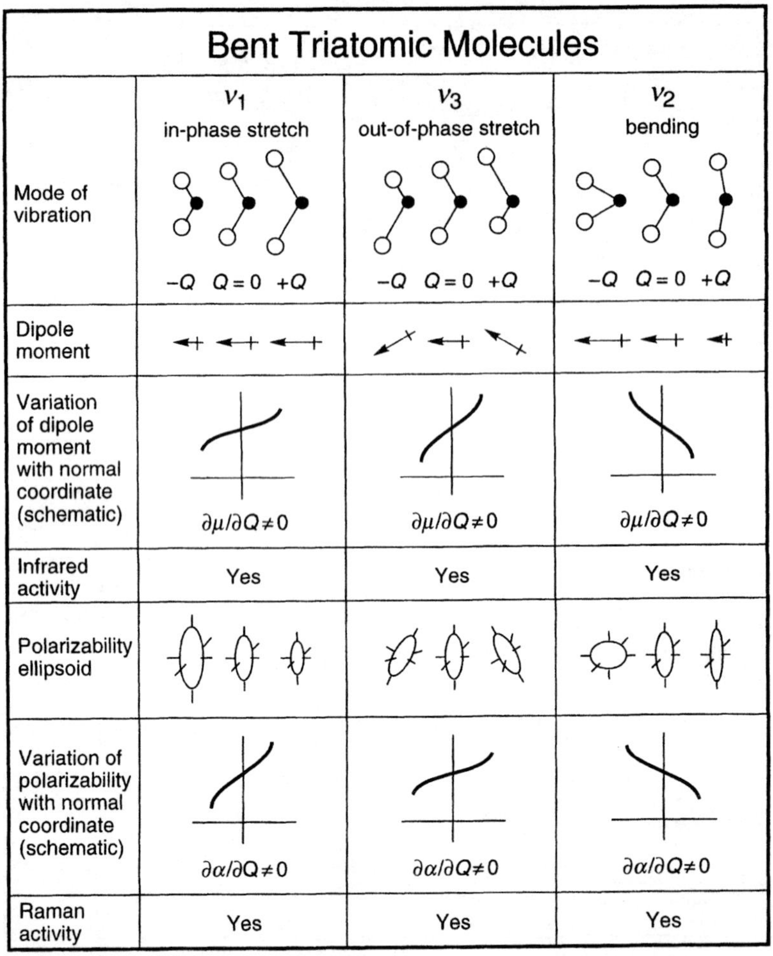

Figure 12. Dipole moment (μ) and polarizability (α) variations during normal vibrations for nonlinear symmetrical triatomic molecules (e.g. SO_2) and the IR and Raman activities. The center column in each panel (ν_i) shows the equilibrium position ($Q = 0$) of the molecule while the right ($+Q$) and left ($-Q$) are the extremes of each vibration.

and of the polarizability tensor are

$$[\alpha]_k = \langle \upsilon'_k | \alpha_{xx} | \upsilon_k \rangle + \langle \upsilon'_k | \alpha_{xy} | \upsilon_k \rangle + \langle \upsilon'_k | \alpha_{xz} | \upsilon_k \rangle + \cdots + \langle \upsilon'_k | \alpha_{zz} | \upsilon_k \rangle \qquad (18)$$

Each quantity in the integrals of (17) and (18) has a definite behavior with respect to any symmetry operation, and so the condition that the matrix elements $[\mu]_k$ and $[\alpha]_k$ shall not vanish must be the same for all transitions between vibrational states of two particular symmetries, regardless of the particular molecule involved. If the symmetry species (irreducible representations) of $|\upsilon_k\rangle$ and $|\upsilon'_k\rangle$, and the operator components μ_ρ and $\alpha_{\rho\sigma}$ (ρ and $\sigma = x,\ y$, or z) are $\Gamma(\upsilon_k),\ \Gamma(\upsilon_k'),\ \Gamma(\mu_\rho)$, and $\Gamma(\alpha_{\rho\sigma})$, this condition is

$$\Gamma(\upsilon_k') \times \Gamma(\mu_\rho) \times \Gamma(\upsilon_k) \supseteq A \tag{19}$$

for infrared absorption and

$$\Gamma(\upsilon_k') \times \Gamma(\alpha_{\rho\sigma}) \times \Gamma(\upsilon_k) \supseteq A \tag{20}$$

for Raman scattering, where A denotes the totally symmetric species of the molecular *point group* and the symbol $\supseteq$ means "equals to" (nondegenerate states) or "contains" (degenerate states).

If we take $\Gamma(\upsilon_k) = A$ (which is always true for the ground vibrational state of any molecule) and $\Gamma(\upsilon_k') = \Gamma(Q_k)$ (which is also true since the first excited vibrational state always transforms as the normal vibration itself), we find that the $\Gamma(Q_k)$ must be the same as $\Gamma(\mu_\rho)$ and $\Gamma(\alpha_{\rho\sigma})$ in order to satisfy Equations (19) and (20), respectively. In other words, the molecular vibration is active in infrared absorption if it belongs to the same representation as at least one of the dipole moment components ($\mu_x,\ \mu_y,\ \mu_z$) or, since the dipole moment is a vector, as one of the Cartesian coordinates ($x,\ y,\ z$). In contrast, the molecular vibration is active in Raman scattering if it belongs to the same representation as at least one of the polarizability components ($\alpha_{xx},\ \alpha_{xy}$, etc.) or, since the polarizability is a tensor, as one of the binary products of Cartesian coordinates ($x^2,\ xy$, etc.) or their linear combinations ($x^2 + y^2, x^2 - y^2$, etc.).

The symmetry species of the Cartesian coordinates and their binary products are customarily given in the point group *character tables*. Infrared and Raman activities for vibrational modes of any symmetry species can therefore be simply read off from these tables. Figure 13 is an example of a character table for the point group C_{4u}. It follows from that table that $\Gamma(z) = A_1$ and $\Gamma(x, y) = E$, and hence fundamental vibrations of type A_1 and E are *allowed* and of type $A_2,\ B_1$, and B_2 are *forbidden* in the infrared spectrum for any molecule that belongs to the C_{4v} point group. The vibrations of type A_2 are also *forbidden* in the Raman spectrum of C_{4v} molecules but those of type $A_1,\ B_1,\ B_2$, and E are all Raman *allowed* because $\Gamma(x^2 + y^2),\ \Gamma(z^2) = A_1,\ \Gamma(x^2 - y^2) = B_1,\ \Gamma(xy) = B_2$, and $\Gamma(xz,\ yz) = E$.

2.6 Polarization of Raman Scattering

The incident and scattered photons in Raman scattering both involve a polarization, and their mutual orientation is important for the scattering intensity, even if the sample is unordered. Because of the two polarization directions, the polarizability is a tensor quantity with nine elements, $\alpha_{xx},\ \alpha_{xy},\ \alpha_{xz}$, etc. If the molecules in the sample were all aligned in a given direction, then the scattering intensity would be determined by α_{zz} if the incident photon vector were aligned with the molecular z axis and the scattered light was analyzed in the z direction. Only those vibrations having nonzero α_{zz} would be represented in this spectrum. On the other hand, if the scat-

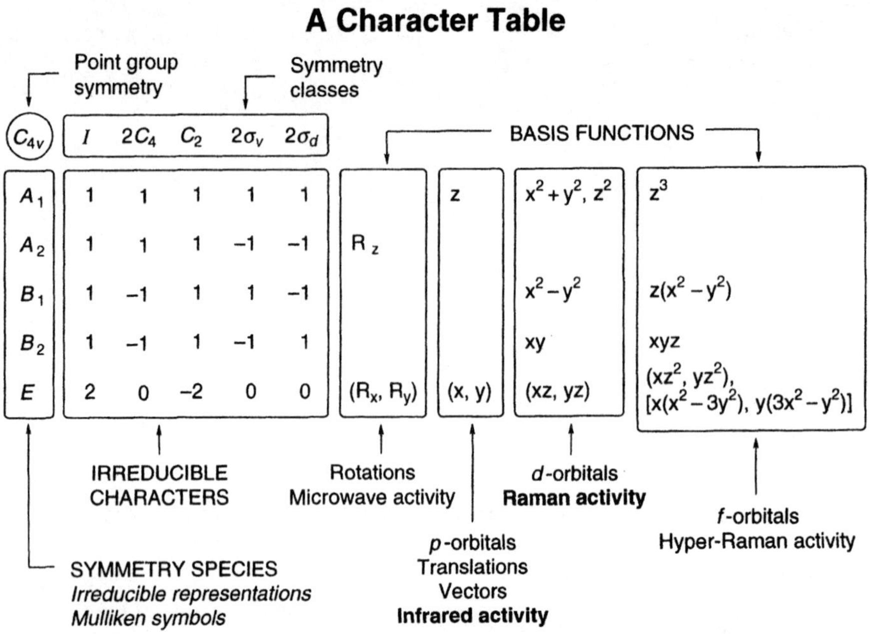

Figure 13. The structure of the character table for the point group C_{4v}.

tered light were analyzed in the x direction then vibrations with nonzero α_{xz} would be represented. This situation is illustrated graphically in Figure 14 for the usual scattering geometry in which scattered light is collected at 90° to the incident beam.

When the sample is unoriented the direction of the incident polarization is unrelated to the molecular axes, but the scattered light can be analyzed parallel (‖) or perpendicular (⊥) to the incident light. Experimentally, this is accomplished by inserting a polarization analyzer between the sample (as neat liquid or in solution) and the entrance slit of the Raman instrument, which upon rotation by 90° allows the perpendicular or parallel polarized light to pass through to the detector (Figure 15). The ratio of the two intensities ($I_\perp/I_\parallel$) is called the *depolarization ratio* ρ, and it can be expressed in terms of the rotational invariants of the scattering tensor, $\overline{\alpha}$ (trace), γ_s (anisotropy), and γ_{as} (antisymmetry):

$$\rho = \frac{3\overline{\alpha} + 5\gamma_{as}^2}{45\overline{\alpha}^2 + 4\gamma_s^2} \tag{21}$$

where

$$\begin{aligned} \overline{\alpha} &= \tfrac{1}{3}\sum \alpha_{\rho\sigma} \\ \gamma_s^2 &= \tfrac{1}{3}\sum(\alpha_{\rho\rho} - \alpha_{\sigma\sigma})^2 + \tfrac{3}{4}\sum(\alpha_{\rho\sigma} - \alpha_{\sigma\rho})^2 \\ \gamma_{as}^2 &= \tfrac{3}{4}\sum(\alpha_{\rho\sigma} - \alpha_{\sigma\rho})^2 \end{aligned} \tag{22}$$

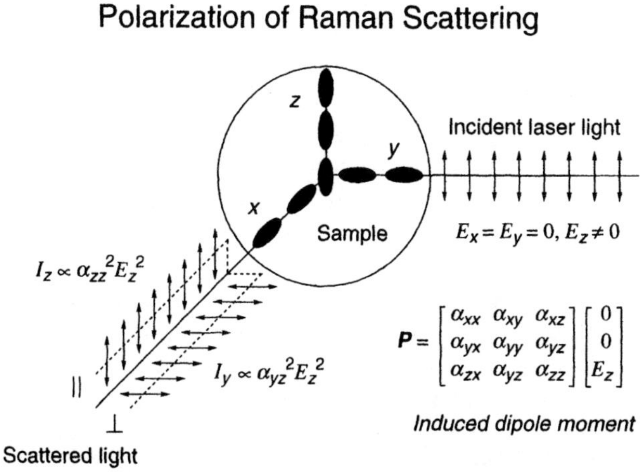

Figure 14. Polarization properties of Raman scattering. The polarizability change associated with the vibration is a tensor quantity and symmetry information can be obtained by analyzing the scattered light into components parallel (||) and perpendicular (⊥) to the incident light vector, even for randomly oriented molecules. This information is contained in the vibrational invariants of the tensor, which determine the depolarization ratio (see text).

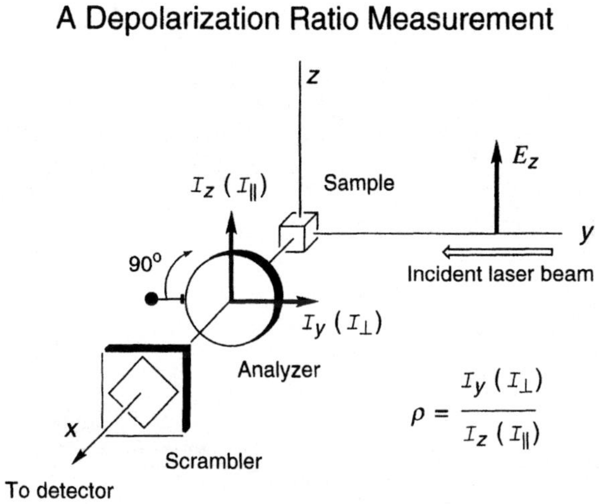

Figure 15. Schematic layout of a typical 90° Raman depolarization experiment showing the positions of the polarization analyzer and scrambler. The analyzer may simply be a Polaroid sheet which can be rotated by 90 degrees in order to allow the parallel (||) and perpendicular (⊥) components of the scattered light to pass through to the detector. The function of a polarization scrambler (or an optical wedge) is to change linear into circular polarization of the light entering the Raman spectrometer slit in order to avoid measurement errors due to the variable spectrometer transmittance of the light polarized in different directions.

Under most circumstances the scattering tensor is symmetric, i.e., $\alpha_{\rho\sigma} = \alpha_{\sigma\rho}$, and consequently $\gamma_{as} = 0$. In that case the depolarization ratio must be 3/4 (when $\overline{\alpha} = 0$) or less ($\overline{\alpha} \neq 0$). Only for vibrations which preserve the symmetry of the molecule (totally symmetric vibrations) are diagonal polarizability tensor elements nonzero ($\overline{\alpha} \neq 0$); these modes have $\rho < 3/4$, and are said to be *polarized.* The higher the symmetry of a molecule, the closer to zero is the value of ρ for a totally symmetric vibration. It should also be pointed out that the polarized bands are usually the strongest features in the Raman spectrum. Non-totally symmetric vibrations ($\overline{\alpha} = 0$) ordinarily give rise to *depolarized bands* ($\rho = 3/4$). In the polarized Raman spectrum of CCl_4 (T_d symmetry) shown in Figure 16, the dominant band at 459 cm^{-1} loses nearly all intensity in the perpendicular scattering ($\rho = 0.006$) and a number of weaker bands at 218, 314, 759, and 788 cm^{-1} are only slightly weakened ($\rho = 0.75$). The band at 459 cm^{-1} is therefore immediately assignable to a totally symmetric vibration of the molecule and the ones at 218, 314, 759, and 788 cm^{-1} cannot arise from totally symmetric modes.

In special cases involving the resonance Raman effect, the scattering tensors are not symmetric, $\alpha_{\sigma\rho} \neq \alpha_{\rho\sigma}$, in which case $\gamma_{as} \neq 0$, and $\rho > 3/4$, a situation described as *anomalous polarization.* If the symmetry is high enough, the scattering tensor may be antisymmetric, and $\alpha_{\sigma\rho} = -\alpha_{\rho\sigma}$, in which case $\gamma_s = 0$ and $\rho = \infty$

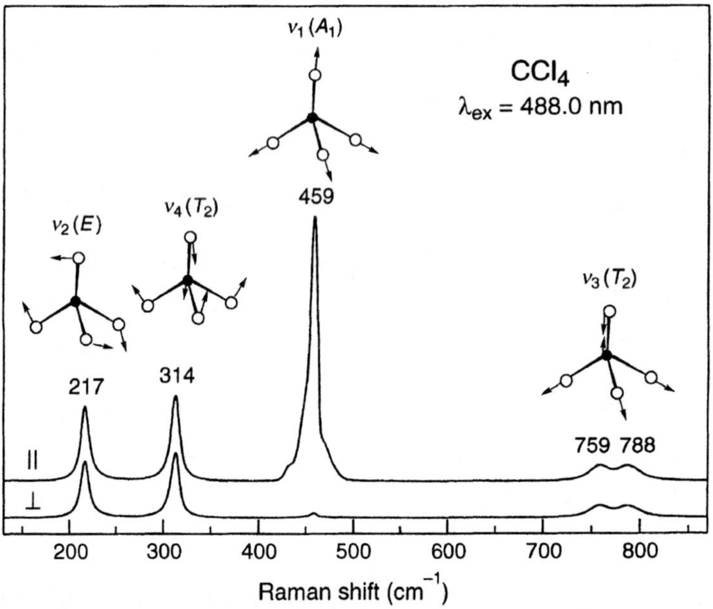

Figure 16. Parallel (||) and perpendicular (⊥) scattered Raman spectra from liquid CCl_4 showing the polarized (ν_1) and depolarized (ν_2, ν_3, ν_4) bands associated with totally symmetric (A_1) and nontotally symmetric (E, T_2) vibrations, respectively. Experimental conditions: 488.0 nm excitation wavelength, 90° scattering from capillary, 100 mW laser power, and 5 cm^{-1} slit widths.

(the band intensity in parallel scattering is zero), a situation termed *inverse polarization*. Metalloporphyrins exemplify the antisymmetric vibrational scattering, and heme proteins provided the first instance of anomalously polarized Raman bands.[24] Figure 17 depicts spectra of nickel(II) porphine (NiP) and indeed the strongest bands show exactly the opposite to what is observed for CCl_4 (Figure 16), i.e., the perpendicular scattered radiation ($\perp$) is dramatically more intense than that with parallel radiation ($\|$). NiP has D_{4h} symmetry and these bands arise from the A_{2g} vibrational modes that have rotational symmetry themselves ($\Gamma(R_z) = A_{2g}$ in D_{4h} point group). Modes of this type are normally Raman forbidden but can be activated in the resonance Raman spectrum by vibronic mixing of excited electronic states (see Section 3.3).

The depolarization ratio in Raman scattering has no true counterpart in infrared absorption. However, the dipole moment is a vector quantity, and the absorption of infrared light depends on the relative orientation of the molecule and the photon electric vector. If all the molecules are aligned, as in a crystal or a stretched film, and the photon vector points along a molecular axis, e.g. z, then absorption occurs for those vibrations which displace the dipole along z. Vibrations which are purely x or y polarized would be absent. Thus extra information about the character of the vibration is available from infrared absorption for oriented samples. If the sample is unordered, all polarization information is lost.

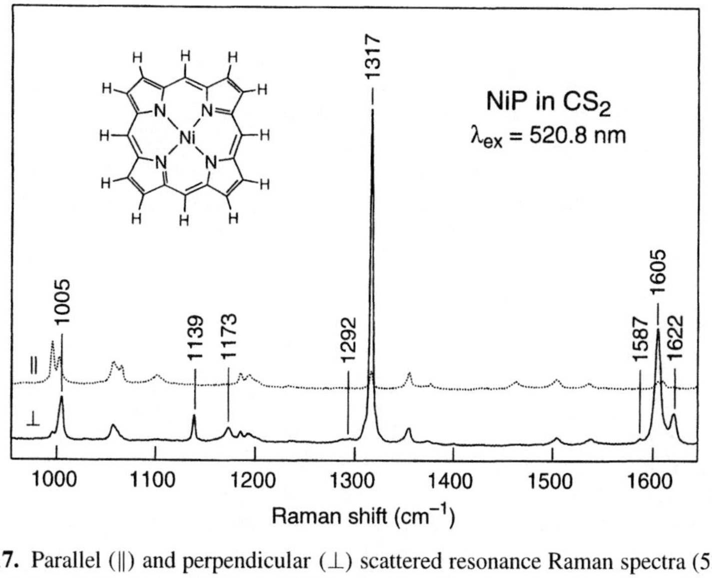

Figure 17. Parallel ($\|$) and perpendicular ($\perp$) scattered resonance Raman spectra (530.9 nm excitation) from NiP (P = porphine) in CS_2 showing the anomalously polarized bands (labeled) associated with the ring A_{2g} vibrations, which have antisymmetric scattering tensors. Experimental conditions: 530.9 nm excitation wavelength (Kr^+ ion laser), 135° backscattering from spinning NMR tube,[56] 150 mW laser power, and 3 cm^{-1} slit widths.

3 RESONANCE RAMAN SPECTROSCOPY

3.1 Enhancement of Raman Scattering

The intensity of Raman scattering depends strongly on the excitation frequency. Figure 18 diagrams what happens if, for a given molecular system, we change laser excitation frequency so that it moves from the transparent region (*a*) to the vicinity of an allowed electronic absorption band (*b–e*). As discussed before, excitation in

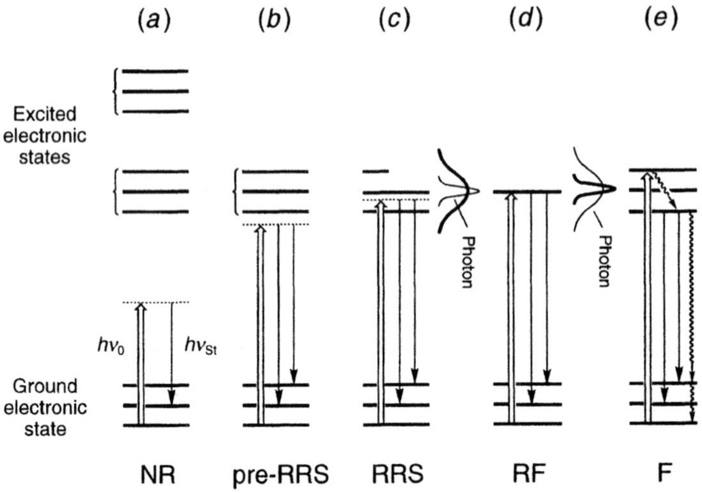

Figure 18. Diagram of Stokes transitions ($h\nu_{St}$) of Raman scattering (*a–c*) and fluorescent emission (*d*, *e*) as a function of excitation energy ($h\nu_0$). (*a*) Normal Raman scattering occurs upon excitation in the transparent region of the molecule. The molecular polarizability creates a nonstationary (virtual) state, which does not resemble any particular molecular state. Many excited electronic states contribute to this state and the scattering process is nonselective in that all vibrations (usually fundamental) that can change the polarizability may be observed. (*b*) Preresonance Raman scattering in which $h\nu_0$ approaches the energy of an electronic transition. The interacting state is than dominated by this electronic state, which produces some enhancement of Raman scattering (10–100-fold) and possible appearance of overtones. (*c*) Resonance Raman scattering in which the interacting state is dominated by a few vibronic levels in the interior of the excited electronic state. Many Raman bands are then attenuated by the absorption, but those associated with normal coordinates that can carry the molecule into its excited state geometry may be greatly enhanced (10^3–10^6 times). (*d*) Resonance fluorescence occurs when $h\nu_0$ coincides with a single sharp level of the electronic manifold (small molecules in gas phase). The distinction between RRS and RF is that the RF involves actual population of the upper state (its width is smaller than that of the $h\nu_0$ pulse), the emission decaying with the excited state lifetime, whereas RRS process is simultaneous (actually $<10^{-14}$ s) with the excitation pulse. Off exact resonance with the upper state, only RR scattering occurs. (*e*) Relaxed fluorescence in which emission to the ground state is preceded by prior radiationless relaxation to the lowest vibrational level of the excited electronic state. This produces a red-shifted broad envelope of emission on the 10^{-6} to 10^{-8} s time scale.

the transparent region of a scattering system leads to a normal (or off-resonance) Raman effect. In this circumstance, the intensity of Raman bands is described by a ν^4 law (2), and some advantage can be taken of this dependence by using the higher-frequency irradiation (e.g., the blue and green lines of the argon laser, rather than the red line of the helium–neon laser). However, when the frequency of the exciting laser line is close to (preresonance) or equal to (resonance) the frequency of an allowed electronic absorption of a molecule, as in (*b*) and (*c*), the intensities of certain Raman bands are greatly increased (up to 10^6-fold) above their off-resonance values. This happens because the polarizability, and its dependence on the molecular motions, is enhanced via the electronic transition. Not all Raman bands intensify, however, but only those due to vibrational transitions that couple with the electronic transition. This is illustrated in Figure 19 which shows a dramatic increase in intensity of the Cu–SS(Cys) vibrational modes near 400 cm^{-1} as the excitation wavelength (647.1 nm) approaches that of the (Cys)S→Cu(II) charge-transfer electronic

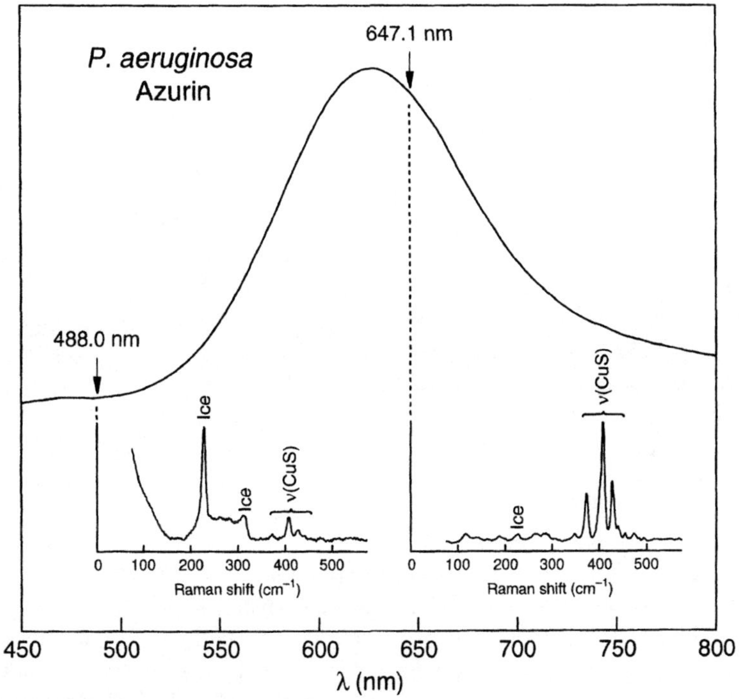

Figure 19. Vibrational enhancement selectivity available from resonance Raman spectroscopy. The UV-visible spectrum of a *P. aeruginosa* azurin is shown together with two different Raman spectra (frozen solution at 77 K[56]) that derive from laser excitation within the S(Cys)→Cu(II) charge-transfer absorption band at 625 nm (647.1 nm) and away from the absorption (488.0 nm). Excitation within resonance leads to dramatically increased Raman scattering from the Cu active site, whereas off-resonance excitation produces a spectrum dominated by bands of nonchromophoric ice.

transition (~600 nm) in the blue copper protein, azurin.[25] Another example of resonance Raman (RR) scattering was seen in Figure 17, where stretching modes of the porphyrin π bonds are enhanced because of their coupling with the π–π^* resonant transitions of NiP.[26,27] Thus the correct identification of the vibrational modes showing RR enhancement will aid in the assignment of the resonant electronic transition and vice versa.

The laser frequency is an important consideration in Raman applications because resonance enhancement greatly increases both the sensitivity and the selectivity of the technique. Nonresonance Raman scattering is weak enough that sample concentrations in the molar range are often required to obtain good quality spectra, but resonance enhancement can lower the required concentration to the mM to μM ranges. Selectivity is an equally important feature since vibrational spectroscopy frequently suffers from crowding and overlap of bands when complex molecules or mixtures are investigated. Because only Raman modes capable of coupling with electronic transitions of a molecule are enhanced, the RR spectrum is simplified and one can be sure that the enhanced bands are associated with the chromophore; interferences from other parts of the molecule or other constituents of the mixture are eliminated. Of course this selectivity means a loss of information if the non-chromophoric parts of the sample are of interest. Wide tunability of the laser source is desirable so that different chromophores in the sample (or different excited states of the same chromophore, which may enhance different Raman bands) can be accessed.

Another advantage of wide laser tunability is the ability to minimize undesirable side effects of laser irradiation, namely photodegradation and fluorescence. It is often possible to reduce photodegradation by tuning the laser to the long-wavelength region of a molecule's absorption spectrum, while interference from fluorescence can be reduced by tuning to the blue side of the fluorescence envelope. RR scattering and fluorescence are intimately connected since in both phenomena the emitted photon results from excitation in an absorption band (see Figure 18 *c*, *e*). Excitation into the absorption band (upward arrow) is followed by emission processes that produce the Rayleigh and Raman peaks and in certain molecular systems a broad fluorescence envelope. This broad envelope, properly called *relaxed fluorescence*, follows rapid vibrational relaxation of the photoexcited molecule to the lowest level of the excited state, from which emission occurs to many levels in the ground state. If vibrational relaxation is not rapid, a situation that can be encountered in small molecules with widely spaced vibrational levels, then it is possible to produce *unrelaxed or resonance fluorescence*, at the same position as the Raman peak, provided the laser is in exact resonance with one of the excited state vibrational levels (Figure 18*d*). Time resolution of the emission can distinguish the two processes,[28] since resonance fluorescence decays with the lifetime of the excited state while RR scattering is instantaneous, coinciding with the excitation pulse. Of course if the laser is tuned away from the resonant level (outside its bandwidth) the resonance fluorescence disappears, while the RR peak is still observed, at the same frequency shift. In complex molecules vibrational relaxation is generally rapid, and resonance fluorescence is rarely seen. Indeed it is common for vibrational relaxation to continue from the bottom of the excited state potential to high vibrational levels of the ground state and then to the ground vibrational level (wavy arrows in the diagram), thereby quenching the fluorescence (*radiationless relaxation*). Relatively few molecules fluoresce strongly, and this is what makes RR spectroscopy possible, since RR scattering, even though enhanced, is intrinsically much weaker than fluorescence.

High backgrounds due to fluorescence are a common frustration for Raman spectroscopists. As mentioned above, the interference can sometimes be reduced by decreasing the laser wavelength, and external heavy-atom quenching agents can sometimes be added to the sample (a KI salt is often quite effective in this regard). Coherent Raman techniques[6,10,14] and Fourier transform (FT) Raman spectroscopy in the near infrared[13] can be applied to fluorescent samples. Coherent Raman techniques are effective because the directional character of the signal lends itself to spatial filtering from the isotropic fluorescence; these techniques are technically difficult to implement however. In FT-Raman instruments the sample is irradiated with a laser line at 1064 nm, that is, in the spectral region remote from the fluorescence excitation of virtually every material for the fluorescence to interfere with the Raman signal. Often the most effective remedy is to purify the sample, since in many cases the interference is not from the intrinsic fluorescence of the molecule under study but from strongly fluorescing impurities.

3.2 Enhancement Mechanisms and Requirements for RR Scattering

The theory of the resonance Raman effect distinguishes two major intensity enhancement mechanisms that play a dominant role in the RR spectra: (1) *Franck–Condon scattering*, which involves displacement of the potential minima of the ground and excited electronic states along a vibrational normal coordinate Q_k; and (2) *Herzberg–Teller vibronic coupling*, which involves a transfer of the transition moment between different excited electronic states induced by a vibrational excitation.[29] This means that by moving into the resonance region, those fundamentals which reflect the change in geometry when converting the molecule from its ground to excited state (Franck–Condon allowed) or those which are able to vibronically couple the resonant excited state to some other electronic state with a different transition moment (Herzberg–Teller allowed) will be strongly enhanced. In general, the former effect is most pronounced for *totally symmetric modes* and *allowed resonant electronic transitions*, whereas the latter is pronounced for *non-totally symmetric modes* and *weakly allowed or forbidden resonant electronic transitions*.

To determine the scattered radiation spectrum of an oscillating molecule under conditions of resonance excitation, we again must look at the matrix elements of the polarizability transition moment $[\alpha]_k$ (12); however, this time we must consider how the polarizability varies not only with normal modes of vibration (13) but also with frequency of the incident radiation that excites them. If we consider a molecule in a molecular state $|g\rangle$ (initial) perturbed by the electromagnetic wave of frequency ν_0 so that it passes into a molecular state $|f\rangle$ (final) while scattering light of frequency $\nu_0 \pm \nu_k$ ($\nu_k = \nu_f - \nu_g$), then the matrix elements of α for the transition k, $[\alpha_{\rho\sigma}]_k$, are given by the Kramers–Heisenberg dispersion equation:[29,30]

$$[\alpha_{\rho\sigma}]_k = \langle f|(\alpha_{\rho\sigma})_k|g\rangle = \frac{1}{h}\sum_r \left(\frac{\langle f|M_\rho|r\rangle\langle r|M_\sigma|g\rangle}{\nu_r - \nu_g - \nu_0 + \mathrm{i}\Gamma_r} + \frac{\langle f|M_\sigma|r\rangle\langle r|M_\rho|g\rangle}{\nu_r - \nu_g + \nu_0 + \mathrm{i}\Gamma_r} \right) \quad (23)$$

where $\rho, \sigma = x, y, z$, which independently refer to the molecule-fixed nonrotating Cartesian coordinate system and represent the polarizations of the incident (σ) and scattered light (ρ), $|g\rangle$ and $|f\rangle$ are the initial and final states of the molecule, and the summation is over all the intermediate states $|r\rangle$. The integrals $\langle f|M_\sigma|r\rangle$ and $\langle r|M_\rho|g\rangle$ are the electric dipole transition moments, along the ρ and σ directions; M is the electron position operator; $h\nu_0$ is the energy of the incident radiation; and Γ_r is a damping factor (which prevents the denominator at resonance from reaching zero), reflecting the finite natural lifetime and sharpness of the intermediate state, $|r\rangle$.

The intermediate states, $|r\rangle$, are the electronically excited molecular eigenstates (Section 2.3), each eigenstate being weighted by an energy denominator according to its nearness to resonance. Far from resonance ($\nu_r - \nu_g \gg \nu_0$) the magnitude of $[\alpha_{\rho\sigma}]_k$ is independent of ν_0; the number of excited electronic states contributing to the polarizability is large. In this case all the energy denominators in the expression for the element $\alpha_{\rho\sigma}$ of the scattering tensor α_k are large, and hence the weighting factors are small and nonselective (normal Raman or Raman scattering). As a result, the intermediate state has no well-defined symmetry and only symmetric components of α_k, $\alpha_{\rho\sigma} = \alpha_{\sigma\rho}$, make any significant contribution to the Raman scattering. This also explains the general weakness or absence of overtones in normal Raman scattering.

The situation changes under conditions of resonance ($\nu_r - \nu_g \approx \nu_0$) with a particular electronic state, $|e\rangle$, because this state now dominates the sum over the other states in the scattering tensor α_k, provided it has a large enough transition dipole moment from the initial (ground) state. Furthermore, the first term of (23) becomes dominant and the scattered intensity is expected to increase drastically when the frequency of the incident light is tuned to the frequency of the electronic transition, ν_{eg}. (The second term of (23) is nonresonant and produces a negligible, slowly varying background contribution to the large resonant part when ν_0 is in resonance region.) One of the energy denominators in α_k becomes small, leading to a large weighting factor for the resonant eigenstate, $|e\rangle$. The intermediate state, $|r\rangle$, in the RR scattering process assumes the symmetry and geometry of this dominant excited state, often leading to overtone progressions and sometimes to contributions from antisymmetric tensor components. Also, while frequencies of the RR scattered photons are still a function of the electronic ground state, their intensities are determined by the properties of the electronic excited state.

To obtain information as to which vibrations are subject to resonance enhancement, we apply (following Albrecht[29]) the adiabatic Born–Oppenheimer approximation of separability of electronic and vibrational wavefunctions for $|g\rangle$, $|r\rangle$, and $|f\rangle$ states in (23), since the nuclei move much slower than the electrons. Thus, by taking the initial and final states to belong to the ground electronic state and the intermediate state to the resonant excited electronic state, and dropping the nonresonant term, (23) can be rewritten as:

$$[\alpha_{\rho\sigma}]_k = \frac{1}{h}\sum_{\upsilon} \frac{\langle j|(\mu_\rho)_e|\upsilon\rangle \langle \upsilon|(\mu_\sigma)_e|i\rangle}{\nu_{\upsilon i} - \nu_0 + \mathrm{i}\Gamma_\upsilon} \tag{24}$$

where $(\mu_\rho)_e = \langle g|(\mu_\rho)_e|e\rangle$ and $(\mu_\sigma)_e = \langle e|(\mu_\sigma)_e|g\rangle$ are the pure electronic transition moments, along ρ and σ directions, for the resonant excited state e; of which υ is a particular vibrational level of bandwidth Γ_υ; $\nu_{\upsilon i}$ is the transition frequency from the ground vibrational level i to the level υ; $|i\rangle$, $|j\rangle$ and $|\upsilon\rangle$ represent vibrational states of a given normal coordinate, Q_k; and the summation is over all excited state vibrational levels, υ. The dependence of $(\mu_{\rho,\sigma})_e$ on Q_k is small if the Born–Oppenheimer approximation is valid[31] and therefore it can be expanded as a Taylor series, analogously to (6) and (13):[30]

$$\mu_e = \mu_e^0 + \sum_k \mu_e' Q_k + \cdots \tag{25}$$

where $\mu_e' = (\partial\mu_e/\partial Q_k)_0$. (In writing (25) we have dropped the polarization subscripts.) When the electronic resonant transition is weakly allowed, μ_e' can be of the same magnitude as μ_e^0, or even exceed it, if the excited state $|e\rangle$ can gain absorption strength from other excited states by vibronic mixing via the coordinate Q_k (called *vibronic coupling*). In the Herzberg–Teller formalism for the vibronic coupling:[29–31]

$$\mu_e' = \frac{\mu_s^0\langle s|(\partial H_e/\partial Q_k)_0|e\rangle}{\nu_s - \nu_e} \tag{26}$$

where $|s\rangle$ is another excited state that can be mixed in to $|e\rangle$ by Q_k; ν_s and μ_s are the frequency and transition dipole moment of the mixing electronic state, $|s\rangle$; and $(\partial H_e/\partial Q_k)_0$ is the vibronic coupling operator that connects two excited states $|e\rangle$ and $|s\rangle$, with H_e being the Hamiltonian for the total electronic energy of the molecule. Thus, the stronger the transition to the state $|s\rangle$, and the closer it is in energy to $|e\rangle$, the larger μ_e' will be.

Substitution of (26) and (25) into (24) yields an expression with many terms, of which the first three terms, called A-, B-, and C-term, are the dominant mechanisms in resonance Raman scattering.

$$[\alpha_k] = A + B + C + \cdots \tag{27}$$

where

$$A = (\mu_e^0)^2 h^{-1} \sum_{\upsilon} F_\upsilon(\nu_{\upsilon i} - \nu_0 + \mathrm{i}\Gamma_\upsilon)^{-1}; \quad F_\upsilon = \langle j|\upsilon\rangle \langle \upsilon|i\rangle \tag{28}$$

$$B = \mu_e^0 \mu_e' h^{-1} \sum_{\upsilon} F_\upsilon' (\nu_{\upsilon i} - \nu_0 + \mathrm{i}\Gamma_\upsilon)^{-1};$$

$$F_\upsilon' = \langle j|Q_k|\upsilon\rangle \langle \upsilon|i\rangle + \langle j|\upsilon\rangle \langle \upsilon|Q_k|i\rangle \tag{29}$$

$$C = (\mu_e')^2 h^{-1} \sum_{\upsilon} F_\upsilon'' (\nu_{\upsilon i} - \nu_0 + \mathrm{i}\Gamma_\upsilon)^{-1}; \quad F_\upsilon'' = \langle j|Q_k|\upsilon\rangle \langle \upsilon|Q_k|i\rangle \tag{30}$$

Each term of (28–30) is factored into parts that can be related to the electronic and vibrational wavefunctions and excitation frequency. The electronic contribution (matrix elements of μ_e^0 and μ_e'), weighted with the differential energy denominators $(\nu_{\upsilon i} - \nu_0 + \mathrm{i}\Gamma_\upsilon)$, determines the *total* enhancement of all the normal vibrations for a given electronic transition and a given ν_0. Since μ_e^0 and μ_e' in α_k also determine the optical absorption spectrum, structural information often emerges from the interpretation of both the RR and UV-visible spectra. The vibrational part described by Franck–Condon factors and $(\partial H_e/\partial Q_k)_0$ (vibronic coupling operator) governs the intensity distribution among the enhanced modes, and is thus responsible for the selectivity of the vibrational mode enhancement. The enhanced modes contain information about the vibronic nature of the electronic transitions as well as the molecular distortion and dynamics at the electronic excited state(s).

The A-term is the leading RR scattering mechanism encountered in practice and involves vibrational interactions with a single excited electronic state, $|e\rangle$, by way of Franck–Condon overlap integrals, $\langle j|\upsilon\rangle$ and $\langle \upsilon|i\rangle$ (FC scattering); in order for this term not to vanish it follows from (28) that both $\mu_e^0 \neq 0$ and $F_\upsilon \neq 0$. Consequently, the A-term is most pronounced for strongly allowed electronic transitions, with large values of $(\mu_e^0)^2$ and substantial nonorthogonalities of the FC factors $\langle \upsilon|i\rangle$ and $\langle j|\upsilon\rangle$ upon electronic excitation, such as electric-dipole allowed σ–σ^*, π–π^*, and charge-transfer transitions. Excitations into a weak electronic transitions, such as forbidden π–π^*, ligand field d–d, and spin forbidden transitions, will not produce significant Franck–Condon scattering.

The integrals $\langle \upsilon|i\rangle$ and $\langle j|\upsilon\rangle$ are overlap integrals between the vibrational wavefunctions of the vibrational level υ of the resonant excited state e with the initial and final vibrational levels i and j, which are the 0 and 1 vibrational levels of the ground electronic state for the fundamental Raman transitions. They become nonzero if there is a difference in the vibrational frequency between ground (g) and excited (e) electronic states (i.e., $\nu_k^e - \nu_k^g \neq 0$) and, more importantly, a shift in the excited state potential along the vibrational coordinate, $\Delta_k^e = Q_k^e - Q_k^g$, as illustrated in Figure 20. For negligible change in vibrational frequency (i.e., $\nu_k^e \approx \nu_k^g = \nu_k$), the Franck–Condon factors for each of the $k = 3N - 6$ ($3N - 5$ for linear molecules) normal modes of vibration can be shown[32] to be only a function of their displacements, Δ_k^e, in the excited electronic state, with the property that:

$$\lim_{\Delta_k^e \to 0} \left[\langle j|\upsilon\rangle \langle \upsilon|i\rangle\right] = 0 \tag{31}$$

The result of this relation is that, if $\Delta_k^e = 0$, then according to (28) the corresponding frequency ν_k will be absent in the RR spectrum, and vice versa; if Δ_k^e is large,

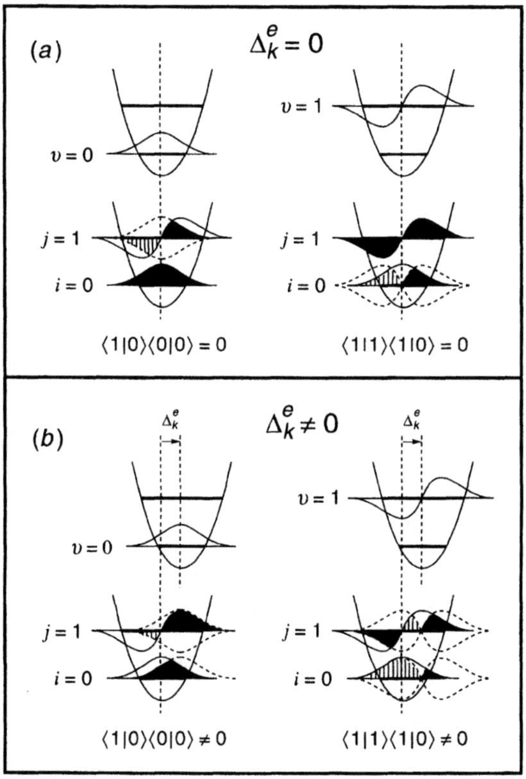

Figure 20. Schematic representation of the Franck–Condon overlap products (opposite signs indicated by different shading patterns) involved in A-term RR scattering of a vibrational fundamental, from the ground vibrational level ($i = 0,\ j = 0$) via the first two vibrational levels of the excited state ($\upsilon = 0,\ 1$). If there is no shift of the excited state potential (a), then $\langle 0|0\rangle = \langle 1|1\rangle = 1$, because of exact overlap, but $\langle 1|0\rangle = 0$ because the $\upsilon = 0$ and 1 wavefunctions are orthogonal and positive and negative overlaps cancel. If there is a shift (b) this cancellation is no longer exact and $\langle 0|0\rangle \neq 0$, so that the product is nonzero.

then this vibration will appear strongly in the RR spectrum (often forming a progression of the corresponding overtones). In general, totally symmetric modes have substantial displacements Δ_k^e and $\nu_k^e \neq \nu_k^g$, whereas by symmetry non-totally symmetric modes can only have different frequencies. Therefore nonorthogonality of the vibrational wavefunctions in the ground (initial and final) state and an excited state is much stronger for totally symmetric than for non-totally symmetric modes, and the Franck–Condon principle yields much higher Raman intensities for the former than for the latter.[33,34]

On the other hand, B-term scattering arises from the vibronic coupling of the resonant state, $|e\rangle$, to other excited state, $|s\rangle$, of the appropriate symmetry (HT scattering). Because the magnitude of μ'_e increases with the oscillator strength of μ_s^0 and decreasing frequency separation between the $|e\rangle$ and $|s\rangle$ states (26), the B-term

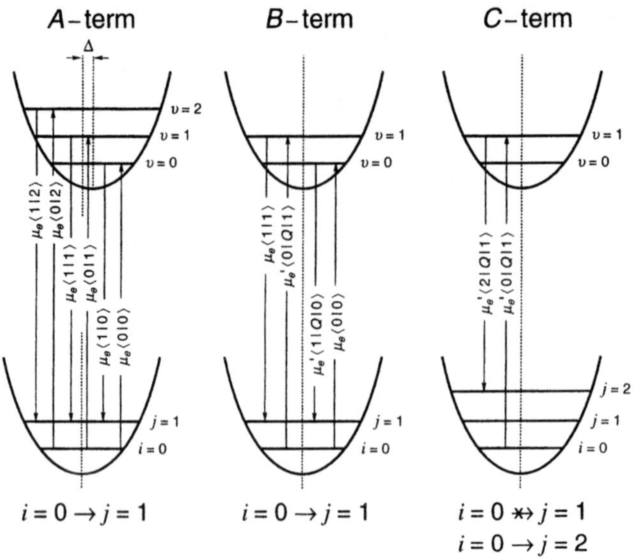

Figure 21. Diagrammatic representation of Raman processes responsible for *A*-, *B*-, and *C*-term resonance scattering for fundamentals (*A* and *B*) and overtones (*C*). The applicable electronic transition moment and vibrational overlap integral are indicated for each contributing transition for $\upsilon = 0,\ 1$, and 2 intermediate levels of the resonant excited state.

scattering is most pronounced when exciting frequency is tuned to a weakly allowed electronic transition that is mixed vibronically with a nearby strong one ($\mu_e' > \mu_e^0$) in this circumstance). If these two states are of the same symmetry, they may be coupled via the totally symmetric modes (the allowed symmetry of the active vibration is contained in the direct product representations of the two electronic states, $\Gamma_e \times \Gamma_s$). In such a case the *B*-term may produce resonance enhancement of totally symmetric vibrations if the FC contribution (*A*-term) to these modes is small.[34] (*A*-term scattering may be small if $\Delta_k^e \approx 0$ upon excitation, leading to an $F_\upsilon \approx 0$.)

The Herzberg–Teller vibronic coupling mechanism, however, is a prime cause of RR activity for the non-totally symmetric modes. This is because, in nondegenerate states, the Franck–Condon (*A*-term) contribution to their Raman intensity will vanish since $\Delta_k^e = 0$ by symmetry for such modes ($F_\upsilon = 0$), while the Q_k-dependent function F_υ' in the *B*-term does not vanish even if $\Delta_k^e = 0$; in this circumstance, the matrix elements of Q_k, $\langle j|Q_k|\upsilon\rangle \sim \delta(j, \upsilon - 1) \neq 0$ and $\langle \upsilon|Q_k|i\rangle \sim \delta(\upsilon, i + 1) \neq 0$ (selection rules for harmonic oscillator), as well as the diagonal FC overlaps, $\langle \upsilon|i\rangle \sim \delta(\upsilon, i) = 1$ and $\langle j|\upsilon\rangle \sim \delta(j, \upsilon) = 1$, survive, as illustrated diagrammatically in Figure 21. Consequently, $F_\upsilon' \neq 0$ in (29) and the *B*-term > 0 for non-totally symmetric vibrations, even if there is no excited state shift of the potential upon electronic excitation. Complications may arise in the case of degenerate excited electronic states, due to the Ranner–Teller and Jahn–Teller effects of potential deformation and level splitting along an appropriate non-totally symmetric coordinate.[34] These effects may lead to significant Franck–Condon overlaps and an *A*-term contribution, which could be comparable to that of the *B*-term.

Nevertheless, striking enhancements can be, and usually are observed for non-totally symmetric modes which are effective in mixing two electronic states of different symmetry or two different components of two degenerate states of the same symmetry. The more efficient the vibronic coupling of the two excited electronic states, the more effectively B-term enhancement can be observed. Moreover, under non-resonance conditions, when the energy denominator $(\nu_{\upsilon i} - \nu_0 + i\Gamma_\upsilon)$ can be factored out, the sum over F'_υ reduces to $\langle j|Q_k|i\rangle$, which is nonzero for $j = i \pm 1$, i.e., fundamental modes. Thus off-resonance Raman scattering is preserved by the B-term (although under these conditions any specific resonance effects are lost in the sum over all excited states of (23)).

This selective enhancement is one of the most important and valuable aspects of the RR spectroscopy, since it leads to a considerable simplification of the observed spectra; the RR spectra consist primarily of bands arising from either totally or non-totally symmetric fundamentals depending upon the nature of the resonant electronic transitions.

Let us now apply this vibrational selectivity to an actual example.

3.3 An Example: *A*- and *B*-Term RR Enhancement Pattern in NiEtio-III

Resonance Raman scattering from metalloporphyrins, whose low-lying intense π–π^* electronic transitions are conveniently excited with visible lasers, provides an excellent example of the differential A- and B-term vibrational enhancement pattern. Their first two excited states are well described by Gouterman's four-orbital model,[35] presented in Figure 22, along with the electronic absorption spectrum of nickel(II) etioporphyrin-III (NiEtio-III)[36] which occurs naturally in crude oils and marine sediments. The lowest-lying unoccupied π^* orbitals are degenerate having e_g symmetry (in the idealized D_{4h} point group), while the two highest filled π orbitals of a_{1u} and a_{2u} symmetries are nearly degenerate. Consequently, the $a_{1u} \rightarrow e_g^*$ and $a_{2u} \rightarrow e_g^*$ excitations (both of $a_{1u} \times e_g = a_{2u} \times e_g = E_u$ symmetry) undergo strong configuration interactions resulting in well separated states, with energies in the violet and yellow-green regions of the spectrum. The transition dipole moments add up for the higher energy transition to produce the very strong absorption B (or Soret) band at 391 nm, but nearly cancel for the lower energy transition leading to a much weaker Q_0 (or α) band at 551 nm. There is also a composite side band at 515 nm that results from vibronic mixing between the Q_0 and B transitions, referred to as the Q_1 or β band. Its maximum vibronic intensity is shifted $\approx$1300 cm^{-1} above Q_0, which corresponds to an approximate mean frequency of the most effective mixing vibrations of the porphyrin skeleton. As a result, metalloporphyrin RR spectra display striking differences in the vibrational enhancement patterns when the laser excitation wavelength falls in one of these absorption bands, as briefly discussed below.

Figure 23 shows the parallel ($\|$) and perpendicular ($\perp$) scattered RR spectra from NiEtio-III in CH_2Cl_2 solution obtained with excitation wavelength at 406.7 nm, while Figure 24 shows the spectra of the same solution excited at 530.9 (top) and 568.2 nm (bottom), in near-resonance with the Soret and Q absorption bands, respectively.[36] Similar to other metalloporphyrins, the in-plane stretching and deformation

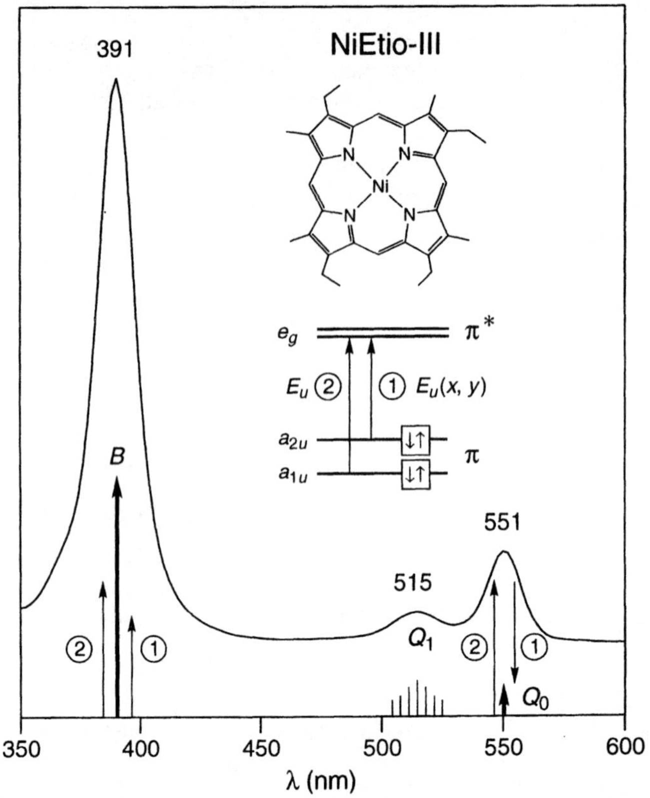

Figure 22. Absorption spectrum of NiEtio-III (Etio-III = etioporphyrin-III) in CH_2Cl_2 and its interpretation via Gouterman's four orbital model. The $e_g \leftarrow a_{2u}, a_{1u}$ orbital excitations (labeled 1 and 2), being of the same symmetry (E_u) and nearly degenerate, interact strongly, the transition dipoles adding up for the intense B transition at 391 nm, and nearly canceling for the weaker Q_0 transition at 551 nm. About 10% of the B band intensity is borrowed by the Q transition producing the Q_1 vibronic side band 515 nm, above Q_0 by ≈ 1300 cm^{-1}, the average frequency of the vibronically effective modes.

vibrations of the Etio macrocycle (labeled ν_i) exhibit the strongest increase in the intensity; this is due to the interactions with the π–π^* resonant transitions[26,27] that are also polarized in the porphyrin plane. Table 2 gives the in-plane skeletal mode frequencies and assignments.[36]

The porphyrin chromophore multiple resonant state conditions lead to further differentiation within the set of the porphyrin skeletal modes. Different bands are enhanced depending upon whether the excitation wavelength approaches those of the visible (α and β) or near-UV (Soret) bands. Excitation in the vicinity of the Soret absorption band (406.7 nm excitation, Figure 23) produces spectra that are dominated by polarized (p) peaks arising from the totally symmetric vibrations, A_{1g} (labeled ν_{2-9}). (The depolarization ratio, $\rho = I_\perp/I_\parallel$, is close to the expected value, 1/8, for

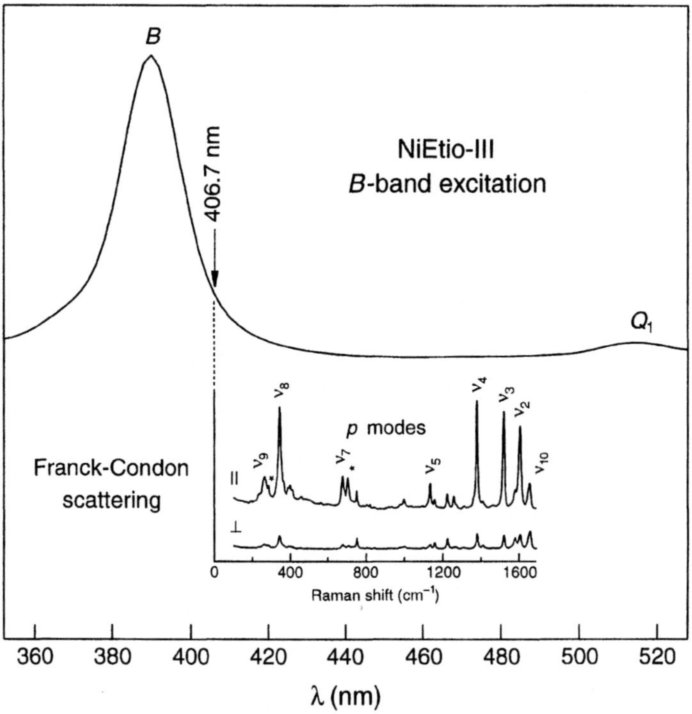

Figure 23. Soret-resonant excitation and Franck–Condon (A-term) scattering in metalloporphyrins. The Soret absorption band (B) is shown for NiEtio-III (CH_2Cl_2 solution) together with the RR spectra in parallel (∥) and perpendicular (⊥) scattering excited at 406.7 nm. Resonance with the Soret absorption band enhances mainly polarized bands (p), arising from totally symmetric modes, A_{1g}. The asterisks indicate solvent bands (CH_2Cl_2).

resonance with in-plane electronic transitions in D_{4h} symmetry.) Their dominance reflects that only totally symmetric vibrations, that is, those capable of shifting the excited state potential surface, are subject to the Franck–Condon (A-term) enhancement mechanism for allowed electronic transitions (28). Structure sensitive modes ν_2, ν_3, ν_4, and ν_8 are especially strong, indicating large geometric displacements in the excited state along these vibrational coordinates. Since the Soret excited state is degenerate (E_u), several Jahn–Teller active modes, those of B_{1g} and B_{2g} symmetries, also appear in the 406.7 nm spectrum as weak depolarized (dp) bands, with $\rho \approx 3/4$ (see for example ν_{10}).[37]

Because the α absorption band shows moderate strength (Figure 22), polarized peaks are also enhanced via FC scattering in the Q-resonant spectra (530.9 and 568.2 nm excitations, Figure 24), albeit their intensity pattern is different from that of the Soret-resonant spectrum (Figure 23). Thus, the high-frequency bands arising from ν_2, ν_3, and ν_4 modes dominate the 406.7 nm-excited spectrum but are absent or have negligible intensity in the 530.9 and 568.2 nm-excited spectra, where

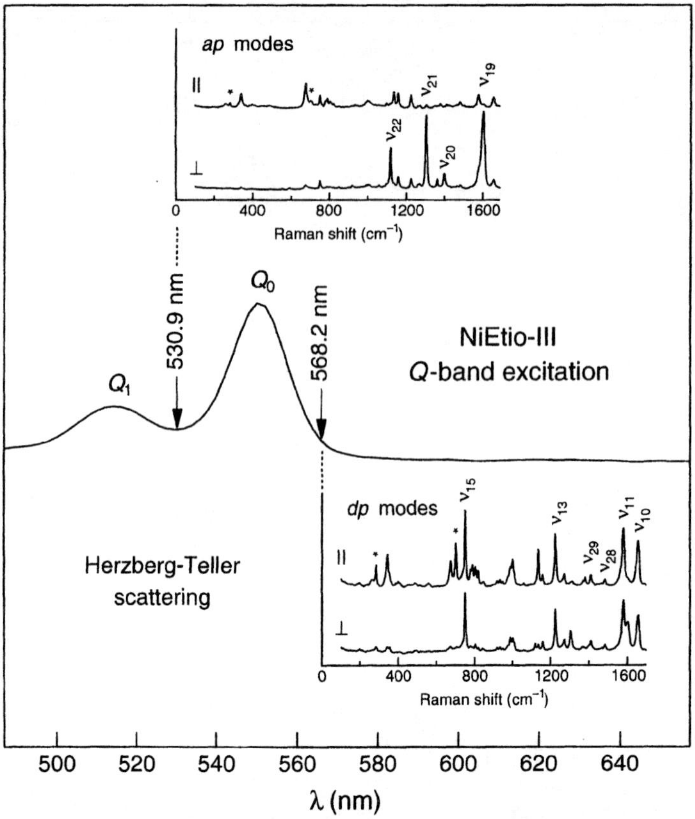

Figure 24. Q-resonant excitation and Herzberg–Teller (B-term) scattering in metalloporphyrins. The Q absorption bands (Q_0 and Q_1) are shown for NiEtio-III (CH_2Cl_2 solution) together with the RR spectra in parallel (||) and perpendicular (⊥) scattering excited at 530.9 (*top*) and 568.2 nm (*bottom*). Resonance in the Q band region enhances depolarized (B_{1g}, B_{2g}) and anomalously polarized (A_{2g}) modes that are effective in vibronic mixing of the Q and Soret electronic transitions. Due to interference effects the A_{2g} modes are brought out most strongly via excitation between Q_0 and Q_1 (530.9 nm), while the B_{1g} and B_{2g} modes are brought out more strongly with excitation outside the Q bands (568.2 nm). The asterisks indicate solvent bands (CH_2Cl_2).

ν_5, ν_7, and ν_8 are the strongest polarized peaks. This difference implies quite different shapes for the Soret and Q excited state potentials.

When the laser wavelength is near resonance with the α band (568.2 nm, Figure 24), however, the spectrum is dominated by the B-term (Herzberg–Teller scattering) enhancement via α-Soret vibronic coupling (which is also responsible for the β absorption sideband). Depolarized Raman bands arising from B_{1g} (labeled ν_{10-17}) and B_{2g} (labeled ν_{28-34}) non-totally symmetric vibrations are dominant. The vibronic coupling strength of the B_{1g} vibrations is greater than B_{2g}, and the B_{1g} RR bands are more intense by factors of 5–20.[36]

Table 2 Allocation of NiEtio-III In-Plane Skeletal RR Frequencies (cm^{-1}) to Local Coordinates[a]

Local Coordinate[b]	A_{1g} (p)[c]		B_{1g} (dp)		A_{2g} (ap)		B_{2g} (dp)	
$\nu(C_mH)$	ν_1	3041[d]					ν_{27}	3040[d]
$\nu(C_\alpha C_m)_{asym}$			ν_{10}	1656	ν_{19}	1603		
$\nu(C_\beta C_\beta)$	ν_2	1605	ν_{11}	1579				
$\nu(C_\alpha C_m)_{sym}$	ν_3	1520					ν_{28}	1483
ν(Pyr. quarter-ring)					ν_{20}	1399	ν_{29}	1411
ν(Pyr. half-ring)$_{sym}$	ν_4	1380	ν_{12}	1324[e]				
$\delta(C_mH)$			ν_{13}	1226	ν_{21}	1304		
$\nu(C_\beta C_1)_{sym}$	ν_5	821	ν_{14}	1129				
ν(Pyr. half-ring)$_{asym}$					ν_{22}	1119	ν_{30}	1159
$\nu(C_\beta C_1)_{asym}$					ν_{23}	1060	ν_{31}	1009
δ(Pyr. def.)$_{asym}$					ν_{24}	591	ν_{32}	948
ν(Pyr. breathing)	ν_6	675	ν_{15}	751				
δ(Pyr. def.)$_{sym}$	ν_7	731	ν_{16}	753				
δ(Pyr. rot.)					ν_{25}	553	ν_{33}	490
ν(NiN)	ν_8	342	ν_{18}	168[f]				
$\delta(C_\beta C_1)_{asym}$					ν_{26}	243[d]	ν_{34}	197
$\delta(C_\beta C_1)_{sym}$	ν_9	260	ν_{17}	305[f]				
δ(Pyr. transl.)							ν_{35}	144[f]

[a] Mode frequencies from CS_2 solution RR spectra at room temperature.[36]
[b] See Figure 50 (Sec. 4.5) for local mode definitions.[39]
[c] P = polarized, dp = depolarized, and ap = anomalously polarized RR bands.
[d] Calculated freguency for NiOEP; not observed.[38]
[e] Observed only in the meso-d_4 isotopomer.[38]
[f] Values from NiOEP RR spectra.[40]

The A_{2g} vibrations, normally Raman inactive, are enhanced as well at 568.2 nm (see for example ν_{19}, ν_{20}, and ν_{21}); they are Herzberg–Teller active since $E_u \times E_u = A_{1g} + A_{2g} + B_{1g} + B_{2g}$. They also have antisymmetric scattering tensors, $\alpha_{xy} = -\alpha_{yx}$, leading to unique *anomalous polarization* (ap) that is characterized by greater intensity in the perpendicular ($\perp$) than in the parallel ($\|$) scattering component ($\rho \gg 3/4$) (Section 2.4). The A_{2g} modes have rotational symmetry, e.g., they involve alternating stretching and contraction of equivalent bonds around the ring (Figure 25), and therefore they can rotate the transition moments, giving an antisymmetric tensor. The *ap* bands become predominate when the excitation wavelength is moved to 530.9 nm, in between the α and β absorptions (Figure 24, top). This additional enhancement selectivity among the $A_{2g}/B_{1g}(B_{2g})$ non-totally symmetric modes occurs because the Q-band resonance is also subject to an interference effect between α and β vibronic excitations that produces frequency dependent contributions to the Raman scattering tensor.[26] The two terms in F'_v of the B-term (29) add for B_{1g} and B_{2g} modes, but subtract for A_{2g} modes, because of the tensor symmetries. As a result, the interference is constructive for A_{2g} but destructive for B_{1g} (B_{2g}) modes at excitation wavelengths falling in between the two transitions (where the energy denominators for the α and β reso-

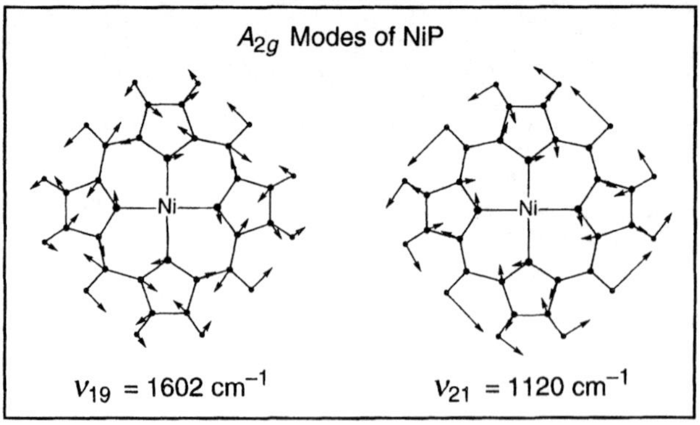

Figure 25. Eigenvectors of the A_{2g} modes, ν_{19} and ν_{21}, of NiP (P = porphine),[39] which can rotate the transition moments, giving antisymmetric scattering tensors; note, e.g., a counterclockwise motion of the H atoms but clockwise motions of the N, C_α, and C_β atoms in ν_{21}.

nances are opposite in sign), while the opposite happens for wavelengths above β or below α.[33]

Vibronic activity is also expected in the Soret-resonant RR spectra (due to Soret–Q mixing) but is quantitatively much less important than the Franck–Condon activity because of the very large Soret absorption strength. The insets of Figure 26 show that *ap* bands (which are activated vibronically but not via the Jahn–Teller effect, in contrast with the *dp* bands which can be activated by both mechanisms) can be detected at high magnification in Soret-resonant spectra.[38]

The mode selectivity afforded by wavelength tuning is extremely useful in sorting out the numerous overlapping metalloporphyrin vibrational modes, especially when combined with isotopic substitution. Figure 27 illustrates how five modes which are crowded into a 16 cm^{-1} region of NiTPP (TPP = tetraphenylporphyrin) RR spectrum can be disentangled in this way.[39]

As a result of the different RR scattering mechanisms and extensive isotope frequency shifts, it has been possible to obtain an unusually complete assignment of the porphyrin vibrational modes, and to derive a reliable force field.[38–40] These assignments are strongly supported by recent density functional theory-scaled quantum mechanical (DFT-SQM) force field calculations.[41]

3.4 *C*-Term RR Scattering, Overtones, and Combination Bands

C-term RR scattering is a dominant mechanism when the exciting frequency is in resonance with a vibronic side band of a forbidden or weakly allowed 0–0 electronic transition, where the side band is allowed ($\mu_s^0 \neq 0$) due to the vibronic coupling between the 0–0 transition and another strongly allowed transition ($|g\rangle \rightarrow |s\rangle$). Under such conditions, the vibronic transition moment μ_e' described by (26) is much greater than μ_e^0 and since the electronic factor is $(\mu_e')^2$ in (30), the C-term can out-

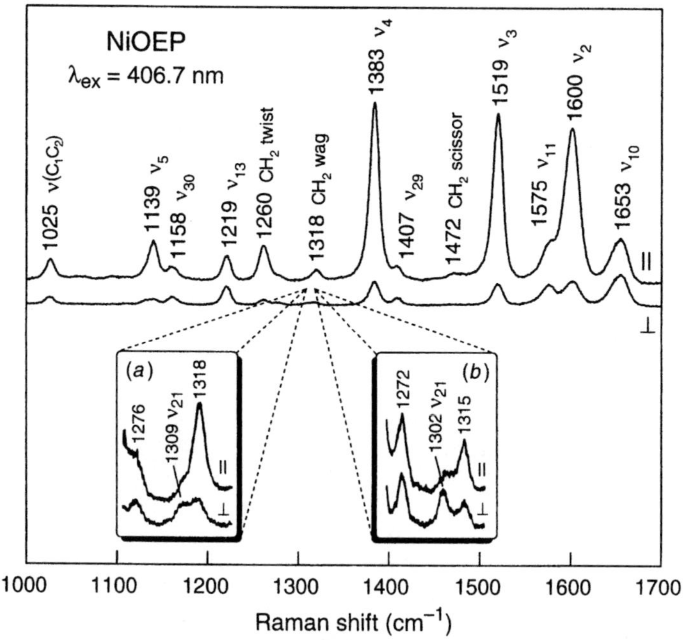

Figure 26. The Soret-resonant scattering from NiOEP (OEP = octaethylporphyrin) in CH_2Cl_2 excited at 406.7 nm. Insets are exploded views of the 1250–1350 cm^{-1} region in CH_2Cl_2 solution (*a*) and tetragonal crystals (*b*) showing enhancement of the *ap* ν_{21} (A_{2g}) skeletal mode via Soret–Q vibronic coupling.[38]

weigh the B-term. The special feature of the C-term RR scattering is that it gives rise to first overtones ($k = l$) and binary combination tones ($k \neq l$) of those vibrational modes that are effective in coupling the two electronic states, instead of the fundamentals. This is because F''_{υ} function of the C-term (30) contains products of two Q-dependent integrals, $\langle j|Q_k|\upsilon\rangle$ and $\langle \upsilon|Q_l|i\rangle$. Since each of these connects vibrational states differing by one quantum, the final state, $|j\rangle$, must differ from $|i\rangle$ by two quanta, i.e., $j = i \pm 2$, as diagrammatically shown in Figure 21. Consequently, only first overtones ($k = l$) and binary combinations ($k \neq l$) are enhanced by the C-term. An example is shown in Figure 28, where overtone and combination bands involving the high-frequency *ap* and *dp* modes $\nu_{19}(A_{2g})$, $\nu_{10}(B_{1g})$, $\nu_{11}(B_{1g})$, $\nu_{20}(A_{2g})$, and $\nu_{29}(B_{2g})$ in the β-resonant spectrum of vanadyl tetraphenylporphyrin, (VO)TPP, are seen to be dramatically stronger than the fundamentals.[27,42] The α absorption band shows no discernible intensity in this case (Figure 28*a*), indicating accidental degeneracy of the a_{1u} and a_{2u} porphyrin π orbitals, while the β band retains $\approx$10% of the intensity of the Soret band. As a result, $\mu_e^0 \approx 0$ but $\mu_e' \neq 0$ and $B \ll C$. The C-term mechanism also manifests itself in the β-excited RR spectrum of vanadyl octaethylporphyrin, (VO)OEP (Figure 29), despite the fact that (VO)OEP exhibits classical metalloporphyrin UV-visible spectrum (see Figure 22) with the α absorption band being markedly stronger than the β band (Figure 29*a*). Overtone and combination

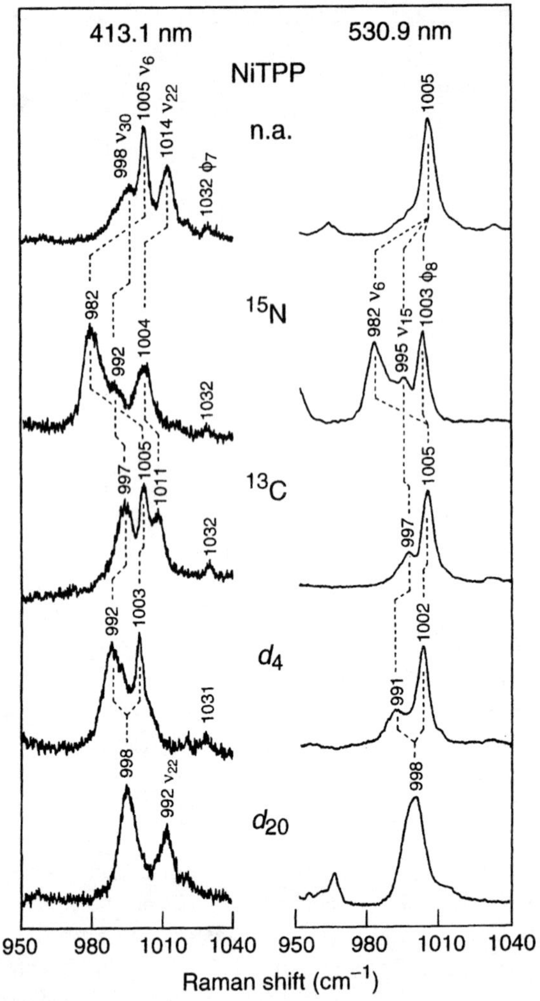

Figure 27. Details of the 950–1050 cm^{-1} RR spectra of NiTPP (TPP = tetraphenylporphyrin) in CS_2 excited at 413.1 nm (*left panel*) and 530.9 nm (*right panel*). The five closely spaced modes, ν_6, ν_{15}, ν_{22}, ν_{30}, and ϕ_8 can be distinguished by isotope substitution, combined with variable wavelength excitation.[39]

bands involving the OEP $\nu_{10}(B_{1g})$ and $\nu_{19}(A_{2g})$ modes are seen to have comparable strengths to those of the fundamentals. However, unlike the *ap* and *dp* RR fundamentals of (VO)TPP, which are very weak or missing (Figure 28), those of (VO)OEP are intense (Figure 29) because both $\mu_e^0 \neq 0$ and $\mu_e' \neq 0$ for (VO)OEP; consequently, the RR scattered intensity for β-band excitation of (VO)OEP is dominated by the B-term contribution in the fundamental region but by the C-term contribution in the overtone region. An extreme case of C-term dominance is found in benzene,[43] whose first two electronic transitions, B_{2u} and B_{1u}, are forbidden by symmetry. However,

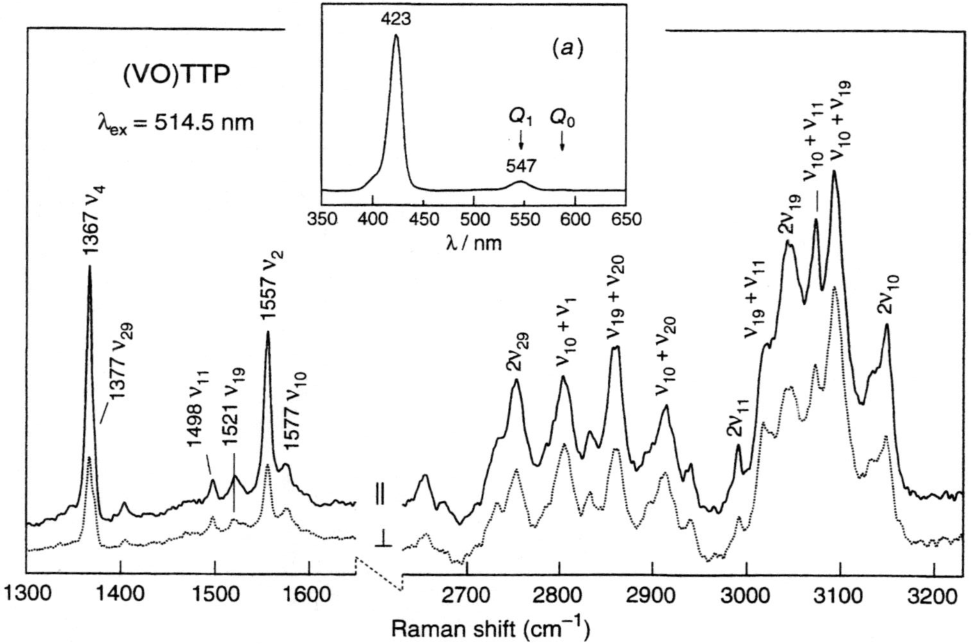

Figure 28. The RR spectrum of solid (VO)TPP (rotating KCl pellet[23]) with 514.5 nm (Q_1) excitation showing strong enhancement of overtone and combination modes involving the vibronically active ν_{19}, ν_{10}, ν_{11}, and ν_{29}, which themselves are very weak. The Q_0 absorption band (*a*) is virtually invisible in this case, giving very large overtone/fundamental ratios via the much larger *C*-term than *B*-term scattering.

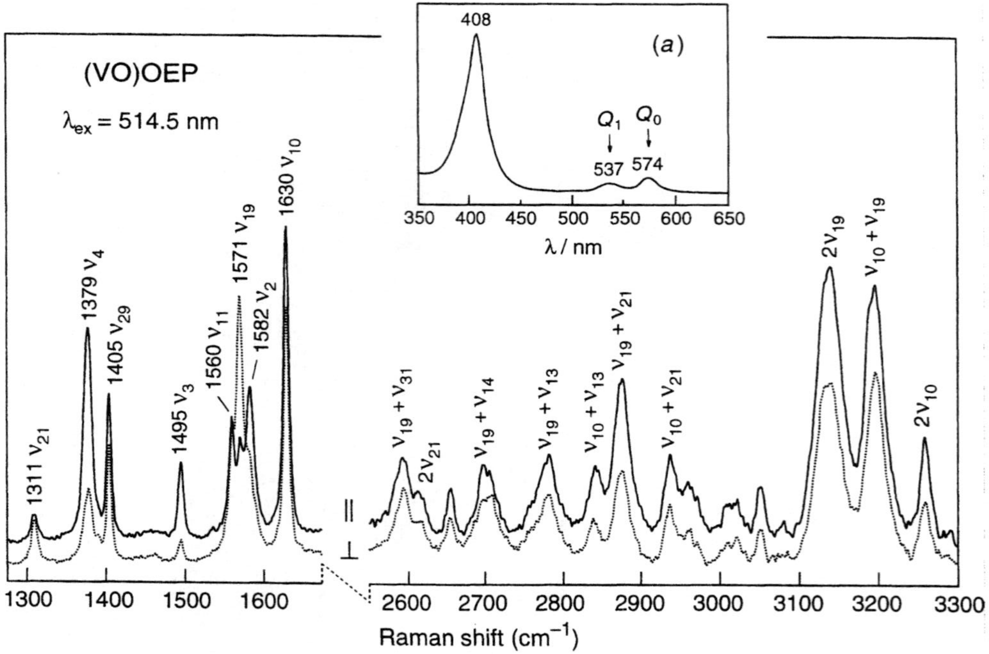

Figure 29. The RR spectrum of solid (VO)OEP (rotating KCl pellet[23]) with 514.5 nm (Q_1) excitation showing strong enhancement of the vibronically active fundamentals ν_{19}, ν_{10}, ν_{11}, and ν_{29}, as well as their first overtones and binary combination modes. The Q_0 absorption band (*a*) is stronger than Q_1 in this case, giving similar overtone and fundamental intensities via the comparable contributions from the *C*-(overtones) and *B*-term (fundamentals) scattering at near-Q_1 excitation.

the B_{1u} transition is strongly mixed vibronically with the nearby allowed transition, E_{1u}. Overtones of the mixing modes are observed in B_{1u}-resonant RR spectra, but the fundamentals are missing since μ_e^0, and therefore the B-term, is zero by symmetry.[43]

Overtones can also be enhanced by the A-term, via the Franck–Condon factors, when in resonance with strongly allowed transitions. In fact, Franck–Condon overlap integrals in (28) may possess appreciable magnitudes for up to several quanta of j if Δ_k^e is sufficiently large in the excited state. Intense overtone progression $n_k \nu_k$ of totally symmetric modes ν_k can then be seen in the RR spectrum. Clark and coworkers have carried out extensive work in this area on coordination compounds.[44,45] The longest progressions observed thus far are for the diatomic I_2 (up to $n_1 = 25$), the linear-chain $\{[Pt(tn)_2][Pt(tn)_2Br_2]\}^{4-}$ (up to $n_1 = 18$), tetrahedral SnI_4 (up to $n_1 = 15$), and μ-oxo-bridged $[Ru_2OCl_{10}]^{4-}$ (up to $n_1 = 12$) for which the respective $\nu_1 < 400$ cm^{-1}.[44,45]

Figure 30 is a scan of 488.0-nm excited RR spectrum out to 7000 cm^{-1} for a solid "red" form of bis(dithiooxalato)nickel(II), $K_2[Ni(dto)_2]$, in resonance with the Ni(II) $\rightarrow$ $S_2C_2O_2^{2-}$ metal-to-ligand charge-transfer (MLCT) transition.[46] Two progressions are seen: $n_1\nu_1$ extending to $n_1 = 6$, and $n_1\nu_1 + \nu_2$ extending to $n_1 = 4$, where $\nu_1 = 1085$ cm^{-1} is a combination of CC and CS stretching coordinates and $\nu_2 = 1602$ cm^{-1} is a combination of CO and CC stretching coordinates. The wavenumbers of each member of these progressions are given in Table 3. This spectrum is unusual in the high fundamental frequencies seen in the overtone progres-

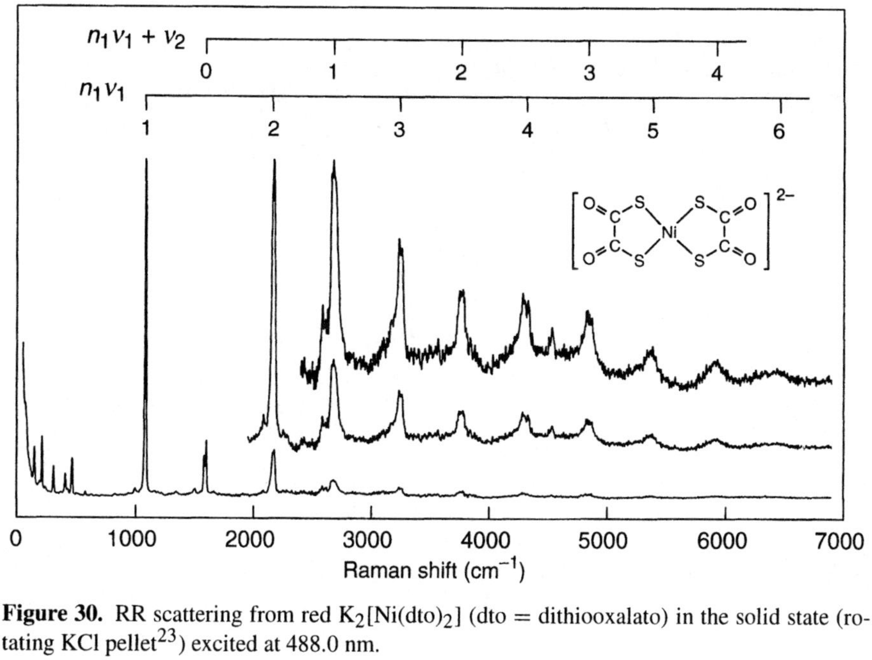

Figure 30. RR scattering from red $K_2[Ni(dto)_2]$ (dto = dithiooxalato) in the solid state (rotating KCl pellet[23]) excited at 488.0 nm.

Table 3 Wavenumbers of the $n_1\nu_1$ and $n_1\nu_1 + \nu_2$ Progressions in the Resonance Raman Spectrum of $K_2[Ni(dto)_2]$ (488.0-nm Excitation)[a]

$n_1\nu_1$ progression		$n_1\nu_1 + \nu_2$ progression	
band	$\tilde{\nu}$, cm^{-1}	band	$\tilde{\nu}$, cm^{-1}
ν_1	1085	$\nu_1 + \nu_2$	2685
$2\nu_1$	2167	$2\nu_1 + \nu_2$	3766
$3\nu_1$	3245	$3\nu_1 + \nu_2$	4842
$4\nu_1$	4318	$4\nu_1 + \nu_2$	5924
$5\nu_1$	5390		
$6\nu_1$	6460		

[a]Data from Czernuszewicz et al.[46]

sions. Most MLCT-resonant RR spectra of other planar four-coordinate complexes lack bands attributable to overtones.[45] Clearly, the NiS_2C_2 ring of $K_2[Ni(dto)_2]$ expands significantly upon MLCT electronic excitation.

The observation of a number of overtones of the ν_1 fundamental in the RR spectrum of $K_2[Ni(dto)_2]$ makes it possible to determine the harmonic wavenumber, ω_1, and anharmonicity constant, X_{11}. In general, the observed wavenumber $\nu(n_1)$, of any overtone of the ν_1 fundamental, is given by the expression[1]

$$\nu(n_1) = G(n_1) - G(0) = n_1\omega_1 + X_{11}\big(n_1^2 + n_1 d_1\big) + \cdots \tag{32}$$

where $G(n_1)$ is the term value of the n_1th vibrational level, and n_1 and d_1 are the vibrational quantum number and degeneracy, respectively, of this fundamental. With the simplification that $d_1 = 1$ (since ν_1 is totally symmetric and consequently nondegenerate), a plot of $\nu(n_1)/n_1$ vs. n_1 will then give a straight line with a slope equal to X_{11} and an intercept equal to $\omega_1 + X_{11}$.

It is also possible to analyze a progression of the type $n_1\nu_1 + \nu_2$ (i.e., the progression in a totally symmetric fundamental, ν_1, that is based on one quantum of another fundamental, ν_2), using the expression[1]

$$\begin{aligned}\nu(n_1\nu_1 + \nu_2) &= G(n_1, n_2) - G(0, n_2 - 1)\\ &= n_1\omega_1 + \omega_2 + X_{11}\big(n_1^2 + n_1 d_1\big) + X_{22}(2n_2 - 1 + d_2)\\ &\quad + X_{12}\big[n_1 n_2 + n_1(d_1/2) + (d_2/2)\big] + \cdots\end{aligned} \tag{33}$$

Since the observed wavenumber of the ν_2 fundamental is given by the expression

$$\begin{aligned}\nu_2 &= G(n_1, n_2) - G(n_1, n_2 - 1)\\ &= \omega_2 + X_{22}(2n_2 - 1 + d_2) + X_{12}\big[n_1 + (d_1/2)\big] + \cdots\end{aligned} \tag{34}$$

we obtain the following equation:

$$\begin{aligned}&\nu(n_1\nu_1 + \nu_2) - \nu_2\\ &= n_1\omega_1 + X_{11}\big(n_1^2 + n_1 d_1\big) + n_1 X_{12}\big[n_2 - 1 + (d_2/2)\big] + \cdots\end{aligned} \tag{35}$$

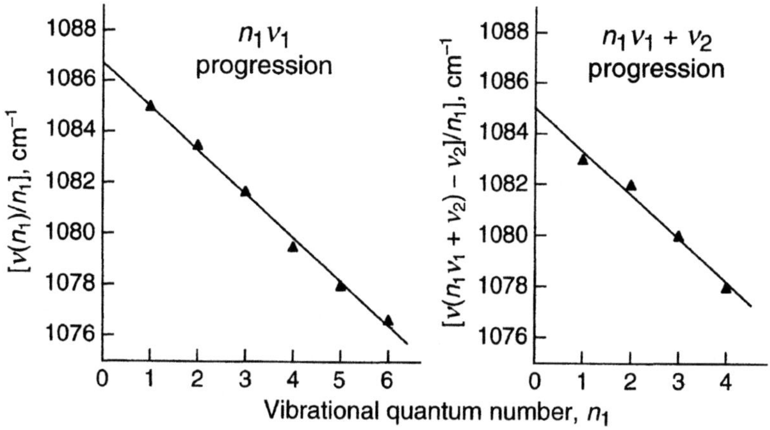

Figure 31. Plots of $\nu(n_1)/n_1$ versus n_1 for $n_1\nu_1$ and $n_1\nu_1 + \nu_2$ progressions for red $K_2[Ni(dto)_2]$. Data from Czernuszewicz et al.[46]

A plot of $[\nu(n_1\nu_1+\nu_2)-\nu_2]/n_1$ vs. n_1 gives the same slope, X_{11}, as that of $\nu(n_1)/n_1$ vs. n_1, but it differs in its intercept precisely by the quantity X_{12}. Figure 31 shows that the $n_1\nu_1$ and $n_1\nu_1+\nu_2$ progressions are well-behaved, a plot of frequency against vibrational quantum number being accurately linear for both progressions, and giving the same anharmonicity constant, $X_{11} = -1.74 \pm 0.05$ cm^{-1} from the $n_1\nu_1$ progression and $X_{11} = -1.70 \pm 0.06$ cm^{-1} from the $n_1\nu_1 + \nu_2$ progression.[46] However, the value of the harmonic frequency $\omega_1 = 1088.5 \pm 0.8$ cm^{-1} derived from the $n_1\nu_1$ progression is 1.7 cm^{-1} larger than that derived from the $n_1\nu_1+\nu_2$ progression, $\omega_1 = 1086.7 \pm 0.9$ cm^{-1}. This difference gives the cross-term X_{12} in (34). Thus X_{12} is similar in magnitude to X_{11}.

Occasionally, the difference combination tones, $\nu_1 - \nu_2$, can also be enhanced in the RR spectrum; such instances have recently been confirmed experimentally for some metalloporphyrin species,[42] and for the μ-oxo-bridged dinuclear iron(III) center in the hemerythrin model complex $[Fe_2O(O_2CCH_3)_2(HBpz_3)_2]$ ($HBpz_3$ = hydrotris(1-pyrazolyl)borate anion).[47] Figure 32 compares the 75–875 cm^{-1} RR spectra of the latter compound and its μ-^{18}O isotopomer excited at 406.7 nm wavelength, which is resonant with μ-O $\rightarrow$ Fe(III) CT transition.[47] Such an electronic transition results in a significant lengthening of the Fe-oxo bonds; thus the RR spectra are dominated by an intense Fe–O–Fe symmetric stretching mode, ν_s (530 cm^{-1}, ^{16}O; 513 cm^{-1}, ^{18}O), and a long progression of $n\nu_s$ overtones, out to $n = 5$ (not shown). Several other ^{18}O-sensitive features are detected at high magnification in the fundamental region, including the antisymmetric Fe–O–Fe stretch, ν_{as} at 754 cm^{-1} (719 cm^{-1}, ^{18}O), and the Fe–O–Fe angle deformation mode, δ_d at 104 cm^{-1} (103 cm^{-1}, ^{18}O). The δ_d mode forms a combination band at 633 cm^{-1} with ν_s ($\nu_s + \delta_d = 634$ cm^{-1}) and also a difference band at 428 cm^{-1} ($\nu_s - \delta_d = 426$ cm^{-1}). Identification of the latter band is confirmed via its ^{18}O shift (18 cm^{-1}, which is essentially the same as that seen for the 530 cm^{-1} ν_s

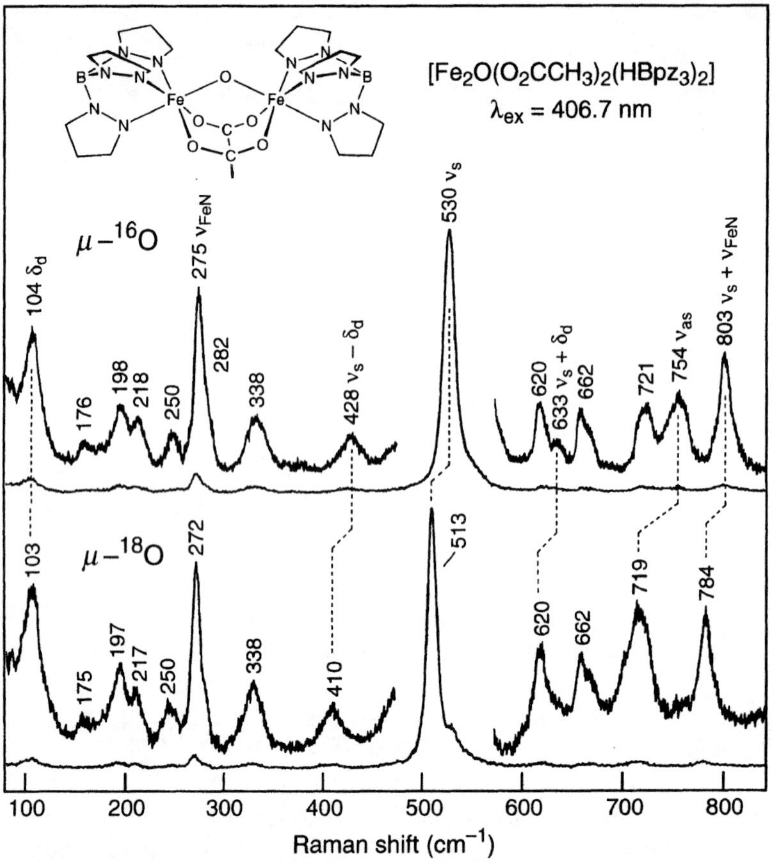

Figure 32. Low-temperature (77 K) RR spectra of crystalline $[Fe_2O(O_2CCH_3)_2(HBpz_3)_2]$ and its μ-^{18}O isotopomer obtained with 406.7 nm excitation from a KCl pellet attached to a liquid-N_2 cell cold finger.[47]

mode) and temperature-dependent intensity. If the 428-cm^{-1} band is due to a difference mode, then its intensity should diminish as the temperature is lowered and the initial level (104 cm^{-1}) is depopulated. Figure 33 shows that this does indeed happen; as the temperature is lowered, the 428-cm^{-1} band steadily loses intensity relative to its 338- and 275-cm^{-1} neighbors.[47]

The A-term scattering can also give rise to the even quanta overtones (which are themselves symmetric) of non-totally symmetric vibrations, provided there is a difference of vibrational frequency ($\nu_k^e \neq \nu_k^g$), i.e., a change in force constant of the mode in the excited state. This produces a change in the shape of the potential energy along the normal coordinate, Q_k, and leads to nonzero Franck–Condon products for the overtone transition. Thus overtones can sometimes be stronger than fundamentals, even in resonance with allowed transitions; such instances, for example, have recently been reported for the phenyl ring of phenylalanine,[48] and for sev-

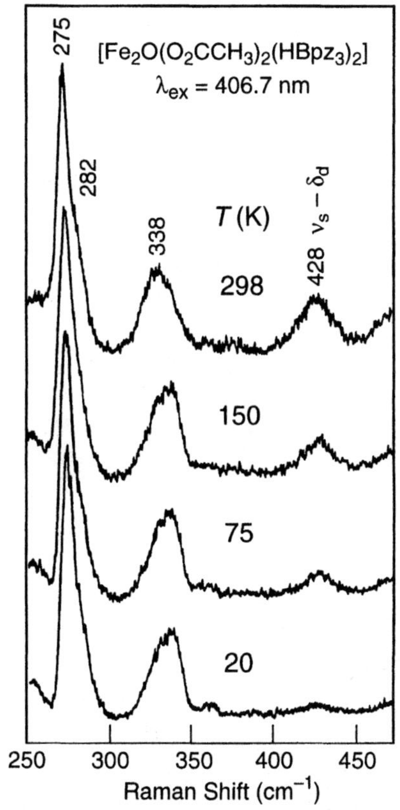

Figure 33. Intensity temperature dependence of the 428 cm^{-1} $\nu_s(FeOFe) - \delta_d(FeOFe)$ difference Raman band for $[Fe_2O(O_2CCH_3)_2(HBpz_3)_2]$. All spectra were obtained with 406.7 nm excitation from a KCl pellet attached to a liquid-He closed-cycle refrigerator cold finger.[47]

eral μ-oxo-bridged dinuclear iron(III),[47,49] manganese(III),[50] and vanadium(III)[51,52] complexes. Figure 34 is the RR scattering by the (μ-oxo) divanadium(III) core of polycrystalline $[V_2O(O_2CCH_3)_2(HBpz_3)_2]$ and its μ-^{18}O isotopomer obtained at low temperature (77 K) in the 100–1400 cm^{-1} region with 676.4 nm excitation, near resonance with the μ-oxo $\rightarrow$ V(III) CT transition.[51] Besides a strong enhancement of the V–O–V symmetric stretch, ν_s (536 cm^{-1}, ^{16}O; 520 cm^{-1}, ^{18}O), and appearance of $2\nu_s$ (1070 cm^{-1}, ^{16}O; 1042 cm^{-1}, ^{18}O), Figure 34 shows that the first overtone, $2\nu_{as}$ (1346 cm^{-1}, ^{16}O; 1283 cm^{-1}, ^{18}O), of the corresponding antisymmetric V–O–V stretch, ν_{as} (685 cm^{-1}, ^{16}O; 653 cm^{-1}, ^{18}O), is substantially more intense than the fundamental and $2\nu_s$. In this case, the fundamental is of B_2 symmetry (under the effective C_{2v} point group of the bent V–O–V bridge) and requires a vibronic mechanism (generally much weaker than a Franck–Condon A-term mechanism) for activation, whereas its overtone is of $B_2 \times B_2 = A_1$ symmetry and is Franck–Condon allowed.

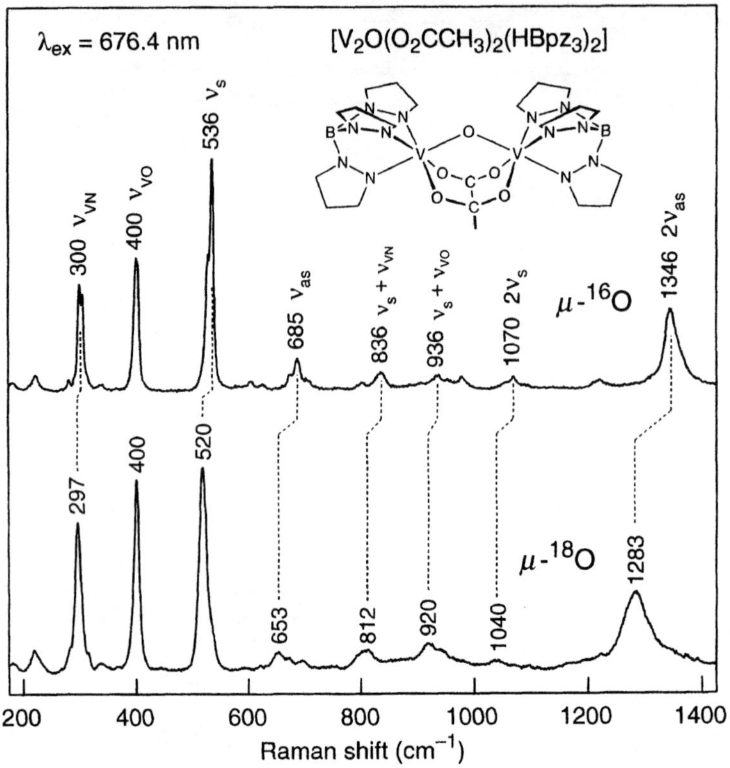

Figure 34. Low-temperature (77 K) RR spectra of crystalline $[V_2O(O_2CCH_3)_2(HBpz_3)_2]$ and its μ-^{18}O isotopomer obtained with 676.4 nm excitation from a KCl pellet attached to a liquid-N_2 cell cold finger.[51]

4 MOLECULAR VIBRATIONS AND ANALYSIS OF RR SPECTRA

Understanding in detail the nature of vibrational modes being monitored by RR spectroscopy is the key in applying it as a probe of electronic and molecular structure. Insofar as the positions of peaks in the RR spectrum are ground-state vibrational frequencies (Figure 18), the principles of interpretation are, of course, the same as those for IR and normal Raman (i.e., far from resonance) spectra.[2–4,53–55] Thus the frequencies of the RR scattered light are also interpreted in terms of atomic masses, molecular geometry (bond distances and bond angles), force constants (bond strength), and local environmental influences (e.g., solvent and conformational effects, pH changes, H-bonding, spin and redox states, etc.). However, resonance excitation is unique in that the enhancement is restricted to Raman lines arising from electrons making up the chromophoric bonds. Often they are the only features observable. For example, resonance with π–π^* electronic transitions, which give rise to strong absorption bands of polyene and aromatic chromophores, enhances predominantly stretching modes of the π bonds[26,27] (Table 2 and Figures 23–24). Similarly, resonance with ligand → metal or metal → ligand CT transitions, which are present in coordination compounds and proteins and enzymes containing transition metals at their active site, enhances modes in which the metal-ligand and internal ligand bonds are stretched[25,56] (Figures 19, 30, 32, 34). Consequently, the task of ascribing the RR enhanced bands to a particular structural element is greatly simplified as compared to the IR and normal Raman spectra.

The fundamentals of vibrational motion and spectra of molecules in terms of their chemical structures and properties have been extensively treated in the literature.[1,2,57–61] In the following sections we shall concentrate on general considerations only. The main points emerge from the vibrational characteristics of diatomic and triatomic oscillators, which can be considered as building blocks for the vibrations of more complex molecules.

4.1 Vibrations of a Diatomic Molecule

To a first approximation the frequency of the stretching vibration of a diatomic molecule, X–Y, is described by Hooke's law

$$\begin{aligned} \nu(\mathrm{Hz}) &= (1/2\pi)\sqrt{F(1/m_X + 1/m_Y)} \quad \text{or} \\ \widetilde{\nu}(\mathrm{cm}^{-1}) &= \nu/c = 1303.1\sqrt{F(\mu_X + \mu_Y)} \end{aligned} \tag{36}$$

in which μ_X and μ_Y are reciprocals of the masses of the atoms and F is the spring constant for the bond between them. Vibrational spectroscopists use the second form of the equation, with frequency expressed as wavenumbers (cm^{-1}), the reciprocal of the wavelength; the factor 1303.1 gives ν in cm^{-1}, when F is expressed in millidyne per Ångström (mdyne $\mathrm{Å}^{-1}$) and masses of the atoms are expressed in amu. The stiffer the spring and the lighter the atoms, the higher the bond stretching frequency. Because the reciprocals of the masses are added to produce the *reduced mass* of the diatom, the lighter atom dominates the frequency. Figure 35 shows that the vibrational frequencies of the hydrogen halides are in direct proportion to the square root of the force constant (a), since the reduced mass is essentially unity for all of them (b) (*force constant effect*). Between HF and HI the frequency falls by about $\sqrt{3}$, corresponding to the ~3-fold decrease in the force constant. In contrast the halogens, which also show a factor of ~3 variation in force constant (Figure 35*a*), have a much wider frequency range (Figure 36), reflecting the much greater reduced mass of I_2 than F_2 (Figure 35*b*) (*force constant effect and mass effect*).

Hooke's law describes a *linear harmonic oscillator*. Its vibrational potential energy curve, $V(Q)$, arising from the variation of total electron energy with internuclear displacement coordinate $Q = R - R_e$, is approximated by

$$V(Q) = \tfrac{1}{2}FQ^2 \tag{37}$$

which gives a parabolic well (Figure 37, dashed curve) with force constant $F = (\partial^2 V/\partial Q^2)_0$ equal to the curvature at the bottom of the potential. This parabolic

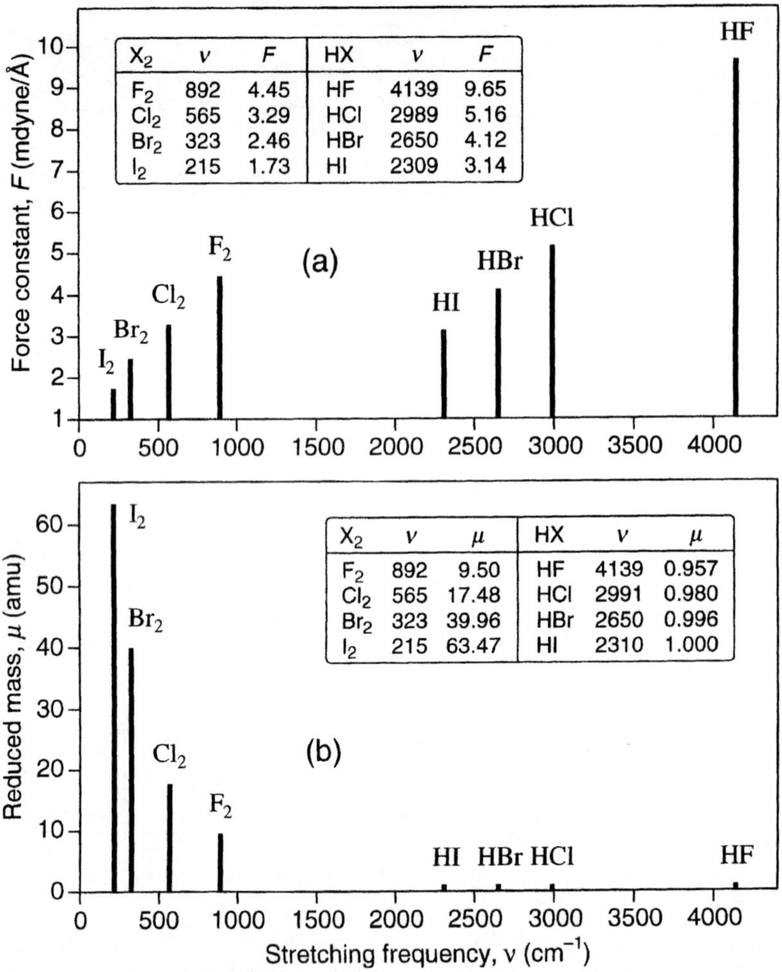

Figure 35. The effects of changing the force constant and mass on the stretching frequency of diatoms. Hydrogen halides vibrate at much higher frequencies than halogens because $\mu(\mathrm{HX}) \approx 1 \ll \mu(\mathrm{X_2})$. Since $\mu(\mathrm{HX})$ changes very little (b), the $\nu(\mathrm{HX})$ trend, HF > HCl > HBr > HI, is linear in $\sqrt{F}$ (a) (*force constant effect*). This is not the case for halogens, however, where both *mass effect and force constant* are operative (a, b). Data from Nakamoto.[2]

potential curve is derived from a Taylor expansion in the neighborhood of a stable equilibrium position ($Q = 0$),

$$V(Q) = V_0 + \left(\frac{\partial V}{\partial Q}\right)_0 Q + \frac{1}{2!}\left(\frac{\partial^2 V}{\partial Q^2}\right)_0 Q^2 + \frac{1}{3!}\left(\frac{\partial^3 V}{\partial Q^3}\right)_0 Q^3 + \cdots \quad (38)$$

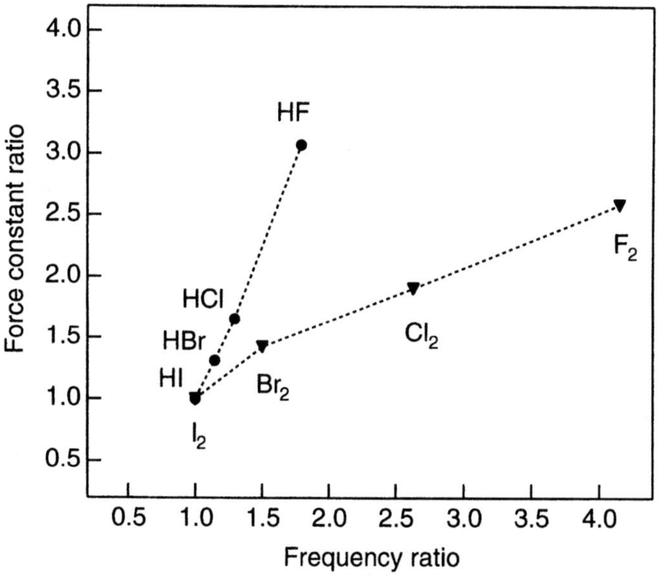

Figure 36. Plots of normalized force constants (F_{HX}/F_{HI}, F_{XX}/F_{II}) versus normalized frequencies (ν_{HX}/ν_{HI}, ν_{XX}/ν_{II}) for stretching vibrations of hydrogen halides and halogens, showing a larger spread of frequencies along the halogen series due to the combined effects of force constants and reduced masses. Data from Nakamoto.[2]

and acts as the potential for the Schrödinger wave equation governing vibratory motion of the nuclei,

$$-\frac{h^2}{8\pi^2\mu}\frac{d^2\Psi(Q)}{dQ^2} + V(Q)\Psi(Q) = E_\upsilon\Psi(Q) \tag{39}$$

The first two terms in (38) vanish at the equilibrium position because $Q = 0$ is the vibrational coordinate origin and V_0 is the minimum of the energy scale, so that $V_0 = 0$ and $(\partial V/\partial Q)_0 = 0$. If we further neglect the cubic and higher order terms (which is reasonable only for sufficiently small bond length distortions, i.e., for small values of Q) and equate $(\partial^2 V/\partial Q^2)_0$ with a force constant F, then (38) becomes identical with (37), the first (harmonic) approximation to $V(Q)$. The eigenvalues of (39), with (37), are then the harmonic stationary states on the parabola with the quantized energies

$$\widetilde{E}_\upsilon(\text{cm}^{-1}) = \widetilde{\nu}_e(\upsilon + \tfrac{1}{2}), \quad \upsilon = 0, 1, 2, 3, \ldots \tag{40}$$

and an uniform spacing between these levels that corresponds to the classical Hooke's law of frequency (36), $\Delta\widetilde{E} = \widetilde{\nu}_e = 1303.1\sqrt{F(\mu_X + \mu_Y)}$. As a result, the allowed energy transitions between vibrational levels of a harmonic oscillator are $\Delta\upsilon = 1$ for infrared absorption and $\Delta\upsilon = \pm 1$ for Raman scattering. The transition

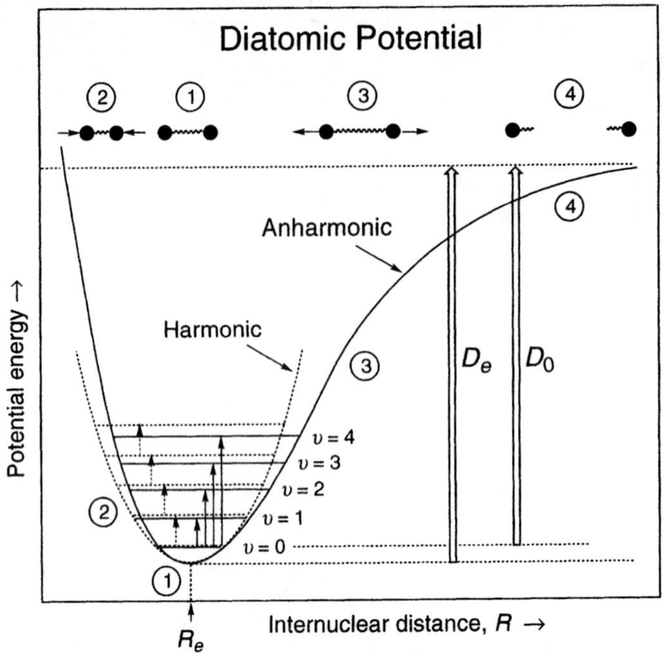

Figure 37. The harmonic (Hooke's law) and anharmonic (Morse-type) potentials for a diatomic oscillator. At (1), the balance between the repulsion of the two positively charged nuclei and the attraction of the electrons gives the minimum energy of the system (equilibrium position, R_e), and the molecule is stable. If the bond is compressed (2) or stretched (3), however, the potential energy of this system rapidly increases by the amount of energy required to restore equilibrium. The allowed energy is quantized, and this is represented by the horizontal lines that are labeled with the vibrational quantum number $\upsilon = 0, 1, 2, \ldots$. These vibrational levels are equally spaced in the harmonic model, and the only upward transitions that can take place are those in which $\Delta\upsilon = +1$ (fundamental and its hot bands, at identical frequencies). In a real molecule (anharmonic oscillator), the level spacing decrease as υ increases, and collapse at the dissociation energy (4); the selection rule is also relaxed, and transitions of $\Delta\upsilon = +2, +3, \ldots$ (overtones) are observed.

from the $\upsilon = 0$ to the $\upsilon' = 1$ vibrational level is the fundamental transition, while the transitions $\upsilon' = 1 \rightarrow \upsilon' = 2$, $\upsilon' = 2 \rightarrow \upsilon' = 3$, etc., are the "hot bands" that have the same frequency as the fundamental (Figure 37, broken arrows); transitions from $\upsilon = 0$ to higher levels (harmonics, or overtones) are forbidden, however.

Although harmonic oscillator can adequately explain most of the features in the vibrational spectrum and can provide a remarkably good fit for the experimental frequencies (Figure 38), some of the finer details such as increased frequency separation with increasing vibrational energy (Figure 39) and appearance of overtones indicate that its potential is inadequate to represent a real molecule. In a real molecular oscillator, the restoring force $(-FQ)$ is not directly proportional to the displacement Q; i.e., Hooke's law is not obeyed because F progressively decreases for the succes-

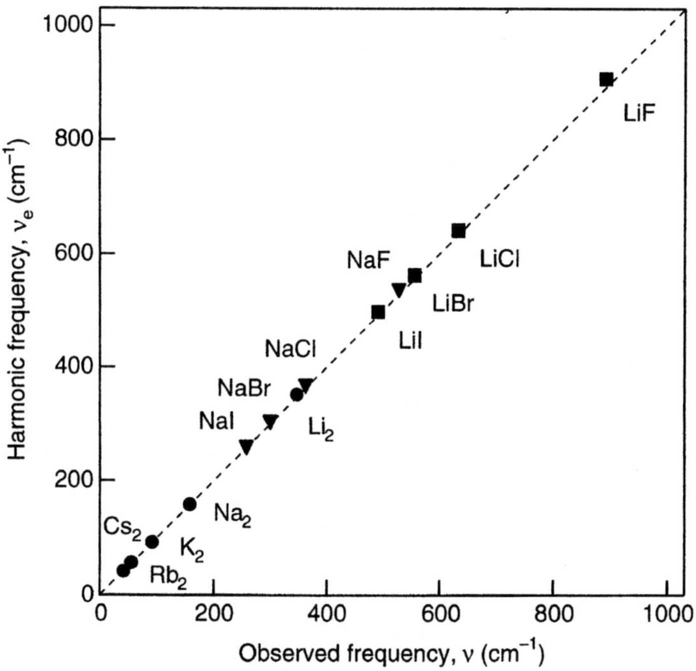

Figure 38. At lower vibrational energies a bond in a molecule is almost elastic; the potential is closely approximated by a parabola, and the atoms vibrate with simple harmonic motion in accordance with Hooke's law. Data from Nakamoto.[2]

sive vibrational levels due to a weakening of the attraction of the electronic binding when the atoms move farther and farther away from each other (Figure 37). A more realistic description is given by (38) with cubic and higher terms retained, or by an empirical function called the *Morse potential* (Figure 37, solid curve) that incorporates a dissociation limit, the energy at which the atoms fly apart. The Morse potential is

$$V(Q) = D_e\left[1 - \exp(-\beta Q)\right]^2 \tag{41}$$

where D_e is the spectroscopic dissociation energy and β is the spreading parameter of the curve. The solution to the Schrödinger equation (39), with $V(Q)$ expressed by (38) or (41), is the *anharmonic oscillator*, which has the allowed vibrational energy levels

$$\widetilde{E}_\upsilon(\text{cm}^{-1}) = \widetilde{\nu}_e\left(\upsilon + \tfrac{1}{2}\right) - \chi_e\widetilde{\nu}_e\left(\upsilon + \tfrac{1}{2}\right)^2 + \cdots \tag{42}$$

that, unlike those of the harmonic oscillator, are no longer equally spaced and become closer and closer as $E_\upsilon \rightarrow D_e$ (Figure 37). Anharmonicity has several important consequences: (1) The actual vibrational frequencies are *always* lower than the har-

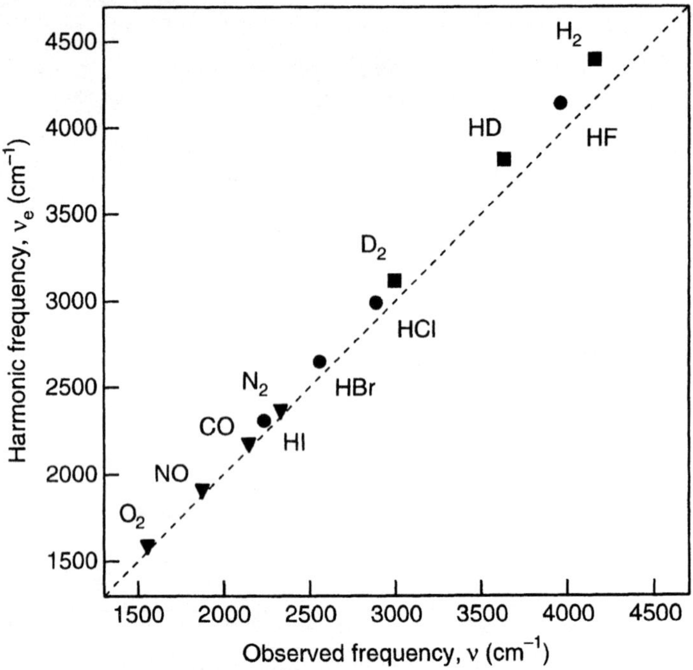

Figure 39. At higher vibrational energies it becomes more difficult to compress the molecule than to stretch it; the potential curve departs from the parabola, and the vibration becomes anharmonic. Data from Nakamoto.[2]

monic ones, and the higher the frequency, the larger will be the discrepancy between the two values (Figures 38, 39). This is because at higher vibrational energies, it becomes more difficult to compress the molecule than to extended it ($|-Q| < |+Q|$), and the actual vibration departs from a simple harmonic motion ($|-Q| = |+Q|$). (2) As we have seen for $K_2Ni(dto)_2$, the spacing between the vibrational energy levels becomes smaller as υ increases (Table 3) and, if we can measure the progressive diminution in spacing, we can obtain a value for χ_e ($X_{11} = -\chi_e\widetilde{\nu}_e$) and from this a value for D_e ($D_e(\text{cm}^{-1}) \approx \widetilde{\nu}_e^2/(4\chi_e\widetilde{\nu}_e)$ if the Morse potential is assumed). (3) The hot bands ($\upsilon' = 1 \rightarrow \upsilon' = 2$, etc.) will have slightly lower frequencies than the fundamental ($\upsilon = 0 \rightarrow \upsilon' = 1$) band, affecting the width and the shape of the fundamental band. (4) Transitions from the ground to higher vibrational levels ($\upsilon = 0 \rightarrow \upsilon' = 2, 3, 4, \ldots$) are no longer forbidden in an anharmonic oscillator (Figure 37, solid arrows), allowing for the overtone bands to be observed (see Figure 30), at frequencies that are lower than $2, 3, 4, \ldots,$ times the frequency of the fundamental (Table 3). Finally (5), in polyatomic molecules, anharmonicity causes even further transitions that involve simultaneous excitation of two or more modes of vibration, such as combination bands (see Figures 28–34), difference bands (Figure 33), and Fermi doublets (Figure 6). The presence of these additional bands results in a complex vibrational spectrum even for triatomic oscillators (Figure 32).

Table 4 Relationship Between Various Molecular Parameters for Dioxygen and Its Ions[a]

Species	Bond order	Bond distance (Å)	Bond energy (kJ/mol)	$\nu(O_2)$ (cm^{-1})	Force constant (mdyne/Å)
O_2^+	2.5	1.123	625.1	1876	16.59
O_2	2.0	1.207	490.4	1580	11.76
O_2^-	1.5	1.280	—	1094	5.67
O_2^{2-}	1.0	1.49	204.2	791/736	2.76

[a]Data from Nakamoto.[2]

The Morse potential force constant (curvature at the bottom of the potential well) is given by

$$F = \left(\frac{\partial^2 V}{\partial Q^2}\right)_{Q=0} = 2D_e\beta^2 \tag{43}$$

which provides a theoretical basis for our intuitive feeling that the force constant should be directly related to the bond energy. As an example, Table 4 shows that experimental O–O stretching force constants for a series of dioxygen species O_2^+, O_2, O_2^-, and O_2^{2-} do vary almost linearly with the O–O bond order, bond distance, and bond dissociation energy. These systematics have proven very useful in characterizing the nature, bonding, and geometries of the dioxygen ligands in a wide range of coordination compounds and metalloproteins that bind molecular O_2, using their characteristic $\nu(O_2)$ frequencies.[2] There are a variety of empirical relationships among force constant, bond energy and bond distance, which have been critically reviewed by Burgi and Dunitz.[62] The most useful is Badger's rule,[63] or its more current variants,[64,65] which relates the force constant to bond distance with empirical parameters that depend on the row of the periodic table containing the connected atoms. These relations can be used to obtain starting approximations of force constants from structural data, or to constrain the ratios of force constants of bonds connecting similar atoms, in vibrational analyses of complex molecules.

Although the spreading parameter β in (43) is not a constant,[7] usually a large force constant does indicate a stronger and, thereby, more stable bond if the series consists of molecules having a similar type of bonding. A particularly revealing comparison is provided by the high-valent oxometal porphyrins of the five adjacent members of the third-row transition metal ions Ti^{IV}, V^{IV}, Cr^{IV}, Mn^{IV}, and Fe^{IV}. Titanyl, vanadyl, and chromyl porphyrins are stable entities, while manganyl and ferryl decompose readily. This reactivity correlates directly with the M–oxo bond strength as revealed by the metal–oxo stretching frequencies, ν_{MO}, in the RR spectra. Figure 40 shows that as the number of d electrons increases from 0 (Ti^{IV}) to 4 (Fe^{IV}), the observed ν_{MO} frequency increases monotonically from 981 cm^{-1} (Ti^{IV})[42] to 1007 cm^{-1} (V^{IV})[66] and 1025 cm^{-1} (Cr^{IV}),[42] then suddenly drops to 757 cm^{-1} (Mn^{IV}),[67] and increases again but only to 842 cm^{-1} (Fe^{IV}),[68] even though Mn occupies a position between

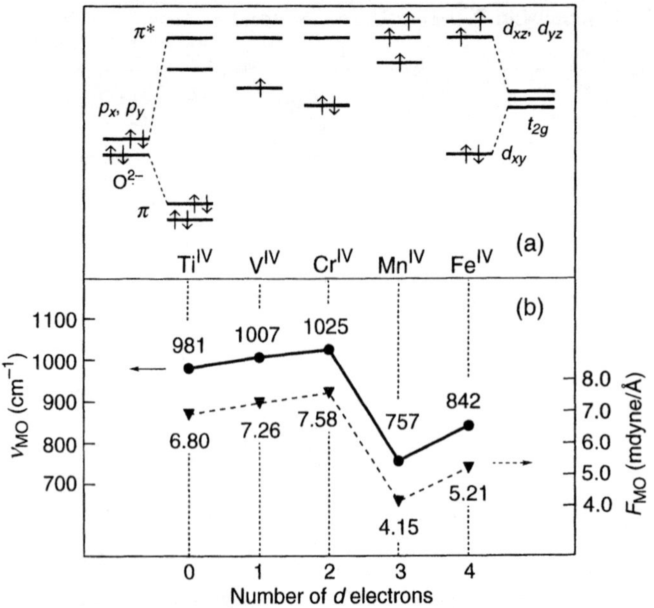

Figure 40. (a) Bonding diagram showing the O^{2-} $p\pi$–$M^{IV}d\pi$ interactions that modulate the M–O strength. (b) Comparison of the M–O stretching force constants, F_{MO}, (calculated for a diatomic oscillator) and RR frequencies, ν_{MO}, for Ti^{IV}, V^{IV}, Cr^{IV}, Mn^{IV}, and Fe^{IV} porphyrins; (Ti^{IV}O)OEP in hexane (this work), (V^{IV}O)OEP in hexane,[66] (Cr^{IV}O)TPP in hexane (this work), (Mn^{IV}O)TMP in CH_3CN,[67] (Fe^{IV}O)TMP in CH_2Cl_2[68].

Cr and Fe in this series. This peculiar behavior can be explained in terms of the force constants calculated on the diatomic oscillator approximation (36), the d electron count, and the special stability and low polarizability of the high-spin half-filled t_{2g} subshell that come into play for the d^3 configuration of Mn^{IV}.[67] Figure 40b provides a qualitative bonding diagram. The O^{2-} ligand donates both σ (omitted from the diagram) and π electrons to the M^{4+} ions. The filled p orbitals of the ligand (p_x and p_y, with z as the M–O direction) interact with the M^{4+} d_π orbitals (d_{xz}, d_{yz}), which become antibonding. The d_{xy} orbital is nonbonding and its energy decreases with increasing effective nuclear charge from Ti^{IV} (d^0) to Fe^{IV} (d^4), with the exception of Mn^{IV} (d^3), where it is brought close to (d_{xz}, d_{yz}) by the exchange interactions in the high-spin half-filled t_{2g} subshell. This orbital has 0, 1, and 2 d electrons for Ti^{IV}, V^{IV}, and Cr^{IV}, respectively, allowing unimpeded O → M π interactions. Since there are two π orbital overlaps, the result is formally a M≡O triple bond in all three oxides. The high bond order is reflected in the elevated force constants, 6.8 (Ti^{IV}), 7.26 (V^{IV}), and 7.58 (Cr^{IV}) mdyne/Å, with the F_{MO} trend, $Cr^{IV} > V^{IV} > Ti^{IV}$, that parallels the trend in effective nuclear charge. However, two electrons enter the π-antibonding d_{xz} and d_{yz} in the Mn^{IV} (d^3) and Fe^{IV} (d^4) species, reducing the bond order by one unit to M=O in both cases. This is reflected in the ~33% decrease in the force constants, to 4.15 (Mn^{IV}) and 5.21 (Fe^{IV}) mdyne/Å. The extra lowering of

F_{MnO} and, thereby, the anomalous weakness of the $\mathrm{Mn^{IV}}$–oxo bond is attributable to the contraction of the d orbitals in the high-spin half-filled shell (Figure 40b), which reduces their availability for π bonding.[67]

4.2 Vibrations of a Triatomic Molecule

We next consider the stretching vibrations of a symmetrical linear triatomic molecule, Y–X–Y (see Section 2.5). There are two such vibrations, one for each of the bonds, and it is clear from the diagram in Figure 41 that the two bond stretching motions interact strongly. When the two bonds are stretched in-phase, the displacements are in opposition and motion of the central X atom is canceled. The frequency for this vibration is given by

$$\widetilde{\nu}_{\mathrm{s}}(\mathrm{cm}^{-1}) = 1303.1\sqrt{F_{\mathrm{s}}\left[\mu_{\mathrm{Y}} + \mu_{\mathrm{X}}(1+\cos\alpha)\right]} \tag{44}$$

where the subscript "s" stands for symmetric (the motion retains the molecular symmetry), $\mu_{\mathrm{Y}} = 1/m_{\mathrm{Y}}$ and $\mu_{\mathrm{Y}} = 1/m_{\mathrm{Y}}$, and α is the Y–X–Y angle. Since $\alpha = 180°$ for a linear triatom, (44) reduces to

$$\widetilde{\nu}_{\mathrm{s}} = 1303.1\sqrt{F_{\mathrm{s}}\mu_{\mathrm{Y}}} \tag{45}$$

and the frequency for the symmetric stretching vibration does not depend on the mass of the central atom; the reduced mass for this motion is just the reciprocal of the mass of Y.

When the two bonds stretch out-of-phase (one stretches while the other contracts), the displacements reinforce one another and the X atom moves twice as far as it did in the X–Y diatom (assuming the same force constant). The frequency for this motion is given by

$$\widetilde{\nu}_{\mathrm{as}}(\mathrm{cm}^{-1}) = 1303.1\sqrt{F_{\mathrm{as}}\left[\mu_{\mathrm{Y}} + \mu_{\mathrm{X}}(1-\cos\alpha)\right]} \tag{46}$$

or, for $\alpha = 180°$, simply by

$$\widetilde{\nu}_{\mathrm{as}} = 1303.1\sqrt{F_{\mathrm{as}}(\mu_{\mathrm{Y}} + 2\mu_{\mathrm{X}})} \tag{47}$$

(The subscript "as" stands for antisymmetric; the motion destroys the symmetry of the molecule.) The X atom contribution to the reduced mass is doubled. If X and Y are the same atoms, and if the force constant is the same for the in- and out-of-phase stretches (i.e., $F_{\mathrm{s}} = F_{\mathrm{as}}$), their frequencies are predicted to be in the ratio $1{:}\sqrt{3}$, while the X–Y diatomic frequency for the same force constant would scale as $\sqrt{2}$. As an example, the C=C stretching frequency for ethylene is 1623 cm^{-1} while the in- and out-of-phase C=C frequencies of allene are 1071 and 1980 cm^{-1}.[4] These values are not too far from the expected $1{:}\sqrt{2}{:}\sqrt{3}$ intervals; closer agreement cannot

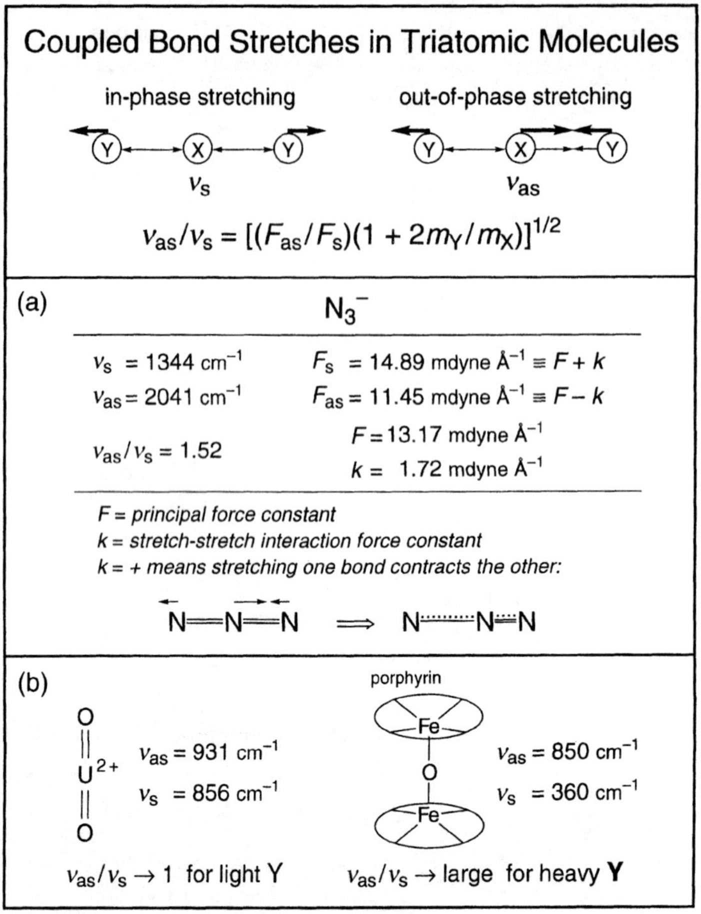

Figure 41. In- and out-of-phase stretching modes of a linear Y–X–Y triatomic molecule, showing how the stretching motions oppose or reinforce one another, leading to a higher frequency for the out-of-phase mode due to the kinematic coupling. (a) If all three masses and the two force constants are the same, the ratio of the frequencies should be $\sqrt{3}$. The ratio is less than this for the azide ion[69] because the electronic structure lowers the energy of the out-of-phase motion, giving a positive stretch-stretch interaction force constant. (b) If the mass of the central atom is much greater than that of the outer atoms, e.g., UO_2^{2+},[70] the two frequencies approach one another (mass uncoupling), while in the reverse situation, e.g., $(Fe^{III}porphyrin)_2O$,[71] the ratio becomes very large.

be expected in view of the fact that the motions of the H atoms also contribute somewhat to these frequencies. The frequencies of the azide ion, N_3^-,[69] can be treated exactly in our simple scheme (Figure 41). The in- and out-of-phase frequency ratio, 1.52, is significantly less than $\sqrt{3}$, showing that the force constant does not have the same value for the two motions. This is due to an electronic interaction between the two bond displacements, which introduces a cross term in the potential energy

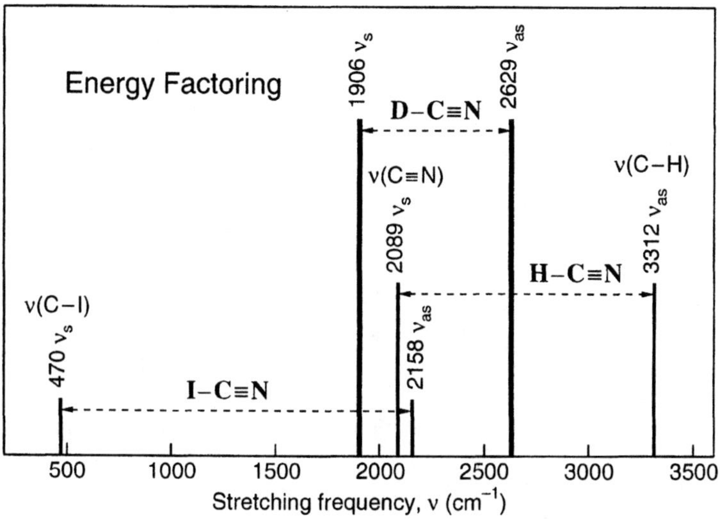

Figure 42. Uncoupling due to energy separation: For Y–C≡N molecules the two stretching modes can be classified as in-phase (lower frequency) and out-of-phase (higher frequency). But the interaction is strong only when the frequencies are not too far apart, as in DCN. For HCN the higher frequency (3312 cm^{-1}) involves stretching of the C–H bond, exclusively, while the lower frequency (2089 cm^{-1}) is mostly C≡N stretching. Conversely, for ICN the higher frequency (2158 cm^{-1}) is mostly C≡N stretching, while the lower frequency (470 cm^{-1}) is mostly C–I stretching. Data from Colthup et al.[3]

second derivative, called an interaction force constant, k. This term adds to the bond stretching constant (the principal valence force constant) for the in-phase motion ν_s but subtracts from it for the out-of-phase motion ν_{as}. The positive interaction constant for $N_{3}-$, can be understood on the basis of its electronic structure. Since there are two double bonds in line, stretching one of them makes it harder to stretch but easier to contract the other one, because π electron density can be transferred from the first to the second bond. Thus the frequency is raised for the in-phase but lowered for the out-of-phase motion. Interaction force constants are generally smaller than 10% of the principal valence force constant, but can be somewhat larger for interacting π bonds.

The ratio of the atomic masses also affects the in- and out-of-phase frequency separation. The frequencies approach one another when the outer atoms are much lighter than the central atom, which then makes little contribution to either motion. Conversely very large separations are seen when the outer atoms are much heavier than the central atoms. These cases are illustrated in Figure 41 with UO_2^{2+} [70] and the μ-oxo dimer of Fe^{III} porphyrin.[71,72]

The in- and out-of-phase motions interact less for unsymmetrical linear triatoms if the force constants of the two bonds or the masses of the two outer atoms are very different. This is illustrated in Figure 42 for a series of Y–C≡N (Y = H, D, I) molecules. The natural frequency for the C≡N triple bond is about 2100 cm^{-1}. One of the two stretching frequencies is close to this value for HCN ($\nu_s = 2089$ cm^{-1}) as well as ICN ($\nu_{as} = 2158$ cm^{-1}). Calculation of the atomic motions[3] shows that in these cases C≡N stretching is a good approximate description of the motion, whether it is in- or out-of-phase, because the remaining bond is displaced very little. Only when the two frequencies approach one another, as in DCN, do the two bond displacements interact strongly, producing shifts in the natural frequencies and thoroughly mixed motions. As the frequencies move apart the interaction of the displacements decreases, an effect known as *energy factoring*. Approximate energy factoring is widely encountered in complex molecules and is of great assistance in permitting qualitative band assignments.

Uncoupling of adjacent bond stretches is also encountered if the triatom (or triatomic fragment of a more complex molecule) is allowed to bend, as illustrated in Figure 43. The mutual opposition or reinforcement of the two bond displacements is now modified by the cosine of the bond angle. Thus coupling is maximal when

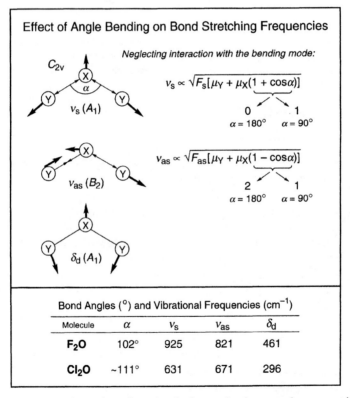

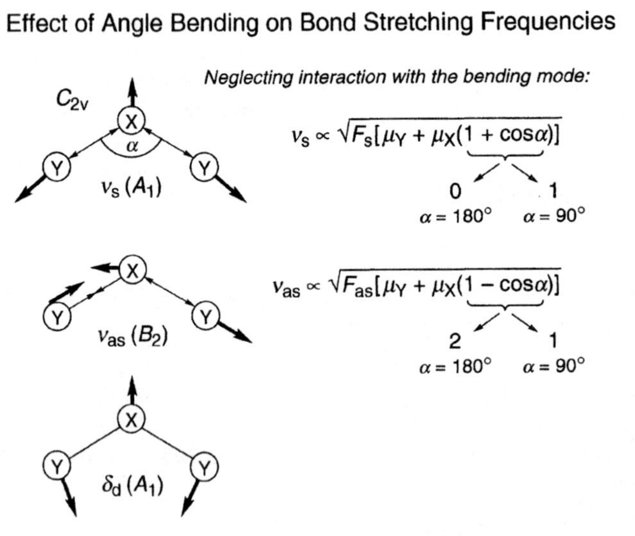

Bond Angles (°) and Vibrational Frequencies (cm^{-1})

Molecule	α	ν_s	ν_{as}	δ_d
F_2O	102°	925	821	461
Cl_2O	~111°	631	671	296

Figure 43. Modes of a bent triatomic molecule. Interaction between the symmetric and antisymmetric stretch decreases to zero as the bond angle α decreases to 90°. The symmetric stretch interacts with the bend, but this is a small effect since the bending frequency is typically about one half of the stretching frequency. For angles near 90°, however, this effect is sufficient to raise ν_s above ν_{as} (F_2O).[73]

the bond angle is 180° but vanishes when it is 90°. The equations given in Figure 43 predict equal frequencies at 90° for the in- and out-of-phase stretches if the force constants are the same (negligible interaction constant), and the same frequency as for an X–Y molecule with the same force constant. However, the equations do not account for mixing of the in-phase stretch with the angle bending motion of the molecule. These motions each contribute to the two symmetric modes of the molecule, ν_s and δ_d, but not to the one antisymmetric mode, ν_{as} (the out-of-phase stretch); modes of different symmetry cannot mix. (This is why the bending mode could be left out of the discussion of the bond stretches of a linear triatom; mixing is precluded because angle bending destroys the symmetry axis of the molecule, while the bond stretches preserve it (see Figure 11).) The effect of mixing angle bending into the in-phase stretching mode of a bent triatom is not large, because the bending frequency is typically only about half that of the stretching frequency (energy factoring). The table in Figure 43 shows that the effect is sufficient to push the in-phase above the out-of-phase mode for bond angles near 90°, as in the case of F_2O (α = 102°).[73] For Cl_2O the angle is slightly larger ($\alpha = \sim 111°$) and the in-phase frequency is below the out-of-phase frequency.[74]

Expressions (44) and (46) further tell us that knowledge of the ν_s and ν_{as} frequencies should lead to the prediction of the bond angle. Indeed, excellent agreement between the observed (X-ray crystallography and/or EXAFS spectroscopy) and calculated angles is often possible if isotope shifts are obtained for ν_s and ν_{as}. Table 5 gives typical results of such analyses for the oxo-bridged units in dinuclear complexes of Fe^{III}, Mn^{III}, and V^{III}, which display a wide range (~120–180°) of the M–O–M angles. These units are of special interest in that many Fe and

Table 5 Bond Angles (°) and Vibrational Frequencies (cm^{-1}) of Oxo-Bridged Compounds[a]

		∠M–O–M		ν(M–O–M)		
Compound[c]	Structure	obsd	calcd	ν_s (Δ^{18}O)	ν_{as} (Δ^{18}O)	Ref.
$[V_2O(l\text{-his})_4]$ (*aq*)	linear	—	180	—	734 (−35)	52
$[Fe_2O(TPP)]$	linear	175	180	363 [+6][b]	885	71
$[Fe_2O(hedta)_2]^{2-}$	monobridged	165	167	409 (−2)	838	49
$[V_2O(l\text{-his})_4]\cdot 2H_2O$ (*s*)	monobridged	154	153	—	720 (−35)	52
$[Fe_2O(phen)_4(H_2O)_2]^{4+}$	monobridged	155	156	395 (−5)	827 (−39)	49
$[V_2O(OAc)_2(HBpz_3)_2]$	tribridged	144	142	499 (−11)	699 (−37)	51
$[V_2O(O_2P(OPh)_2)_2(HBpz_3)_2]$	tribridged	134	132	536 (−16)	685 (−35)	51
$[Fe_2O(hdp)_2(OBz)]^+$	dibridged	129	124	494 (−17)	763 (−43)	49
$[Fe_2O(OAc)_2(HBpz_3)_2]$	tribridged	124	127	530 (−17)	754 (−35)	47
$[Mn_2O(OAc)_2(HBpz_3)_2]$	tribridged	125	128	554 (−18)	710 (−37)	50

[a]Observed ∠M–O–M values from X-ray crystal structures. Calculated ∠M–O–M values based on secular equation (44) for ν_s(MOM) with ^{16}O and ^{18}O, or ^{56}Fe and ^{54}Fe (TPP complex).
[b][Δ^{54}Fe].
[c]Ligands: *l*-his = *l*-histidine, TPP = tetraphenylporphyrin, hedta = N-hydroxyethylenediamine-acetate, phen = 1,10-phenanthroline, $HBpz_3$ = hydrotris(1-pyrazolyl)borate, OAc = acetate, OBz = benzoate, $O_2P(OPh)_2$ = diphenylphosphate.

Mn proteins and enzymes also incorporate oxo-bridged structural motifs at their actives sites to carry out a variety of biological functions, while the V^{III}–O–V^{III} dimers have been implicated in the biochemistry of tunicates. Figure 44 shows how changing the V–O–V angle affected the ν_s(VOV) and ν_{as}(VOV) stretches in the RR spectra of $[V_2O(O_2CCH_3)_2(HBpz_3)_2]$ and its μ-^{18}O isotopomer discussed above (Figure 34). The ν_s(VOV) and ν_{as}(VOV) occur at 536 and 685 cm^{-1}, respectively, in the acetate-bridged complex, but at 499 (ν_s) and 699 (ν_{as}) cm^{-1} in $[V_2O(O_2P(OPh)_2)_2(HBpz_3)_2]$, i.e., when the cobridging carboxylato ligands were

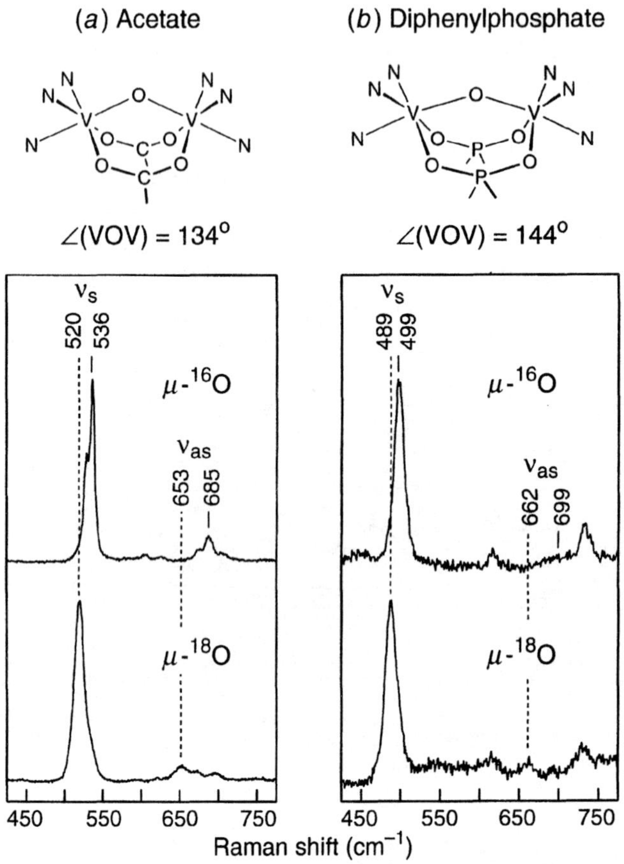

Figure 44. RR scattering (676.4 nm excitation) in the 450–750 cm^{-1} spectral region from crystalline (*a*) $[V_2O(O_2CCH_3)_2(HBpz_3)_2]$ and (*b*) $[V_2O(O_2P(OPh)_2)_2(HBpz_3)_2]$ and their μ-^{18}O isotopomers, showing the effects of changing the V–O–V angle on the characteristic symmetric (ν_s) and antisymmetric (ν_{as}) stretching frequencies of the supported μ-oxo divanadium(III) bridge. These frequencies conform to the values expected for the V–O–V angles seen in the crystal structures of the two dimers, 134 and 144° (Table 5);[51] the ν_s and ν_{as} are expected to shift downward and upward, respectively, upon widening of the μ-oxo bridge angle caused by replacement of acetate with diphenylposphate.

replaced with diphenylphosphates.[51] The magnitude of the isotopic shift upon labeling the μ-oxo group with ^{18}O changed as well; it decreased for the symmetric stretch, from 16 cm^{-1} (acetate) to 11 cm^{-1} (phosphate), but increased for the antisymmetric one, from 32 cm^{-1} (acetate) to 37 cm^{-1} (phosphate). This pattern of vibrational frequencies and isotopic shifts (lower frequency and smaller ^{18}O shift for ν_s but higher frequency and larger ^{18}O shift for ν_{as}) in the phosphate analog is consistent with the larger V–O–V angle found for this complex (Table 5). Using either the force constants derived from (44) and (46) for $[V_2O(O_2CCH_3)_2(HBpz_3)_2]$ ($F_s = 4.46$ and $F_{as} = 2.20$ mdyne/Å, with $\alpha = 135°$), or the ^{16}O and ^{18}O RR frequencies observed for $[V_2O(O_2P(OPh)_2)_2(HBpz_3)_2]$, a simple triatom calculation gives a value of 142° for the V–O–V angle in the phosphate analog that agrees well with the crystallographically determined angle of 144°.[51]

An even more interesting case is a monobridged $[V_2O(l\text{-his})_4]$ (l-his = l-histidine) complex; its originally reported bridging V–O–V angle (153.9°)[75] appeared to lie well outside the range of values for either supported (130–145°) or unsupported (165–180°) μ-oxo dimers. As seen in Figure 45, this unusual bending distortion of $[V_2O(l\text{-his})_4]$ is caused by strong intramolecular hydrogen bonding between the two coordinated histidines in the solid state which is lost in solution, resulting in a more linear V–O–V bond. Evidence for these features was provided by a new X-ray crystal structure, and by analyzing the differences between the solid state and solution RR frequencies of the ν_{as}(VOV) stretch and its first overtone, $2\nu_{as}$(VOV), using H_2O, D_2O, and $H_2{}^{18}O$ for band identification (Figure 46).[52] Although no corresponding ν_s(VOV) stretching vibration is observed for this molecule (*vide infra*), the bridge ν_{as} and $2\nu_{as}$ modes are strongly enhanced at 720 and 1434 cm^{-1} in the solid state and at 734 and 1454 cm^{-1} in H_2O solution (middle traces); this assignment is secured by the large, ~35 and ~70 cm^{-1}, shifts to lower frequency upon $H_2{}^{18}O$ exchange (bottom traces). Furthermore, solid samples of $[V_2O(l\text{-his})_4]$ prepared from D_2O revealed deuterium-dependent positions for ν_{as} (725 cm^{-1}) and $2\nu_{as}$ (1446 cm^{-1}) while in solution no differences between H_2O and D_2O were noted (top traces). We may thus conclude that $\angle$(VOV) $\rightarrow$ 180° upon dissolution due to a breaking of the intramolecular hydrogen bonding that stabilizes the acute V–O–V angle in the solid (~154°). Breaking the hydrogen bonds between the histidine ligands in solution should result in a linear μ-oxo V^{III} dimer, the expected structure for an unsupported V^{III}–O–V^{III} bridge. The incorporation of $F_{as} = 2.20$ mdyne/Å derived above into (46) accurately reproduces the solid state ν_{as}(VOV) frequencies with $\angle$(VOV) = 154° found experimentally in solid $[V_2O(l\text{-his})_4]\cdot 2H_2O$, and also predicts that the ν_{as}(VOV) mode should occur at 735 (^{18}O) and 699 (^{18}O) cm^{-1} for $\angle$(VOV) = 180°. These frequencies are in nearly perfect agreement with those observed for ν_{as}(VOV) in H_2O and $H_2{}^{18}O$ (Figure 46), and, consequently, the V–O–V structure of $[V_2O(l\text{-his})_4]$ in aqueous solution must be very close to linear. This force constant also reveals that an upward shift in ν_{as} frequency from 720 (hydrogen bonded species) to 724 cm^{-1} (deuterium bonded) in the solid state corresponds to an opening of the V–O–V angle from 154° to ~160° upon H/D exchange,

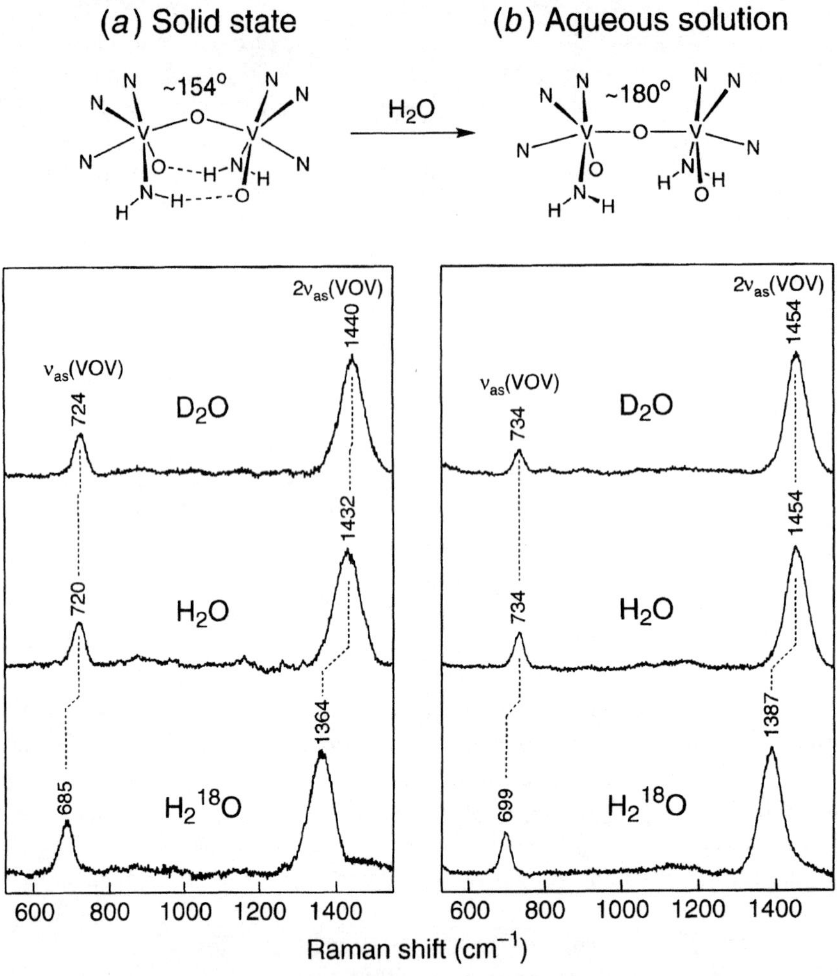

Figure 45. Comparison of RR spectra (488.0 nm excitation) of $[V_2O(l\text{-his})_4]$ in (*a*) solid and (*b*) aqueous solution reveals[52] that the unusual bending distortion in $[V_2O(\text{l-his})_4]$ is caused by strong intramolecular hydrogen bonding in the solid state (deuterium-dependent spectrum) which is lost in solution (D_2O-insensitive spectrum), resulting in a more linear V–O–V bond (upshifted ν_{as} and $2\nu_{as}$ bands).

in accord with weaker deuterium bonding, as opposed to hydrogen bonding, between the histidine ligands. Thus, the positions of the enhanced V–O–V Raman bands can be used to monitor the ground-state structure and dynamics.[51,52]

An interesting question is why the characteristic RR features of $[V_2O(l\text{-his})_4]$ are only a moderately enhanced ν_{as}(VOV) and a very intense $2\nu_{as}$(VOV) while the corresponding ν_s(VOV) and its harmonics are not observed at all. In triatomic M^{III}–O–M^{III} complexes, the dominant feature usually arises from the symmetric stretch and the antisymmetric stretch is usually weak or unobserved (see Figure 32).

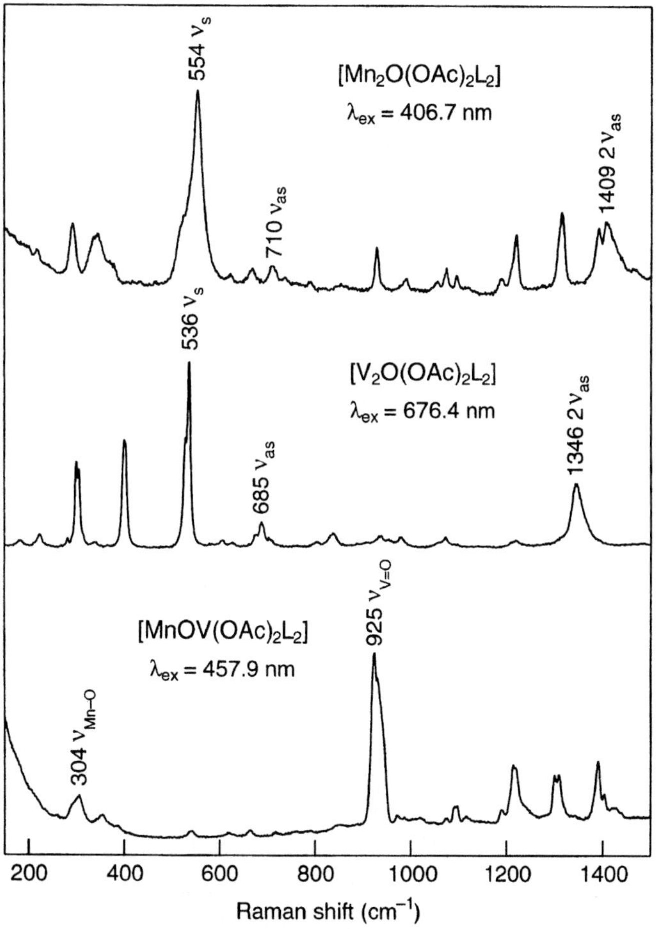

Figure 46. RR spectra of solid [LV(μ-O)(μ-OAc)$_2$MnL] (*bottom*), [V$_2$(μ-O)(μ-OAc)$_2$L$_2$] (*middle*), and [Mn$_2$(μ-O)(μ-OAc)$_2$L$_2$] (*top*) (where L = HBpz$_3^-$ ligand and OAc = acetate ion).[76]

This also applies to the V–O–V dimers but apparently only to those in which $\angle$(VOV) < $\sim$150° (Figures 34 and 45). Such observations are in line with expectations for excitation into the strong μ-O $\rightarrow$ M CT absorption bands in the visible region that these systems exhibit, since the M–O bonds are weakened in the CT excited state. If a μ-oxo bridge does not undergo a change in symmetry upon excitation, which is the more common occurrence, then both M–O bonds equally elongate in the excited state; as a result, the RR active modes are the in-phase stretches, ν_s(MOM). Exceptions arise when (1) the M–O–M bridge exhibits some inherent asymmetry in the ground electronic state and/or (2) when there is dynamic stabilization of the μ-O $\rightarrow$ M CT excited state that is asymmetrical due to differential electronic in-

Another example is the identification of the HO–Mn^{III}–OH and HO–Mn^{IV}=O bonds, formed by hydrolysis of $[Mn^{III}TMPyP]^+$ (TMPyP = tetrakis(methylpiridinium)porphyrin) in pH 14 aqueous solution and after chemical or electrochemical oxidation of this solution, using $H_2{}^{18}O$ and D_2O.[67] Figure 48 shows how the course of electrooxidation was monitored in 1 M NaOH aqueous solution by *in situ* RR spectroelectrochemistry, while the effects of $H_2{}^{18}O$ and D_2O substitutions on the RR spectra obtained before and after conversion of Mn^{III} to Mn^{IV}TMPyP with potassium ferricyanide are displayed in Figure 49. As the potential was raised through the

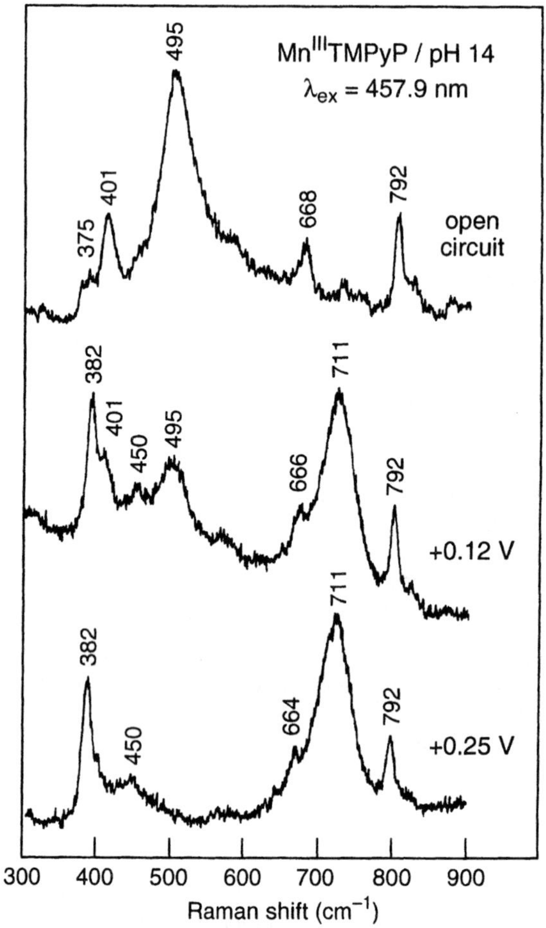

Figure 48. RR spectra (457.9 nm excitation) of manganese TMPyP in 1 M NaOH at various oxidation potentials, illustrating how RR scattering was used[67] to follow chemical and electrochemical reactions; as the potential was raised through the Mn^{IV}/Mn^{III} redox potential, the 495 cm^{-1} $Mn(OH)_2$ symmetric stretching band of the chemically formed $(HO)_2Mn^{III}$TMPyP is replaced by a Mn^{IV}=O stretching band at 711 cm^{-1} of the electrogenerated (HO) (Mn^{IV}O)TMPyP.

Mn^{IV}/Mn^{III} redox potential in the electrochemical cell or by addition of excess ferricyanide, the prominent broad band at 495 cm^{-1} was replaced by an equally broad and strong band at ~710 cm^{-1}. The Mn^{III}TMPyP species in 1 M NaOH is shown to be a *trans*-dihydroxide adduct by virtue of a 495 cm^{-1} RR band identified, on the basis of its 26 cm^{-1} $H_2{}^{18}O$ and 14 cm^{-1} D_2O downshifts (Figure 49, left panel), as the symmetric $Mn–(OH)_2$ stretch. A simple linear-triatomic oscillator calculation with point mass (17 amu) OH ligands reproduces this frequency with an $F_s = 2.46$ mdyne/Å and predicts $H_2{}^{18}O$ and D_2O shifts (point mass 19 and 18 ligands) of 27 and 14 cm^{-1}, in excellent agreement with experiment. The accuracy of this calcula-

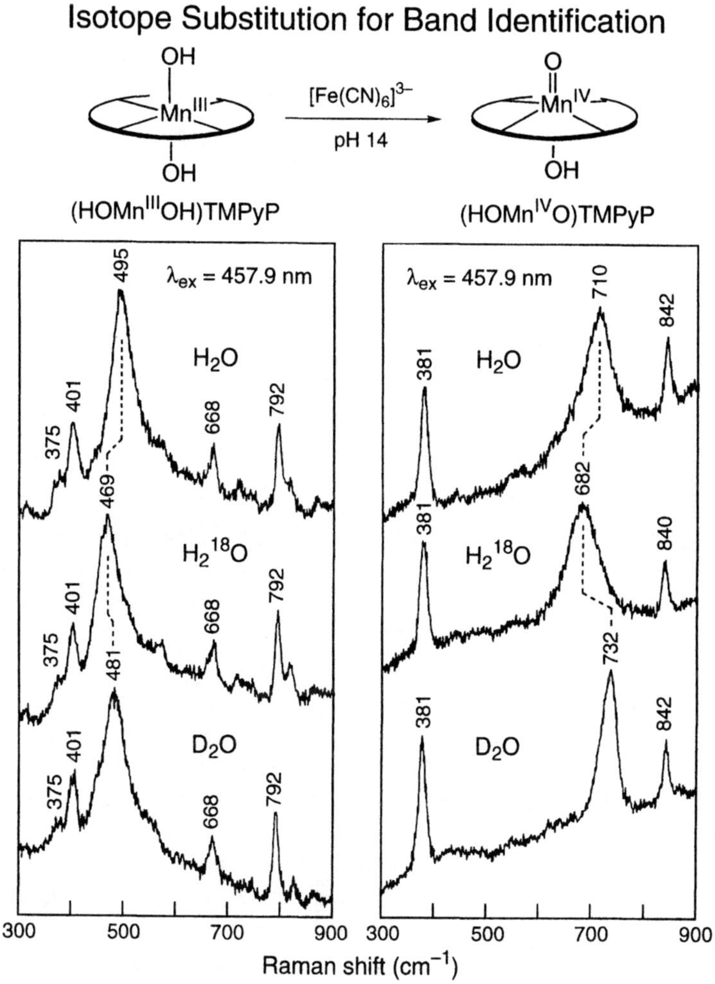

Figure 49. RR spectra (457.9 nm excitation) of $(HO)_2Mn^{III}$TMPyP (*left panel*) and its ferricyanide oxidized $(HO)(Mn^{IV}O)$TMPyP product (*right panel*) in 1 M NaOH (*top*), 1 M $Na^{18}OH$ (*middle*), and 1 M NaOD (*bottom*) aqueous solution.[67]

Table 6 Normal Mode Calculations for a HO–Mn=O Model[a]

Mode character	ν (cm^{-1})		PED (%)[b]			
	obsd	calcd	ν_{MnO}	ν_{MnOH}	δ_{MnO}	ν_{OH}
			$H^{16}O–Mn=^{16}O$			
$\nu_{Mn^{16}O}$	711	710	75	9	16	
$\nu_{Mn^{16}OH}$		534	10	74	8	
$\delta_{Mn^{16}OH}$		785	7	17	76	
$\nu_{^{16}OH}$		3591				100
			$H^{18}O–Mn=^{18}O$			
$\nu_{Mn^{18}O}$	684	686	75	14	10	
$\nu_{Mn^{18}OH}$		508	21	73	5	
$\delta_{Mn^{18}OH}$		774	4	12	84	
$\nu_{^{18}OH}$		3579				100
			$D^{16}O–Mn=^{16}O$			
$\nu_{Mn^{16}O}$	732	731	64	30	5	
$\nu_{Mn^{16}OD}$		475	30	45	25	
$\delta_{Mn^{16}OD}$		620	6	24	70	
$\nu_{^{16}OD}$		2623				100

[a]Only the modes in the Mn–O–H plane were calculated. The assumed geometry was Mn=O = 1.85 Å, Mn–OH = 2.00 Å, OH = 1.00 Å, ∠O–Mn=O = 105°. The following stretching (K) and bending (H) force constants (mdyne $Å^{-1}$) were used: $K_{MnO} = 3.43$, $K_{MnOH} = 2.60$, $K_{OH} = 7.20$, and $H_{MnOH} = 0.30$. $H_{O–Mn=O}$ was set to zero, for convenience; the stretching frequencies are not affected by its magnitude.

[b]Potential energy distribution (PED); % contribution from the indicated internal coordinates. All data are from Czernuszewicz et al.[67]

tion implies that the Mn–$(OH)_2$ mode does not couple significantly to the porphyrin vibrations; the lack of coupling is because the Mn–OH and Mn–N(porphyrin) bands are orthogonal. If instead a Mn–OH diatomic oscillator is assumed, modeling a five-coordinate complex, then the required force stretching constant is 1.87 mdyne/Å and the predicted $H_2{}^{18}O$ and D_2O shifts are 20 and 10 cm^{-1}, significantly smaller than observed. Thus a *trans*-dihydroxo complex is indicated by the vibrational data.[67]

On the other hand, the 710 cm^{-1} ^{18}O-sensitive RR band of the oxidized species has been identified as ν_{MnO} of the $Mn^{IV}=O$ moiety when there is a *trans* axial OH^- ligand.[67] Proof that this band is associated with a *trans* OH^- adduct is provided by its anomalous *upshift*, to 732 cm^{-1}, in D_2O (Figure 49, right panel). Although some sensitivity to D_2O associated with H-bond differences might be expected, the upshift is much too large to be attributed to a difference in D- versus H-bonding. It must instead reflect differential coupling with an internal coordinate involving H or D atoms such as the Mn–O–H bending coordinate, δ_{MnOH}, of a *trans* axial OH^- ligand. If the natural frequency for δ_{MnOH} were ~750 cm^{-1}, it could depress the $Mn^{IV}=O$ stretching frequency by a near resonant interaction. This interaction would be relieved in D_2O because of the large drop in the δ_{MnOD} frequency, resulting in a rise of the ν_{MnO} frequency. Any calculations based on the HO–Mn^{IV}=O triatom model (assuming a point mass OH^- ligand) cannot predict such a coupling nor it can explain the observed D_2O upshift; clearly, an explicit hydrogenic oscillator (δ_{MnOH}) that mixes with ν_{MnO} must be included in the model calculation of ν_{MnO} frequency in this case. Indeed, with the use of normal coordinate analysis (NCA) techniques, which will be discussed in the next section, it was possible to calculate the observed 21 cm^{-1} upshift with a four-atom H–O–Mn^{IV}=O model using physically reasonable force constants and structure parameters, as given in Table 6.[67] This calculation also predicted the correct ^{18}O frequency shift, which landed further confidence in the proposed *trans* HO–Mn^{IV}=O structure.

4.4 Normal Mode Analysis

A molecule containing N atoms has $3N$ degrees of freedom. Six of these (five for linear molecules) are the translations and rotations of the molecule as a whole, leaving $3N-6$ ($3N-5$) internal degrees of freedom that comprise the molecular vibrations. These can be calculated by solving Newton's equations of motion, provided that the geometry of the molecule, the masses of the atoms, and the force constants (collectively called the force field) are all known. The result of this calculation is a set of $3N-6$ eigenvalues, which are the vibrational frequencies, and the corresponding eigenvectors, which describe the motion. A proper solution leads to periodic motions, in which the atoms move along recurring paths. These are called the normal coordinates, Q, which describe the normal modes. For example, the arrows in Figure 43 represent the eigenvectors for the three normal modes of a typical bent triatom. To the extent that the harmonic approximation is valid, each normal mode involves motion on a separate parabolic potential, like the diatomic potential in Figure 37, but with the normal coordinate as the abscissa. For methods of carrying out the calculations the reader is referred to Herzberg,[1] Wilson et al.,[57] Gans,[61], Diem,[6] and Nakamoto.[2]

The main problem in normal coordinate analysis (NCA) is that the force field is not known in advance. Moreover it cannot be determined unambiguously from the vibrational spectra because, except for diatoms, the number of force constants always exceeds the number of modes. Even for a bent Y–X–Y triatom there are three modes but four valence force constants (two principal constants, stretch and bend, and two interaction constants, stretch-stretch and stretch-bend), and a fifth constant (called a Urey–Bradley constant) should be added to take into account the nonbonded repulsion between the terminal atoms. Thus the problem is overdetermined.

Historically, vibrational spectroscopists have adopted an empirical approach, using force constants transferred from small molecules having bonding features of the larger molecule under study. These force constants can then be adjusted incrementally in order to minimize the discrepancy between calculated and experimental frequencies. While this procedure cannot lead to a unique solution, it is possible to constrain the force constants to physically reasonable values, on the basis of the extensive body of empirical analyses. The purpose of the exercise is to provide a reliable guide to the assignment of the vibrational spectra. In this situation it is important to have isotopic data, since the calculation of isotope shifts is a stringent test of the adequacy of the force field, as well as the assignments. An eigenvector is accurately calculated if the isotope shifts of the atoms participating in the motion are reproduced, whether or not the force field is unique. An example of a simple NCA calculation has been given in Table 6 for the four-atom HO–Mn^{IV}=O unit of ($HOMn^{IV}O$)TMPyP (Section 4.3).

4.5 *Ab Initio* Calculation of Spectra

The alternative approach is to calculate the force constants from a potential surface obtained via an *ab initio* electronic structure calculation.[78] Until recently sufficiently accurate *ab initio* potentials have been beyond reach computationally for all but the simplest molecules. However, this limitation has been rapidly receding, thanks to the power of current computers, and especially thanks to improvements in electronic structure methodology. The currently favored approach, Density Functional Theory [DFT] has the advantage of including the effects of electron correlation in a computationally efficient manner.[79,80] The latest versions of DFT give remarkably accurate calculations for quite complex molecules, as illustrated in Table 7 for nickel porphine.[41] Scaling of the force constants is required, since *ab initio* potentials systematically overestimate the vibrational frequencies,[81] but only four scale factors are required for the in-plane NiP modes, one for each type of bond, and the values are the same as for free-base porphine.[82–84] The root-mean-square deviations for 43 observed frequencies is only 0.8%, compared to 2.1% for the best available empirical force field from NCA.[41]

Table 7 is arranged according to the local coordinate, as defined in Figure 50. Although the normal mode eigenvectors[38,39] are complex (see Figure 25), they can be decomposed into local coordinate contributions, and modes of different symmetry but similar compositions have similar frequencies. For example, modes involving out-of-phase stretching of adjacent $C_\alpha C_m$ methine bridge bonds, have frequencies near 1600 cm^{-1}, while modes involving in-phase stretching of the same bonds are about 50 cm^{-1} lower, on average. This difference mainly reflects the restricted motion of the central atom of a triatomic unit with >90° bond angle when the two bonds stretch in-phase (see Section 4.2). However, other factors are also important, including the positive interaction constant between adjacent bonds in an aromatic ring, and the fact that the two types of stretches contribute to different symmetry blocks, and mix with different coordinates. In particular, the $C_\beta C_\beta$ stretch mixes with the in-phase $C_\alpha C_m$ stretch in the A_{1g} block, so that ν_2 and ν_3 are mixtures of both coordinates. However, in the B_{1g} block the $C_\beta C_\beta$ stretch mixes with the out-of-phase $C_\alpha C_m$ stretch, producing the ν_{10} and ν_{11} pair. The $C_\beta C_\beta$ stretch does not contribute to the A_{2g} or B_{2g} blocks, so that the modes ν_{19} and ν_{28} are relatively pure out-of-phase and in-phase $C_\alpha C_m$ stretches, respectively.

The DFT calculation produces all the normal modes, out-of-plane (oop) as well as in-plane, whereas the empirical approach (NCA) is severely limited for oop modes, because their spectroscopic activity is severely limited by the D_{4h} selection rules.[40]

Table 7 **Observed and Calculated Vibrational Frequencies (cm^{-1}) for Nickel Porphine and Its Isotopomers[a,b]**

Local coordinate	A_{1g}	B_{1g}	A_{2g}	B_{2g}	E_u
$\nu(C_mH)$	ν_1			ν_{27}	ν_{36}
	—			—	—
	3097 (0, 778, 0, 778)			*3041 (0, 772, 0, 788)*	*3042 (0, 772, 0, 772)*
	3073 (0, 801, 0, 801)			3073 (0, 802, 0, 802)	3073 (0, 801, 0, 802)
$\nu(C_\alpha H_m)_{asym}$		ν_{10}	ν_{19}		ν_{37}
		1650 (—, 8, 4, 14)	**1611 (—, 13, 6, 19)**		**1592 (—, —, —, —)**
		1649 (0, 8, 3, 11)	*1602 (2, 9, 1, 10)*		*1628 (1, 8, 2, 10)*
		1659 (0, 11, 3, 13)	1615 (0, 16, 2, 18)		1595 (0, 15, 5, 22)
$\nu(C_\beta C_\beta)$	ν_2	ν_{11}			ν_{38}
	1574 (—, 8, 22, 28)	**1505 (—, 1, 51, 51)**			**1547 (—, 4, 20, 27)**
	1564 (0, 6, 21, 31)	*1529 (0, 0, 41, 42)*			*1550 (1, 5, 27, 35)*
	1585 (0, 7, 22, 31)	1521 (0, 2, 51, 53)			1558 (1, 6, 23, 31)
$\nu(C_\alpha C_m)_{sym}$	ν_3			ν_{28}	ν_{39}
	—			**1505 (—, 28, 18, 24)**	**1462 (—, 4, 42, 42)**
	1459 (0, 13, 19, 30)			*1492 (6, 12, 9, 22)*	*1472 (5, 12, 20, 30)*
	1469 (0, 6, 28, 32)			1504 (3, 7, 15, 23)	1467 (0, 2, 41, 44)
ν(Pyr quarter-ring)			ν_{20}	ν_{29}	ν_{40}
			1354 (—, 7, 41, 76)	**1368 (—, 0, 44, 53)**	**1385 (—, 12, —, —)**
			1351 (—, 0, 30, 53)	*1381 (0, 0, 55, 55)*	*1379 (12, 1, 37, 39)*
			1356 (0, 5, 30, 72)	1365 (3, 3, 36, 38)	1397 (7, 44, 4, 45)
ν(Pyr half-ring)$_{sym}$	ν_4	ν_{12}			ν_{41}
	1376 (—, 2, 9, 9)	**1383 (—, 62, 59, 68)**			**1319 (—, 4, 54, 54)**
	1384 (2, 2, 23, 23)	*1319 (11, 10, 10, 26)*			*1338 (10, 7, 23, 39)*
	1384 (8, 1, 9, 9)	1394 (6, 63, 66, 65)			1322 (2, 1, 45, 52)
$\delta(C_mH)$		ν_{13}	ν_{21}		ν_{42}
		1185 (—, 247, 6, 237)	**1139 (—, 229, 117, 291)**		**1150 (—, —, 54, 204)**
		1197 (1, 260, 10, 267)	*1120 (1, 255, −111, 272)*		*1138 (2, −96, 235, 247)*
		1193 (2, 247, 5, 241)	1142 (7, 227, −116, 296)		1151 (7, 182, 57, 203)

Table 7 (*Continued*)

Local coordinate	A_{1g}	B_{1g}	A_{2g}	B_{2g}	E_u
$\nu(C_\beta H)_{sym}$	ν_5	ν_{14}			ν_{43}
	—	—			—
	3097 (0, 0, 778, 778)	*3097 (0, 0, 779, 779)*			*3097 (0, 0, 778, 778)*
	3121 (0, 0, 791, 791)	3120 (0, 0, 791, 791)			3120 (0, 0, 791, 791)
$\nu(\text{Pyr half-ring})_{asym}$			ν_{22}	ν_{30}	ν_{44}
			1005 (—, **−7**, **−90**, **−184**)	**1062** (—, —, —, —)	**1037** (—, **10**, **126**, —)
			992 (4, −24, −78, −195)	*1036 (19, 19, −14, 2)*	*1012 (13, −17, −53, −122)*
			1006 (13, −7, −93, 186)	1062 (20, 14, 96, −10)	1037 (16, 11, 126, 142)
$\nu(C_\beta H)_{asym}$			ν_{23}	ν_{31}	ν_{45}
			—	—	—
			3087 (0, 0, 789, 789)	*3088 (0, 0, 789, 789)*	*3088 (0, 0, 790, 790)*
			3101 (0, 0, 811, 811)	3101 (0, 0, 812,812)	3101 (0, 0, 812, 812)
$\delta(\text{Pyr def})_{asym}$			ν_{24}	ν_{32}	ν_{46}
			806 (—, **23**, **18**, **39**)	**819** (—, **4**, **20**, **39**)	**806** (—, —, —, —)
			796 (0, 34, 19, 46)	*836 (0, 8, 49, 54)*	*820 (1, 13, 26, 37)*
			808 (0, 23, 19, 36)	825 (0, 8, 25, 60)	806 (1, 11, 37, 42)
$\nu(\text{Pyr breathing})$	ν_6	ν_{15}			ν_{47}
	995 (—, **3**, **8**, **10**)	**1003** (—, **−17**, —, —)			**995** (**0**, **−4**, **12**, **−3**)
	1020 (20, 3, 6, 9)	*1015 (14, −27, 19, −6)*			*1003 (20, 5, 5, 12)*
	998 (19, 2, 8, 10)	999 (15, −21, 15, −24)			993 (15, −5, 9, 3)
$\delta(\text{Pyr def})_{sym}$	ν_7	ν_{16}			ν_{48}
	732 (—, **21**, **12**, **23**)	**732** (—, **67**, **27**, **77**)			**745** (—, **46**, **45**, **67**)
	725 (0, 20, 4, 24)	*710 (3, 76, 1, 76)*			*722 (1, 58, 18, 68)*
	732 (0, 22, 6, 25)	740 (3, 69, 32, 78)			746 (1, 52, 35, 68)
$\delta(\text{Pyr rot.})$			ν_{25}	ν_{33}	ν_{49}
			429 (—, **10**, **25**, **32**)	**435** (—, **3**, **36**, **36**)	**366** (—, —, —, —)
			416 (1, 4, 17, 24)	*424 (0, 0, 35, 35)*	**352** (**3**, **3**, **16**, **19**)
			437 (4, 10, 23, 32)	427 (0, 0, 35, 35)	371 (3, 6, 5, 11)

Table 7 (*Continued*)

Local coordinate	A_{1g}	B_{1g}	A_{2g}	B_{2g}	E_u
ν(NiN)	ν_8	ν_{18}			ν_{50}
	369 (—, **2**, **9**, **10**)	**237** (—, **0**, **5**, **5**)			**420** (—, —, —, —)
	378 (2, 1, 7, 8)	*245 (1, 0, 4, 4)*			*413 (2, 2, 10, 12)*
	362 (3, 1, 10, 9)	233 (2, 0, 4, 4)			411 (1, 1, 19, 20)
$\delta(C_\beta H)_{asym}$			ν_{26}	ν_{34}	ν_{51}
			1317 (—, **68**, **466**, **405**)	**1195** (—, **0**, —, **245**)	**1250** (—, —, —, —)
			1319 (8, 40, 464, 467)	*1183 (0, 1, 218, 224)*	*1272 (2, 372, 63, 369)*
			1329 (5, 72, 476, 415)	1193 (7, 1, 109, 233)	1263 (6, 40, 38, 121)
$\delta(C_\beta H)_{sym}$	ν_9	ν_{17}			ν_{52}
	1066 (—, **1**, **291**, **293**)	**1058** (—, **−22**, **292**, **292**)			**1064** (0, **4**, **270**, **300**)
	1079 (1, 2, 297, 298)	*1064 (0, −2, 286, 287)*			*1074 (1, 0, 295, 299)*
	1067 (2, 0, 294, 297)	1062 (0, −18, 292, 294)			1061 (1, −4, 267, 292)
δ(Pyr trans)				ν_{35}	ν_{53}
					289 (—, —, —, —)
				202 (2, 2, 1, 3)	*281 (0, 1, 4, 5)*
				232 (2, 3, 2, 5)	293 (0, 0, 9, 9)

[a] ν_i ← vibrational mode (see Figure 50 for definitions of local coordinates).[39]
Frequency (Δ^{15}**N**, Δ**d**$_4$, Δ**d**$_8$, Δ**d**$_{12}$) ← observed.[39]
Frequency ($\Delta^{15}N$, Δd_4, Δd_8, Δd_{12}) ← calculated with valence force field.[39]
Frequency (Δ^{15}N, Δd$_4$, Δd$_8$, Δd$_{12}$) ← calculated with SQM force field.[41]
[b] Reassigned experimental frequencies are underlined.[41]

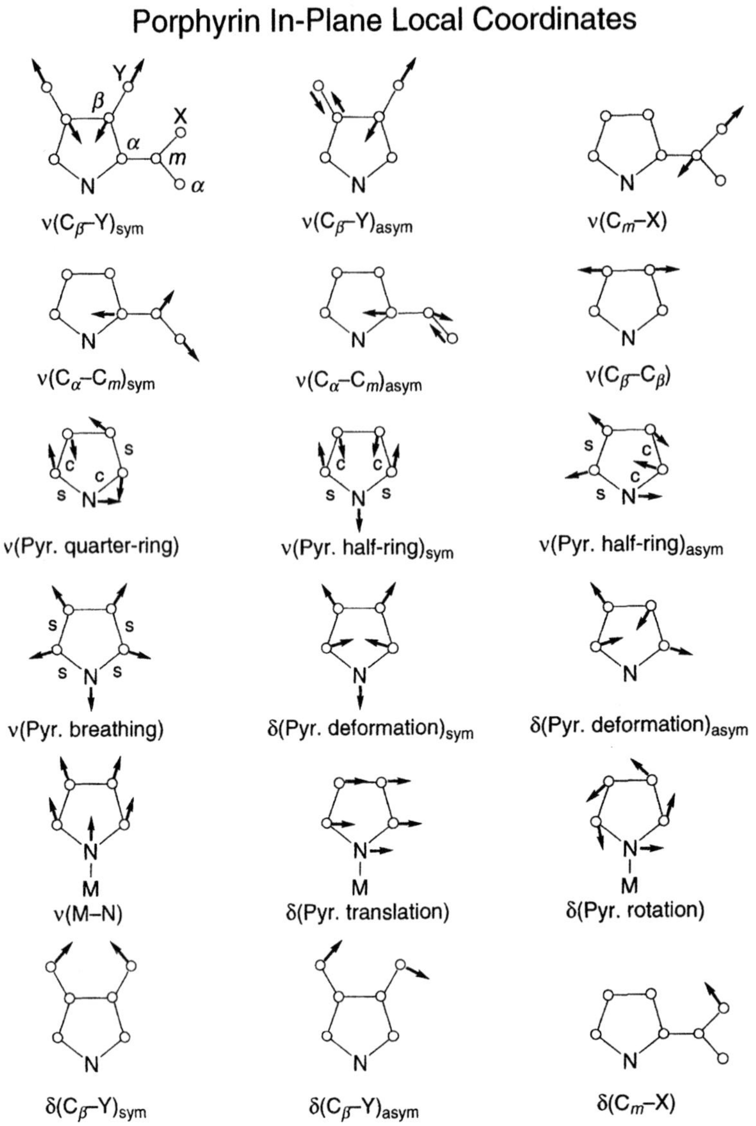

Figure 50. Illustration of the local coordinates used in classifying the in-plane porphyrin skeletal modes.[39] The stretching motion of the $C_\alpha N$ and $C_\alpha C_\beta$ pyrrole bonds are considered collectively, and classified according to the phases of the individual bonds indicated by "s" (stretching) and "c" (contraction).

The DFT frequencies are in good agreement with experiment for the A_{2u} modes, which are active in the IR spectrum (Table 7).

Moreover, the DFT calculation is informative about intrinsic distortions of the molecule under study, because distortion modes produce imaginary frequencies.

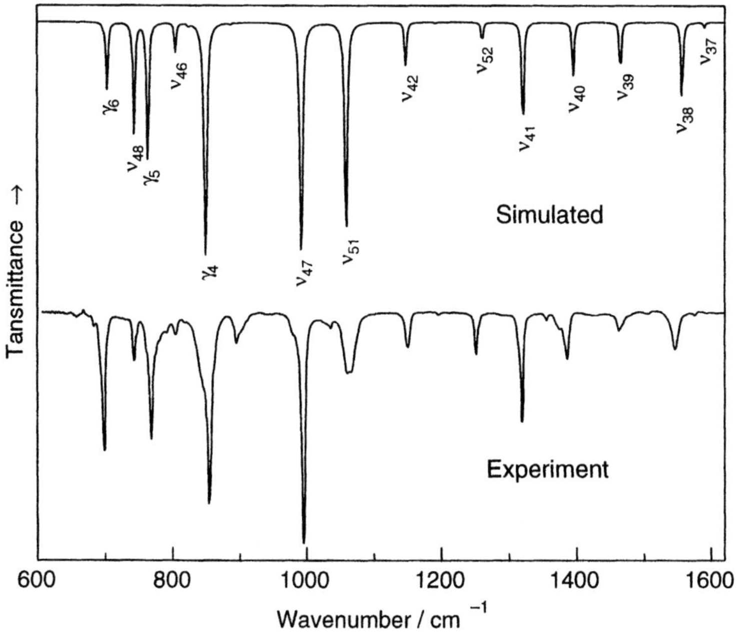

Figure 51. Comparison of the experimental IR spectrum (600–1600 cm^{-1}) of NiP with the SQM/DFT simulated spectrum using a Lorentzian line profile of 4 cm^{-1} half-width.

Thus, when the NiP was constrained to a planar geometry (D_{4h} symmetry), an imaginary frequency was calculated for γ_{14}, an oop mode that leads to a ruffling of the porphyrin (D_{2d} distortion).[85] When the planarity constraint was relaxed, the DFT energy was minimized for a slightly ruffled structure. This result provides striking theoretical support for Hoard's explanation of the interplay between metal–$N_{pyrrole}$ bond length and the porphyrin planarity.[86] When this bond is significantly shorter than the natural radius of the porphyrin cavity (~2.00 Å), then the pyrrole rings are drawn toward the center through a ruffling distortion of the macrocycle. Since the planar geometry maximizes the π overlap, the equilibrium structure finds a balance between the energy gain from short bonds and the energy penalty to the π system from ruffling. The Ni–$N_{pyrrole}$ distance (~1.96 Å in the D_{4h} molecule) is just small enough to induce ruffling (~1.93 Å in the D_{2d} molecule); the DFT predicts an energy gain of only 0.1 kcal/mol for the ruffled structure, and this energy is easily overcome by intermolecular forces in the crystal, which contains an essentially planar structure.[87] Interestingly, more pronounced ruffling is observed for nickel octaethylporphyrin (NiOEP), in one of its polymorphs,[88] presumably because the steric clashes of the substituent groups amplify the ruffling effect of the metal. This effect is even more pronounced in sterically crowded porphyrins such as nickel octaethyltetraphenylporphyrin (NiOETPP), which show large out-of-plane distortions.[89–91]

Finally, *ab initio* calculations readily yield IR and Raman intensities. For NiP, these produced excellent replicas of the experimental spectra (Figures 52 and 53),

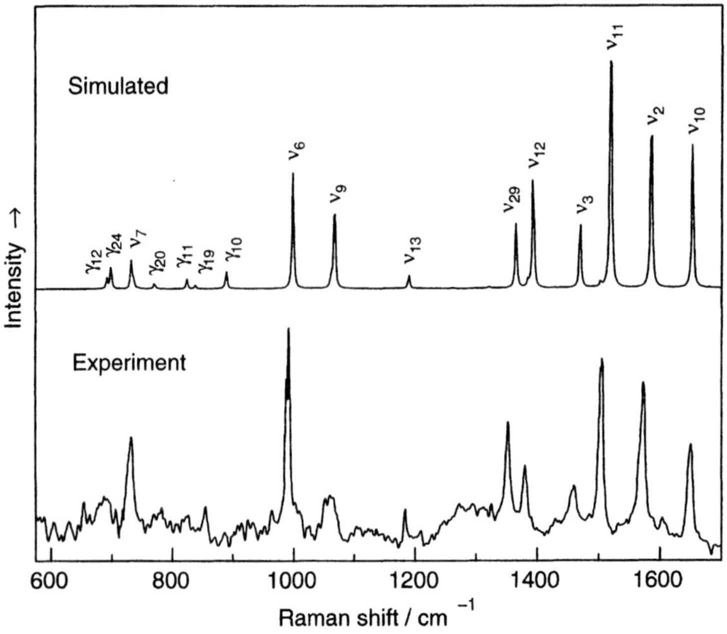

Figure 52. Comparison of the experimental off-resonance Raman spectrum of NiP (600–1700 cm^{-1}) with the SQM/DFT simulated spectrum using a Lorentzian line profile of 4 cm^{-1} half-width and the ν^4 correction.

leaving no doubt about the mode assignments. To evaluate intensities, the Raman spectra were obtained off-resonance, using 1.064 μm laser excitation and Fourier-transform spectrometry. A striking observation is the high intensity of several non-totally symmetric B_{1g} modes, and this effect is fully captured in the calculated spectra (Figure 52). Normally, non-resonant Raman spectra are dominated by totally symmetric modes, which have the largest effect on the molecular polarizability derivative (see Section 2.4). In porphyrin, however, rectangular B_{1g} modes can strongly alter the polarizability if the eigenvectors are composed of rectangular distortions of the local coordinates (e.g., the pyrrole rings), as is the case for the strong FT-Raman bands.[41]

The porphyrin ruffling relaxes the selection rules, and this effect too was accurately captured by the DFT calculation. Predicted intensities were too small for detection for most of the newly-activated modes, consistent with experiment. But, one E_g mode, γ_{22}, was predicted to have significant IR intensity, and is indeed observed in the FTIR spectrum (Figure 51). This mode involves a swiveling motion of the pyrrole rings, and generates a significant dipole in the ruffled structure.

Calculation of resonance Raman intensities is also desirable for spectral interpretation, but this requires evaluating the geometrical distortion in the excited state. *Ab initio* methods are not yet up to this task, although new developments are on the horizon. In the meantime, the semiempirical INDO method, has been found to give

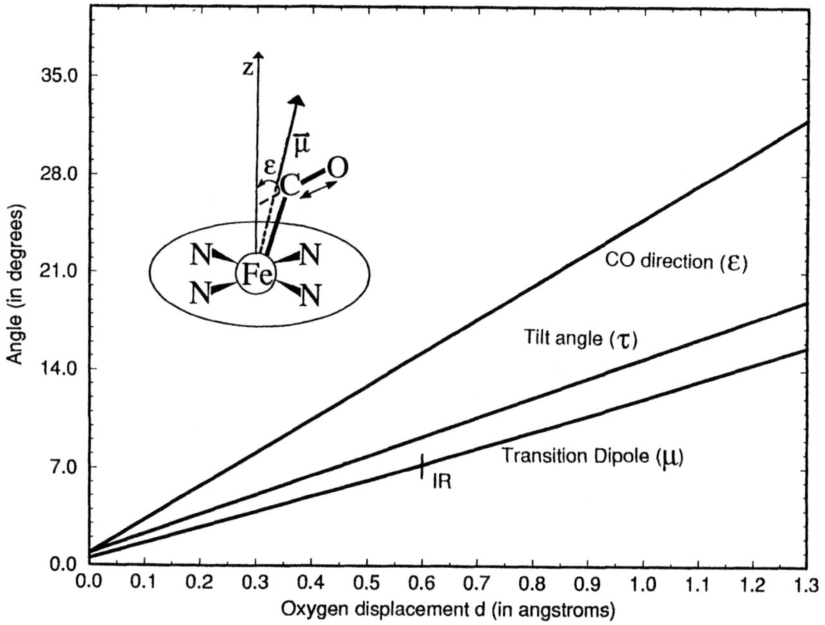

Figure 53. Calculated C–O stretching transition dipole direction (μ) along the DFT minimum energy distortion path.

Table 8 Energy Difference Between Planar and Nonplanar Structure of Nickel Porphine[41]

Method/Basis Set	Energy (a.u.)	ΔE (kcal/mol)[a]
DFT[b]/3–21G	−2483.975 448	
DFT[c]/3–21G	−2483.979 384	2.470
DFT[b]/6–311G	−2496.694 781	
DFT[c]/6–311G	−2496.694 980	0.125
DFT[b]/6–31G(d),VTZ[d]	−2496.776 288	
DFT[c]/6–31G(d),VTZ[d]	−2496.776 290	0.002
DFT[b]/6–311G(d)	−2496.953 711	
DFT[c]/6–311G(d)	−2496.953 879	0.105

[a] $\Delta E = E(D_{2d}) - E(D_{4h})$.
[b] Minimum corresponding to D_{4h} symmetry.
[c] Minimum corresponding to D_{2d} symmetry.
[d] 6–31G(d) for H, C, and N; VTZ for Ni.

excellent results, in conjunction with a DFT force field in the ground state.[92,93] INDO gives reliable excited state energies, but is not appropriate for calculating geometries. However, the geometric distortion can be gauged with sufficient accuracy by converting INDO-derived bond order changes to geometry.[92] An even better method is to find the slope of the excited state potential by calculating the INDO energy at a series of fixed geometries along each normal coordinate.[93] In this manner, both Soret-enhanced (A-term)[92] and Q-band-enhanced (B-term)[93] RR spectra of NiP have been successfully modeled.

DFT can also be used to model distortions that may result from macromolecular interactions, as in a protein. For example, the effect of steric distortion of carbon monoxide bound to a heme was investigated by constraining the CO to distort along a minimum energy path in a heme–CO model that was otherwise optimized via DFT.[94–96] The distortion potential was found to be surprisingly soft because of an electronic interaction between CO bending and tilting coordinates. In addition, the transition dipole moment for the CO stretching mode was found to remain much closer to the heme normal than the CO bond direction, because π bonding delocalizes the oscillating electrons even if the nuclear motions are localized to the CO unit (Figure 53).[96] This finding resolved an apparent discrepancy between X-ray crystallography and polarized IR spectroscopy with respect to the geometry of myoglobin–CO. Crystallography has consistently produced bent structures,[97,98] while IR spectroscopy[99,100] gives an angle of $<7°$ from the heme normal. However, the intuitively appealing assumption in the IR method that the measured angle corresponds to the CO direction is shown by the DFT calculations to be invalid. A 7° transition dipole was found to be consistent with the crystallographic geometries. However, because this distortion requires only a small energy, it is unlikely to play an important role in controlling CO binding, as has been widely supposed.[101,102]

4.6 Excited States

A plot of RR intensity against the excitation wavelength is called an *excitation profile* (EP). Such plots track the absorption spectrum in a general way, but differ from it in detail, frequently showing more structure. There are three basic reasons for such differences: (1) the absorption dipole change depends on the vibrational overlap integrals for the transition from the ground to the excited state, but the Raman polarizability change depends on the product of overlap integrals for the up and down transitions (Figure 21); the latter can have a distinctly different wavelength dependence. (2) The absorption oscillator strength depends on the sum of the squares of the overlap integrals for different modes, whereas the Raman intensity is proportional to the square of the polarizability change, so that the sum over the the terms in Eq. (27) is formed before squaring. Consequently, the scattering contributions from the individual vibrational levels of the excited state can either add or subtract (the denominators change signs as the laser frequency is tuned through the local resonance, and the numerators can be positive or negative); contributions from nearby excited states can also interfere negatively as well as positively. (3) The EP emphasizes electronic factors that are particularly sensitive to the mode under consideration, while the absorption shows no such selectivity. Consequently, the EP's can provide information about the excited state that may be hidden in the absorption spectrum.

The quantitative treatment of RR intensities and EP's, as well as of the absorption spectrum, is of considerable current interest, because it can provide interesting information about the resonant excited states. If the system is well behaved, it is possible to calculate EP's from the absorption spectrum, and vice versa, with only the coupling strengths of the individual normal modes as adjustable parameters, using transform theory.[103,104] This methodology has been applied extensively to heme proteins by Champion.[105]

Another approach is to model both the EP's and the absorption spectrum using the sum-over-states formalism introduced above. The equations as written are inadequate for this purpose, however, since they tacitly assume that only a single normal mode contributes to the intensity of a given RR band. In reality the vibrational wavefunctions are multidimensional, and each one contains all the normal modes in some quantum state. Only one normal mode at a time is excited in the vibrational fundamentals, but the numerators of the terms in Eq. (27) should all be multiplied by Franck–Condon products $\Pi \langle i|v\rangle\langle v|i\rangle$, one for each normal mode that is not excited. These products are close to unity only if the excited state is not significantly displaced along the mode in question. Otherwise they depart from unity and become significant determinants of the RR intensity and the EP. Multimode effects considerably complicate the sum-over-states calculation,[106] although computer programs are available that can handle a substantial number of modes.

Another alternative for the investigation of excited state properties from RR intensities is the time-dependent theory developed by Heller and coworkers.[107–109] In this approach, the sum-over-states expressions are transformed from a frequency to a time basis, and the resulting equations can be interpreted as reflecting the evolution of the overlap of a wave-packet moving along the excited state potential surface with the initial and final vibrational wavefunctions of the ground state. This approach has the conceptual advantage of providing a dynamical description of the excited state, and also has computational advantages for systems with many active modes, or when there are large distortions in the excited state. A full description of applications to RR intensities and EP's has been given by Myers and Mathies.[106]

5 CONCLUDING REMARKS

Vibrational spectroscopy is a powerful method for structural analysis, because its features are sensitive to the details of bonding and conformation. The vibrational frequencies depend on the strengths of bonds and the geometrical arrangement of the bonded atoms. Intensities depend on molecular symmetry and on the electronic structure. IR and Raman spectra serve as reporters for these fundamental properties of molecules, and can sensitively monitor changes resulting from chemical reactions and intermolecular forces.

The utility of vibrational spectroscopy has been greatly increased by recent technical advances:

1. The large dynamic range of FTIR instrumentation permits the acquisition of highly accurate difference spectra, so that small changes in complex samples, such as single residue alterations in proteins, are readily detected.

2. The wide wavelength range of available lasers allows the resonance Raman effect to be exploited for multiple chromophores. In the context of metalloproteins, Raman modes can be enhanced both for the metal coordination group and for aromatic groups elsewhere in the protein, by appropriate tuning of the laser in the visible and UV regions.
3. Reliable methods have become available for the calculation of vibrational frequencies and intensities from accurate computations of force fields and of electronic structure.

Thus, the use of vibrational spectroscopy will spread increasingly into the mainstream of structural chemistry. Its exploitation requires a clear understanding of the relationships between molecular vibrations and structure, so that the experiment can be properly designed to test specific structural hypotheses. We hope that the principles and examples discussed in this chapter serve to illuminate these relationships.

REFERENCES

1. Herzberg, G. *Molecular Spectra and Molecular Structure. II. Infrared and Raman Spectra of Polyatomic Molecules*; Van Nonstrand Reinhold: New York, 1945.
2. Nakamoto, K. *Infrared and Raman Spectra of Inorganic and Coordination Compounds*, 5th ed.; John Wiley & Sons, Inc.: New York, 1997; Vol. B.
3. Colthup, N.B.; Daley, L.H.; Wiberley, S.E. *Introduction to Infrared and Raman Spectroscopy*, 3rd ed.; Academic Press: New York, 1990.
4. Doolish, F.R.; Fateley, W.G.; Bentley, F.F. *Characteristic Raman Frequencies of Organic Compounds*, 2 ed.; Academic Press: New York, 1974.
5. Parker, F.S. *Application of Infrared, Raman and Resonance Raman Spectroscopy in Biochemistry*; Plenum Press: New York, 1983.
6. Diem, M. *Introduction to Modern Vibrational Spectroscopy*; John Wiley & Sons: New York, 1993.
7. Harris, D.C.; Bertolucci, M.D. *Symmetry and Spectroscopy*; Oxford University Press: New York, 1978.
8. Fadini, A.; Schnepel, F.-M. *Vibrational Spectroscopy: Methods and Applications*; Ellis Horwood: Chichester, 1989.
9. Carey, P.R. *Biochemical Applications of Raman and Resonance Raman Spectroscopies*; Academic Press: New York, 1982.
10. Ferraro, J.R.; Nakamoto, K. *Introductry Raman Spectroscopy*; Academic Press: New York, 1994.
11. *Practical Raman Spectroscopy*; Gardiner, D.J.; Graves, P.R., Eds.; Springer-Verlag: Berlin, 1989.
12. *Biological Applications of Raman Spectroscopy*; Spiro, T.G., Ed.; Wiley-Interscience: New York, 1987; Vol. 1.
13. *Fourier-Transform Raman Spectroscopy. From Concept to Experiment*; Chase, D.B.; Rabolt, J.F., Eds.; Academic Press: New York, 1994.

14. *Modern Techniques in Raman Spectroscopy*; Laserna, J.J., Ed.; John Wiley & Sons: New York, 1996.
15. *Surface Enhanced Raman Scattering*; Chang, R.K.; Furtak, T.H., Eds.; Plenum Press: New York, 1982.
16. Gerwert, K.; Souvignier, G.; Hess, B. *Proc. Natl. Acad. Sci. USA* **1990**, *87*, 9774.
17. Long, D.A. *Raman Spectroscopy*; McGraw-Hill: New York, 1977.
18. Nakamoto, K.; Czernuszewicz, R.S. *Meth. Enzym.* **1993**, *226*, 259.
19. Raman, C.V.; Krishnan, K.S. *Nature* **1928**, *121*, 501.
20. Raman, C.V.; Krishnan, K.S. *Nature* **1928**, *121*, 619.
21. Landsberg, G.S.; Mandeltsam, L.J. *Naturwissenschaften* **1928**, *16*, 557.
22. Czernuszewicz, R.S.; Johnson, M.K. *Appl. Spectrosc.* **1983**, *37*, 297.
23. Czernuszewicz, R.S. *Appl. Spectrosc.* **1986**, *40*, 571.
24. Spiro, T.G.; Strekas, T.C. *Proc. Natl. Acad. Sci. USA* **1972**, *69*, 2622.
25. Spiro, T.G.; Czernuszewicz, R.S. *Meth. Enzym.* **1995**, *246*, 416.
26. Spiro, T.G.; Li, X.-Y. In *Biological Applications of Raman Spectroscopy*; Spiro, T.G., Ed.; Wiley-Interscience: New York, 1988; Vol. 3; pp. 1–38.
27. Spiro, T.G.; Czernuszewicz, R.S.; Li, X.-Y. *Coord. Chem. Rev.* **1990**, *100*, 514.
28. Rousseau, D.L.; Williams, D.F. *J. Chem. Phys.* **1976**, *64*, 3519.
29. Albrecht, A.C. *J. Chem. Phys.* **1961**, *34*, 1476.
30. Tang, J.; Albrecht, A.C. *J. Phys. Chem.* **1968**, *49*, 1144.
31. Albrecht, A.C. *J. Chem. Phys.* **1960**, *33*, 156.
32. Siebrand, W.; Zgierski, M.Z. *J. Chem. Phys.* **1975**, *71*, 3561.
33. Spiro, T.G.; Stein, P. *Annu. Rev. Phys. Chem.* **1977**, *28*, 501.
34. Siebrand, W.; Zgierski, M.Z. In *Excited States*; Lim, E.C., Ed.; Academic Press: New York, 1979; Vol. 4; pp. 1–136.
35. Gouterman, M. In *Porphyrins*; Dolphin, D., Ed.; Academic Press: New York, 1979; Vol. 3; pp. 1–156.
36. Rankin, J.G.; Czernuszewicz, R.S. *Org. Geochem.* **1993**, *20*, 521.
37. Shelnutt, J.A.; Cheung, L.D.; Chang, R.C.C.; Yu, N.-T.; Felton, R.H. *J. Chem. Phys.* **1977**, *66*, 3387.
38. Li, X.-Y.; Czernuszewicz, R.S.; Kincaid, J.R.; Stein, P.; Spiro, T.G. *J. Phys. Chem.* **1990**, *94*, 47.
39. Li, X.-Y.; Czernuszewicz, R.S.; Kincaid, J.R.; Su, Y.O.; Spiro, T.G. *J. Phys. Chem.* **1990**, *94*, 31.
40. Li, X.-Y.; Czernuszewicz, R.S.; Kincaid, J.R.; Spiro, T.G. *J. Am. Chem. Soc.* **1989**, *111*, 7012.
41. Kozlowski, P.M.; Rush III, T.S.; Jarzecki, A.A.; Zgierski, M.Z.; Chase, B.; Piffat, C.; Ye, B.-H.; Li, X.-Y.; Pulay, P.; Spiro, T.G. *J. Phys. Chem.* **1999**, in press.
42. Yan, Q. Ph.D. Thesis. University of Houston, 1996.
43. Ziegler, L.D.; Hudson, B. *J. Chem. Phys.* **1981**, *74*, 982.
44. Clark, R.J.H.; Stewart, B. *Struct. Bonding* **1979**, *36*, 1.
45. Clark, R.J.H.; Dines, T.J.D. *Angew. Chem. Int. Ed. Engl.* **1986**, *25*, 131.
46. Czernuszewicz, R.S.; Nakamoto, K.; Strommen, D.P. *J. Am. Chem. Soc.* **1982**, *104*, 1515.

47. Czernuszewicz, R.S.; Sheats, J.E.; Spiro, T.G. *Inorg. Chem.* **1987**, *26*, 2063.
48. Fodor, S.P.A.; Copeland, R.A.; Grygon, C.A.; Spiro, T.G. *J. Am. Chem. Soc.* **1989**, *111*, 5509.
49. Sanders-Loehr, J.; Wheeler, W.D.; Shiemke, A.K.; Averill, B.A.; Loehr, T.M. *J. Am. Chem. Soc.* **1989**, *111*, 8084.
50. Sheats, J.E.; Czernuszewicz, R.S.; Dismukes, G.C.; Rheingold, A.L.; Petrouleas, V.; Stubbe, J.-A.; Armstrong, W.H.; Beer, R.H.; Lippard, S.J. *J. Am. Chem. Soc.* **1987**, *109*, 1435.
51. Bond, M.R.; Czernuszewicz, R.S.; Dave, B.C.; Yan, Q.; Mohan, M.; Verastque, R.; Carrano, C.J. *Inorg. Chem.* **1995**, *34*, 5857.
52. Czernuszewicz, R.S.; Yan, Q.; Bond, M.R.; Carrano, C.J. *Inorg. Chem.* **1994**, *33*, 6116.
53. Adams, D.M. *Metal-Ligand and Related Vibrations*; Edward Arnold: London, 1967.
54. Ferraro, J.R. *Low Frequency Vibrations of Inorganic and Coordination Compounds*; Plenum Press: New York, 1971.
55. Szymanski, H.A. *Interpreted Infrared Spectra*; Plenum Press: New York, 1964, 1966, 1967; Vol. 1–3.
56. Czernuszewicz, R.S. In *Methods in Molecular Biology*; Jones, C., Mulloy, B., Thomas, A.H., Eds.; Humana Press: Totowa, NJ, 1993; Vol. 17; p. 345.
57. Wilson, E.B.; Decius, J.C.; Gross, P.C. *Molecular Vibrations: The Theory of Infrared and Raman Spectra*; McGraw-Hill: New York, 1955.
58. Steele, D. *Theory of Vibrational Spectroscopy*; W.B. Saunders: Philadelphia, 1971.
59. Woodward, L.A. *Introduction to the Theory of Molecular Vibrations and Vibrational Spectroscopy*; Oxford University Press: London, 1972.
60. Cyvin, S.J. *Molecular Structure and Vibrations*; Elsevier: Amsterdam, 1972.
61. Gans, P. *Vibrating Molecules*; Chapman and Hall: London, 1971.
62. Burgi, H.-B.; Dunitz, J.D. *J. Am. Chem. Soc.* **1987**, *109*, 2924.
63. Badger, R.M. *J. Chem. Phys.* **1935**, *3*, 710.
64. Conradson, S.P.; Sattelberger, A.P.; Woodruff, W.M. *J. Am. Chem. Soc.* **1988**, *110*, 1309.
65. Herschbach, D.R.; Laurie, V.W. *J. Chem. Phys.* **1961**, *35*, 458.
66. Su, Y.O.; Czernuszewicz, R.S.; Miller, L.A.; Spiro, T.G. *J. Am. Chem. Soc.* **1988**, *110*, 4150.
67. Czernuszewicz, R.S.; Su, Y.O.; Stern, M.K.; Macor, K.A.; Kim, D.; Groves, J.T.; Spiro, T.G. *J. Am. Chem. Soc.* **1988**, *110*, 4158.
68. Czernuszewicz, R.S.; Macor, K.A. *J. Raman Spectrosc.* **1988**, *19*, 553.
69. Gray, P.; Waddington, T.C. *Trans. Faraday Soc.* **1956**, *53*, 901.
70. Jones, L.H. *J. Chem. Phys.* **1955**, *23*, 2105.
71. Burke, J.M.; Kincaid, J.R.; Spiro, T.G. *J. Am. Chem. Soc.* **1978**.
72. Fleischer, E.; Srivastava, T.S. *J. Am. Chem. Soc.* **1969**, *91*, 2403.
73. Ogden, J.S.; Turner, J.J. *J. Chem. Soc. A* **1967**, 1483.
74. Rochkind, M.M.; Pimentel, G.C. *J. Chem. Phys.* **1965**, *42*, 1361.
75. Kanamori, K.; Teraoka, M.; Maeda, H.; Okamoto, K. *Chem. Lett.* **1993**, 1731.
76. Dean, N.S.; Cooper, J.K.; Czernuszewicz, R.S.; Ji, D.; Carrano, C. *Inorg. Chem.* **1997**, *36*, 2760.

77. Mohan, N.; Müller, A.; Nakamoto, K. In *Advances in Infrared and Raman Spectroscopy*; Clark, R.J.H., Hester, R.E., Eds.; Heyden: London, 1970; Vol. 1; pp. 173–226.
78. Fogarasi, G.; Pulay, P. *Ann. Rev. Phys. Chem.* **1984**, *35*, 191.
79. Labanowski, J.; Andzelm, J. *Density Functional Methods in Chemistry*; Springer: New York, 1991.
80. Parr, R.G.; Yang, W. *Density Functional Theory of Atoms and Molecules*; Oxford University Press: New York, 1989.
81. Pulay, P.; Fogarasi, G.; Pongor, G.; Boggs, J.E.; Vargha, A. *J. Am. Chem. Soc.* **1983**, *105*, 7073.
82. Kozlowski, P.M.; Zgierski, M.Z.; Pulay, P. *Chem. Phys. Lett.* **1995**, *247*, 379.
83. Kozlowski, P.M.; Jarzecki, A.A.; Pulay, P. *J. Phys. Chem.* **1996**, *100*, 7007.
84. Kozlowski, P.M.; Jarzecki, A.A.; Pulay, P.; Li, X.-Y.; Zgierski, M.Z. *J. Phys. Chem.* **1996**, *100*, 13985.
85. Scheidt, W.R.; Lee, Y.J. *Struct. Bonding* **1987**, *64*, 2.
86. Hoard, J.L. *Science* **1971**, *174*, 1295.
87. Jentzen, W.; Turowska-Tyrk, I.; Scheidt, W.R.; Shelnutt, J.A. *Inorg. Chem.* **1996**, *35*, 3559.
88. Cullen, D.L.; Meyer, E.F. *J. Am. Chem. Soc.* **1974**, *96*, 2095.
89. Brennan, T.D.; Scheidt, W.R.; Shelnutt, J.A. *J. Am. Chem. Soc.* **1988**, *110*, 3919.
90. Meyer, E.F.; Jr. *Acta Crystallogr. Sect. B* **1972**, *B28*, 2162.
91. Piffat, C.; Melamed, D.; Spiro, T.G. *J. Phys. Chem.* **1993**, *97*, 7441.
92. Rush III, T.; Kumble, R.; Mukherjee, A.; M.E. Blackwood, J.; Spiro, T.G. *J. Phys. Chem.* **1996**, *100*, 12076.
93. Kumble, R.; Rush III, T.S.; M.E. Blackwood, J.; Kozlowski, P.M.; Spiro, T.G. *J. Phys. Chem. B* **1998**, *102*, 7280.
94. Ghosh, A.; Bocian, D.F. *J. Phys. Chem.* **1996**, *100*, 16.
95. Spiro, T.G.; Kozlowski, P.M. *J. Biol. Inorg. Chem.* **1997**, *2*, 516.
96. Spiro, T.G.; Kozlowski, P.M. *J. Am. Chem. Soc.* **1998**, *120*, 4524.
97. Kuriyan, J.; Wilz, S.; Karplus, M.; Petsko, G.A. *J. Mol. Biol.* **1986**, *192*, 133.
98. Quillin, M.L.; Arduini, R.M.; Olson, J.S.; G.N. Phillips, J. *J. Mol. Biol.* **1993**, *234*, 140.
99. Lim, M.; Jackson, T.A.; Antinrud, P.A. *Science* **1995**, *269*, 1995.
100. Ivanov, D.; Sage, J.T.; Keim, M.; Powell, J.R.; Asher, S.A.; Champion, P.M. *J. Am. Chem. Soc.* **1994**, *116*, 4139.
101. Collman, J.P.; Brauman, J.I.; Halbert, T.R.; Suslick, K.S. *Proc. Natl. Acad. Sci. USA* **1976**, *73*, 3333.
102. Stryer, L. *Biochemistry*, 3rd ed.; Freeman & Co: New York, 1988.
103. Tonks, D.L.; Page, J.B. *Chem. Phys. Lett.* **1979**, *66*, 449.
104. Page, J.B.; Tonks, D.L. *J. Chem. Phys.* **1981**, *75*, 5694.
105. Champion, P.M. In *Biological Applications of Raman Spectroscopy*; Spiro, T.G., Ed.; Wiley-Interscience: New York, 1988; Vol. 3; pp. 249–292.
106. Meyers, A.B.; Mathies, R.A. In *Biological Applications of Raman Spectroscopy*; Spiro, T.G., Ed.; Wiley-Interscience: New York, 1988; Vol. 2; pp. 1–58.
107. Lee, S.-Y.; Heller, E.J. *J. Chem. Phys.* **1979**, *71*, 4777.
108. Heller, E.J.; Sundberg, R.L.; Tannor, D. *J. Phys. Chem.* **1982**, *86*, 1822.
109. Tannor, D.J.; Heller, E.J. *J. Chem. Phys.* **1982**, *77*, 202.

8 Photoelectron Spectra of Inorganic and Organometallic Molecules in the Gas Phase Using Synchrotron Radiation

G.M. BANCROFT AND Y.F. HU
Department of Chemistry
University of Western Ontario
London, Ontario, Canada N6A 5B7
E-mail: scigmb@uwoadmin.uwo.ca
yhu@julian.uwo.ca

Inorganic Electronic Structure and Spectroscopy, Volume I: Methodology.
Edited by E. I. Solomon and A. B. P. Lever.
ISBN 0-471-15406-7.

1 INTRODUCTION

1.1 Basic Principles

Although the first photoelectron spectrum of Au metal was recorded photographically by Robinson as early as 1925,[1] photoelectron spectroscopy first developed in the 1960's with the pioneering studies of gases by Siegbahn[2,3] and Turner[4,5] using X-rays (1253.6 eV (9.9 Å) and 1486.6 eV (8.3 Å)) and ultraviolet photons (21.22 eV (584 Å)), respectively. These studies resulted in the two areas called Electron Spectroscopy for Chemical Analysis (ESCA, or X-ray photoelectron spectroscopy (XPS))[2,3] and molecular photoelectron spectroscopy (PES, or ultraviolet photoelectron spectroscopy (UPS))[4,5], respectively.

When an energetic photon beam irradiates a molecule, ionization can occur and electrons (so-called photoelectrons) are ejected:

$$M + h\nu = M^+ + e \tag{1}$$

The photon energy ($h\nu$) has to be larger than the ionization potential (IP) or binding energy (BE) of the bound valence or core electrons. The photon energy is transferred completely to the electron, and $h\nu$ is partitioned between the BE and the kinetic energy (E_k) of the ejected electron, as given by Einstein's photoelectric equation:[6]

$$h\nu = BE + E_k \tag{2}$$

The kinetic energy is measured very accurately with an electrostatic analyzer.[7,8] If the photon energy is known, and is large enough, Eq. 2 enables the determination of the *BE*'s of all electrons (valence, inner valence and core) in a molecule.

A photoelectron spectrum consists of a plot of number of electrons with a given IP versus E_k or *BE* (Figure 1). According to convention, the spectra are plotted with

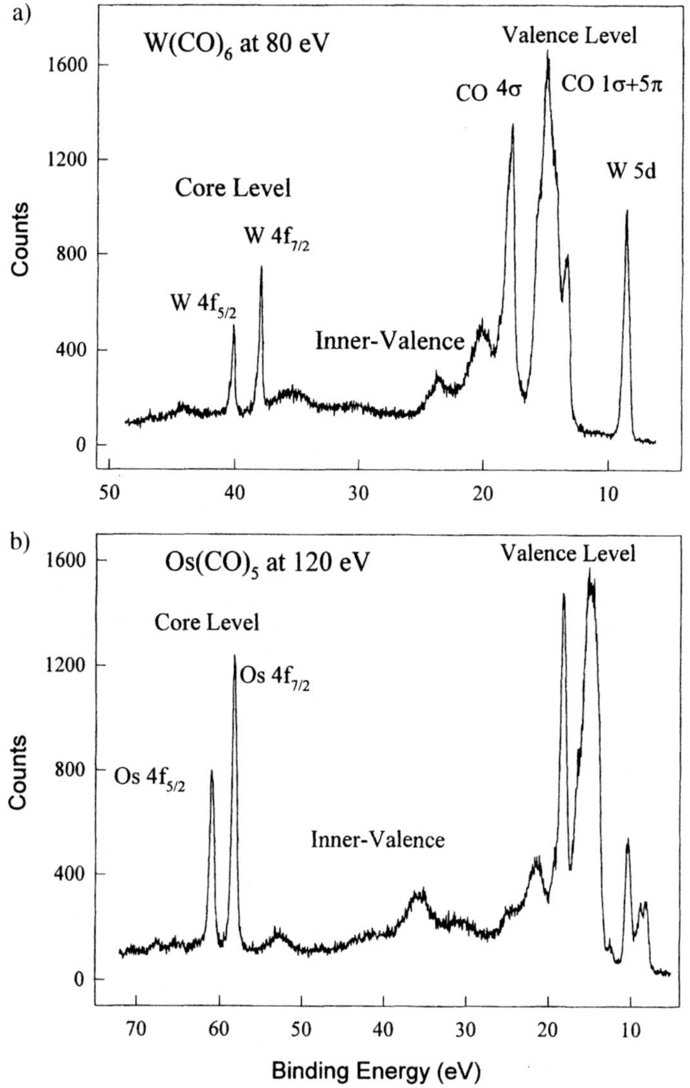

Figure 1. Broadscan spectra of a) $W(CO)_6$ and b) $Os(CO)_5$ in the gas phase taken at 80 and 120 eV with monochromatized synchrotron radiation at instrumental resolution of <0.2 eV.[9,10] Note the intense valence band peaks at low BE, the broad inner valence peaks, and the W $4f$ core level peaks at high BE.

E_k increasing from left to right, and *BE* increasing from right to left (Figure 1). In its simplest form, one molecular or atomic orbital gives rise to one peak, and gives us the best way of obtaining molecular and atomic orbital *BE*'s. For example, in Figure 1a, the lowest *BE* peak (at ~8 eV) arises from the ejection of the valence W 5*d* electrons.[9] The next peaks between ~14 eV and 20 eV arise from the ejection of electrons in CO molecular orbitals. The broad peaks between ~20 and ~37 eV can be assigned to inner valence electrons from C 2*s* and O 2*s* orbitals; and the narrow peaks at ~40 eV can be assigned to W 4*f* core electrons. It is immediately obvious in Figure 1 that more than one peak arises from one orbital (e.g. two W 4*f* peaks), the widths of peaks are very different (the narrow W 4*f* and the broad inner valence peaks), and the relative intensities of the peaks are very different (e.g. the Os 4*f* peaks in Figure 1b[10] are relatively much more intense than the W 4*f* peaks in Figure 1a). The effects that give rise to peak splittings, broadenings, and intensity changes will be discussed in some detail in this chapter.

In addition, it is important to emphasize that photoelectron spectroscopy is only useful because there is little or no multiple ionization under normal conditions. The ion M^+ decays or relaxes to neutral M very quickly ($<10^{-12}$ sec) before the next photon ejects the next electron (a very intense photon source has 10^{12} photon/sec in a mm^2 area). Thus, we are studying the ionization of neutral M, not M^+, M^{++}, etc. This relaxation process, in core levels, occurs by Auger or fluorescence decay (Figure 2). The decay rate determines the inherent linewidth of the level using the Heisenberg uncertainty principle:

$$\Gamma_i \tau = \hbar \tag{3}$$

where Γ_i is the inherent width of the photoelectron line, $\tau (= t_{1/2}/0.693)$ is the mean lifetime of the hole state and $\hbar (= h/2\pi)$ is Planck's constant.

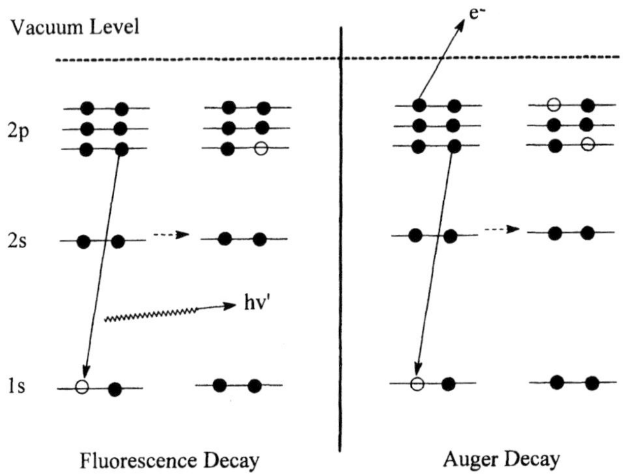

Figure 2. The fluorescence and Auger decay processes after injection of a 1*s* electron in Ne.

Taking $h = 6.626 \times 10^{-34}$ joule·sec and converting to eV, we find that:

$$\Gamma_i(eV) = \frac{4.56 \times 10^{-16}}{t_{1/2}} \tag{4}$$

Thus if $t_{1/2} = 10^{-12}$ sec (e.g. valence hole), Γ_i is less than 1 meV; while if $t_{1/2} = 4 \times 10^{-15}$ sec (e.g. outer core electron), Γ_i is ~0.1 eV.

The total width of an observed photoelectron peak (Γ_{obs}) is determined by the instrumental linewidth Γ_{inst} and Γ_i. Γ_{inst} is given by the source photon width (Γ_{ph}) and the resolving power of the electron analyzer (Γ_e):

$$\Gamma_{inst} = \sqrt{\Gamma_{ph}^2 + \Gamma_e^2} \tag{5}$$

Γ_{ph} usually is the most important contributing factor to Γ_{inst} and Γ_{obs}, and a lot of effort has been made to decrease Γ_{ph} for higher energy photons.

1.2 Conventional Photoelectron Spectroscopy Using Laboratory Sources

As mentioned previously, as a result of the pioneering studies by Siegbahn and Turner, photoelectron spectroscopy in the 1960's, 1970's, and into the 1980's, was divided into two areas: XPS and UPS. These two areas had quite different characteristics. UPS, using HeI (21.2 eV) and HeII (40.8 eV) radiations, studied valence bands at high resolution ($\leqslant$20 meV).[4,5,8] The inherent linewidths of valence levels are long (~10^{-12} sec), the photon linewidths are very narrow (~1 meV), electron resolutions are small for low kinetic energy electrons (a few meV), and ionization cross sections are very large.[11] All these result in intense narrow lines, although photoelectron spectra of large inorganic and organometallic molecules are usually broad ($\geqslant$0.3 eV) due to unresolved vibrational structure. Excellent reviews summarize the extensive literature of spectra of inorganic and organometallic compounds up to 1980.[12,13]

In contrast, XPS, using either Mg Kα or Al Kα sources (1253.6 and 1486.6 eV, respectively) were used to study core levels mainly at relatively low resolution (~1 eV total linewidths).[2,3,14] Of course, valence and inner valence levels, could also be studied; but ionization cross sections are very small at these photon energies (notice the rapid decrease in cross section with increasing photon energy (Figure 3)), and the resolution is much poorer than in UPS. Because the dominating factor in the linewidths was the photon linewidth (Γ_i is often $\leqslant$0.15 eV, and Γ_e is $\ll$0.1 eV), a great deal of effort was made to monochromatize the Al Kα source to give a photon width of $\leqslant$0.4 eV which gave substantially narrower linewidths of 0.3 to 0.5 eV.[15–17] With laboratory sources, it has not been possible to get narrow photon widths with good intensity above 40.8 eV (HeII radiation) and very few core levels can be studied with 40.8 eV photons.

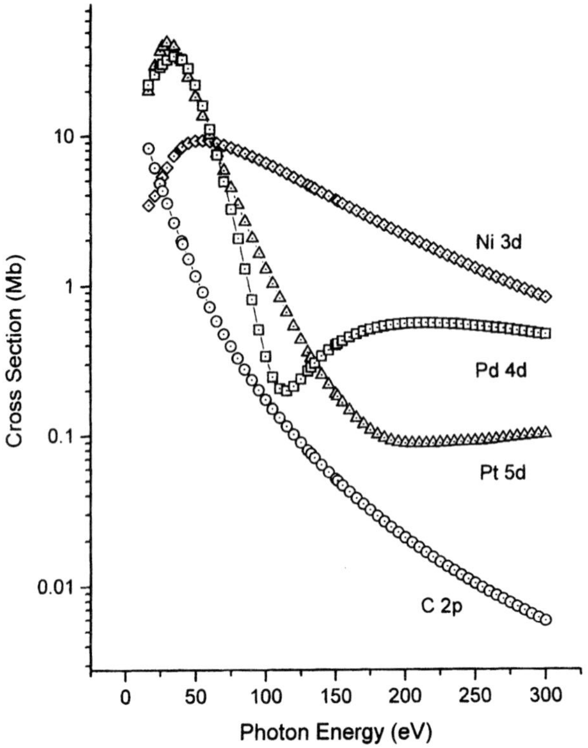

Figure 3. Photoionization cross section for the C $2p$, Ni $3d$, Pd $4d$, and Pt $5d$ orbitals.[11]

1.3 Synchrotron Radiation—Properties and Uses

In the last ten years, monochromatized synchrotron radiation (SR)[18] has been used as a photon source for gas phase photoelectron studies; and this source has given the whole field of electron spectroscopy (photoelectron, Auger, resonance Auger) new impetus and direction. SR yields all photon energies in the UV and soft X-ray region (Figure 4) at far higher intensities than laboratory UV or X-ray sources. Many spectra of core levels have been taken at total resolutions (Γ_{inst}) of $\leqslant$0.1 eV;[19,20] and as seen in Figure 5, this difference in resolution enables us to resolve features that could not be seen with either nonmonochromatic or monochromatic X-ray sources.[19,21,22] With recent advances in SR technology, electron spectra in the last year have been taken at resolutions of ~10 meV,[23,24] and there will be many new advances in the field in the next ten years because of this excellent resolution. Moreover, because the energy can be varied very easily, cross section variations can be readily studied for both gases and solids.[25,26] These different photon energies result in very large differences in relative intensity (Figure 6),[27] which are extremely useful for assigning the spectra together with the atomic valence cross sections (Figure 3).

For gas phase, solid state, and indeed liquid samples, SR has blurred the distinction between UPS and XPS; we can now obtain high resolution spectra of valence

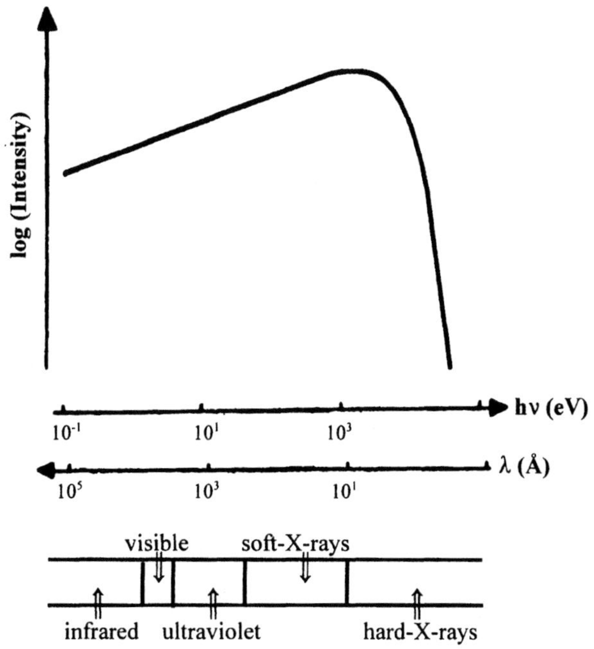

Figure 4. *Top*: The spectral distribution, on a logarithmic scale, of the radiation emitted by a typical SR source. The emission covers extremely wide range of photon energies and wavelengths. *Bottom*: The different parts of the electromagnetic spectrum included in this range. In particular, the ultraviolet region at energies larger than ~10 eV is called "vacuum ultraviolet."[18]

and core levels of gas phase molecules and obtain these spectra at a wide variety of photon energies (e.g. Figure 1a at 80 eV; Figure 1b at 120 eV).

Because of the importance of SR to this field, it is important to describe briefly the origin and use of SR.[18] SR is the electromagnetic radiation emitted by very high energy (relativistic) electrons bent by a magnetic field in a synchrotron (Figure 7). High energy electrons are injected in high vacuum (I, Figure 7) into a ring structure that has a periodic array of magnets. The electrons are bent by three dipole magnets at each corner of the ring, and SR is emitted in a collimated beam from two or three ports after each magnet. This radiation has some remarkable properties (Figure 4). First, it spans a very broad range of photon energies continuously from the far-infrared to the ultraviolet and X-ray regions. Second, the intensity and brightness (photons per unit area) can be orders of magnitude higher than laboratory photoelectron sources. Third, SR is linearly polarized. To increase the intensity and brightness of the radiation, multiple magnetic insertion devices (Figure 7) called undulators (U, Figure 7) or wigglers[18] are now being used in all SR sources. Indeed, the latest SR sources (so-called third generation sources) utilize these insertion devices most efficiently, and yield unprecedented brightness. Results from these synchrotrons (such as the Advanced Light Source at Berkeley) are just appearing;[24]

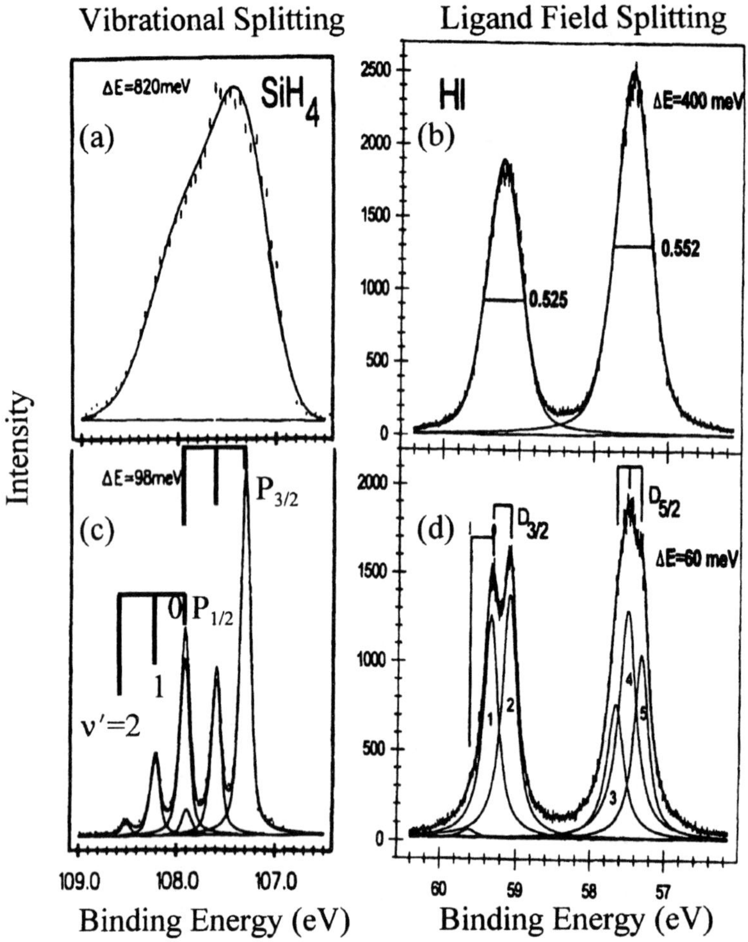

Figure 5. The Si $2p$ and I $4d$ core level spectra of SiH_4 and HI: a) SiH_4 at ~800 meV resolution, typical of a nonmonochromatized Mg Kα source; b) HI at ~400 meV resolution, typical of a monochromatized Al Kα source; c) SiH_4 at ~100 meV instrumental resolution taken at 130 eV with monochromatized SR; d) HI at <100 meV resolution taken at 90 eV with monochromatized SR (Refs. 19–22).

and the effect on the quality and resolution of photoelectron spectra is sometimes remarkable.

How is the continuous SR spectrum (Figure 4) utilized for high resolution experiments? Beamlines have to be constructed to select, and vary, the photon energy; and also focus the beam to the interaction region of the photoelectron spectrometer (Figure 7). The most important device in the beamline is the monochromator (G and DCM in Figure 7). In the far UV and soft X-ray region, grating monochromators are used; whereas in the X-ray region, crystal monochromators are used. Reflection gratings consist of a reflecting surface with a periodic array of lines. Diffraction of

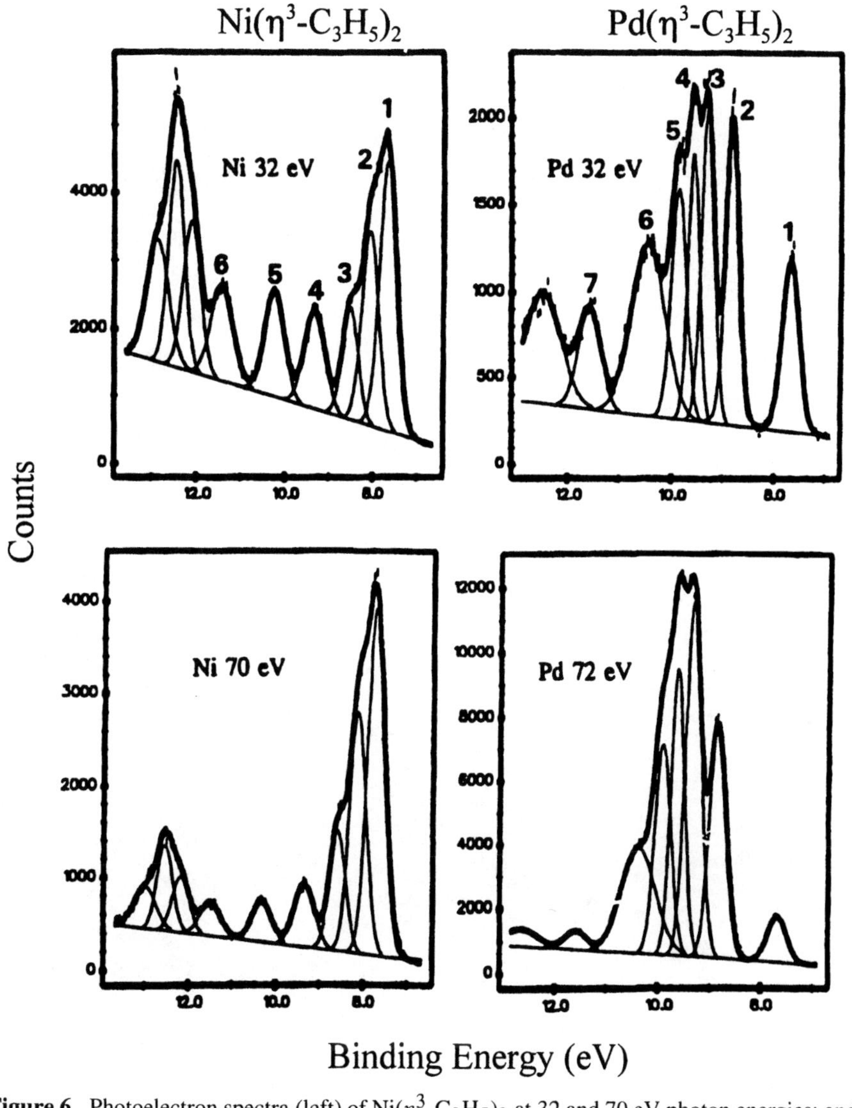

Figure 6. Photoelectron spectra (left) of Ni(η^3-C_3H_5)$_2$ at 32 and 70 eV photon energies; and photoelectron spectra (right) of Pd(η^3-C_3H_5)$_2$ at 32 and 72 eV photon energies.[27]

different wavelengths (in first-order) is given by the well-known equation:

$$d(\sin\theta_i + \sin\theta_d) = \lambda \tag{6}$$

where d is the distance between two adjacent lines, θ_i and θ_d are the incidence and diffraction angles measured perpendicularly from the measured surface. The resolution improves linearly with decrease in d. Many ingenious monochromators have

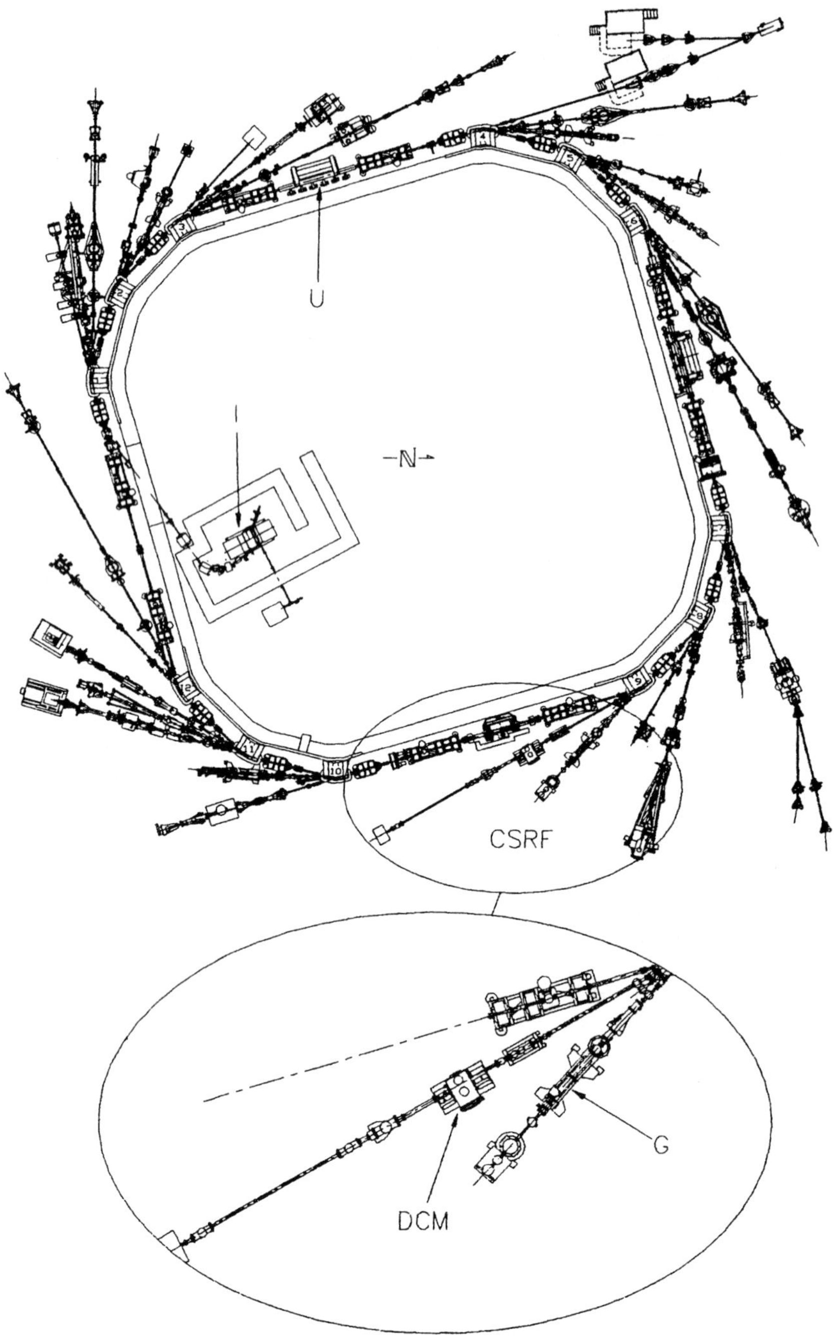

Figure 7. The Aladdin synchrotron with (bottom) two Canadian beamlines. I: Injector, U: Undulator, G: Grasshopper grating incidence monochromator, DCM: Double crystal monochromator. All our photoelectron spectra with SR were taken with the Grasshopper beamline.

been designed and built successfully—usually with a fixed exit slit for easy use. The "Grasshopper" monochromator[28] (G in Figure 7) provides good resolution over a very wide spectral range of 20 to ~1000 eV with a wavelength resolution $\Delta\lambda$ of ~0.03 Å for an 1800 grooves/mm grating ($d = 0.55\ \mu$m). Since $\Delta\lambda/\lambda \approx \Delta h\nu/h\nu$, the resolution $\Delta h\nu$ at 100 eV can be 30 meV; but $\Delta h\nu$ increases greatly at higher $h\nu$. More recent spherical grating and plane grating monochromators give much better resolutions, especially above 100 eV photon energy; and resolutions of <0.1 eV can still be obtained at >500 eV photon energy.

The operation of crystal monochromators (DCM, Figure 7) for X-ray use,[29] is based on Bragg's law which describes diffraction of photons from a crystal:

$$\lambda = 2d \sin \theta_d \tag{7}$$

where d is the spacing of lattice planes in the crystal. The resolution of crystal monochromators (as also used in laboratory monochromatized Al Kα sources) is >0.3 eV, and is not sufficient for high resolution experiments. Moreover, the majority of narrow (small Γ_i) and intense core photolines have *BE*'s $\ll$ 1000 eV; so that crystal monochromators will not be important for high resolution gas phase photoelectron experiments.

1.4 Scope of this Review

It is the purpose of this chapter to review the present and future applications of SR to photoelectron spectroscopy of inorganic and organometallic gas phase molecules. However, for a pedagogical treatment, we must discuss, at least briefly, the important parameters and ideas of photoelectron spectroscopy. Most of these were formulated and discussed in the 1960's and 1970's—e.g. Koopmans' theorem, vibrational structure, chemical shifts, core equivalent model, spin-orbit splitting, ligand-field splitting, shake-up satellites, and cross section variations. Then in the next sections, we will show the power of monochromatized SR to: obtain (often simultaneously) high resolution core, valence and inner valence spectra, and then use the high resolution to resolve effects on core levels. We will then show how these effects can be utilized for structure and bonding studies, and for assigning valence bands from variable energy spectra. Because intensities and resolutions are still improving with SR sources, we speculate in the concluding sections on the uses of SR for inorganic molecules in the future.

2 PRINCIPLES OF PHOTOELECTRON SPECTROSCOPY

2.1 Koopmans' Theorem, Molecular Orbitals, and Vibrational Structure

Perhaps the most important use of photoelectron spectroscopy is to obtain *BE*'s of atomic and molecular orbitals using Eq. 2. The *BE* for a given AO or MO is actually the difference between the *total* energies (T) of the final state (the molecular ion) and the initial state (the neutral molecule):

$$BE = T_f - T_i \tag{8}$$

These energies (and the adiabatic *BE*) can be calculated, but with great difficulty for most molecules. Instead, the calculated energies (or eigenvalues) for the MO's of the *neutral* molecule are used, along with Koopmans' theorem[30] which states:

> *The vertical ionization energy for removal of an electron from an MO is equal to minus (a stable orbital has a negative eigenvalue) the corresponding eigenvalue obtained from a Hartree–Fock self-consistent field calculation for the* neutral *molecule.*

$$BE_j = -\varepsilon_j \tag{9}$$

where ε_j is the eigenvalue for a given orbital. Koopmans' theorem implies that all electrons in the molecule are frozen in their initial orbitals when the photoelectron is ejected. In reality, the other electrons relax (their energies change) as a result of the created vacancy. The relaxation energy is given by:

$$E_R = -\varepsilon - [T_f - T_i] \tag{10}$$

The Koopmans' *BE* ε_j is always *larger* than the relaxed true *BE*. While E_R can be over 10 eV for core levels, E_R is normally only 1 eV for valence levels (Ref. 7, p. 75). In the interpretation of differences in *BE* for both core and valence levels, the ground state approach is normally used, E_R is neglected, and we compare measured IP's with computed MO energies using Eq. 9. For first row transition metal complexes, this is controversial (see later, and Volume II, Chapter 9, by Gruhn and Lichtenberger).

In the above section, the terms adiabatic and vertical IP's are used. To explain these terms, we have to look at vibrational structure. Transitions from the molecule to the ion, go from ground state $\nu = 0$ (for most molecules) to a series of vibrational states in the ion $\nu' = 0, 1, 2$, etc. governed by Franck–Condon factors (Figure 8),[7] and usually more than one peak is observed for one MO for both valence bands (Figure 9) *and* core levels (Figure 5, SiH_4). Figure 9 shows the correspondence of the He I photoelectron spectrum (in eV) with the three valence MO's of N_2 ($\sigma_g\ 2p$, $\pi_u\ 2p$ and $\sigma_u\ 2s$, note small σ, π) with the three final ion states $^2\Sigma_g$, $^2\Pi_u$ and $^2\Sigma_u$ (note the capital Σ, Π).[31] Each band has a different number of vibrational peaks with different vibrational spacings than in neutral N_2 (2345 cm^{-1}). The adiabatic IP is the energy corresponding to the transition from $\nu_0 \rightarrow \nu'_0$ (as for $^2\Sigma_g$ and $^2\Sigma_u$ states); whereas the vertical IP is the energy corresponding to the $\nu_0 \rightarrow \nu'_1$ transition (the most probable transition) for the $^2\Pi_u$ state. The spacing of the vibrational levels is governed by the energy separation of the vibrational levels ν' in the *final* ion state.

In the simplistic early interpretation of vibrational splittings of valence bands, it was thought that removal of a nonbonding electron would yield an ion with about the

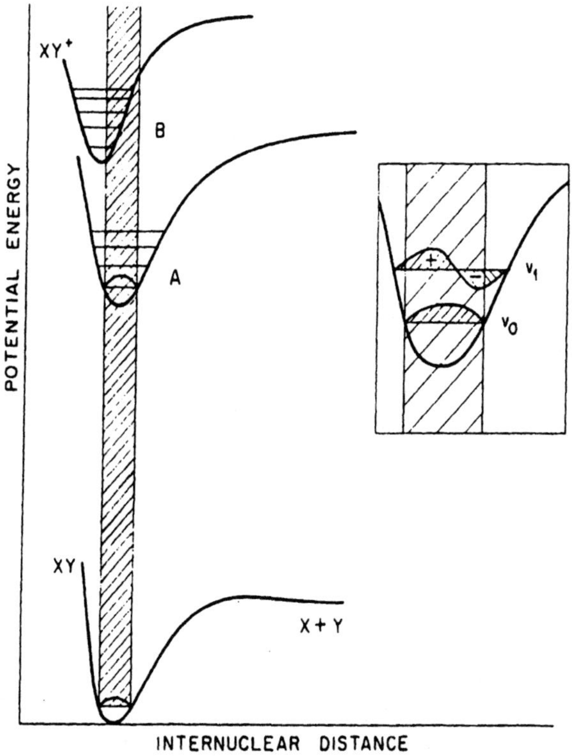

Figure 8. Schematic of Franck–Condon transitions in a diatomic molecule XY from $\nu = 0$ in the ground state to different ion states.[7] In ion state A, the internuclear distance is very similar to the ground state, and the overlap of $\nu = 0$ wavefunction and the $\nu' = 0$ wavefunction is very large indeed (insert), while the positive and negative portions of the $\nu' = 1$ wavefunction cancel almost completely with the $\nu = 0$ wavefunction. One intense ($\nu = 0 \rightarrow \nu' = 0$) and one very weak ($\nu = 0 \rightarrow \nu' = 1$) photoelectron peak arises. For ion state B, there is a large decrease in internuclear distance, and the most intense transitions are $\nu = 0$ to $\nu' \neq 0$, and a large envelope of peaks is seen.

same internuclear distance as the ground state and yield almost the same vibrational splitting as the ground state.[8] In contrast, removal of a bonding electron, would yield an ion with a *longer* internuclear distance, and give a lot of vibrational peaks with a *lower* vibrational frequency than the neutral molecule because of the weakened bond. Finally, removal of an antibonding electron yields an ion with a *shorter* internuclear distance, and gives a lot of vibrational structure with *higher* vibrational frequency than the neutral molecule. In the N_2 spectrum, these ideas would suggest that the $\sigma_g 2p$ electron is weakly bonding, the $\pi_u 2p$ electron is more strongly bonding, and the $\sigma_u 2s$ electron is weakly antibonding.

Vibrational structure on large organometallic molecules is normally not well resolved because of the small M–C vibrational frequencies. For example in $W(CO)_6$, it is possible on the W 5*d* levels to resolve the CO vibrational frequency and also the

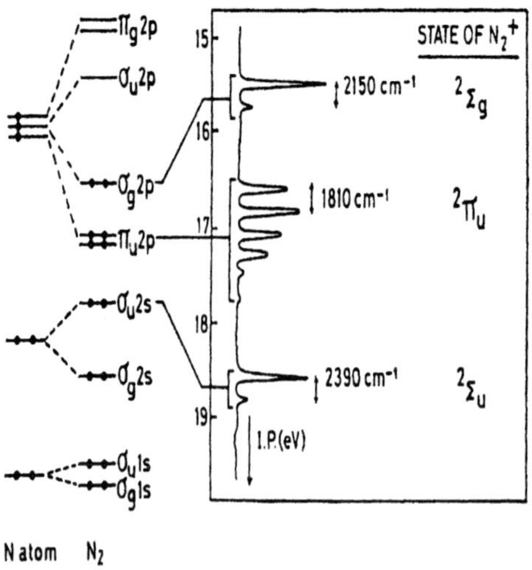

Figure 9. Left: the MO diagram for N_2; middle: the photoelectron spectrum of N_2; right: the ion states of N_2^+.[31]

W–C frequency of 50 meV (Figure 10).[32] As a result, unless the resolution is very high indeed, only broad (~0.3 eV) peaks are seen for most organometallics.

2.2 Vibrational Splitting on Core Levels and the Core Equivalent Model

However, the discovery of vibrational structure on the C1*s* core spectrum of CH_4[17,18] and then on Si 2*p* spectra of silanes[20] (Figure 5) show that the above generalizations can be qualitative at best. To rationalize the splitting in core levels, it is best to consider the core equivalent model* which states when a core electron is removed from an atom or molecule, the valence electrons relax as if the nuclear charge of the atom had increased by one unit.[33,34] As far as the valence electrons are concerned, the *removal* of a core electron is the same as *adding* a proton to the nucleus. Thus, the core equivalent of ionized C^*H_4 (with a C 1*s* electron removed) is NH_4^+; the core equivalent of Si^*H_4 (with a Si 1*s*, 2*s*, or 2*p* electron removed) is PH_4^+. The decrease in internuclear distance is 0.06 ± 0.01 Å in both cases, and three resolvable peaks with similar relative intensities (Franck–Condon factors) are seen for both cases. Similarly, the core equivalent of a series of F molecules is Ne^+ ($CF_4 \rightarrow CF_3Ne^+$), and large F1*s* vibrational structure is observed which can be rationalized by the core

* For the purpose of this chapter we shall define the term "core equivalent" as a molecule in which the core-ionized atom, with nuclear charge Z, is replaced by an atom with nuclear charge $Z + 1$. Although confusing, the precedent is already established in the literature to use the term "equivalent cores" to describe molecules such as disilane (SiH_3-SiH_3) which, from symmetry considerations, have silicon cores in equivalent geometric positions; these types of molecules yield photoelectron spectra with unique characteristics. We will use these two terms precisely as written earlier (core equivalent or equivalent core) to express their assigned definitions.

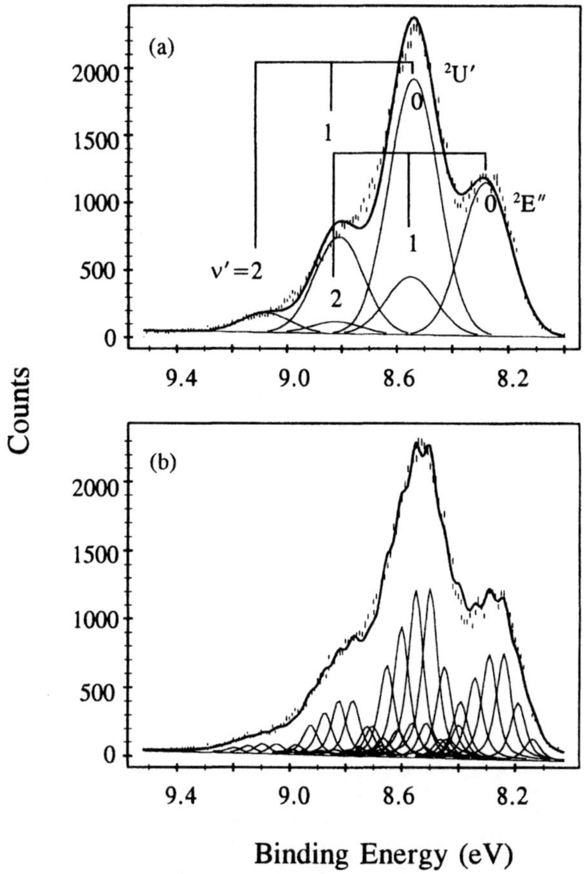

Figure 10. The high resolution W 5d photoelectron spectra of gas phase $W(CO)_6$. In (a), the spectrum is fitted considering just spin-orbit splitting, and a ν_1(C–O) vibrational component, (b) fits a ν_2(W–C) vibrational envelope as well.[32]

equivalent model.[35] On the other hand, the core equivalent for ionized HI is HXe^+, the internuclear distances are virtually identical, and little or no vibrational structure is observed in the HI spectrum.

Indeed, the core equivalent model is seen to be useful for rationalizing the very extensive W–C vibrational structure in the $W(CO)_6$ spectrum (Figure 10). The core equivalent for W ionized $W(CO)_6$ is $Re(CO)_6^+$, and the M–C bond length for $Re(CO)_6^+$ is 2.01 Å compared to 2.07 Å for W–C in $W(CO)_6^+$. Both W–C and CO vibrations are immediately obvious on the W 5d spectrum; the W–C vibrations broaden the W 4f peak, and the CO vibrational structure is obvious.

The core equivalent model is very useful for interpreting other photoelectron parameters, such as the chemical shift and shake-up. This will be discussed in the following sections.

2.3 Spin-Orbit and Jahn–Teller Splitting

As seen in previous figures, many peaks are split into two, and the major reason for this splitting is due to the coupling of spin and orbital angular momentum (so-called spin-orbit splitting). Removal of an electron from a filled p, d, or f orbital *always* gives a doublet in the photoelectron spectrum (e.g. the W $4f$ and Os $4f$ doublets in Figure 1, the Si $2p$ and I $4d$ doublets in Figure 5), although for valence and outer core electrons of light elements, the resolution has to be good enough to resolve this splitting (e.g. Figure 5a). Removal of a Si $2p$ electron from the filled p^6 configuration, yields a p^5 configuration with the same total orbital (L) and total spin (S) angular momentum as for a p^1 configuration ($L = 1$ and $S = 1/2$). These two momenta can couple to give $J = L \pm S$, yielding $J = 3/2$ and $1/2$. The two ion states are labeled $^2P_{3/2}$ and $^2P_{1/2}$ ($2 = 2S + 1$; P signifies $L = 1$; and $3/2$ and $1/2$ are the J values). The higher J value is at lower *BE*. In the spirit of Koopmans' theorem, these ion states are often labeled $p_{3/2}$ and $p_{1/2}$; but of course, these configurations do not exist in the neutral atom or molecule.

The spin-orbit interaction is usually expressed by the Hamiltonian, $H_{S.O}$:

$$H_{S.O} = \lambda[1/2(L_+S_- + L_-S_+) + L_ZS_Z] \tag{11}$$

Operation of this Hamiltonian on the p wavefunctions yields the Hamiltonian matrix and the two energies corresponding to the $P_{3/2}$ and $P_{1/2}$ states. For a filled core d level, we obtain $D_{5/2}$ and $D_{3/2}$ states (Figure 5d); while for the filled core f level, $F_{7/2}$ and $F_{5/2}$ states are obtained (Figure 1, the W $4f$ and Os $4f$ levels). Note again that the lower case letters are often used.

The intensity ratio of the two spin-orbit peaks is, at high photon energy, given by the m_j multiplicity of the levels. Thus, $P_{3/2}$ has four m_j levels ($+3/2$, $+1/2$, $-1/2$, $-3/2$), while $P_{1/2}$ has two such levels ($+1/2$, $-1/2$), yielding an intensity ratio $I_{3/2} : I_{1/2} = 2 : 1$ (Figure 5b). The intensity ratio for the D states is $I_{5/2} : I_{3/2} = 3 : 2$ (Figure 5c, 5d); while the ratio for the F states is $I_{7/2} : I_{5/2} = 1.33 : 1$ (Figure 1, the ratio here is quite a bit bigger than this).

It is important to emphasize that for core levels, the spin-orbit splitting is chemically insensitive. Thus, for all Si compounds (gases, solids, surfaces, liquids) the Si $2p$ spin-orbit splitting is 0.61 eV. For a given core level, the spin-orbit splitting increases with increase in atomic number (e.g. spin-orbit splitting for P $2p$ and S $2p$ is 0.89 and 1.20 eV, respectively); while the spin-orbit splitting for deeper core levels increases dramatically (the spin-orbit splitting from the Xe $5p$, $4d$, $3d$, $3p$, and $2p$ levels is 1.30, 1.98, 12.6, 61.5, and 319.9 eV, respectively).

Spin-orbit splitting is often seen on valence levels as well (e.g. the W $5d$ levels in $W(CO)_6$, Figure 10). Nearly all degenerate valence states in molecules give spin-orbit splitting.[8]

In addition, Jahn–Teller splitting can be important in yielding extra peaks in the valence band. The Jahn–Teller theorem states that a nonlinear molecule in a degenerate electronic state is unstable towards distortion which removes the degeneracy. A good example of Jahn–Teller splitting occurs in the ionized t_2 valence orbital of SiH_4. Often, Jahn–Teller splitting further splits a spin-orbit split state (e.g. in CBr_4^+

and $Pb(CH_3)_4^+$).[8] Jahn–Teller splitting is generally not of importance in the following discussion of valence band spectra of the xenon fluorides and organometallics, so we refer the interested reader to reference 8 for further details.

2.4 The Chemical Shift and Core Equivalents

Siegbahn and his coworkers[2,3] found, very early in their work, that the core *BE* for a given element varied a great deal from an atom (A) such as Xe to a molecule (M) such as XeF_2.

$$\Delta E = BE(M) - BE(A) \tag{12}$$

ΔE was, quite naturally, termed the chemical shift. The xenon fluorides (XeF_2, XeF_4, and XeF_6) were one of the first series of inorganic molecules studied. The initial Xe 3*d* spectra (width >1 eV) showed chemical shifts of 2.9, 5.4, and 7.7 eV, respectively.[36,37] The Xe 4*d* spectra were later recorded with the monochromatized Al Kα source at 0.5 eV resolution (Figure 11); and the Xe 4*d* chemical shifts were similar to the Xe 3*d* chemical shifts for XeF_2 and XeF_4 (2.84 and 5.30 eV, respectively).[38] One of the major objectives of the early gas phase core level work was to use the chemical shifts to provide information on the charge distribution in molecules, and on bond characters.[2,3,36,37]

Before examining the origin of these shifts, it is important to comment on aspects of the core level spectra already presented (W 4*f* and Os 4*f* in Figure 1; Si 2*p* and I 4*d* in Figure 5; and the Xe 4*d* in Figure 11). First, the above chemical shifts for the Xe 3*d* and Xe 4*d* levels show that the chemical shifts for all core levels are very similar. However, for best chemical sensitivity, the narrowest core level is studied for a given element; and it was soon found that the linewidths were smallest for the lowest *BE* 1*s*, 2*p*, 3*d*, 4*d*, and 4*f* levels for a given element. Thus, the 1*s* level is recorded for light elements such as B, C, N, and O (*BE* is between 200 to 540 eV); the 2*p* level is recorded for elements such as Si, P, S, and the first row transition elements (*BE* is between 100 to 1000 eV); the 3*d* level is recorded for elements such as As and the second row transition elements (*BE* is between 40 to 400 eV); the 4*d* (and sometimes the 3*d*) level is recorded for elements such as In, Sn, I, and Xe ($BE < 100$ eV); and the 4*f* level is recorded for heavy elements such as the third row transition metals and actinides (*BE* is between 20 to 400 eV). All of these levels have inherent linewidths of $\leqslant 0.3$ eV;[39,40] and often $\leqslant 0.1$ eV (e.g. C 1*s*, Si 2*p*, Xe 4*d*, and W 4*f*). The hole states for these levels have the longest $t_{1/2}$ values ($>3 \times 10^{-15}$ sec) and the narrowest linewidths (Eq. 4). Other levels such as the Si 2*s* or Si 1*s*, and the Xe 4*p*, 4*s*, 3*p* have much broader lines (>1 eV) mainly due (with the exception of Si 1*s*) to so-called Coster Kronig Auger transitions—Auger transitions within the same *n* values (i.e. $n = 3$ or 4 in Xe)[7] which decrease the $t_{1/2}$ value greatly to $\sim 10^{-16}$ sec.

Second, the chemical shift between Xe and XeF_2 and Xe and XeF_4 are typical (<10 eV) to those observed for most elements, and they are comparable to chemical *BE*'s (1 eV = 23 Kcal).

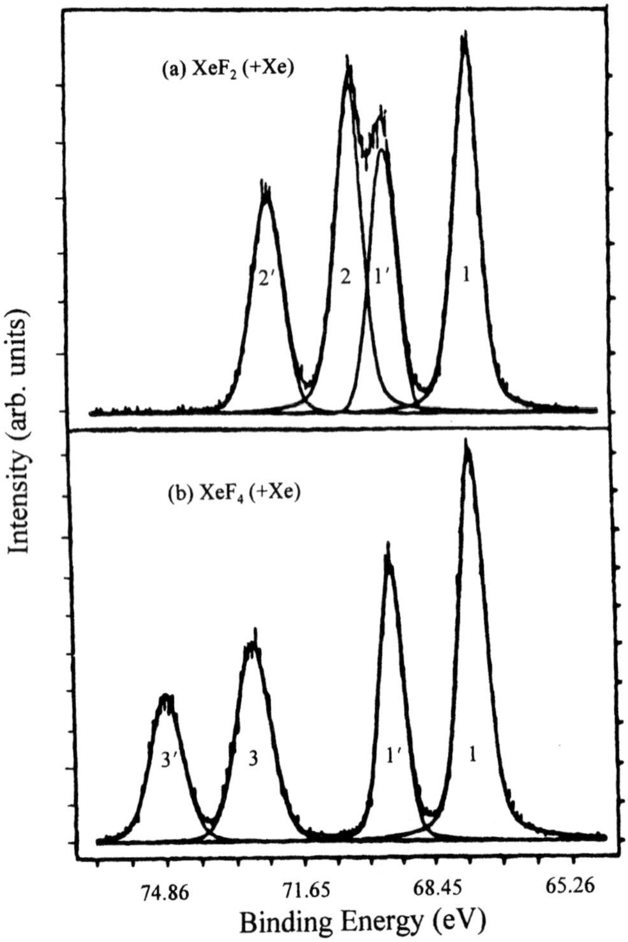

Figure 11. The Xe $4d_{5/2}$ and $4d_{3/2}$ photoelectron spectra taken at a total resolution of ~0.5 eV with a monochromatized Al Kα source. a) XeF_2 (2 and 2′) with Xe (1 and 1′); b) XeF_4 (3 and 3′) and Xe (1 and 1′).[38]

Third, it is apparent in Figure 11 that the XeF_2 and XeF_4 4*d* linewidths are broader than the Xe 4*d* linewidths in Xe gas. This broadening could be due to vibrational splitting, or ligand field (or molecular field) splitting. SR enabled us to resolve peaks in the XeF_2 and XeF_4 spectra; and this ligand field splitting will be discussed in detail in the next sections of this chapter.

How do these shifts arise? Qualitatively, the *BE* of a core electron increases from Xe to XeF_2 because the electronegative F withdraws Xe 5*p* valence electron density (see later). The Xe 5*p* electrons shield (to a very small extent) the Xe 4*d* (or Xe 3*d*) electrons from the nucleus. Thus, removal of Xe 5*p* electron density, results in less shielding for the Xe core electrons, the Xe 4*d* electrons are drawn closer to the nucleus, and their *BE* increases.

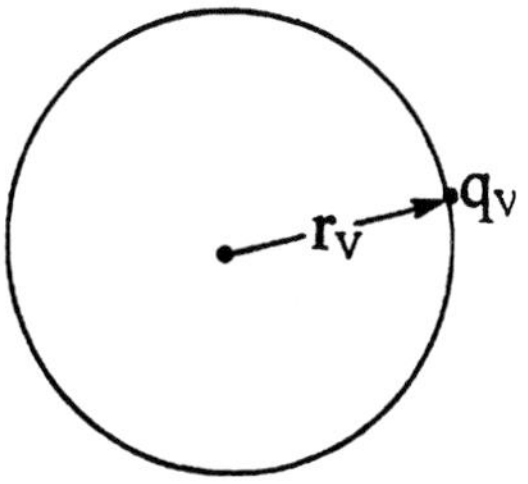

More quantitatively, Siegbahn and coworkers, very early on,[2,3] formulated a simple electrostatic model to rationalize chemical shifts. If we neglect differences in relaxation energies (see later), the major contributing terms to the chemical shift are the V_i and V_{ij} terms—the so-called Madeling or point charge potentials. Much of the chemical shift arises because rearrangement of the valence electrons (radius r_V, charge q_V) changes the potential felt by the core electrons inside this hollow electrostative sphere (above). The potential felt by core electrons (V_i) inside this sphere is $q_V e^2/r_V$. The chemical shift is then given by:

$$\Delta V_i = \frac{q_V e^2}{r_V}(M) - \frac{q_V e^2}{r_V}(A) = \frac{\Delta q_V e^2}{r_V} \tag{13}$$

assuming that $1/r_V$ is a constant for a given element. Carlson and coworkers have tabulated $\langle 1/r_V \rangle$ values;[7] and $e^2/r_V (= k)$ values (the chemical shift for a change in charge Δq_V of 1). For example, the average $\langle 1/r_V \rangle$ value for a Xe $5p$ valence electron is $0.597\ a_0^{-1} = 1.13 \times 10^8\ \text{cm}^{-1}$ $[0.597 \times (0.529 \times 10^{-8})^{-1}]$; $e^2(1/r) = (4.8 \times 10^{-10})^2 \times 1.13 \times 10^8 = 2.60 \times 10^{-11}$ ergs $= 16.2$ eV (1 erg $= 6.24 \times 10^{11}$ eV). Thus, removal of a Xe $5p$ electron should lead to a chemical shift of 16.2 eV on the Xe $4d$ or Xe $3d$ level.

The e^2/r_V values for valence electrons of all elements have been listed (Table 5.7 in Ref. 7), and they vary from ~10 eV to ~40 eV. Chemical shifts are never this large, because we have to look at the effects of negative charges on the F atoms in XeF_2 which *decrease* the shift. If each F has a net charge q_j acting as a point charge on the core electron in atom i of interest, then:

$$V_{ij} = \sum q_j e^2 / R_{ij} \tag{14}$$

where R_{ij} is the distance between atoms i and j. Then the total change in potential seen by the Xe core electrons between XeF_2 and Xe can be written (summing Eqs. 13 and 14)

$$\Delta V = k \Delta q_V - V_{ij} \tag{15}$$

However, ΔV is *not* the chemical shift ΔE in Eq. 12. The relaxation energy difference must be included:[41–43]

$$\Delta E = \Delta V - \Delta R \tag{16}$$

where ΔR is the difference in relaxation energy between Xe and the compound XeF_2. This ΔR can be obtained by determining the shifts in Auger lines, together with the *BE* shift.[43] ΔR is 0.6 eV, so that ΔV is 3.5 eV from Eq. 16. This ΔR is 20% of ΔV, and must be considered, although it is usually not determined. As a result, the ΔE values are normally used to determine charges; and we illustrate this for XeF_2. We take $k_{\mathrm{Xe}} = 16.2$ eV,[7] $R_{\mathrm{Xe-F}} = 2.00$ Å, and remember that for electroneutrality:

$$2q_F + q_{\mathrm{Xe}} = 0 \tag{17}$$

Substituting into Eq. 15:

$$\Delta V \approx \Delta E = 2.87 = 16.2q_{\mathrm{Xe}} + 2e^2 q_F / R_{\mathrm{Xe-F}} \tag{18}$$

$e^2/R_{ij} = 7.17$ eV, as determined above from k.

Thus:

$$2.87 = 16.2q_{\mathrm{Xe}} + 14.3q_{\mathrm{F}} \tag{19}$$

From Eq. 17, $q_{\mathrm{Xe}} = -2q_{\mathrm{F}}$, and solving we obtain $q_{\mathrm{F}} = -0.16$e and $q_{\mathrm{Xe}} = +0.32$e. The first calculation used a k of 13.0 eV, yielding a $q_{\mathrm{F}} = -0.24$e and $q_{\mathrm{Xe}} = +0.48$e.[37] Including ΔR, and using $k = 10.8$ eV, Aksela et al.[43] calculated $q_{\mathrm{F}} = -0.48$e and $q_{\mathrm{Xe}} = +0.96$e. Obviously, a reliable k is of crucial importance for obtaining reliable q's, and the ΔR value must be known. Finally, since both ΔV and ΔR increase linearly with the number of fluorines, Aksela et al.[43] assumed with Carroll et al.[37] that the F charge is constant for all compounds and found that a $k_{\mathrm{Xe}} = 11.8$ eV and $q_{\mathrm{F}} = -0.38$e gives the best fit to all the chemical shift data for the three compounds.

Many different theoretical methods (e.g. CNDO, *ab initio*, Xα) have been applied, often successfully, to calculate charges and chemical shifts.[3,44] However, conceptually, there are two very simple methods that give semiquantitative to quantitative agreement with experiment. Siegbahn et al.[3] developed the simple Pauling electronegativity concept to determine the charges on atoms, and then used Eq. 15 to predict ΔV. To illustrate this method, we do a simple calculation. The C 1s chemical shift difference between CH_4 and CBr_4 is 4.0 eV.[3,14] Take $k_{\mathrm{C2p}} = 22.0$ eV, $R_{\mathrm{C-H}} = 1.0$ Å, $R_{\mathrm{C-Br}} = 2.0$ Å, and electronegativities for H = 2.20, C = 2.45, and Br = 3.05.[45] To obtain the charge on the carbon, we use the Pauling formula for calculating the formal charge:

$$I_C = \frac{\chi_A - \chi_B}{|\chi_A - \chi_B|}\left(1 - e^{[-0.25(\chi_A - \chi_B)^2]}\right) \tag{20}$$

where χ_B and χ_A are the electronegativities of C and either H or Br, respectively and $I =$ the partial ionic character or the partial charge on the C atom.

Then $q_C = \sum I$ for the four C–H or C–Br bonds.

	Bond	$\chi_A - \chi_B$	I_i	No. of Bonds	$\sum I_i$
CH_4	C–H	2.20-2.45	−0.02	4	−0.08
CBr_4	C–Br	3.05-2.45	+0.086	4	+0.34

Based on just the first term in the chemical shift equation, we would expect a chemical shift difference of 0.424 × 22.0 = 9.3 eV. This is far too large and this is because we have neglected the charges on the H and Br. Applying Eq. 16 to both CH_4 and CBr_4: $\Delta V_{CH_4} = -0.08(22.0) + 0.08(14.34) = -0.61$ eV; $\Delta V_{CBr_4} = (+0.344)22.0 - (0.344)7.17 = +5.11$ eV; where 14.34 is e^2/R_{C-H} and $7.17 = e^2/R_{C-Br}$. Thus, the calculated shift between CH_4 and CBr_4 is +5.7 eV, which is in the same ballpark as the observed 4.0 eV considering the simplicity of the model and the neglect of ΔR.

Another useful method for predicting and rationalizing changes in *BE* uses the core equivalent model mentioned earlier:[33,34] the removal of a core electron is chemically equivalent to an increase of nuclear charge by one: that is, the total thermochemical energy E_T for N^*H_3 (N 1*s* core ionized) is approximately equal to that for H_3O^+.

The photoionization of a N 1*s* electron from the NH_3 and N_2 molecules in the gas phase can be represented as:

$$NH_3 \rightarrow N^*H_3^+ + e \qquad BE(NH_3) \tag{21}$$

$$N_2 \rightarrow N^*N^+ + e \qquad BE(N_2) \tag{22}$$

Thermodynamic data cannot be used to calculate the energy change in the above equations because the heat of formations of N^*H_3 and N^*N are not known. From equivalent cores: $N^*N^+ \equiv NO^+$ and $N^*H_3^+ \equiv OH_3^+$ and heats of formation of NO^+ and OH_3^+ can be obtained. Substituting into Eqs. 21 and 22, and subtracting Eq. 21 from Eq. 22 yields:

$$N_2 + OH_3^+ \rightarrow NO^+ + NH_3 \qquad \Delta BE = \Delta E_T \tag{23}$$

For this reaction, ΔE_B rom N 1*s* photoelectron spectra is 4.35 eV; while ΔE_T is calculated from heats of formation to be 3.5 eV, in semiquantitative agreement with ΔE_B.

Similarly, for Xe and XeF_2, we can write:

$$Xe \rightarrow Xe^{*+} + e, \qquad XeF_2 \rightarrow Xe^*F_2^+ + e \tag{24}$$

Taking the core equivalents of $Xe^{*+} \equiv Cs^+$ and $Xe^*F_2^+ \equiv CsF_2^+$, we can write:

$$XeF_2 + Cs^+ \rightarrow CsF_2^+ + Xe \qquad \Delta BE \equiv \Delta E_T \tag{25}$$

Assuming that ΔH_f^0 for CsF_2^+ is the same as that for $Cs^+ + 2F$, the ΔE_T of 2.7 eV compares favorably with the ΔBE of 2.9 eV mentioned earlier. For a large number of molecules, there is a good correlation between the ΔBE and ΔE_T values.

For large inorganic and organometallic molecules, the above methods usually cannot readily be used. However, the chemical shift is still very useful for discussing relative bonding characteristics in these molecules. For example, the *BE* of a metal normally increases by 1 eV for an increase in oxidation state.[14]

BE shifts in both valence and core levels in organometallic compounds[46–53] can be very useful in studying the σ donor and π^* acceptor properties of ligands, and the additivity of the σ and π^* effects at a metal center.[46–53] The ligand additivity model[46] proposes that valence metal orbital ionizations are shifted in a linear way as the number of CO in $Mo(CO)_6$ are replaced by PMe_3 ligands. Both $Mo(CO)_{6-n}(PMe_3)_n$ and $W(CO)_{6-n}(PMe_3)_n$ molecules ($n = 0$–3) have been studied. Both the Mo 3*d* and valence t_{2g} 4*d* electrons (Mo^0 has a d^6 configuration with the six *d* electrons in the t_{2g} orbital) show a linear decrease in *BE* for both the 3*d* and the valence levels as *n* increases (Figure 12). This shift in *BE* (ΔBE) from the parent $M(CO)_6$ molecule can be described by the following equation:

$$\Delta BE = n\Delta E_Q^i \tag{26}$$

where *n* is the total number of PMe_3 ligands, and ΔE_Q^i is the change in charge potential talked about earlier (Eq. 15). In Figure 12, the shift is 0.50 ± 0.04 eV per phosphine for the valence 4*d* levels, and 0.65 ± 0.01 eV for Mo 3*d* core levels. The decrease in both *BE* is due to the better donor and poorer π acceptor nature of PMe_3 relative to the CO ligand. For both σ and π effects, Mo becomes more positive, decreasing the *BE* ratios.

In addition, the different π acceptor abilities of the two ligands, splits the t_{2g} levels into two. This splitting of between 0.2–0.5 eV is expressed by an extra term, ΔE_S in Eq. 26. The splitting is proportional to the difference in π abilities of the CO and PMe_3. Indeed, this splitting for a series of $LM(CO)_5$ complexes (M = Cr, Mo, W; L = PEt_3, PMe_3, $P(NMe_2)_3$, $P(OEt)_3$, $P(OMe)_3$, PF_3) decreases in the above order, showing that the π acceptor ability increases in the above order.[50]

Why is the core level shift larger than the valence shift? As shown above, core *BE*'s are determined by the charge distributions and the relaxation energies. Valence *BE*'s respond to these factors (usually smaller ΔR values) but are also sensitive to bonding effects. The principle of core valence ionization correlation states that the ΔBE of a nonbonding valence orbital should be 0.8 of the ΔBE of the core orbital in the same two molecules.[47,52,53] A valence/core ratio of >0.8 indicates a contribution of bonding to the valence shift. Although transition metal complexes never have the strictly nonbonding lone pair, the ratio of valence/core *BE* is 0.74 ± 0.06 in Figure 12, in reasonable agreement with Jolly's predictions of 0.8 ± 0.1.

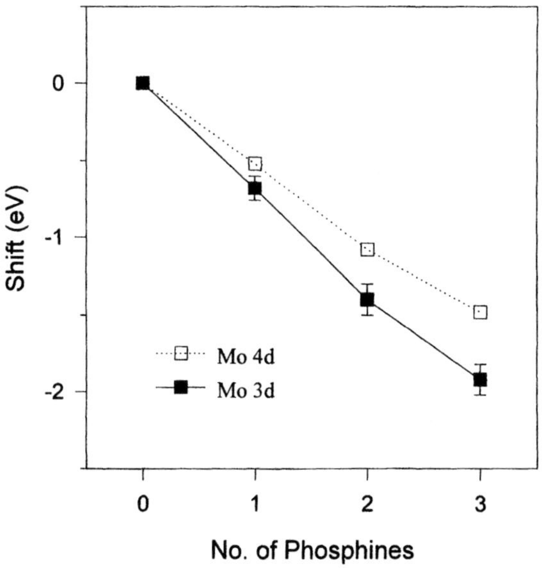

Figure 12. The changes in chemical shift in $Mo(CO)_{4-X}(PMe_3)_X$ ($X = 0, 1, 2, 3$) compounds for the Mo $3d$ core level and the Mo $4d$ valence level. The ratio of the valence to core level shifts is 0.74 ± 0.06.[47]

2.5 Ligand Field (or Molecular Field) Splitting on Main Group Elements

2.5.1 ***Ligand Field Splittings on d Levels*** In Figure 5, the I $4d$ orbitals are split into five levels, a $4d_{3/2}$ doublet and a $4d_{5/2}$ triplet. What causes this splitting, and what can it be used for? A summary of early work is given in reference 54.

The splitting of partially filled metal *valence d* orbitals in transition metal complexes has been studied extensively for many decades, and has been extremely important for characterizing structure and bonding in hundreds of transition metal complexes.[54–57] Crystal field and ligand field theories have been used to interpret the spectra. To give one example, in Cu^{2+} complexes, the ^2D state (d^9 configuration) is split into two states in octahedral symmetry and into four with a tetragonal distortion (Figure 13).[54] In linear $CuCl_2$ ($D_{\infty h}$ or $C_{\infty v}$ symmetries) in the gas phase (Figure 13), the ^2D state is split into three states, and the $^2\Delta_g$ and $^2\Pi_g$ states are split by spin-orbit splitting. The order of states $\Delta > \Pi > \Sigma$ corresponds to the opposite one-electron order $\sigma(d_{z^2}) > \pi(d_{xz}, d_{yz}) > \delta(d_{xy}, d_{x^2-y^2})$, as expected on the basis of simple electrostatics. For this linear species, the required crystal field potential is:[58]

$$V = A_0^0 Y_0^0 + A_2^0 r^2 Y_2^0 + A_4^0 r^4 Y_4^0 \quad (27)$$

where the A's are expansion coefficients for the spherical harmonics Y_l^m.

The $A_0^0 Y_0^0$ term has no angular dependence and shifts all d orbitals by the same amount (the chemical shift). The A_2^0 and A_4^0 terms give rise to the splitting in Fig-

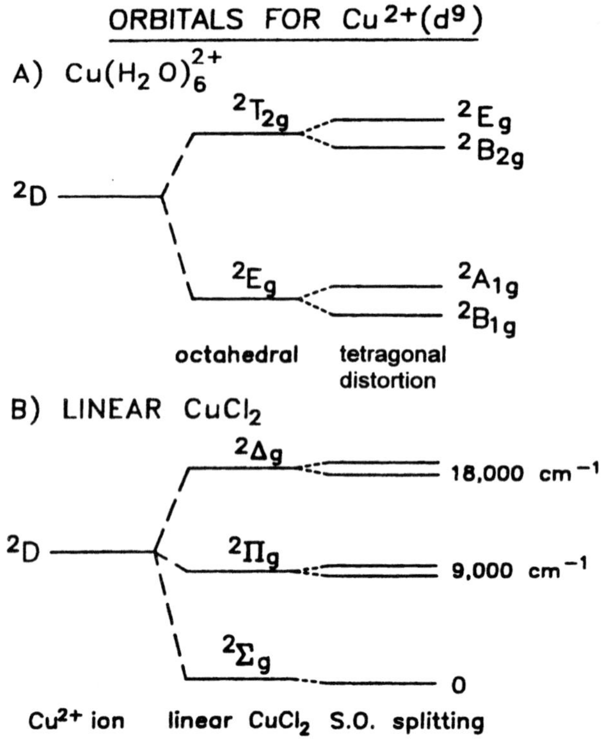

Figure 13. Crystal field splitting of Cu $3d$ orbitals in Cu^{2+} ($3d^9$) complexes. a) $Cu(H_2O)_6^{2+}$ and b) linear gas phase $CuCl_2$. The ^2D state splits in linear $CuCl_2$ into three states $^2\Delta_g$, $^2\Pi_g$, and $^2\Sigma_g$, and spin-orbit splitting of $^2\Delta_g$ and $^2\Pi_g$ yields a total of five states.[54]

ure 5d. The method of operator equivalents[59] is conveniently used for calculation of the matrix elements of V, and H becomes (neglecting the $A_0^0 Y_0^0$ term):

$$H = C_2^0\left[3L_z^2 - L(L+1)\right] + C_4^0\left[35L_z^4 - 30L(L+1)L_z^2 - 25L_z^2 - 6L(L+1) + 3L^2(L+1)^2\right] \quad (28)$$

Considering only the C_2^0 term in the Hamiltonian (Hc_2^0), the energy levels can be obtained easily for $m_L = \pm 2(d_{xy}, d_{x^2-y^2})$, $\pm 1(d_{xz}, d_{yz})$ and $0(d_{z^2})$ for the d^1 case. Thus, $\langle 2|Hc_2^0|2\rangle = 6C_2^0$, $\langle 1|Hc_2^0|1\rangle = -3C_2^0$, and $\langle 0|Hc_2^0|0\rangle = -6C_2^0$. The diagonal elements for the C_4^0 term are obtained similarly. Using the hole-particle analogy, the signs for Cu^{2+} (d^9) are reversed, and energy levels are obtained:

$$E_\Sigma = +6C_2^0 + 72C_4^0; \qquad E_\Pi = +3C_2^0 - 48C_4^0; \qquad E_\Delta = -6C_2^0 + 12C_4^0; \quad (29)$$

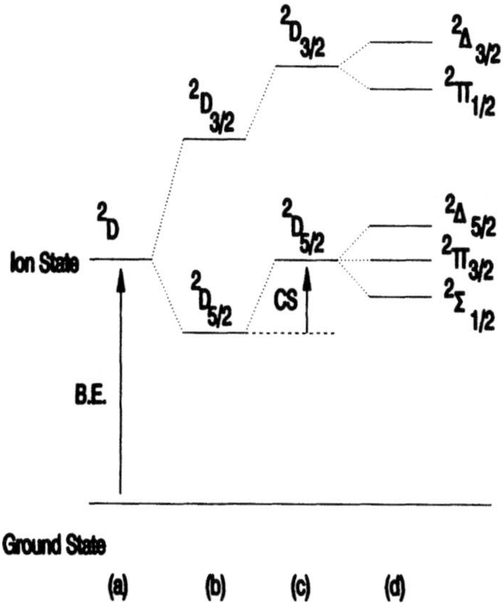

Figure 14. Chemical effects on a ^2D ion state (d^9 configuration) after a) photoionization of a main group d^{10} core level, b) spin-orbit splitting, c) chemical shift, and d) ligand field splitting.[54]

These are just the expressions obtained by Hougen et al.[58] using A_2 and A_4 parameters taking $C_2^0 = 2/3A_2$ and $C_4^0 = 1/6A_4$. From the two observed UV transitions at 9,000 cm^{-1} and 18,000 cm^{-1}, and knowing from the above electrostatic arguments that $E_\Delta > E_\Pi > E_\Sigma$, we can readily solve for $C_4^0 = -43$ cm^{-1} (0.005 eV) and $C_2^0 = -1286$ cm^{-1} (0.159 eV). Notice that both signs are negative and $C_2^0/C_4^0 \approx 30$. The non-cubic C_2^0 term dominates. Spin-orbit splitting (described by the Hamiltonian in Eq. 11) further splits the $^2\Pi$ and $^2\Delta$ states. For first row transition metals, this is much smaller than the ligand field splitting.

As we have derived in Section 2.3., the photoionization of a d^{10} configuration in a main group element such as I results in a d^9 configuration—a ^2D state just like the Cu $3d^9$ example discussed above (Figure 14). This ^2D state should be split by spin-orbit splitting and ligand field (or molecular field) splitting into five states in a linear molecule such as HI, just as five states are observed for linear $CuCl_2$. However, now the spin-orbit splitting is larger than the ligand field spitting. Note that we end up with the same Δ, Π, and Σ states as for $CuCl_2$.

It turns out that the ligand field splittings are small (<0.3 eV), and high resolution spectra are required to observe and characterize this splitting. These were first observed and characterized on the narrow low *BE d* levels of Zn and Cd using narrow HeI and HeII sources (Figure 15): at *BE*'s between 16 and 19 eV. Note that the splittings are $\leqslant$0.2 eV. Apart from small impurity and/or vibrational peaks, the spectra[54,60] consist of a $D_{5/2}$ triplet and a $D_{3/2}$ doublet as indicated in Figure 14. Combining the crystal field Hamiltonian for these linear molecules (Eq. 20) with the

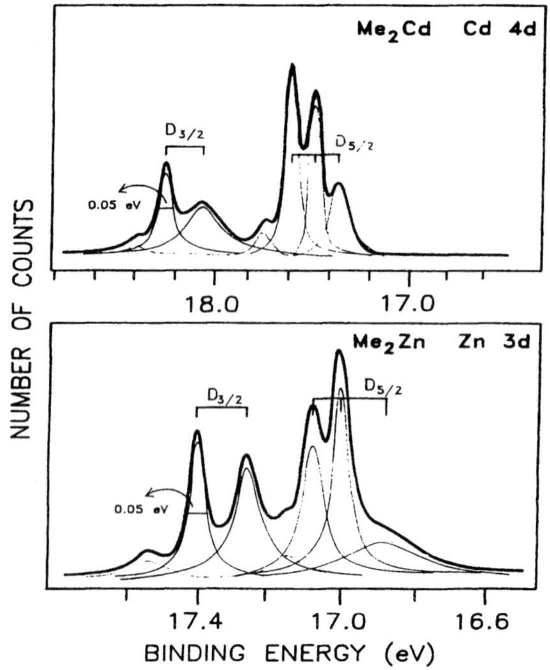

Figure 15. Cd 4*d* and Zn 3*d* photoelectron spectra of Me_2Cd and Me_2Zn, respectively taken with HeI radiation.[60] Note the $D_{3/2}$ doublet and $D_{5/2}$ triplet in both cases.

spin-orbit interaction (Eq. 11), it is still rather easy to diagonalize the Hamiltonian matrix, and the resulting equations for the five energies are (with the approximate term symbols):[60]

$$E(^2\Delta_{5/2}) = E_{4d} - 6C_2^0 + 12C_4^0 - \lambda \tag{30a}$$

$$E(^2\Delta_{3/2}) = E_{4d} - \frac{3}{2}C_2^0 - 18C_4^0 + \frac{1}{4}\lambda + \frac{1}{2}\sqrt{81C_2^{02} - 1080C_2^0C_4^0 - 27C_2^0\lambda + 3600C_4^{02} + 180C_4^0\lambda + \frac{25}{4}\lambda^2} \tag{30b}$$

$$E(^2\Pi_{3/2}) = E_{4d} - \frac{3}{2}C_2^0 - 18C_4^0 + \frac{1}{4}\lambda - \frac{1}{2}\sqrt{81C_2^{02} - 1080C_2^0C_4^0 - 27C_2^0\lambda + 3600C_4^{02} + 180C_4^0\lambda + \frac{25}{4}\lambda^2} \tag{30c}$$

$$E(^2\Pi_{1/2}) = E_{4d} + \frac{9}{2}C_2^0 + 12C_4^0 + \frac{1}{4}\lambda + \frac{1}{2}\sqrt{9C_2^{02} + 72C_2^0C_4^0 - 3C_2^0\lambda + 14400C_4^{02} - 120C_4^0\lambda + \frac{25}{4}\lambda^2} \tag{30d}$$

Table 1 E_{4d}, C_2^0, C_4^0 and λ for Me_2Cd and Et_2Cd[a, 60]

	Me_2Cd	Et_2Cd
$C_2{}^0$	−0.0225 (8)	−0.023 (2)
$C_4{}^0$	−0.0008 (1)	−0.0012 (4)
λ	+0.685 (10)	+0.688 (20)
E_{4d}	17.747	17.497

[a]Estimated error in parentheses.

Table 2 One-Electron Eigenfunctions and Energies for the Cd 4*d* Level in Me_2Cd[60]

			E (eV)	
Approximate M_i	Term	Eigenfunction	Expt.	Calc.
1/2	$^2\Sigma_{1/2}$	$0.54d_\pi\beta + 0.84d_\sigma\alpha$	17.349	17.344
3/2	$^2\Pi_{3/2}$	$0.38d_\delta\beta + 0.93d_\pi\alpha$	17.483	17.475
5/2	$^2\Delta_{5/2}$	$d_\delta\alpha$	17.589	17.601
1/2	$^2\Pi_{1/2}$	$0.84d_\pi\beta - 0.54d_\sigma\alpha$	18.053	18.064
3/2	$^2\Delta_{3/2}$	$0.93d_\delta\beta - 0.38d_\pi\alpha$	18.262	18.252

$$E(^2\Sigma_{1/2}) = E_{4d} + \frac{9}{2}C_2^0 + 12C_4^0 + \frac{1}{4}\lambda$$
$$- \frac{1}{2}\sqrt{9C_2^{02} + 720C_2^0C_4^0 - 3C_2^0\lambda + 14400C_4^{02} - 120C_4^0\lambda + \frac{25}{4}\lambda^2} \qquad (30e)$$

With five peak positions, and five equations, the four unknowns (E_{4d}, C_2^0, C_4^0 and λ) can be readily obtained, and these are given in Table 1 for Me_2Cd and Et_2Cd. Using these parameters, the agreement between calculated and observed peak positions for Me_2Cd (Table 2) is really quite remarkable—within 0.012 eV for all peak positions. The ordering of the energy states in Table 2, $^2\Delta_{3/2} > {}^2\Pi_{1/2} > {}^2\Delta_{5/2} > {}^2\Pi_{3/2} > {}^2\Sigma_{1/2}$, is the same as that observed, $\Delta > \Pi > \Sigma$, for $CuCl_2$ without spin-orbit splitting. This sequence corresponds to the orbital energy ordering $\sigma > \pi > \delta$ which is just the order mentioned earlier for $CuCl_2$ for a purely electrostatic perturbation of the metal *d* orbitals. Several other comparisons between the Me_2Cd and $CuCl_2$ spectra should be made. First, the signs of C_2^0 and C_4^0 are both negative as for $CuCl_2$. This, of course, follows from the same energy ordering above. Second, the ratios of C_2^0/C_4^0 of 20 for Me_2Cd and 19 for Et_2Cd are both large and rather similar to the C_2^0/C_4^0 ratio of 30 obtained for $CuCl_2$. However, the magnitudes of C_2^0 and C_4^0 are close to an order of magnitude smaller than for $CuCl_2$ while λ for Cd^{2+} is much larger than that for Cu^{2+}. Similar results are obtained for the Zn 3*d* peaks in Me_2Zn. The Zn ion state ($3d^9$) is isoelectronic with the Cu $3d^9$ state in $CuCl_2$.

For accurate values of the chemical shift, it is obvious that E_{4d} rather than $4d_{5/2}$ or $4d_{3/2}$ energies should really be used. When the ligand field splitting is not resolved, the "average" $4d_{5/2}$ energy could be misleading by as much as 0.1 eV because of the different intensities of the $4d_{5/2}$ triplet and the asymmetric positions for the $d_{5/2}$ triplet.

The C_2^0 values have been interpreted using an atomic theoretical framework.[40] The electric field which contributes to the C_2^0 term at the electronic site can be written as a sum of terms from the p valence electrons (so-called $C^0_{2\,\text{valence}}$) and a point charge crystal field term, $C^0_{2\,\text{lattice}}$. Taking Me_2Zn as an example, the linear C-Zn-C linkage involves $4s$ and $4p_z$ Zn bonding orbitals. The dominant contribution to the observed Zn $3d$ C_2^0 value comes from the $C^0_{2\,\text{valence}}$ term, the electrostatic interaction between the Zn $4p_z$ electrons and the core Zn $3d$ orbitals. Considering this pd interaction, atomic F^2, G^1, and G^3 Slater–Condon integrals are of importance, but F^2 is usually dominant. We can then write:

$$C^0_{2\,\text{valence}} \propto F^2(\Delta\rho) \tag{31}$$

where $\Delta\rho = n_{p_x} - (1/2)(n_{p_z} + n_{p_y})$ (n's are the orbital populations). For Me_2Zn, $\Delta\rho$ is very close to n_{p_z}, and $\Delta\rho \approx n_{p_z}$.

The point charge contribution can be readily calculated from crystal field theory. For example, one point charge (Ze) and a distance R from the Zn nucleus gives:

$$C^0_{2\,\text{lattice}} = \frac{1}{21}\langle r^2\rangle_{3d} Ze^2/R^3 \tag{32}$$

where $\langle r^2\rangle_{3d}$ is the mean square radius of the Zn $3d$ orbital. The $C^0_{2\,\text{lattice}}$ term is expected to be smaller than $C^0_{2\,\text{valence}}$ for Me_2Zn, mainly because the substantial Zn $4p_z$ electron density is much closer to the Zn $3d$ electrons than the ligand charges. However, the $C^0_{2\,\text{lattice}}$ term will dominate the observed C_2^0 for more ionic compounds, such as the alkali halides.

For low-lying core levels with $BE < 25$ eV, it was possible to resolve ligand field splittings on the outer metal d levels with HeII radiation. For a variety of Zn, Cd, Tl, Ga, and In compounds, the crystal field Hamiltonian can be used for the common symmetries such as $D_{\infty h}$, $C_{\infty v}$, D_{3h}, and C_{3v}; and the splitting should be 0 for T_d and O_h symmetries. As indicated by Eq. 31, the splitting should be sensitive to bonding and structure. First, $|C_2^0|$ for the Zn and Cd compounds, respectively, increase in the order Cl < Br < I < Me < Et < nPr < Me_3SiCH_2. This increase can be readily attributed to an increase in valence p_z ($4p_z$ for Zn, and $5p_z$ for Cd) electron density in the above order. This order is consistent with the accepted order of increasing donor strengths for these ligands. There are also correlations between chemical shift and C_2^0 values which are consistent with this order. Second, gas phase studies of InI_3 show that it gives a narrow featureless In $4d$ spectra,[61] although the In $4d$ inherent linewidths (~0.15 eV) are much greater than the Cd $4d$ linewidths. This lack of splitting shows that the InI_3 is a dimer with close to tetrahedral local symmetry about the In, rather than planar C_{3v} manometric species such as Me_3In which give splittings similar to Me_2Cd.[62]

A few other comments on these spectra seem warranted. First, the extra small peaks are probably due to vibrational splitting; but this is a very small effect in most of the $3d$, $4d$, and $5d$ recorded to date. However, as shown in Figure 5, and in later spectra, vibrational splitting on the $2p$ levels is large; and P $2p$ and S $2p$ spectra of nontetrahedral compounds such as PF_3 or COS give very complex spectra with *both* vibrational and ligand field splitting. Second, these splittings can be readily calculated. For example, Me_2Zn has a substantially lower C_2^0 (16.5 meV) than Me_2Cd (22.5 meV, Table 1). This is mainly due to the lower F^2 value for Zn (0.0715 Ry) compared to Cd (0.0830 Ry). Also, the $\Delta\rho$ values for Me_2Zn and Me_2Cd from *ab initio* calculations are 0.39 and 0.43e, respectively.[54] The theoretical ratio of C_2^0 (Zn)/C_2^0 (Cd) is 0.78, and this is in excellent agreement with the experimental ratio of 0.75. Third, the linear R_2Cd or R_2Zn species give a negative C_2^0 value, while Me_3M (M = Ga, In) have positive C_2^0 values corresponding to an excess of valence charge density in p_x and p_y. Linear XeF_2 has a positive C_2^0 because of withdrawal of $5p_z$ electron density from the Xe $5p^6$ configuration by the electronegative fluorines, while planar XeF_4 has a negative C_2^0 value (see next section).

2.5.2 ***Ligand Field Splittings on p Levels*** Ligand field splitting should also split p and f levels; and in early studies of manometric alkali halides this splitting was observed on the valence band I $5p$ orbitals of manometric gas phase NaI and LiI.[63,64] For p orbitals, the C_4 terms vanish, and the Hamiltonian simplifies to:

$$H = \Delta BE + C_2^0[3L_z^2 - L(L+1)] + \lambda[(1/2)(L_+S_- + L_-S_+) + L_zS_z] \quad (33)$$

where ΔBE is the shift of the free I^- ion $5p$ BE by the Na^+ point charge (from 3.51 to 8.85 eV, a shift ΔBE of 5.3 eV). The ligand field splitting just splits the $p_{3/2}$ level into two while leaving the $p_{1/2}$ level unsplit (but shifted). The energies for the three levels including the BE are:

$$E_{\Pi_{3/2}} = BE - C_2^0 - 1/2\lambda \quad (34a)$$

$$E_{\Pi_{1/2}} = BE + \frac{1}{2}\left[C_2^0 + \frac{1}{2}\lambda - \sqrt{(3\lambda/2)^2 - 3\lambda C_2^0 + 9(C_2^0)^2}\right] \quad (34b)$$

$$E_{\Sigma_{1/2}} = BE + \frac{1}{2}\left[C_2^0 + \frac{1}{2}\lambda + \sqrt{(3\lambda/2)^2 - 3\lambda C_2^0 + 9(C_2^0)^2}\right] \quad (34c)$$

The $^2P_{3/2}$ level splits into two—the $\Pi_{1/2}$ and $\Pi_{3/2}$ states. Taking the quoted peak positions given by Potts el al.[64] for these levels gives $\Sigma_{1/2} - \Pi_{1/2} = 0.96$ eV, and $\Pi_{1/2} - \Pi_{3/2} = 1.18$ eV. Solving for λ and C_2^0 in Eqs. 34 gives $(C_2^0)_{\text{obs}} = +0.13$ eV and $\lambda = 0.68$ eV. Taking the possible errors into account, $(C_2^0)_{\text{obs}}$ could be as large as 0.19 eV.

Although Berkowitz and coworkers[63] have used MO theory to calculate the ΔBE and C_2^0 value, it is instructive to use the simple point charge formulation to calculate both. We take $R_{\text{Na-I}} = 2.715$ Å, and can now calculate the ΔBE from I^- to be just $e^2/R = (4.00 \times 10^{-10})^2/2.71 \times 10^{-8} = 8.50 \times 10^{-12}$ ergs $= 5.30$ eV (1 erg $= 6.24 \times 10^{11}$ eV), in excellent agreement with the observed value. This is perhaps more easily obtained using atomic units (1 a.u. $(r) = 0.529$ Å; 1 a.u. $(E) =$ 27.21 eV). Thus 2.715 Å $= 5.13$ a.u. and $1/R = 0.195$ a.u. $= 5.30$ eV. C_2^0 is given by Eq. 32, taking $\langle r^2 \rangle = 7.201$ a.u. and $C_2^0 = 0.20 \times 1 \times 7.201/(5.13)^3 = 0.0107$ a.u. which when multiplied by 27.21 becomes 0.291 eV. This is double the experimental value of 0.13 eV. Better agreement (0.22 eV) is obtained by considering Sternheimer shielding factors or covalency effects.[40] Obviously, Eq. 32 shows that C_2^0 will increase with decreasing R, and this is exactly what is seen in the alkali halides.

2.5.3 *Correlation of Ligand Field Splitting and Nuclear Field Gradients* It is interesting to note that the C_2^0 in Eqs. 28 and 33 transforms like the nuclear quadrupole Hamiltonian (neglecting the η (or C_2^0) term).[65–67]

$$H_Q = \frac{e^2 q_n Q}{4I(2I-1)}[3I_{z^2} - I(I+1)] \tag{35}$$

where eQ is the nuclear quadrupole moment of the nucleus, eq_n is the so-called field gradient set up by the ligands, and I and I_z are the nuclear spin momentum and momentum operators. For $I = 3/2$, and an axially symmetric molecule;

$$\Delta E_Q = 1/2 e^2 q_n Q \tag{36}$$

Comparing $H_{C_2^0}$ (the first term of Eq. 28) with Eq. 35, it is immediately obvious that $C_2^0 \propto q_n$. For this reason, we initially termed the Cd and Zn d splitting, the electric field gradient splitting. Jørgenson pointed out that the electric field gradient could only be measured at the nucleus. However, it is the same asymmetric ligand field that is measured at the electronic site and at the nucleus, and the ligand-field splitting offers us a marvelous opportunity for observing the transmission of the ligand field through the atom to the nucleus for main group elements.[66,67]

As we have done for the ligand field splitting, the nuclear field gradient (eq_n) is divided into a valence and point charge term. These are both multiplied by so-called Sternheimer factors $(1 - R)$ and $(1 - \delta_\infty)$, respectively.[67]

$$q_n = (1 - R)q_{\text{valence}} + (1 - \delta)q_{\text{lattice}} \tag{37}$$

where,

$$q_{\text{valence}} = K_\rho \Delta\rho \tag{38}$$

and

$$q_{\text{lattice}} = \sum_i \frac{Z_i(3\cos^2\theta_i - 1)}{R_i^3} \tag{39}$$

where K_ρ is a constant. Obviously, Eqs. 38 and 39 have identical form to Eqs. 31 and 32, respectively, discussed earlier for C_2^0. The eq_n values can be readily calculated, and are sensitive to both structure and bonding.[65–67] Obviously, C_2^0 should correlate with eq_n, and we will show this in Section 3, using SR studies of I 4d and Xe 4d levels.

2.6 Shake-Up and Other Effects

Extra peaks are sometimes seen on the low kinetic energy (high BE) side of core level, or even valence level peaks (Figure 16).[9,68] These extra peaks—so-called shake-up satellites—are broader than the main peak and between 5 to 20 eV of the main peak. They are often useful in distinguishing different oxidation states.

Why do these occur? As a result of a sudden change in the central potential of an atom (due to core ionization), an electron in an occupied valence orbital may be

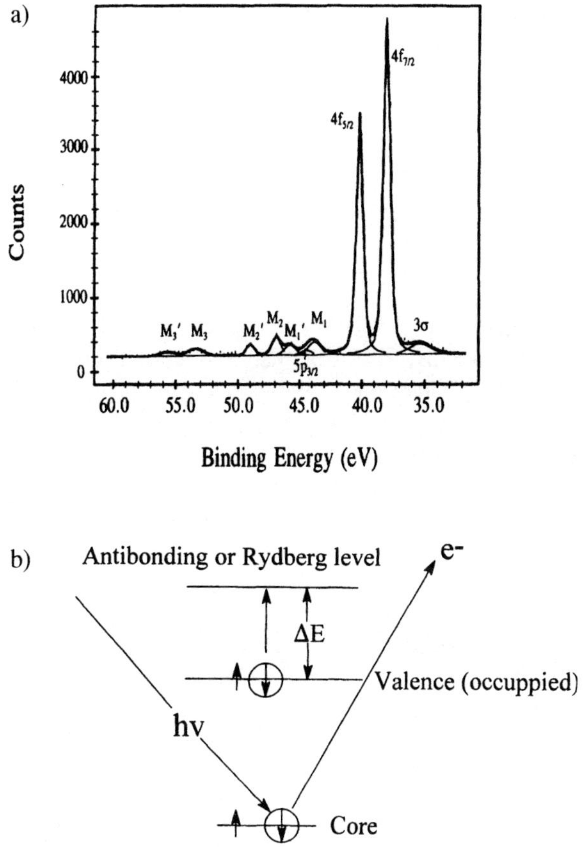

Figure 16. Shakeup on the W 4f level on $W(CO)_6$. a) W 4f spectrum taken at 150 eV photon energy,[9,68] b) the shakeup process.

excited into an unoccupied state (shake-up) or continuum state (shake-off)—see Figure 16. One way of thinking of this process is that, after core ejection, the remaining electron state is not an eigenfunction of the core state, but can be expressed as a *combination* of relaxed states which result in a "main line", and shake-up satellites.

First, the kinetic energy of the outgoing electron is *decreased* by the amount ΔE, the shake-up energy. As in Figure 16, the satellites are normally weak. Second, there are no directional characteristics to the perturbing potential. The selection rules are $\Delta l = \Delta s = \Delta j = 0$—a monopole selection process (see below). For example, in Figure 16, the low energy satellites are due to transition of a t_{2g} electron in the filled $2t_{2g}$ W $5d$ orbitals[69] [$W^0 \equiv (t_{2g})^6$] into the empty $3t_{2g}$ antibonding orbitals of high CO π^* character.[69] The energy of this transition is best estimated using the core equivalent model. For example, the electronic spectrum of $W(CO)_6$ places the $2t_{2g} \rightarrow 3t_{2g}$ transition at 4.53 eV.[69] However, for the core equivalent species $Re(CO)_6^+$, the energy of this transition increases to 5.83 eV, in good agreement with the shake-up energy of 5.7 eV.

Shake-up peaks are present in valence band spectra as well. For example, in Figure 1, the peak at ~20 eV is a very intense shake-up peak from the CO valence peak between 13 and 15 eV.

The probabilities for electron shake-up can be calculated using the sudden approximation.[70] This approximation is based on the assumption that the initial and final states are well defined, and that the change in the Hamiltonian is instantaneous. The probability for an electron in a given single electron orbital (n, l, j) going from an initial to final state is:

$$(P_{i \to f})_{n,l,j} = |\phi_f^* \phi_i d\tau|^2 \tag{40}$$

where ϕ_i is the single-electron wavefunction of a given orbital of the ground state before the event that will give rise to a change in central potential, and ϕ_f^* is the single-electron wave function of one of the vacant orbitals into which the electron is excited. Eq. 36 of course indicates the monopole nature of the transition.

Other effects, such as multiplet splitting and configuration interaction, also give rise to multiple peaks in core level photoelectron spectra.[7] These effects are rarely seen, and will only be discussed briefly. For example, multiplet splitting occurs on the $3s$ levels of transition metal ions with unpaired d electrons (such as $Cr^{3+} \equiv 3d^3$ or $Fe^{3+} \equiv 3d^5$). In the hole state, the $3s$ electron can have the same, or opposite spin, as the valence electrons. Two $3s$ peaks result. Configuration interaction occurs, for example, in the $4p$ or $5p$ orbitals of Xe or Th, respectively.[38,71] In Xe, the $4d$ *BE* is ~70 eV and the $4p$ *BE* is ~140 eV. Ionization/excitation of *two* Xe $4d$ electrons gives rise to a final state (s) with the same total angular momentum and parity of the one hole Xe $4p_{1/2}$ state, and similar total energies. Several peaks can arise.

2.7 Intensity Variations With Photon Energy: Cross Sections

In Section 1.2., we mentioned that the ionization cross sections decrease greatly from the UV to X-ray energies (Figure 3). These figures also show that different levels have very different cross section shapes between threshold and 100 eV photon energy. For example, the C $2p$ cross section decreases rapidly from threshold to 100 eV, while the Ni $3d$ cross section increases from threshold to ~50 eV before decreasing.

Price[4,72] showed very early on that the relative intensities of valence peaks varied greatly between 21.2 and 40.8 eV. These intensity changes were widely used to assign metal d orbitals in transition metal compounds:[12,13] the intensities of peaks from metal d ionizations increase dramatically from HeI to HeII radiation, relative to those of ligand-based peaks. The variable energy SR source has been extremely useful in studying cross section changes.[25,26] Here we briefly outline some of the general theory used when considering photoelectron intensities. Reference 25 is an excellent place to turn for more information.

The probability of photoionization to an ion state is normally designated by the photoionization or photoelectron cross section, σ. The photoionization flux is angle dependent, and also dependent on the polarization of the radiation.[73] However, the majority of the measurements on inorganic and organometallic compounds have been performed at the magic angle ($\theta_x = \theta_y = \theta_z = 54.7°$) or the pseudo magic angle where the band intensity is just proportional to the cross section. Carlson has done some fine work on inorganic molecules—analyzing the photoelectron flux at different angles.[74] The so-called β value (the angular parameter) can be derived and this can be a useful fingerprint for atomic and molecular orbitals.

At the magic angle, σ and the intensities I are proportional to the square of the dipole matrix element for excitation of an electron:[7]

$$I \propto \left[\int \phi_i \underline{r} \phi_f^* d\tau \right]^2 \tag{41}$$

where ϕ_i in Eq. 41 is the initial electron in its atomic or molecular orbital, ϕ_f is the outgoing free electron and $\underline{r}$ is the dipole operator. In the dipole approximation, the allowed transitions are $\Delta l = \pm 1$; in other words a p electron is excited to an s or d type free electron. The $\Delta l = +1$ transition usually dominates.

The above dipole integral can, to a first approximation, be regarded as a direct overlap between the initial p wavefunction and the outgoing d wave (Figure 17).[75] Consider the cross sections of the Ne $2p$ and Ar $3p$ orbitals. The radial parts of the Ne $2p$ and Ar $3p$ wavefunctions are shown along with the outgoing d wavefunction for a) zero kinetic energy, and b) 10 eV kinetic energy. In the latter case, the deBroglie wavelength ($\lambda_D = h/mv$) can be readily calculated as follows:

$$E_k = 10\, eV = 1/2mv^2 \tag{42}$$

Since 1 eV $= 1.602\times10^{-12}$ ergs, and $m = 9.11\times10^{-28}$ g, then $v = 1.9\times10^{8}$ cm/sec. Using $\lambda = h/mv$, we get $\lambda = 3.9 \times 10^{-8}$ cm $= 3.9$ Å. Similarly, λ_D for E_k of 1, 100, and 1000 eV are 12, 1.2, and 0.4 Å, respectively.

For Ne, the overlap between ϕ_{2p} and ϕ_d is initially small, and it increases to a maximum near ~10 eV kinetic energy ($\lambda_D = 4$ Å, Figure 17). The interaction between ϕ_{2p} and ϕ_d is the largest with the initial orbital and outgoing wave are

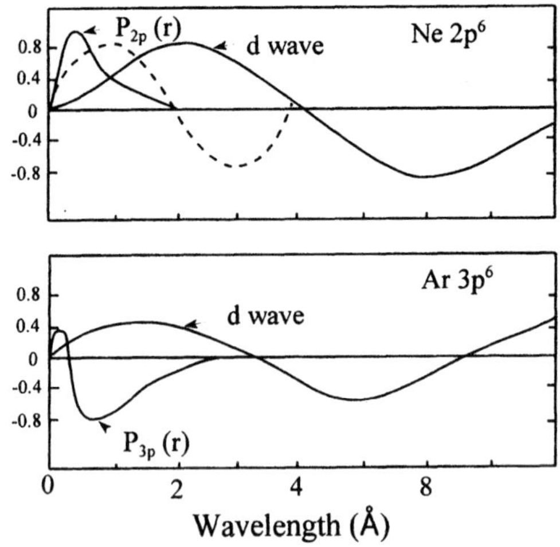

Figure 17. The radial wavefunctions for the Ne $2p$ and Ar $3p$ orbitals (solid lines, left side). The radial wavefunction for an outgoing d wave with zero KE is also shown as a solid line. The radial wave function for a KE of 10 eV is shown in a dashed line as well for Ne.[75]

in the same spatial region. At $E_k > 15$ eV, the negative lobe begins to overlap with the $2p$ wavefunction, and the positive and negative parts of the dipole matrix element cancel each other—and I begins to decrease. When $E_k = 100$ eV ($\lambda_D = 1.2$ Å), positive and negative contributions begin to approach each other, and the cross section decreases by two orders of magnitude.

For the Ar $3p$ cross section, ϕ_{3p} and ϕ_d have opposite signs when $E_k \sim 0$; but the close spatial match gives the largest cross section when E_k approaches 0. At 25 eV kinetic energy ($\lambda_D = 2.5$ Å), the negative lobe of the d wave contributes as much to the integral as the positive lobe. The cross section goes formally to zero—the so-called Cooper minimum; and then increases before decreasing gradually. The cross section does not go to zero in practice, because although the $l \rightarrow l + 1$ cross section goes to zero, the $l \rightarrow l - 1$ does not.

Note that atomic subshells whose radial wavefunctions have no nodes ($1s$, $2p$, $3d$, and $4f$) do not give Cooper minima, while all others should (number of nodes $= n - l - 1$). Thus in Figure 3, the C $2p$ and Ni $3d$ cross sections do not show a Cooper minimum, but the Pd $4d$ cross section does.

For d and f orbitals, the cross sections often have a maximum some way above threshold. For example, the Ni $3d$ cross section in Figure 3 has a maximum at about 50 eV photon energy, above which the cross section falls off rapidly. This is in contrast to s and p orbitals, where any maxima are within a few eV of threshold. For a d or f electron, a double well potential is produced with a potential or centrifugal barrier which inhibits the interaction between the d or f electron and the outgoing f or g wave respectively. The interaction improves at higher photon energies, giving a cross section maximum before decreasing rapidly due to the oscillatory overlap effects discussed above.

In molecules, the cross section for a MO is determined usually by the atomic orbital components of that molecular orbital. For organometallic molecules, the cross section variation of nonbonding metal *d* and ligand carbon $2p$ orbitals should behave like their atomic counterparts, and those of the bonding MO's behave in an intermediate manner. This statement is quantified by the so-called Gelius model.[16,38] The intensity of the jth MO, I_j^{MO} is proportional to (neglecting the angular parameter β):

$$I_j^{\mathrm{MO}} \propto \sum P_{A\lambda}\sigma_{A\lambda}^{AO} \tag{43}$$

$\sigma_{A\lambda}^{AO}$ are the atomic cross sections[16] for all orbitals λ of atoms A in an MO, and $P_{A\lambda}$ is the probability of finding in the jth MO an electron belonging to the atomic $A\lambda$ orbital. These probabilities are given from the population analysis from an MO treatment, and the $\sigma_{A\lambda}$ are theoretical values from such tabulations as given in Yeh and Lindau.[11] Xα calculations for large molecules have been especially useful for obtaining the $P_{A\lambda}$ values.[26]

This model was supposed to be valid only at high photon energy; and indeed at X-ray energies, good agreement between theoretical and experimental relative intensities has been obtained.[16,38] It is not useful at HeI energies; but the SR radiation results (discussed later) show that it works rather well for many molecules above ~40 eV photon energy. Indeed, qualitatively, the spectral changes between HeI and HeII radiation have been widely, and usually successfully, used to assign metal *d* and ligand orbitals using this model.[12,13] However, SR has revolutionized the use of the model, because spectra can be obtained above 40 eV at continuous photon energies. At low photon energies, dramatic changes in relative intensity are sometimes seen (Figure 18) which are due to types of resonances which will be discussed later. These resonances are often useful for assignment purposes as well.

3 APPLICATIONS OF SYNCHROTRON RADIATION TO CORE LEVELS

3.1 Introduction

As has been seen in a few of the spectra shown in the beginning of this article (Figures 1, 5, and 6), it has been possible to obtain both core and valence level spectra at instrumental resolutions of $\leqslant$0.1 eV. This resolution has enabled several groups now to resolve vibrational and ligand field effects on a number of core levels, and study these effects theoretically. Of course, chemical shifts and shake-up peaks are much better resolved than with a laboratory source. For example, the Si $2p$ spectra of organo silicon compounds, $H_XSi^{(1)}[Si^{(2)}(CH_3]_{4-X}$, with two inequivalent Si atoms illustrate this point very well (Figure 19).[77] Because of vibrational broadening (see Figure 5 and next section), the total linewidths are ~0.4 eV; with a laboratory source, the linewidths would be much larger, and the two inequivalent Si $2p$ doublets could not be resolved. Addition of H to $Si^{(1)}$ increases the Si $2p$ *BE* by 0.3 eV for each additional H, in good agreement with the Si $2p$ shift on Si metal as one to four H are added. Reviews of this work are seen in two conference overviews.[20,77(b)]

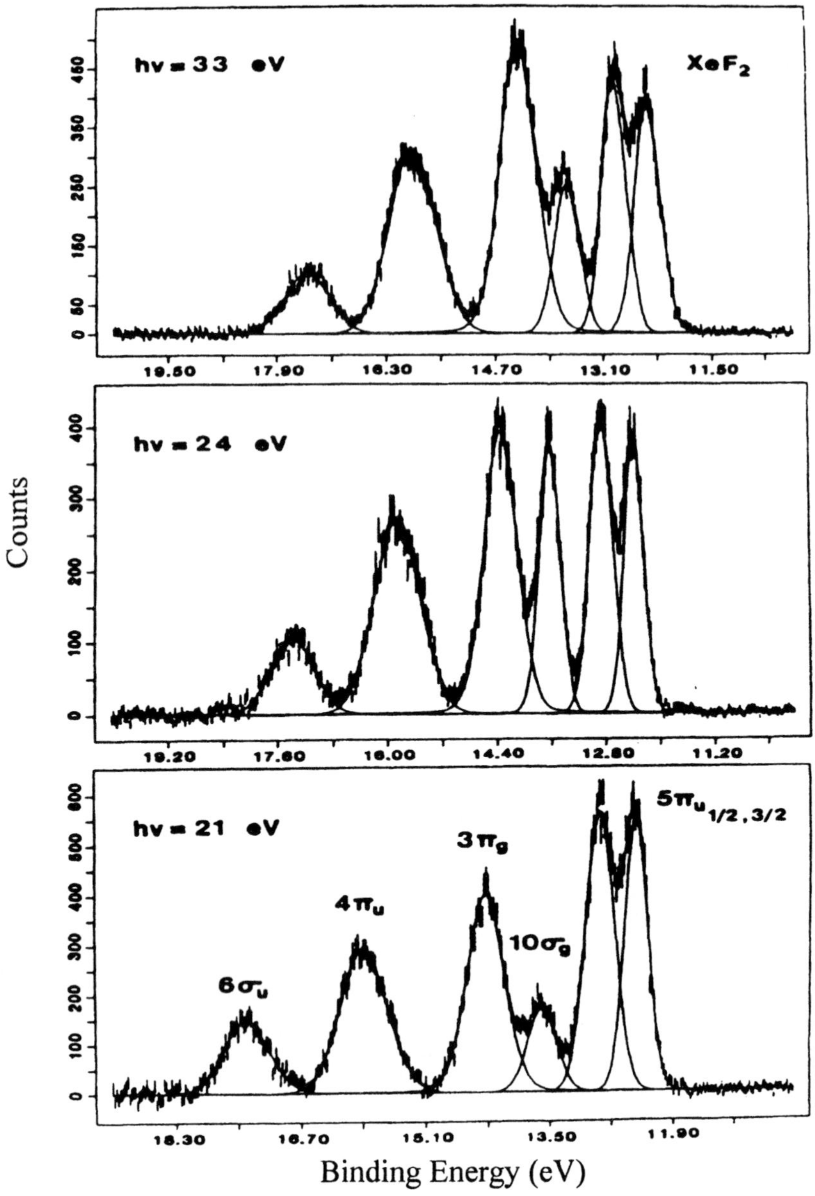

Figure 18. Valence band photoelectron spectra of XeF_2 at 21, 24 and 33 eV photon energies taken with SR. The molecular orbital assignment is given in the bottom plot. Reprinted with permission from Ref. 76. Copyright 1986 American Institute of Physics.

In the following sections, we will review the high resolution core level work; and indicate the use of these spectra to obtain structural and bonding information in inorganic and organometallic molecules.

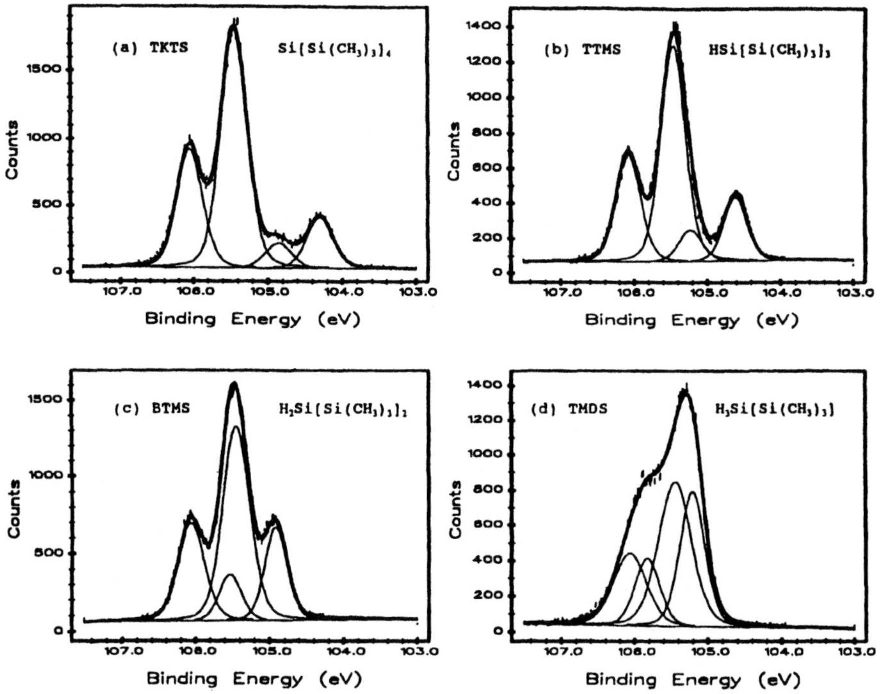

Figure 19. The high resolution Si $2p$ photoelectron spectra of Si molecules taken at 130 eV with SR: a) TKTS, b) TTMS, c) BTMS, and d) TMDS. Reprinted with permission from Ref. 77. Copyright 1992 American Institute of Physics.

3.2 Ligand Field Splittings: Bonding and Structure

3.2.1 ***Core d Levels*** Ligand field (or molecular field) splittings have been resolved on high resolution synchrotron based photoelectron spectra of many molecular core d levels (I $4d$[79,80], Xe $4d$[81], Br $3d$[77,82,83], Sn $4d$[84], and Ge $4d$[84]), and core p levels (S $2p$[78,85,86], Cl $2p$[87] and P $2p$[88]). These splittings are present in other core level spectroscopies using SR (such as photoabsorption spectra[77,83,87,89–91] of many of these levels, and also Auger spectroscopy[22,77,92]). Here, we will give a brief overview of the uses of the photoelectron spectra to obtain bonding and structural information on inorganic molecules in the gas phase.

High resolution I $4d$ core level spectra of ICl, IBr, HI and I_2 are shown in Figure 20.[20,79,80] As for the HI spectrum shown in Figure 5, the spectra consist of a $D_{5/2}$ triplet and a $D_{3/2}$ doublet with very little noticeable vibrational splitting. This is fortunate because vibrational effects would severely complicate these spectra, and make it very difficult to resolve the ligand field splitting. It is perhaps surprising that there is little or no vibrational splitting; because the core equivalent of HI is HXe^+. However, it will be shown that the bond lengths for HI and HXe^+ are within 0.01 Å, and thus it is no longer surprising that no observable vibrational splitting is observed. There are chemical trends in these spectra which are obvious. First, as expected from the discussion in Section 2, the I $4d$ BE increases in the order of I_2 < IBr < ICl as the

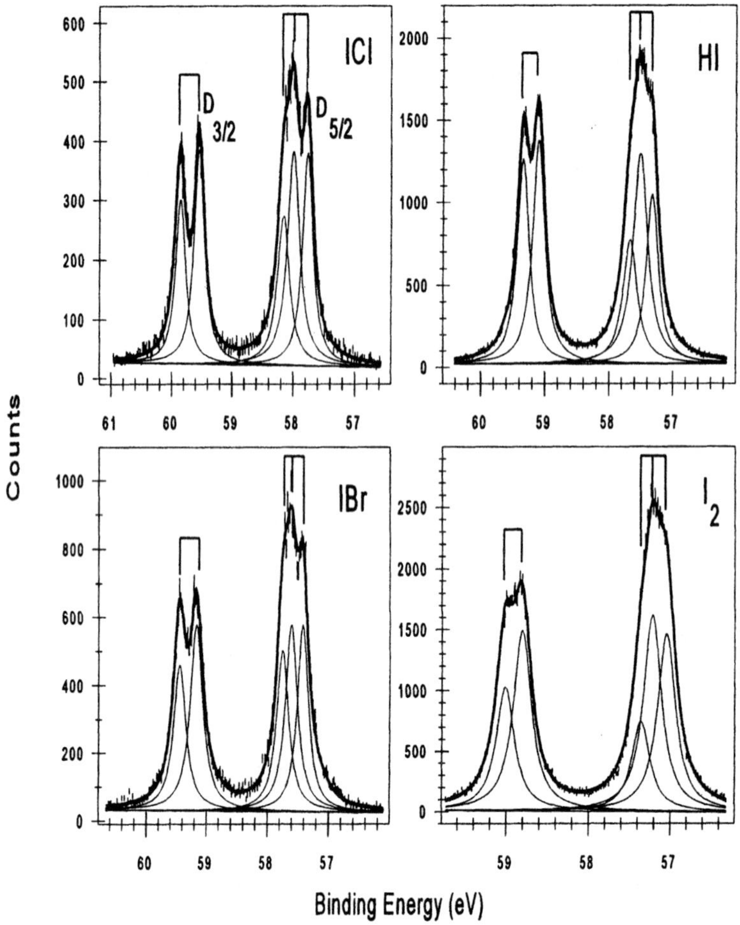

Figure 20. The I 4*d* photoelectron spectra of ICl, IBr, HI, and I_2 taken with 90 eV SR (near the maximum of the I 4*d* cross section and <100 meV resolution). The $D_{3/2}$ doublet and $D_{5/2}$ triplet is evident in each case.[80]

electronegativity of the ligand increases. Second, the ligand field splitting increases in the same order; and third, the observed linewidths (~0.2 eV, inherent linewidth ~0.15 eV) increase in the opposite order, ICl < IBr < I_2.

Using the five equations in Eq. 30, E_{4d}, λ, and C_2^0 are readily obtained. The spin-orbit splitting, $5/2\lambda$, is a constant at 1.74 eV. The trends in C_2^0 and E_{4d} are nicely illustrated in Figure 21, with a good linear correlation between E_{4d} and C_2^0 for seven molecules. This is not unexpected considering the origins of C_2^0 (Eqs. 31, 32) and *BE* (Eqs. 15, 16). $C_2^0 \propto \Delta\rho$ (Eq. 31), and $\Delta\rho = n_{p_z} - 1/2(n_{p_x} + n_{p_y})$. For I_2, the electron configuration on one I is formally $5p^5$ ($n_{p_x} = n_{p_y} = 2$; $n_{p_z} = 1$). There is a deficiency of electron density along the molecular axis and C_2^0 is posi-

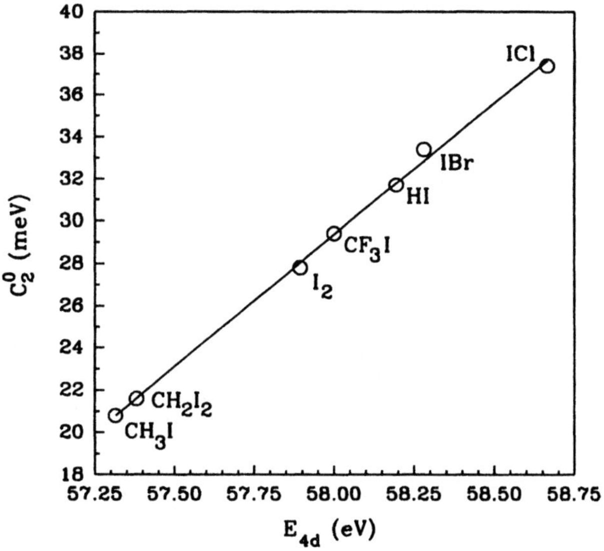

Figure 21. Plot of the asymmetric component of the crystal field, C_2^0 against E_{4d}. The line of best fit is $C_2^0 = 12.49(E_{4d}) - 694.95$ ($R = 0.999$).[80]

tive. As the electronegativity of the ligand increases from I to Cl, n_{p_z} decreases, and C_2^0 increases. As the electronegativity increases from I to Cl, the charge q on the I increases, and the *BE* also increases, because the first *BE* term (Eq. 15), kq, dominates over the counteracting e^2/R term. The correlation between C_2^0 and *BE* is thus expected. However, the position of HI in this plot is probably unexpected. From electronegativity arguments, we would expect both E_{4d} and C_2^0 for HI to be smaller than I_2. Relaxation effects, are probably responsible for this effect,[93] because H (and F) are known to give very different ΔR values (Eq. 16) than other common substituents.

The linewidth trend, ICl $<$ IBr $<$ I_2,[79,80] has been explained as being due to a change in Auger relaxation rate because of a smaller $5p$ electron density in ICl than in I_2, a smaller NVV Auger rate in ICl than in I_2, and a longer lifetime in ICl than in I_2. This would lead to a narrower line for ICl than I_2. Other lindwidths—for CH_3I, CH_2I_2 and CF_3I are dominated by vibrational broadening. It is often difficult to distinguish lifetime broadening from vibrational broadening.

Figure 22 shows a good linear correlation between ^{129}I nuclear field gradients and C_2^0 for seven iodine and two Xe compounds, as expected from the discussion in Section 2.5.3. We can now use additive treatments[65–67] discussed for the ^{57}Fe nuclear quadrupole splitting (Eq. 36) to obtain structural information on XeF_6 from ligand field splittings. Figure 23 shows that XeF_2 yields a resolved splitting; but in XeF_6 although no splitting is resolved, the Xe $4d$ linewidths of >0.36 eV are about double the linewidths in Xe gas, XeF_2, and XeF_4. The major assumption in these additivity treatments is that a ligand or substituent bonded to a given element always gives the same contribution to the C_2^0 or *BE* regardless of the other substituents bonded

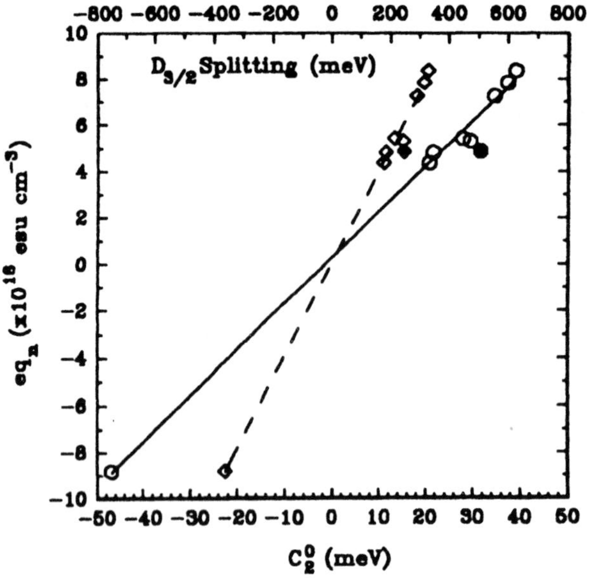

Figure 22. A plot of eq_n vs C_2^0 and $\Delta D_{3/2}$. ○ represents C_2^0 and ◇ represents $\Delta D_{3/2}$. The compounds are XeF_4, CH_3I, CH_2I_2, I_2, CF_3I, IBr, ICl, and XeF_2. The • and ◆ refer to the position of HI which shows evidence of final state valence electron relaxation. Excluding HI, the line of best fit for C_2^0 is $eq_n = 0.196(C_2^0) + 0.299$ ($R = 0.998$) and for $\Delta D_{3/2}$ is $eq_n = 0.0246(\Delta D_{3/2}) + 0.02111$ ($R = 0.998$).[80]

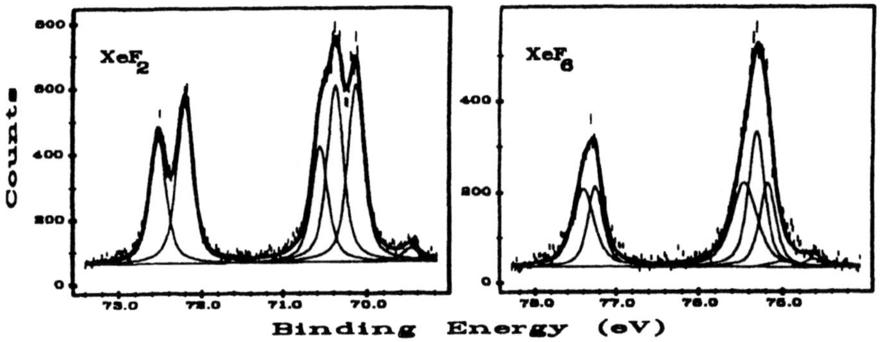

Figure 23. Xe 4*d* photoelectron spectra taken with 94 eV SR, of XeF_2, and XeF_6 at experimental resolution of 0.13 eV.[81]

to that element. The linear increase in Xe 3*d* and 4*d* chemical shifts in the XeF_x ($x = 2, 4, 6$) series discussed in Section 2.4. showed that indeed the charge on the F was very close to a constant in all these molecules.

The additive treatment can be formulated in either a point charge (Eq. 32) or valence (Eq. 31) formalism.[67,94,95] It is convenient to use the point charge formalism.

For *cis* and *trans*-FeA_2B_4 isomers of Fe(II) such as $FeCl_2(ArNC)_4$, we can write from Eqs. 36 and 39:

$$\Delta E_Q = \sum PQS(3\cos^2\theta - 1) \tag{44}$$

where *PQS* is the partial quadrupole splitting $[1/2e^2Q(Z_i/R_i^3)]$, and θ is the angle of the bond of interest from molecular Z axis. For *trans*-FeA_2B_4, $(3\cos^2\theta - 1)$ for each A is $+2$; and $(3\cos^2\theta - 1)$ for each B is -1; for *cis*-FeA_2B_4, $(3\cos^2\theta - 1)$ for each A is -1 (the Z axis is along B-Fe-B!). For an octahedral FeA_6 compound, it is obvious that $\Delta E_Q = 0$. ΔE_Q and C_2^0 measure the distortion from cubic symmetry. Thus:

$$\begin{aligned} \Delta E_Q(trans) &= 4(PQS)_A - 4(PQS)_B; \\ \Delta E_Q(cis) &= -2(PQS)_A + 2(PQS)_B \end{aligned} \tag{45}$$

The predicted ratio is $\Delta E_Q = 2{:}-1$; and the experimental ratios are in close agreement.[67] If one *PQS* is calculated, a table of *PQS* values can be calculated for many ligands bonded to Sn and Fe; and these values have been very useful for predicating and explaining the ΔE_Q values for more complexes isomers of six-coordinate Fe and Sn, and four and five coordinate compounds of Sn.[67,94,95] Also, in Me_2SnL_4 compounds, in which the Sn–C bonds dominate the ΔE_Q value, a simple equation from Eq. 41 has been successfully applied to determine the C-Sn-C bond angle in many Me_2Sn compounds.[96]

We can now use the same additive treatments for C_2^0 values. For XeF_2 and XeF_4, the Xe 4*d* spectra (Figure 23) show that the ligand field splitting can indeed be resolved using high resolution SR (although the total resolution here is >0.1 eV).[80] The XeF_2 spectrum should be compared with the unresolved spectrum (Figure 11) taken with a monochromatized Al Kα source.[38] Both XeF_2 and XeF_4 give quite large ligand field splittings ($D_{3/2}$ splittings of 0.35 and 0.37 eV, respectively) as expected for the linear and square planar structures, respectively (the largest possible distortions from octahedral symmetry).

For C_2^0, we can now write the equivalent equations to Eq. 44 for ΔE_Q:

$$C_2^0 = (PLFS)_F \sum (3\cos^2\theta - 1) \tag{46}$$

where *PLFS* is the partial ligand field splitting. For linear XeF_2, $\theta = 0$ or $180°$, $(3\cos^2\theta - 1) = 2$ for both F; and for square planar XeF_4, $(3\cos^2\theta - 1) = -1$ for each F. Thus:

$$C_2^0(XeF_2) = -C_2^0(XeF_4) = 4(PLFS)_F \tag{47}$$

The observed C_2^0 values are: +39.1 meV for XeF_2 and −46.8 meV for XeF_4 (the different signs are evident from the different splittings of the $D_{5/2}$ triplet). These values (and the ^{129}Xe e^2qQ values for XeF_2 and XeF_4) are within 20% of each other, and indicate that the additive treatment is working reasonably well. The average $(PLFS)_F$ of +10.8 meV is obtained from these two C_2^0 values (+9.8 meV from XeF_2; +11.7 meV from XeF_4). We can now use this value to obtain the structure of XeF_6. The Xe 4*d* widths are almost 0.20 eV broader than for Xe, XeF_2, and XeF_4. The broad XeF_6 peaks show that C_2^0 cannot be zero as expected for an octahedral structure (Eqs. 44, 45). The broad Xe 4*d* peaks of XeF_6 were fitted to a doublet and triplet, and the C_2^0 value was determined from the $4d_{3/2}$ splitting, since $C_2^0 \propto \Delta 4d_{3/2}$. The C_2^0 value is $\pm 18 \pm 2$ meV.

The literature indicates that the most likely structure of gas phase XeF_6 is a C_{3v} structure (Figure 24) with $\theta_{1,2,3} < \theta_{4,5,6}$.[97–99] Assuming this C_{3v} structure (it could also be C_{2v}), we can write:

$$C_2^0 = 3(PLFS)_F(3\cos^2\theta_{1,2,3} - 1) + 3(PLFS)_F(3\cos^2\theta_{4,5,6} - 1) \qquad (48)$$

It was not possible to solve $\theta_{1,2,3}$ and $\theta_{4,5,6}$ concurrently. Therefore to solve Eq. 48, the angle $\theta_{1,2,3}$ was held constant at different fixed angles (48°, 50°, 52°,

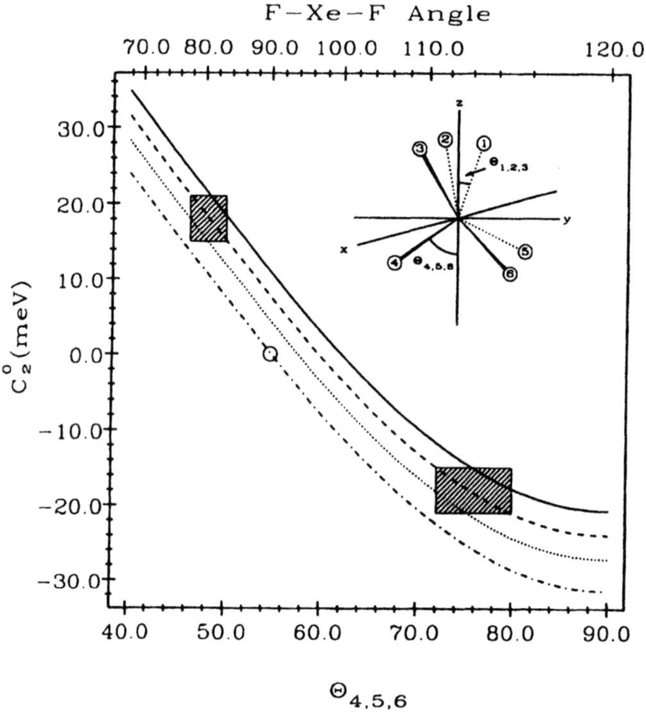

Figure 24. The structure of XeF_6 is in the upper right corner. Plot of C_2^0 versus $\Theta_{4,5,6}$ and F-Xe-F bond angle for different $\Theta_{1,2,3}$ values (48°, 50° --, 52° ··· 54.7° -.-). (o) represents the position of XeF_6 if it is perfectly octahedral (The F-Xe-F angles are 90° and Θ_{1-6} is 54.7°). In the case of $\Theta_{1,2,3}$ being 50°, the errors on C_2^0 and $\Theta_{4,5,6}$ are given by the dimensions of the box.[81]

and 54.7°) and $\theta_{4,5,6}$ was allowed to float. Figure 24 is a plot of C_2^0 versus $\theta_{4,5,6}$ while holding $\theta_{1,2,3}$ constant. From the electron diffraction and latest theoretical values[97–99] for $\theta_{1,2,3}$ of 49° and 52° respectively, we fix $\theta_{1,2,3}$ at 50°. The two boxed-in regions refer to the experimentally measured C_2^0 and the size of the boxes gives the estimated error, which we take as 50% greater than the standard deviation in C_2^0, due to deficiencies in the model.

Since it is not possible to directly determine the sign of C_2^0 for XeF_6, two possible angles for $\theta_{4,5,6}$ are possible. If C_2^0 is positive, the F-Xe-F angle falls between 77 and 84° indicating that the angle defined by $\theta_{4,5,6}$ is closing up and the lone pair of electrons are acting like a ring of charge in the x-y plane forcing the ring to close. In the negative C_2^0 case, the F-Xe-F angle is 111–117° thereby forcing $\theta_{4,5,6}$ to open up, indicating that the lone pair is directed toward the face of the cone.

The majority of the experimental and latest theoretical results suggest strongly that the latter distortion takes place, and thus C_2^0 is negative. This gives estimates for $\theta_{1,2,3}$ and $\theta_{4,5,6}$ of $50 \pm 2°$ and $76 \pm 4°$, respectively. These angles are in good agreement with the latest theoretical values of 49° and 76°, respectively,[99] but not in as good agreement with the electron diffraction results of 52° and 67°, respectively.[97,98]

Ligand field splitting has also been resolved on the Br 3d spectra of HBr and Br_2,[77,82,83] and this splitting is very comparable to that for HI and I_2. Again, in these spectra, very little vibrational splitting is observed. In contrast, vibrational splitting (broadening) is very important for many elements (e.g. Si 2p (see next section), Ge 3d and Sn 4d). Because of this vibrational structure—and the broader natural linewidths for Ge 3d, Sn 4d (and other d levels such as As 3d and Sb 4d), it will not be possible to obtain accurate C_2^0 values for Ge and Sn compounds of the type $(CH_3)_xMX_{4-x}$ ($x = 0$–4, M = Ge, Sn, X = Cl, Br, I).[83]

3.2.2 ***Core p and f Levels*** As might be expected from the treatment in Section 2.5.2, ligand field splittings have recently been resolved on the S 2p level of compounds such as H_2S (Figure 25),[78] COS, CS_2,[85,86] the Cl 2p level of HCl,[87] and the P 2p level of PF_3.[88] The vibrational effects are, like HBr and HI, rather small in H_2S (Figure 25) and HCl making it easy to obtain the $P_{3/2}$ splitting (110 meV in H_2S; 80 meV in HCl), and use Eqs. 34 to derive the C_2^0, λ and BE values. The C_2^0 values are 41 meV for HCl, and 53 meV for H_2S (note that for H_2S, the H-S-H bond angle is very close to 90° (actually 92°), and the Z axis for a 90° bond angle is perpendicular to the H-S-H plane, analogous to the case for *cis*-FeA_2B_4 discussed earlier).[100] These C_2^0 values are larger but comparable to the C_2^0 values for HBr and HI, as expected from the similar bonding and similar F^2 values (Eq. 31).[40] Moreover, the relative $|C_2^0|$ values for HCl and H_2S can be rationalized. C_2^0 for HCl and H_2Cl^+ (H-Cl-H = 90°, Z axis perpendicular to the H-Cl-H plane) should be equal but opposite in sign from Eq. 44. Because Cl is more electronegative than S, ρ will be smaller for HCl than H_2S; and C_2^0 for HCl would be expected to be smaller for HCl than H_2S in agreement with the observed values.

Ligand field splitting should also be present on f levels of the heavy elements.[40] The $F_{7/2}$ level would split into four levels, and the $F_{5/2}$ levels would split into three levels. Ligand field broadening has been observed on the 4f levels of Os and W in

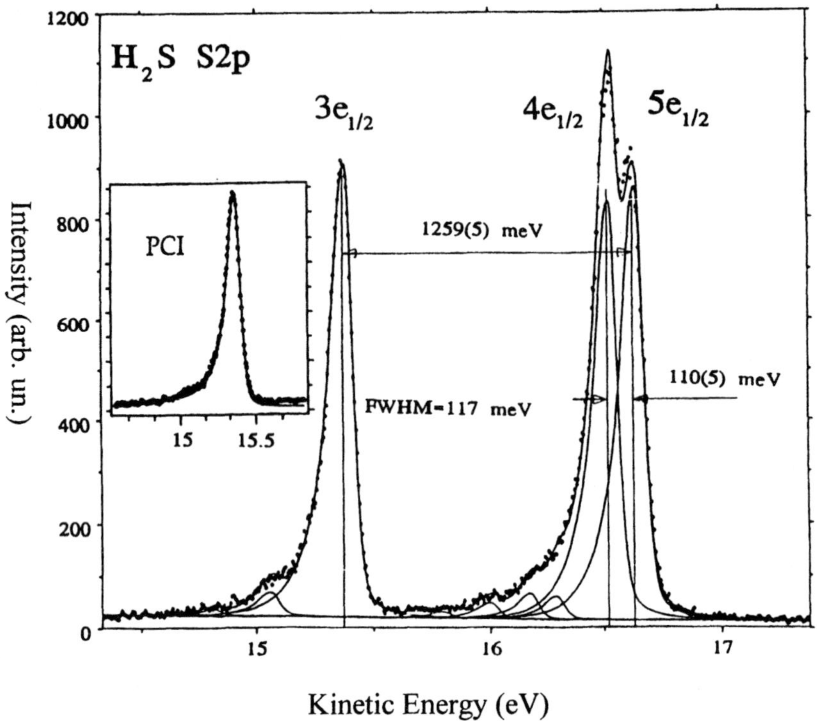

Figure 25. The S 2p core photoelectron spectrum of H_2S. The fitted curves give accurate positions for the main (0 → 0)$3e_{1/2}$, $4e_{1/2}$, and $5e_{1/2}$ molecular field split relativistic levels. The insert shows the result of a fit of a PCI line profile to the $3e_{1/2}$ core photoelectron line. The lifetime obtained from this fit was 70(10) meV.[81]

molecules such as $Os(CO)_5^9$ and $W(CO)_{6-x}(PMe_3)_x$,[101] but it is unlikely that these splittings will be resolved for quantitative studies.

3.2.3 ***Theoretical Calculations and Other Core Level Spectra*** Theoretical studies of these splittings are now beginning to appear after the initial theoretical studies on Me_2Cd, and Me_2Zn.[54,60] Multi configuration SCF calculations on H_2S, HCl, and HBr molecules[100] using the CASSCF approximation[102] and the SIRIUS program[103] gives theoretical C_2^0 values within 15% of the experimental values for these three molecules. A configuration interaction-base method[104–106] using large basis sets gives good agreement with experiment for the S 2p splitting on H_2S, OCS, CS_2, and SO_2.[85,86]

Finally in this Section, ligand field splittings have to be taken into account when analyzing the high resolution photoabsorption and Auger spectra involving p and d levels.[22,77,82] Indeed, the first analysis of ligand field splitting on a core level was performed in the early 1970's by Comes et al. on XeF_2 and XeF_4.[89] High resolution photoabsorption spectra of the 2p level of H_2S[90] and the Br 3d level of HBr[83,91] reveal both core level and Rydberg level ligand field splittings to give remarkably complex overlapping spectra. The Auger spectra are also very interesting because vibrational splitting becomes more important in the double hole state; and the HI and HBr NVV and MVV Auger spectra[22,82] show a very complicated overlap of vibrational and ligand field splittings. Somewhat amazingly, in the H_2S LVV Auger spectra,[78,92] vibrational splitting dominates, and one ligand field level from the $2p^5$ hole state is barely evident.

3.3 Vibrational Splitting

As shown in Section 2.2 previously, extensive vibrational splitting can be observed on core levels—such as the C $1s$ and Si $2p$ levels.[17–20,107,108] SR from undulator sources is just now giving better photon resolutions so that better resolved spectra will still be obtained in the future. For example, recent high resolution results on the C $1s$ levels of CO and CH_4,[109,110] show better resolved spectra than previously obtained with monochromatized Al Kα radiation or bending magnet SR.

From an inorganic chemist's viewpoint, we want to review here what we know now about the systematics of vibrational splittings: the variations in vibrational patterns on the Si $2p$ level for different substituents,[76,107,108] and the periodic trends for a series of congeric and isoelectronic molecules.[111,112] As outlined in Section 2.2, the core equivalent model is often very useful for rationalizing the number of vibrational peaks and the vibrational frequencies. We will emphasize that resolution of these splittings is important to obtain accurate *BE* (and accurate relative *BE*'s), and that these vibrational patterns can be readily calculated.[85,111–113]

Figure 26 shows the high resolution Si $2p$ photoelectron spectra of SiH_4, SiD_4, and Si_2H_6.[20,77] These four spectra illustrate the different types of vibrational splitting that are seen in Si molecules with total linewidths usually ~130–150 meV. Since the Si $2p$ inherent linewidths are probably only ~40 meV, it is apparent that even higher resolution spectra should be obtainable with undulator sources on third generation synchrotrons.

In SiH_4, the two largest peaks correspond to the adiabatic $\nu = 0$ to $\nu' = 0$ transition for the Si $2p_{3/2}$ and Si $2p_{1/2}$ peaks separated by the Si $2p$ spin-orbit splitting of 0.613 eV. These two spin-orbit lines are split into a vibrational manifold of three observable peaks (a fourth is apparent at very high resolutions) from ($\nu' = 0, 1, 2$) with an energy spacing of 0.295 eV (2379 cm^{-1}) corresponding to the totally symmetric ν' vibrational mode of a_1 symmetry (Table 3). In CH_4, a very similar spectrum is seen on the C $1s$ level, with a vibrational frequency of 0.396 eV.[19,109] As discussed in Section 2.2., these vibrational lines arise because the Si-H and C-H bond lengths are substantially shorter (~0.06 Å) in the ion state than in the ground state. Moreover, the vibrational frequency of 2379 cm^{-1} is much larger than the ground state frequency, ν, of 2180 cm^{-1}, and rather close to the core equivalent ν in PH_4^+ (2295 cm^{-1}) or PH_3 (2327 cm^{-1}) (Table 3).[19,20,77]

At higher resolution (Figure 26b), the SiH_4 spectrum shows additional shoulders on all peaks indicating the presence of a second vibrational progression. The vibrational frequency of 0.11 eV corresponds to the asymmetric ν_2 bending frequency of

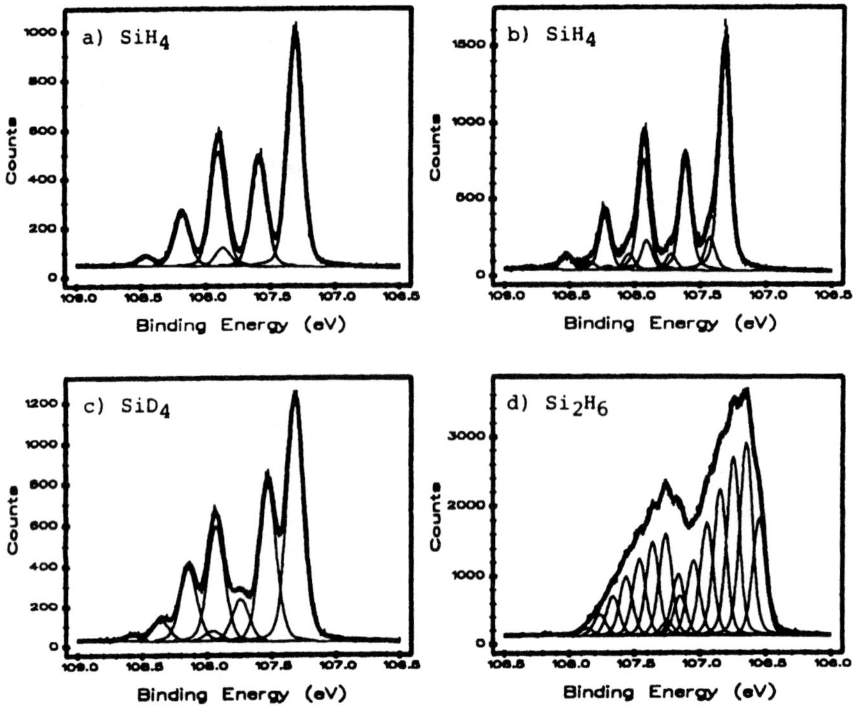

Figure 26. High resolution Si $2p$ photoelectron spectra of SiH_4, SiD_4, and Si_2H_6.[20,77] In a), c), d), the total linewidth of the lines is $\geqslant$130 meV, whereas in b) the total linewidth is 95 meV. The vibrational frequencies are given in Table 3.

Table 3 Vibrational Frequencies from Si 2*p* and C 1*s* Photoelectron Spectra,[20] and Core Equivalent Values

		Vibrational Frequency (cm^{-1})			
Molecule	Bond	Ground State	Core Equiv.	Experimental	Vibrational Mode
SiH_4	Si-H	2180	2327	2379	ν_1
	Si-H	970,910	—	887	$\nu_{2,4}$
SiD_4	Si-D	1545	1694	1710	ν_1
Si_2H_6	Si-H	844	—	823	ν_6
SiF_4	Si-F	801	893	847	ν_1
CH_4	C-H	2917	3335	3162	ν_1

SiH_4.[20] Both ν_1 and ν_2 are seen at better resolution on the Si $2p$ photoabsorption spectrum of SiH_4.[114]

In SiD_4, there are four vibrational peaks corresponding to $v' = 0, 1, 2, 3$, with an energy spacing of 0.212 eV (1710 cm^{-1}). This value is exactly that expected from the square root of the ratio of the reduced masses (μ) of Si-H and Si-D.

$$\frac{E(\mathrm{SiH_4})}{E(\mathrm{SiD_4})} = \sqrt{\frac{\mu(\mathrm{SiD_4})}{\mu(\mathrm{SiH_4})}} = 1.39 \tag{49}$$

The experimental ratio (0.295/0.212) equals 1.39! The vibrational frequency is again very close to that of the core equivalent species, PD_4^+ and PD_3 of 1654 cm^{-1} and 1694 cm^{-1}, respectively (Table 3).

The Si $2p$ spectrum of Si_2H_6 (Figure 26d) is rather surprising. Qualitatively, it is apparent that neither the symmetric Si–H nor Si–Si vibrations (~0.3 eV and 0.06 eV, respectively) are excited in this molecule. It is apparent that there exists vibrational fine structure with an energy spacing of ~0.1 eV. By far the best fit results from a single vibrational progression corresponding to $\nu' = 0$–7 with an energy spacing of 0.102 eV (823 cm^{-1}). This is the same vibrational frequency observed for the small peaks in Figure 5b in SiH_4, and in the Si $2p$ absorption spectrum of Si_2H_6.[114] Obviously, the vertical *BE* or IP is at least 0.1 eV higher than the adiabatic IP in Si_2H_6.

This is the first core level photoelectron spectrum to show vibrational structure from non-totally symmetric vibrational modes. In molecules with degenerate core orbitals from two or more equivalent cores, ionization of core electrons may result in allowed excitations of non-totally symmetric vibrational modes due to vibronic coupling between electronic and vibrational states.[115–117] This theory has been tested on the C $1s$ photoabsorption spectra of several hydrocarbons such as ethylene[116] and acetylene.[117]

For other Si molecules such as $Si(CH_3)_4$, $Si[N(CH_3)_2]_4$, $Si(OCH_3)_4$ and SiF_4, very extensive vibrational splitting (or broadening) is seen.[77] Indeed in the $Si(OCH_3)_4$ spectrum (Figure 27b), the Si $2p$ spin-orbit splitting is not even resolved. In both SiF_4 and $Si(OCH_3)_4$, there is enough resolution to be reasonably confident that the fits (Figure 27b) have semiquantitative significance, but better resolution will be essential to completely characterize the splitting. In both cases, the totally symmetric vibrational frequency dominates the spectra. These fits are made with very few variables—a constant spin-orbit splitting, a constant vibrational splitting from the core equivalent species (e.g. for $Si(OCH_3)_4$, the P–O frequency from $P(OCH_3)_3$ of 1012 cm^{-1} was used), and the same relative intensities (Franck–Condon factors) for both Si $2p_{3/2}$ and $2p_{1/2}$ envelopes. Ten peaks for each of the spin-orbit components gives a good fit to the $Si(OCH_3)_4$ spectrum, while the narrower linewidth for $Si(CH_3)_4$ yields seven vibrational peaks (P–C frequency in $P(CH_3)_3$ of 652 cm^{-1}).[77]

In the series of molecules SiH_4, SiD_4, $Si(CH_3)_4$, $Si[N(CH_3)_2]_4$, $Si(OCH_3)_4$, and SiF_4, the vibrational manifold increases from H(3) to D(4), C(7), N(>7, <10), O(10), and F(11).[77] This obviously leads to very different overall widths of the Si $2p_{3/2}$ and Si $2p_{1/2}$ peaks (Figure 27 and Table 4), different adiabatic and vertical *BE* for each compound, and differences in ΔBE (vertical-adiabatic) depending on whether or not the adiabatic or vertical IP is used. For example, the ΔBE for $Si(CH_3)_4$ is 0.18 eV; while for $Si(OCH_3)_4$ ΔBE is 0.57 eV. Better resolution is still needed to quantify

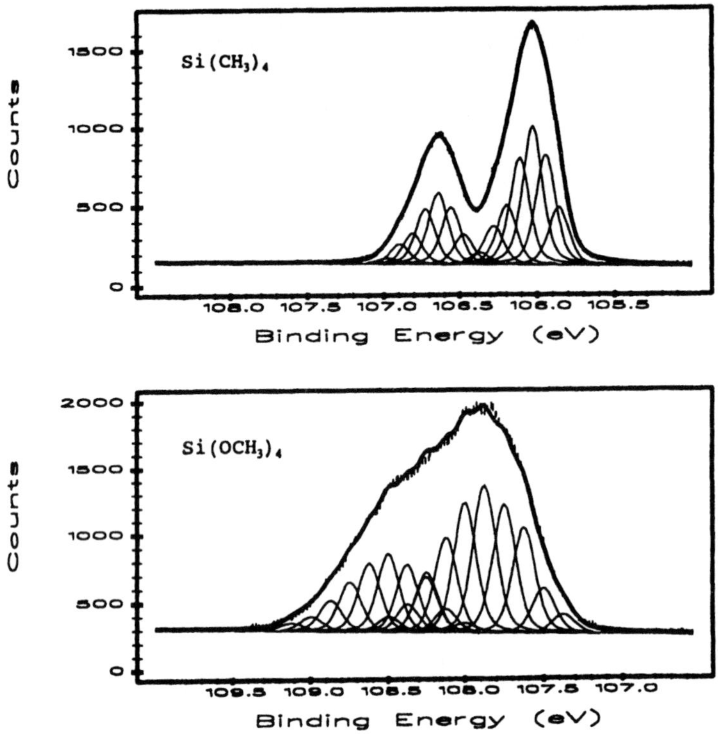

Figure 27. Si $2p$ photoelectron spectra of TMS (upper) and TMOS (lower). The vibrational energy spacing was chosen from that of the core-equivalent species (see text). Reprinted with permission from Ref. 77. Copyright 1992 American Institute of Physics.

Table 4 Linewidths, Vertical BE, and Adiabatic BE for the Eight Compounds in the Methyl/Methoxy and the Methyl/Dimethyl Amino Series. The Linewidths Quoted from the Two-Peak Fit are Accurate to within ± 20 meV.[77]

Molecule	Linewidth (meV)	Vertical BE (eV)	Adiabatic BE (eV)
$Si(CH_3)_4$	355	106.04 (1)	105.86 (2)
$Si(CH_3)_3(OCH_3)$	449	106.59 (1)	106.3 (1)
$Si(CH_3)_2(OCH_3)_2$	514	107.02 (1)	106.6 (1)
$Si(CH_3)(OCH_3)_3$	605	107.41 (1)	107.0 (1)
$Si(OCH_3)_4$	647	107.89 (1)	107.42 (2)
$Si(CH_3)_3[N(CH_3)_2]$	414	106.09 (1)	105.7 (1)
$Si(CH_3)_2[N(CH_3)_2]_2$	467	106.14 (1)	105.7 (1)
$Si(CH_3)[N(CH_3)_2]_3$	498	106.18 (1)	105.8 (1)

these numbers; and to resolve and characterize the vibrational envolopes in mixed ligand Si species (Table 4). Note that the overall Si $2p$ linewidths increase almost linearly in the series $Si(CH_3)_{4-X}(OCH_3)_X$ ($X = 0$–3) and $Si(CH_3)_{4-X}[N(CH_3)_2]_X$ ($X = 0$–3) as X increases.

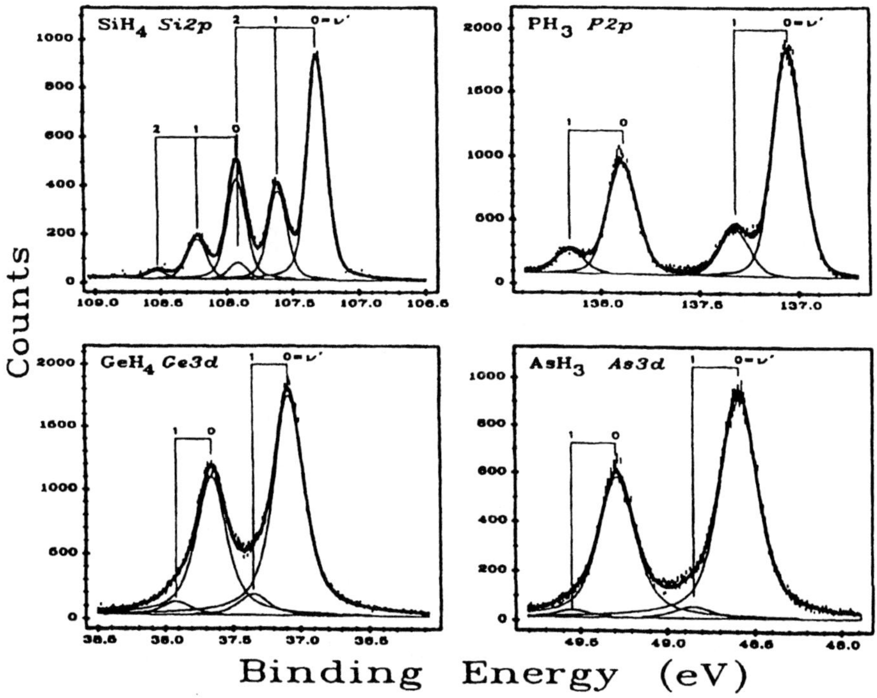

Figure 28. Experimental photoelectron spectra of SiH_4, Si $2p$; PH_3, P $2p$; GeH_4, Ge $3d$; and AsH_3, As $3d$.[111]

So far, systematic studies have been obtained only on the Si $2p$ levels of silicon molecules. However, we have shown both experimentally and theoretically that core level vibrational structure decreases down a group and across a row.[111,112] First, Figure 28 shows that there is little vibrational structure on the Ge $3d$ level of GeH_4 compared to the Si $2p$ level of SiH_4.[111] Also, the $\nu' = 1$ vibrational intensity is much smaller on the As $3d$ level of AsH_3 than on the P $2p$ level of PH_3. To illustrate the second trend across a row, it is apparent from the $2p$ spectra of SiH_4, PH_3 (Figure 28),[111] H_2S[78] (Figure 25), and HCl[87] that the vibrational splitting decreases greatly from SiH_4 to HCl. This trend is also apparent on the $3d$ spectra of GeH_4, AsH_3,[111] and HBr,[77,82] although the $\nu' = 1$ peaks are rather weak and not resolved in these cases. These qualitative trends can be rationalized using the bond lengths of ground state and core equivalent species in Table 5.[118–120] Thus, as noted above, the bond length in PH_4^+ is 0.06 Å smaller than in SiH_4, and a number of vibrational peaks are expected. In contrast, the GeH_4 and AsH_3 bond lengths are within 0.005 Å, and little or no vibrational splitting is expected. For the hydrogen halides, the very similar bond lengths for HX (X = Cl, Br, I) and HY^+ (Y = Ar, Kr, Xe) (within 0.007 Å) suggest that little or no vibrational splitting should be observed, consistent with the experimental results. However, the large bond length increase (0.07 Å) from HF to the core equivalent HNe^+ (Table 5) should lead to a large vibrational progres-

Table 5 Bond Lengths (in Å). Most data are from *Structure Data of Free Polyatomic Molecules*, edited by Hellwege, K.H. and Hellwege, A.M. (Springer, Berlin, 1987), unless noted otherwise.[111]

CH_4	NH_3	NH_4^+	HF	NeH^+
1.094	1.030	1.034	0.917[a]	0.991[b]
SiH_4	PH_3	PH_4^+	HCl	ArH^+
1.481	1.420	1.42	1.275[a]	1.280[b]
GeH_4	AsH_3		HBr	KrH^+
1.525	1.520		1.414[a]	1.421[b]
SnH_4	SbH_3		HI	XeH^+
1.7108	1.7039		1.609[a]	1.603[c]

[a]Ref. 119; [b]Ref. 120; [c]Ref. 121.

sion in the F 1*s* photoelectron line. Indeed, a large vibrational F 1*s* broadening in HF was seen earlier using Mg Kα radiation,[35] consistent with this large bond length change.

More quantitatively, multiconfiguration self consistent field (MCSCF) *ab initio* and local density-functional studies on SiH_4 and GeH_4[111,112] show that the experimental Franck–Condon factors are well reproduced for the Si 2*p* and Ge 3*d* levels. Recent vibrational calculations on the S 2*p* spectra of S molecules such as COS and H_2S[85,86] also give good agreement with experiment.

Do these periodic trends hold for other ligands such as the halides and methyl? Ge 3*d* and Sn 4*d* spectra of $M(CH_3)_xX_{4-x}$ (M = Ge, Sn; X = Cl, F; x = 0–4) compounds certainly give less vibrational broadening than for the Si analogues.[84] For example, the Ge 3*d* linewidths for the Ge 3*d* spin-orbit doublet in $GeCl_4$ and GeF_4 of 0.28 and 0.31 eV are only slightly broader than the widths in GeH_4 of 0.26 eV. All of these widths seem to indicate an inherent Ge 3*d* linewidth of $\geqslant$0.15 eV, which will make it impossible to ever resolve these splittings and quantify them.

Finally, can vibrational broadening/splitting be observed on core levels of transition metal compounds? Indeed, in $W(CO)_6$ (Figure 29), the CO vibrational splitting is seen clearly on the W 4*f* levels. Such splitting/broadening is also present in the Re 4*f* levels of $Re(CO)_5X$ (X = Cl, Br, I) complexes[9] and the Os 4*f* level of $Os(CO)_5$.[10] In addition, the rather broad linewidth for each W 4*f* component suggests there is a substantial W–C vibrational series on the 4*f* levels, just as there is on the W 5*d* valence band levels (Figure 10). Because of the inherent linewidth of the W 4*f* level is ~0.1 eV, the W–C series (ν = 0.05 eV, 400 cm^{-1}) will never be resolved.[32]

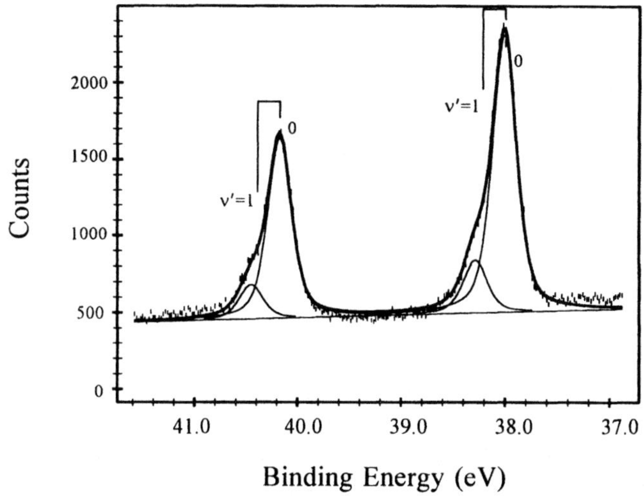

Figure 29. The high resolution W $4f$ photoelectron spectrum of $W(CO)_6$.[32] The spectrum is fitted just considering spin-orbit splitting and a ν_1(C–O) vibrational component, as in Fig. 10.

4 APPLICATIONS OF SYNCHROTRON RADIATION TO VALENCE BAND STUDIES

4.1 Introduction

We have already shown in Figure 3 that atomic cross sections, and the photon energy dependence of these cross sections, varies widely for different atomic levels (e.g. C $2p$ versus Ni $3d$, Figure 3). SR has enabled researchers to vary the photon energy routinely between 20 and 200 eV photon energy, and photon energies up to 1000 eV at high resolution will shortly be used routinely. Using Koopmans' theorem (Eq. 9) as usual, and the Gelius intensity model (Eq. 41 and Section 2.7.) for molecular orbitals, the molecular orbital ordering, in a host of inorganic and organometallic molecules, can be readily obtained with the aid of molecular orbital calculations. Based on energies alone, it is not possible to be confident of the MO assignments, even for small molecules such as SiF_4 or $Ni(\eta^3\text{-}C_3H_5)_2$, based on even the most sophisticated MO calculations. The variation of photoelectron intensities with photon energy provides a more powerful method for assigning the MO ordering.

In addition to the use of the Gelius model, resonance features (e.g. core level induced resonances and shape resonances[25]) can be very useful for assignment purposes for gas phase spectra. In the following sections, we will look at the use of the Gelius model, and resonances, for assignment of spectra of $M(\eta^3\text{-}C_3H_5)_2$ (M = Ni, Pd, Pt), $TiCl_4$, SiF_4, and XeF_2. Several similar cross section studies have been made on solid state inorganic species (e.g. Refs. 26, 122–124), but these are beyond the scope of this article.

4.2 Assignment of the Photoelectron Spectra of Organometallics

4.2.1 ***The Gelius Model*** Even for relatively simple organometallic compounds such as $Ni(\eta^3\text{-}C_3H_5)_2$, there has been a long debate about the assignment of the photoelectron spectrum (Figure 6); and the validity of Koopmans' theorem for these molecules.[27] Many different MO calculations along with HeI and HeII spectra led to a host of different assignments (see references in Refs. 27, 124). The controversy centered mainly on the assignment of the low *BE* peak at 7.8 eV (peak 1 in Figure 6). Despite experimental inconsistencies, nearly all the theoretical and experimental studies, between 1969 and 1993, agreed that the peak at 7.8 eV arises from the ligand nonbonding $7a_u$ orbital of high C $2p$ character (see Figure 30 for our ground state $X\alpha$ results[27]) rather than the mainly Ni $3d$ orbitals $13a_g$, $12a_g$, $6b_g$, $11a_g$, and $5b_g$. Note, in Figure 30, that our MO calculation also gives $7a_u$ as the HOMO.

However, the spectra at 32 eV and 70 eV photon energies (Figure 6) show immediately that, for $Ni(\eta^3\text{-}C_3H_5)_2$, peaks 1 and 2 *increase* in intensity from 32 to 70 eV; while for the Pd analogue, the intensity of peak 1 *decreases* sharply from 32 to 70 eV.

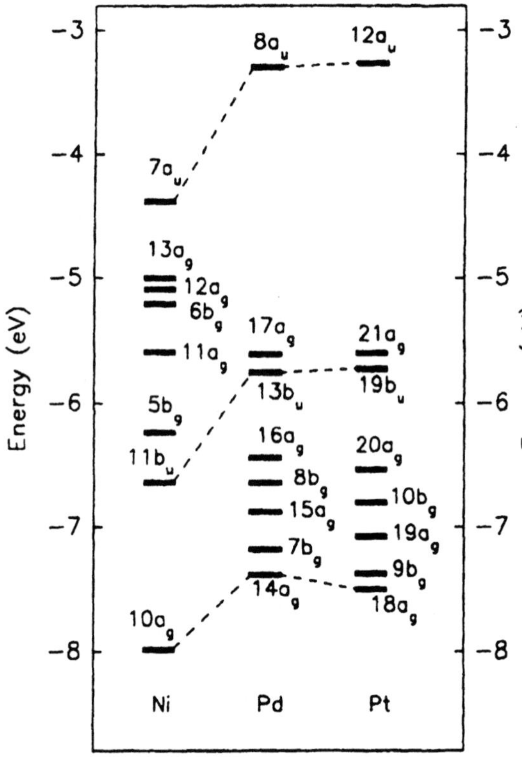

Figure 30. The $X\alpha$-SW MO ordering for the three molecules $M(\eta^3\text{-}C_3H_5)_2$ (M = Ni, Pd, and Pt). The ligand a_u orbitals are the HOMO's in all cases.[27]

From the cross section plots in Figure 3, these four spectra show immediately that peaks 1 and 2 for Ni(η^3-C_3H_5)$_2$ must arise mainly from Ni 3*d* orbitals; while peak 1 for Pd(η^3-C_3H_5)$_2$ must arise from the 8a$_u$ orbital of mainly C 2*p* character (Figure 30). Immediately, the two Ni spectra suggest strongly that the conclusions from over twenty previous experimental and theoretical studies (including our theoretical calculation) were incorrect.

However, the assignments are more complex. First, from Figure 30, we have eight MO's in the region of interest, and only six peaks in Ni(η^3-C_3H_5)$_2$, seven peaks in Pd(η^3-C_3H_5)$_2$, and eight peaks in the Pt analogue (Figure 31). The large width of peak 6 in the Pd spectra (Figures 6, 31) suggests strongly that this peak arises from ionization from two MO's. However, it is certainly not obvious how to assign the six peaks to the eight MO's in the Ni(η^3-C_3H_5)$_2$ spectrum. A detailed analysis of the intensities as a function of photon energy is obviously required.

In addition to the valence band peaks from ~7 eV to ~12 eV *BE*, spectra of both valence and inner valence regions (Figure 31) were recorded. As for W(CO)$_6$ and Os(CO)$_5$ (Figure 1) inner valence peaks A to F are very broad and rather weak. Also, in Figure 31, it is apparent that the energies and intensities of peaks A to F are very similar for all three molecules. In contrast, the pattern of peaks in the valence region is very different for Ni(η^3-C_3H_5)$_2$ compared to the Pd and Pt analogues. In other late transition series such as η^5-C_5H_5M(CO)$_2$ (M = Co, Rh, Ir),[125] again the Co spectra are quite different from those of the Rh and Ir analogues. All of the valence and inner valence peaks have been assigned with the aid of an $X\alpha$-SW calculation;[27]

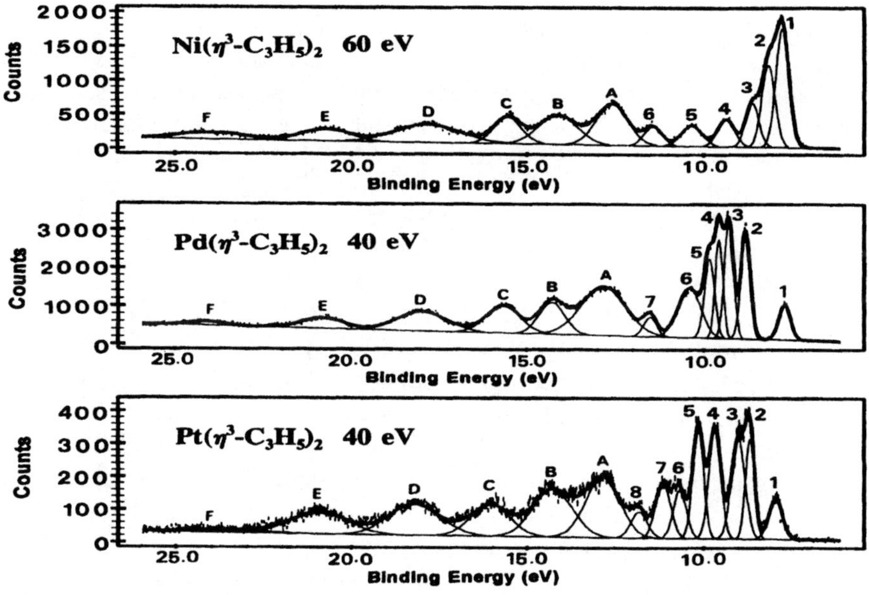

Figure 31. Broadscan spectra of the three M(η^3-C_3H_5)$_2$ molecules showing both the valence and inner valence levels.[27]

and the calculations for the valence levels give an orbital sequence very similar to the sequences from previous $X\alpha$ calculations.[126,127]

For the detailed analysis of the valence band region (Figure 30) spectra were taken every few eV from 20 eV to ~100 eV. Branching ratios were calculated for each peak from the peak areas ($BR_i = A_i / \sum A_i$), and compared with theoretical BR ($BR_i = \sigma_i / \sum \sigma_i$) using both the Gelius method (Eq. 43), and the $X\alpha$-SW method using Davenport's program.[128]

The orbital populations, $P_{A\lambda}$ (Eq. 43) were obtained from $X\alpha$ orbital compositions, and the atomic $\sigma_{A\lambda}^{AO}$ values were obtained from Yeh and Lindau's tables.[11] The experimental and theoretical branching ratios in Figure 32 reflect the atomic cross sections, and provide a powerful way of assigning the peaks to the MO's. Five photoelectron bands were expected in the region of bands 1–3 (Figure 30), and the challenge is to match the observed and calculated BR to make the assignment (Figure 32). Thus, band 1 must be assigned to two orbitals of very high Ni 3*d* character because of the very large increase in BR at higher photon energies, and the good semiquantitative agreement with *both* Gelius and $X\alpha$-SW cross sections (Figure 32).[27] Band 2 is assigned to a Ni 3*d* orbital ($6b_g$) plus the ligand $7a_u$ orbital. The slight increase in BR towards high photon energy reflects the increase in Ni 3*d* cross section and the decrease in C 2*p* cross section. Bands 3 and 4 show intermediate behavior because these orbitals have 93% Ni 3*d* ($11a_g$) and 69% Ni 3*d* ($5b_g$) character. Bands 5 and 6 have BR's which decrease sharply, and they can be assigned readily to the majority ligand orbitals $11b_u$ and $10a_g$, respectively.

Thus these spectra show that peak 1 arises from *two* Ni 3*d* orbitals, and not the *one* ligand $7a_u$ orbital from all previous studies. This assignment is not in agreement with either the $X\alpha$-SW ground state orbital energies (Figure 30) or the transition state energies which are supposed to take relaxation energies into account.

A similar treatment for the Pd and Pt compounds enables assignments based on semiquantitative agreement between experimental and theoretical cross sections.[27] The Pd 4*d* Cooper minimum (Figure 3) is very helpful here in confirming the assignment of the Pd 4*d* orbitals. Obviously, the $8a_u$ orbital is assigned to peak 1 in the Pd and Pt analogues.

Many other organometallic molecules have been studied in the last ten years using SR based photoelectron spectroscopy including $M(CO)_6$ (M = Cr, Mo, W),[129] $M(\eta^5\text{-}C_5H_5)_2$ (M = Fe, Ru, Os),[130] $U(\eta^8\text{-}C_8H_8)_2$,[131] $(\eta^5\text{-}C_5H_5)PtMe_3$,[132] $(\eta^7\text{-}C_7H_7)Ta(\eta^5\text{-}C_5H_4Me)$,[133] $(\eta^7\text{-}C_7H_7)M(\eta^5\text{-}C_5H_5)$ (M = Ti, Nb, Mo),[133] $Cr(\eta^6\text{-}C_6H_6)_2$,[134] $Mo(\eta^6\text{-}C_6H_5Me)_2$,[134] $M(\eta^5\text{-}C_5H_5)_2$ (M = V, Cr, Co, Ni),[135] $(\eta^5\text{-}C_5H_5)_2M(CO)_2$ (M = Co, Rh, Ir),[125] $[(\eta^5\text{-}C_5H_4^iPr)MoS]_4$,[136] $(\eta^5\text{-}C_5H_5)M(\eta^3\text{-}C_3H_5)$ (M = Ni, Pd),[137] and (hfac)$MPMe_3$ (M = Cu, Ag).[138] Recent work has focused largely on using core level induced valence band resonances to help valence band assignments;[25] and our work has focused on the variation of metal *d* orbital energies across and down the transition metal series. These two aspects will be discussed in the next two sections.

Before looking at the above two topics, it is important to emphasize here that, even using variable energy spectra, it is not possible to be confident about assignments for some compounds; and indeed for a few molecules it is not possible to use MO calcu-

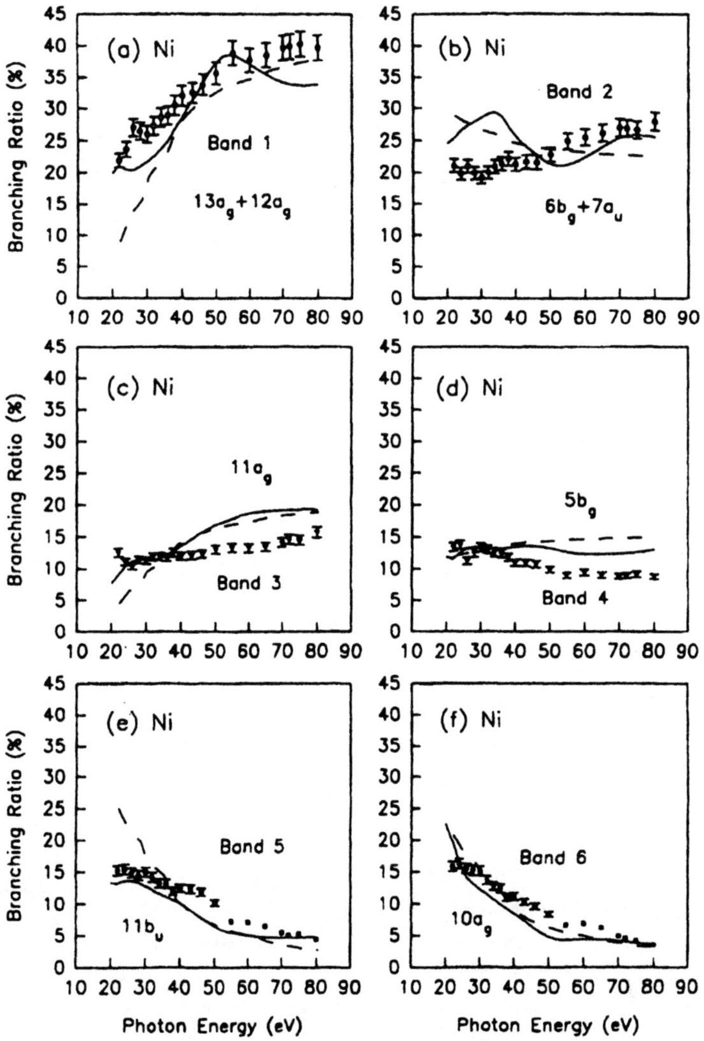

Figure 32. Comparison of experimental branching ratios (circle, triangles, and squares) with $X\alpha$ ones (solid lines) and ones from the Gelius model (dashed lines) for the six valence bands of $Ni(\eta^3\text{-}C_3H_5)_2$ in the photon energy range of 20 to 80 eV. The band assignments are indicated on each plot.[27]

lations and Koopmans' theorem to assign spectra. The HeI photoelectron spectrum of CpNiNO was first published in 1974.[139] The spectrum was assigned based on SCF calculations.[140] Other more recent theoretical calculation using INDO[141] and MS-$X\alpha$ method[142] agreed with the original assignment: the first four bands were assigned to $5e_1 < 7a_1 < 3e_2 < 4e_1$ in order of increasing *BE*. However, the first two bands are separated by only 0.23 eV (1855 cm^{-1}), which is very close to the NO vibrational frequency. Two very recent detailed SR studies[143,144] have yielded two different assignments: $5e_1 < 7a_1 < 3e_2 < 4e_1$,[143] in agreement with the original assignments; and $5e_1 < 7a_1$, $3e_2 < 4e_1$, where the first *and* second bands are due to $5e_1$ orbital, split by the NO vibrational frequency.[144] Because the splitting of the first two bands is ~0.2 eV, and the linewidth of band 2 is >0.2 eV, it is not possible to obtain accurate BR for bands 1 and 2 separately. The theoretical energies in both papers tend to favor the second assignment; while the peak widths and overall BR tend to favor the first assignment. A recent private communication[145] indicates that because of strong correlation between one-electron states, there is a complete breakdown of the one-electron picture, and it is not possible to assign peaks to individual MO's. Clearly, higher resolution spectra and more calculations are required to solve this interesting controversy.

4.2.2 ***Resonant Features*** Often, valence band cross sections can be heavily influenced by resonant features. In these cases, the Gelius model does not apply. There are two types of common resonances:[25,26] core level enhanced resonances and shape resonances.

Valence band cross sections can be enhanced when the photon energy is tuned to the excitation or ionization of a metal or ligand core electron.[25] This effect is most dramatic for a transition metal *d* electron when the metal core *p* electron is resonantly excited or ionized. For example, for the W 5*d* electrons in $W(CO)_6$ (Figures 1 and 10), the W 5*d* intensity is dramatically enhanced close to the W $5p_{3/2}$ *BE* (43.85 eV, small broad peak in Figure 1) and W $5p_{1/2}$ *BE* at 53 eV (Figure 33). This is a consequence of the interaction of two ionizing channels:

1) The "normal" channel: $5p^6 5d^6 \rightarrow 5p^6 5d^5 + e$
2) The resonant channel: $5p^6 5d^6 \rightarrow 5p^5 5d^7 \rightarrow 5p^6 5d^5 + e$

Note the final states are the same (Figure 33) via both channels.

A very dramatic enhancement of the metal *d* intensity is often seen due to the second channel involving excitation and super Coster–Kronig Auger decay. The degree of enhancement can also be used for determining covalency in molecules, but there are several cases now where this effect leads to ambiguous interpretations: for example, in $Ru(\eta^5\text{-}C_5H_5)_2$,[130] enhancements of orbitals with no *d* content occurs. This has been explained by interchannel coupling whereby final state ionization channels mix and borrow intensity.[25]

Shape resonances[146–148] are also important, and are not accounted for by the Gelius model—although they are accounted for in the $X\alpha$ program of Davenport.[128] These shape resonances occur at low photoelectron energies (<20 eV kinetic energy) and are normally accounted for by two theories. In an MO-type interpretation, the electron is resonantly excited into an antibonding orbital in the continuum. In the second model, a centrifugal potential or electronegative ligand creates an effective potential which traps the photoelectron. Alternatively, Tse has shown[149] that these resonances can be accounted for by an EXAFS type treatment, where scattering off the neighboring atoms gives enhancement of the outgoing electron wave. Although,

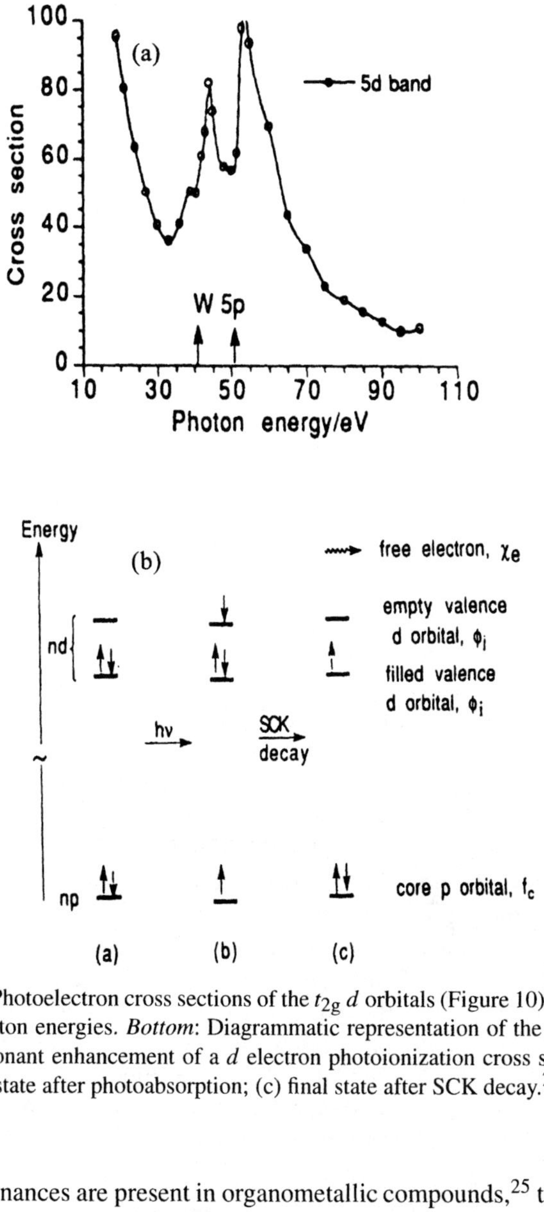

Figure 33. *Top*: Photoelectron cross sections of the t_{2g} d orbitals (Figure 10) of $W(CO)_6$ from 20 to 100 eV photon energies. *Bottom*: Diagrammatic representation of the processes underlying $p \rightarrow d$ resonant enhancement of a d electron photoionization cross section: (a) initial state; (b) excited state after photoabsorption; (c) final state after SCK decay.[25]

these shape resonances are present in organometallic compounds,[25] the most intense resonances come in inorganic compounds with electronegative ligands, such as F or Cl. These will be discussed in Section 4.3.

4.2.3 ***Periodic Differences in the Transition Metal d Binding Energies*** From the spectra in Figures 6 and 31, it is immediately obvious that the average *BE* for the 10 d electrons is over 1 eV larger for the Pd and Pt analogues than in $Ni(\eta^3\text{-}C_3H_5)_2$. From other spectra in the literature, it is apparent that the average *BE* difference

Table 6 Differences in "Metal *d*" Ionization Potentials (ΔIP) between the Second- and First-Row-Metal Organometallic Compounds[125]

Compound	ΔIP
$M(CO)_6$ (M = Cr, Mo)	0.10
$CpM(CO)_3$ (M = Mn, Re)[a]	0.34
$M(\eta^5\text{-}C_5H_5)_2$ (M = Fe, Ru)	0.59
$CpM(CO)_2$ (M = Co, Rh)	0.71
$M(\eta^3\text{-}C_3H_5)_2$ (M = Ni, Pd)	1.17
$(hfac)MP(CH_3)_3$ (M = Cu, Ag)[b]	1.9

[a] ΔIP is between the third- and first-row metals.
[b] hfac = $CF_3C(O)CHC(O)CF_3$.

(ΔBE) between metal 3*d* and 4*d* is much larger for the late transition metals than the early transition metals (Table 6).[125]

What is the reason for this trend, in which the energy difference between 3*d* and 4*d* or 5*d* orbitals of analogous compounds appears to increase for the later transition metals?

Lichtenberger et al.[47,150,151] attributed the *d BE* separation in $CpM(CO)_2$ (M = Cr, Rh) mainly to the larger relaxation energy associated with first row complexes. The orbital relaxation effect is mainly due to the orbital contraction on ionization. This effect makes the ion state relatively more stable, and lowers the *BE*. On the other hand, Ziegler et al.[152] proposed that trends in the thermal stability and kinetic lability of the metal–carbonyl bond in $M(CO)_6$ (M = Cr, Mo, W), $M(CO)_5$ (M = Fe, Ru, Os), and $M(CO)_4$ (M = Ni, Pd, Pt) were due to the ground state energy differences between the first and second or third row transition metals. They suggested that, in the ground state, the 4*d* and 5*d* metal orbitals are lower in energy than the 3*d*, since *d*–*d* repulsions are smaller for the diffuse 4*d* and 5*d* orbitals than for the contracted 3*d* orbitals, and that this difference should increase from the early to the late transition metals. Due to the lanthanide contraction, the energies of 4*d* and 5*d* metal orbitals are relatively close. The $X\alpha$ calculations carried out for $CpM(CO)_2$ (M = Co, Rh, Ir)[125] and $M(\eta^3\text{-}C_3H_5)_2$ (M = Ni, Pd, Pt)[27] support Ziegler's interpretation in terms of the difference in ground state d orbital energies. It is also supported by the difference in slopes among experimental curves of relative band intensity variation (BR) for the HOMO's of $CpM(CO)_2$ (M = Co, Rh, Ir).[125]

4.3 The Assignment of Photoelectron Spectra of Inorganic Molecules—Shape Resonances

Even for relatively simple prototype inorganic molecules such as $TiCl_4$ and SiF_4, there has been considerable controversy over the assignment of the photoelectron spectra and the MO ordering. The MO diagram for a d^0 tetrahedral molecule such as $TiCl_4$ is shown in Figure 34.[153] Other nontransition metal analogues such as CX_4

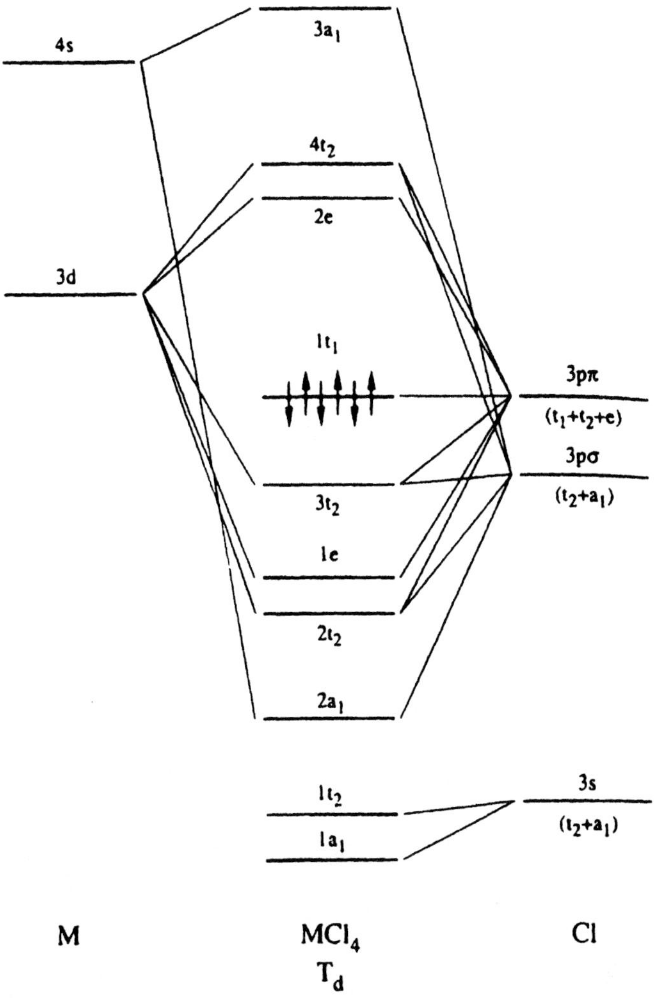

Figure 34. Qualitative molecular orbital diagram for a first-row transition metal in tetrahedral Cl_4 ligand field.[153]

and SiX_4 (X = Cl, Br, I) have the same basic MO diagram with the C $2p$ and Si $3p$ orbitals involved in the bonding.[154]

All of these tetrahedral molecules have the five MO's in the valence region (Figure 34). Some of these molecules, such CF_4, gives five well-resolved peaks labeled A–E;[155] while $TiCl_4$ gives a strong overlap of peaks C and D (Figure 35),[153,155,156] and SiF_4 gives a strong overlap of peaks B and C.[154,157] For CF_4, the assignment of the spectrum follows the MO ordering in Figure 34: $t_1 < t_2 < e < t_2 < a_1$ in order of increasing *BE*. However, INDO calculations[157] for SiF_4 gave the assignment: $1t_1 < 5t_2 < 5a_1 < 1e < 4t_2$; and in $TiCl_4$, there has been a longstanding controversy which has only recently been resolved with a detailed SR study.[153]

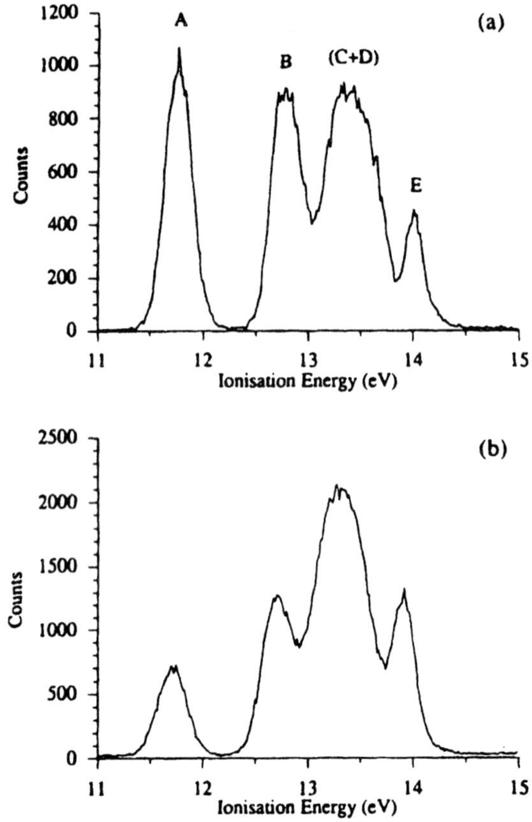

Figure 35. Photoelectron spectra of $TiCl_4$, acquired with SR at (a) 24 and (b) 40 eV.[153]

In $TiCl_4$, there have been at least seven HeI studies, two HeII studies, and a large number of calculations including INDO, SCF-$X\alpha$ and *ab initio* calculations (references in 153, 156). For example, $X\alpha$ calculations give the ordering $1e < 2a_1 < 2t_2$ while *ab initio* calculations give the ordering $1e < 2t_2 < 2a_1$. Cowley in 1979[12] stated that some of the bands in $TiCl_4$ cannot be assigned unequivocally based on HeI and HeII spectra and the many different calculations.

In 1982, the original ordering given in Figure 34 was questioned by our HeI and HeII study along with $X\alpha$ calculation.[156] The spectra taken at 24 and 40 eV with SR (Figure 35)[153] show the band intensity trends from our HeI, HeII study:[157] peaks A and B decrease relatively in intensity from ~20 to 40 eV, while the intensity of peak E increases relatively at 40 eV. The intensity decreases for peaks A and B are expected; ionization cross sections for the Cl 3*p* band orbitals $1t_1$ and $3t_2$ drop sharply from ~20 to 40 eV photon energy (see Figure 3 for the analogous C 2*p* cross section trend). However, the increase in relative intensity for peak E is not expected if it arises from ionization of the $2a_1$ electrons. This orbital cannot have any Ti 3*d* character (it has some Ti 4*s* character), whereas all calculations indicate that both

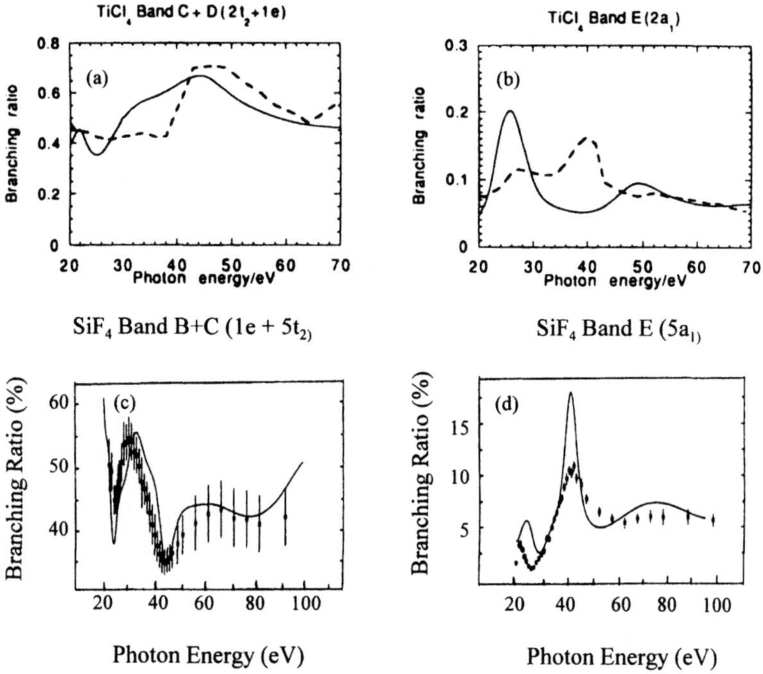

Figure 36. *Top*: Comparison of experimental (dashed line) and MS-$X\alpha$ theoretical (solid line) BR for bands C + D and E for $TiCl_4$.[153] *Bottom*: Comparison of experimental (dashed line) and MS-$X\alpha$ theoretical (solid line) BR for bands B + C and E for SiF_4. Adapted with permission from Ref. 154. Copyright 1985 American Institute of Physics.

the 1e and $2t_2$ orbitals have substantial Ti 3*d* character. Because the 3*d* character is normally the important factor for enhancing the band intensity from HeI to HeII radiation, peak E should be assigned to the 1e or $2t_2$ orbitals.[156] Combined with the $X\alpha$ energies, peak E was assigned to the $2t_2$ orbital, and not the $2a_1$ orbital.

The synchrotron study combined with a DV-$X\alpha$ calculation[158] showed that at about 40 eV photon energy, there was a quite narrow shape resonance (Figure 36) which enhanced the BR for peak E by almost a factor of two. A resonance also occurs for band C + D at about 45 eV photon energy. There is reasonable agreement between experiment and the MS-$X\alpha$ theoretical results (Figure 36a, b) only if band E is assigned to the $2a_1$ orbital. The calculation gives the resonance at ~50 eV rather than 40 eV, but the agreement above 50 eV photon energies in both plots is excellent. Also the DV-$X\alpha$ transition state calculation gave quantitative agreement with the experimental *BE* (within 0.16 eV for all bands). Bursten et al.[153] have suggested that the resonance at 40–50 eV is a multiple scattering resonance of the type proposed earlier by Tse.[149] The predicated kinetic energy of such resonance is E (eV) $= 151/R^2$. The theoretical E for $TiCl_4$ is 35 eV, yielding $R = 2.1$ Å compared to R_{Ti-Cl} of 2.185 Å.

Similarly for SiF_4, peak E also shows a strong shape resonance at ~40 eV photon energy. Again the agreement between experiment and theory in Figure 36d

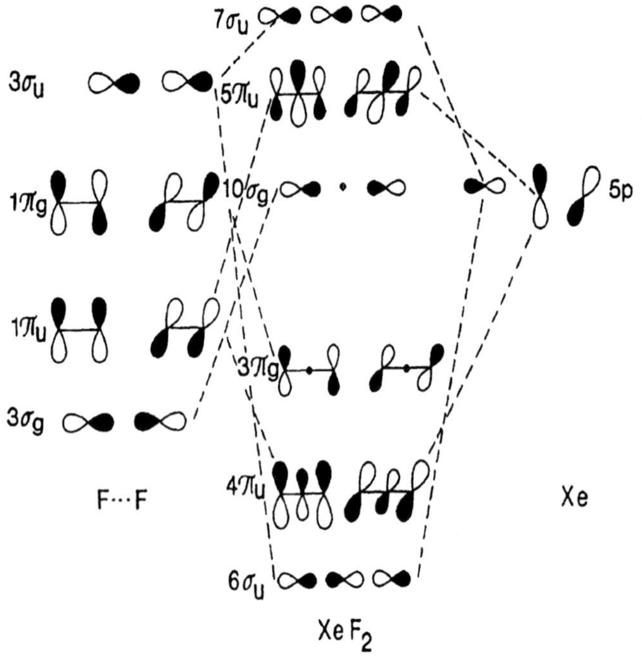

Figure 37. The molecular orbital ordering for XeF_2.[149]

shows that peak E arises from the $5a_1$ orbital, and the assignment is the same as for CF_4 and $TiCl_4$, and in disagreement with the INDO calculation mentioned above.

Obviously, SR is essential to assign these spectra. The intensity changes at two (or even a few) photon energies cannot be used for assignment purposes using the Gelius-type treatment. Spectra have to be taken at many photon energies especially in the 20–50 eV range.

Finally, we return to the valence band spectra of XeF_2 (Figure 18).[76] Once again there were large discrepancies in predicted MO assignments from *ab initio*[159,160] and $X\alpha$[161] calculations. The $X\alpha$ calculations gave the ordering in Figure 37 ($5\pi_u$ is the HOMO); whereas the *ab initio* calculations reversed the $10\sigma_g$ and π_u ordering. The spin-orbit splitting of the degenerate $5\pi_u$ level (Figure 18) shows immediately that the former ordering is correct. A detailed SR study combined with MS-$X\alpha$ calculations confirmed this assignment. In Figure 18, there is obviously a very large, very sharp resonance feature for the $10\sigma_g$ ionization centered at ~24 eV photon energy. This resonance is reproduced in the $X\alpha$ calculations (Figure 38). Obviously, the very different cross section behavior for the two orbitals and the agreement with theory, confirms the calculations. Tse has suggested that this resonance might be due to a similar, but sharper, delayed onset resonance seen for Ni 3*d* (Figure 3) and Xe 4*d*. The apparent broad resonance at ~45 eV photon energy is probably due to a scattering resonance off the F as discussed above for $TiCl_4$.

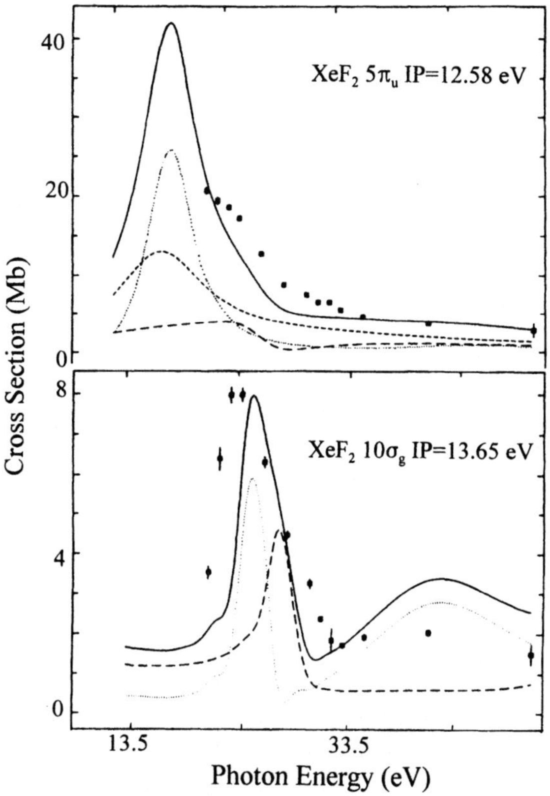

Figure 38. *Top*: Plot shows the experimental and theoretical MS-$X\alpha$ (solid line) photoionization cross section (in Mb) of the $5\pi_u$ valence orbital of XeF_2; plus the partial-channel cross sections (dashed lines). *Bottom*: Plot shows the experimental and theoretical MS-$X\alpha$ (solid line) photoionizaiton cross section (in Mb) of the σ_g valence ortibal of XeF_2; plus the partial-channel cross sections (dashed lines). Reprinted with permission from Ref. 76. Copyright 1986 American Institute of Physics.

5 FUTURE STUDIES

It is important to emphasize that most of the results discussed here can still be improved in the future with the higher intensity and resolution from intense undulator radiation, and third generation SR sources such as the ALS at Berkeley and MAX II at Lund, and even higher resolution photoelectron spectrometers.[162] Thus, it should be possible to resolve the Si $2p$ spectra better; and resolve better both ligand field and vibrational effects on many narrow core levels. The work at ALS on the P $2p$, S $2p$, and Cl $2p$, levels[86,88] gives total linewidths of <100 meV, enabling resolution of *both* ligand field and vibrational splittings on a number of P and S molecules.[86,88]

In the valence band area, higher intensity and resolution will certainly enable resolution of many more vibrational features in these spectra, and these higher resolution spectra should aid assignments in several cases. Core/valence resonances will be studied at higher photon energies, because high intensity/resolution has just being obtained recently on undulator beamlines from third generation synchrotron. A better understanding of these resonances is still required. Also, valence bands at low photon energies (<20 eV) need to be studied to characterize shape resonances. To obtain much higher resolutions, the technique of zero kinetic energy (ZEKE) photoelectron spectroscopy has been developed.[163–165] Lasers have been used mostly as the photon source at <20 eV, but high resolution SR will be important in this area in the future.

In addition, high resolution will enable resolution of the extensive splittings in photoabsorption and Auger spectra. But perhaps more exciting, the discovery of the Auger resonance Raman effect (Auger spectra after resonance excitation into Rydberg orbitals) gives the opportunity of beating the lifetime linewidth on core level Auger spectra.[23,24] Indeed, total linewidths of <20 meV have been seen on the resonance Auger Kr MVV and Xe NVV spectra.[23,24] The only molecule studied so far is HBr.[166] In addition, studies of photodissociation of molecules either by looking at the Auger spectra[167] or electron ion coincidence is a powerful way of studying selective fragmentation near core level ionization potentials; and much more work will be done in this area in the near future on larger molecules.[168,169]

ACKNOWLEDGMENTS

We would like to dedicate this article to the memory of Ednor M. Rowe, former director and associate director of the Tantalus and Aladdin synchrotron facilities (or SRC) at the University of Wisconsin. Without his encouragement and guidance, the Canadian Synchrotron Radiation Facility (CSRF) would not have existed, and most of the results in this article would not have been generated.

Many collaborators were extremely important in generating many of the ideas and results discussed in this work. G.M.B. would like to begin by thanking his collaborators during his 1975–1976 sabbatical: W. Gudat, D.E. Eastman, U. Gelius, S. Svensson, P.A. Malmquist, and K. Siegbahn. That year generated many of the ideas which led to CSRF and the gas phase results. Then, considering graduate students (with their Ph.D. degree date in parenthesis), we would like to thank D.K. Creber (1978), J.S. Tse (1980), T.C.S. Chan (1982), D.J. Bristow (1982), B.W. Yates (1986), L. Dignard-Bailey (1986), J.E. Bice (Forrest) (1987), D.S. Yang (1989), J.D. Bozek (1991), J.N. Cutler (1993), D.G.J. Sutherland (1993), Z.F. Liu (1993), X. Li (1995), Y.F. Hu (1996), J. Johnson and J.Wu. Postdoctorate fellows/research associates R.P. Gupta, L.L. Coatsworth and E. Pellach are also gratefully acknowledged for their important contributions. My senior collaborators R.J. Puddephatt, G. Schrobilgen, R. Pomeroy, J.S. Tse, R.G. Cavell, H. Aksela and S. Aksela provided compounds, equipment, theoretical calculations, and many new ideas.

Bob Lazier did yeomen service operating our laboratory ESCA-36 from 1972–1978. But Kim Tan in Madison has make the largest contribution of any individual to this work for the last 17 years; he built and assembled the Grasshopper beamline (1980–1983), he has kept this beamline operating routinely for 14 years; and he was always an important helper and initiator with all the graduate students who traveled to Madison.

We also have to thank the staff at Tantalus and Aladdin for their excellent help. We are also grateful for the continuous financial support from NSERC and NRC, and Norman Sherman of NRC for his continuous administrative help. NSF must also be thanked for their support to Tantalus and Aladdin, NSF Grant NO. DMR-9212658 to the Synchrotron Radiation Center.

REFERENCES

1. Robinson, H.R. *Phil. Mag.* **1925**, *50*, 241.
2. Siegbahn, K.; Nordling, C.; Fahlman, A.; Nordberg, R.; Hamrin, K.; Hedman, J.; Johansson, G.; Bergmark, T.; Karlsson, S.-E.; Lindgren, I.; Lindberg, B.J. *ESCA-Atomic, Molecular and Solid State Structure Studied by Means of Electron Spectroscoppy*; Nova Acta Regiae Soc. Sci. Upsaliensis Ser. IV; **1967**, *Vol. 20.*
3. Siegbahn, K.; Nordling, C.; Johansson, G.; Hedman, J.; Heden, P.F.; Hamrin, K.; Gelius, U.; Bergmark, T.; Werme, L.O.; Manne, R.; Baer, Y. *ESCA Applied to Free Molecules*; North Holland: Amsterdam, 1969.
4. Price, W.C.; Turner, D.W. *Phil. Trans. Roy. Soc. London*, **1970**, *A268.*
5. Turner, D.W.; Baker, C.; Baker, A.D.; Brundle, C.R. *Molecular Photoelectron Spectroscopy*; Wiley: New York, 1970.
6. Einstein, A. *Ann. Phys.* **1905**, *17*, 32.
7. Carlson, T.A. *Photoelectron and Auger Spectroscopy*; Plenum: New York, 1975.
8. Eland, J.H.D. *Photoelectron Spectroscopy*; Wiley: New York, 1984.
9. Hu, Y.F.; Bancroft, G.M.; Liu, Z.; Tan, K.H. *Inorg. Chem.* **1995**, *34*, 3716.
10. Hu, Y.F.; Bancroft, G.M.; Davis, H.B.; Male, J.I.; Pomeroy, R.K.; Tse, J.S.; Tan, K.H. *Organometallics* **1996**, *15*, 4493.
11. Yeh, J.J.; Lindau, I. *At. Data Nucl. Data Tables* **1985**, *32*, 1.
12. Cowley, A.H. *Prog. Inorg. Chem.* **1979**, *26*, 46.
13. Green, J.C. *Struct. Bonding (Berlin)* **1981**, *43*, 37.
14. Jolly, W.L.; Bomben, K.D.; Eyermann, C.J. *At. Data Nucl. Data Tables* **1984**, *31*, 433.
15. Siegbahn, K. *J. Electron Spectrosc. Relat. Phenom.* **1974**, *5*, 3.
16. Gelius, U. *J. Electron Spectrosc. Relat. Phenom.* **1974**, *5*, 985.
17. Gelius, U.; Asplund, L.; Basilier, E.; Hedman, S.; Helenelund, K.; Siegbahn, K. *Nucl. Inst. Methods B* **1984**, *229*, 85.
18. Margaritondo, G. *Introduction to Synchrotron Radiation*; Oxford University Press: New York, 1988.
19. Bozek, J.D.; Bancroft, G.M.; Cutler, J.N.; Tan, K.H. *Phys. Rev. Lett.* **1990**, *65*, 2757.
20. Bancroft, G.M.; Bozek, J.D.; Cutler, J.N.; Hu, Y.F.; Liu, Z.F.; Sutherland, D.G.; Tan, K.H. in *The Proceeding of the Tenth International Conference on Vacuum Ultraviolet Radiation Physics*; Wuilleumier, F.J.; Petrofee, Y.; Nenner, I. Eds., World Scientific: Singapore, 1993, p. 191.
21. Karlsson, L.; Svensson, S.; Carlsson-Gothe, M.; Keane, M.P.; de Brito, A.N.; Correia, N.; Wannberg, B. *J. Phys. B:* **1989**, *22*, 3001.

22. Cutler, J.N.; Bancroft, G.M.; Tan, K.H. *J. Phys B:* **1991**, *24*, 4897.
23. Aksela, S. *J. Electron Spect. Relat. Phenom.* **1996**, *79*, 247.
24. Langer, B.; Berrah, N.; Farhat, A.; Hemmers, O.; Bozek, J.D. *Phys. Rev. A* **1996**, *53*, R 1946.
25. Green, J.C. *Acc. Chem. Res.* **1994**, *27*, 131.
26. Didziulis, S.V.; Cohen, S.L.; Gewirth, A.E.; Solomon, E.I. *J. Am. Chem. Soc.* **1988**, *110*, 250.
27. Li, X.; Bancroft, G.M.; Puddephatt, R.J.; Liu, Z.F.; Hu, Y.F.; Tan, K.H. *J. Am. Chem. Soc.* **1994**, *116*, 9543.
28. (a) Brown, F.C.; Bach, R.Z.; Lien, N. *Nucl. Inst. Methods* **1978**, *152*, 73. (b) Tan, K.H.; Bancroft, G.M.; Coatsworth, L.L.; Yates, B.W. *Can. J. Phys.* **1983**, *60*, 131.
29. Yang, B.X.; Middleton, F.H.; Olsson, B.G.; Bancroft, G.M.; Chen, J.M.; Sham, T.K.; Tan, K.H.; Wallace, D. *Nucl. Inst. Methods A* **1992**, *316*, 422.
30. Koopmans, T. *Physica* **1933**, *1*, 104.
31. Baker, A.D.; Betteridge, D. *Photoelectron Spectroscopy, Chemical and Analytical Aspects*; Pergamon Press: New York, 1972.
32. Hu, Y.F.; Bancroft, G.M.; Bozek, J.D.; Liu, Z.F.; Sutherland, D.G.J.; Tan, K.H. *J. Chem. Soc., Chem. Com.* **1992**, 1276.
33. Hollander, J.M.; Jolly, W.L. *Act. Chem. Res.* **1970**, *3*, 193.
34. Jolly, W.L.; Henderickson, D.N. *J. Am. Chem. Soc.* **1970**, *92*, 1863.
35. Bristow, D.J.; Bancroft, G.M. *J. Am. Chem. Soc.* **1983**, *105*, 5634.
36. Karlsson, S.E.; Sieghabn, K.; Bartlett, N. in Ref. 3, p. 133.
37. Carroll, T.X.; Shaw, Jr., R.W.; Thomas, T.D.; Kindle, C.; Bartlett, N. *J. Am. Chem. Soc.* **1974**, *96*, 1989.
38. Bancroft, G.M.; Malmquist, P.-A.; Svensson, S.; Basilier, E.; Gelius, U.; Sieghahn, K. *Inorg. Chem.* **1978**, *17*, 1595.
39. Krause, M.O. *J. Phys. Chem. Ref. Data* **1979**, *8*, 307.
40. Gupta, R.P.; Tse, J.S.; Bancroft, G.M. *Phil. Trans. Royal Soc.* **1980**, *293*, 535, and references therein.
41. Shirley, D.A. *Phys. Rev. A* **1973**, *7*, 1520.
42. Thomas, T.D. *J. Elect. Spect. Relat. Phemon.* **1980**, *20*, 117.
43. Aksela, S.; Bancroft, G.M.; Bristow, D.J.; Aksela, H.; Schrobilgen, G.J. *J. Chem. Phys.* **1985**, *82*, 4809.
44. Russ, S.; Grodzicki, M. *Phys. Scr.* **1990**, *42*, 58.
45. Huheey, J.E.; Keiter, E.A.; Keiter, R.L. *Inorganic Chemistry: Principles of Structure and Reactivity*; Harper Collins College: New York, 1993, p. 188.
46. Bursten, B.E. *J. Am. Chem. Soc.* **1982**, *104*, 1299.
47. Lichtenberger, D.L.; Kellogg, G.E. *Acc. Chem. Res.* **1987**, *20*, 379.
48. Lichtenberger, D.L.; Kellogg, G.E.; Landis, G.H. *J. Chem. Phys.* **1985**, *83*, 2759.
49. Bursten, B.E.; Darensbourg, D.J.; Kellogg, G.E.; Lichtenberger, D.L. *Inorg. Chem.* **1984**, *23*, 4361.
50. Yarbrough, L.W., II; Hall, M.B. *Inorg. Chem.* **1978**, *17*, 2269.
51. Bancroft, G.M.; Dignard-Bailey, L.; Puddephatt, R.J. *Inorg. Chem.* **1984**, *23*, 2369.
52. Jolly, W.L. *Acc. Chem. Res.* **1983**, *16*, 370.

53. Beach, D.B.; Jolly, W.L. *Inorg. Chem.* **1986**, *25*, 875.

54. Bancroft, G.M.; Tse, J.S. *Comments Inorg. Chem.* **1986**, *5*, 89.

55. Jorgensen, C.K. *Modern Aspects of Ligand Field Theory*; North-Holland: Amsterdam, 1971.

56. Ballhausen, C.J. *Introduction to Ligand Field Theory*; McGraw-Hill: New York, 1962.

57. Cotton, F.A.; Wilkinson, G. *Advanced Inorganic Chemistry, Fifth Edition*; John Wiley & Sons: New York, 1988.

58. Hougen, J.T.; Leroi, G.E.; James, T.C. *J. Chem. Phys.* **1961**, *34*, 1670.

59. Bleaney, B.; Stevens, K.W.H. *Report Prog. Phys.* **1953**, *16*, 108.

60. (a) Bancroft, G.M.; Creber, D.K.; Basch, H. *J. Chem. Phys.* **1977**, *67*, 4891. (b) Bancroft, G.M.; Creber, D.K.; Ratner, M.A.; Moskowitz, J.M.; Topiol, S. *Chem. Phys. Lett.* **1977**, *50*, 233.

61. Forrest, J.E.; Bancroft, G.M.; Coatsworth, L.L. *Inorg. Chem.* **1986**, *25*, 2181.

62. Bancroft, G.M.; Coatsworth, L.L.; Creber, D.K.; Tse, J.S. *Phys. Scr.* **1977**, *16*, 217.

63. Berkowitz, J.; Dehmer, J.L.; Walker, T.E.H. *J. Chem. Phys.* **1973**, *59*, 3645.

64. Potts, A.W.; Williams, T.A.; Price, W.C. *Proc. Roy. Soc. (London) A* **1974**, *341*, 147.

65. See the Mössbauser Chapter by P. Gutlich in this volume.

66. Lucken, E.A.C. *Nuclear Quadrupole Coupling Constants*; Academic Press: New York, 1969.

67. Bancroft, G.M. *Mossbauer Spectroscopy: An Introduction for Inorganic Chemists and Geochemists*; McGraw-Hill: London, 1973.

68. Bancroft, G.M.; Boyd, B.D.; Creber, D.K. *Inorg. Chem.* **1978**, *17*, 1008.

69. Beach, N.A.; Gray, H.B. *J. Am. Chem. Soc.* **1968**, *90*, 5713.

70. Carlson, T.A.; Krause, M.O. *Phys. Rev.* **1965**, *137*, A1655.

71. Bancroft, G.M.; Sham, T.K.; Larsson, S. *Chem. Phys. Lett.* **1977**, *46*, 551.

72. Price, W.C.; Potts, A.W.; Streets, D.G. in *Electron Spectroscopy*; Shirley, D.A., Ed.; North-Holland: Amsterdam, 1972.

73. Carlson, T.A.; Fahlman, A.; Krause, M.O.; Keller, P.R.; Taylor, J.W.; Whitley, T.A.; Gimm, F.A. *J. Chem. Phys.* **1984**, *80*, 3521 and references therein.

74. Carlson, T.A.; Fahlman, A.; Svensson, W.A.; Krause, M.O.; Whitley, T.A.; Grimm, F.A.; Piancastelli, W.N.; Taylor, J.W. *J. Chem. Phys.* **1984**, *81*, 3828.

75. Berkowitz, J. *Photoabsorption, Photoionization and Photoelectron Spectroscopy*; Academic Press: New York, 1979.

76. Yates, B.W.; Tan, K.H.; Bancroft, G.M.; Coatsworth, L.L.; Tse, J.S.; Schrobilgen, G.J. *J. Chem. Phys.* **1986**, *84*, 3603.

77. (a) Sutherland, D.G.J.; Bancroft, G.M.; Tan, K.H. *J. Chem. Phys.* **1992**, *97*, 7918. (b) Sutherland, D.G.J.; Liu, Z.F.; Bancroft, G.M.; Tan, K.H. *Nucl. Inst. Methods B* **1994**, *87*, 183.

78. Svensson, S.; Ausmees, A.; Osborne, S.J.; Bray, G.; Gel'mukhanov, F.; Agren, H.; Naves de Brito, A.; Sairanen, O.P.; Kivimaki, A.; Nommiste, E.; Aksela, H.; Aksela, S. *Phys. Rev. Lett.* **1994**, *72*, 3021.

79. Cutler, J.N.; Bancroft, G.M.; Sutherland, D.G.; Tan, K.H. *Phys. Rev. Lett.* **1991**, *67*, 1531.

80. Cutler, J.N.; Bancroft, G.M.; Tan, K.H. *J. Chem. Phys.* **1992**, *97*, 7932.

81. Cutler, J.N.; Bancroft, G.M.; Bozek, J.D.; Tan, K.H.; Schrobilgen, G.J. *J. Am. Chem. Soc.* **1991**, *113*, 9125.

82. Liu, Z.F.; Bancroft, G.M.; Tan, K.H.; Schachter, M. *J. Electron Spect. Relat. Phenom.* **1994**, *67*, 299.

83. Johnson, J.; Cutler, J.N.; Bancroft, G.M.; Hu, Y.F.; Tan, K.H. submitted to *J. Phys. B*.

84. Cutler, J.N.; Bancroft, G.M.; Tan, K.H. *Chem. Phys.* **1994**, *181*, 461.

85. Siggel, M.R.F.; Field, C.; Saethre, L.F.; Borve, K.J.; Thomas, T.D. *J. Chem. Phys.* **1996**, *105*, 9035; Borve, K.J. *Chem.Phys. Lett.* **1996**, *262*, 801.

86. Bozek, J.D.; Thomas, T.D. personal communication.

87. Aksela, H.; Kukk, E.; Aksela, S.; Sairanen, O.P.; Kivimaki, A.; Nommiste, E.; Ausmees, A.; Osborne, S.J.; Svensson, S. *J. Phys. B* **1995**, *28*, 4259.

88. Bozek, J.D. personal communication.

89. Comes, F.J.; Haensel, R.; Nielsen, U.; Schwarz, W.H.E. *J. Chem. Phys.* **1973**, *58*, 516.

90. Hudson, E.; Shirley, D.A.; Domke, M.; Remmers, G.; Kaindl, G. *Phys. Rev. A* **1994**, *49*, 161.

91. Puttner, R.; Domke, M.; Schulz, S.; Gutierrez, A.; Kaindl, G. *J. Phys. B* **1995**, *28*, 2425.

92. Gel'mukhanov, F.; Agren, H.; Svensson, S.; Aksela, H.; Aksela, S. *Phys. Rev. A* **1996**, *53*, 1379.

93. Aitken, E.J.; Bahl, M.K.; Bomben, K.D.; Gimzewski, J.K.; Nolan, G.S.; Thomas, T.D. *J. Am. Chem. Soc.* **1980**, *102*, 4873.

94. Bancroft, G.M. *Coord. Chem. Rev.* **1973**, *11*, 247.

95. Bancroft, G.M.; Platt, R.H. *Adv. Inorg. Radiochem.* **1972**, *15*, 59.

96. Sham, T.K.; Bancroft, G.M. *Inorg. Chem.* **1975**, *14*, 2281.

97. Gavin, R.M., Jr.; Bartell, L.S. *J. Chem. Phys.* **1968**, *48*, 2460.

98. Pitzer, K.S.; Bernstein, L.S. *J. Chem. Phys.* **1975**, *63*, 3849.

99. Klobukowski, M.; Huzinaga, S.; Seijo, L.; Barandiaran, Z. *Theor. Chim. Acta* **1987**, *71*, 237.

100. Johnson, J.; Liu, Z.F.; Bancroft, G.M.; Cutler, J.N. to be published.

101. Wu, J.; Bancroft, G.M.; Puddephatt, R.J.; Hu, Y.; Li, X.; Tan, K.H. to be published.

102. Roos, B.O. *Adv. Chem. Phys.* **1987**, *69*, 399.

103. Jenssen, H.J.A.; Agren, H.; Olsen, J. in Modern Techniques in Computational Chemistry: *MOTECC-90*; Clementi, E., Ed., ESCOM: London, **1990**.

104. Pierloot, K.; Dumez, B.; Widmark, P.O.; Roos, B.O. *Theor. Chim. Acta.* **1995**, *90*, 87.

105. Chong, D.P.; Langhoff, S.R. *J. Chem. Phys.* **1986**, *84*, 5606.

106. Siegbahn, P.E.M.; Blomberg, M.R.A.; Pettersson, G.M.; Ross, B.O.; Almlof, J. *Stockholm*; University of Stockholm: Sweden, 1995.

107. Bozek, J.D.; Bancroft, G.M.; Tan, K.H. *Phys. Rev. A* **1991**, *43*, 3597, **1991**.

108. Sutherland, D.G.J.; Bancroft, G.M.; Tan, K.H. *Surf. Sci.* **1992**, *262*, L96.

109. Koppe, H.M.; Itchkawitz, B.; Kilcoyne, A.L.D.; Feldhaus, J.; Kempgens, B.; Kivimaki, A.; Neeb, M.; Bradshaw, A.M. *Phys. Rev. A* **1996**, *53*, 4120.

110. Koppe, H.M.; Kilcoyne, A.L.D.; Feldhaus, J.; Bradshaw, A.M. *J. Electron Spect. Relat. Phenom.* **1995**, *75*, 97.

111. Liu, Z.F.; Bancroft, G.M.; Cutler, J.N.; Sutherland, D.G.J.; Tan, K.H.; Tse, J.S.; Cavell, R.G. *Phys. Rev. A* **1992**, *46*, 1688.

112. Liu, Z.F.; Bancroft, G.M.; Tse, J.S.; Agren, H. *Phys. Rev. A* **1995**, *51*, 439.

113. Asplund, L.; Gelius, U.; Hedman, S.; Helenelund, K.; Siegbahn, K.; Siegbahn, P.E.M. *J. Phys. B* **1985**, *18*, 1569.

114. Sutherland, D.G.J.; Bozek, J.D.; Bancroft, G.M.; Tan, K.H. *Chem. Phys. Lett.* **1992**, *199*, 341.

115. Cederbaum, L.S.; Domcke, W. *J. Chem. Phys.* **1977**, *66*, 5084.

116. Gadea, F.A.; Kopper, H.; Schirmer, J.; Cederbaum, L.S.; Randall, K.J.; Bradshaw, A.M.; Ma, Y.; Sette, F.; Chen, C.T. *Phys. Rev. Lett.* **1991**, *66*, 883.

117. Ma, Y.; Chen, C.T.; Meigs, G.; Randall, K.J.; Sette, F. *Phys. Rev. A* **1991**, *44*, 1848.

118. Huber, K.P.; Herzberg, G. *Constants of Diatomic Molecules*; Van Nostrand Reinhold: New York, 1979.

119. Klein, R.; Rosmus, P. *Z. Naturforsch. A* **1984**, *39*, 349.

120. Rogers, S.A.; Brazier, C.R.; Bernath, P.F. *J. Chem. Phys.* **1987**, *87*, 159.

121. Butcher, K.D.; Didziulis, S.V.; Briat, B.; Solomon, E.I. *J. Am. Chem. Soc.* **1990**, *112*, 2231.

122. Didziulis, S.V.; Cohen, S.L.; Butcher, K.D.; Solomon, E.I. *Inorg. Chem.* **1988**, *27*, 2238.

123. Lichtenberger, D.L.; Ray, C.D.; Stepniak, F.; Chen, F.; Weaver, J.H. *J. Am. Chem. Soc.* **1992**, *114*, 10492.

124. Moncrieff, D.; Hillier, I.H.; Saunders, V.R.; von Niessen, W. *Chem. Phys. Lett.* **1986**, *131*, 545.

125. Li, X.; Bancroft, G.M.; Puddephatt, R.J.; Hu, Y.F.; Tan, K.H. *Organometallics* **1996**, *15*, 2890.

126. Hancock, G.C.; Kostic, N.M.; Fenske, R.F. *Organometallics* **1983**, 2, 1089.

127. Guerra, M.; Jones, D.; Distefano, G.; Torroni, S.; Foffani, A.; Modelli, A. *Organometallics* **1993**, *12*, 2203.

128. (a) Davenport, J.W. Ph.D. Dissertation, University of Pennsylvania, Philadelphia, PA, **1976**. (b) Davenport, J.W. *Phys. Rev. Lett.* **1976**, *36*, 945.

129. Cooper, G.; Green, J.C.; Payne, M.P.; Dobson, B.R.; Hillier, I.H. *J. Am. Chem. Soc.* **1987**, *109*, 3836.

130. Cooper, G.; Green, J.C.; Payne, M.P. *Mol. Phys.* **1988**, *63*, 1031.

131. Brennan, J.G.; Green, J.C.; Redfern, C.M. *J. Am. Chem. Soc.* **1989**, *111*, 2372.

132. Yang, D.S.; Bancroft, G.M.; Puddephatt, R.J.; Tan, K.H.; Cutler, J.N.; Bozek, J.D. *Inorg. Chem.* **1990**, *29*, 4956.

133. Green, J.C.; Kaltsoyannis, N.; Sze, K.H.; MacDonald, M. *J. Am. Chem. Soc.* **1994**, *116*, **1994**.

134. Brennan, J.G.; Cooper, G.; Green, J.C.; Kaltsoyannis, N.; MacDonald, M.A.; Payne, M.P.; Redfern, C.M.; Sze, K.H. *Chem. Phys.* **1992**, *164*, 271.

135. Brennan, J.; Cooper, G.; Green, J.C.; Payne, M.P.; Redfern, C.M. *J. Electron Spect. Relat. Phenom.* **1993**, *66*, 101.

136. Davis, C.E.; Green, J.C.; Kaltsoyannis, N.; MacDonald, M.A.; Qin, J.; Rauchfuss, T.B.; Redfern, C.M.; Stringer, G.H.; Woolhouse, M.G. *Inorg. Chem.* **1992**, *31*, 3779.

137. Li, X.; Tse, J.S.; Bancroft, G.M.; Puddephatt, R.J. *Organometallics* **1995**, *14*, 4513.

138. Li, X.; Bancroft, G.M.; Puddephatt, R.J.; Yuan, Z.; Tan, K.H. *Inorg. Chem.* **1996**, *35*, 5040.

70, 417.

140. Hillier, I.H.; Saunders, V.R. *Mol. Phys.* **1972**, *23*, 449.
141. Bohm, M.C. *Z. Naturforsch.* **1981**, *36A*, 1361.
142. Modellie, A.; Foffani, A.; Scagnolari, F.; Torroni, S.; Guerra, M.; Jones, D. *J. Am. Chem. Soc.* **1989**, *111*, 6040.
143. Li, X.; Tse, J.S.; Bancroft, G.M.; Puddephatt, R.J.; Tan, K.H. *Inorg. Chem.* **1996**, *35*, 2515.
144. Field, C.N.; Green, J.C.; Mayer, M.; Nasluzov, V.A.; Rosch, N.; Siggel, M.R.F. *Inorg. Chem.* **1996**, *35*, 2504.
145. Tse, J.S.; von Niessen, W. private communication.
146. Dehmer, J.L. in *Resonances in Electron-Molecule Scattering, van der Waals Complexes, and Reactive Dynamics*; Truhlar, D.G.; Ed.; American Chemistry Society: Washington, DC, 1984; Vol. 263.
147. Dehmer, J.L. *J. Chem. Phys.* **1972**, *56*, 4496.
148. Langhoff, P.W. in *Resonances in Electron-Molecule Scattering, van der Waals Complexes, and Reactive Dynamics*; Truhlar, D.G.; Ed.; American Chemistry Society: Washington, DC, 1984; Vol. 263.
149. Tse, J.S. *J. Chem. Phys.* **1988**, *89*, 920.
150. Lichtenberger, D.L.; Calabro, D.C.; Kellogg, G.E. *Organometallics* **1984**, *3*, 1623.
151. Calabro, D.C.; Lichtenberger, D.L. *Inorg. Chem.* **1980**, *19*, 1732.
152. Ziegler, T.; Tschinke, V.; Ursenbach, C. *J. Am. Chem. Soc.* **1987**, *109*, 4825.
153. Bursten, B.E.; Green, J.C.; Kaltsoyannis, N.; MacDonald, M.A.; Sze, K.H.; Tse, J.S. *Inorg. Chem.* **1994**, *33*, 5086.
154. Yates, B.W.; Tan, K.H.; Bancroft, G.M.; Coatsworth, L.L.; Tse, J.S. *J. Chem. Phys.* **1985**, *83*, 4906.
155. Green, J.C.; Green, M.L.H.; Joachim, P.J.; Orchard, A.F.; Turner, D.W. *Philos. Trans. R. Soc. London A* **1970**, *268*, 111.
156. Bancroft, G.M.; Pellach, E.; Tse, J.S. *Inorg. Chem.* **1982**, *21*, 2950 and references therein.
157. Jonas, A.E.; Schweitzer, G.K.; Grimm, F.A.; Carlsson, T.A. *J. Electron Spect. Relat. Phenom.* **1972/73**, *1*, 29.
158. Ellis, D.E. *J. Phys. B* **1977**, *10*, 1.
159. Basch, H.; Moskowitz, J.W.; Hollister, C.; Hankin, D. *J. Chem. Phys.* **1971**, *55*, 1922.
160. Bartell, L.S.; Rothman, M.J.; Evig, C.S.; Van Wazer, J.R. *J. Chem. Phys.* **1980**, *73*, 367.
161. Rosen, A.; Ellis, D.E. *Chem. Phys. Lett.* **1974**, *27*, 595.
162. Eland, J.H.D.; Baltzer, P.; Lundquist, M.; Wannberg, B.; Karlsson, L. *Chem. Phys.* **1996**, *212*, 457.
163. Grant, E.R.; White, M.G. *Nature* **1991**, *354*, 249.
164. Muller-Dethlefs, K.; Schlag, E.W. *Ann. Rev. Phys. Chem.* **1991**, *42*, 109.
165. Wang, K.; McKoy, V. *Ann. Rev. Phys. Chem.* **1995**, *46*, 275.
166. Liu, Z.F.; Bancroft, G.M.; Tan, K.H.; Schacter, M. *Phys. Rev. Lett.* **1994**, *72*, 621.
167. Liu, Z.F.; Bancroft, G.M.; Tan, K.H.; Schacter, M. *Phys. Rev. A* **1993**, *48*, 4019.
168. Lee, K.; Hulbert, S.L.; Kuiper, P.; Ji, D.; Hanson, D. *Nucl. Inst. Methods A* **1994**, *347*, 446.
169. Hitchcock, A.P.; Cavell, R.G. private communication.

9 X-Ray Absorption Spectroscopy and EXAFS Analysis: The Multiple-Scattering Method and Applications in Inorganic and Bioinorganic Chemistry

HUA HOLLY ZHANG

Department of Chemistry
Stanford University
Stanford, CA 94305, USA

BRITT HEDMAN AND KEITH O. HODGSON

Stanford Synchrotron Radiation Laboratory, SLAC
Stanford University
Stanford, CA 94309, USA
E-mail: hedman@ssrl.slac.stanford.edu
hodgson@ssrl.slac.stanford.edu

Inorganic Electronic Structure and Spectroscopy, Volume I: Methodology.
Edited by E. I. Solomon and A. B. P. Lever.
ISBN 0-471-15406-7.

1 OVERVIEW

Extended X-ray absorption fine structure (EXAFS) refers to the region of an X-ray absorption spectrum where the absorption coefficient exhibits an oscillatory pattern as a function of photon energy. In the EXAFS region, a core electron absorbs sufficiently large photon energy to overcome its ionization energy and, with the excess photon energy, escape into the continuum where the ejected photoelectron can be visualized as an outgoing wave originating and propagating away from the absorbing atom (or absorber). For an isolated atom, the EXAFS absorption coefficient decreases monotonically with increasing photon energy. However, if the absorbing atom is surrounded by neighboring atoms, the outgoing photoelectron wave will be backscattered by these atoms (or scatterers), thus producing an incoming photoelectron wave. The interaction between an outgoing wave and an incoming wave creates a pattern of constructive and destructive interference as the photon energy is increased. A constructive interference between the two waves results in a local maximum whereas a destructive interference gives a local minimum (Figure 1). These

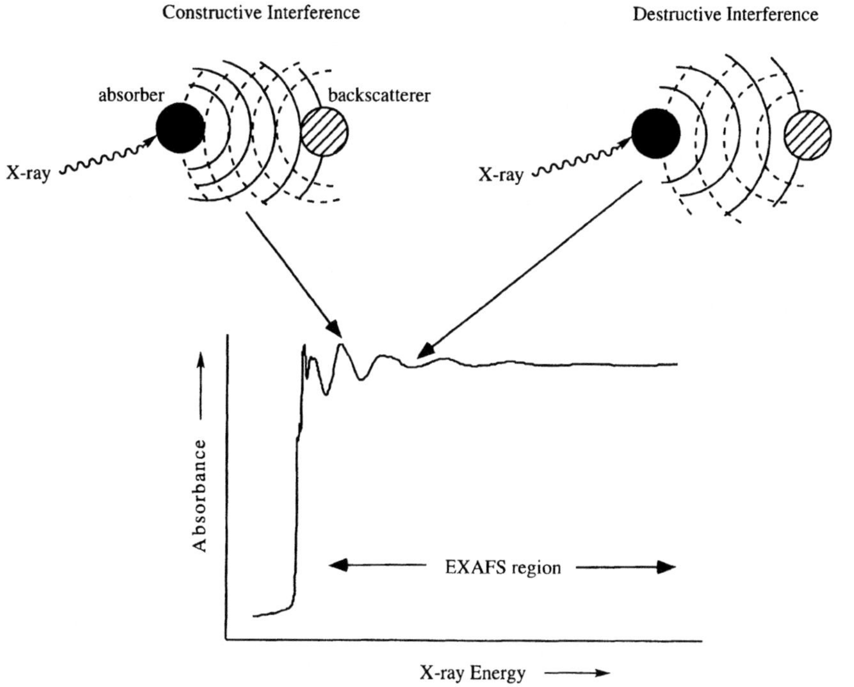

Figure 1. Pictorial illustration of the constructive and destructive interference of outgoing and incoming photoelectron waves that give rise to EXAFS.

interferences therefore raise or lower the absorption coefficient relative to that for a free atom, producing the oscillation or fine structure, known as EXAFS.

Because the oscillation of the X-ray absorption coefficient is a direct consequence of the interactions between the photoabsorbing atom and its surrounding environment, EXAFS provides information leading to structural characterization of the photoabsorber and its local environment. Specifically, it allows quantitative determination of distances between the photoabsorber and scatterers, the types and numbers of the scatterers, the thermal and static disorder of the system through Debye–Waller (DW) factors, and when strong multiple-scattering (MS) effects are present, the angular arrangement of ligands around the photoabsorber.

EXAFS, as a tool for structural determination, occupies a unique niche. Because EXAFS does not depend on long-range order, the technique can be used to obtain structural information for samples of any physical state, such as gases, liquids, solutions, amorphous solids, crystalline solids, etc. This is a particularly attractive feature for biological systems (on which this chapter will focus) given the fact that in many cases they are difficult to crystallize and therefore inaccessible for X-ray crystallography. With the availability of synchrotron radiation sources that provide X-ray intensities many orders-of-magnitude higher than conventional X-ray tubes, EXAFS data of good quality can be collected on dilute samples, an added advantage for proteins that are hard to obtain at high concentrations. In addition, because the EXAFS data are measured for a particular absorption edge that is unique for an atom of specific atomic number, the technique is highly element specific. In other words, it allows one to probe the immediate environment of the particular absorbing species when the photon energy matches its absorption edge, without interference of atoms of other element types in the sample. Further, by tuning the photon energy, one can perform EXAFS data collection at more than one absorption edge for the same sample if it contains several different absorbing species. Typically the onset of absorption K-edges for transition metals of similar atomic number are separated by more than several hundred electron volts (eV), so there is a considerable energy range from which useful information about a particular absorbing species can be extracted. In the cases where the active site of a protein contains more than one type of metal atom (*e.g.* Cu and Zn in Cu/Zn superoxide dismutase,[1] Cu and Fe in cytochrome *c* oxidase[2–4]), complementary information about the active site can thus be obtained by analyzing EXAFS data from these different edges (see for example ref 5 for Cu/Zn superoxide dismutase and ref 6 for cytochrome *c* oxidase).

The application of EXAFS to biological problems is generally used in the following two areas: to refine (or correct) metal active site structures previously characterized by protein crystallography; and to elucidate structures of active sites with an unknown metal environment.

While protein crystallography provides a three-dimensional structure of an entire protein molecule, including the location of the metal centers and their ligands, detailed information about the active site, especially accurate metal-ligand distances and the presence of small molecules bound to the active site, may not be available. The accuracies of distances, typically only to a few tenths of an Ångström, are often limited by the resolution of the data and the size of the structure. EXAFS as a technique for local structural determination, however, typically gives much higher accuracy for metal-ligand distances. In particular, first-shell distances can be determined to as accurate as 0.01–0.02 Å. With such high accuracy, subtle structural changes under different conditions (*e.g.* a change in the oxidation state) can be detected. In addition, EXAFS is very useful in defining structural details for proteins in the presence of substrates and inhibitors as well as in investigating transient enzyme intermediates in a catalytic cycle using time-resolved techniques. Such species are often not attainable as stable crystals, but can be trapped and kept as frozen solutions at low temperature (*e.g.* 4 K) for EXAFS measurements. Examples of complementarity between EXAFS and crystallography in determining protein structures include applications to rubredoxin and plastocyanin, and these studies are described in more detail in refs 7,8. The second application of the EXAFS technique can be considerably more challenging, depending upon the complexity of the structure. Results from other spectroscopies such as EPR, Mössbauer, MCD, and resonance Raman may be used to infer a postulated structure which EXAFS analysis can build upon.

Application of the EXAFS method to bioinorganic systems covers a wide variety of systems with quite different functions, ranging from structural characterization to mechanistic study. For example, EXAFS has been used to detect metal–metal (M–M) interactions,[9] to identify bridging ligation,[10] to confirm a high oxidation state of a metal evidenced by a short M-O bond,[11] and to examine structural change upon change of the oxidation state[12] or binding of substrates.[13]

As described in Chapter 1 of volume II, EXAFS is typically defined as the energy region ~40–1000 eV above an absorption edge in an X-ray absorption spectrum. The near-edge region,[14] loosely defined between the absorption edge and the beginning of EXAFS, is sometimes included in the EXAFS region, but it requires a slightly different approach to its analysis and MS theory is typically used to explain the structure in this region (see discussion below). The other two regions of the spectrum are the so-called pre-edge and edge regions where the incident photon energy is below the ionization threshold. These two regions contain transitions of core electrons to any bound state valence level. For example, an Fe K-edge corresponds to the transition of $1s \rightarrow 4p$, whereas its pre-edge feature is a result of a $1s \rightarrow 3d$ transition. Analyzing the edge and pre-edge features gives information about the electronic and the geometric structure of a photoabsorbing atom and its interactions with its ligands. In particular, the shape and position of an edge are indicative of the oxidation state of the photoabsorber and are dependent on the geometry of the coordination sphere.[15] The pre-edge feature can be correlated to, for example, ligand charge transfer, symmetry, and spin state of the absorbing site.[16] These two regions have been introduced in Chapter 1, whereas this chapter will focus mainly on the EXAFS region of the X-ray absorption spectrum. For a discussion of the properties of synchrotron radiation, the reader is referred to Chapter 8.

This chapter will introduce basic EXAFS theory and data analysis methods, and present recent advances in the application of theoretical developments using the GNXAS program suite as an example. MS treatment of the EXAFS data for selected inorganic systems will be discussed in detail using GNXAS analysis to demonstrate the contributions from MS effects and their utilization in the determination of structural parameters. This will be followed by an application of the method to a specific biological system of unknown structure at the active site. The problem and challenges currently encountered in data analysis will be discussed last. The focus of this chapter is on MS aspects of the EXAFS technique, illustrating its effects on data analysis and its applications in bioinorganic systems. Readers are referred to ref 17 for review of EXAFS in general, refs 18–20 for reviews of EXAFS theory including MS theory, and refs 8,21–24 for biological applications.

2 EXAFS THEORY

2.1 Single-Scattering (SS) Theory

2.1.1 ***The EXAFS Equation (Plane-Wave Approximation)*** SS refers to the case where the outgoing photoelectron wave is scattered only once before returning to the photoabsorber. The photoelectron wave is formed when the incident photon energy is large enough to excite a core electron into the continuum. The kinetic energy of the photoelectron can be determined by the free electron relation

$$\frac{p^2}{2m_e} = E - E_0 \tag{1}$$

where p and m_e are the momentum and the mass of the photoelectron, respectively, E is the incident photon energy, and E_0 is the threshold energy for ionization of the core electron. The wavelength λ of the photoelectron is given by the deBroglie equation

$$\lambda = \frac{h}{p} \tag{2}$$

where h is Planck's constant. The outgoing photoelectron wave from the absorbing atom is scattered by the surrounding atoms, forming an incoming photoelectron wave with which the outgoing wave interacts. As suggested by Equations (1) and (2), the wavelength of the photoelectron λ decreases when the photon energy E increases. Therefore upon scanning of the photon energy through the EXAFS region, there will be a smooth and continuous interference pattern between the outgoing and backscattered wave, giving rise to the oscillation. The peak of the oscillation occurs when the backscattered photoelectron wave is in phase with the outgoing wave at the absorbing atom while the minimum of the oscillation corresponds to the situation where the two waves are 180° out of phase.

When the energy of the photoelectron is sufficiently high, and therefore its wavelength λ is sufficiently short, the curvature of the photoelectron wave can be ignored and a plane-wave approximation can be made. Based on this approximation, SS theory[25–27] arrives at the following expression to describe the EXAFS behavior:

$$\chi(k) = \sum_j \underbrace{\frac{N_j A_j(k) S_0^2}{kR_j^2} \exp(-2\sigma_j^2 k^2) \exp\left(\frac{-2R_j}{\lambda(k)}\right)}_{\textit{Total Amplitude Function}} \times \underbrace{\sin[2kR_j + \phi_j(k)]}_{\textit{Total Phase Function}} \tag{3}$$

In this expression, the EXAFS amplitude $\chi(k)$ is defined as the modulation of the absorption coefficient μ of a particular atom relative to the smooth background absorption coefficient μ_s, normalized by the absorption coefficient that would be observed for a free atom μ_0:

$$\chi(k) = \frac{\mu - \mu_s}{\mu_0} \tag{4}$$

The photon energy is conveniently expressed in terms of the photoelectron wave vector:

$$k = \sqrt{\frac{2m_e}{\hbar}(E - E_0)} \tag{5}$$

for simpler mathematical manipulation of structural parameters. Expressing the photon energy using k also results in an approximately evenly-spaced oscillation. In Equation (5), $\hbar$ is Planck's constant h divided by 2π.

Equation (3) can be broken down into two components, the amplitude and the phase functions. The total amplitude function is made up of several contributions. It is inversely related to the distance R_j of the jth atom away from the absorber, directly proportional to the backscattering amplitude $A_j(k)$ from each of the N_j scatterers and the total number of equivalent scatterers N_j, and modified by an amplitude reduction factor S_0^2 (due to many-body effects such as shake-up and shake-off processes in which other electrons at the absorbing atom are excited or ionized along with the photoelectron which, as a result, has less energy).[19] In addition, the EXAFS is dampened by two exponential terms:

$$\exp(-2\sigma_j^2 k^2) \quad \text{and} \quad \exp\left(\frac{-2R_j}{\lambda(k)}\right)$$

The former is the so-called DW-factor term that accounts for the thermal vibration (assuming harmonic vibration) and static disorder (assuming Gaussian pair distribution) of the system. σ_j^2 is in fact the mean-square deviation of R_j. The latter term accounts for the inelastic loss in the scattering process where $\lambda(k)$ is the photoelectron mean free path (explaining the finite core hole lifetime). The oscillation of the photoelectron wave is defined by the total phase function in Equation (3), where $\phi_j(k)$ is the total phase shift experienced by the photoelectron during the scattering process, twice from the absorber (going out and coming back) and once from the scatterer. The frequency of the oscillation is determined by R_j. The EXAFS for any absorber-scatterer pair is therefore represented as a damped sine wave with the amplitude, frequency and phase shift characteristic of the atoms involved. The total EXAFS is the sum of the individual sine waves describing each absorber-scatterer pair interaction.

In practice, an EXAFS spectrum is typically k^3-weighted to enhance the high-k region, because the total amplitude is weaker at high k due to the $1/k$ dependence and the damping effect of the DW factor in Equation (3). In addition, when multiple shells of scattering atoms exist in the system of interest, with different distances away from the photoabsorber, it is useful to study individual contributions separately. This can be done by performing a Fourier transform (FT) of the EXAFS data in k space which separates the waves of different distances into unique peaks in the conjugate R space.[28] In other words, the FT provides a radial distribution of scatterers from the absorber as a function of their distances. It should be noted that the FT peaks center at distances that are shorter than the corresponding R_j, and the difference (*ca.* 0.4 Å) is due to the presence of the total phase shift $\phi_j(k)$ in Equation (3). It should also be noted that this phase shift has a Z-dependence, and in all but the simplest cases, this means that FTs are not useful for obtaining accurate distance information.

2.1.2 ***Structural Information*** There are two types of parameters present in Equation (3): parameters that are necessary to account for the scattering process (*e.g.* S_0^2

and $\lambda(k)$) and parameters that bear structural information (*e.g.* N_j, R_j and σ_j^2). From a bioinorganic chemistry point of view, the structural parameters are the ones relevant to understanding the metal sites in bioinorganic systems. Table 1 summarizes the structural information that can be extracted from an EXAFS spectrum. Figure 2a–2d illustrates the effects of the structural parameters on a k^3-weighted EXAFS signal, using both the EXAFS data and its FT. In general, a shorter interatomic distance, a larger coordination number, a larger atomic number (stronger scatterer), or a smaller DW factor results in larger EXAFS amplitude and FT magnitude, with R_j and Z having additional effects on the frequency and amplitude envelope, respectively.

Interatomic distance: Because the frequency of each sine wave is a function of R_j, analyzing the frequency can provide information about interatomic distances (Figure 2a). In general, the frequency is not severely affected by the noise level of the data, and can typically be well determined from an EXAFS spectrum, so R_j can be determined with an accuracy level of within 0.02 Å for first-shell distances[21] and on average 0.04 Å for outer shell distances.[29] However, because R_j is inversely related to the EXAFS amplitude, shorter distances between the absorber and the scatterer result in stronger EXAFS signals than those from longer distances. This means that inner-shell distances from the absorber can normally be determined with higher accuracy than those from the outer-shells. In addition, the uncertainty of distance determination is complicated by the strong correlation between R_j and E_0 (which has an effect on the phase of the oscillation).[21] (More discussion is found in Section 5.2).

Another issue worth noting is the resolvability of two shells of similar distances from the absorber. Whether these two shells can be separated depends on the data range used for data analysis. Equation (6) gives the resolution limit of two shells of similar atomic number[30,31]:

$$\Delta R = \frac{\pi}{2\Delta k} \tag{6}$$

where Δk is the range of the data analyzed in k-space. For a typical data range of 4–15 Å^{-1} ($\Delta k = 11$ Å^{-1}), shells that differ by more than ~0.14 Å can be separated by EXAFS analysis. It should be noted, however, that Equation (6) gives an upper limit of resolving two shells. In reality, high noise level may substantially lower the resolution.[32]

Table 1 Summary of Structural Information Obtainable From EXAFS

From Each Sine Wave	Information Revealed
Frequency	Interatomic distance R_j (between absorber and scatterer)
Overall magnitude	Coordination number N_j (number of scatterers)
Phase shift and amplitude envelope	Atomic number Z (scatterer identification)
Damping effect	Structural disorder σ_j^2 (Debye–Waller factor)

Interatomic Distance (R)

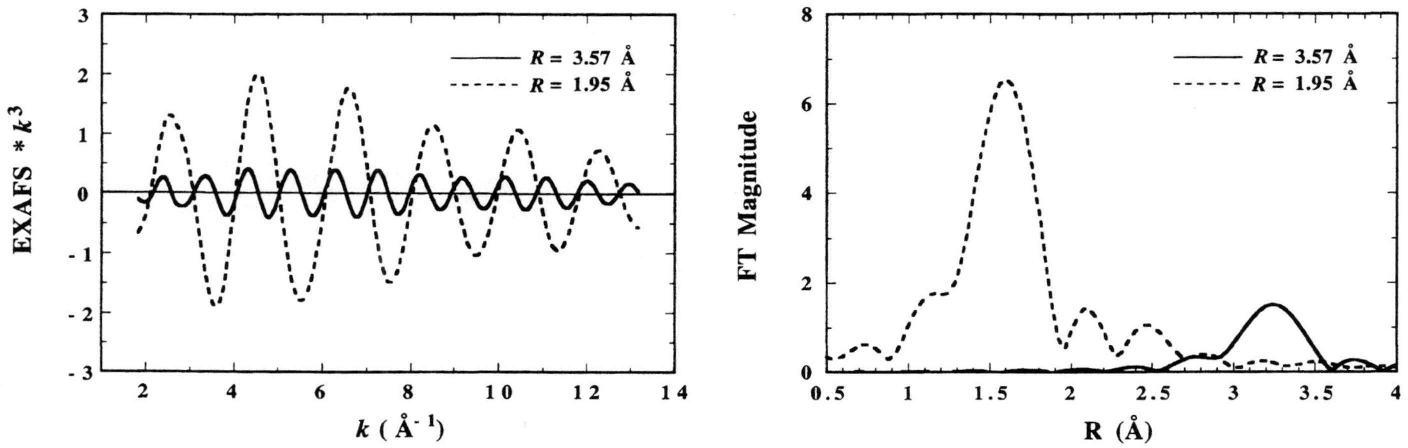

Figure 2a. Effects of the interatomic distance (R) on an EXAFS signal (left) and its FT (right). R affects both the frequency and the amplitude. The inverse relationship between R and the amplitude of an EXAFS signal means that a longer distance ($R = 3.57$ Å) gives a weaker signal, which in turn generates a FT peak of smaller magnitude that is further away from the origin. As part of the phase function ($\sin[2kR + \phi(k)]$), R affects the frequency of the sine wave, with a higher frequency resulting from a longer distance. If two frequencies are different enough, the two corresponding R's can be resolved (see discussion in text).

Coordination Number (CN)

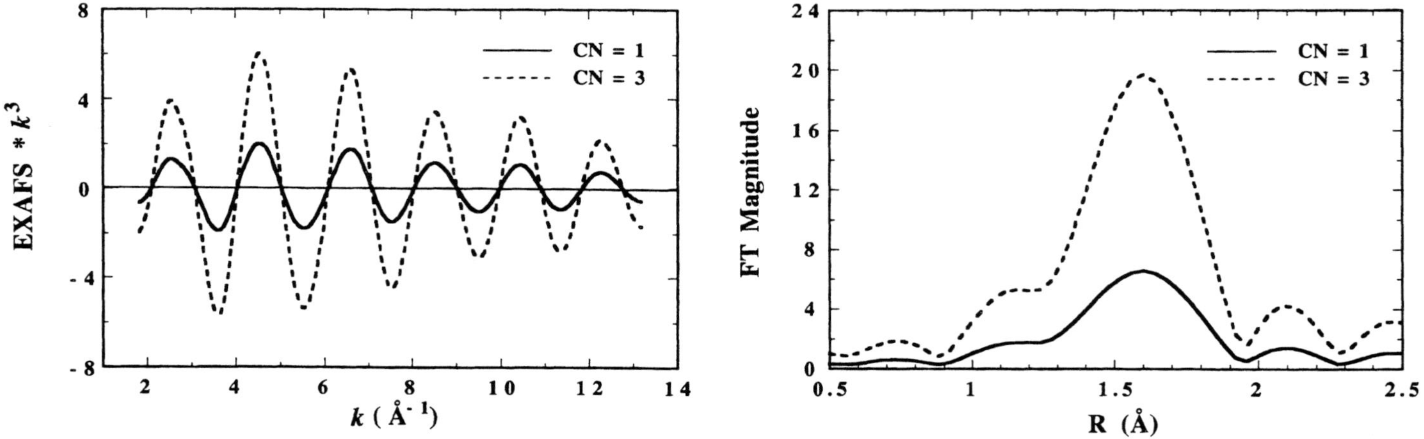

Figure 2b. Effects of the coordination number (CN) on an EXAFS signal (left) and its FT (right). CN acts as a scaling factor to the amplitude. A larger CN (CN = 3) generates a stronger EXAFS signal and a FT of larger magnitude.

Atomic Number (Z)

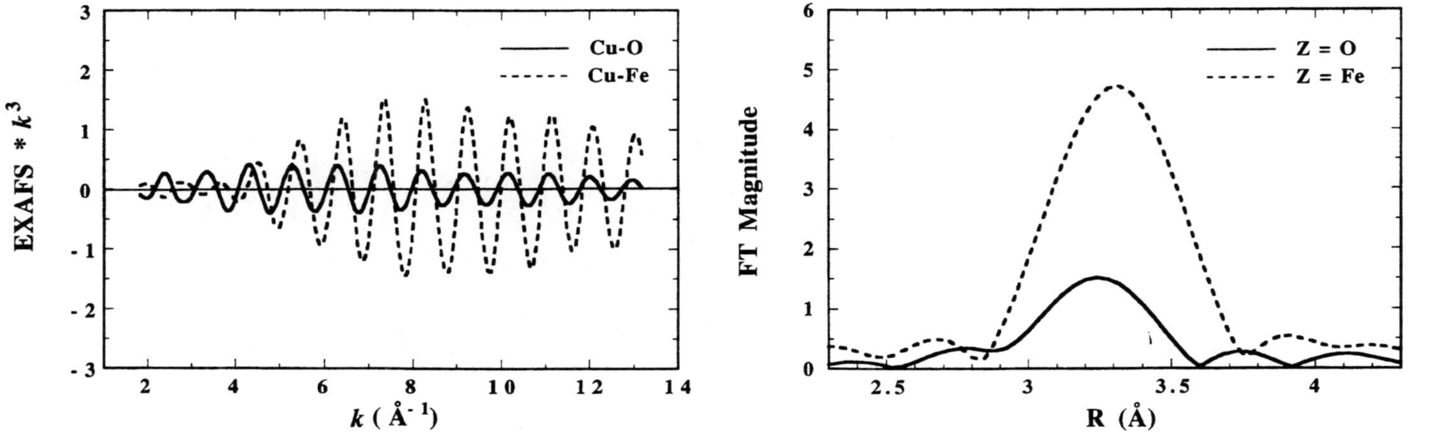

Figure 2c. Effects of the atomic number (Z) on an EXAFS signal (left) and its FT (right). Z affects both the phase and amplitude envelope. Atoms with larger Z (Fe *vs* O) are stronger backscatterers and they give rise to stronger EXAFS signals and FTs. In addition, EXAFS signals from low-Z backscatterers are attenuated more rapidly towards high k than those from high-Z backscatterers, especially metal atoms, which typically have an amplitude envelope peaking at high k.

Debye-Waller Factor (σ^2)

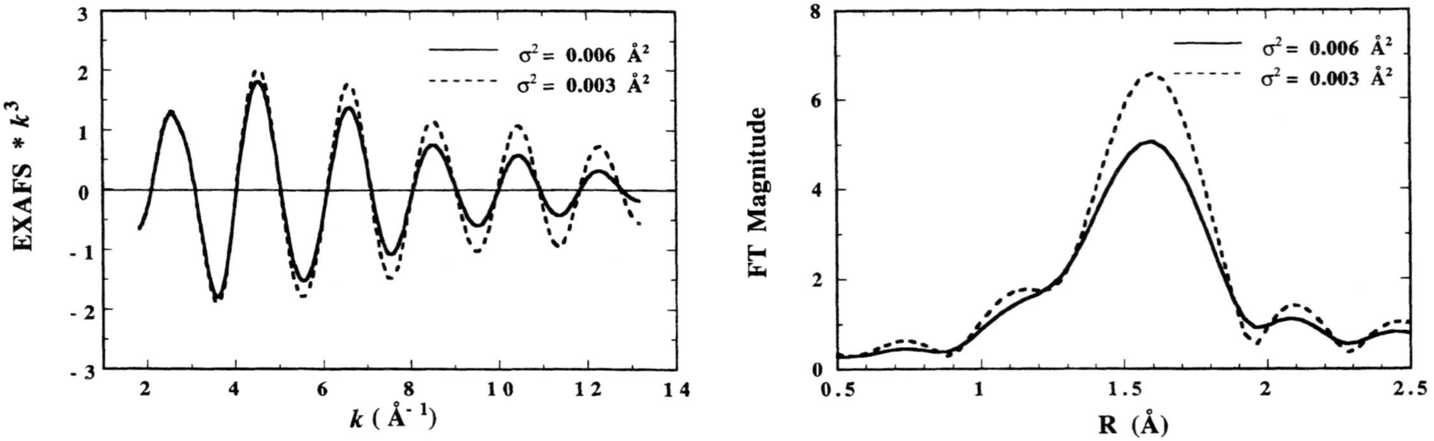

Figure 2d. Effects of the Debye–Waller factor (σ^2) on an EXAFS signal (left) and its FT (right). The DW factor dampens out the amplitude. The bigger its value, the larger its effect. The damping effect is usually larger at higher k and it can significantly affects the useful data range (see discussion in text).

Coordination number: The amplitude of an EXAFS wave reveals information about coordination numbers (Figure 2b). However, a number of parameters contribute to the amplitude function, especially the DW factor that can have a large damping effect on the amplitude. The correlation between the coordination number and the DW factor, as well as uncertainties in other parameters contributing to the amplitude function, typically limits the determination of the coordination number to be within ±25%.[21] In more favorable cases ±10% can be achieved. It might be noted that it is also sometimes possible to use information from the edge region to place further limits on the coordination numbers (*e.g.* distinguishing tetrahedral from octahedral coordination). In other cases, comparison of bond distances can be used as a way of identifying the coordination geometry.[33]

Atomic number: As indicated in Equation (3), both the amplitude and the phase functions are j dependent. This means that the amplitude envelope and the phase shift can be used to identify scatterers of different atomic number Z. The amplitude envelope is especially useful when the scatterers have large difference in Z. In biological applications, the amplitude envelope is often used to help identify a heavy scatterer (*e.g.* a metal atom). Figure 2c compares the EXAFS wave of a Cu-O pair with that of a Cu-Fe pair. For a heavy scatterer such as Fe, the maximum amplitude occurs at higher k than for a low-Z scatterer. With a k^3-weighting scheme, the differences at the high-k region are even more apparent. This distinguishing feature of the amplitude envelope of a heavy scatterer has been used as a way of detecting a M···M interaction in the EXAFS analysis of metalloproteins.[34,35] However, care needs to be taken when claiming the detection of a M···M interaction outside 3Å. Metal–carbon signals often interfere in this region and unique identification of a M···M signal can be problematic.[32,36,37]

The amplitude functions, however, vary relatively slowly with Z, and thus distinguishing scatterers of similar Z is often not feasible. This gives rise to a difficulty in uniquely characterizing metal centers in proteins where many biologically relevant ligands, such as C, N, and O, have a similar Z. In such cases, other spectroscopic and structural techniques are needed to obtain additional information.

There is another challenge associated with low-Z scatterers. The backscattering power usually decreases with decreasing Z. In the high-k region, the EXAFS signal of a low-Z scatterer decreases rather rapidly. This implies that, in biological applications, low-Z ligands are more difficult to identify than their metal counterparts. An improved data quality, that is, a better signal-to-noise ratio, is needed to expand the useful k range. In addition, extending the data analysis to near-edge region (a region where MS effects strongly contribute) will also significantly increase the possibility of detecting low-Z atoms (see further discussion below).

DW factor: The damping effects relate to a certain level of disorder in the system (Figure 2d). In the DW term, σ_j^2 is the mean-square displacement from the equilibrium separation point of the scatterer from the absorber. It has contributions from two effects, a vibrational effect and a static effect:

$$\sigma_j^2 = \sigma_{vib}^2 + \sigma_{stat}^2 \tag{7}$$

σ_{vib}^2 results from the vibrational motion of the two atoms, which is temperature dependent. It can be estimated using a harmonic model. Often σ_{vib}^2 is used to measure bond strength for closely related systems. However, in reality, precise determination of this term using EXAFS analysis is not easily achievable. It should be noted that for systems with large disorder, the approximation of a harmonic vibration mode can lead to error in the determination of structure and thermal parameters.[38,39] The outer-shell structure is especially affected by anharmonicity.[40] In such cases, non-Gaussian corrections are needed.[41] σ_{stat}^2 originates from the static disorder of the system, that is, a structural distribution of the interatomic distances, and it is temperature independent. These two terms therefore can be separated by a temperature-dependent EXAFS measurement. Because the DW factor is overall temperature sensitive, a stronger EXAFS signal can be obtained by lowering the temperature to reduce the thermal vibration, resulting in enhanced amplitude, particularly at high k, and thus improving the signal-to-noise level.

2.1.3 ***Limitations of Plane-Wave SS Theory*** The SS equation described above uses the plane-wave approximation in which the curvature in the photoelectron wave is ignored and the photoelectron is treated as a plane wave. This approximation greatly simplifies the theory and is a reasonable approximation when the photoelectron energy is sufficiently high (short photoelectron wavelength). However, the theory breaks down in the low-k region, especially in the near-edge region ($k < 3$ Å^{-1}) where the approximation no longer holds. It is this low-k region that is especially relevant for many biological investigations. As described in the previous section, the behavior of low-Z backscattering atoms manifests itself more strongly in the low-k region, and many biological ligands have a greater chance of being identified if these data were reliably interpreted.

While the SS theory is adequate for analyzing first coordination shells, it fails to take into account the long-range MS interactions (>3 Å) which can be especially significant in the near-edge region. Some ligands commonly found in bioinorganic systems, such as imidazole and porphyrin, have been shown to generate strong MS effects. In addition, long-range interactions have been observed in bridge units through one or more bridging ligands (see discussion in Section 4). For these systems to be analyzed in detail by EXAFS, a spherical wave MS theory is required.

2.2 Multiple-Scattering (MS) Theory

Multiple-scattering effects refer to the case where the photoelectron is scattered by more than one atom before returning to the absorber. Figure 3 schematically shows some of the pathways a photoelectron travels through a three-atom (or three-body) configuration, where B is the intervening atom between the absorber A and the scatterer C. A MS effect occurs when the photoelectron is scattered via B in this particular configuration (*i.e.* pathway (II) and (III)). (I) is the direct two-atom (or two-body) backscattering pathway. The EXAFS signal of a three-body system, therefore, has contributions from all of these processes, with relative strengths depending on how

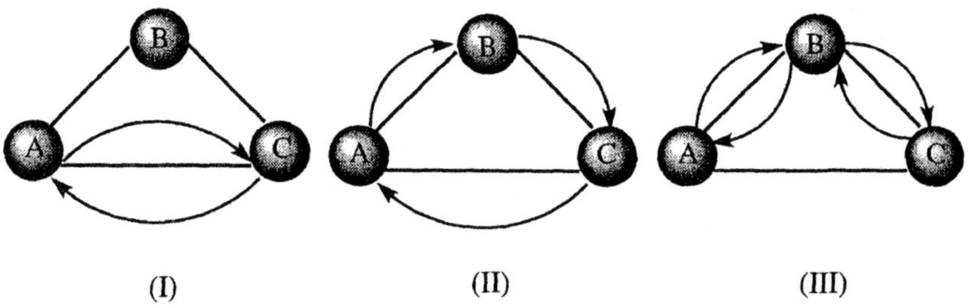

Figure 3. Typical scattering pathways in a three-body configuration. (I) is the single-scattering pathway, whereas (II) and (III) are two possible multiple-scattering pathways.

strong the MS effects are in the system. The net effect of traveling through the intervening atom is that the amplitude and the phase of the photoelectron wave are modified. When such modifications are large, MS effects can no longer be ignored, as serious errors in coordination numbers and interatomic distances may otherwise result.

The strength of MS effects depends on the scattering angle of the forward scattering atom. In other words, the magnitude of MS effects is highly geometry sensitive. For systems that have a $\angle ABC$ of $< \sim 150°$, MS effects are generally weak and can therefore be ignored.[42–44] On the other hand, if $\angle ABC$ is $> \sim 150°$, MS effects are significant and can be as much as an order of magnitude higher than the SS effects only. The strongest MS effects occur when $\angle ABC \cong 180°$, that is, A-B-C is arranged in a linear fashion. In such cases, the outgoing photoelectron wave is strongly forward-scattered by the intervening atom B, resulting in a significant amplitude enhancement. MS can also be quite prominent for certain rigid ligands and can be of such magnitude that they dominate over SS signals even in structures that are not collinear. Examples of such cases as well as a comparison of MS and SS signals and angle dependence of MS effects are given in Section 4.

In bioinorganic systems, typical MS effects can be observed from some functional groups, such as M–C≡N and M–C=O ligand arrangements[45,46] and in some coordinated ring structures such as imidazole and porphyrin.[47] Strong MS effects can also be found when two metal atoms are bridged by small atoms such as oxygen in a nearly linear configuration[43] thus helping detect the presence of M–M interactions in metalloproteins in cases where the bridge angle is $> \sim 150°$.

With MS analysis, structural information beyond 3 Å can be obtained in cases when MS effects are strong. Also, while SS EXAFS yields only radial distribution of the neighboring atoms around a metal center, MS analysis may provide some insights into the geometry of a metal center; the angular dependence of the MS effects means that it is possible to obtain direct information about bond angles, in addition to distance information. Moreover, with the availability of MS theory and analysis methods, the useful data range can be extended to lower k, where more information can be extracted and the ratio of structure variables to observables can be improved in least squares determination of the metrics (see Section 5).

3 DATA ANALYSIS

3.1 Data Reduction

Before any useful metrical information can be extracted, the raw experimental data must be first processed to isolate the EXAFS data that can then be studied using the theoretical formulation described above. This part of the analysis is well developed and relatively standardized. Figure 4 outlines the steps in data reduction. For a detailed description of the data reduction procedures, the reader is referred to refs 21,22,48.

Briefly, during the data reduction, a pre-edge subtraction must be performed to remove the background absorption due to lower-Z elements or the scatter background. A spline fit to the data above the edge is then performed to remove the smoothly varying portion of the data corresponding to the absorption coefficient of an isolated atom. This spline function normally consists of a set of cubic polynomials. In addition, the EXAFS data must be normalized to the edge jump and a function approaching the experimental atomic X-ray absorption falloff after the edge. If glitches (*i.e.* sharp changes or spikes in background caused by diffraction effects at discrete monochromator energies) exist in the data, it is often necessary to remove them before spline subtraction so as not to distort the spline and also limit the useful range of the data.

Fourier filtering is used to isolate the EXAFS frequency in a well-defined transform peak, and a Gaussian window with width of 0.1 $Å^{-1}$ is usually used. This filtered FT peak is backtransformed into k space to generate the wave for that particular frequency, which can then be fit separately, thus simplifying the analysis. Problems can arise, however, if the FT peaks are not well separated, which can lead to truncation errors due to artifacts introduced by separating the overlapping peaks.

3.2 Curve-Fitting

Curve-fitting involves the comparison of experimental EXAFS data with a hypothetical structure, either theoretically calculated or built up from a known structure. The hypothetical model is then refined, by varying such structural parameters as distance and DW factor, to minimize its difference with the experimental spectrum. Such refinement is typically performed by a nonlinear least-squares fitting algorithm, and the quality of the refinement is judged by a goodness-of-fit parameter that indicates the level of agreement between the experimental and the hypothetical spectra and gives quantitative comparisons among different models used to fit a given experimental data set. A smaller goodness-of-fit value indicates a better fit.

A key element of this data analysis is the determination of the backscattering amplitude and phase functions. There are two main approaches to obtaining these two functions: empirical analysis and *ab initio* calculation. The first approach involves the use of pairwise amplitude and phase functions that have been extracted from EXAFS data of suitable models of known structure.[21,22,49] The advantage of using the empirical method is that it is relatively easy to use, and amplitude and phase functions are easy to obtain if the models are available. However, this method assumes that

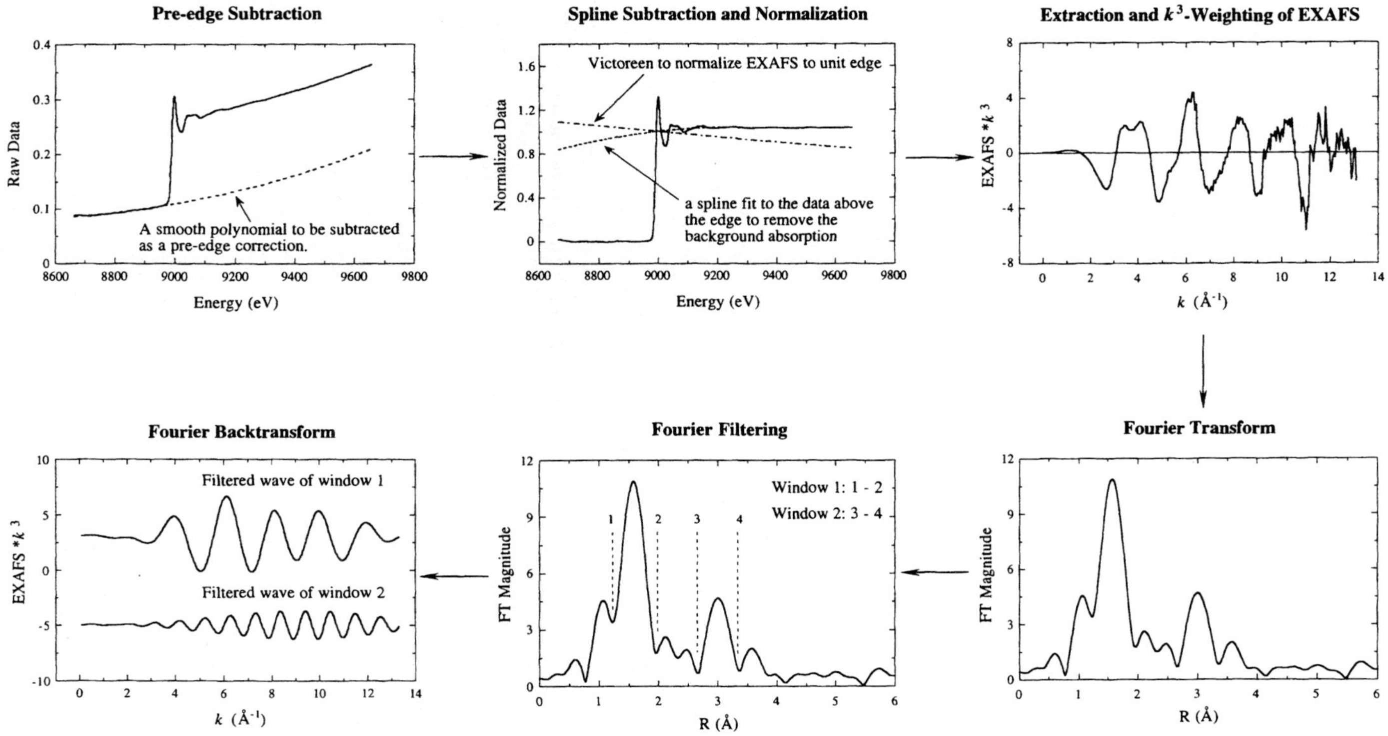

Figure 4. Steps in EXAFS data reduction and extraction of EXAFS signals. Details are given in the text.

both the amplitude and phase are transferable from the model to the unknown. This implies that the chemical environment of the model and the unknown need to bear a reasonable similarity to each other in order for the assumption to hold. This is especially critical for absorber-scatterer distances. Ideally, the model should have all the scatterers of interest occurring at the same distance from the absorbing atom, with no other atoms occurring at similar distances. Furthermore, because this method cannot easily account for MS effects, the transferability typically breaks down for scatterers beyond ~3 Å away from the absorber. Other limitations of the method include truncation effects of Fourier filtering and unknown DW factors.

Rather than being determined empirically, the phase and amplitude functions can be obtained from tables containing parameters that have been calculated *ab initio*. Teo and Lee[50] were among the first to attempt such calculations, but with the plane-wave approximation, their method was limited to use with data at $k \geqslant 4$ Å^{-1}. At low k, they gave relatively poor description of the EXAFS due to inaccurate scattering amplitude functions. McKale and coworkers improved the method by using a curved-wave SS formula.[51] This method allows the EXAFS analysis to be extended to a somewhat lower k region and has been shown to give more accurate structural information.[32] However, this method provides no estimation of the electron mean free path and is less accurate for very high-Z elements. A recent study shows that an addition of an energy-dependent mean free path to the table improved the accuracy by 15–25%.[52]

The next evolution in EXAFS data analysis involves theoretical simulations of EXAFS signals including those from MS contributions. The development of theoretical methods has been the focus of many efforts in recent years. The problems in developing an improved MS data analysis method include a correct account of MS contributions and an efficient algorithm for MS calculations. Significant progress has been made in theoretical methods and program suites based on these theoretical approaches have now become available. Among them, the three widely used data analysis programs are EXCURVE[53,54], FEFF[55,56], and GNXAS.[57] All of them employ curved-wave theory and have the capability of MS calculations. These programs differ mainly in the way MS effects are introduced and in the treatment of configurational average effects.[58]

The following section will provide a relatively detailed description of GNXAS to give readers a good picture of how a theoretical simulation and fitting of an EXAFS signal works and how a MS approach like GNXAS can be used to analyze rather complex bioinorganic systems.

4 GNXAS MS ANALYSIS

4.1 Introduction

The GNXAS approach is an MS-based analysis. Because MS effects are the result of interactions of several scattering atoms with the photoabsorber, they actually probe the n-body distribution function. The theory thus centers on the connection between MS signals and n-body distribution functions, as its name suggests. In the GNXAS notation, g_n stands for the n-body distribution function and XAS stands for X-ray absorption spectroscopy.

This MS-based data analysis program suite has the following features. As a theory, it incorporates the inelastic losses through a complex potential, the so-called Hedin–Lundqvist potential.[59,60] It treats SS and MS signals according to the proper n-body distribution function with correct treatment of the configurational average of MS signals. This treatment of MS signals provides an opportunity to determine correlated distances and DW factors in an n-body configuration, which in biological applications would most commonly occur as a two-, three-, or four-body structural unit. As an advanced fitting program, GNXAS simplifies the conventional data reduction steps by directly fitting the model spectrum to the experimental raw data, eliminating the tedious procedure of manual spline-background removal, and the spline refinement is performed simultaneously with the fitting. In addition, GNXAS can account for complex multi-electron excitation features in the background, and has the capability of simultaneously fitting a number of spectra containing multiple edges and performing a rigorous statistical error analysis using contour plots (see discussion in Section 5.2).

The GNXAS data analysis method has been successfully applied to a large number of systems, from simple inorganic molecules[61–63] to complicated bioinorganic systems[29,46,64–66], and proteins.[67,68] A comprehensive description of the theory, data analysis and applications of GNXAS can be found in refs 44,58,69. The following section briefly summarizes the GNXAS data analysis method.

4.2 Fitting Method

GNXAS data analysis first builds a theoretical EXAFS spectrum using a generated model, either from a known structure or from a hypothetical one. Typically the crystallographic fractional coordinates or Cartesian coordinates of the model are input to generate a model cluster up to a certain cutoff distance away from the photoabsorber. The n-body ($n = 2, 3, 4$) configurations, built around the photoabsorber, within the cluster are identified as the relevant peaks in the g_n (pair, triplet, and quadruplet) distribution functions. Each atom is associated with an appropriate phase shift calculated using muffin-tin approximation.[44] EXAFS signals, $\gamma^{(2)}$, $\gamma^{(3)}$, and $\gamma^{(4)}$ associated with each two-, three-, and four-body configuration, are combined along with an appropriate background to form the theoretical EXAFS spectrum. The model absorption coefficient is defined in energy (E) space as

$$\alpha_{\text{mod}}(E) = J\alpha_0(E)[1 + \chi(E)] + \beta(E) \tag{8}$$

and is composed of an atomic-like background modeled within the hydrogenic approximation $\alpha_0(E)$, a structural $\chi(E)$ term, and a background $\beta(E)$ that can include double-electron excitations. J is the absorption coefficient jump taking into account the thickness and density of the photoabsorber.

This theoretical EXAFS spectrum is then refined against the experimental absorption data using a least-squares minimization procedure that varies structural and non-structural parameters. It should be noted that GNXAS refines the spline simultaneously with other parameters, with the reasoning that EXAFS background cannot be defined exactly and that the relatively smooth background variation will not correlate significantly with the EXAFS oscillations.

The structural parameters varied depend on the n-body configuration involved. For a two-body signal, the parameters required are the bond distance (R) and the bond variance σ_R^2, a DW-like factor. The structure of a three-body signal is slightly more complicated; it is defined by two bond distances (R_1 and R_2) and the angle (θ) between the two bonds. Thermal vibration and static disorder of the structure are described by terms belonging to a symmetric covariance matrix and they are bond variances ($\sigma_{R_1}^2$ and $\sigma_{R_2}^2$), angle variance (σ_θ^2), bond-bond correlation (ρ_{R_1,R_2}) and bond-angle correlations ($\rho_{R_1,\theta}$ and $\rho_{R_2,\theta}$). Therefore, a total of 9 parameters is required in a fit for a three-body configuration. A four-body configuration is even more complicated and requires up to a total 27 parameters to fully describe. In many cases, however, this number can be significantly lowered.[66]

The fit of the theoretical spectrum to the experimental data also requires the inclusion of a few non-structural parameters. E_0, S_0^2, Γ_e (effective mean-free path parameter), and E_r (experimental resolution) are typically varied along with the structural parameters. The coordination numbers are typically either fixed at known crystallographic values or systematically stepped through but not refined.

The quality of a fit is determined by inspection of the EXAFS residual and their FT along with the goodness-of-fit value $\mathcal{R}$, which is defined as

$$\mathcal{R}_{N-n}(x_1, x_2, \ldots, x_n) = \frac{N}{N-n} \frac{\sum_{i=1}^{N} [\alpha_{\text{exp}}(k_i) - \alpha_{\text{mod}}(k_i; x_1, x_2, \ldots, x_n)]^2 k_i^p}{\sum_{i=1}^{N} [\alpha_{\text{exp}}(k_i)]^2 k_i^p} \tag{9}$$

This equation is a χ-squared-like statistical function dependent on the structural and background parameters ($x_1, x_2, \ldots, x_n$) and on the noise level. In the equation, α_{exp} and α_{mod} are the experimental and the theoretical EXAFS signals, respectively, N is the number of experimental data points, and n is the number of fitting parameters (variables). This goodness-of-fit value thus takes into account the number of parameters used in the refinement and is a way of comparing fits with different number of fitting parameters (see further discussion in Section 5.2).

4.3 GNXAS Applications to Inorganic Systems

4.3.1 ***A Heterometal Cuboidal Cluster: GNXAS SS Treatment*** The geometry of cuboidal clusters is typically such that all internal bond angles are significantly below 150°. The EXAFS analysis of this type of structure is therefore relatively simple, because MS effects are generally weak and only SS treatment of the EXAFS data is required. One such example is $MoFe_4S_6(PEt_3)_4Cl$ (Figure 5)[29] that has been shown to be among the best currently available models of the local environments in the iron-molybdenum cofactor of the enzyme nitrogenase. The cluster has a crystallographically imposed C_3 symmetry, with three structurally identical iron atoms (Fe(2)), and

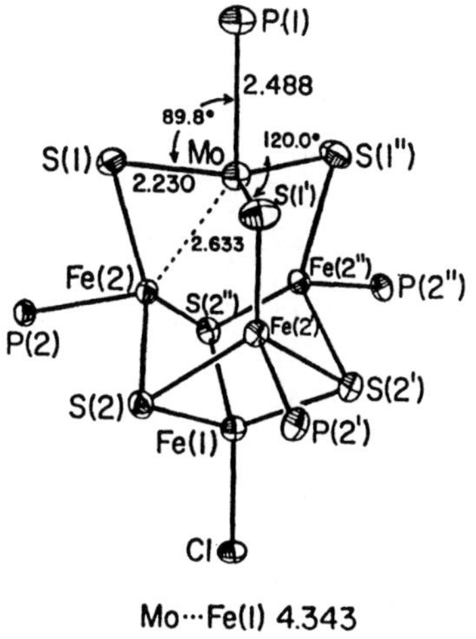

Figure 5. Crystal structure of the heterometal cuboidal cluster $MoFe_4S_6(PEt_3)Cl$.

one additional iron (Fe(1)) on the symmetry axis. This Fe(1) has a terminal chloride ligand and is located in a distorted tetrahedral site with ∠Cl-Fe(1)-S = 115° and ∠S-Fe(1)-S = 103°. Each Fe(2) has a terminal Et_3P ligand and exhibits unusual distorted trigonal pyramidal coordination with a phosphorus atom at the apex and three sulfur atoms forming the base. The site of the Mo atom is trigonal planar with ∠P(1)-Mo-S(1) = 89–90° and ∠S(1)-Mo-S(1) = 120°. The Mo···Fe(2) separation is 2.633 Å, which indicates the likelihood of direct M···M interactions.

Figure 6a shows the FT of the Mo K-edge EXAFS data for the compound. Intensity in the FT can be seen as far out as ~4.3 Å. Because all the angles involving Fe or Mo in the cuboidal-type structure are less than 150°, one would not expect MS effects to be strong. In fact, the GNXAS fitting results show that the EXAFS data could be well explained by 8 two-body SS waves (Figure 6b). The two major FT peaks were fit by three waves from strong backscatterers S(1) at 2.23 Å, P(1) at 2.49 Å, and Fe(2) at 2.63 Å.

As would be expected, contributions from the further shells are smaller in magnitude and contribute less significantly to the total EXAFS signal. Among them, the strongest contributions come from the heavier scatters, S(2), Fe(1), and P(2). The Mo···Fe(1) wave was shown to be the main contributor in the FT region at ~3.9 Å (Figure 6a). When this wave was omitted, this region was not at all well fit. Therefore, although the Mo···Fe(1) distance is as long as 4.3 Å, it contributes a small, yet significant, amount to the total EXAFS signal, and the SS treatment is sufficient to account for the long-range Mo–Fe(1) interaction.

To evaluate the importance of MS effects in this system, theoretical three-body pathways were calculated and their magnitudes were compared with those of the two-body signals. It was shown that most of them would not make significant contributions. Even when the five most intense signals (Mo-P(1)-C(1), Mo-S(1)-S(1), Mo-Fe(2)-S(2), Mo-Fe(2)-Fe(1), and Mo-Fe(2)-P(2)) were added to the 8 two-body

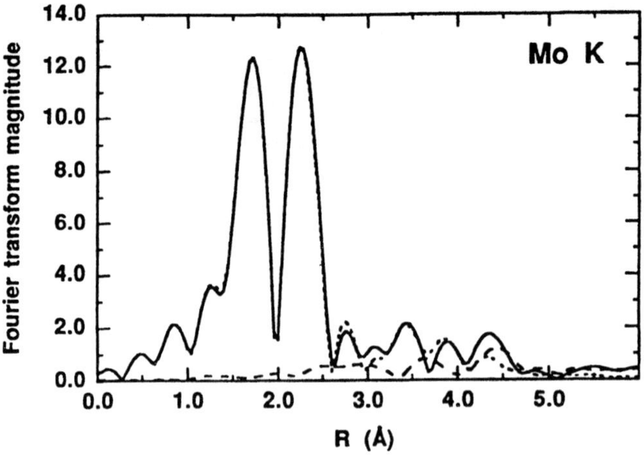

(a)

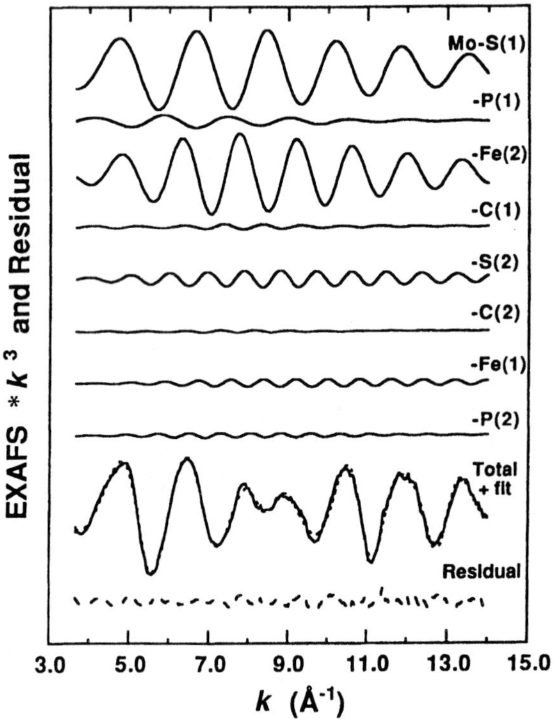

(b)

Figure 6. GNXAS analysis results for Mo K-edge EXAFS data of $MoFe_4S_6(PEt_3)Cl$. (a): FTs of the experimental data (—) and fit to the data (· · · ·). Also shown is the FT of the residual (- - -). (b): Individual EXAFS signals in the final fits along with the total signal (···) compared with the experimental data (—). (The ordinate scale is 8 between two consecutive longer marks.)

pathways, the fit resulted in insignificant changes in the two-body parameters, and the $\mathcal{R}$ value (Equation (9)) increased from 1.46×10^{-6} to 1.68×10^{-6}. The inclusion of three-body pathways, therefore, did not improve the quality of the fit, indicating that three-body MS signals are not important in this cluster.

The GNXAS calculation gave excellent structural agreement with the crystallographic data. The distances for the three first-shell atoms deviated by less than 0.01 Å while the outer-shell distances were determined to be within 0.04 Å of the crystallographic values (due to weak signals and static disorder in the crystal structure). Similar results and conclusions were obtained for the Fe K-edge data.

4.3.2 *Three-Body Inorganic Systems: GNXAS MS Treatment*

4.3.2.1 *Comparison of MS and SS Contributions* The GNXAS analysis in Section 4.3.1 demonstrated that MS effects are not significant for clusters with internal bond angles significantly below 150°, and that SS contributions are sufficient to accurately describe the total signal. However, when a three-body configuration has an angle greater than ~150°, MS effects can be large, and an MS EXAFS treatment of the data may be necessary. In other words, for such a three-body system, its effective EXAFS signal can have contributions from both the two-body signal $\gamma^{(2)}$ involving only the two end atoms (direct backscattering) and the three-body signal $\gamma^{(3)}$ involving all of the three atoms: the absorbing, intervening, and scattering atoms (backscattering via the intervening atom). The relative contributions of $\gamma^{(2)}$ and $\gamma^{(3)}$ vary as a function of the bond angle. Figure 7 compares the theoretical $\gamma^{(2)}$ and $\gamma^{(3)}$signals corresponding to the three-body Fe-C-N configuration with an Fe-C-N angle varying from 90° to 180°.[44] For an Fe-C-N angle of < 150°, (*i.e.* 90° or 120°), the $\gamma^{(2)}$ signal greatly outweighs the $\gamma^{(3)}$ signal over a rather extended region in k space, particularly beyond $k > \sim 5$ Å^{-1}. In such cases, the MS interaction $\gamma^{(3)}$ can be ignored during the refinement. For an Fe-C-N angle of $\geqslant$ 150°, however, the situation is reversed. Considerable amplitude enhancement of $\gamma^{(3)}$ signal can be observed when the Fe-C-N angle is beyond the threshold of 150°. In these cases, the $\gamma^{(3)}$ signal is dominant over $\gamma^{(2)}$ and a satisfactory fit can only be obtained when the Fe-C-N $\gamma^{(3)}$ MS contribution is included.

4.3.2.2 *Angle-Dependence of MS: Direct Probe of Bond Angles* The angular dependence of MS effects suggests that direct angular information can be obtained if MS effects are sufficiently strong. Therefore, one can determine bond angles in a system of unknown structure. A published study on angle determination involves a series of $\{FeNO\}^7$ complexes that have various Fe-N-O angles.[64] These complexes have been described as having different electronic structures for different geometric structures. Detailed understanding of these electronic and geometric structures is required in order to probe the electron distribution related to oxygen activation in nitrosyl-bound non-heme iron enzymes. These enzyme-NO complexes serve as analogs of the possible dioxygen intermediates of a large number of enzymes involved in the binding and activation of dioxygen.

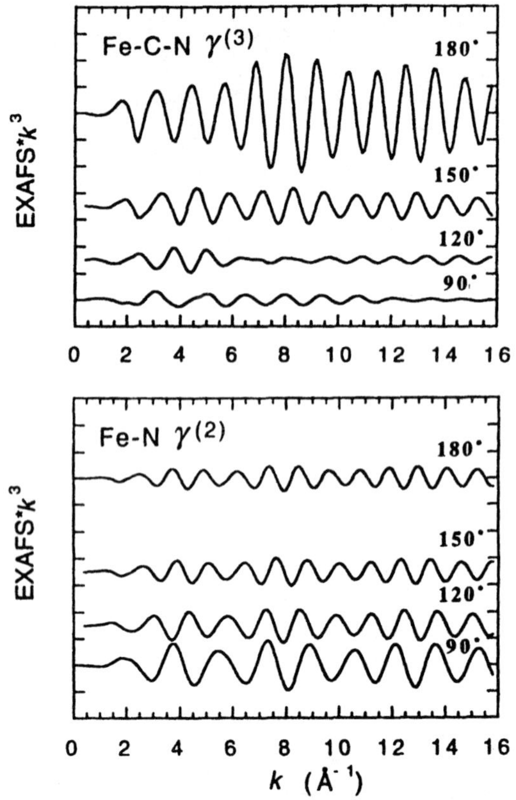

Figure 7. Comparison of the EXAFS three-body MS signal ($\gamma^{(3)}$) of Fe-C-N with the two-body SS signal ($\gamma^{(2)}$) corresponding to the long bond Fe–N when the Fe-C-N angle equals 180°, 150°, 120°, and 90°. Note the differences in the EXAFS amplitudes of $\gamma^{(2)}$ and $\gamma^{(3)}$ at each Fe-C-N angle, and the large amplitude enhancement of $\gamma^{(3)}$ at Fe-C-N angles above 150°. (The ordinate scale is 10 between two consecutive marks.)

The $\{FeNO\}^7$ complexes used in the study had been crystallographically characterized with an Fe-N-O angle of 177.5(5)°, 156(1)°, 147(5)°, or 127(6)°. This series of complexes with a varying Fe-N-O angle therefore allows a systematic GNXAS study from a nearly linear to a bent structure. As would be predicted, the best fits for most of the complexes were obtained when the Fe-N-O angle was near to the crystallographically determined values. However, the sensitivity of the fit to the Fe-N-O angle varied largely depending on the actual bond angle. The sensitivity study was performed by fixing all the distances, angles, and other non-structural parameters using the values from the best fit and calculating a theoretical EXAFS spectrum with Fe-N-O angles ranging from 90° to 180°. For the nearly linear Fe-N-O unit (177.5°), the $\mathcal{R}$ value increased dramatically in the calculations when the Fe-N-O angle in the fits was constrained below 170°. When the Fe-N-O angle was around 150°, relatively high $\mathcal{R}$ values were obtained for fits above 155° but the $\mathcal{R}$ values for all the fits below 155°

were very similar, suggesting that an upper limit of 155° could be set for the Fe-N-O angle. For a crystallographically determined Fe-N-O angle of 127°, because of its weak signal, $\mathcal{R}$ values were insensitive to the variation of the Fe-N-O angle in the fits. This sensitivity study shows that a signal must be a significant component in the total EXAFS signal for accurate angle determination by GNXAS. In addition, the GNXAS analysis is usually capable of differentiating a linear from a bent three-body structural unit, with the analysis being very sensitive when the angle is between 150° and 180°.

These observations were used to investigate the Fe-N-O angle in a structurally unknown complex, FeEDTA-NO. The crystallographically-characterized $[Fe(OH_2)EDTA]^-$ complex was chosen as the initial structural model because the first-shell empirical fits for both FeEDTA-NO and $[Fe(OH_2)EDTA]^-$ gave similar Fe-O and Fe-N distances, and their edge spectra also look very similar.[70] The $[Fe(OH_2)EDTA]^-$ structural model was modified to include a short Fe-N bond at ~1.8 Å and an N-O distance at 1.1 Å. By systematically varying the Fe-N-O angle from 90° to 180°, the refinement arrived at the best fit with an Fe-N-O angle of 156°. Further, the Fe-N and N-O distances obtained from the best fit were 1.78 and 1.10 Å, respectively, consistent with those in the $\{FeNO\}^7$ series, giving confidence that a chemically reasonable structure, especially the bond angle, had been determined.

4.3.2.3 *M···M Interaction in a Heterometallic System* The importance and usefulness of MS effects can also be seen in the structural definition of linear and nonlinear homo- and heterometallic M-X-M′ bridges that may occur in synthetic and biological bridged assemblies. Examples include Fe-O-Fe complexes that model the active site of an oxygen-transport protein, hemerythrin,[43] and Mn-O-Mn complexes that are the likely structural elements of the Photosystem II manganese complex.[71] The example presented here is the heme-based molecular complexes containing the bridge unit $[Fe^{III}\text{-}X\text{-}Cu^{II}]$ with X = O^{2-} [72–74] and OH^- [73,74]. These complexes are intended as actual or potential analogues of the oxidized binuclear heme a_3-Cu_B site in the superfamily of heme-copper respiratory oxidases.[2–4]

The two oxygen-bridged complexes, $[(OEP)Fe^{III}\text{-}O\text{-}Cu^{II}(Me_6tren)]^{1+}$ ("oxo") and $[(OEP)Fe^{III}\text{-}(OH)\text{-}Cu^{II}(Me_5tren)(OClO_3)]^{1+}$ ("hydroxo"), present a significant challenge for MS EXAFS analysis.[65,73] The crystal structures of the oxo[72] and hydroxo[73] complexes reveal that the two have similar environments around the Fe and Cu centers, but they differ significantly in the Fe-O-Cu bridge angle which is practically linear (175.2(3)°) in the oxo complex and markedly bent (157.0(2)°) in the hydroxo complex. Consequently, the oxo/hydroxo pair forms a unique set with which to examine the MS effect as a function of the bridge angle and its effects on the Fe–Cu interaction.

We have shown in the previous section that the strongest MS effects occur when the intervening atom is placed linearly between the photoabsorber and the backscatterer, resulting in a significant amplitude enhancement in the EXAFS signal. This effect is clearly demonstrated in the FTs of the Fe K-edge data of the linear oxo complex (Figure 8a). The unusually large FT feature at ~3.1 Å is particularly striking. The GNXAS analysis shows that this feature results from a strong Fe···Cu interaction via the Fe-O-Cu linear MS pathway. The FT of the hydroxo complex (Figure 8b),

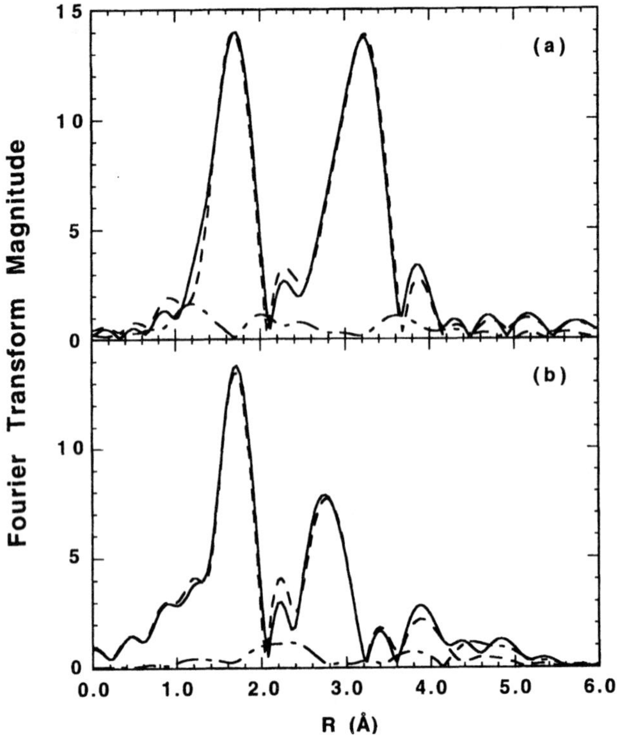

Figure 8. Comparison of the FTs of the Fe K-edge EXAFS data for (a) the linear and (b) the bent Fe-O-Cu bridge in $[(OEP)Fe\text{-}O\text{-}Cu(Me_6tren)]^{1+}$ and $[(OEP)Fe\text{-}OH\text{-}Cu(Me_5tren)(OClO_3)]^{1+}$, respectively. Experimental: (—), fit to the data: (- - -) and residual: (–·–). The unusually high FT peak at ~3.1 Å in (a) derives from the strong MS effects of a linear Fe-O-Cu configuration.

however, is rather different. It has the more normal appearance one observes from the falloff in amplitude with increasing R. Its corresponding FT feature at 3.4 Å is much smaller, reflecting a weaker MS contribution from the bent structure. Similar MS effects and impact on the Fe···Cu interaction are observed for Cu K-edge data.

The strong Fe···Cu interaction via the linear oxo bridge allowed an accurate determination of the bridge structure. The calculated distance for Fe···Cu from various fits (including multiple-edge fits) was consistently within 0.01 Å of the crystallographic distance of 3.57 Å, and the Fe-O-Cu angle deviated less than 2° from the crystal structure. The relatively weak Fe···Cu interaction in the hydroxo complex, on the other hand, made the determination of the structural parameters less accurate. The calculated Fe···Cu distance fluctuated from fit to fit with a deviation between 0.01 Å and 0.03 Å from the crystallographic value of 3.80 Å, and the average Fe-O(H)-Cu angle was determined to be 153°, deviating by 4° from the crystallographic value. The Fe···Cu interaction, however, is significant. The inclusion of the Fe-O(H)-Cu contribution in fits decreased the $\mathcal{R}$ value by ~25% in Fe and ~65% in Cu K-edge fits.

A similar pair of oxo/hydroxo complexes have been studied[75] using EXCURVE, in which the structurally characterized $[(F_8\text{-TPP})Fe^{III}\text{-O-}Cu^{II}(TMPA)]^+$ was used to derive $[(F_8\text{-TPP})Fe^{III}\text{-OH-}Cu^{II}(TMPA)]^+$ whose X-ray crystal structure was not known. These two complexes are related by protonation and deprotonation. The oxo bridge has a crystallographically determined angle of 178.2(4)° whereas the angle of the hydroxo bridge was determined from EXAFS to be 157(5)°. In this study, the Fe···Cu MS interaction could be clearly observed in the oxo complex. For the hydroxo complex, however, only a SS interaction was used in the Fe K-edge fit and no Fe-Cu interaction was observed from the Cu K-edge.

4.3.2.4 *MS Effects in Rigid Inorganic Systems* MS effects can sometimes become significant even when bond angles of three-body configurations are below 150°. This situation is typically seen in metal-coordinated rigid systems such as imidazole and porphyrin where bond angles involving the metal atom are in the range of ~100° to ~120°. The first quantitative examination of the MS effects in an imidazole ring was performed on a series of tetrakis(imidazole)copper(II) compounds.[47] It showed that major improvement to the fit can be obtained when MS effects in the ring are included, and that strong MS contributions are present in the EXAFS over an extended range above the edge. By including MS in the EXAFS analysis, it is possible to extend the low-energy fitting range to below $k = 3$ Å^{-1}. Further, MS analysis can sometimes allow the determination of the number of histidine ligands in a mixed-ligand complex.[47]

From a GNXAS point of view, the regularity of a ring system results in symmetry-related multiples of a three- or four-body pathway; therefore an MS signal is amplified rather than canceled out. In a porphyrin structure, eight of the Fe-N-C_β pathways with an Fe-N-C_β angle of around 125° can generate an EXAFS signal that is stronger than one linear Fe-C-N pathway[66] (Figure 9). Therefore, it is essential to include MS contributions from rigid structures like these in fits.

4.3.3 *Four-Body Inorganic System: GNXAS MS Treatment*

4.3.3.1 *Long-Range M–M Interaction (~5 Å) in a Linear System* As with a three-body system, a linear four-body unit can also generate strong MS effects which provide an opportunity to study long-range interactions out to 5 Å, especially M···M interactions in a [M-X-Y-M′] structural unit. EXAFS analysis of such linear or nearly linear four-body configuration may require treatment of a correlated *four-body* MS pathway, and an efficient description of four-body configurations is thus required. Both GNXAS and FEFF MS analysis have been successfully applied to four-body structural units in a number of systems. The FEFF applications include cyanocuprate reagents with a linear Cu-C≡N-Cu structure[76] and a Mo complex with a linear Mo-N=N-Mo configuration.[77] In both cases, a M···M interaction was detected at ~4.9 Å.

The first application of GNXAS four-body analysis to molecular inorganic complexes was to a series of Fe^{III}-CN-Cu^{II} bridged complexes.[66] These complexes are considered potential analogues of the cyanide-inhibited heme-copper oxidases.[78–81] This study provided an excellent example of examining the long-range interactions

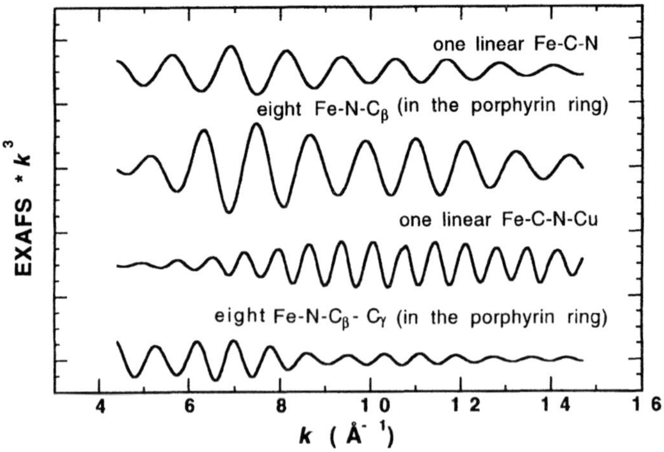

Figure 9. Comparison of three-body and four-body MS EXAFS signals from linear configurations involving a cyanide ligand, Fe-C-N and Fe-C-N-Cu, with those from the nonlinear structure units in a porphyrin ring, Fe-N-C_β and Fe-N-C_β-C_γ. The Fe-C-N three-body and Fe-C-N-Cu four-body pathways generate strong EXAFS signals due to their linear configurations. On the other hand, the regularity of the porphyrin structure results in multiples of the three- and four-body nonlinear pathways that give rise to equally strong (or even stronger) EXAFS signals, *e.g.* eight Fe-N-C_β pathways in the porphyrin with an Fe-N-C_β angle of ~125°. (The ordinate scale is 4 between two consecutive longer marks.)

between two metals through a four-body configuration, and the extent to which these interactions are significant from the EXAFS point of view.

In the study, two structurally defined cyanide-bridged complexes, $[(py)(OEP)Fe\text{-}CN\text{-}Cu(Me_6tren)]^{2+}$ (**1**) and $[(py)(OEP)Fe\text{-}CN\text{-}Cu(TIM)]^{2+}$ (**2**), were analyzed. The two complexes have similar bridge structures at the Fe end, with a linear Fe-C-N unit and a porphyrin ligation to Fe. They are, however, quite different from the Cu end of the bridge. **1** has a nearly linear Cu-N-C unit with a bridge angle of 174° and a Cu-N distance of 1.90 Å while **2** has a Cu-N-C angle of 147° and a Cu-N distance of 2.17 Å:

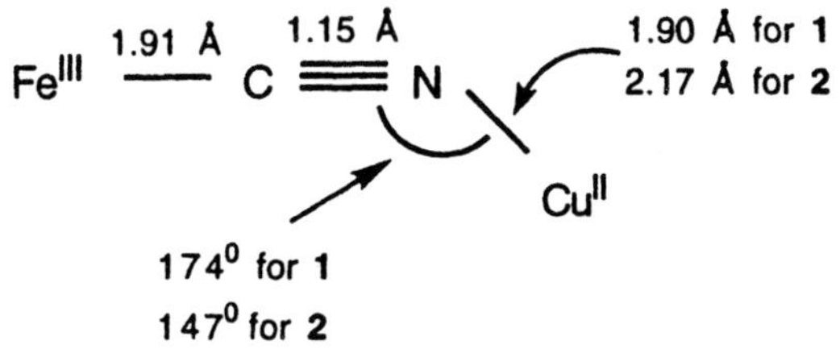

The large difference in bridge geometry therefore affords the unique opportunity to investigate the angular sensitivity of the four-body MS effects and to compare the results with those from the three-body oxo- and hydroxo-bridged cases.[65]

Because the Fe environments of the two complexes are very similar, illustrations will be made for the Fe K-edge data. Figure 10 shows the FTs of the data and fit for the two complexes. A careful inspection of the FTs shows strong similarity between

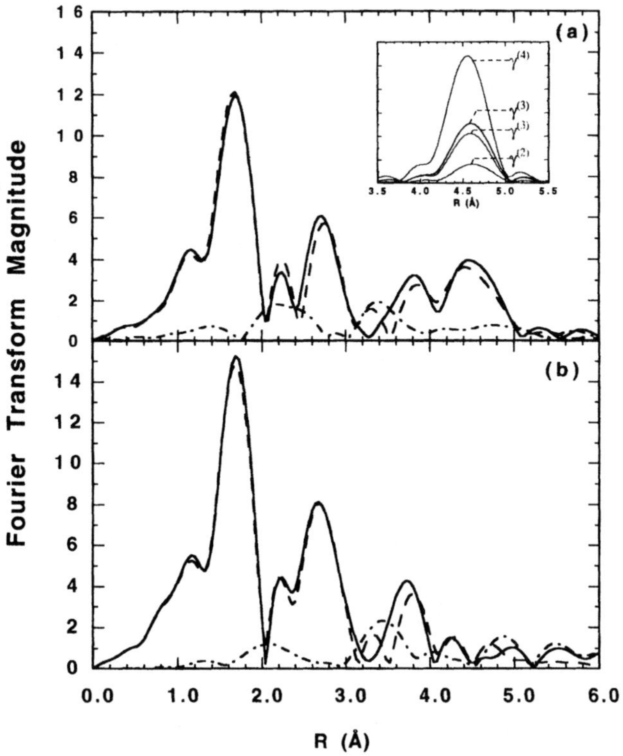

Figure 10. Comparison of the FTs of the Fe K-edge EXAFS data for (a) the linear and (b) the bent Fe-C-N-Cu bridge in $[(py)(OEP)Fe\text{-}CN\text{-}Cu(Me_6tren)]^{2+}$ (**1**) and $[(py)(OEP)Fe\text{-}CN\text{-}Cu(TIM)]^{2+}$ (**2**), respectively. Experimental: (—), fit to the data: (- - -) and residual: (-·-). Note the 4.6-Å peak in the linear bridge case due to a long-range Fe···Cu interaction through the strong MS effects of a linear Fe-C-N-Cu bridge. This peak is absent in the bent bridge case. Inset: Components of the 4.6-Å FT feature. See text for details.

the two complexes up to 4 Å. Beyond this point, **2** is essentially featureless whereas **1** has a significant peak at ~4.6 Å and its EXAFS signal exhibits more features, especially in the k region above 9 $Å^{-1}$ (Figure 11a compared with Figure 11c). This difference between **1** and **2** was attributed to the difference in the long-range Fe–Cu interactions. The 4.6-Å FT feature of **1** could be well explained by a linear four-body Fe-C-N-Cu contribution, whose strong EXAFS signal is shown in Figure 9. The bent structure at the Cu end of the bridge in **2**, on the other hand, results in substantial decrease in the Fe–Cu MS contributions. Its Fe-C-N-Cu EXAFS is so weak that it does not contribute in a statistically meaningful fashion. As a result, the Fe–Cu distance of 5.02 Å known from the crystal structure could not be fit to the data.

4.3.3.2 *Comparison of Three-Body and Four-Body M–M Interactions* Both the three-body and four-body structures have demonstrated strong MS effects in a linear arrangement and the angle-dependence of the MS effects. However, the three-body

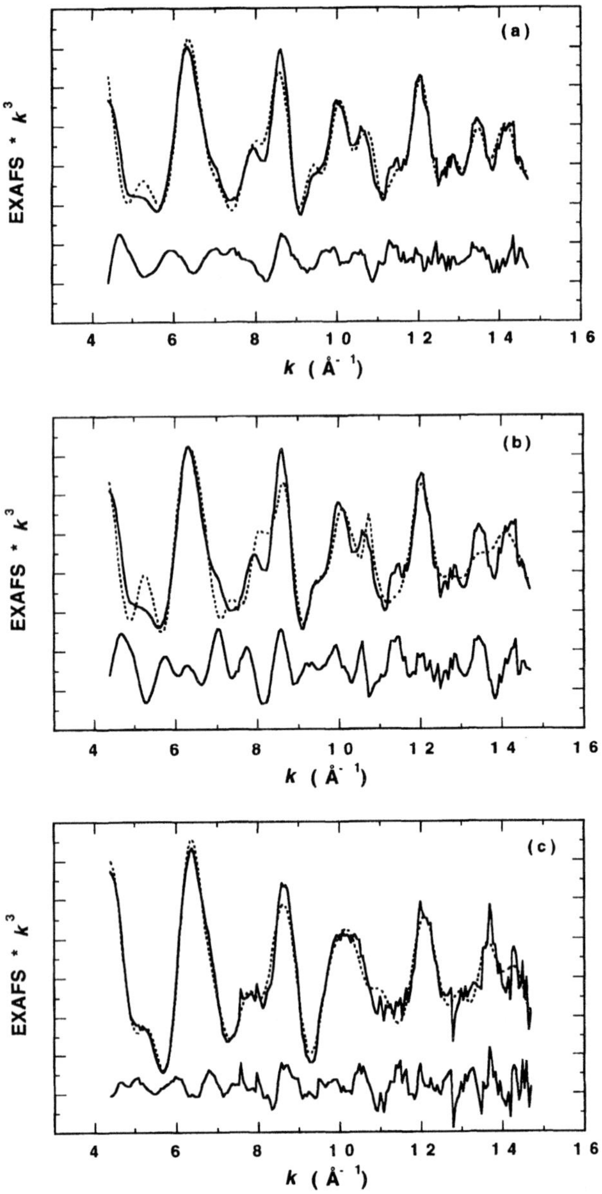

Figure 11. The Fe K-edge experimental data (—) *vs* the fit signal (···) for (a) the linear Fe-C-N-Cu bridged complex (**1**) with the $\gamma^{(4)}$ MS contribution, (b) the linear Fe-C-N-Cu bridged complex (**1**) without the $\gamma^{(4)}$ MS contribution, and (c) the bent Fe-C-N-Cu bridged complex (**2**). The fit residual is shown below the total signal. Compared to the EXAFS signal in (c), the signal in (a) exhibits more features or high-frequency components. When $\gamma^{(4)}$, the linear Fe-C-N-Cu four-body MS contribution, is not included in the fit, a high-frequency pattern in the residual is clearly seen. (The ordinate scale is 2.5 between two consecutive longer marks.)

MS effects of Fe-O-Cu described in Section 4.3.2.3 are relatively short-range, compared to the Fe-C-N-Cu MS effects in **1**. The Fe···Cu MS interaction in the oxo case is in the range of 3.5–3.8 Å while its corresponding interaction in **1** has been extended out to ~5.0 Å. The situation in the two bent structures, however, are quite different. In the hydroxo case, the three-body Fe-O(H)-Cu EXAFS signal was weak yet significant to the fit, indicating that its MS effects were still relatively strong. On the other hand, the bent structure of **2**, with a similar bridge angle around 150°, required no Fe-C-N-Cu contribution to the fit, suggesting that the MS as well as SS Fe···Cu interaction becomes insignificant when the Fe···Cu distance is longer (5.0 Å *vs.* 3.8 Å).

This comparison demonstrates that when a structural unit deviates significantly from linearity, MS effects, similar to SS effects, fall off fast and the long-range interaction becomes less observable when the distance becomes longer.

4.3.3.3 *MS vs. SS Contributions in a Linear Four-Body Fe-C-N-Cu Signal* As with the EXAFS of a three-body system, the total four-body Fe-C-N-Cu EXAFS signal[66] has contributions from four components, one two-body SS signal $\gamma^{(2)}$ Fe–Cu, two three-body MS signals $\gamma^{(3)}$ involving Fe-C···Cu and Fe···N-Cu, and one four-body MS signal $\gamma^{(4)}$ Fe-C-N-Cu. A breakdown of the 4.6-Å peak in **1** into its four components (Figure 10 inset) allows a closer look at this four-body MS effects. A simple calculation indicated that the $\gamma^{(2)}$ accounts for 7% of the total intensity, the two $\gamma^{(3)}$, Fe···N-Cu and Fe-C···Cu, account for 20% and 23%, respectively, and $\gamma^{(4)}$ accounts for 50%. Figure 12a displays the relative strengths of the EXAFS signals

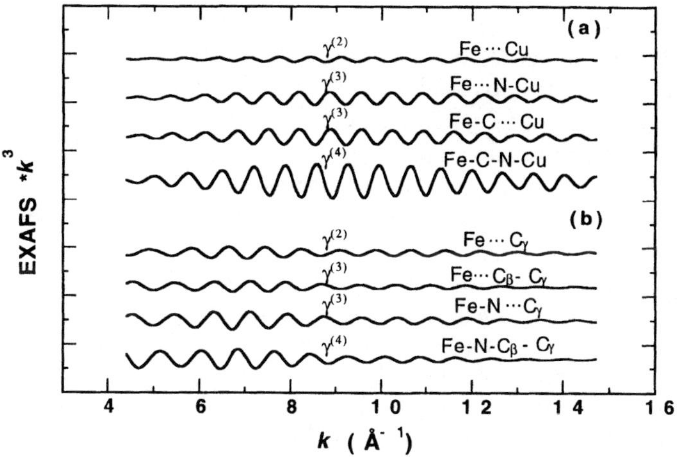

Figure 12. Relative Fe K-edge EXAFS strengths of the four contributing pathways in a four-body configuration for (a) the linear Fe-C-N-Cu bridge and (b) the Fe-N-C_β-C_γ pathway in the porphyrin of $[(py)(OEP)Fe\text{-}CN\text{-}Cu(Me_6tren)]^{2+}$ (**1**). While the four contributions in (b) are relatively equal in strength, $\gamma^{(4)}$ is the major contribution in (a), suggesting the importance of the long-range linear MS effects in the total EXAFS signal. (The ordinate scale is 10 between two consecutive longer marks.)

of the four components. It is evident that the linear four-body interaction $\gamma^{(4)}$ is the main contributor to the intense outer-shell feature. A fit without $\gamma^{(4)}$ increased the $\mathcal{R}$ value from 0.411×10^{-8} to 1.17×10^{-8}, almost a factor of three, and a high-frequency wave dominated in the EXAFS residual (see Figure 11b).

4.3.3.4 *A Linear Four-Body Fe-C-N-Cu Configuration vs. the Four-Body Fe-N-C_β-C_γ Pathway in Porphyrins* Figure 12 compares the relative EXAFS strengths of the four contributing pathways in a four-body linear bridge Fe-C-N-Cu and Fe-N-C_β-C_γ in the porphyrin of **1**.[66] It is clear that these two types of four-body signal receive a different set of contributions from its four components. While the linear four-body signal has a dominant $\gamma^{(4)}$character and less than 10% $\gamma^{(2)}$ character, the porphyrin four-body signal has a more even distribution among $\gamma^{(2)}$, $\gamma^{(3)}$, and $\gamma^{(4)}$ (Figure 12). This clearly demonstrates the mechanism by which the two types of the four-body configurations contribute to the total EXAFS for these complex systems (see discussions in Sections 4.3.2.4 and 4.3.3.1 for the Fe-N-C_β-C_γ and the Fe-C-N-Cu configurations, respectively).

4.4 Implication of Using the GNXAS MS Approach for Study of Biological Systems

The model EXAFS studies using GNXAS in the previous sections showed the importance of MS effects in some of the EXAFS data. Its behaviors and its impact on data analysis were clearly demonstrated. In addition, these studies established the utility and reliability of the GNXAS approach and provided a reliable means to determine additional information for structures of chemical interest from EXAFS analysis, especially angular information and long-range interactions. This section gives an example in which GNXAS MS analysis was applied to an unknown biological system, with the knowledge and experience obtained from the model studies.

The nitrogenase enzyme system catalyzes the biological reduction of dinitrogen to ammonia. The molybdenum-containing nitrogenase consists of a MoFe protein and an Fe protein that is the electron donor in catalysis.[82] The MoFe protein in turn contains two Fe-S P-clusters and two iron-molybdenum cofactors, the so-called FeMoco, which is believed to be the possible site of substrate binding and reduction.[83,84]

EXAFS studies of the Mo site in the MoFe protein, with FeMoco native[85] and extracted as an "isolated" form,[86] provided the first definitive structural evidence that the Mo was a part of a polynuclear cluster containing Fe and S, and that the basic features of the Mo environment remained unchanged in isolated FeMoco. Recent crystal structure studies[87,88] have generated complete structural details about the MoFe protein including the structure of the FeMoco (Figure 13) within the protein. However, questions about the detailed structure of the isolated FeMoco and reactivity of the FeMoco site remain unanswered.

EXAFS studies have thus been continued to probe various states as well as mutants of this protein. One of the aspects of the investigation has been the study of the interaction of substrates and other exogenous ligands with the system. This example presents Mo K-edge EXAFS studies of isolated FeMoco and MoFe protein

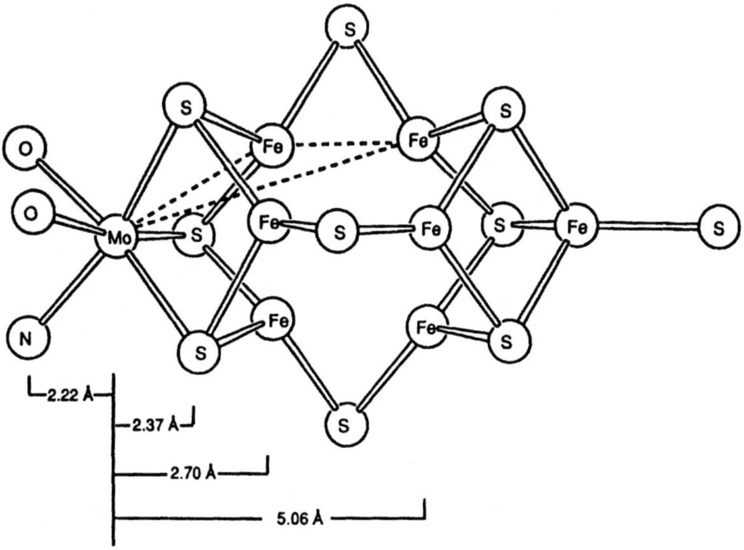

Figure 13. Structural model of the FeMoco site in nitrogenase. The three-body Mo-Fe-Fe(long) MS pathway through which a long-range Mo···Fe(long) interaction takes place is indicated by the dotted line.

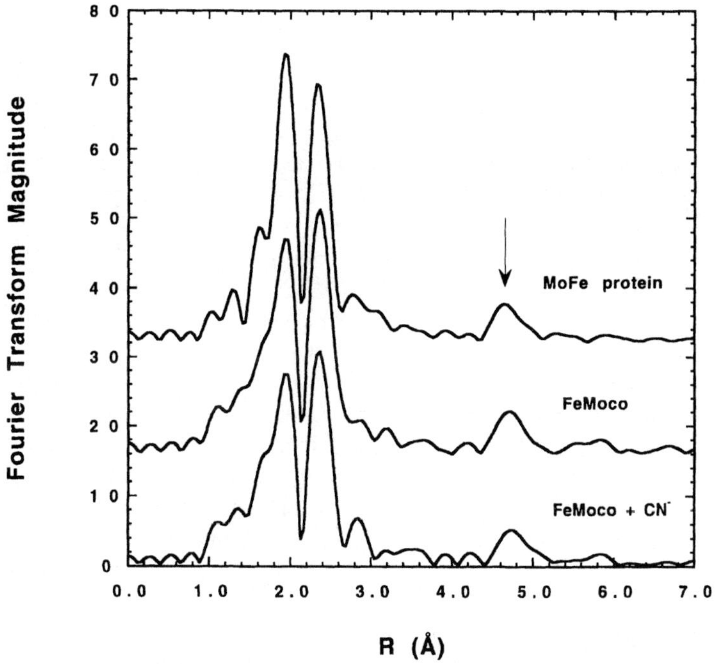

Figure 14. FTs of the Mo K-edge EXAFS for MoFe protein, isolated FeMoco and FeMoco + CN^-. The FT peak at ~4.8 Å is present in all three cases and is attributed to the SS and MS contributions from Mo···Fe(long).

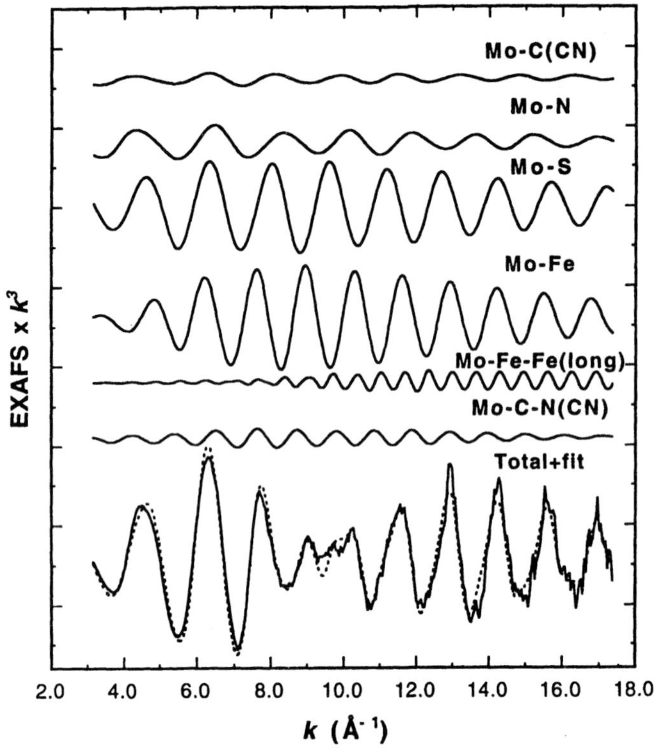

Figure 15. Individual contributions to the EXAFS signal in the final fit for Mo K-edge data for FeMoco + CN^-. The total signal is also shown (· · · ·) compared with the experimental data (—). The Mo-Fe-Fe(long) signal gives rise to the FT peak at ~4.8 Å shown in Figure 14. Also note the Mo-C-N signal that gains significant intensity through the linear MS effect. (The ordinate scale is 10 between two consecutive longer marks.)

using the GNXAS MS method.[68] Figure 14 shows the FTs of the Mo K-edge data from isolated FeMoco with and without added CN^- as well as the dithionite-reduced MoFe protein. All the FTs show a significant asymmetric peak at the high R region between 4.5 Å and 5.2 Å. Using GNXAS analysis, it was shown that this peak derives from a long-range Mo···Fe interaction (referred to later as Fe(long)) through a combination of SS and MS contributions. Figure 15 shows the individual EXAFS contributions of the fit and it is clear from the relative strength of the EXAFS signals that the Mo···Fe(long) interaction is very strong given such a long distance. A careful analysis further showed that the Mo-Fe-Fe(long) $\gamma^{(3)}$ interaction contributed 40% of the total signal while the Mo···Fe(long) SS $\gamma^{(2)}$ contributed 60%. The presence of this Mo···Fe(long) interaction in both FeMoco and the MoFe protein in the FT around $R = 4.8$ Å demonstrates that the long-range order in the FeMoco cluster is preserved when FeMoco is extracted from the MoFe protein.

The FeMoco with and without CN^- give very similar FTs, except for the small but reproducible feature at $R = 2.9$ Å for the CN^- added FeMoco (Figure 14). The MS analysis attributed this to a second-shell contribution from CN^-. The theoretical

Mo-C-N signal in Figure 15 shows that this linear unit makes a significant MS contribution due to the linear configuration. This FT feature thus provides direct evidence that the added CN^- coordinates to Mo in the isolated FeMoco.

Finally, the GNXAS analysis gives accurate metrical details of the FeMoco sites. These results differed initially from those of FeMoco from the X-ray crystal structure of the MoFe protein, but crystal structure refinement to higher resolution has yielded metric values closer to those of the GNXAS results.[89]

This study gives confidence in studying solution conditions where inhibitors and substrates interact. From a MS EXAFS point of view, it demonstrates the sensitivity of the GNXAS MS method; small molecule ligation can be studied even in complex clusters when MS effects enhance the visibility of binuclear ligands through linear or near-linear coordination.

5 CHALLENGES IN DATA ANALYSIS

5.1 Number of Justifiable Fitting Parameters

In the curve-fitting method described above, a number of parameters are refined to optimize the theoretical spectrum. Obviously, the fit result will improve if an increased number of parameters are allowed to vary. However, whether the inclusion of additional parameters in a fit is justifiable depends on the data range used. Specifically, the information content of an EXAFS spectrum places an upper limit on the number of parameters that can be varied in a fit. This limit is the number of independent data points N_{idp} of the particular data range. Stern[90] suggested using the following equation:

$$N_{idp} = \frac{2\Delta k \Delta R}{\pi} + 2 \tag{10}$$

where Δk is the data range analyzed in k space and ΔR is the FT interval used. If the number of parameters varied in a fit is more than N_{idp}, the fit is underdetermined, indicating that a large number of solutions are possible, with no clear defined minimum in the refinement space.

This restriction has the following consequences:

(1) Within the limit of N_{idp}, additional parameters may be included in a fit. Because a larger number of fitting parameters always results in an apparent better fit, a valid comparison between two fits should take into account the total number of fitting parameters, as suggested by Equation (9). Therefore, any increase in the number of these parameters should be reflected in the goodness-of-fit value.
(2) Because the data range Δk typically available is rather limited (*e.g.* k = 3–15 Å^{-1}), N_{idp} sometimes can be quite small depending on ΔR. In the case of a GNXAS fit where ΔR is the entire FT range of the useful data (*e.g.* 5 Å), N_{idp} can be as large as 35. However, since all shells within this FT range have to be included in order to obtain reasonable fits, the number of fitting parameters

to account for all these shells can be large and this large N_{idp} may still seem small. This problem can potentially be quite severe when analyzing data for complicated biological systems, especially when considering MS contributions. In GNXAS, for example, one additional three-body MS signal introduces up to 9 additional parameters.

To solve this problem, it is necessary to increase the amount of information available. One of the techniques developed to date is a group fitting of rigid structures such as imidazole. Co *et al.*[91] exploited the fact that the internal planar geometry of the imidazole group is always preserved, such that the coordinates of all the atoms in the structure can be determined by the metal-to-first-shell nitrogen (from the imidazole group) distance. Binsted and coworkers[92] further developed the technique, and their restrained and constrained refinement have so far been widely used. In this method, a well defined chemical structure is refined as one unit, with its internal structure treated as fixed. In the case of an imidazole ring, the only parameters needed to include this structure in the fit are the metal-to-first-shell distance, at most three angular parameters defining the orientation of the structure, and their corresponding bond/angle variances. This treatment results in a significant decrease in the fitting parameters, especially those from imidazole MS contributions. In addition, this method has the capability of allowing for the structural variability in these rigid structures. Information from other structural techniques is used as additional observation in the data analysis. The slight structural differences are weighed and then contribute to the fit index being minimized.

In cases where an EXAFS sample contains more than one photoabsorbing species, increasing the amount of information can be achieved by simultaneously analyzing the EXAFS spectra of all the edges, the so-called multiple-edge EXAFS refinement.[93] Common structural units from these edges share a subset of parameters and thus only one set is needed for all of them in the refinements. In the model complex of cytochrome *c* oxidase,[65] (see Section 4) the Fe-O-Cu subunit is shared by both Fe and Cu, so the Fe-O and Cu-O distances and the Fe-O-Cu angle are constrained to be the same for both the Fe and the Cu K-edge EXAFS data. This multiple-edge method not only reduces the number of fitting parameters, but also increases the accuracy and consistency of the structural results.

5.2 Statistical Analysis of Fitting Parameters

In determining structural parameters from EXAFS curve-fitting, it is important to assess the accuracy of fitting results. Error estimates can be measured by examining how much a given variable must change to produce a given decrease in the $\mathcal{R}$-value for a particular noise level of the experimental data. In general, the uncertainties in EXAFS determinations are dominated by systematic errors (from data collection and intrinsic limitations in the theory) and correlation effects. A global estimate of uncertainties comes from the comparison of the difference between the fitting results and their corresponding crystallographic values. Statistical errors can be further evaluated from correlations between parameters.

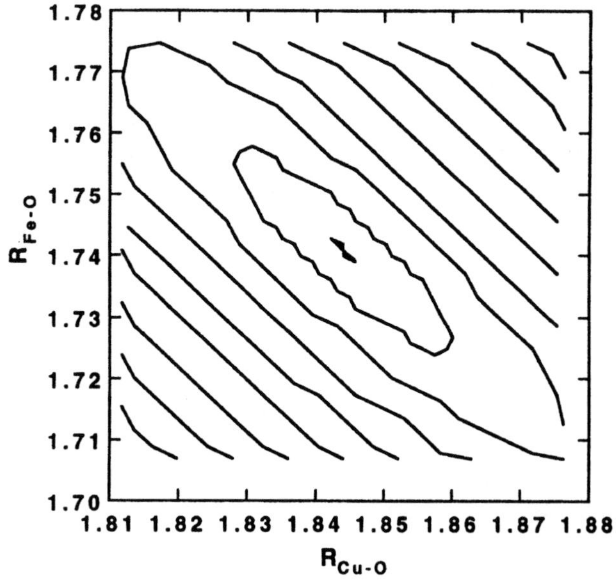

Figure 16. A representative contour plot, $R_{\text{Fe-O}}$ *vs* $R_{\text{Cu-O}}$, showing the statistical correlation between the two distances. The innermost curve corresponds to the 95% confidence interval from which the statistical errors are determined.

Due to the limited number of parameters that can be fit for a given EXAFS data set, refinements are often underdetermined. As a result, fitting parameters become highly correlated. Correlation can greatly increase the uncertainties on the fitting parameters. For example, the two parameters that affect the phase of an EXAFS wave, R_j and E_0, are strongly correlated. Every 2–3 eV increase in E_0 results in *ca.* 0.01 Å decrease in R_j. The uncertainty in E_0 therefore may introduce additional uncertainty on distance determination. The amplitude of an EXAFS wave is determined partially by the strongly correlated pair σ_j^2 and N_j, and their correlation makes the determination of coordination number less accurate. The presence of correlation means that a proper statistical analysis of a fit is necessary.

A rigorous way of analyzing the correlation effects and evaluating statistical uncertainties is through multidimensional contour plots in parameter space.[65,69,94] For example, a two-dimensional contour plot shows the interrelation between two parameters whose statistical errors can be obtained from the plot. A two-dimensional contour plot typically has an approximately ellipsoidal shape that is tilted proportionally to the degree of the correlation between the two parameters. Figure 16 shows a typical contour plot, in this case the correlation between two distances ($R_{\text{Fe-O}}$ and $R_{\text{Cu-O}}$).[65] High statistical correlation (and therefore large statistical errors) between the two can be seen from the plot. Therefore, determination of one parameter is expected to have a large effect on the accuracy of the other. From the maximum and minimum values relative to both axes given by the innermost contour (95% confidence level), the statistical errors of the two distances were determined to be within ±0.02 Å for both $R_{\text{Fe-O}} = 1.74$ Å and $R_{\text{Cu-O}} = 1.84$ Å.

5.3 Limited Data Range in Data Analysis

As discussed in the above sections, the available range of EXAFS data determines the amount of information that is available and has significant impact on the analysis and fitting results. In Section 2.1.2, it was discussed that the k-range available for analysis limits the resolution in bond length, making it difficult to separate closely spaced shells of similar atomic numbers. In addition, an FT of shorter k-range generally gives broader peaks which may worsen the peak overlapping, making it difficult or impossible to perform Fourier filtering. A limited k-range also has profound impacts on the interpretation of protein structure. For example, an existing M···M interaction in the structure may not be observed from EXAFS given only a short data range for analysis.[33]

For EXAFS data of biological systems, the problem of limited k-range is compounded by relatively poor data quality (as compared to model data), increased number of ligand types and higher disorder of the system. All of these factors contribute to further limit the information content of the EXAFS data.

Furthermore, the limitations described in the two sections above (Sections 5.1 and 5.2) are directly related to the data range used. Equation (10) clearly shows the effect of Δk on the number of fitting parameters allowed. A larger number of parameters may be justified when the information content of the EXAFS spectrum is increased by increasing the k-range. It is also shown that extending the number of independent points in the fitting procedure can greatly reduce the correlation effects and consequently reduce uncertainties on the fitting parameters.

Extending the k-range is therefore highly desirable. However, challenges currently encountered involve both theoretical and experimental aspects. To extend the data to lower k-range requires an improved theoretical understanding of MS scattering effects and the development of sophisticated data analysis packages that are capable of performing efficient MS calculations. To extend the data to higher k-range, on the other hand, involves the improvement of experimental techniques, such as higher intensity of synchrotron radiation, that will provide a better signal-to-noise ratio and a better quality of data.

6 CONCLUSIONS

X-ray absorption spectroscopy, and in particular EXAFS, has over the past 25 years evolved into a powerful and frequently applied method for local structure determination in many scientific fields, including inorganic and bioinorganic chemistry. This is largely due to the availability of synchrotron radiation, its continued improvement in flux and brightness, advancements in optical systems, and the availability of increasingly sophisticated detectors and other instrumentation. It is also due to the continued development of theory and its conversion into advanced computer codes for functional calculation and data analysis. With the understanding of MS effects and the more recent development and improvement of MS theory and codes, EXAFS has become increasingly powerful, and is being applied to even larger and more complicated systems. Its dimension has been extended to include the study of spatial orientation in addition to that of radial distribution of atoms around an absorbing center. In particular, information regarding long range (~5 Å) interactions and bond angles can be obtained reliably in certain cases along with other structural information.

As EXAFS has become a standard technique for determining local environments, a variety of more refined and more difficult experiments can be performed, such as single-crystal and time-resolved EXAFS studies, to selectively explore specific bonding aspects of structures and to probe reaction mechanisms at various stages. Meanwhile, experiment and theory continue to advance and address problems such as those discussed in Section 5.3. Significant improvements will be made in the future in the data quality and available data range for EXAFS analysis, and will result in improved accuracy and resolution.

ACKNOWLEDGMENT

Most of the data described in Section 4.3 were measured at the Stanford Synchrotron Radiation Laboratory, which is supported by the U.S. Department of Energy, Office of Basic Energy Sciences, Divisions of Chemical and Materials Sciences, and in part by the National Institutes of Health, National Center for Research Resources, Biomedical Technology Program, and by DOE's Office of Biological and Environmental Research.

REFERENCES

1. Bordo, D.; Djinovic, K.; Bolognesi, M. *J. Mol. Biol.* **1994**, *238*, 366 and references therein.
2. Babcock, G.T.; Wikström, M. *Nature* **1992**, *356*, 301.
3. Malmström, B. *Acc. Chem. Res.* **1993**, *26*, 332.
4. Calhoun, M.W.; Thomas, J.W.; Gennis, R.B. *Trends. Biol. Sci.* **1994**, *19*, 325.
5. Murphy, L.M.; Strange, R.W.; Hasnain, S.S. *Structure* **1997**, *5*, 371.
6. Henkel, G.; Müller, A.; Weissgräber, S.; Buse, G.; Soulimane, T.; Steffens, G.C.M.; Nolting, H. *Angew. Chem. Int. Ed. Engl.* **1995**, *34*, 1488.
7. Garner, C.D. *Adv. Inorg. Chem.* **1991**, *36*, 303, and references therein.
8. Ellis, P.J. Ph.D. thesis, University of Sydney, 1995.
9. Blackburn, N.J.; Barr, M.E.; Woodruff, W.H.; van der Ooost, J.; de Vries, S. *Biochemistry* **1994**, *33*, 10401.
10. Powers, L.; Lauraeus, M.; Reddy, K.S.; Chance, B.; Wikström, M. *Biochim. Biophys. Acta.* **1994**, *1183*, 504.
11. Penner-Hahn, J.E.; Eble, K.S.; McMurry, T.J.; Renner, M.; Balch, A.L.; Groves, J.T.; Dawson, J.H.; Hodgson, K.O. *J. Am. Chem. Soc.* **1986**, *108*, 7819.

12. Gu, Z.; Dong, J.; Allan, C.B.; Choudhury, S.B.; Franco, R.; Moura, J.J.G.; Moura, I.; LeGall, J.; Przybyla, A.E.; Roseboom, W.; Albracht, S.P.J.; Axley, M.J.; Scott, R.A.; Maroney, M.J. *J. Am. Chem. Soc.* **1996**, *118*, 11155.

13. Wang, X.; Randall, C.R.; True, A.E.; Que, L.Jr. *Biochemistry* **1996**, *35*, 13946.

14. Durham, P.J. in *X-Ray Absorption: Principles, Applications, Techniques of EXAFS, SEXAFS and XANES*; Koningsberger, D.C. and Prins, R., Eds.; John Wiley & Sons: New York, 1988; p. 53.

15. Kau, L.-S.; Spira-Solomon, D.J.; Penner-Hahn, J.E.; Hodgson, K.O.; Solomon, E.I. *J. Am. Chem. Soc.* **1987**, *109*, 6433.

16. Westre, T.E.; Kennepohl, P.; DeWitt, J.G.; Hedman, B.; Hodgson, K.O.; Solomon, E.I. *J. Am. Chem. Soc.* **1997**, *119*, 6297.

17. Teo, B.K. *EXAFS: Basic Principles and Data Analysis*; Springer-Verlag, New York, 1986.

18. Stern, E.A. in *X-Ray Absorption: Principles, Applications, Techniques of EXAFS, SEXAFS and XANES*; Koningsberger, D.C. and Prins, R., Eds.; John Wiley & Sons: New York, 1988; p. 3.

19. Gurman, S.G. in *Synchrotron Radiation and Biophysics*; Hasnain, S.S., Ed.; Ellis Horwood Limited: Chichester, U.K., 1990; p. 9.

20. Fonda, L. *J. Phys. Condens. Matter* **1992**, *4*, 8269.

21. Cramer, S.P. and Hodgson, K.O. *Prog. Inorg. Chem.* **1979**, *25*, 1.

22. Scott, R.A. *Methods Enzymol.* **1985**, *117*, 414.

23. Cramer, S.P. in *X-Ray Absorption: Principles, Applications, Techniques of EXAFS, SEXAFS and XANES*; Koningsberger, D.C. and Prins, R., Eds.; John Wiley & Sons: New York, 1988; p. 257.

24. Yachandra, V.K. *Methods Enzymol.* **1995**, *246*, 638.

25. Stern, E.A. *Phys. Rev. B.* **1974**, *10*, 3027.

26. Ashley, C.A.; Doniach, S. *Phys. Rev. B* **1975**, *11*, 1279.

27. Lee, P.A.; Pendry, J.B. *Phys. Rev. B* **1975**, *11*, 2795.

28. Stern, E.A.; Sayers, D.E.; Lytle, F.W. *Phys. Rev. B* **1975**, *11*, 4836.

29. Nordlander, E.; Lee, S.C.; Cen, W.; Wu, Z.Y.; Natoli, C.R.; Di Cicco, A.; Filipponi, A.; Hedman, B.; Hodgson, K.O.; Holm, R.H. *J. Am. Chem. Soc.* **1993**, *115*, 5549.

30. Bunker, G. *Nucl. Instrum. Methods* **1983**, *207*, 437.

31. Lee, P.A.; Citrin, P.H.; Eisenberger, P.; Kincaid, B.M. *Rev. Mod. Phys.* **1981**, *53*, 769.

32. Riggs-Gelasco, P.J.; Stemmler, T.L.; Penner-Hahn, J.E. *Coord. Chem. Rev.* **1995**, *144*, 245.

33. Pickering, I.J.; George, G.N.; Dameron, C.T.; Kurz, B.; Winge, D.R.; Dance, I.G. *J. Am. Chem. Soc.* **1993**, *115*, 9498.

34. Shu, L.; Liu, Y.; Lipscomb, J.D.; Que, L.Jr. *J. Biol. Inorg. Chem.* **1996**, *1*, 297.

35. Shin, W.; Sundaram, U.M.; Cole, J.L.; Zhang, H.H.; Hedman, B.; Hodgson, K.O.; Solomon, E.I. *J. Am. Chem. Soc.* **1996**, *118*, 3202.

36. Scott, R.A.; Eidsness, M.K. *Comm. Inorg. Chem.* **1988**, *7*, 235.

37. DeWitt, J.G.; Bentsen, J.G.; Rosenzweig, A.C.; Hedman, B.; Green, J.; Pilkington, S.; Papaefthymiou, G.C.; Dalton, H.; Hodgson, K.O.; Lippard, S.J. *J. Am. Chem. Soc.* **1991**, *113*, 9219.

38. Eisenberger, P.; Brown, G.S. *Solid State Commun.* **1979**, *29*, 481.

39. Crozier, E.D.; Seary, A. *Can. J. Phys.* **1980**, *58*, 1388.

40. Dalba, G.; Diop, D.; Fornasini, P.; Rocca, F. *J. Phys. Condens. Matter* **1994**, *6*, 3599.

41. Crozier, E.D.; Rehr, J.J.; Ingalls, R. in *X-Ray Absorption: Principles, Applications, Techniques of EXAFS, SEXAFS and XANES*; Koningsberger, D.C. and Prins, R., Eds.; John Wiley & Sons: New York, 1988; p. 373.

42. Teo, B.K. *J. Am. Chem. Soc.* **1981**, *103*, 3990.

43. Co, M.S.; Hendrickson, W.A.; Hodgson, K.O.; Doniach, S. *J. Am. Chem. Soc.* **1983**, *105*, 1144.

44. Westre, T.E.; Di Cicco, A.; Filipponi, A.; Natoli, C.R.; Hedman, B.; Solomon, E.I.; Hodgson, K.O. *J. Am. Chem. Soc.* **1995**, *117*, 1566.

45. Binsted, N.; Cook, S.L.; Evans, J.; Greaves, G.N.; Price, R.J. *J. Am. Chem. Soc.* **1987**, *109*, 3669.

46. Filipponi, A.; Di Cicco, A.; Zanoni, R.; Bellatreccia, M.; Sessa, V.; Dossi, C.; Psaro, R. *Chem. Phys. Lett.* **1991**, *184*, 485.

47. Strange, R.W.; Blackburn, N.J.; Knowles, P.F.; Hasnain, S.S. *J. Am. Chem. Soc.* **1987**, *109*, 7157.

48. Sayers, D.E.; Bunker, B.A. in *X-Ray Absorption: Principles, Applications, Techniques of EXAFS, SEXAFS and XANES*; Koningsberger, D.C. and Prins, R., Eds.; John Wiley & Sons: New York, 1988; p. 211.

49. Cramer, S.P.; Hodgson, K.O.; Stiefel, E.I.; Newton, W.E. *J. Am. Chem. Soc.* **1978**, *100*, 2748.

50. Teo, B.K.; Lee, P.A. *J. Am. Chem. Soc.* **1979**, *101*, 2815.

51. McKale, A.G.; Veal, B.W.; Paulikas, A.P.; Chan, S.K.; Knapp, G.S. *J. Am. Chem. Soc.* **1988**, *110*, 3763.

52. Vaarkamp, M.; Dring, I.; Oldman, R.J.; Stern, E.A.; Koningsberger, D.C. *Phys. Rev. B.* **1994**, *50*, 7872.

53. Gurman, S.J.; Binsted, N.; Ross, I. *J. Phys. C* **1984**, *17*, 143.

54. Gurman, S.J.; Binsted, N.; Ross, I. *J. Phys. C* **1986**, *19*, 1845.

55. Rehr, J.J.; Mustre de Leon, J.; Zabinsky, S.I.; Albers, R.C. *J. Am. Chem. Soc.* **1991**, *113*, 5135.

56. Mustre de Leon, J.; Rehr, J.J.; Zabinsky, S.I.; Albers, R.C. *Phys. Rev. B* **1991**, *44*, 4146.

57. Filipponi, A.; Di Cicco, A.; Tyson, T.A.; Natoli, C.R. *Solid State Commun.* **1991**, *78*, 265.

58. Filipponi, A.; Di Cicco, A.; Natoli, C.R. *Phys. Rev. B* **1995**, *52*, 15122.

59. Hedin, L.; Lundqvist, S. *Solid State Phys.* **1969**, *23*, 1.

60. Lundqvist, B.I. *Phys. Kondens. Mater.* **1967**, *6*, 193.

61. Di Cicco, A.; Stizza, S.; Filipponi, A.; Boscherini, F.; Mobilio, S. *J. Phys. B* **1992**, *25*, 2309.

62. D'Angelo, P.; Di Cicco, A.; Filipponi, A.; Pavel, N.V. *Phys. Rev. A* **1993**, *47*, 2055.

63. Burattini, E.; D'Angelo, P.; Di Cicco, A.; Filipponi, A.; Pavel, N.V. *J. Phys. Chem.* **1993**, *97*, 5486.

64. Westre, T.E.; Di Cicco, A.; Filipponi, A.; Natoli, C.R.; Hedman, B.; Solomon, E.I.; Hodgson, K.O. *J. Am. Chem. Soc.* **1994**, *116*, 6757.

65. Zhang, H.H.; Filipponi, A.; Di Cicco, A.; Lee, S.C.; Scott, M.J.; Holm, R.H.; Hedman, B.; Hodgson, K.O. *Inorg. Chem.* **1996**, *35*, 4819.

66. Zhang, H.H.; Filipponi, A.; Di Cicco, A.; Scott, M.J.; Holm, R.H.; Hedman, B.; Hodgson, K.O. *J. Am. Chem. Soc.* **1997**, *119*, 2470.
67. Conradson, S.D.; Burgess, B.K.; Newton, W.E.; Di Cicco, A.; Filipponi, A.; Wu, Z.Y.; Natoli, C.R.; Hedman, B.; Hodgson, K.O. *Proc. Natl. Acad. Sci. USA* **1994**, *91*, 1290.
68. Liu, H.I.; Filipponi, A.; Gavini, N.; Burgess, B.K.; Hedman, B.; Di Cicco, A.; Natoli, C.R.; Hodgson, K.O. *J. Am. Chem. Soc* **1994**, *116*, 2418.
69. Filipponi, A.; Di Cicco, A. *Phys. Rev. B* **1995**, *52*, 15135.
70. Zhang, Y.; Pavlosky, M.A.; Brown, C.A.; Westre, T.E.; Hedman, B.; Hodgson, K.O.; Solomon, E.I. *J. Am. Chem. Soc.* **1992**, *114*, 9189.
71. Dittmer, J.; Dau, H. *Ber. Bunsenges. Phys. Chem.* **1996**, *100*, 1993.
72. Lee, S.C.; Holm, R.H. *J. Am. Chem. Soc.* **1993**, *115*, 5833, 11789.
73. Scott, M.J.; Zhang, H.H.; Lee, S.C.; Hedman, B.; Hodgson, K.O.; Holm, R.H. *J. Am. Chem. Soc.* **1995**, *117*, 568.
74. Kauffmann, K.E.; Goddard, C.A.; Zang, Y.; Holm, R.H.; Münck, E. *Inorg. Chem.* **1997**, *36*, 985.
75. Fox, S.; Nanthakumar, A.; Wikström, M.; Karlin, K.D.; Blackburn, N.J. *J. Am. Chem. Soc.* **1996**, *118*, 24.
76. Stemmler, T.L.; Barnhart, T.M.; Penner-Hahn, J.E.; Tucker, C.E.; Knochel, P.; Böhme, M.; Frenking, G. *J. Am. Chem. Soc.* **1995**, *117*, 12489.
77. Laplaza, C.E.; Johnson, M.J.A.; Peters, J.C.; Odom, A.L.; Kim, E.; Cummins, C.C.; George, G.N.; Pickering, I.J. *J. Am. Chem. Soc.* **1996**, *118*, 8623.
78. Lee, S.C.; Scott, M.J.; Kauffmann, K.; Münck, E.; Holm, R.H. *J. Am. Chem. Soc.* **1994**, *116*, 401.
79. Scott, M.J.; Lee, S.C.; Holm, R.H. *Inorg. Chem.* **1994**, *33*, 4651.
80. Scott, M.J.; Holm, R.H. *J. Am. Chem. Soc.* **1994**, *116*, 11357.
81. Gardner, M.T.; Deinum, G.; Kim, Y.; Babcock, G.T.; Scott, M.J.; Holm, R.H. *Inorg. Chem.* **1996**, *35*, 6878.
82. Burgess, B.K. *Chem. Rev.* **1990**, *90*, 1377, and references therein.
83. Hawkes, T.R.; McLean, P.A.; Smith, B.E. *Biochem. J.* **1984**, *217*, 317.
84. Scott, D.J.; May, H.D.; Newton, W.E.; Brigle, K.E.; Dean, D.R. *Nature (London)* **1990**, *343*, 188.
85. Cramer, S.P.; Hodgson, K.O.; Gillum, W.O.; Mortenson, L.E. *J. Am. Chem. Soc.* **1978**, *100*, 3398.
86. Cramer, S.P.; Gillum, W.O.; Hodgson, K.O.; Mortenson, L.E.; Stiefel, E.I.; Chisnell, J.R.; Brill, W.J.; Shah, V.K. *J. Am. Chem. Soc.* **1978**, *100*, 3814.
87. Kim, J.; Rees, D.C. *Science* **1992**, *257*, 1677.
88. Bolin, J.T.; Ronco, A.E.; Morgan, T.V.; Mortenson, L.E.; Xuong, N.H. *Proc. Natl. Acad. Sci. U. S. A.* **1993**, *90*, 1078.
89. Peters, J.W.; Stowell, M.H.B.; Soltis, M.; Finnegan, M.G.; Johnson, M.K.; Rees, D.C. *Biochemistry* **1997**, *36*, 1181.
90. Stern, E.A. *Phys. Rev. B* **1993**, *48*, 9825.
91. Co, M.S.; Scott, R.A.; Hodgson, K.O. *J. Am. Chem. Soc.* **1981**, *103*, 986.
92. Binsted, N.; Strange, R.W.; Hasnain, S.S. *Biochemistry* **1992**, *31*, 12117.
93. Di Cicco, A. *Phys. Rev. B* **1996**, *53*, 6174.
94. Di Cicco, A. *Phys. B* **1995**, *208&209*, 125.

10 Electronic Structure Calculations on Transition Metal Complexes: *Ab-Initio* and Approximate Models

CHARLES H. MARTIN AND MICHAEL C. ZERNER

Quantum Theory Project
Departments of Chemistry and Physics
University of Florida
Gainesville, FL 32611, USA
E-mail: zerner@qtp.ufl.edu

Inorganic Electronic Structure and Spectroscopy, Volume I: Methodology.
Edited by E. I. Solomon and A. B. P. Lever.
ISBN 0-471-15406-7.

1 INTRODUCTION

There has been an amazing amount of progress in the last 25 years in quantum chemistry. Although the principles of quantum mechanics that were necessary for a theoretical understanding of chemistry in terms of nuclei and electrons were laid out over 50 years ago, much was required in the development of computers, coding algorithms, matrix manipulation, integral evaluation, and, simply, experience. Experience determined what worked well, and what did not. Quantum chemistry did not become a serious contender with experiment until the introduction of gradient methods that allowed for the generation of molecular structure in a more or less automatic fashion[1–6] and in the developments in the theory that introduced corrections (electron correlation) to the independent particle theory (molecular orbital theory) that allowed calculations to vie with experiment for accuracy. For nearly twenty years semiempirical methods, such as π-electron theory[7] and INDO/S,[8–13] served the need to predict and/or interpret electronic spectra, and only within the past five years have *ab-initio* methods appeared that can compete in accuracy, if not in efficiency, with these simpler theories. Today there are many commercially available program packages of the semiempirical,[7–13] density functional,[14–17,29] and *ab-initio*[17–23] types that can be used in a more or less automatic fashion to study a great variety of chemical problems, including molecular structure, reactivity, spectra, and numerous electronic properties. Even today, however, computational methods can not easily treat molecules containing transition metal atoms, especially those with multiple open-shell electrons, with nearly the ease or utility with which they can treat molecules containing hydrogen and main group elements.[24–28]

For many years, transition metals seemed to be the purview of the very successful Extended Hückel Theories (EHT), described later in this chapter, and Density Functional Theory (DFT), as described elsewhere in this book. The EHT models have been used for many years to explain many phenomena in a qualitative sense, and they yield frontier orbitals which seem to reproduce well the properties of the underlying many-electron states. The DFT methods are more quantitative. Indeed, a number of methods, such as the Xα method, provide not only ground states but also excited state information.[29] DFT avoids many of the convergence problems which traditional SCF procedures encounter, as discussed below, both through the use of Kohn Sham orbitals and local exchange functionals. However, DFT has problems treating molecules with numerous open shells because it is not appropriate for multiplet structure in general, and, if nonintegral values of occupation numbers are required to simulate degeneracies, the applicability of the model is open to debate. In contrast, both semiempirical and *ab-initio* methods offer a more rigorous approach to treating such systems, but not without their computational and theoretical limitations, some of which are discussed in this chapter.

It is interesting here to speculate on the usefulness of classical models (molecular mechanics, MM) for studying transition metal systems because these methods have had a large impact in studying the structure, vibrational properties and thermodynamics of rather large systems.[30,31] These methods model molecules simply as balls and springs, and contain parametrized potential energy expressions specific for each bond stretch, bond angle bend, and dihedral angle twist. Nonbonded interactions are represented by Coulomb electrostatic interactions and potentials such as Lennard–Jones potentials, i.e., $\varepsilon_{i,j}[A_{i,j}/r_{i,j}^{12} - B_{i,j}/r_{i,j}^{6}]$, that model exchange repulsion and dispersive attractions. These expressions require parameters specific for each individual atom, atom pairs, triplets of atoms, and even quartets of atom types, in addition to the Lennard Jones parameters, $\varepsilon_{i,j}$, $A_{i,j}$, and $B_{i,j}$. For example, in the simplest theory, a halide atom would define one type, but a carbon atom requires at *least* three types, according to sp, sp^2, or sp^3 hybridization.

Even given the complex hybridization, valence, and spin states of carbon atom, a typical transition metal offers an order of magnitude more difficulty, and parametrizing a useful MM model for even a single transition metal atom will contain as many adjustable parameters as all of the elements of the first row, many of which have still not been successfully parametrized. For example, an iron atom can not just be considered a single atom type because it can bond between one to seven neighbors, it may span any number of valence states (Fe(0) to Fe(IV)), and it may form a wide range of spin-states (high, low, or intermediate). Worse, these characteristics of the iron atom will change with rather subtle changes in molecular bond length and bond angle. Each of these different situations necessitates defining a different atom type, making for a very complex MM model. Nevertheless, some MM models do include force field parameters for transition metals, but these characteristics must not change in order for these models to work well. Therefore the metal atom can not be directly involved with any of the actual chemistry (i.e., electron transfer, bond breaking, etc.). Where these models do perform well is in predicting geometries in highly coordinated ionicly bonded complexes.[30,31]

Turning to the more robust and rigorous *ab-initio* and semiempirical methods, we should like to understand how well the currently available quantum chemical techniques, such as Hartree Fock theory and simple excited state methods, perform for transition metal complexes, and, furthermore, to determine the accuracy, applicability, and limitations of the most sophisticated methods when applied to transition metal complexes. First, extensive computational research demonstrates that many of the problems once believed to arise in the study of transition metals complexes, such as the need for highly correlated, multi-reference configuration treatments,[32,33] seems less problematic for simple transition metal complexes. For example, *ab-initio* theory remains unable to treat the chromium dimer. However, this case is more anomalous than ubiquitous; *ab-initio* methods perform much better for other metal dimers. A number of transition metal systems, especially those containing only one or two transition metals, seem well within the grasp of semiempirical SCF and SCF-CIS treatments. For greater accuracy, one can apply highly correlated *ab-initio* methods, such as the Symmetry Adapted Cluster Configuration Interaction (SAC-CI) theory[34–46] or the Complete Active Space plus Second Order Perturbation Theory (CASPT2) approach[23,47–53] which provide both ground and electronically excited states with near chemical accuracy (2 kcal/mol). The first row transition metals incur the most difficulties because of the highly local nature of the d orbitals.[54] Such problems include difficulties with converging SCF calculations and the need for large amounts of electron correlation corrections. For the heavier elements, one might expect that relativistic effects would complicate the calculations, but, as we shall see, most of these effects, such as the relativistic contraction of the s orbitals and the spin-orbit coupling, can be correctly approximated with simple model potentials. Many of the difficulties in treating transition metal complexes results from two interrelated complications that arise in certain systems that contain a large number of open-shell electrons distributed among many localized d and/or f orbitals.

Many transition metal complexes contain numerous *open-shells* (colloquial for electrons in partially-filled orbitals) and many nearly degenerate electronic states.

Consequently, standard self consistent field (SCF) calculations, originally designed, tested, and optimized for closed-shell molecules, converge quite poorly, and require sophisticated theoretical methods and extensive human interaction. If the system is open-shell, then most often the user must assign electrons to the resultant molecular orbitals, MO's, and predict in advance the spin state, and even forcibly guide the calculation to the energy minimum. Predictions frequently rely upon simple ligand field theory or Heisenberg (Ising) spin models. Even if one can guess the proper spin state and orbital occupancies, the most general open-shell SCF procedure, the Restricted Open-Shell Hartree–Fock (ROHF) method, converges very slowly, if at all, for complexes with numerous open shells. In principle, the orbital occupancy and spin state may be determined with a complete active space self consistent field (CASSCF) calculation,[33,55,56] but this requires diagonalizing a very large matrix, and remains limited to systems with fourteen or fewer valence molecular orbitals, such as transition metal dimers and small organometallic complexes Alternative SCF approaches include the Unrestricted Hartree–Fock (UHF) theory, the older Hyper Hartree–Fock approach, and the recently developed Configuration Averaged Hartree Fock technique (CAHF).

The second general problem is also associated with the very local nature of d- or f-orbitals. The atomic nature of transition metal atoms or ions in molecular complexes is most often preserved, and it is this observation that makes crystal and ligand field theories, based essentially on perturbation theory, so useful. But this is at odds with the basic philosophy of molecular orbital theory that ignores atomic structure, which solves a molecular problem determined by the positions of the nuclei that describe the complex, and then assign electrons to orbitals based upon some type of *aufbau* (building up) principle. The atomic nature of the problem is lost, and this must be restored through corrections to the Hartree–Fock (independent particle) model by including electron correlation, as described below. Hartree–Fock theory is usually an excellent theory for molecules of main group elements, and its refinements either through perturbation theory,[5,24–27] coupled cluster theory[22,25,27,28] or configuration interaction[20,21,25,27] are reasonably straightforward. A Hartree–Fock calculation, by itself, may be reasonable for a closed-shell transition metal system, but any calculation on transition metal containing systems, even those on closed-shell systems, usually requires a post Hartree–Fock refinement before it is robust enough to really trust.

2 SOME QUANTUM MECHANICS

2.1 Introduction

The starting point of nearly all calculations today is the time independent, fixed nuclei (Born–Oppenheimer[24–27]), non-relativistic Schrödinger equation:

$$\mathbf{H}(1, 2, \ldots)\Psi_J(1, 2, \ldots) = E_J\Psi_J(1, 2 \ldots) \tag{1}$$

with **H** the many-electron fixed-nuclei Coulombic Hamiltonian,

$$\mathbf{H}(1,2,\ldots) = \frac{1}{2m}\sum_{i}\mathbf{p}_i^2 - \sum_{(A,i)} Z_A e^2/R_{A,i} + \sum_{(i<j)} e^2/r_{i,j} + \sum_{(A<B)} Z_A Z_B e^2/R_{A,B} \tag{2a}$$

$$= -\frac{1}{2}\sum_{i}\nabla_i^2 - \sum_{(A,i)} Z_A/R_{A,i} + \sum_{(i<j)} 1/r_{i,j} + \sum_{(A<B)} Z_A Z_B/R_{A,B} \tag{2b}$$

and E_J the energy of the state described by the many-electron wavefunction Ψ_J. In Eqn. 2a $\mathbf{p}_i$ is the momentum of the i'th electron, and the first term represents the kinetic energy of all the electrons, the second term is the Coulomb attraction between electrons of charge e and all nuclei A of charge Z_A, the third term is the electron-electron repulsion, and the last term is the nuclear-nuclear repulsion. This last term *must* be considered as part of the electronic energy in order to generate potential energy curves for the nuclei. We have also adapted the most common units used in quantum mechanics, the atomic unit, in Eqn. 2b. In this equation the mass of the electron $m = 1$, the charge on the electron $e = 1$, and Planck's constant h is set to 2π. The energy unit is called an a.u., or *Hartree*, h, and $1h = 27.2114$ eV $= 627.5$ kcal/mol. The unit of distance is the *Bohr* $= 0.52917$ Å. We use bold symbols for operators (or matrices, see below), so that they are not confused with multiplicative factors whenever confusion might occur. The symbol $r_{i,j}$ is shorthand for $|r_i - r_j|$, the scaler distance between particle i and j. Finally a symbol such as $\sum_{(i<j)}$ is shorthand for the double summation $\sum_i \sum_{(j>i)}$. Note that since the nuclear-electronic term and the nuclear-nuclear term are functions of nuclear positions which we fix for this problem, E_J and $\Psi_J(1,2,\ldots)$ are *parametric* functions of the nuclear position; that is, although there is no explicit dependence of E or Ψ on the set of nuclear positions $\{R_A\}$, the value and the function change with $\{R_A\}$.

Eqn. 2 is nonrelativistic, and seems to be quite adequate for most problems, especially for ground state problems dealing with the lighter atoms. Relativistic effects will be necessary for treating the heavier transition metal atoms, and for detailed spectroscopy even of the lighter transition metal series, but they may be included in a nonrelativistic formalism using simple model potentials and/or perturbation theory, as described in Section 5 below.

In principle, we would like to solve the Hamiltonian eigenvalue problem as accurately as possible. But there is a trade-off between accuracy and expediency due to the enormous computational requirements required to actually solve the molecular Schrödinger equation. *Solving* this problem requires expanding the **H** matrix representation of Eqn. 1 in an infinite many-electron basis, which is referred to as the full CI solution.[25–27] Although *completely* spanning the orbital (basis set) space and then the many particle space is likely not required in most cases for usable results, the dimension of the matrices required grows exponentially with the number of electrons in the molecule, and even a small molecule like water can require billions of many-electron functions. The goal of modern quantum chemistry is to find ways around diagonalizing large matrices. The *dressed* (scaled) integrals of semiempirical theory and the approximation methods of electron correlation discussed below are attempts to avoid diagonalizing large matrices.

2.2 Introduction to the Hartree–Fock Method

Nearly all electronic structure calculations today start with the solution to a one-electron effective operator, an operator generated by assuming that the many-electron wavefunction is a product of one-electron functions. Such functions give rise to many-electron Hamiltonians that are the sums of one-electron operators. The coupling terms in such a problem are the two-electron repulsions, and these terms are large enough to make suspect the assumption that this approach will be successful, even if corrected subsequently by perturbation theory. The essence of the Hartree–Fock approach is to minimize the differences between an effective one electron operator $\mathbf{v}(i)$ that averages over the energy of electron i in the field of all other $N-1$ electrons, and the explicit repulsion between this electron and all others, $\sum_{(j)} e^2/r_{i,j}$.

The steps in this approximation are:

Step 1. The molecular orbital (mo) approximation:

$$\Psi_J(1, 2, \ldots) = \mathbf{A}\mathbf{O}_S(\phi_1(1)\phi_2(2)\ldots) \tag{3a}$$

in which $\mathbf{A}$ is the antisymmetrizer, $\mathbf{O}_S$ is the spin projection operator, and ϕ_i is a molecular orbital, and the assignment of electrons in the product $(\phi_1(1)\phi_2(2)\ldots)$ is called an electronic configuration. In the closed-shell case, $\mathbf{O}_S$ is simply the unit operator, and

$$\begin{aligned}\Psi_J(1, 2\ldots) &= \mathbf{A}(\phi_1(1)\phi_2(2)\ldots)\\ &= (1/\sqrt{2n})\det\begin{vmatrix}\phi_1(1)\ \phi_1(2)\ldots & \phi_1(2n-1) & \phi_1(2n)\\ \phi_2(1)\ \phi_2(2)\ldots & & \ldots\phi_2(2n)\\ \ldots & \phi_n(2n-1)\ \ldots & \\ \phi_n(1)\ \phi_n(2)\ldots & \ldots\phi_n(2n-1) & \phi_n(2n)\end{vmatrix}\\ &= |\phi_1(1)\phi_2(2)\ldots\phi_n(2n)|\\ &= |\phi_1\phi_2\ldots\ \ldots\phi_n|\end{aligned} \tag{3b}$$

is a Slater determinant,[26] ensuring that electrons are Fermions (i.e., that the wavefunction is antisymmetric with respect to the exchange of the coordinates of any two electrons), and that the Pauli exclusion principle is obeyed (If two rows of a determinant are equal, the determinant is zero, and the probability for such a system to exist, $\Psi^*\Psi$, is also zero). In the open-shell case, $\mathbf{O}_S$ creates a linear combination of Slater determinants that ensures that the resulting wave function is an eigenfunction of the spin operator, that is, $\mathbf{S}^2\Psi = s(s+1)\Psi$, in such a fashion that we can also label the state as a spin eigenfunction, ${}^{2s+1}\Psi$, where $2s+1$ is the *multiplicity*, or spin degeneracy. Such linear combinations of determinants that are eigenfunctions of spin are often referred to as *configuration* state *functions*, CSFs.[58]

Note in the above that the subscript J in Ψ_J will designate the configuration and the multiplicity should this be important.

Step 2. Linear Combination of Atomic (-like) Orbitals (LCAO-MO):

This step, systematized by Roothaan[59] and Hall,[60] allows for the conversion of a many-dimensional differential equation into a matrix problem, and in many ways recognizes the fact that an atom in a molecule is essentially an atom, and it behooves us to get the energy of atoms more or less correct before beginning the molecular problem. Molecular orbitals are expanded as

$$\phi_i = \sum \chi_\mu C_{\mu_i} \tag{4}$$

or

$$\phi = (\phi_1, \phi_2, \ldots, \phi_n) = (\chi_1, \chi_2, \ldots)(\mathbf{C}_1, \mathbf{C}_2, \ldots) = \mathbf{XC}$$

in which χ_μ is an atomic orbital (ao), and the C_{μ_i}'s are the molecular orbital coefficients. The atomic orbitals, $\mathbf{X}$, a row matrix, solve an atomic problem similar to the molecular problem as outlined here. The $\mathbf{C}_j$ are the molecular orbital coefficients of ϕ_j, as a column matrix.

As we will examine later, the expansion of the molecular orbitals in terms of atomic-like orbitals can be made more and more accurate, until the LCAO-MO approximation reaches its limit, the *Hartree–Fock* limit. The choice of the set $\mathbf{X}$ defines the basis set. The ultimate accuracy of a calculation depends on this choice.

Step 3. The Variational Principle:

The variational principle states that the expectation value of the Hamiltonian $\mathbf{H}$ representing any system will be an upper bound to the true energy of the system, the only constraints being the spin and space symmetry of the assumed wavefunction.[25,26]

$$W_J(\Psi) \equiv \langle \Psi_J | \mathbf{H} | \Psi_J \rangle / \langle \Psi_J | \Psi_J \rangle \geqslant E_J \tag{5a}$$

$$\delta W_J / \delta \Psi_J = 0 \Rightarrow (\mathbf{F} - \varepsilon_j \Delta)\mathbf{C}_j = 0 \tag{5b}$$

or

$$\mathbf{f}(i)\phi_j(j)\varepsilon_j\phi_j(i) \tag{5c}$$

with $\mathbf{f}(i)$, the Fock operator for electron i, $\mathbf{F}$ the Fock, or energy, matrix, with matrix elements

$$F_{\mu,\nu} = \langle \chi_\mu | \mathbf{f} | \chi_\nu \rangle$$

and ε_j is the molecular orbital energy. In obtaining the expressions for the Fock operator, most often the mo's are constrained to be *orthonormal*,

$$\langle \phi_j | \phi_k \rangle = \int d\tau(1)\phi_j(1)\phi_k(1) = \delta_{j,k} \tag{5d}$$

where $d\tau(1)$ represents an integration over the volume coordinates of electron 1, and $\delta_{j,k} = 1$ if $j = k$, and 0 if $j \neq k$ (the *Kroenicker delta*). Δ is the basis set metric, or *overlap* matrix with elements

$$\Delta_{\mu,\nu} = \langle \chi_\mu | \chi_\nu \rangle$$

Step 4. The Self Consistent Field (SCF):

The one-electron Fock operator **f**, the result of the variational principle, treats each electron within the average field of other electrons, placing each electron in a molecular orbital which itself depends nonlinearly on the other orbitals. Unlike the original many-electron Hamiltonian, which is a linear operator, the Fock operator requires an iterative, self-consistent solution which involves an initial guess for every orbital, some algorithm for varying the molecular orbitals, and a convergence criterion. The initial guess orbitals, represented by the matrix of coefficients $\mathbf{C}^0$, typically arise from a simple (projected) extended Hückel calculation, from a semi-empirical calculation (for *ab-initio* procedures), or through a diagonalization of the one-electron operator (which contains only the kinetic energy and nuclear-electron potential. This is the **h** matrix below). The $\mathbf{C}^0$ matrix defines the initial Fock matrix (or in the open-shell cases, see below, matrices), $\mathbf{F}(\mathbf{C}^0)$, which is then solved (Eqn. 4) for $\mathbf{C}^1$. Then $\mathbf{C}^0$ and $\mathbf{C}^1$ are used to estimate new MO's, $\underline{\mathbf{C}}^1$, that are used to form a new Fock matrix, and this procedure repeated until the electrostatic field assumed to form the Fock matrix agrees with that obtained within some accepted tolerance. Although this procedure is not guaranteed to converge to the lowest energy state, or even a proper state of the system, there are ways to check this that usually involve configuration interaction of some type. The SCF method is summarized below:

$$\mathbf{F}(\mathbf{C}^0) \to \mathbf{C}^1 \to \underline{\mathbf{C}}^1 \to \mathbf{F}(\underline{\mathbf{C}}^1) \to \mathbf{C}^2 \to \underline{\mathbf{C}}^2 \to \text{ until convergence} \tag{6}$$

This iterative procedure seldom proceeds efficiently, or at all, if one merely uses $\mathbf{C}^n$ to calculate $\mathbf{F}(\mathbf{C}^n)$ to obtain $\mathbf{C}^{n+1}$. One uses such schemes as damping,[60] dynamic damping,[61] or level-shifting[62] to extrapolate $\underline{\mathbf{C}}^n$. It is often necessary in calculations on transition metal systems, especially for open-shell systems, to estimate the *d*-orbital occupation and force this occupation, either through orbital symmetry, or through examining principally *d* orbital MO's. Without this, especially in the early stages of an SCF calculation, the *d* character of the MO's and the occupancy of the molecular orbitals of *d* AO's, will oscillate strongly between empty and fully occupied. Such biased calculations may incorrectly converge to an excited state, if at all, and the user must beware.

As a result of our formal training, we, as chemists, often give the greatest significance to the orbitals and orbital energies. Yet this significance applies only insofar as the orbitals reflect the symmetry and nodal structure of states, and insofar as orbital energies reflect the energies of ionization processes. Only the many-electron wavefunctions Ψ_J and the state energies E_J actually solve the electronic structure problem: ϕ_i and ε_i are mathematical conveniences along the way, and they can prove misleading. Consider, for example, Figure 1, which shows an orbital diagram for porphinato-Fe(II) , Fe(II)P.[64] Note that the d orbitals are not the highest

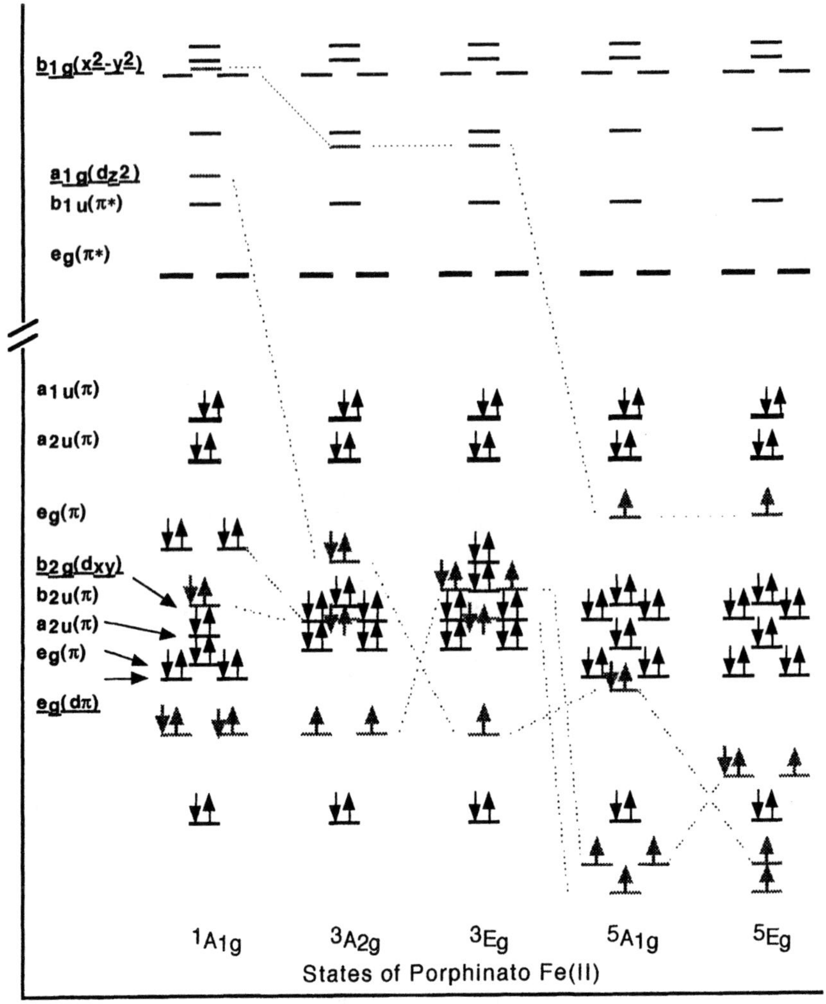

Figure 1. The orbital diagram for the low-lying states of different multiplicities for Fe(II)Porphyrine. Note that the d orbitals are not the highest occupied, and the occurance of partially occupied d orbitals with orbital energies less than those of doubly occupied ligand-based orbitals, see text. Shaded arrows designate d-electrons.

occupied, even in the $^1A_{1g}$ case. Yet calculations on the positive ion clearly demonstrate that an electron detaches from a d-like MO. Note that in the open-shell cases some partially-occupied d-like MO's lie lower in orbital energy than do the doubly-occupied heterocyclic orbitals. Yet removing an electron from the HOMO (highest occupied molecular orbital) represents a charge transfer excitation, and, if we were to actually perform this calculation, would create a state of considerably higher energy. Although it is the differences in *state* energies, not orbital energies, that are observable, how can we rationalize the fact that the low-lying d orbitals ionize easier than the HOMO does? The electrons repel each other more (via Coulomb repulsion) when residing in the localized d orbitals than when occupying the delocalized porphyrin orbitals (i.e. the HOMO), therefore the differences in the electron-electron Coulomb repulsion compensate for the inverted ordering of the d-orbital and HOMO orbital energies.

2.3 The Hartree–Fock Problem for Open-Shell Systems

The traditional Hartree–Fock (HF) equations apply to closed-shell molecules, and, in many transition metal complexes, the HF method provides a reasonable approximation to the ground state wavefunction. However, complexes with more than one transition metal atom, and, indeed, transition metal clusters, often present a huge challenge to the computational chemist because of the large number of low-lying, nearly degenerate spin states that confound the SCF convergence procedures. Several computational methods generalize the HF equations for open shells. These include the Unrestricted Hartree–Fock (UHF) method,[65] the Restricted Open-Shell Hartree–Fock (ROHF) method[68–71] the Hyper-Hartree–Fock method,[72] and the recently developed Configuration Averaged Hartree–Fock (CAHF) method.[78,80]

The UHF theory works well for high-spin systems, but this approach breaks the symmetry between the alpha and beta orbitals, thus complicating more extensive calculations (although perturbation theory and coupled cluster calculations with a UHF starting point work well for ground states, see below). The ROHF approach requires the user to specify *a priori* the occupancy of the open-shell orbitals and the electronic spin state, precluding its use as an exploratory tool. The CAHF method avoids the convergence problems and serves as an approximate starting point for more extensive configuration interaction (CI) calculations.

Let us examine each of these approaches in detail, starting again with the Fe(II)P example.

The $^1A_{1g}$ state of Fe(II)P, Figure 1, presents a typical molecular orbital diagram for a closed-shell system. The Restricted Hartree–Fock (RHF) wavefunction is given by

$$^1\Psi_0 = |\Phi_1\alpha\Phi_1\beta\ldots\phi_n\alpha\Phi_n\beta| \tag{7}$$

Applying the variational principle to

$$\mathbf{H}\Psi_\kappa = E_\kappa\Psi_\kappa \tag{8}$$

yields the molecular orbital Fock equations for MO Φ_i and orbital energy ε_i, with the matrix elements for **f** given by

$$\langle \chi_\mu | \mathbf{f} | \chi_\nu \rangle = \langle \mu | \mathbf{f} | \nu \rangle = h_{\mu,\nu} + \sum_{\sigma,\lambda} P_{\sigma,\lambda} \big[\langle \mu, \sigma | \nu, \lambda \rangle - \langle \mu, \nu | \sigma, \lambda \rangle / 2 \big] \tag{9}$$

In Eqn. 9, **h** is the one-electron operator of Eqn. 2b,

$$h_{\mu,\nu} = \langle \mu | \mathbf{h} | \nu \rangle = \langle \mu | - \tfrac{1}{2} \sum_i \nabla_i^2 - \sum_{(A,i)} Z_A / R_{A,i} + V | \nu \rangle \tag{10}$$

and V is an effective core potential, if some of the inner-shell orbitals have been neglected. The matrix **P**, the *first order density matrix*, one-matrix, or Fock–Dirac density matrix is given by

$$P_{\sigma,\lambda} = \sum_a n_a C_{\lambda,a} C_{\sigma,a} \tag{11}$$

with n_a the occupation of MO Φ_a, 2 or 0 in the closed-shell case. The first term in the sum of Eqn. 9 is the Coulomb term, the second, the exchange term. The two-electron integrals are given by

$$\begin{aligned} \langle \mu, \sigma | \nu, \lambda \rangle &= \int \delta\tau(1) \int \delta\tau(2) \chi_\mu(1) \chi_\nu(1)\, {}^{1/r}1,2 \chi_\sigma(2) \chi_\lambda(2) \\ &= (\mu, \nu | \sigma, \lambda) \end{aligned} \tag{12}$$

The last equality gives the relationship between the many body representation of the two-electron integral and the chemist's or Mulliken, $(\mu, \nu | \sigma, \lambda)$, notation.

This prescription yields an eigenfunction of the spin operator $\mathbf{S}^2$, with eigenvalue $s(s+1)$.

$$\begin{aligned} \mathbf{S}^2 \Psi_\kappa &= s(s+1)\Psi_\kappa = 0(1)\Psi_\kappa, \\ {}^{2s+1}\Psi_\kappa &= {}^1\Psi_\kappa. \end{aligned} \tag{13}$$

2.3.1 ***Unrestricted Hartree–Fock (ROHF)*** Figure 2 pictures a high-spin Unrestricted Hartree–Fock (UHF) wavefunction,[65] which, in this case, is given by

$$\begin{aligned} {}^{\text{uhf}}\Psi_0 = &\, |\Phi_1^\alpha \alpha(1) \Phi_1^\beta \beta(2) \ldots \Phi_p^\alpha \alpha(2p-1) \Phi_p^\beta \beta(2p) \\ &\times \Phi_{p+1}^\alpha \alpha(2p+1) \Phi_{p+2}^\alpha \alpha(2p+2) \Phi_{p+3}^\alpha (2p+3)| \end{aligned} \tag{14}$$

The electron labels above indicate the $2p+3$ α spin electrons and $2p$ β spin electrons. As mentioned previously, we can neglect the electron labels if we maintain lexiconical order and do not forget about the three additional electrons (traditionally and arbitrarily assigned α spin).

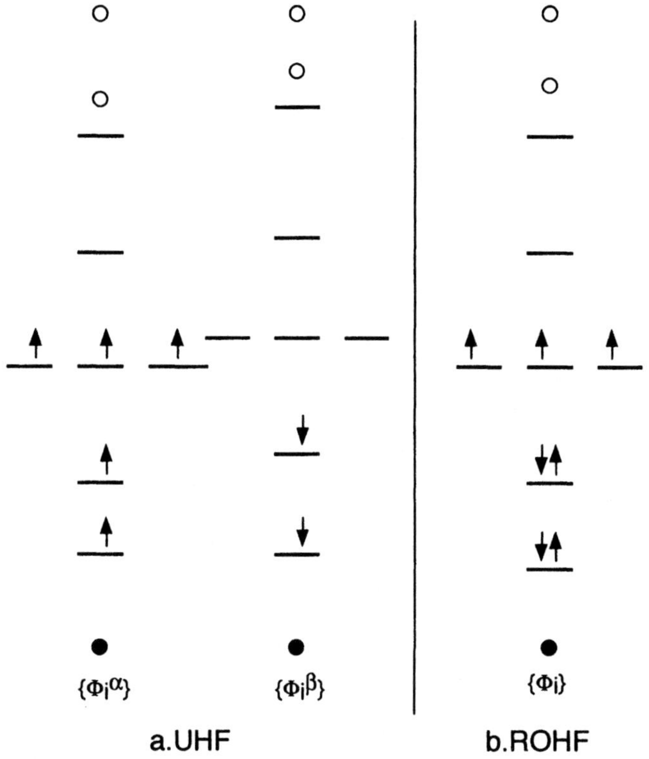

Figure 2. A schematic of (*a*) an open-shell UHF quartet, and (*b*) an open-shell ROHF quartet.

The UHF theory defines different spatial orbitals for the alpha spin $\{\Phi_i^\alpha\}$ and for the beta spin, $\{\Phi_i^\beta\}$, Different Orbitals for Different Spins, DODS.[74] The alpha spin-orbitals are orthonormal, as are the beta spin-orbitals, but they are not *biorthogonal*,

$$\langle \Phi_i^\alpha | \Phi_j^\alpha \rangle = \delta_{ij}, \quad \langle \Phi_i^\beta | \Phi_j^\beta \rangle = \delta_{ij}, \quad \text{but} \quad \langle \Phi_i^\alpha | \Phi_j^\beta \rangle \neq \delta_{ij}. \tag{15}$$

Invoking the variational principle for this wavefunction yields two Fock equations, one for the alpha spin orbitals and one for the beta,

$$\mathbf{f}^\alpha(j)\Phi_j^\alpha(j) = \varepsilon_j^\alpha \Phi_j^\alpha(j). \tag{16a}$$

$$\mathbf{f}^\beta(i)\Phi_j^\beta(i) = \varepsilon_j^\beta \Phi_j^\beta(i). \tag{16b}$$

The Fock operators differ from the closed-shell case as they contain different exchange operators: that is, only electrons of like spin can exchange:

$$\langle \mu | \mathbf{f}^\alpha | \nu \rangle = h_{\mu,\nu} + \sum_{\sigma,\lambda} \left[P_{\sigma,\lambda}^T \langle \mu, \sigma | \nu, \lambda \rangle - P_{\sigma,\lambda}^\alpha \langle \mu, \nu | \sigma, \lambda \rangle \right] \tag{17}$$

$$P^{\alpha}_{\sigma,\lambda} = \sum_{a} n^{\alpha}_{a} C^{\alpha}_{\lambda,a} C^{\alpha}_{\sigma,a} \tag{18}$$

with similar expressions for $\mathbf{P}^{\beta}$. In Eqn. 18, n^{α} (and n^{β} in the analogous equation for $\mathbf{P}^{\beta}$), the occupation numbers, are 1 or 0. The total density is

$$\mathbf{P}^{T} = \mathbf{P}^{\alpha} + \mathbf{P}^{\beta}, \tag{19}$$

and the spin density is defined as

$$\rho = \mathbf{P}^{\alpha} - \mathbf{P}^{\beta}. \tag{20}$$

The UHF wave function *breaks* spin symmetry; that is, the UHF wave function is not an eigenfunction of the $\mathbf{S}^2$ operator

$$\mathbf{S}^{2\,\mathrm{uhf}}\Psi_{\kappa} \neq s(s+1)\,{}^{\mathrm{uhf}}\Psi_{\kappa}, \tag{21}$$

but it can be expanded in terms of spin eigenfunctions that span the same space,

$${}^{\mathrm{uhf}}\Psi_{Sz} = \sum_{S=Sz}^{N/2} {}^{2S+1}\Psi_{Sz} C_S, \tag{22}$$

where the C_S are the expansion coefficients, and the $\{{}^{2S+1}\Psi_{Sz}\}$ functions represent spin eigenfunctions. The functions $\{{}^{2S+1}\Psi_{Sz}\}$ should not be viewed as simple RHF functions obtained from the RHF procedure, or the Restricted Open-Shell Hartree–Fock (ROHF) procedure described below. Rather, they represent linear combinations of the functions obtained through the UHF variational method, and, as such, are akin to the CI functions discussed in the following section. The mixing of other spinmultiplicities into the UHF wavefunction that does not relate to the $\mathbf{S}^2 = Sz(Sz+1)$ value is referred to as *spin contamination*.

A simple annihilation scheme might be invoked to lead to better spin properties, one in which the next highest component of spin is annihilated[66,67,77]

$$\begin{aligned} A_s &= \left[\mathbf{S}^2 - (S+1)(S+2)\right] \\ A_s\,{}^{\mathrm{uhf}}\Psi_{Sz} &= \left[-2(Sz+1)\right]{}^{2Sz+1}\Psi_{Sz}C_{Sz} + [0]^{2Sz+3}\Psi_{Sz}C_{Sz+1} + \\ &\quad + \left[(sz+1)(sz+2)\right]{}^{2Sz+5}\Psi_{Sz}\,C_{Sz+2} + \ldots \end{aligned} \tag{23}$$

Another somewhat more complex approach, the PUHF approach, simply *projects* the desired spin state from the UHF wavefunction[78,79]

$$\theta_S\,{}^{\mathrm{uhf}}\Psi_{Sz} = {}^{2S+1}\Psi_{Sz}. \tag{24}$$

Inevitably the UHF method leads to lower energies than does the RHF or ROHF method, Figure 3, and projecting or annihilating components of higher multiplicities

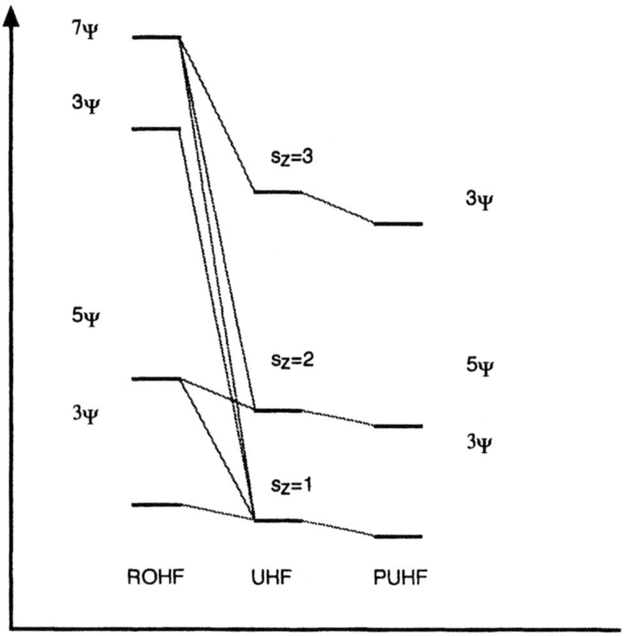

Figure 3. The energy of the UHF wavefunction compared to ROHF and projected UHF.

general yields even lower energies. The UHF method is variational, and it is a Fock-type method, but it does contain some of the correlation energy, by definition. That is, the correlation energy is *defined* as the difference between the exact non-relativistic Born Oppenheimer Energy, and that obtained by the RHF or ROHF procedure.[25,26]

We examine the case of the 2-Fe model ferrodoxin illustrated in the figure at the top of Table 1.[80] The $S_z = 5$ UHF calculation, corresponding to all ten electrons in the ten d orbitals all of α spin, or *ferromagnetically* (FM) coupled (all open-shell orbitals on all atoms have the same spin), is nearly pure undectet ($s = 5$, $2s + 1 = 11$), as can be seen from the weight of the pure undectet in the UHF wavefunction of 0.9993, or from the expectation value $\langle \mathbf{S}^2 \rangle = 30.008$ (were this a pure undectet, $2s + 1 = 11$, then this value would be $s(s + 1) = 30$.) The lowest energy for this system that we calculate using the INDO model (see below) is, however, for a singlet. This is obtained from projecting the UHF $S_z = 0$ anti-ferromagnetic (AF) case, with all up-spin electrons on one Fe(II) ion, and all down-spin on the other. Note that this UHF wavefunction is mostly triplet. The AF case cannot be considered a pure spin state, nor does it even approximate one. The expectation value $\mathbf{S}^2$ of a pure AF coupling would be 5, close to the 4.72 value actually calculated for the UHF wavefunction. In general, the UHF energy is most reliable for those states of highest multiplicity for a given number of electrons in a set number of open-shell orbitals. Note that the calculated $\langle \mathbf{S}^2 \rangle$ value deviates more from the $S_z(S_z + 1)$ value the smaller S_z becomes. As shown in Table 2, the prediction is that the ground state of this model is diamagnetic, or singlet. Note further that the observation that the UHF

Table 1 Spin Components ("Weights") for 2-Fe Model Ferredoxin Obtained from a UHF Calculation

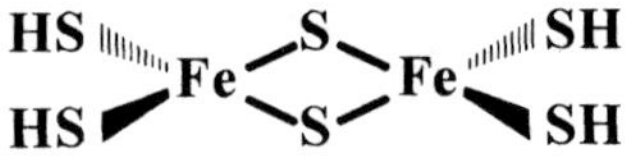

$2S+1/Sz =$	5	4	3	2	1	0
1	—	—	—	—	—	*0.18*
3	—	—	—	—	*0.49*	0.37
5	—	—	—	*0.81*	0.36	0.29
7	—	—	*0.88*	0.18	0.13	0.13
9	—	*0.93*	0.11	0.01	0.02	0.03
11	*0.9993*	0.07	0.00	0.00	0.00	0.00
13	0.0007	0.00	0.00	0.00	0.00	0.00
$\langle S^2 \rangle$	30.008	20.66	12.95	7.26	5.11	4.72
$S_z(S_z+1)$	30	20	12	6	2	0
$\langle S^2 \rangle$AF =						5
AF = 0.160 (singlet) +0.357 (triplet) +0.298 (quintet) +						
relative energy						
SCF (kJ/mol)	0.00	109	140	230	282	−35.2

energy for the highest multiplicity is generally a good estimate (the projection only lowering this energy by 30 cm^{-1}) does not necessarily imply that the wavefunction is *good*; the $\langle \mathbf{S}^2 \rangle$ value does suggest this. There are two additional observations from this table. The first is that the tridectet is of much higher energy, as it requires uncoupling of spins on the sulfur ligands. Secondly, only the AF (minimum multiplicity) and FM (maximum multiplicity) seem consistent in their prediction. This is because all other multiplicities involve the loss of one-center exchange, and this raises the energy considerably. The prediction that the FM case is higher in energy than the AF case by 4,840 cm^{-1} is verified by the FM calculation which produces a value 3,540 cm^{-1}. Finally note that, in spite of all the spin contamination in the AF case, the spin projection only lowers the projected singlet by 630 cm^{-1}. This is due to the fact that the higher multiplicities that mix into this wavefunction are the weakly coupled ones, as described below, and delineated in Table 2.

2.3.2 ***Restricted Open-Shell Hartree–Fock (ROHF)*** For first row elements, open-shell calculations generally begin with the very reliable Restricted Open-Shell Hartree–Fock (ROHF) procedures,[69,70] which construct a single configuration, open shell, SCF wavefunction that is an eigenfunction of $\mathbf{S}^2$. Consider the quartet of Figure 2, ${}^4\Psi_0$. The ROHF wavefunction is given by

$$\begin{aligned} {}^4\Psi_0 = \big| & \Phi_1\alpha(1)\Phi_1\beta(2)\ldots\Phi_p\alpha(2p-1)\Phi_p\beta(2p) \\ & \times \Phi_{p+1}\alpha(2p+1)\Phi_{p+2}\alpha(2p+2)\Phi_{p+3}\alpha(2p+3)\big| \end{aligned} \tag{25}$$

Table 2 Relative Energies in 1000 cm^{-1} for 2-Fe Model Ferredoxin: 1000 cm^{-1} = 2.86 kcal/mol

$2S+1$	11	9	7	5	3	1
1	—	—	—	—	—	**0.00**
3	—	—	—	-	26.1	0.25
5	—	—	—	22.5	27.6	0.76
7	—	—	13.5	22.9	30.5	1.60
9	—	5.76	27.7	25.2	36.1	2.86
11	*3.54*	24.6	45.6	33.2	79.6	*4.84*
13	56.7	82.1	105.0	97.7	126.2	64.8
SCF	3.57	6.48	15.3	22.8	27.2	0.63
$\langle S^2 \rangle$	30.008	20.66	12.95	7.26	5.11	4.72
$S_z(S_z+1)$	30	20	12	6	2	0

Table 3 Numerical Hartree–Fock Results for Some Transition Metal Atoms, from Davidson, Reference 73, in 1000 cm^{-1}. The Numbers in Parenthesis Are Experimental Values

$d^n s^2 \rightarrow$	$d^{n+1}s$	d^{n+2}	$d^n sp$	n
Sc	8.1 (11.5)	36.1 (33.8)	7.7 (15.7)	1
Cr	−10.2 (−7.7)	46.4 (27.6)	7.5 (17.2)	2
Fe	14.5 (6.9)	60.2 (32.9)	8.8 (19.4)	6
Ni	10.2 (0.2)	44.1 (14.7)	15.5 (25.8)	8

and the common case of the two-orbital two-electron open-shell singlet with $2n$ electrons as

$$ {}^1\Psi_0 = \left\{ |\Phi_1\alpha\Phi_1\beta\ldots\Phi_n\alpha\Phi_{n+1}\beta| + |\Phi_1\alpha\Phi_1\beta\ldots\Phi_{n+1}\alpha\Phi_n\beta| \right\}/\sqrt{2} \tag{26} $$

These functions are eigenfunctions of the operator $\mathbf{S}^2$ by construction. Although the resulting procedure is somewhat more complex, it can be used directly to calculate the states of any system in which the total energy can be expressed as

$$ \begin{aligned} E = {} & 2\sum h_i + \sum (2J_{i,j} - K_{i,j}) && i \text{ closed}, \ j \text{ closed} \\ & + \sum n_m h_m + \sum n_m (2J_{i,m} - K_{i,m}) && i \text{ closed}, \ m \text{ open} \\ & + \sum n_m n_0 (2a_{m,o} J_{m,o} - b_{m,o} K_{m,o})/4 && m \text{ open}, \ o \text{ open} \end{aligned} \tag{27} $$

In the above, h_i is the one electron energy of orbital Φ_i,

$$ h_i = \langle \Phi_i | h | \Phi_j \rangle \tag{28a} $$

$J_{i,j}$ is the Coulomb integral between MO's (contrast with Eqn. 9 and 12)

$$J_{i,j} = \langle \phi_i(1)\phi_j(2)|1/r_{1,2}|\phi_i(1)\phi_j(2)\rangle \tag{28b}$$

and $K_{i,j}$ is the molecular exchange integral,

$$K_{i,j} = \langle \phi_i(1)\phi_j(2)|1/r_{1,2}|\phi_j(1)\phi_i(2)\rangle, \tag{28c}$$

The a and b coefficients of Eqn. 27 are called the *vector-coupling coefficients*, that are specific to the situation being treated. There are tables of vector coupling coefficients in reference 80. For the open-shell quartet depicted in Figure 2, the energy is

$$\begin{aligned}
E = {} & 2h_1 + 2h_2 + \cdots + 2h_p + 2J_{1,1} + 2J_{1,2} + \cdots \\
& + 2J_{p,p} - K_{1,1} - K_{1,2} - \cdots - K_{p,p} && \text{closed-closed} \\
& + h_{p+1} + h_{p+2} + h_{p+3} + 2J_{1,p+1} + 2J_{2,p+1} + \cdots \\
& + 2J_{p,p+3} - K_{1,p+1} - \cdots - K_{p,p+3} && \text{closed-open} \\
& + J_{p+1,p+2} + J_{p+1,p+3} + J_{p+2,p+3} - K_{p+1,p+2} \\
& - K_{p+1,p+3} - K_{p+2,p+3} && \text{open-open}
\end{aligned}$$

This can be obtained directly from the figure, or from the wavefunction given in Eqn. 25. (To read this from the figure note that each electron in spatial orbital ϕ_i contributes an h_i and there is a Coulomb repulsion contribution J for each electron-electron repulsion, but only electrons of like spin can lead to an exchange term K, and K has negative sign.) Comparison with the general equation for E (Eqn. 27) yields $n_{p+1} = n_{p+2} = n_{p+3} = 1$; $a_{p+1,p+2} = a_{p+1,p+3} = a_{p+2,p+3} = 1$ and $b_{p+1,p+2} = b_{p+1,p+3} = b_{p+2,p+3} = 1$. The vector coupling coefficients are written as $\mathbf{a} = (1, 1, 1, 1, 1, 1)$ and $\mathbf{b} = (2, 2, 2, 2, 2, 2)$. (That is, as a *vector*) as these matrices are always of a symmetric form, and only the lower triangle need be given, e.g.,

$$\mathbf{a} = \begin{vmatrix} 1, & 1, & 1 \\ 1, & 1, & 1 \\ 1, & 1, & 1 \end{vmatrix} \qquad \mathbf{b} = \begin{vmatrix} 2, & 2, & 2 \\ 2, & 2, & 2 \\ 2, & 2, & 2 \end{vmatrix} \tag{29}$$

These matrices are of the form (see Eqn. 27)

$$\begin{array}{lcccccl}
 & \text{open MO 1} & \text{open MO 2} & \text{open MO 3} & \ldots & & \\
\mathbf{a} = & \left| \begin{array}{c} a_{1,1} \\ a_{2,1} \\ a_{3,1} \\ \ldots \end{array} \right. & \begin{array}{c} a_{1,2} \\ a_{2,2} \\ a_{3,2} \\ \ldots \end{array} & \begin{array}{c} a_{1,3} \\ a_{2,3} \\ a_{3,3} \\ \ldots \end{array} & \left. \begin{array}{c} \ldots \\ \ldots \\ \ldots \\ \ldots \end{array} \right| & \begin{array}{l} \text{open MO 1} \\ \text{open MO 2} \\ \text{open MO 3} \\ \end{array}
\end{array}$$

and if they *both* can be *simultaneously* arranged such that both **a** and **b** have equal elements arranged in blocks, such as

$$\mathbf{a} = \left|\begin{array}{cc:cc} a_{1,1} & a_{1,1} & a_{1,3} & \dots \\ a_{1,1} & a_{1,1} & a_{1,3} & \dots \\ \hdashline a_{3,1} & a_{3,1} & a_{3,3} & \dots \\ \dots & \dots & \dots & \dots \end{array}\right| \begin{array}{l} \text{open MO 1} \\ \text{open MO 2} \\ \text{open MO 3} \\ \end{array}$$

then the MO's in each block can be arranged in shells that make the calculation far simpler.

$$\begin{array}{c} \phantom{\mathbf{a} =} \begin{array}{cc} \text{shell 1} & \text{shell 2} \end{array} \\ \mathbf{a} = \left|\begin{array}{ccc} \mathbf{a}_{1,1} & \mathbf{a}_{1,2} & \dots \\ \mathbf{a}_{1,2} & \mathbf{a}_{2,2} & \dots \end{array}\right| \begin{array}{l} \text{shell 1} \\ \text{shell 2} \end{array} \end{array}$$

In the case of Eqn. 29, there is only one open-shell operator, $\mathbf{a} = 1$ and $\mathbf{b} = 2$. This is always the situation for the case of the highest multiplicity for a given open-shell structure, i.e. $\mathbf{a} = 1$ and $\mathbf{b} = 2$. For the common case of the open-shell singlet, Eqn. 26, the energy is:

$$\begin{aligned} E = {} & 2h_1 + 2h_2 + \cdots + 2h_p + 2J_{1,1} + 2J_{1,2} + \cdots \\ & + 2J_{p,p} - K_{1,1} - K_{1,2} - \cdots - K_{p,p} && \text{closed-closed} \\ & + h_{p+1} + h_{p+2} + 2J_{1,p+1} + 2J_{2,p+1} + \cdots \\ & + 2J_{p,p+2} - K_{1,p+1} - \cdots - K_{p,p+2} && \text{closed-open} \\ & + J_{p+1,p+2} + K_{p+1,p+2} && \text{open-open} \end{aligned}$$

The exchange term $+K_{p+1,p+2}$ comes from the cross-term in the wavefunction, and should serve as a caution that only when a single determinant describes the state can the energy of the spin-state be read from the MO diagram. Comparison with Eqn. 27 leads to $2a_{p,p} - b_{p,p} = 0$ (there are no $J_{p,p}$, and $J_{p,p} = K_{p,p}$), $2a_{p+1,p+1} - b_{p+1,p+1} = 0$, $a_{p,p+1} = 1$, and $b_{p,p+1} = -2$. Note that no choice of $a_{p,p}$ can be made that will lead to a single open-shell operator. Setting $a_{p,p} = 1$ leads to $b_{p,p} = 2$, and

$$\mathbf{a} = \left|\begin{array}{cc} 1, & 1 \\ 1, & 1 \end{array}\right| \qquad \mathbf{b} = \left|\begin{array}{cc} 2, & -2 \\ -2, & 2 \end{array}\right| \tag{30}$$

or $\mathbf{a} = (1, 1, 1)$ and $\mathbf{b} = (2, -2, 2)$. The choice of the diagonal elements are arbitrary in this case as long as $2a_{p,p} - b_{p,p} = 0$ and $2a_{p+1,p+1} - b_{p+1,p+1} = 0$. For the corresponding triplet function change the sign in the wavefunction of Eqn. 26 then the cross term gives $-K_{p+1,p}$, and $b_{p,p+1} = +2$. Now the choice of $a_{p,p}$ and $a_{p+1,p+1}$ does matter. Choosing $a_{p,p} = 1$ and $a_{p+1,p+1} = 1$, leads to $\mathbf{a} = (1, 1, 1)$ and $\mathbf{b} = (2, 2, 2)$, and a single operator will represent the open-shell, $\mathbf{a} = (1)$ and

b = (2), in agreement with the general case of the greatest multiplicity for the given open-shell, Eqn. 29.

In general, the UHF energy is lower than is the ROHF energy, as it allows for two sets of MO's, and a greater degree of variational freedom. The projected UHF energies are usually lower still, as the contaminating spin multiplicities generally lie higher in energy. Therefore, eliminating these higher energy components and renormalizing the wave function lowers the energy *for the highest multiplicity cases*. This situation is summarized in Figures 3 and 4.

The ROHF method routinely fails for large transition metal complexes with many low lying nearly degenerate electronic states. Consider the two weakly coupled d^5 systems of Tables 1 and 2. There are available for 10 electrons in 20 spin orbitals (10 spatial d orbitals) some

$$20!/10!10! = 184{,}756$$

states, NOT including possibilities that involve possible low-lying d^4s configurations. Even so, some of these 184,756 states are very highly excited—those that might involve charge transfer, for example. Ignoring charge transfer, there are still

$$(10!/5!5!)^2 = 63{,}504 \text{ states}$$

$$^{uhf}\Psi_{SZ} = \sum^{N/2} {}^{2s+1}\Psi_{SZ}\, C_S,$$

$$3\text{-}^{uhf}\Psi_1 = {}^3\Psi_1\, C_1 + {}^5\Psi_1\, C_5 + {}^7\Psi_1 C_7 +$$

not a simple ROHF function

$$\underline{\qquad}\quad {}^7\psi$$

$$\underline{\qquad}\quad {}^5\psi$$

$$\underline{\qquad}\quad {}^3\Psi(\text{ROHF})$$

$$\underline{\qquad}\quad 3\text{-}^{uhf}\psi$$

$$\underline{\qquad}\quad {}^3\Psi(\text{PUHF}) = {}^3\Psi$$

$$E(\text{PUHF}) \leq E(\text{UHF}) \leq E(\text{ROHF})$$

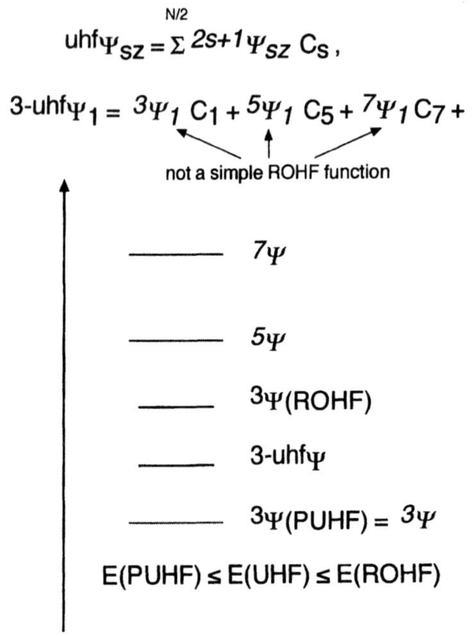

Figure 4. The relationship between UHF, RHF, and Projected-UHF for the case of three open-shell orbitals.

available to the system. Proceeding further, if we further limit the possibilities to those in which each *d* orbital has but one electron, this yields,

$$2^{10} = 1024 \text{ states.}$$

And finally we might consider only the weak (Heisenberg-type) coupling case of two transition metal ions each with spin 5/2 (each a sextet). This yields states of multiplicities 11 through 1, a total of 36 spin-states. No matter how we arrange the possibilities, we can not immediately identify a single ground state wavefunction as we can for compounds of the main group elements! And the final answer depends on the details of the system being considered.

Very often, especially if we know in advance that the calculations must include extensive electron correlation, all that is really required are orbitals that are *good enough* to start the correlated treatment. Several schemes provide such orbitals simply by averaging over the possible spin-states including the hyper Hartree–

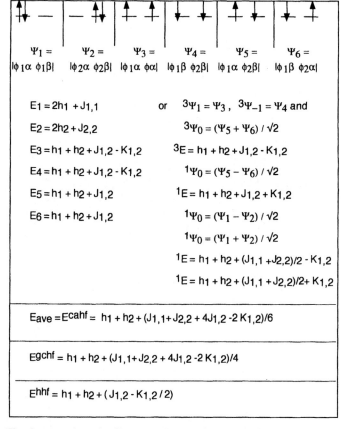

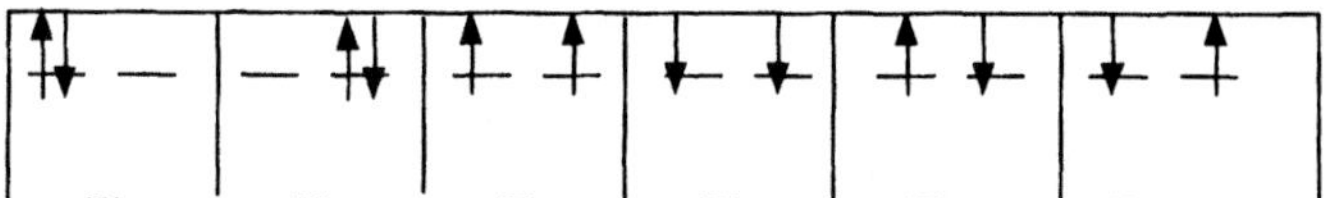

$\Psi_1 = |\phi_1\alpha\ \phi_1\beta|$ $\Psi_2 = |\phi_2\alpha\ \phi_2\beta|$ $\Psi_3 = |\phi_1\alpha\ \phi\alpha|$ $\Psi_4 = |\phi_1\beta\ \phi_2\beta|$ $\Psi_5 = |\phi_1\alpha\ \phi_2\beta|$ $\Psi_6 = |\phi_1\beta\ \phi_2\alpha|$

$E_1 = 2h_1 + J_{1,1}$	or	${}^3\Psi_1 = \Psi_3,\ {}^3\Psi_{-1} = \Psi_4$ and
$E_2 = 2h_2 + J_{2,2}$		${}^3\Psi_0 = (\Psi_5 + \Psi_6)/\sqrt{2}$
$E_3 = h_1 + h_2 + J_{1,2} - K_{1,2}$		${}^3E = h_1 + h_2 + J_{1,2} - K_{1,2}$
$E_4 = h_1 + h_2 + J_{1,2} - K_{1,2}$		${}^1\Psi_0 = (\Psi_5 - \Psi_6)/\sqrt{2}$
$E_5 = h_1 + h_2 + J_{1,2}$		${}^1E = h_1 + h_2 + J_{1,2} + K_{1,2}$
$E_6 = h_1 + h_2 + J_{1,2}$		${}^1\Psi_0 = (\Psi_1 - \Psi_2)/\sqrt{2}$
		${}^1\Psi_0 = (\Psi_1 + \Psi_2)/\sqrt{2}$
		${}^1E = h_1 + h_2 + (J_{1,1} + J_{2,2})/2 - K_{1,2}$
		${}^1E = h_1 + h_2 + (J_{1,1} + J_{2,2})/2 + K_{1,2}$

$$E_{ave} = E^{cahf} = h_1 + h_2 + (J_{1,1} + J_{2,2} + 4J_{1,2} - 2K_{1,2})/6$$

$$E^{gchf} = h_1 + h_2 + (J_{1,1} + J_{2,2} + 4J_{1,2} - 2K_{1,2})/4$$

$$E^{hhf} = h_1 + h_2 + (J_{1,2} - K_{1,2}/2)$$

Figure 5. The six states that arise from two electrons in two orbitals, and the energies obtained from the various averaging schemes.

Fock method,[72] the grand-canonical Hartree–Fock method,[73] and the configuration-averaged Hartree–Fock scheme[78,80] briefly discussed below. Figure 5 demonstrates these averaging schemes for two electrons in two orbitals. We develop the CAHF scheme below.

2.3.3 ***Configuration Averaged Hartree–Fock (CAHF)*** Consider N electrons distributed in $2m$ spin-orbitals (m spatial orbitals) in all possible ways. The number of possible ways N is given by

$$N = 2m!/\big[N!(N-2m)!\big]. \tag{31}$$

We first consider how many times $J_{i,i}$ occurs in these N determinants, $\eta(J_{i,i})$. Fixing one electron in $\phi_i\alpha$ and one in $\phi_i\beta$ leaves $N-2$ electrons in the remaining $2m-2$ spin orbitals, then

$$\begin{array}{ccccccccc} \underline{x} & \underline{x} & \underline{x} & \underline{x} & \_ & \underline{x} & \underline{x} & \_ & \ldots \eta(J_{i,i}) = \frac{(2m-2)!}{[(2m-N)!(N-2)!]} \\ \phi_i\alpha & \phi_i\beta & \phi_j\alpha & \phi_j\beta & \phi_k\alpha & \phi_k\beta & \phi_l\alpha & \phi_l\beta & \ldots \end{array} \tag{32}$$

The number of $J_{i,j}$ that can occur is then four times this number, for the electrons in ϕ_i and ϕ_j can be α, α or α, β or β, α or β, β; i.e.,

$$\eta(J_{ij}) = 4(2m-2)!/\big[(2m-N)!(N-2)\big]. \tag{33}$$

There are, or course, no exchange integrals between two electrons both in ϕ_i and only half as many exchanges as Coulomb interactions in the pair ϕ_i, ϕ_j:

$$\eta(K_{ij}) = 2(2m-2)!/\big[(2m-N)!(N-2)!\big] \tag{34}$$

From this information, the average number of Coulomb and exchange integrals considering N electrons in $2m$ spin-orbitals can be evaluated by dividing the number of integrals $\eta(K_{ij})$ and $\eta(J_{ij})$ by the number of possible states, N of Eqn. 31. Then from the occupancy of all the active orbitals $n = N/m$, and comparing these values with Eqn. 27 yields[61,63]

$$\begin{aligned} N(N-1)/\big[2m(2m-1)\big]J_{i,i} &= N^2/m^2(2a_{i,i}-b_{i,i})J_{i,i}/4 \\ 2m(N-1)/\big[N(2m-1)\big] &= 2a_{i,i}-b_{i,i} \\ 2m(N-1)/\big[N(2m-1)\big] &= a_{i,j} \\ -2m(N-1)/\big[N(2m-1)\big] &= -b_{i,j} \end{aligned}$$

and, thus,

$$\mathbf{a} = \mathbf{b} = 2m(N-1)/\big[N(2m-1)\big] \tag{35}$$

and a single open-shell ROHF operator can be used to calculate the average energy of all states generated by considering all possible arrangements of N electrons in $2m$

spin orbitals. Should there be two shells that need to be considered averaged, one of average occupancy N_i/m_i and N_j/m_j, then a and b are given by

$$\begin{aligned} a_{1,1} &= b_{1,1} = 2m_1(N_1 - 1)/[N_1(2m_1 - 1)] \quad \text{and} \\ a_{2,2} &= b_{2,2} = 2m_2(N_2 - 1)/[N_2(2m_2 - 1)] \quad \text{and} \\ a_{1,2} &= b_{1,2} = 1 \end{aligned}$$

As an example, 6 states arise when considering two electrons in two orbitals of Figure 5: three singlets, $|\Phi_1\alpha\Phi_1\beta|$, $|\Phi_2\alpha\Phi_2\beta|$ and $(|\Phi_1\alpha\Phi_2\beta| + |\Phi_2\alpha\Phi_1\beta|)/\sqrt{2}$, and the three components of the triplet $|\Phi_1\alpha\Phi_2\alpha|$, $|\Phi_1\beta\Phi_2\beta|$, $(|\Phi_1\alpha\Phi_2\beta| - |\Phi_2\alpha\Phi_1\beta|)/\sqrt{2}$. This is the type of average considered in the CAHF procedure above[78] with energy

$$E^{\text{cahf}} = h_1 + h_2 + (J_{1,1} + J_{2,2} + 4J_{1,2} - 2K_{1,2})/6 \tag{36a}$$

and

$$\mathbf{a} = \mathbf{b} = 2/3, \quad n = 1.$$

The grand canonical Hartree–Fock scheme considers $N/2m$ electrons of α and β spin in each orbital. In the case above, $1/2$ electron of α spin and $1/2$ electron of β spin in ϕ_1 and ϕ_2, and the total energy can be written as

$$E^{\text{gchf}} = h_1 + h_2 + (J_{1,1} + J_{2,2} + 4J_{1,2} - 2K_{1,2})/4 \tag{36b}$$

It is clear that in this case there is also one open-shell operator with

$$\mathbf{a} = \mathbf{b} = 1, \quad n = 1.$$

Hyper Hartree–Fock differs significantly from configuration averaged Hartree–Fock and grand canonical Hartree–Fock. The starting point here is to write the average energy as a weighted sum of pair energies,

$$E^{\text{hhf}} = \sum n_i(i) + \tfrac{1}{2}\sum n_i(n_i - 1)(i,i) + \sum n_i n_j(i,j) \tag{37}$$

where (i) is the one-electron energy of an electron in orbital ϕ_i (this is the same as h_i), (i,i) is the average two-electron energy of a pair of electrons in a set of orbitals considered degenerate, and (i,j) the average two-electron energy of a pair of orbitals, one in ϕ_i and the other in ϕ_j. If we again consider two electrons in a set of two degenerate orbitals ϕ_1 and ϕ_2, then

$$\begin{aligned} (1,1) &= J_{1,1} \\ (2,2) &= J_{2,2} \\ (1,2) &= J_{1,2} - K_{1,2}/2 \end{aligned}$$

In this particular case, $n_1 = n_2 = 1$ (there are six determinants as before. One of these determinants has two electrons in ϕ_1, one has no electrons in ϕ_1, and the remaining 4 have one electron in this orbital. The average occupancy is 1, and

$$E^{\text{hhf}} = h_1 + h_2 + J_{1,2} - K_{1,2}/2. \tag{38}$$

This case can also be handled by the open-shell theory described above (i.e. Eqn. 27) but not as a single open-shell operator. Comparing Eqn. 27 with Eqn. 38 yields $2a_{1,1} - b_{1,1} = 2a_{2,2} - b_{2,2} = 0$, $a_{1,2} = b_{1,2} = 1$. Although $a_{i,i}$ and $b_{i,i}$ are not determined uniquely from these conditions, a choice of vector coupling coefficients of

$$\mathbf{a} = (1, 1, 1), \qquad \mathbf{b} = (2, 1, 2)$$

is appropriate. Note that in all the above open-shell models, the one-electron part will always be the same. The Coulomb integrals $J_{i,j}$ will differ as will the exchange part.

Figure 6 shows schematically how the CAHF procedure might be used to advantage in a situation where little is known about the system at the onset, and neither RHF or UHF looks sensible, or converges. On physical grounds we can generally eliminate $2m'$ electrons in m' MO's that are sufficiently low-lying as to not be problematic in their doubly occupation (for example, doubly occupy all inner-shell orbitals, and perhaps leave only the d-like orbitals and electrons as open). The remaining N electrons are averaged using the CAHF procedure, panel a. of

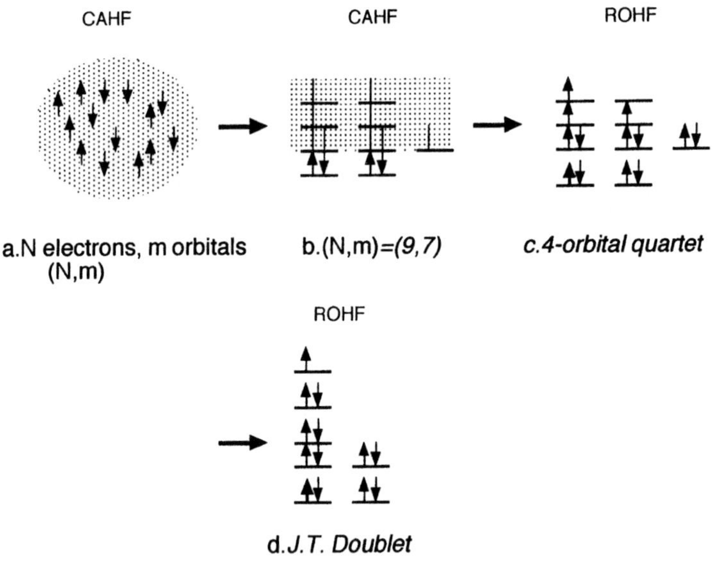

Figure 6. The use of the CAHF procedure to force SCF when little is known about the system in advance.

the figure. These MO's are saved, and used in another CAHF calculation with perhaps 9 electrons in 7 orbitals, and assignment suggested by using the results of the first calculation as a guide. If the results converge well, and the spatial symmetry of the system is maintained in an appropriate fashion, the resulting orbitals from

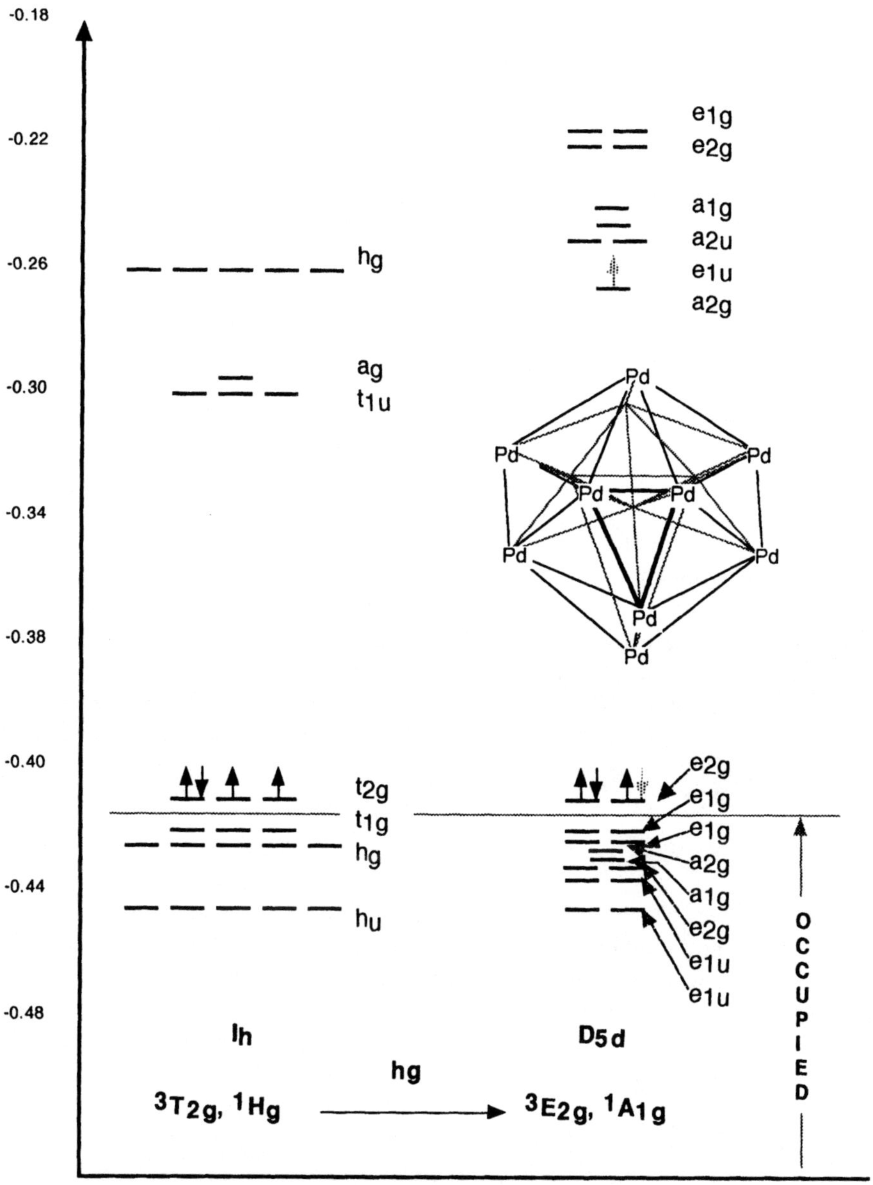

Figure 7. The molecular orbital energies (a.u.) of PD_{13}, in I_h and D_{5d} symmetries. $\uparrow$ indicates the assignment for ${}^3E_{2g}$, $\downarrow$ for ${}^1A_{1g}$.

panel b can be used for an ROHF calculation suggested from this result, panel c. In the figure, the results of ROHF calculation suggest a doublet should be examined, and this would then yield a Jahn–Teller distortion, as suggested in panel d. *Jahn–Teller distortions* arise in nonlinear molecules whenever the state of interests is spatially degenerate, a consequence of an unequal occupation of degenerate orbitals.

As a specific example of how to run sequential CAHF and ROHF calculations, consider the case of a Pd_{13} cluster with I_h symmetry,[81–84] given in Figure 7. The great symmetry of this case leads to orbitals of closely lying energy, and of high degeneracy. Because of this, the SCF procedure will not converge unless "coaxed". Examining the orbitals from an extended Hückel calculation, 130 electrons are averaged over 78 orbitals to yield a set of orbitals that maintain the icosahedral symmetry. A first SCF calculation produces orbitals resulting from averaging over 3.4×10^{30} states. These orbitals serve as the guess for a second SCF calculation that averages 130 electrons over 66 near lying orbitals, representing 8646 states. The final SCF calculation (see Figure 7) uses this second set of orbitals for a regular ROHF calculation of four electrons in three orbitals that represents a proper $^3T_{2g}$ state in I_h symmetry. This state is then allowed to distort, and it is found to distort to a low spin structure of D_{5d} symmetry. This energy is lower than any of those obtained through serendipity from the original I_h calculations which broke symmetry through the uneven occupation of degenerate orbitals, and that did not yield results of any reliability.

In Section 3, we will concentrate mostly on the calculation of the excited states of transition metal complexes and clusters. By far the most systematic way to do this is through a configuration interaction procedure. For this reason calculations of the ROHF type are the most useful, for it is relatively easy to perform post-Hartree–Fock CI on such a wavefunction. The orbitals that result from CAHF calculations also form the basis of a CI expansion, but these orbitals will not be as good as will be the ROHF solution unless a reasonable high *level* of correlation can be explored. By *level* we refer to a type of theory, as SCF, CI, etc.

2.4 The Basis Set Problem

2.4.1 ***Introduction*** In principle, the Hamiltonian **H** has an infinite number of solutions, including the bound molecular electronic states, the quasi-bound excited states, and the electronic continuum. Most quantum chemistry studies only examine the electronic ground state and perhaps a few low-lying excited states and properties, and therefore can employ successfully a very finite basis set, constructed usually from atomic-like orbital basis functions. A typical basis set includes numerous atomic orbitals on each atom, although some calculations employ effective potentials to reduce the number of core electrons and orbitals (see below).

Basis sets for atomic orbitals may be defined over Slater-type orbitals (STO's)[85]

$$\chi_A(\zeta, l, m, n) = (2\zeta/(2n)!)^{1/2} Y_l{}^m(\theta, \Phi) r_A^{n-1} e^{-\zeta r_A} \tag{39a}$$

or Gaussian type orbitals (GTO's)[85–90]

$$\chi_A(\alpha, l, m, n) = (2l-1)!!(2m-1)!!(2n-1)!!x_A^l y_A^m z_A^n e^{-\alpha r_A^2} \tag{39b}$$

ζ in Eqn. 39a, and α in 39b, the pre-exponential factors, are often referred to as the exponents. $Y_l{}^m(\theta, \Phi)$ are the spherical harmonics that give rise to the familiar s, p, d, and f orbitals of atomic theory. (The !! in the normalizing factor of Eqn. 39b denotes the odd number factorial, e.g., $7!! = 7 \times 5 \times 3$.) Although GTO's can also be expanded in terms of the spherical harmonics, it is more common to use Cartesian Gaussian, defined as in Eqn. 39b. The subscript A in Eqn. 39 designates the center (atom) from which the electronic coordinates are measured.

Gaussians have the wrong behavior (derivative) at the origin ($r_A = 0$), the so-called cusp condition, and they fall too rapidly with increasing r_A. Nevertheless, they are the atomic orbitals of choice as it is relatively easy to evaluate the three- and four-center integrals that arise in the molecular electronic structure problem. Most often, to improve the speed of calculation, *contractions* (fixed linear combinations) of *primitive* Gaussians are used.

$$\chi_A(l, m, n) = \sum a_i \chi_A(\alpha_i, l, m, n) \tag{39c}$$

in which the a_i are set to reproduce STO's, or through variational calculations on atoms, or molecules, as described below. In such a case, when specific basis sets are used, the original literature is cited, and the notation (8, 6, 3/8, 4/2) [4, 3, 2/3, 2/2] might be used to indicate that on the transition metal there are 8 primitive s orbitals (e.g., $\chi_A(\alpha_i, 0, 0, 0)$. $i = 1 - 8$, on all transition metal centers A), 6 primitive p orbitals, and 3 primitive d orbitals (8, 6, 3/ contracted to four s symmetry basis functions, three p symmetry basis functions and two d symmetry functions [4, 3, 2/, then the first row elements are represented by 8 primitive s functions and 4 primitive p functions, /8, 4/, contracted to three s and two p functions, /3, 2/, and finally, hydrogen atoms are represented by two primitive GTO's.

2.4.2 *Common Basis Sets*

Common choices today are:[85–90]

(i) MBS — minimum basis set. In this case one contracted function is chosen for each atomic orbital usually occupied in the ground state of the atom. As an example, one function represents the $1s$ atomic orbital of hydrogen, one $1s$, one $2s$, and three $2p$ functions are used for carbon, and one $1s$, one $2s$, three $2p$, one $3s$, three $3p$, five $3d$, one $4s$ and three $4p$ functions (18 in all) are used for the iron atom. Often six d functions are included, rather than five, a consequence of using Cartesian Gaussian functions (see below). A minimum basis set is not, in general, capable of good accuracy. It is, however, convenient to get the lay of the land, to probe initial geometries, and often to detect potential problems that will arise in more demanding calculations. Examples include Slater type orbitals, STO's, or STOnG in which n Gaussian functions are fixed (contracted) to reproduce the behavior of a Slater type orbital: i.e., $(4n, 3n, n/2n, n/n)[4, 3, 1/2, 1/1]$.

In calculations which employ 6 d functions, rather than 5, the extra function combines to generate an additional function of s symmetry. Care must be taken in interpreting the participation of this orbital in the bonding.

(ii) VMBS — valence minimum basis sets. In the above example this would include the $1s$ of the H atom, the $2s$ and the three $2p$ functions of the C atom, and the five $3d$, one $4s$ and three $4p$ (large core) or three $3p$, five $3d$, one $4s$ and three $4p$ (small core) for the Fe atom. The inner-shell orbitals are accounted for by effective core potentials, described below. Generally the large core basis is used for semi-empirical work, with the deficiencies of this small basis accounted for through empirical parameters. This ought not to be a choice of basis for *ab-initio* calculations.

(iii) DZ — double zeta. Two functions are used to represent each orbital generally occupied in the ground state of the atom. This allows for a given function to expand or contract depending on the bonding situation.

(iv) VDZ — valence double zeta. This generally implies MBS for the inner-shell and double zeta for the valence-shell. An example, 6-31G, suggests a contraction of 6 Gaussian functions to represent the $1s$ orbital, and a contraction of 3 Gaussians to represent one valence function and 1 to represent a second valence function, $[10, 4/4](3, 2/2)$.

(v) DZP — double zeta (plus) polarization. This is a double zeta basis set augmented, in general, with functions with angular momentum one greater than that found normally in the ground states of atoms. For example, one would include p functions on hydrogen, d functions on carbon, f functions on transition metals. 6-31G* "polarizes" second row elements, whereas 6-31G** polarizes hydrogen and main group elements. Polarization functions allow the electronic distribution around an atom or ion to polarize off the nucleus. This is a reasonably good basis set, capable of reliably predicting most molecular properties.

(vi) TZP — triple zeta (plus) polarization.

(vii) TZP+ — triple zeta plus polarization plus diffuse functions. Diffuse functions are needed to study negative ions and the spectroscopy of smaller molecules. For example, Rydberg states of molecules can only be represented if these diffuse functions are present.

(viii) ANO — Atomic natural orbitals. These are orbitals tuned to reproduce correlated calculations.[89,90] Most basis sets are tuned to yielding good results at the Hartree–Fock level, and it is assumed, most often correctly, that these orbitals are also good for correlated treatments.

(ix) cc — correlation consistent.[87,88] This is a systematic approach to attempt to reach basis set limits. For example:

cc-pvdz — correlation consistent polarized valence double-zeta
$[9, 5, 1/4, 1](3, 2, 1/2, 1)$

cc-pvtz — correlation consistent polarized triple-zeta
[10, 5, 2, 1/4, 2, 1](4, 3, 2, 1/3, 2, 1)
cc-pvqz — correlation consistent polarized quadruple-zeta
[11, 6, 3, 2, 1/6, 3, 2, 1](5, 4, 3, 2, 1/4, 3, 2, 1)

The assumption here is that as another uncontracted orbital is added to the variational problem, the polarization is also increased. Such systematically improved basis sets do not yet exist for transition metal atoms.

2.4.3 ***Effective Core Potentials*** One seldom considers the $1s$ electrons of the carbon atom when discussing the chemistry of benzene, or the $1s$, $2s$ and $2p$ electrons (The Ne core) of chlorine when considering the chemistry of chlorine compounds. How important are the $1s$, $2s$, $2p$, $3s$ and even $3p$ *electrons* (the electrons in these orbitals) in the chemistry of copper complexes?

If these orbitals, and the electrons they contain, could be ignored, the calculations would save a considerable amount of time. Formally, the Hartree–Fock problem increases in computational time as n^4, with n the basis set size. Reducing n therefore reduces the computational time substantially. For example, the simple $[CuCl_4]^{2-}$ complex contains 108 basis functions at the DZ level, but only 50 basis orbitals excluding the core. Although the smaller basis set is reduced only by a factor of 2, the computational time reduces by $(108/50)^4 \sim 22$, a factor of 22. And, of course, even greater savings result for complexes of the third transition series or complexes of the actinides.

It is not possible to naively *ignore* the inner-shell orbitals and their electrons. Doing so for first row elements, for example causes the $2s$ orbital to mimic the missing $1s$, which contains so much more energy than does the valence shell. A method which ignores the inner-shell orbitals and electrons must effectively project the valence electrons out of the core region[91–98] and simulate the repulsive forces of the core electrons to prevent this variational collapse of valence electrons into the core region.

The form of an effective core potential can be derived by applying standard matrix partitioning theory to the Fock operator, and then expanding the formal one-electron effective potential using perturbation theory.[99] Consider the MO's of an arbitrary molecule, and factor the basis into core and valence orbitals:

$$\phi_i = \sum \chi_\mu C_{\mu i} = \sum_{\text{(core)}} \chi_\alpha C_{\alpha i} + \sum_{\text{(valence)}} \chi_\mu C_{\mu i}$$

$$\phi = (\mathbf{X}_C \mathbf{X}_V)(\mathbf{C}_C \mathbf{C}_V)$$

This factorization permits the Fock operator to be expressed in block matrix form,

$$(\mathbf{F} - \varepsilon\mathbf{\Delta})\mathbf{C} = \begin{vmatrix} \mathbf{F}_{CC} - \varepsilon\Delta_{CC} & \mathbf{F}_{CV} - \varepsilon\Delta_{CV} \\ \mathbf{F}_{VC} - \varepsilon\mathbf{\Delta}_{VC} & \mathbf{F}_{VV} - \varepsilon\mathbf{\Delta}_{VV} \end{vmatrix} \begin{vmatrix} \mathbf{C}_C \\ \mathbf{C}_V \end{vmatrix} = 0 \qquad \text{(40a)}$$

which then leads to two matrix equations,

$$(\mathbf{F}_{CC} - \varepsilon\Delta_{CC})\mathbf{C}_C + (\mathbf{F}_{CV} - \varepsilon\Delta_{CV})\mathbf{C}_V = \mathbf{0} \tag{40b}$$

and

$$(\mathbf{F}_{VC} - \varepsilon\Delta_{VC})\mathbf{C}_C + (\mathbf{F}_{VV} - \varepsilon\Delta_{VV})\mathbf{C}_V = \mathbf{0} \tag{40c}$$

Eqn. 40b is then solved for $\mathbf{C}_C$, the coefficients of the core orbitals, e.g.,

$$\mathbf{C}_c = -(\mathbf{F}_{CC} - \varepsilon\Delta_{CC})^{-1}(\mathbf{F}_{CV} - \varepsilon\Delta_{CV})\mathbf{C}_V \tag{40d}$$

and $\mathbf{C}_C$ is eliminated from Eqn. 40c,

$$(\mathbf{F}_{VV} + \mathbf{V}_{VV} - \varepsilon\Delta_{VV})\mathbf{C}_V = 0 \tag{41a}$$

with the effective potential defined as

$$\mathbf{V}_{VV} = -(\mathbf{F}_{VC} - \varepsilon\Delta_{VC})(\mathbf{F}_{CC} - \varepsilon\Delta_{CC})^{-1}(\mathbf{F}_{CV} - \varepsilon\Delta_{CV}) \tag{41b}$$

Eqns 41 are exact and constitute the partitioning technique (of any matrix problem), but offer by themselves very little relief from the basis set problem. To accomplish this, the inverse of Eqn. 41b is expanded as

$$\begin{aligned}\mathbf{M} &= (\mathbf{A}+\mathbf{B})^{-1} = \mathbf{A}^{-1} - \mathbf{A}^{-1}\mathbf{B}(\mathbf{A}+\mathbf{B})^{-1}\\ &= \mathbf{A}^{-1} - \mathbf{A}^{-1}\mathbf{B}\mathbf{A}^{-1} + \mathbf{A}^{-1}\mathbf{B}\mathbf{A}^{-1}\mathbf{B}(\mathbf{A}+\mathbf{B})^{-1}\end{aligned}$$

where **A** is the diagonal and **B** the off-diagonal of the matrix **M**. This expansion gives rise to the *Brillouin–Wigner* perturbation theory.

$\mathbf{V}_{VV}$ is a *pseudo potential*, meaning that it depends explicitly on the orbital energy, itself, and if utilized in this form would give rise to a different Fock operator $\mathbf{F}' = \mathbf{F}_{VV} + \mathbf{V}_{VV}$ for each orbital. To remove the energy dependence, we make use of the observation that the core orbitals $(\chi_\alpha, \chi_\beta)$ of any molecule remain fairly unchanged, independent of the molecule, and that

$$\mathbf{F}\chi_\alpha \sim \varepsilon_\alpha \chi_\alpha. \tag{42}$$

Then, writing

$$\Delta_{CC} = \mathbf{1} \text{ (orthonormal core orbitals)}$$

and

$$\langle \chi_\mu | \mathbf{F} | \chi_\alpha \rangle \sim \varepsilon_\alpha \langle \chi_\alpha | \chi_\mu \rangle = \varepsilon_\alpha \Delta_{\alpha\mu}$$

one may express $\mathbf{V}_{VV}$ as a series expansion (χ_μ, χ_ν are valence orbitals)

$$\begin{aligned}
\mathbf{V}_{VV} &= -(\mathbf{F}_{VC} - \varepsilon\Delta_{VC})(\mathbf{F}_{CC} - \varepsilon^1)^{-1}(\mathbf{F}_{CV} - \varepsilon\Delta_{CV}) = \mathbf{V}^{(1)} + \mathbf{V}^{(2)} + \cdots \\
V_{\mu\nu}^{(1)} &= -\sum_\alpha (\mathbf{F}_{VC} - \varepsilon\Delta_{VC})_{\mu\alpha}(\varepsilon_\alpha - \varepsilon)^{-1}(\mathbf{F}_{CV} - \varepsilon\Delta_{CV})_{\alpha\nu} \\
&= -\sum_\alpha \Delta_{\mu\alpha}(\varepsilon_\alpha - \varepsilon)^{-1}\Delta_{\alpha\nu} \qquad (43)
\end{aligned}$$

In practice, the effective potential includes only first order terms in the series, and, additionally, assumes $(\varepsilon_\alpha - \varepsilon)^{-1}$ to be independent of the MO eigenvalue ε. Although ε_α can be taken from atomic information, it is far more accurate to take ε_α as an empirical parameter to be fit from accurate calculations in order to compensate for the deficiencies of this approximate development. More common today in the more accurate effective core potentials is the assumption of a functional form for ε_α as

$$\varepsilon_\alpha = \sum_q A_q/r^q,$$

where the A_q are empirical constants depending on the atom, the core orbital angular momenta and the shell. Note that an equation similar to Eqn. 43 can also be derived simply by Schmidt orthogonalizing of the valence orbitals to the core,

$$\begin{aligned}
|\chi_\mu{}^*\rangle &= |\chi_\mu\rangle - \sum_\alpha |\chi_\alpha\rangle\Delta_{\alpha\mu} = \Big(1 - \sum \alpha|\chi_\alpha\rangle\langle\chi_\alpha|\Big)|\chi_\mu\rangle \\
&= \big(1 - |\phi_c\rangle\langle\phi_c|\big)|\chi_\mu\rangle \qquad (44)
\end{aligned}$$

assuming Eqn. 42. This simpler formulation hides some of the further corrections.

Effective core potentials (as opposed to pseudo-potentials) not only save computational time, but they can be used to include relativistic effects that become important for the inner-shell orbitals of the heavier atoms.[91–98,103–105] Relativistic corrections, discussed in more detail in Section 5, become important in heavier atoms because the large nuclear charge leads to electrons in the inner shells with velocities comparable to the speed of light. Such corrections include mass-velocity corrections, *Darwin* corrections, *spin-orbit* interactions, *Breit interactions*, Coreolis couplings, hyperfine interactions, and other effects of smaller magnitude.[91,92,98,100–105] Except for the heaviest of atoms, these effects may be approximately incorporated in non-relativistic corrections using a combination of perturbative methods and/or effective potentials and operators.

When specifying the basis set of heavier atoms, it is customary to include Relativistic Effective Core Potentials (RECP)[91–96,103–105] which both mimic the relativistic mass-velocity corrections and, additionally, reduce the total number of electrons in the molecule. The mass-velocity corrections contract the innermost s orbitals, and, due to orthogonality, the higher shell s orbitals as well. Consequently, the mass-velocity corrections lower the energy of s-rich electronic configurations in transition metal complexes, as given in Figure 8 for the lanthanides, in which this effect is particularly large. The inner p-orbitals also contract, but the d and f orbitals now expand due to the more complete screening of the tighter s and p electrons. Note that the $4f$ electrons have no radial node, and the $5d$ but one, so that electrons in these orbitals are not often close to the heavy nucleus that would lead to contraction.

The literature contains a number of ECPs.[91–96,103–105] There exist many derivations for ECPs, but the most common procedures follow the development above.[104,105]

To define a *Relativistic* ECP, the core projection operator $|\phi^c\rangle\langle\phi^c|$ must be contracted using orbitals from a relativistic SCF type Hamiltonian $\mathbf{H}_{\mathrm{rel}}$. The RECPs of Hay and Wadt[94–96] were derived as approximations to a Hartree–Fock Hamiltonian of the form:

$$\mathbf{H}_{\mathrm{rel}} = \mathbf{H}_{\mathrm{nr}} + \mathbf{H}_{\mathrm{mv}} + \mathbf{H}_{\mathrm{D}} \tag{45}$$

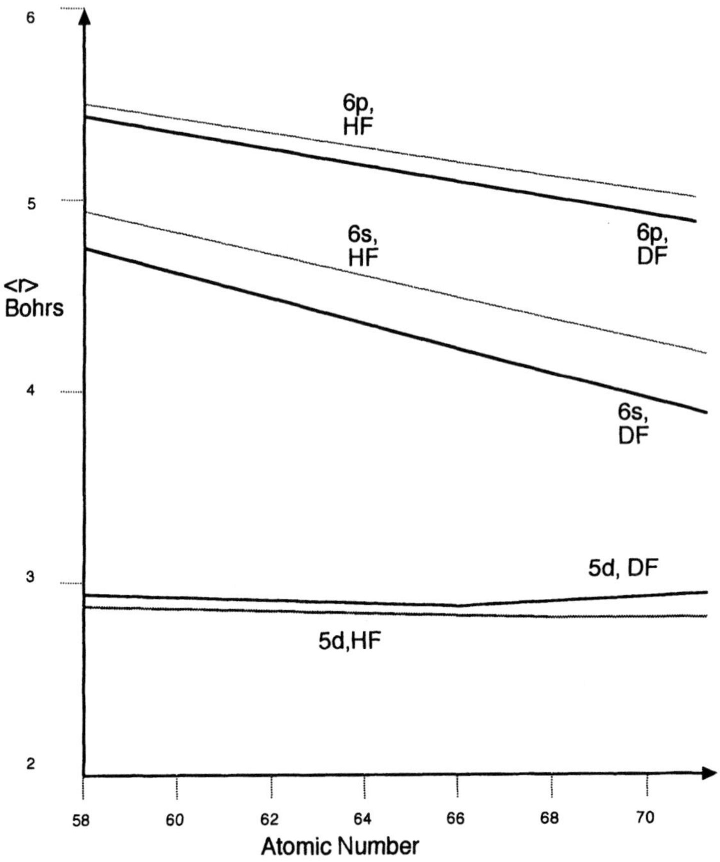

Figure 8. A comparison of the radial extent of nonrelativistic (HF) and relativistic (DF) atomic orbitals for the lanthanides, see text.

where $\mathbf{H}_{nr}$ is the non-relativistic Hartree–Fock Hamiltonian, and $\mathbf{H}_{mv}$ and $\mathbf{H}_D$ include the mass-velocity and Darwin terms, respectively. We discuss these terms further below in Section 5.2.

The resulting RECP $\mathbf{V}(r)$ is decomposed into a finite sum of separate l-dependent potentials $\mathbf{V}_l(r)$ plus a constant term $\mathbf{U}_L(r)$ to account for higher l values:

$$\mathbf{V}(r) = \mathbf{V}_L(r) + \sum_{l=0,L} \mathbf{V}_l(r) - \mathbf{V}_L(r)|l\rangle\langle l| \tag{46}$$

This form depends explicitly on the orbital quantum number l, and is expanded as a sum of Gaussians:

$$\mathbf{V}_l(r) = \sum_k d_k r^{n_k} \exp\left[-\zeta_k r^2\right], \tag{47}$$

where $n_k = [0, 1, 2]$. The parameters d_k, n_k, and $-\zeta_k$ are determined by fitting the RECP to Relativistic HF calculations on the corresponding atoms. There exist a number of other suggested parametric forms for effective core potentials as well.[73–76] (Note that the Gaussian expansion above must have very large exponents because the true mass-velocity correction in the Pauli Hamiltonian, described below in Section 5.2, diverges in regions close to the nucleus. These terms in the operator generate overlap—like terms in the Fock equation.)

Typical RECPs come in two forms depending on which orbitals lie in the core. In particular, for first row transition metal atoms the $3s$ and $3p$ orbitals share similar regions of space as do the $3d$ orbitals. *Small core* effective *core* potentials leave explicit the $3s$, $3p$, $3d$, $4s$, and $4p$ orbitals — large *core* include explicitly only the $3d$, $4s$, and $4p$. It appears that for high accuracy the $3s$ and, especially, the $3p$ must be explicitly included, and this inclusion is even more important for correlated calculations.

2.5 The Correlation Problem

2.5.1 ***General Comments*** The LCAO approximation relies on the idea that an atom in a molecule behaves much as an isolated atom does. But, as already mentioned in the Introduction, atom in molecules lose their atomic spin and angular momenta. This poses no problem for carbon atoms, for example, because the binding energies of a carbon atom in most compounds far exceed the energy separation between distinct atomic carbon electronic configurations. The binding energy of a carbon atom, about 400 kcal/mol = 17.3 eV, greatly exceeds the separation between the ^3P, ^1D, or ^1S states, about 2.5 eV, or indeed, even the separation between the average configuration from s^2p^2 and the average from sp^3, about 5 eV.[70] Therefore it is not important if the molecular ground state is derived from a ^3P, ^1D, or ^1S state, or even from an sp^3 or s^2p^2 configuration. In fact, carbon generally forms four, rather than two, bonds because the gain in energy in forming two additional bonds far exceeds the promotion energy between the sp^3 configuration, available for forming four bonds, and the s^2p^2 configuration, which provide only two. Transition metals, however, often bind with an order of magnitude less energy, and, accordingly, the characteristic atomic spin and spatial angular momenta still persist. This is precisely why crystal field theory,[106] ligand field theory,[107] and the Sugano–Tanabe diagrams[108] prove so very useful.

In many cases we must restore the atomic nature of the transition metal atom to get a good representation of the electronic structure of a complex. This is done through the inclusion of what is known as non-dynamic, or essential, correlation. Exactly the same corrections are needed when examining molecular dissociation. For example, the $^1\Sigma_g^+$ state of H_2 will not dissociate correctly into the ^2S states of the two hydrogen atoms. This kind of correlation becomes most important when two states of differing nature but of the same symmetry approach one another in energy. A common case occurs when the coupling between atoms becomes small, as it does when the two H atoms of Figure 9 become separated and the overlap decreases. The overlap between the d orbitals of a transition metal element and the chelating atoms of a ligand at normal bonding distances is *always* small, suggesting the importance of essential correlation in most bonding situations for transition metal complexes.

Figure 9 also suggests that this kind of correlation will be important when bonds dissociate, as many molecular potential energy surfaces often converge to degenerate atomic or fragment states. Similar situations also arise when electronic states change from covalent at short distance to ionic at large internuclear separation (i.e. in metal halides, such as NaI[109]). Likewise, frequently an excited state may change in character from highly diffuse (Rydberg-like) to highly compact (valence-like) with a relatively small change in geometry.[110] (Note that in situations where two electronic states cross, the Born–Oppenheimer approximation may break down, leading to further, nonadiabatic couplings, or, as in many transition metals, large spin-orbit couplings.)

In addition, the motions of electrons are inherently correlated with the motion of others, as suggested in Figure 9a. The Hartree–Fock model examines the motion of an electron in the average of the other $N - 1$ electrons as discussed previously. Although it does explicitly include a form of electron-electron correlation, through the use of Slater determinants, and thus the Pauli exclusion principle (the wavefunction vanishes, and thus so does the probability of finding two electrons of the same spin in the same region of space at the same time) still the HF approximation allows electrons to get too close, *on average*, and this proximity with the other $N - 1$ electrons leads to an overestimate of the electron-electron repulsion. This type of correlation, referred to as dynamic correlation, effectively reduces the probability that two electrons will be too close, reducing the Coulomb interaction between electrons. The dynamic correlation also includes the electronic polarizability, necessary for treating clusters of heavy transition metal atoms.[111]

A word of caution is perhaps in order here on terminology. CI originally referred to any post-HF model. Thus configuration interaction methods referred to perturbation theories, pair theories, and coupled cluster theories in addition to methods that

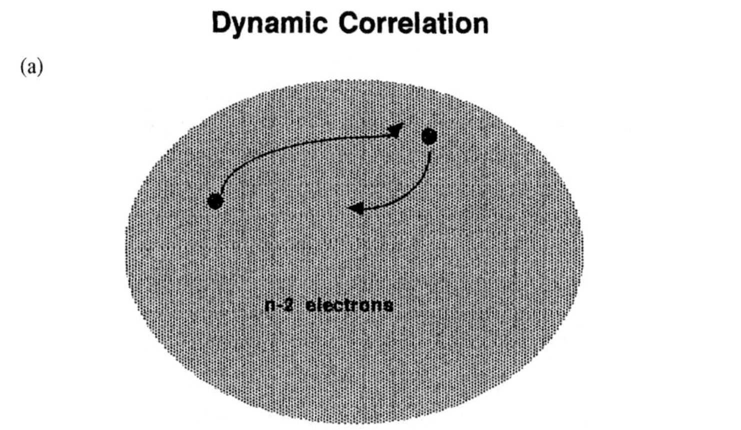

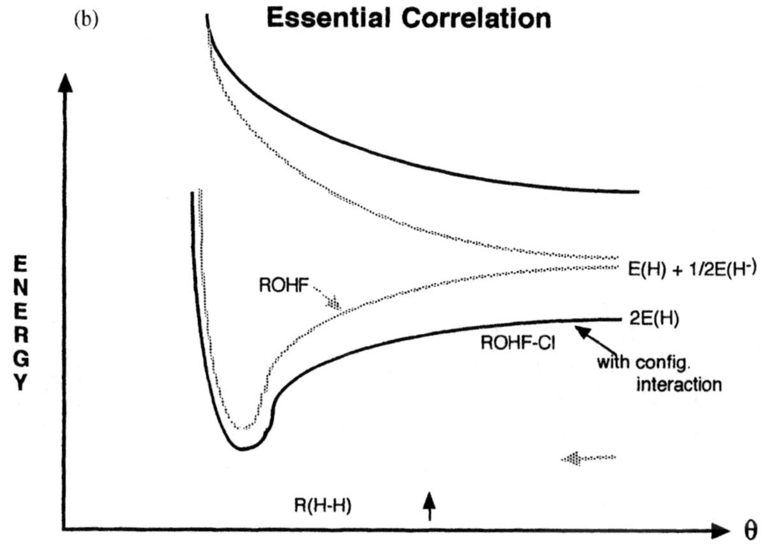

Figure 9. Dynamic and an example of essential correlation.

diagonalize the energy matrix between configurations. More common today CI refers only to the latter—only one technique among many post Hartree–Fock treatments of the correlation problem.

Correlated *ab-initio* calculations on transition metals and transition metal complexes, have revealed a number of unexpected features in transition metals, some which simplify calculations, and others which challenge even the most sophisticated theories.[47,98,100,111,112] First, essential, or static, electron correlation, plays less of a role in second and third row transition metals than those from the first row. This simplification arises because in second and third row transition metals, the outer valence *d* and *s* orbitals are larger. Therefore they bond (overlap) more strongly with chelating ligands than do elements of the first transition series, and can hybridize more easily, reducing the need for significant configuration mixing to restore the atomic character to the metallic ion. This observation, and the lack of near-degeneracy effects in the second row metals allows an SCF determination of molecular geometries, even though the bond energies themselves may require electron correlation. Exceptions to these trends arise in complexes with multiple metal-metal bonds and in metal hydroxides, where the configurations with strong covalent and ionic configurations mix. Especially in the first transition metal series the overlap between the *d* orbitals and the chelating atoms of the ligand is small, and the interaction between them is small. In such cases covalent mixing is often exaggerated by the SCF model because of the near degeneracies between the bonding hybrids of the ligand and the metal *d* orbitals, an event that takes place generally with the stronger ligands. Thus, the need for correlation increases in general with increased covalent bonding between the metal and the ligand valence orbitals. Note that this trend agrees with the nephelauxetic series[48,113,114] [$F^- < H_2O < NH_3 < CN^-$]. The latter members of this series increasingly hide the atomic nature of the metal because of the apparant strong covalent bonding. Restoration of the atomic character of the metal requires the exact same correlation that is required in the H_2 molecule to restore the atomic character to the H atoms when the overlap is similarly small. A more thorough discussion of these effects, for first, second, and third row transition metals, is provided in a recent review by Siegbahn,[112] and a discussion of the covalent bonding is provided by Roos[47] using CASSCF and CASPT2 calculations to demonstrate the effects.

Dynamic correlation is also important in transition metal complexes, and this is especially true in complexes of the latter transition series with strong ligands, where many electrons are in the same region of space. For example in $Cr(CO)_6$ the 6 CO σ-lone pairs (mostly on the chelating C atoms) donate 12 electrons datively to the Cr, and the Cr(0) atom has 6. This is too many electrons in this region of space, and Hartree–Fock calculations of good quality lead to a net repulsion in this complex. Dynamic correlation reduces the electron repulsion, and allows the net bonding.

There is no clear distinction between dynamic and essential correlation, as both are included in a post Hartree–Fock treatment by allowing electronic configurations to interact through the variational principle, or approximations to it. The corrections arise through the one-particle orbitals that are assumed in the initial wavefunction, and both corrections have algebraicly the same appearance. The classification into dynamic or essential rests upon our physical picture of the deficiencies of the independent particle model, and the conceptual and mathematical formulation of our theories. For example, theories which readily incorporate experimental data (ligand field theory, semiempirical theory) may seem quite removed from the numerically driven *ab-initio* models, yet both must somehow take advantage of the chemical inituition collected over years of research. Since transition metal compounds are so much more complex than most other molecules, it is important to elucidate some of these differences so that the user can utilize their chemical inituition in order to design and run effective electronic structure calculations.

It is interesting to note the differences in the assumptions often invoked in *ab-initio* and semiempirical models. Most *ab-initio* methods, particularly state-selective methods (CASPT2, SAC-CI, MRSDCI, etc., see below), first explicitly include the essential correlation for a single state (perhaps by diagonalizing a small matrix), and then add dynamic correlation (using a CI expansion, perturbation theory, etc.). In contrast, semiempirical methods attempt to include the dynamic correlation through empirical parameters at the HF level and then treat the remaining dynamic and essential correlation, if necessary. Hence a semiempirical HF calculation includes some dynamic correlation already, whereas an *ab-initio* HF calculation has none, by definition. There is an exact *ab-initio* theory, effective Hamiltonian theory, which explains how semiempirical parameters may include dynamic correlation.[99,115–121] The effective Hamiltonian theory is developed as a perturbation theory which, due to computational difficulties, has only recently been applied to semiempirical π-electron theory for large polyenes with success.[118–121] Active research in this area continues (See Appendix II for more details on the relationship between semiempirical methods and *ab-initio* theory).

As above, we define the correlation *energy* as the difference in energy between the best Hartree–Fock calculation for a spin-restricted wavefunction (the Hartree–Fock limit) and the exact many-body solution that can be obtained for that given Hamiltonian. Basis set correlation energy refers only to that part of the correlation energy that can be obtained for a given basis set.

2.5.2 ***The Configuration Interaction Method*** The basic assumption of the HF model is the independent particle approximation given in Eqn. 3. Here we have expanded a many-particle function in terms of the product of one-particle functions (orbitals), and this is only approximately so. What is needed is an expansion in terms of many-particle functions. An efficient method for doing this is to expand in terms of determinants, such as those given in Eqn. 3. The rationale for this is (i) that the independent particle approximation is, after all, quite good as far as the total energy is concerned, and (ii) the orbitals that are not occupied in the SCF wavefunction form a convenient *orthonormal* set to use for the expansion. An approximation for the state Ψ_l is then given by

$$\begin{aligned}\Psi_l^{ci} &= \Psi_0 + \sum d_{ia}\Psi_i^a + \sum d_{ijab}\Psi_{ij}^{ab} + \cdots \\ &\quad \text{SCF} + \text{singles} + \text{doubles} + \text{triples} + \cdots \\ &= \mathbf{D}_l\Psi \\ &= (\mathbf{1} + \mathcal{D}_1 + \mathcal{D}_2 + \cdots)\Psi_0 \end{aligned} \tag{48}$$

By singles, or single excitations, we refer to a single electron removed from an orbital occupied in the SCF *reference* function Ψ_0, say ϕ_i, and placed into a virtual orbital, say ϕ_a, generating the electronic configuration Ψ_i^a, etc. The CI expansion provides a convenient framework for deriving and classifying various post-HF approximations depending on the type of reference function Ψ_0 (be it single configurational SCF, multi-configuration SCF, or multireference) and the level of excitations included (singles only, doubles only, singles and doubles, etc.).

Invoking the linear variational principal on the expectation value of Ψ_I^{CI}

$$E_I = \langle \Psi_I | \mathbf{H} | \Psi_I \rangle / \langle \Psi_I | \Psi_I \rangle \tag{49}$$

yields the matrix eigenvalue problem

$$(\mathbf{H}_{\mathrm{CI}} - E_I \mathbf{1}) \mathbf{D}_I = \mathbf{0} \tag{50}$$

in which the CI matrix elements are given by $(H_{\mathrm{CI}})_{IJ} = \langle \Psi_I | H | \Psi_J \rangle$. Solving this matrix equation for a given basis set including all possible excitations is referred to as *full CI*, FCI or CIF, and is equivalent to solving the problem exactly within a given basis set for the given Hamiltonian. If the basis set is a complete basis, this idealized situation is referred to as Complete CI, and solves the problem exactly for any finite dimensional Hamiltonian **H**. This is demonstrated in Figure 10. The third axis, the model axis, emphasizes that often the system of interest, itself, must be reduced in complexity in such a fashion that a calculation of sufficient accuracy is feasible but the chemistry of interest is not compromised. Only a few CIF *ab-initio* calculations exist, and only for molecules of quite modest size.[122]

Less familiar, but more convenient for later discussion, is the development of the CI equations by *projection*. Consider the electronic Schrödinger equation

$$\mathbf{H} \Psi_I^{\mathrm{ci}} = E^{\mathrm{ci}} \Psi_I^{\mathrm{ci}} \tag{51}$$

which is equivalent to Eqn. 49 since Ψ_I^{ci} results from a variational method. Then we can project this equation onto Ψ_0 as

$$\begin{aligned} \langle \Psi_0 | \mathbf{H} | \Psi_I^{\mathrm{ci}} \rangle &= \langle \Psi_0 | \mathbf{H} | (\mathbf{1} + \mathcal{D}_1 + \mathcal{D}_2 + \cdots) | \Psi_0 \rangle \\ &= \langle \Psi_0 | E^{\mathrm{ci}} | (\mathbf{1} + \mathcal{D}_1 + \mathcal{D}_2 + \cdots) | \Psi_0 \rangle = E^{\mathrm{ci}} \end{aligned} \tag{52}$$

since

$$\langle \Psi_0 | \Psi_I^{\mathrm{ci}} | \rangle = \langle \Psi_0 | (\mathbf{1} + \mathcal{D}_1 + \mathcal{D}_2 + \cdots) \Psi_0 | \rangle = 1 \tag{53}$$

since all single, double, etc., excited configurations are orthogonal to the reference state Ψ_0 (this is called *intermediate normalization*). Similarly the right-hand side of this equation simplifies considerably as the Hamiltonian only contains one- and two-electron terms,

$$\begin{aligned} E^{\mathrm{ci}} &= \langle \Psi_0 | \mathbf{H} | (\mathbf{1} + \mathcal{D}_1 + \mathcal{D}_2 + \cdots) | \Psi_0 \rangle \\ &= \langle \Psi_0 | \mathbf{H} | (\mathbf{1} + \mathcal{D}_1 + \mathcal{D}_2) | \Psi_0 \rangle \\ &= E_{\mathrm{SCF}} + \langle \Psi_0 | \mathbf{H} | (\mathcal{D}_1 + \mathcal{D}_2) | \Psi_0 \rangle \end{aligned} \tag{54}$$

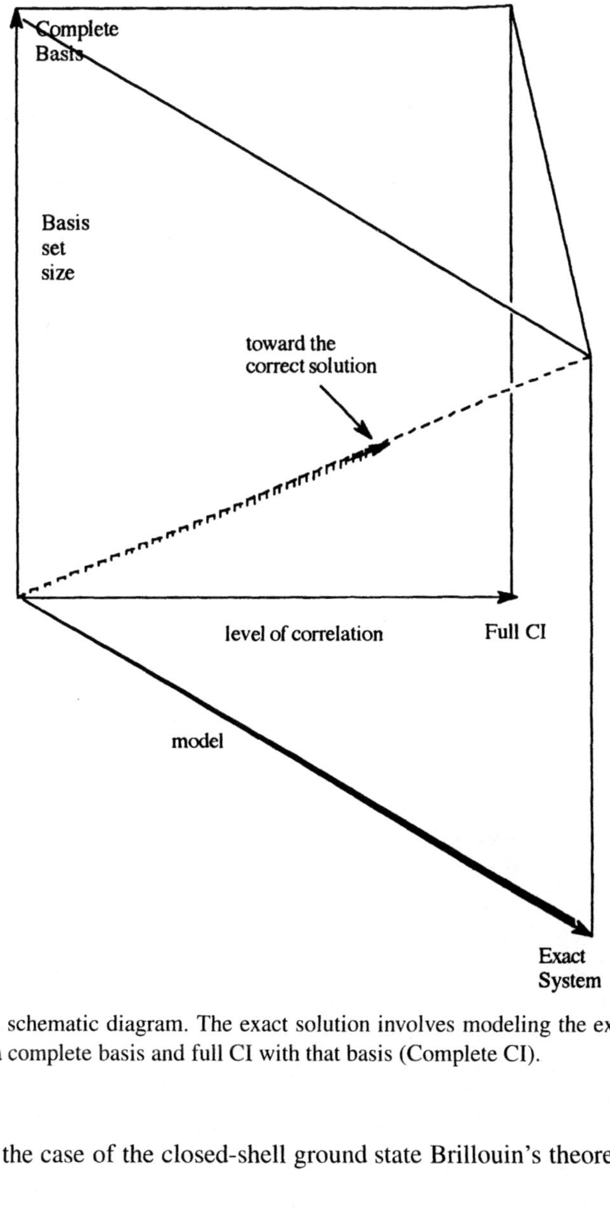

Figure 10. A schematic diagram. The exact solution involves modeling the exact system of interest with a complete basis and full CI with that basis (Complete CI).

Note that in the case of the closed-shell ground state Brillouin's theorem is obeyed and

$$\langle \Psi_0 | \mathbf{H} | \mathcal{D}_1 \Psi_0 \rangle = 0 \tag{55}$$

as discussed in more detail below. Eqn. 48 expresses the CI energy in terms of the d_{ia}, d_{ijab}, etc., coefficients that are obtained iteratively through projection of the Schrödinger Eqn. 51 with $<\Psi_i^{al}$, $<\Psi_{ij}^{ab}I$, etc., rather than through diagonalization. These iterative equations are the basis of such techniques as *direct CI*,[124] and are related to the Davidson diagonalization method.[122,124,125] The idea of projection is developed further below in the description of the coupled cluster method.

It would be ideal to solve this matrix problem, but the size of the problem quickly gets out of hand. If we consider n_o occupied orbitals and n_v virtual orbitals, then the number of single excitations (if the reference ground state is closed-shell) is $n_o n_v$, the number of doubles, $n_o^2 n_v^2$, and so forth. Spatial and spin-symmetry can reduced the dimensions of this problem considerably, but only by factors, not powers, of $n_o n_v$. Furthermore an open-shell reference system complicates the problem considerably.

The singles, doubles, and triples in Eqn. 56a are generally spin-adapted functions, referred to as Configuration State Functions (CSF). Creating spin eigenfunctions and the matrix elements between them is one of the most important areas of quantum chemistry today.[122] It is considerably easier to work in determinants, as the matrix elements are simple and depend only on whether the two determinants differ by one, two, or more spin-orbitals. Although the matrix elements are easier to evaluate, the size of the CI matrix becomes larger, faster, and contains all multiplicities, not only the ones in which we might be interested.

We delineate below several models based on the basic idea of CI as described above.

CIS—Single excitations only, also referred to as MECI (mono-excited CI) or the Tam–Dancoff[27,126–128,141–143] approximation (TDA). This level of theory is often useful for examining spectroscopy and other optical properties, but does not correct errors in the Hartree–Fock solution for the ground state. This is easy to understand. In exciting an electron from a given occupied MO to all virtual MOs, and allowing these singly excited configurations to mix with the ground state determinant, examines the interactions of this electron in the average field of all the other $N-1$ electrons. This is equivalent to the Hartree–Fock method itself. For a well converged SCF, singles do not mix with the Hartree–Fock reference Ψ_0. The formal proof of this is Brillouin's Theorem.[25–27]

There is often confusion on this matter in the case of open-shell Hartree–Fock references. Here there *will* be mixing between singles and the reference state. The energy is slightly lowered, and calculated one-electron properties often improved. To understand this, consider the simple case of the open-shell doublet of Figure 11. There are three classes of single excitations. From doubly occupied (closed) to the open, generating another doublet, from open to virtual, generating a doublet, and then from closed to virtual, generating a quartet and two-doublets. In order to generate two doublets in this latter case, the open-shell orbital must spin-flip, and it is this spin flip that looks like a double excitation, and that allows this configuration to mix with the reference wavefunction. In the figure we have chosen the two-doublets to be orthogonal to ${}^4\Psi_{1/2}$ and to each other. The actual mixing between these doublets ${}^2\Psi_{1/2}(3a)$ and ${}^2\Psi_{1/2}(3b)$ depends upon their mixing through the CI Hamiltonian.

CID—CI-doubles. This level of theory examines the motion of two-electrons in the average of the $N-2$ electrons that remain in the reference determinant. This generally contains the major effects of correlation to the ground state of molecules that are well described by a single CSF, for, as expected, the likelihood of two electrons

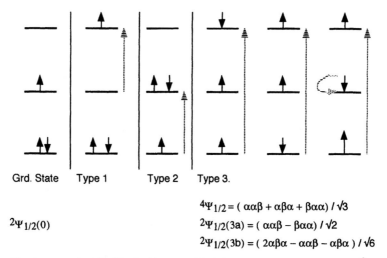

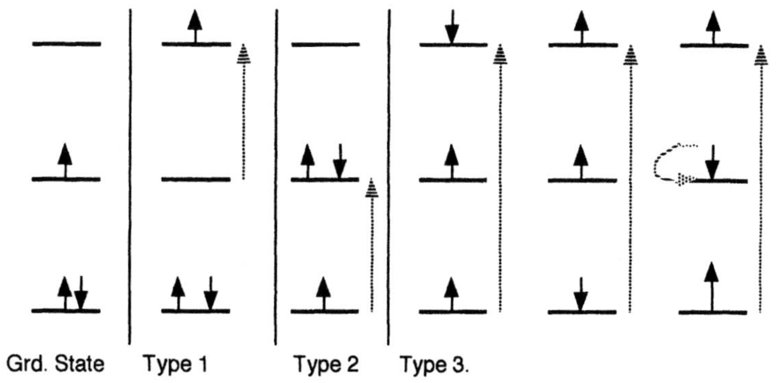

$$^{4}\Psi_{1/2} = (\alpha\alpha\beta + \alpha\beta\alpha + \beta\alpha\alpha)\,/\,\sqrt{3}$$

$$^{2}\Psi_{1/2}(0) \qquad ^{2}\Psi_{1/2}(3a) = (\alpha\alpha\beta - \beta\alpha\alpha)\,/\,\sqrt{2}$$

$$^{2}\Psi_{1/2}(3b) = (2\alpha\beta\alpha - \alpha\alpha\beta - \alpha\beta\alpha)\,/\,\sqrt{6}$$

Figure 11. An example of Brillouin (theorem) Violating Terms. The wavefunction $^{2}\psi_{1/2}(3b)$ requires a spin-flip, and the contribution of this $\alpha\beta\alpha$ determinant allows this combination to interact with the ground state doublet.

being in the same region of space at the same time is far greater than three or more. Second and third order perturbation theory, as discussed below, include doubles.

CISD—CI-singles and doubles only. This is a reasonably good theory for most ground state problems. Although the single excitations do not mix with the reference CSF, they do mix with the doubles, which, in turn, mix with the reference. The singles improve the energy only slightly, but can often have an important effect in improving one-electron properties. CISD is not a good theory for spectroscopy, as it correlates the ground state and improves its characteristics far better than it does most low-lying excited states, which are generally well represented by singles (in physical characteristics—not always in energy). To correlate the singles that are the major components of most low-lying excited states requires triples, and, in particular, those that are doubly excited relative to these singles. CISDT is already a sizable effort for most systems.

MRCI—Multi-reference CI.[32,33] There are situations, especially in transition metal chemistry, in which the ground state requires large amounts of essential correlation, such as when bonds break or at near degeneracies, and, as a result, the ground state wavefunction should include two or more configurations (as determined from physical arguments or a preliminary CI). The multi-configurational ground state serves as the reference to generate singles, singles and doubles, etc., MR-CIS, MR-CISD, etc. MR-CISD is usually an accurate level of theory, but one must acknowledge that the word *accuracy* is subjective! MRD-CISD refers to a multi-reference treatment, where the generating CSFs are doubly related and each of these doubles, as well as the reference CSF, form the basis from which single and double excitations are generated. MRS-CISD is a level of theory that is balanced for spectroscopy.

MC-SCF[28,33,55,56]—Multi-configuration self-consistent field. Such a wavefunction optimizes the MO coefficients **C**, Eqn. 3, simultaneously optimizing the CI coefficients **D**, Eqn. 56. The optimization is performed with respect to energy. This theory is particularly attractive in transition metal calculations, as one can, for example, place into the MC-SCF all those states that involve transitions among the *d* orbitals in an attempt to describe the metal well. An MC-SCF treatment is often set up to contain most of the essential correlation; dynamic correlation is often included by following such a treatment with a CISD, using the MC-SCF states for generating the configurations.

CAS-SCF.[56] This is a special case of an MC-SCF where all the configurations within a given subspace are included (a FCI within this space). For example, a $(m, N) = (6, 5)$ CAS refers to all the CSF that can be generated with $N = 5$ electrons in $m = 6$ orbitals, or $(2m)!/N!(2m - N)! = 792$ states. This becomes much larger than the more controlled MC-SCF or MR-CI, and often includes determinants that have little effect, but the fact that all the configurations within the subspace are considered reduces the complexity of the calculation. The (10, 10) case of Tables 1 and 2 would involve 185,000 CSFs and would represent a state-of-the-art calculation.

2.5.3 ***Perturbation Theory*** The most compact CI expansion (for the ground state) can be derived using perturbation theory,[25–28] and, as one would expect, perturbative methods provide a very competitive alternative to CI expansions. Perturbation theory first partitions the exact Hamiltonian into two components: $\mathbf{H} = \mathbf{H}_0 + \mathbf{V}$, where $\mathbf{H}_0$ is some simple, solvable Hamiltonian, and $\mathbf{V}$ is the perturbation, which is to be treated using a power series expansion. One should take note that the partitioning of $\mathbf{H}$ is arbitrary, that $\mathbf{V}$ may not be small, and series expansions may only converge *asymptotically*. As with the CI expansion, perturbation theory naturally classifies into cases where $\mathbf{H}_0$ describes an SCF or CASSCF reference function, and may proceed using a state-selective (CASPT2[47,48] or Effective Hamiltonian) approach.[48,99,115,131] Unlike the CI approach, however, the reference functions for perturbation theory remain more limited.

Today almost all calculations performed using perturbation theory are of the Møller–Plesset Rayleigh Schrödinger type, designated MPn, where n is the order of perturbation theory used. The zero'th order wavefunction is generally the Hartree–Fock one, and $\mathbf{H}_0$ is defined as the sum of the Fock operators, Eqns. 6 and 9.

$$\mathbf{H}_0 = \sum \mathbf{f}(i) \tag{56a}$$

The zero'th order energy then is

$$E^0 = \sum_i \varepsilon_i n_i \tag{56b}$$

The first order correction is[25]

$$E^{(1)} = \langle \Psi^0 | \mathbf{V} | \Psi^0 \rangle = -\sum_{(i,j)} [2J_{i,j} - K_{i,j}] = -\sum_{(i,j)} \langle i, j | | i, j \rangle \tag{57}$$

with the perturbation, the *fluctuation potential*, defined as

$$\mathbf{V} = \mathbf{H} - \mathbf{H}_0 = \sum_{(i<j)} 1/r_{i,j} - \sum_i \{\mathbf{f}(i) - \mathbf{h}(i)\} \tag{58}$$

V is the difference between the Fock potential, which generates the average field of the other $N-1$ electrons, and the explicit electron-electron repulsion. Note that the self-consistent field (Hartree–Fock) energy is

$$E^{HF} = E^0 + E^{(1)}. \tag{59a}$$

Then

$$\begin{aligned} E^{(2)} &= \langle \Psi^0|\mathbf{V}|\Psi^1\rangle = \sum_{(l\neq 0)} \langle \Psi^0|\mathbf{V}|\Psi^{0_l}\rangle\langle \Psi^{0_l}|\mathbf{V}|\Psi^0\rangle/[E_0^0 - E_i^0] \\ &= \sum_{i<j}\sum_{a<b} \langle i,j|\,|a,b\rangle\langle a,b|\,|i,j\rangle/[\varepsilon_i + \varepsilon_j - \varepsilon_a - \varepsilon_b] \end{aligned} \tag{59b}$$

The integral $\langle i,j|\,|a,b\rangle$ is the *many body* interaction between two electrons

$$\langle i,j|\,|a,b\rangle = \langle i,j|\,|a,b\rangle - \langle i,j|\,|b,a\rangle \tag{60}$$

and the sums of Eqn. 59b are over occupied spin-orbitals ϕ_i and ϕ_j and virtual spin-orbitals ϕ_a and ϕ_b. In deriving $E^{(2)}$ an SCF solution was assumed, and therefore Brillouin's theorem is applicable. Due to the two-electron nature of **V** only double excitations can mix with the ground state at second order. The same observation is true at third order, MP3: only double excitations contribute to the energy if the SCF procedure has converged. Fourth order Møller Plesset theory, MP4, is more complex, and includes the effects of single, triple, and quadruple excitations as well as doubles. A simple way to understand this is to recall the two-electron nature of **V**, and to interpret each dash "-" in the next sentence as an order of perturbation theory representing a matrix element $\langle \Psi_K|\mathbf{V}|\Psi_L\rangle$ in such expressions as

$$\langle \Psi_0|\mathbf{V}|\Psi_I\rangle\langle \Psi_I|\mathbf{V}|\Psi_J\rangle\langle \Psi_J|\mathbf{V}|\Psi_K\rangle\langle \Psi_K|\mathbf{V}|\Psi_0\rangle$$

At fourth order the ground state can mix with - doubles and these doubles can mix with - doubles, singles, triples, or quadruples, and these in turn can mix with - doubles, which can, in turn, mix with (*return to*) - the ground state. Note that if Ψ_K were not a double then $\langle \Psi_K|\mathbf{V}|\Psi_0\rangle$ in the above expression would vanish and the entire expression would vanish. The four - appearing in the sentence represents an order in perturbation theory, and four of them (4th order) are required to include singles, triples, or quadruples.

MPRSPT—(Møller–Plesset, Rayleigh Schrödinger Perturbation theory), or MPPT (or, MBPT = many body perturbation theory) can be shown to be *size extensive*, that is, the quality of the calculation does not worsen as the system studied gets larger. Hartree–Fock, CIS, and CIF are also size extensive, but other truncated levels of CI are not. Most often the error this invokes is not serious for highly correlated calculations, but one should nonetheless be aware of this shortcoming. Correction methods, such as the Davidson method, can be used, but they are approximate.[129]

MPPT can be used for closed-shell reference systems, RHF, and for open-shell UHF reference systems. In the latter case the two-electron integrals must be evaluated over α and β spin orbitals, but the perturbation theory has the same formal appearance. The results are generally good providing the UHF solution is valid. This requirement usually restricts the method to examining states of maximum multiplicity for a given number of open-shell orbitals, as has been discussed previously. There is, at present, no convenient MPPT for a general ROHF reference.

MP4 is a reasonably high level of theory for estimating the correlation energy. It is expensive in terms of computer resources, and the convergence of the perturbation theory itself might be questioned if the 4th order results are not at least a factor of 4 smaller than the second order correction. There is also growing evidence that MPPT is *asymptotically* convergent, meaning that the results approach the correct answer as the order of the perturbation theory increases up to a point, and then the results begin to diverge.

MRPT—There is some confusion in the literature regarding the nomenclature of Multi-Reference Perturbation Theories. The state-selective multi-reference perturbation theories employ a single reference, multi-configuration ground state (i.e., a CASSCF reference function) and apply perturbation theory to the reference wavefunction.[47,48] Section 4.2 discusses a newly developed and highly accurate form of this, CASPT2. Note that the CASPT2 approach first treats the essential correlation with a CASSCF calculation, and then adds the dynamic correlation via second order perturbation theory.

QDPT[99,115,131]—Quasi-Degenerate Perturbation Theory, in contrast to MRPT, employs an effective Hamiltonian approach to electron-electron correlation, in which the perturbation series expansion is made of a small CI matrix, and not a reference wavefunction. This approach first includes dynamic correlation into a small effective Hamiltonian matrix, and then introduces the static correlation by diagonalizing the CI matrix for this effective Hamiltonian. QDPT remains in the hands of the experts; no available program packages include this approach. QDPT may prove necessary, however, for calculations on large transition metal compounds because they have many low-lying nearly degenerate states of varying configurational character. An Effective Hamiltonian version of CASPT2 is said to be under development.[47]

2.5.4 ***Coupled Cluster Theory*** Coupled cluster theory[22,26–28,130,132] takes advantage of the observation that the most important corrections to the independent particle model are of the double excitation type, and that, for example, the most important quadruples are the pair-wise doubles. That is, it is far less likely that four electrons are correlated, simultaneously, in the same region of space, then would be the probability of finding two pairs of two electrons simultaneously in two separate proximities.

Consider a wavefunction of the type

$$\psi^{cc} = e^{\mathbf{T}}\Psi_0 = \left(1 + \mathbf{T} + \tfrac{1}{2!}\mathbf{T}^2 + \tfrac{1}{3!}\mathbf{T}^3 + \cdots\right)\Psi_0 \tag{61a}$$

where $\mathbf{T}$ is an excitation operator generating singles, $\mathbf{T}_1$, doubles, $\mathbf{T}_2$, etc., as do the $\mathcal{D}_i$ in Eqn. 56a: i.e.,

$$\mathbf{T} = \mathbf{T}_1 + \mathbf{T}_2 + \mathbf{T}_3 + \cdots \tag{61b}$$

The lowest level of coupled cluster calculation is CCD, coupled cluster doubles, which results from setting $\mathbf{T} = \mathbf{T}_2$.

From Eqn. 61a, we find

$$\begin{aligned}\psi^{ccd} &= e^{\mathbf{T}}\Psi_0 = \left(1 + \mathbf{T}_2 + \tfrac{1}{2!}\mathbf{T}_2\mathbf{T}_2 + \tfrac{1}{3!}\mathbf{T}_2\mathbf{T}_2\mathbf{T}_2 + \cdots\right)\Psi_0 \\ &= \Psi_0 + \sum_{(i<j,a<b)} t_{ij}^{ab}\Psi_{ij}^{ab} + \tfrac{1}{2!}\sum_{(i<j<k<l,a<b<c<d)} t_{ijkl}^{abcd}\Psi_{ijkl}^{abcd} + \cdots\end{aligned} \tag{61c}$$

Note that this contains quadruples, sextuples, etc., via *coupled* doubles, $\mathbf{T}_2\mathbf{T}_2$, $\mathbf{T}_2\mathbf{T}_2\mathbf{T}_2$, etc. CID does not.

$$\psi^{cid} = (\mathbf{1} + \mathcal{D}_2)\Psi_0 = \Psi_0 + \sum_{(i<j,\,a<b)} d_{ij}^{ab}\Psi_{ij}^{ab}. \tag{61d}$$

These additional terms also guarantee that the coupled cluster methods are size extensive. This is rather easy to demonstrate, and is a characteristic of the exponential construction of the wavefunction. Consider two systems, A and B. Then if the systems are separated,

$$\psi^{cc} = e^{\mathbf{T}}\Psi_0 = e^{\mathbf{T}}\Psi_A\Psi_B = e^{\mathbf{T}_A+\mathbf{T}_B}\Psi_A\Psi_B = e^{\mathbf{T}_A}\Psi_A{}^{\mathbf{T}_B}\Psi_B = \psi_A^{cc}\psi_B^{cc} \tag{62}$$

Ψ_0 can be written as the product $\Psi_A\Psi_B$ if the electrons are sufficiently separated so that exchange between them can be ignored. We will return to this idea later.

The coupled cluster equations are developed from the CC Schrödinger equation

$$\mathbf{H}\psi^{cc} = \mathbf{H}e^{\mathbf{T}}\Psi_0 = Ee^{\mathbf{T}}\Psi_0 = E^{cc}\psi^{cc} \tag{63a}$$

and the *shift* equation

$$(\mathbf{H} - E_0)\psi^{cc} = \mathbf{H}_N\psi^{cc} = (E^{cc} - E_0)\psi^{cc} = \Delta E\psi^{cc} \tag{63b}$$

which gives the correlation energy, ΔE. *Projection* by $\langle\Psi_0|$ and $\langle\Psi_{ij}^{ab}|$ then gives

$$\langle\Psi_0|\mathbf{H}_N|e^{\mathbf{T}}\Psi_0\rangle = \Delta E \tag{63c}$$

$$\langle\Psi_{ij}{}^{ab}|\mathbf{H}_N|e^{\mathbf{T}}\Psi_0\rangle = \Delta E\langle\Psi_{ij}{}^{ab}|e^{\mathbf{T}}\Psi_0\rangle \tag{63d}$$

Expansion of $e^{\mathbf{T}}$ gives equations for the *amplitudes* t_{ij}^{ab} in terms of the other t_{kl}^{cd} which are solved iteratively, usually from a starting guess from peturbation theory (see Eqn. 59b)

$$t_{ij}^{ab} = \langle a, b| \, |i, j\rangle / [\varepsilon_i + \varepsilon_j - \varepsilon_a - \varepsilon b] \tag{63e}$$

It is interesting to note that the first iteration of the CC equations generates third order MPPT if the model is formulated this way, the next iteration, fourth order, etc., providing the cluster operator **T** is consistent to this order. Indeed, CC theory can be constructed by summing certain terms in perturbation theory (i.e., all double excitations, for instance) to infinite order.[25]

A very useful level of coupled cluster theory is CCSD,[28,132] in which $\mathbf{T} = \mathbf{T}_1 + \mathbf{T}_2$. Expanding the exponent in Eqn. 61a shows the inclusion of not only singles and doubles, but triples of the form $\mathbf{T}_1\mathbf{T}_1\mathbf{T}_1$ and $\mathbf{T}_1\mathbf{T}_2$, as well as many other terms. Experience has shown that any reasonable set of orbitals can be used in Ψ_0 and the CCSD procedure is robust enough to yield good results.

CCSD is already a reasonably accurate level of theory for many problems. It is generally applied to closed-shell ground states, but open-shell systems can again be represented with an UHF wavefunction. Although there is growing evidence that high level CC theory can handle cases in which there are near degeneracies, single reference coupled cluster will also fail if the problem is truly one of multi-reference nature. As with MRPT, multi-reference coupled cluster (MRCC) remains underdeveloped.

2.5.5 *Quadratic CI* A popular form of CC theory is the more approximate *Quadratic* CI (QCI) model.[132,134] This was originally motivated to be a size extensive version of CI.

As in Eqn. 62, two separated systems in which both wavefunctions are formed from CID yields

$$\begin{aligned} \Psi^{\text{cid}} &= (1 + \mathcal{D}_2)\Psi_0 = (1 + \mathcal{D}_2)\Psi_A\Psi_B(1 + \mathcal{D}_{2A} + \mathcal{D}_{2B})\Psi_A\Psi_B \\ &\neq (1 + \mathcal{D}_{2A})\Psi_A(1 + \mathcal{D}_{2B})\Psi_B = \Psi^{\text{cid}}{}_A\Psi^{\text{cid}}{}_B \end{aligned} \tag{64a}$$

It is clear what needs to be added to make this true:

$$(1 + \mathcal{D}_{2A})(1 + \mathcal{D}_{2B}) = (1 + \mathcal{D}_{2A} + \mathcal{D}_{2B} + \mathcal{D}_{2A}\mathcal{D}_{2B}) \tag{64b}$$

Of course, in general, when the system is not separated, we cannot really localize electrons to one fragment or the other, and therefore $\mathcal{D}_{2A}\mathcal{D}_{2B}$ is replaced by $\mathcal{D}^2$, and then, not to over count, $\mathcal{D}^2/2$. The wavefunction then becomes

$$\Psi^{\text{qci}} = (1 + \mathcal{D}_2 + \mathcal{D}_2^2/2)\Psi_0 \tag{64c}$$

A comparison with the coupled cluster wavefunction of Eqn. 61c indicates that this is just the first three terms of the CCD wavefunction

$$\Psi^{ccd} = \left(1 + \mathbf{T}_2 + \tfrac{1}{2!}\mathbf{T}_2\mathbf{T}_2 + \tfrac{1}{3!}\mathbf{T}_2\mathbf{T}_2\mathbf{T}_2 + \cdots\right)\Psi_0$$

The quadratic CI model is a popular model for closed-shell reference Ψ_0 because it is readily available with the Gaussian94 package.[17] Extension to including singles is straightforward, either from the original Eqn. 64a, or, more readily, from truncating the CCSD equations appropriately. Modern CC codes are reasonably efficient, and since this model contains a more consistent treatment of correlation, it might be preferred over QCI. Indeed, the less general QCI approach can fail when higher order terms become important. Interesting examples include the Be atom[28] and $CuCH_3$.[134]

2.5.6 ***Response Theories*** So far, the electron correlation methods discussed provide accurate wavefunctions for individual electronic states. These modern methods provide many accurate ground state properties, frequently inaccessible to experimental measurement, such as geometries, transition states, and static properties (see Section 3). Experiments often measure, on the other hand, spectroscopic quantities, such as high resolution absorption spectra, solvent effects on spectra, and nonlinear optical spectra. Indeed, even in time-averaged absorption spectra, one can obtain dynamic information from the shape of the absorption curve.[135] Despite their ready availability, spectroscopic experiments frequently prove difficult to interpret, and therefore demand both qualitative theory and quantitative electronic structure calculations. Rather than compute all ground and excited state wavefunctions directly, given a good ground state wavefunction it may prove easier to compute the *response* of the ground state density to an external perturbation, leading to the so-called response methods.

The response methods have a long history and include a wide class of mathematical methods used in many theoretical areas. Response methods are also referred to as Green's function methods,[25] Equations of Motions (described below in Section 4), and methods formulated directly in Liouville space (the so-called Tetradic methods, discussed in great detail by Mukamel[136]). The Green's function methods stem from techniques used to solve inhomogeneous differential equations,[137] and remain extremely popular in solid state physics, although less so in chemistry. The Liouville space methods may seem more familiar to a physical chemistry student from statistical mechanics,[138,139] but also play a key role in interpreting and designing experiments in nonlinear optical spectroscopy. (Note that virtually every method developed in one area of theory, such as statistical mechanics, can usually find application in electronic structure theory, and vice versa! This is certainly so with the coupled-cluster model. As another example, Fulde has shown how to utilize the Mori and Zwanzig formulation of nonequilibrium statistical mechanics in electronic structure theory, and, additionally, describes its relationship to full CI methods.[140])

The final class of methods, the Equations of Motion (EOM), find the most application in electronic structure calculations, both for computing properties and spectra.[27,141–143] The simplest theory computes the response to Hartree–Fock ground states, and is referred to as the Random Phase Approximation (RPA), or the Time Dependent Hartree–Fock (TDHF) theory, and is formulated in Section 4. The RPA approach describes correlated fluctuations between the external potential and the *ground* state density. When used to compute spectra, one replaces the external potential with the HF fluctuation potential (**V**, Eqn. 58), and thus the same mathematical framework can provide the excited state energies which include contributions from all singly and some doubly excited determinants. In this application, the RPA method conceptually resembles a CIS calculation, and, as shown by Freed, involves a summation of an in finite number of terms from a standard CIS calculation.[141]

More highly correlated EOM calculations employ a correlated ground state. Popular examples include the Multiconfigurational Time Dependent Hartree–Fock (MCTDHF) method,[27,142] which employs an MCSCF or CASSCF reference function, and the Coupled Cluster Linear Response (CCLR),[144] which employs a CC ground state. Two methods with recent application in transition metal spectroscopy include the Symmetry Adapted Cluster CI (SAC-CI) and the Coupled Cluster Equations of Motion (CCEOM), both of which resemble the CCLR method. We delay further discussion to Section 4, where we derive the SAC-CI and CCEOM methods and provide a number of examples of transition metal spectroscopy. These methods are included in this section for completeness, as they are derived as a *property* of the ground state—its response to an external perturbation.

2.5.7 ***Comments on Localization vs. Delocalization*** Complexes or clusters of more than one transition metal atom usually have broken symmetry solutions in which the molecular orbitals obtained do not maintain the spatial symmetry of the system. The results of calculation on such systems can often be difficult to interpret.

RHF or ROHF calculations often yield symmetry adapted molecular orbitals without any additional constraints. This is especially the case for well written computer programs using stable algorithms. For example, in a homonuclear diatomic molecule, the $D_{\infty h}$ symmetry is maintained, and all the resulting MO's can be labeled by the irreducible representations of this group, σ_u, σ_g, π_u, etc. Another example is the complex studied in Tables 1 and 2, that is of D_{2h} symmetry. All MO's contain an equivalent contribution form each symmetry equivalent d AO, and a "correct" wavefunction is obtained. Such solutions often yield good geometries and reasonable properties. However, UHF calculations often break this symmetry leading to localized MO's, either ferromagnetically coupled (FM) with all localized MO's with spin up, or antiferromagnetically coupled (AF), in which each atom has its maximum multiplicity, but with spins alternating, one atom all spin up, the next, all spin down, etc., *even though all the atoms are symmetry equivalent.* The ROHF method can also be made to break spatial symmetry by choosing the appropriate vector coupling coefficients. The energies of these symmetry broken solutions are generally far below the symmetry maintained SCF solutions, but the physical description of the ground state is far poorer.

This situation is easy to understand in the simple case of minimum basis set H_2. The ROHF solution at all geometries is ${}^1\Psi_0 = |\Phi_1\alpha(1)\Phi_1\beta(2)|$, with $\Phi_1 = (\chi_A + \chi_B)\sqrt{2}(1+S)$, where $S = \langle\chi_A|\chi_B\rangle$ is the overlap. The energy of the two electrons in this determinant is

$$\begin{aligned} E_0 &= 2h_{1,1} + J_{1,1} \\ &= 2\big(2U_A + 2h_{A,B}\big)/\big(2(1+S)\big) \\ &\quad + \big[2(a,a|a,a) + 2(a,a|b,b) \\ &\quad + 4(a,b|a,b) + 8(a,a|a,b)\big]/\big(4(1+S)^2\big), \end{aligned} \tag{65a}$$

where $(a, a|a, a)$ is the two-electron integral in the chemists' notation, Eqn. 12, for $J_{a,a}$. As mentioned in the discussion around Figure 9, this is incorrect when S becomes small

$$E_0(R_{AB} \to \infty) = 2U_A + (a,a|a,a)/2 \tag{65b}$$

as the energy should approach that of two H atoms, $2U_A$, Figure 9.

The UHF solution of this problem will break space- and spin-symmetry at sufficiently large R_{AB}, or small overlap S, if not constrained, yielding $\Psi_0 = |\chi_A\alpha(1)\chi_B\beta(2)|$ (or $\Psi_0 = |\chi_B\alpha(1)\chi_A\beta(2)|$), which is not spin or spatially correct, but the energy is clearly correct. The proper symmetry adapted solution is a combination of these two, which is accomplished by the spin projection discussed previously.

The preferred way to treat the general situation is to take the symmetry correct wavefunction, and perform a sufficient CI to generate the correct valence bond description of this situation. In the case of H_2 discussed above, we need to consider the doubly excited CSF ${}^1\Psi_2 = |\Phi_2\alpha(1)\Phi_2\beta(2)|$, with $\Phi_2 = (\chi_A - \chi_B)\sqrt{2}(1 - S)$, and a general CI wavefunction of the form

$${}^1\Psi^{ci} = \cos^2\theta|\Phi_1\alpha(1)\Phi_1\beta(2)| + \sin^2\theta|\Phi_2\alpha(1)\Phi_2\beta(2)| \tag{65c}$$

where θ is a parameter to be determined variationally through the diagonalization of the CI matrix. $1\Psi_2$ has energy

$$\begin{aligned} E_2 &= 2h_{2,2} + J_{2,2} \\ &= 2(2U_A - 2h_{A,B})/\big(2(1-S)\big) \\ &\quad + \big[2(a,a|a,a) + 2(a,a|b,a) \\ &\quad + 4(a,b|a,b) - 8(a,a|a,b)\big]/(4(1-S)^2). \end{aligned} \tag{66a}$$

The coupling matrix element of the CI is

$$\begin{aligned} H_{0,2} &= \langle {}^1\Psi_0|\mathbf{H}|{}^1\Psi_2\rangle = J_{1,2} = \big(2(a,a|a,a) + 2(a,a|b,b) \\ &\quad - 4(a,b|a,b)\big)/4(1-S^2) \end{aligned} \tag{66b}$$

and the CI matrix is now

$$\mathbf{H} = \begin{vmatrix} E_0 & H_{0,2} \\ H_{0,2} & E_2 \end{vmatrix} \qquad (66c)$$

At large distances, or small S, E_2 approaches E_0 (the node in the antibonding orbital Φ_2 is no longer important), and $H_{0,2} \Rightarrow (a, a|a, a)/2$. The diagonalization of this matrix under these conditions yields the correct energy, $E = 2U_A$ and $\theta = -45°$.

The overlap between the d orbitals of different transition metal atoms in polynuclear complexes and clusters as mentioned previously is usually very small. The situation above pertains even at the equilibrium geometries of these materials. The CI involved to correct the situation, however, is no longer as simple as that given above. For example, in the Cr dimer,[47,145,146] a particularly pathological system, the covalent sextuple bond, involving essentially two d^5s^1 atoms gives a nearly correct geometry at about 1.65Å, while the localized AF description has minimum energy at about 3.25Å. At this distance the AF solution is over 4000 kJ/mol lower in energy than is the symmetry constrained sextuple bonded structure. The CI required to properly generate the localized situation from the symmetry correct ROHF reference ${}^1\Psi_0$ includes sextuple excitations, a sizable task still not convincingly performed. Now contemplate the case of transition metal clusters!

Experience must be the guide here. Generally the symmetry adapted solutions describe the essential chemistry reasonably well, including even the relative energies of the multiplets at fixed geometries (when refined with an appropriate CI calculation). The symmetry broken solutions, however, have much lower energies, and, for large enough systems, an energy that cannot be obtained even with state-of-the-art correlated treatments that start with a symmetry correct solution. Our experience, as noted previously, however, is that PUHF calculations at experimental geometries yield very good magnetic predictions.

2.6 Semiempirical Molecular Orbital Theory

Semiempirical methods simplify electronic structure calculations substantially, trading accuracy for reduced computational time, resources, and complexity, see references 7–12,116,147–171. These methods perform efficiently for two reasons. First, they generally employ a simple level of *ab-initio* theory (i.e. as SCF or SCF+CIS) within a minimal basis set of valence orbitals. Second, they avoid computing the cumbersome *ab-initio* one- and two-electron integrals directly, replacing these integrals with parameters fit to experimental data. Despite their simplicity and approximate development, semiempirical methods generally perform far better than the corresponding minimal basis set *ab-initio* theory—no doubt a consequence of the various parameter schemes. Indeed, as discussed above in Section 2.4.3, semiempirical methods actually include dynamic correlation (if not also some nondynamic correlation as well) within their empirical parameters (Appendix II briefly elucidates the formal relationship between *ab-initio* QDPT and semiempirical methods). The targeted experimental information usually consists of molecular geometry, thermochemistry, dipole moments and/or electronic spectroscopy, although there exist specially tailored methods that target other properties.

Semiempirical molecular orbital models come in two formulations: those that are of the Extended Hückel type, and diagonalize a Fock operator of the type (Eqn. 5)

$$\mathbf{FC} = \Delta\mathbf{C}\boldsymbol{\varepsilon} \tag{67a}$$

and those that are of the Zero Differential Overlap type (ZDO), solving

$$\mathbf{FC} = \mathbf{C}\boldsymbol{\varepsilon} \tag{67b}$$

and that have assumed an orthonormal basis set of valence type orbitals,

$$\Delta_{\mu\nu} = \langle \chi_\mu | \chi_\nu \rangle = \delta_{\mu\nu}. \tag{68}$$

2.6.1 ***Extended Hückel Approximations*** The first of these models were introduced by Wolfsberg and Helmholtz,[165] and later refined by Lohr and Libscomb.[166] The most well known applications were made by Hoffmann in his derivation of the Woodward–Hoffmann symmetry rules.[169] Ballhausen and Gray[168] and Zerner and Gouterman[169] then devised models that could be used for transition metal complexes in general.

The simplest version of this model equates the diagonal of the Fock operator to the observed ionization potentials of atoms

$$H_{\mu\mu} = -\mathrm{IP}_\mu \tag{69a}$$

and the off-diagonal elements proportional to the overlap

$$H_{\mu\nu} = (H_{\mu\mu} + H_{\nu\nu})K_{\mu\nu}\Delta_{\mu\nu}/2 \tag{69b}$$

$K_{\mu\nu}$ is a constant. In the simplest form of this theory the constant is independent of orbital type, and a value is chosen between 1.75 to 1.90 to reproduce a specific observable of interest.[161–165] It can be shown that the value of K must be greater than 1.0 or the role of bonding and anti-bonding orbitals will be reversed.

Although this simple theory is widely used for molecules containing main group elements, problems arise when molecules contain very polar groups. This theory, by not being applied in an iterative fashion in which the potential in an atom becomes more electron attractive as it becomes positive, and *vice versa*, allows charge to build up by not resisting it. This shortcoming can render this model useless when applied to complexes containing transition elements which often contain highly charged metal ions.

Several improved extended Hückel theories exist to correct for this charge buildup, including the SCC-EHT model (self-consistent charge extended Hückel theory) or IEHT (iterative extended Hückel theory), and the Fenske–Hall model[164] (below).

A generalization of the SCC-EHT Hamiltonian operator can be written as

$$H_{\mu\mu} = -\mathrm{IP}_\mu + (M_A - Z^c A)\gamma_{AA} + (1 - M_{\mu\mu})\gamma_{AA} + \sum (M_B - Z^c B)\gamma_{AB}\colon\ \mu\varepsilon \text{ atom } A \tag{70}$$

a form that can be derived using the Mulliken approximation for integrals as shown below. The notation $\mu\ \varepsilon$ atom A implies that the AO χ_μ is centered on atom A.

In Eqn. 70 $M_{\mu\mu}$ is the Mulliken population, a measure of the number of electrons *belonging to* orbital χ_μ, and M_A is the number of electrons *belonging to* atom A in a molecule.[172,173]

$$M_{\mu\mu} = \sum_\nu P_{\mu\nu}\Delta_{\nu\mu} \tag{71a}$$

$$M_A = \sum_{\mu\varepsilon A} M_{\mu\mu}. \tag{71b}$$

Note that $\sum_A M_A = N =$ the total number of electrons. The words *belonging to* are in italics, for there is no unique way to assign electrons to an atom in a molecule, and there is no physical observable that corresponds to such a charge. In Eqn. 70 Z_B is the atomic number of atom B, and Z_B^c is the core charge of atom B, equal to the number of valence electrons in neutral atom A (i.e., 4 for carbon, 8 for iron, etc.), and γ_{AB} is the two-electron integral

$$\gamma_{\mathrm{AB}} = \langle \overline{\mu}, \overline{\nu} | \overline{\mu}, \overline{\nu} \rangle = \underline{\gamma_{\mu\nu}} = (\overline{\mu}\,\overline{\mu} | \overline{\nu}\,\overline{\nu}) \tag{72a}$$

where $\overline{\mu} = \chi_\mu$ is an average (s-symmetry) atomic orbital centered on atom A. The semiempirical literature commonly employs the Mulliken notation $(\overline{\mu}\,\overline{\mu}|\overline{\nu}\,\overline{\nu})$.

The first two terms in $H_{\mu\mu}$ in Eqn. 70 form the basis of the SCC-EHT, or IEHT, model. This model has been successfully applied to many problems, but any significant charge buildup is now strongly prevented by the assumption that this charge goes to infinity. Note that from the theory of atoms

$$\gamma_{\mu\mu} = \mathrm{IP}_\mu - \mathrm{EA}_\mu + \text{(small corrections)}, \tag{72b}$$

where EA_μ is the electron affinity, and the first correction in Eqn. 70 (the second term on the right hand side) extrapolates the ionization potential of atom A between the positive and negative ion with charge $Q_A = (M_A - Z_A^c)$.

The third term in Eqn. 70 is derivable from the *ab-initio* Fock operator and corrects for the one-center exchange if one assumes the Mulliken approximation for all integrals. This is generally a small term.

$$I_{\mu\nu} = (I_{\mu\mu} + I_{\nu\nu})\Delta_{\mu\nu}/2 \tag{73}$$

The last term in Eqn. 70, seldom used in EHT models, is the *Madelung term*, derivable from the Mulliken Approximation for integrals and the approximation[161,174]

$$(\mu|R_B^{-1}|\mu) = (\overline{\mu}\,\overline{\mu}|\overline{\nu}\,\overline{\nu}) \tag{74}$$

and takes into account the stabilizing or destabilizing effects of the charge build up on neighboring atoms. The model is now capable of yielding reasonable charge distributions in transition metal complexes.

An interesting variant of the Extended Hückel Theory is the Fenske Hall model.[164] The Fock operator is written as

$$f = -\tfrac{1}{2}\nabla^2 + \sum_B V_B \tag{75a}$$

$$f^A = -\tfrac{1}{2}\nabla^2 + V_A \tag{75b}$$

where V_A is the total potential energy caused by atom A.

If one then *assumes* atomic problem is solved

$$f^A \chi_\mu = \varepsilon_\mu^A \chi_\mu \tag{75c}$$

Then

$$\varepsilon_\mu^A = (\mu|-1/2\nabla^2 + V_A|\mu) = -\mathrm{IP}_\mu. \tag{75d}$$

The last equality *assumes* Koopmans' approximation. (Koopmans' approximation, or *theorem*, states that in the absence of molecular orbital relaxation upon ionization—the *frozen orbital* approximation—the molecular orbital energy of the parent is the negative of the ionization energy corresponding to the loss (or gain) of an electron from that orbital.) We then derive the Fenske–Hall Equations,

$$F_{\mu\mu} = \varepsilon_\mu^A + \sum_{B\neq A} (M_B - Z_B)\gamma_{AB} \tag{76a}$$

$$\begin{aligned} F_{\mu\nu} &= (\mu|f + 1/2\nabla^2 - 1/2\nabla^2|\nu) \\ &= (\mu|-1/2\nabla^2 + V_A|\nu) + (\mu|-1/2\nabla^2 + V_B|\nu) + (\mu|1/2\nabla^2|\nu) \\ &\quad + \sum_{C\neq A,B} (\mu|V_C|\nu) \\ F_{\mu\nu} &= (\varepsilon_\mu^A + \varepsilon_\nu^B)(\mu|\nu) + (\mu|1/2\nabla^2|\nu) \\ &\quad + \tfrac{1}{2}(\mu|\nu) \sum_{C\neq A,B} (\mu|(M_C - Z_C^c)|\nu)[\gamma_{AC} + \gamma_{BC}] \end{aligned} \tag{76b}$$

All of the extended Hückel models yield frontier molecular orbitals that are rich in interpretability. They are *physical* in the sense that the orbital energies are generally close to ionization potentials because of their parametrization, and differences between orbital energies are often related back to electronic spectroscopy. Unlike *ab-initio* models, the highest occupied orbitals (the HOMO's) and the lowest unoccupied (LUMO's) are generally metal d. The Extended Hückel models used today are not in general successful for obtaining structural information, nor detailed molecular spectroscopy. But they certainly are simple; and very powerful in the right hands!

2.6.2 ***Zero Differential Overlap Approximations*** Today models based upon the ZDO approximation are more refined and better parametrized than those of the Extended Hückel type, and they have been applied to a greater variety of problems, including the prediction of geometry and detailed molecular electronic spectra. There are now very many examples of the use of ZDO models for examining transition metal complexes. The Intermediate Neglect of Differential Overlap (INDO) model is the most commonly used for this purpose. Models based on the Neglect of Differential Diatomic Overlap approximation (NDDO) such as AM1 and PM3 are under development.[12,13,148,149]

A few reviews of the INDO model Hamiltonian are available.[11,12,161] This model is based on the original INDO model of Pople, Segal, and Santry[162–165] and is similar to the Complete Neglect of Differential Overlap (CNDO) model adapted for spectroscopy by Jaffé and Delbene.[11] The CNDO model, as outlined below, does not contain the one-center two-electron terms (Slater Condon Factors[114]) and is therefore not useful for examining the spectroscopy of transition metal complexes, or as accurate in reproducing multiplet structure. We return to this point below.

Let **F** be the Fock matrix, **H** the one-electron matrix, and **G** the two-electron matrix, and take the UHF case as an example, with no loss of generality. Then the Hartree–Fock Roothaan method yields[59]

$$\mathbf{F}^{\alpha} = \mathbf{H} + \mathbf{G}^{\alpha} \tag{77a}$$

and (see Eqn. 9)

$$F_{\mu\nu}^{\alpha} = h_{\mu\nu} + \sum_{\sigma\lambda} \{P_{\sigma\lambda}(\mu\nu|\sigma\lambda) - P_{\sigma\lambda}^{\alpha}(\mu\sigma|\nu\lambda)\} \tag{77b}$$

with $\mathbf{P}^{\alpha}$ the first order Fock–Dirac a spin-density, with elements given by (see Eqn. 11)

$$P_{\mu\nu}^{\alpha} = \sum c_{\mu i}^{\alpha} c_{\nu i}^{\alpha} n_{\iota}^{\alpha} \tag{78a}$$

with similar expressions for β spin. n_{ι}^{α} is the occupation of the MO $\phi_{\iota}^{\alpha} = \sum_{\nu} X_{\nu} C_{\nu i}$, either zero or one, and the total density **P** is given by

$$\mathbf{P} = \mathbf{P}^{\alpha} + \mathbf{P}^{\beta}. \tag{78b}$$

An orthogonalized basis is assumed in all ZDO models,

$$\Delta_{\mu,\nu} = \langle \chi_\mu | \chi_\nu \rangle = \delta_{\mu\nu}$$

and the one most often assumed is a symmetric orthogonalized basis (also called a Löwdin orthogonalized basis, LTOs)

$$(\chi_\mu, \chi_\mu, \ldots \chi_\mu) = \mathbf{X} \Leftarrow \mathbf{X}\Delta^{-1/2} \tag{79}$$

Although it is not necessary to justify the model using this expansion, it is often invoked. In general, one-center two-electron integrals are slightly increased, two-center two-electron integrals are slightly decreased, and three- and four-center two-electron integrals are greatly reduced in value.[174]

The NDDO model takes advantage of these observations, and differential overlap between AO's on different atoms is neglected, Figure 12 (to be discussed in greater detail below)

$$\chi_\mu^{A*}(1)\chi_\nu^{B}(1)\delta\tau(1) = \chi_\mu^{A*}(1)\chi_\nu^{A}(1)\delta\tau(1)\delta_{AB}$$

and this leads to NDDO.

$$F_{\mu\mu}^{\alpha} = U_{\mu\mu} - \Sigma_{B\neq A} Z_B(\mu|R_B^{-1}|\mu) + \Sigma_{\sigma\lambda}\{P_{\sigma\lambda}(\mu\mu|\sigma\lambda) - P_{\sigma\lambda}^{\alpha}(\mu\sigma|\mu\lambda)\}$$

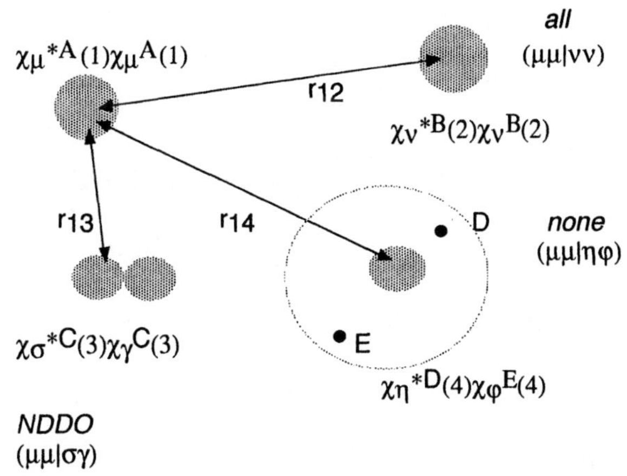

Figure 12. A schematic for the integrals that are included in the various ZDO models. See text for a detailed discussion.

NDDO
diagonal μ on atom A

$$
\begin{aligned}
F^{\alpha}_{\mu\mu} &= [h_{\mu\mu}] + [G^{\alpha}_{\mu\mu}] \\
&= \left[U_{\mu\mu} - \sum_{B \neq A} Z^{c}_{B}(\mu|R^{-1}_{B}|\mu)\right] + \left[\sum_{\sigma\lambda}\{P_{\sigma\lambda}(\mu\mu|\sigma\lambda) - P^{\alpha}_{\sigma\lambda}(\mu\sigma|\mu\lambda)\}\right] \\
&\Rightarrow \left[U_{\mu\mu} - \sum_{B \neq A} Z^{c}_{B}(\mu\mu|\bar{b}\bar{b})\right] + \sum_{B} \sum_{\{\sigma\lambda \text{ both on } B\}} \left\{P_{\sigma\lambda}(\mu\mu|\sigma\lambda)\right. \\
&\quad \left. - \sum_{\{\sigma,\lambda \text{ both on } A\}} P^{\alpha}_{\sigma\lambda}(\mu\sigma|\mu\lambda)\right\}
\end{aligned} \tag{80a}
$$

where an approximation similar to that of Eqn. 74 is again made. An overbar again indicates that only the spherically symmetric part of the integral is maintained.[3,4] $U_{\mu\mu}$ is the core integral, and it is calculated from ionization potentials

$$U_{\mu\mu} = \mathrm{IP}_{\mu} + (Z^{c}_{A} - 1)^{\gamma} AA + \text{higher order Slater–Condon terms} \tag{80b}$$

The one-center two-electron integrals are taken from experimental Slater–Condon radial integrals obtained from atomic spectroscopy

$$(\mu\sigma|\nu\lambda) = f\left(F^{k}, G^{k}, R^{k}\right) \tag{80c}$$

or they are purely empirical.[11,174]

(These effective one- and two-electron integrals can also be derived exactly, in some simple cases, from Effective Hamiltonian Theory, as discussed in Appendix II. Such *ab-initio* calculations explain the need for empirical corrections.)

Off-diagonal ν on A, ν on atom B, A ≠ B

$$
\begin{aligned}
F^{\alpha}_{\mu\nu} &= [h_{\mu\nu}] + [G^{\alpha}_{\mu\nu}] \\
&= \left[\tfrac{1}{2}(\beta^{A}_{\mu} + \beta^{B}_{\nu})\Delta_{\mu\nu}\right] - \sum_{\{\lambda \text{ on } B\}} \sum_{\{\sigma \text{ on } A\}} [P^{\alpha}_{\mu\nu}(\mu\sigma|\nu\lambda)]
\end{aligned} \tag{80d}
$$

where β^{A}_{μ} = the *resonance integral*, and $\Delta_{\mu\nu}$ = weighted overlap. For further details see references.[11]

In the CNDO model, all differential overlap is neglected

$$\chi^{*}_{\mu}(1)\chi_{\nu}(1)\delta\tau(1) = \chi^{*}_{\mu}(1)\chi_{\mu}(1)\delta\tau(1)\,\delta_{\mu\nu} \tag{81a}$$

as are all the higher Slater–Condon integrals.

CNDO
diagonal μ on atom A

$$F^{\alpha}_{\mu\mu} = [h_{\mu\mu}] + [G^{\alpha}_{\mu\mu}]$$

$$\Rightarrow U_{\mu\mu} - \sum_{B \neq A} Z_B \gamma_{AB} + \sum_{\{\sigma \text{ on } B\}} P_{\sigma\sigma}(\bar{\mu}\,\bar{\mu}|\bar{\sigma}\,\bar{\sigma}) - P^{\alpha}_{\mu\mu}(\bar{\mu}\,\bar{\mu}|\bar{\mu}\,\bar{\mu}) \quad (81b)$$

Off-diagonal μ on A, ν on atom B, $A \neq B$

$$F^{\alpha}_{\mu\nu} = [h_{\mu\nu}] + [G^{\alpha}_{\mu\nu}] = \left[\tfrac{1}{2}\left(\beta^A_{\mu} + \beta^B_{\nu}\right)\Delta_{\mu\nu}\right] - P^{\alpha}_{\mu\nu}(\bar{\mu}\,\bar{\mu}|\bar{\nu}\,\bar{\nu}) \quad (81c)$$

Because the INDO model includes the higher Slater-Condon integrals (or Racah coefficients[174]) neglected at the CNDO level, the INDO/S parametrization can describe the numerous low-lying, atomic-like excited states in weakly bound transition metal systems with considerable accuracy. Recall the comments on the electron correlation problem, and the comparison between the splittings of the atomic configurations of carbon vs. transition metals. The s^2p^2 configuration of carbon gives rise to 15 states, of ${}^3P$, ${}^1D$, and ${}^1S$ symmetry, split in energy by 20,000 cm^{-1} = 57 kcal/mol. The CNDO model does not contain this splitting, but rather incorporates an average energy for all s^2p^2 electronic configurations. Hence, CNDO calculations will underestimate the splittings between electronic states of different multiplicities in simple molecules. In contrast, the neglected terms in weakly bonded systems, such as transition metal systems, are as large as 500 kcal/mol, and, consequently, the CNDO scheme offers limited information on the electronic spectroscopy of transition metal complexes. Because the INDO model includes all one-center, two-electron integrals, it performs quite well for many transition metal systems.

INDO

diagonal μ on atom A

$$\begin{aligned} F^{\alpha}_{\mu\mu} &= [h_{\mu\mu}] + [G^{\alpha}_{\mu\mu}] \\ &= \left[U_{\mu\mu} - \sum_{B \neq A} Z_B(\bar{\mu}\,\bar{\mu}|\bar{\nu}\,\bar{\nu})\right] + \left[\sum_{\{\sigma\lambda \text{ both on } A\}} \{P_{\sigma\lambda}(\mu\mu|\sigma\lambda) \right. \\ &\quad \left. - P^{\alpha}_{\sigma\lambda}(\mu\sigma|\mu\lambda)\} + \sum_{\{\sigma \text{ not on} A\}} \{P_{\sigma\sigma}(\bar{\mu}\,\bar{\mu}|\bar{\sigma}\,\bar{\sigma})\}\right] \end{aligned} \quad (83a)$$

Off-diagonal μ on A, ν on atom B, $A \neq B$

$$F^{\alpha}_{\mu\nu} = [h_{\mu\nu}] + [G^{\alpha}_{\mu\nu}] = \left[\tfrac{1}{2}\left(\beta^A_{\mu} + \beta^B_{\nu}\right)\Delta_{\mu\nu}\right] - \left[P^{\alpha}_{\mu\nu}(\bar{\mu}\,\bar{\mu}|\bar{\nu}\,\bar{\nu})\right] \quad (83b)$$

It is interesting to insert Eqn. 56b into the diagonal of any of the ZDO models. This gives rise to

$$\begin{aligned} F_{\mu\mu} &= -\mathrm{IP}_{\mu} + (P_{AA} - Z^c_A)\gamma_{AA} + (1 - P_{\mu\mu})\gamma_{AA} \\ &\quad + \sum (P_{BB} - Z^c_B)\gamma_{AB} + \text{higher order terms} \end{aligned} \quad (84)$$

which can be compared with a similar expression from the EHT model, Eqn. 70. The charges in 70, $(M_B - Z_B^c)$, over Slater-type overlapping orbitals are replaced with those over non-overlapping AO's, $(P_{BB} - Z_B^c)$.

Figure 12 shows the physical meaning behind these integral approximations. Eqn. 84 gives the energy of a probe electron in the distribution $\chi_\mu(1)^*\chi_\mu(1)$. By Koopmans' approximation this would be the IP if there were no other atoms. All the ZDO models contain the classic two-center two-electron Coulomb integral $(\mu\mu|\nu\nu)$ weighted by the number of electrons on orbital χ_ν, $P_{\nu\nu}$. NDDO also contains $(\mu\mu|\sigma\gamma)$ weighted by the number of electrons $P_{\sigma\gamma}$ in the $\chi_\sigma^{*C}(3)\chi_\gamma^C(3)$ distribution. Terms such as $(\mu\mu|\eta\phi)$ are not included. These three-centered integrals over the symmetrically orthogonalized basis are generally quite small,[157] are weighted by small density matrix elements $P_{\eta\phi}$ of varying sign, and tend to cancel in the Fock matrix with corresponding one-electron three-centered nuclear attraction integrals, also neglected.

In all of the ZDO models, the Hamiltonian has an exact form, which, given in second quantized form is

$$H = \sum_{\mu\nu}\langle\mu|h|\nu\rangle\mu^+\nu + \tfrac{1}{4}\sum_{\mu\nu}\sum_{\sigma\lambda}\langle\mu\nu\|\sigma\lambda\rangle\mu^+\nu^+\lambda\sigma \tag{85}$$

The operators μ^+ and ν are Fermion creation and annihilation operators, described in Appendix I. The parameters of this Hamiltonian, $\langle\mu|\mathbf{h}|\nu\rangle$ and $\langle\mu\nu\|\sigma\lambda\rangle$ are calculated in *ab-initio* theory, or are empirically obtained in ZDO models (see Appendix II). This relationship demonstrates that the methods of *ab-initio* quantum chemistry discussed previously and those of semiempirical quantum chemistry are the same.

In the above only integral approximations underlying the models are discussed. The choice of parameters and the details of functional forms for the integrals can be quite complex, and is better examined in the original literature where detailed explanation is given.[11,12] The original CNDO and INDO models (CNDO/1, CNDO/2, INDO/1, INDO/2) were designed to reproduce minimum basis set *ab-initio* calculations.[11,12,160–164] The first useful NDDO models (MNDO, AM1 and PM3) were parameterized directly on experiment to yield ground state geometries, heats of formation, dipole moments and lowest ionization potentials at the SCF level of theory.[11,12,147–149] The CNDO/s and INDO/s models are parameterized at the CIS level to yield experimental molecular electronic spectra.[11,12,157–160]

3 THE GROUND STATE

3.1 General Concepts

One of the most dramatic advances in quantum mechanics in the last two decades, since the pioneering work of Pulay,[175] and of Kormonicki and McIver[5,6] has been the introduction of gradient methods into quantum chemistry codes.[1–5] With these, it is possible in a more or less routine fashion to obtain the shapes of molecules just by specifying the nuclei and the number of electrons in the system, and an educated guess as to what the structure might be. After obtaining the structure, similar techniques can be used to estimate the vibrational spectrum of the model compound, and even examine its reactivity.

There are, of course, some provisos to this optimistic view. The accuracy of the results we obtain depend on the quality of the quantum mechanics we can use. This is clearly a function of the size of the system. Larger systems require more approximate methods, and with these come some uncertainties. In addition, larger systems may have many low-lying minima, and sorting through these multiple minima for the lowest energy one is still an uncertain process.[5] Although chemical intuition is a good guide for a starting geometric structure, often very interesting tautomers can be missed this way—by being trapped in a local minima not easily connected to the global minimum energy structure. Intuition is also a poor guide to uncovering transition states, as we, as yet, have less experience here. Transition state structures do not exist *experimentally*.[6] They are well defined, however, *theoretically* as the minimum energy pass between two stable structures, perhaps the reactant and the product of a chemical reaction.[3]

In this section we briefly discuss the general problem of uncovering extreme points of the potential energy surface, those points in which

$$\mathbf{f}\mathbf{f}^{+} = \sigma = 0 \tag{86a}$$

where $\mathbf{f}$ is a row matrix with elements given by

$$f_i = \partial E(\mathbf{X})/\partial x_i \tag{86b}$$

E is the potential energy surface obtained from the quantum mechanics assuming the fixed-nuclei Born–Oppenheimer approximation,[7] and x_i is one of the $3n$ coordinates describing the location of n atoms. Figure 13 presents a simple potential energy surface. Extreme points include minima (stable structures), transition states (defined below as the minimum energy pass between two stable structures) and higher level *saddles* of less chemical interest.

The Hessian matrix, or matrix of second derivative, $\mathbf{H}$ (the force constant matrix), is given by

$$H_{ij} = \partial^2 E/\partial x_i \partial x_j \tag{87}$$

Minima of the potential energy surface are obtained when the squared gradient norm, $\sigma = 0$, and all the eigenvalues of the Hessian are positive or zero, see Figure 13. If $\mathbf{x}$ are Cartesian coordinates there will be six (or five, for linear molecules) zero eigenvalues of the Hessian corresponding to the three translational and three (or two) rotational degrees of freedom that are without constraint in the fixed nuclei approximation.[6] In practice, an iterative procedure is invoked that will reduce σ to below a set threshold. In such a procedure the rotational degrees of freedom may mix with

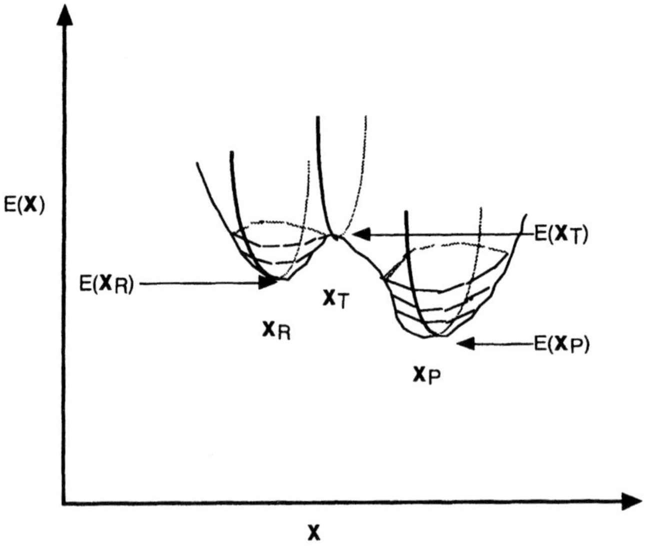

Figure 13. A potential energy surface of two geometric variables, showing two minima, $E(X_R)$ and $E(X_P)$ and a transition state separating them, $E(X_T)$.

the low-lying vibrations yielding small positive eigenvalues for the rotations unless a smaller threshold is set, or the rotational degrees of freedom are projected out. True minima of the potential energy surface should be checked by calculating the second derivative matrix, and ensuring that all eigenvalues of this matrix are, indeed, positive, or zero.

Transition state structures are characterized as extreme points where one and only one eigenvalue of the Hessian is negative.[2] This negative eigenvalue corresponds to the imaginary frequency of transition state theory,[3] and the eigenvector yields the reaction coordinate at the transition state geometry.

3.2 Gradients

Let $E = E(\mathbf{X})$ be the Born–Oppenheimer Energy at the geometry $\mathbf{x}$. We begin by expanding this energy in a Taylor series,

$$E(\mathbf{X}') = E(\mathbf{X}) + \mathbf{f}(\mathbf{X})\mathbf{q}^+ + 1/2\mathbf{q}\mathbf{H}(\mathbf{X})\mathbf{q}^+ + \cdots \tag{88}$$

$$f_i(\mathbf{X}) = (\partial E/\partial X_i)_{\mathbf{x}} : \mathbf{q} = (\mathbf{X}' - \mathbf{X}) : H_{ij} = \left(\partial^2 E/\partial X_i \partial X_j\right)_{\mathbf{x}}$$

gradient — step — Hessian or Force Constant Matrix

Let $\mathbf{X} = \mathbf{X}_e$ be an extreme point on the Born–Oppenheimer surface. Then

$$\mathbf{f}\mathbf{f}^+ = \sigma = 0$$

and, assuming a quadratic surface,

$$E(\mathbf{X}) = E(\mathbf{X}_e) + \tfrac{1}{2}\mathbf{q}\mathbf{H}(\mathbf{X})\mathbf{q}^+ \tag{89}$$

It is easy to see from this that if $\mathbf{H}$ is positive semi-definite, $E(\mathbf{X}_e)$ is a minimum. Similarly, we can expand the gradient,

$$\mathbf{f}(\mathbf{X}') = \mathbf{f}(\mathbf{X}) + \mathbf{q}\mathbf{H}(\mathbf{X}) + \cdots \tag{90}$$

and, again assuming a quadratic function, at $\mathbf{X}' = \mathbf{X}_e$

$$\mathbf{f}(\mathbf{X}) = -\mathbf{q}\mathbf{H}(\mathbf{X}) \tag{91}$$

If $\mathbf{H}^{-1}$ exists, then

$$\mathbf{q} = -\mathbf{f}(\mathbf{X})\mathbf{H}^{-1}(\mathbf{X}) = -\mathbf{f}(\mathbf{X})\mathbf{G}(\mathbf{X}) \tag{92}$$

$\mathbf{q}$ then gives a step toward the minimum. If the surface were quadratic, this one step would reach the minimum, but unfortunately, for real molecules, this is only locally true. Eqn. 92 is a generalization of the Newton–Ralphson[8] procedure into many dimensions.

We might note that the inverse of $\mathbf{H}$ does not, in fact, exist because of the zero eigenvalues corresponding to the rotations and translations discussed previously. In practice this will cause no difficulty, for the procedures we use either work in $(3n-6)$ or $(3n-5)$ internal coordinates, or they project these zero eigenvalue directions from the Hessian, or they never update the approximate Hessian in those directions with zero eigenvalues.

Eqn. 92 is the working equation for all present-day geometry optimization methods. It requires evaluation, or approximation, of the gradients and the Hessian.[3,4]

In general, first derivatives are evaluated using analytic expressions, as this evaluation takes little more time than does the evaluation of the energy *for any wavefunction*. Second derivatives are most often approximated using quasi-Newton update procedures, as described below. The evaluation of second derivatives are far more time consuming than are the first derivatives, generally requiring a factor of n more time, where n is the size of the basis set. For example, for Hartree–Fock theory coupled perturbed Hartree–Fock theory (CPHF)[175] is used, and this adds a step that is n^5.

The quasi-Newton *update* methods update an approximate Hessian based upon the numerical experience gained in examining the potential energy surface. The quasi-Newton condition is

$$-\mathbf{G}^{k+1}\boldsymbol{\gamma}^k = \boldsymbol{\delta}^k \tag{93a}$$

where

$$\mathbf{G}^{k+1} = \left(\mathbf{H}^{k+1}\right)^{-1} \tag{93b}$$

is the inverse Hessian, and $\boldsymbol{\gamma}^k$ is obtained numerically from the differences between the calculated gradients at two different geometries, $\mathbf{X}^{k+1}$ and $\mathbf{X}^k$

$$\boldsymbol{\gamma}^k = \mathbf{f}^{k+1} - \mathbf{f}^k \tag{93c}$$

$$\boldsymbol{\delta}^k = \alpha^k \mathbf{q}^k = \mathbf{X}^{k+1} - \mathbf{X}^k \tag{93d}$$

$\boldsymbol{\delta}^k$ is known as the *step*.

Most methods begin a geometry search with an initial Hessian inverse assumed to be the unit matrix, $\mathbf{G}^0 = \mathbf{1}$ (in which case the first step is steepest descent. The zero eigenvalues of the Hessian never appear as these directions are never updated from the unit initial value) or from the inverse of a diagonal matrix in which common force constants are tabulated (in such a case internal coordinates must be used or the inverse will contain these 6 (or 5) ill defined values). The update then is calculated as $\mathbf{e}$,

$$\mathbf{G}^{k+1} = \mathbf{G}^k + \mathbf{e} \tag{94}$$

There is a hierarchy of methods for updating the inverse Hessian. The most common are based upon:[1–3]

i. Fletcher's method
ii. Davidson–Fletcher–Powell (DFP)
iii. Murtagh–Sargent Method (MS)
iv. Broyden Method
v. Broyden–Fletcher–Goldfarb–Shanno Method (BFGS)

The easiest to understand are the rank 1 methods. Using Eqns. 93a and 94, we get for the k'th update $\mathbf{e}^k$

$$\mathbf{e}^k|\boldsymbol{\gamma}^k\rangle = -|-\boldsymbol{\delta}^k\rangle - \mathbf{G}^k|\boldsymbol{\gamma}^k\rangle \tag{95a}$$

Let

$$\mathbf{e}^k = a|n\rangle\langle n| \tag{95b}$$

where $|n\rangle$ is a column matrix of one element, which defines rank 1. Then

$$a|\mathbf{n}\rangle\langle\mathbf{n}|\boldsymbol{\gamma}^k\rangle = -|\boldsymbol{\delta}^k\rangle - \mathbf{G}^k|\boldsymbol{\gamma}^k\rangle \tag{95c}$$

Since a is a scale, we can set

$$a\langle\mathbf{n}|\boldsymbol{\gamma}^k\rangle = 1 \tag{95d}$$

yielding

$$
\begin{aligned}
|\mathbf{n}\rangle &= -|\boldsymbol{\delta}^k\rangle - \mathbf{G}^k|\boldsymbol{\gamma}^k\rangle \\
&= (|\boldsymbol{\delta}^k\rangle + \mathbf{G}^k|\boldsymbol{\gamma}^k\rangle)(\langle\boldsymbol{\gamma}^k|\mathbf{G}^k + \langle\boldsymbol{\delta}^k|)/(\langle\boldsymbol{\gamma}^k|\mathbf{G}^k|\boldsymbol{\gamma}^k\rangle + \langle\boldsymbol{\delta}^k|\boldsymbol{\gamma}^k\rangle)
\end{aligned} \tag{96}
$$

which is the Davidon–Fletcher–Powell Equation.

In general, the geometry has converged to within specified tolerances before the updated Hessian is really very accurate. We note that the Hessian or its inverse need not be determined so accurately for this search, however, for when we have reached a minimum, or any extreme point, $\mathbf{q} = \mathbf{0}$ when $\mathbf{f} = \mathbf{0}$, regardless of the accuracy of $\mathbf{G}$, Eqn. 92.

3.3 Optimization Strategies

A search strategy consists of four basic parts:

1. A search direction $\mathbf{q}^k$ is chosen.
2. A line search is performed to determine how far along in this direction the move should be. One chooses a value α^k and moves to $\mathbf{X}^{k+1} = \mathbf{X}^k + \alpha^k\mathbf{q}^k$. (In other words, it is useful to know that New York is north of Miami, but it is equally important to estimate how far north this is!) Figure 14 schematically shows the relationship between line search direction $\mathbf{q}^k$, and line search length α^k.
3. A test for convergence is made. If convergence is satisfied, quit. If not,
4. Calculate or estimate a new Hessian $\mathbf{H}^{k+1}$, or its inverse $\mathbf{G}^{k+1}$ and repeat steps 1 through 3.

Line search strategies can be quite involved. The loosest line searches only demand that the energy be reduced step by step in determining minima (or $\mathbf{ff}^+$ be reduced if searching for transition states). The more sophisticated line searches can extrapolate using surface information, contracting the step, or expanding the step, appropriately,[1] see Figure 14.

Convergence is determined on the basis of energy

$$E(\mathbf{X}_n) - E(\mathbf{X}_{n-1}) < \varepsilon \tag{97a}$$

and/or the maximum gradient or the gradient norm

$$\sigma = \mathbf{f}(\mathbf{X}_n)\mathbf{f}(\mathbf{X}_n)^+ < \delta \tag{97b}$$

and/or the step size

$$\mathbf{X}_n\mathbf{X}_n^+ < \delta' \tag{97c}$$

where ε, δ, and δ' are chosen thresholds. Usually ε is chosen to be of the order of 10^{-4} Hartrees (less than 0.1 kcal/mole) and δ between 10^{-3} and 10^{-4} Hartree/Bohr,

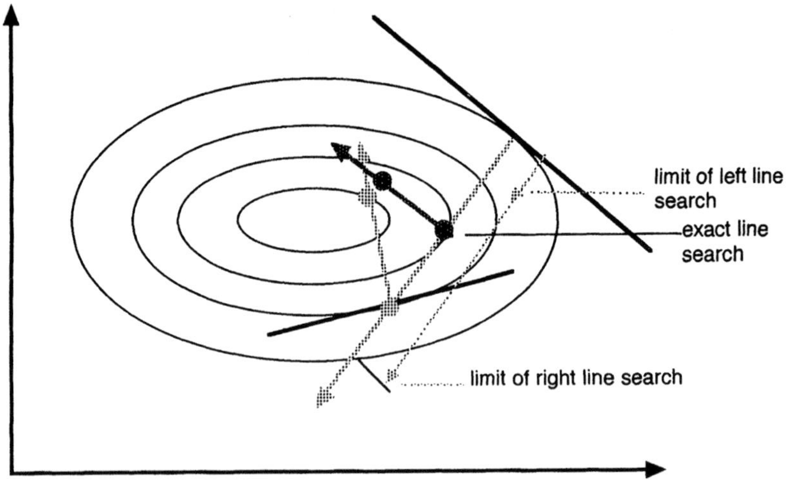

Figure 14. Line search strategies. The darker lines designate an exact line search that minimizes the energy along the search direction, and that can take several calculations. The lighter lines designate a partial line search, and the figure is designed to demonstrate that this is most often a more efficient procedure than is the exact line search. The limits of partial line search are designated.

a value general consistent with the energy criterion, with bond distances good to 0.001Å. The accuracy with respect to angles is less certain, as many dihedrals have very soft potentials, and one must be careful here. For this reason a threshold on δ' should not be used in isolation from the other thresholds.

The progress of a typical geometry search is summarized in Table 4 for the $^1A_1$ state of Fe(II) porphyrin, the situation given in Figure 1. The second optimization shown in this table uses an approximation of the Hessian using first order perturbation theory,[2,175] and to our knowledge, is more successful than any simple update procedure.

3.4 Transition States

Transition states separate reactants from products, and they can be as simple as the planar form of ammonia separating the two pyramidal forms, or as complex as the maximum energy on the path separating two separate reacting species.[176] All transition states are characterized as possessing $\sigma(\mathbf{X}^t) = \mathbf{f}(\mathbf{X}_n^t)\mathbf{f}(\mathbf{X}_n^t)^+ = 0$, and one and only one eigenvalue of the Hessian matrix $\mathbf{H}^t$ is negative (that is, this point of the potential energy surface is uphill in energy in all directions, but downhill only in the reaction direction). In addition, $E(\mathbf{X}^t)$ should be the highest energy on this path separating reactant from product.

Unlike searching for minima, searching for transition states does not always succeed. There are a variety of reasons for this. One, certainly, is the requirement that a transition state has *one and only one* negative eigenvalue of the Hessian, and this

Table 4 The Progress of a Geometry Minimization of the $^1A_1$ State of Fe(II) Porphyrin, Figure 1, Starting from a Drawn Structure Using the INDO/1 Model Hamiltonian. The Energy is in Hartrees, as is the SCF Convergence. The Latter is Examined to Ensure the Accuracy of the Calculated Gradients that Are Evaluated from Analytic Expressions that Assume SCF Convergence for Their Accuracy. The Gradients Are Given in Hartrees/Bohr. Left and Right Fails Refer to the Line Search (There Are None, Corresponding to a Loose Line Search) and Alpha Is Given in Equation 93d

Cycle BFGS	Energy	SCF Conv.	Grad Norm	Max Grad	Alpha	Left Fails	Right Fails
0	−202.485808	0.000000	0.083019	0.017200	0.000000	0	0
1	−202.490553	0.000000	0.049127	0.013900	0.400000	0	0
2	−202.492572	0.000000	0.024816	0.005673	1.000000	0	0
3	−202.485707	0.000000	0.024393	0.005358	1.000000	0	0
4	−202.498135	0.000000	0.024840	0.006173	1.000000	0	0
5	−202.499029	0.000000	0.011568	0.002282	1.000000	0	0
6	−202.499279	0.000000	0.008594	0.001783	1.000000	0	0
7	−202.499362	0.000000	0.004894	0.001446	1.000000	0	0
8	−202.499391	0.000000	0.000562	0.000134	1.000000	0	0
CPU TIME: 0 H 0 MIN 15.79 SEC; REAL TIME: 0 H 0 MIN 16 SEC (98.7%)							
First Order Approximation (ref 173)							
0	−202.485808	0.000000	0.083019	0.017200	0.000000	0	0
1	−202.499234	0.000000	0.005516	0.001176	1.000000	0	0
2	−202.499388	0.000000	0.000673	0.000179	1.000000	0	0
CPU TIME: 0 H 0 MIN 37.84 SEC; REAL TIME: 0 H 0 MIN 43 SEC (88.0%)							

must always be checked, and a second is that the wavefunction employed must be flexible enough to describe the transition state properly.

Obtaining transition states by these methods, as mentioned, is more difficult. Most of the methods utilized use explicitly calculated second derivatives, and, to date, none of the update procedures are accurate enough to guarantee an uphill climb in one direction, yet downhill in all others, see Figure 13 and Figure 15. As the calculation of second derivatives are expensive in both computer time and storage, large molecules are in the hands of specialists.

The *object* function to be minimized in transition state searches is $\sigma(\mathbf{X})$. The general starting point is again a Taylor series.

$$\alpha(\mathbf{X}_{k+1}) = \sigma(\mathbf{X}_k) + \mathbf{q}_{k+1}\mathbf{V}(\mathbf{X_k}) + \tfrac{1}{2}\mathbf{q}_{k+1}\mathbf{T}(\mathbf{X})\mathbf{q}^+{}_{k+1} + \cdots \tag{98}$$

with

$$\mathbf{V}(\mathbf{X_k})_i = (\partial\sigma(\mathbf{X})/\partial X_i)_{\mathbf{X}=\mathbf{X}_k} \tag{99}$$

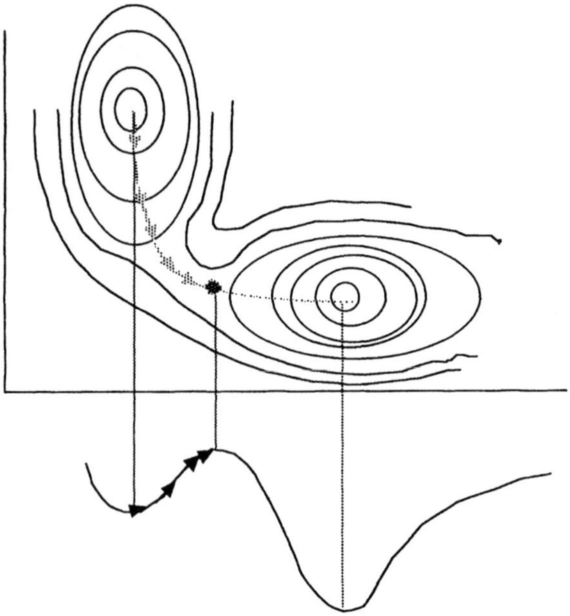

Figure 15. A schematic of a search for a transition state starting from the reactants.

and

$$\mathbf{T}(\mathbf{X_k})_{i,j} = (\partial^2\sigma(\mathbf{X})/\partial X_j \partial X_j)_{\mathbf{X}=\mathbf{X}_k} \tag{100}$$

At the minimum value of σ, $\sigma(\mathbf{X}^t) = 0$, and $\mathbf{V}(\mathbf{X}^t) = \mathbf{0}$. This yields

$$\mathbf{q}_{k+1} = -\mathbf{T}(\mathbf{X_k})^{-1}\mathbf{V}(\mathbf{X_k}). \tag{101}$$

an iterative expression in which $\mathbf{q}_{k+1}$ should approach $\mathbf{q}^t$. Of course, from the above, we are guaranteed only to find an extreme point ($\sigma = 0$), not necessarily a transition state. When σ is below the specified threshold, or the step is below a specified value (see above in searching for a minima) the Hessian must be calculated and the *signature* (the number of positive and negative roots) checked to ensure that only one is negative.

In the above,

$$\mathbf{V}_k = 2\mathbf{H}_k\mathbf{f}_k^+ \tag{102a}$$

$$\mathbf{T}_k = 2(\mathbf{H}_k\mathbf{H}_k^+ + \mathbf{C}_k) \tag{102b}$$

$$(\mathbf{C}_k)_{i,j} = \sum_\alpha f_\alpha\left(\partial^2 f_\alpha/\partial X_i \partial X_j\right) \tag{102c}$$

Table 5 The Anatomy of a Transition State Search Using the ZINDO Program and the Augmented Hessian to Uncover the Region of One Negative Eigenvalue Starting from a Minimum[a] and then Switching to the Newton Method, Equation 103 to Find the Nearest Extreme POint, See Text

IT	Energy (Hartrees)	SCF Conv.	Grad Norm	Max Grad	α	Comments
0	−12.522890	0.000003	0.000057	0.000028	0.000000	Augmented
1	−12.522849	0.000013	0.001337	0.000700	0.000000	Hessian
2	−12.521285	0.000002	0.013850	0.005653	0.000000	Cycles until
3	−12.516034	0.000005	0.021692	0.009547	0.000000	negative root
3	−12.509819	0.000006	0.025393	0.012428	0.000000	Minimize
0	−12.509819	0.000006	0.025393	0.012428	0.000000	the gradient
1	−12.507360	0.000008	0.023395	0.012907	0.325770	norm
2	−12.507139	0.000004	0.008693	0.004908	0.967428	cycles
3	−12.507084	0.000000	0.000900	0.000493	1.000000	stop

[a]Reference 2.

C is the tensor of third derivatives of the energy with respect to nuclear coordinate, which, if really required, would render this method intractable in most cases. Rather near an extreme point $\mathbf{f} \rightarrow \mathbf{0}$ and $\mathbf{T}_k \Rightarrow 2\mathbf{H}_k\mathbf{H}_k^+$. Under such conditions,

$$\mathbf{q}_{k+1} = -\mathbf{T}(\mathbf{X_k})^{-1}\mathbf{V}(\mathbf{X_k}) = -\mathbf{H}_k^{-1}\mathbf{f}_k^+ \tag{103}$$

which is *equivalent* to the Newton–Ralphson search for a minimum, except the function being minimized is σ, not the energy, $E(\mathbf{X})$. This suggests the most common, and often most successful, method, of finding transition states: guessing, and then minimizing the norm or the gradient squared (σ), and hoping to find only one negative eigenvalue.

More elegant ways of searching for transition states start from product or reactant, or both, and then march uphill along either the softest normal mode (obtained by diagonalizing the Hessian), or along that mode with greatest overlap (scalar or dot product) with the vector connecting reactant with product, i.e., $\mathbf{d} = \mathbf{X}_{\text{product}} - \mathbf{X}_{\text{reactant}}$. Such methods are often called *eigenvector following*. Another interesting method is the gradient extremal method.[177] All such methods contain an uphill climb, with energy increasing and σ increasing, see Figure 15, and Table 5, along a soft mode, or projected mode, until a negative root of the Hessian is uncovered. They may then continue following this mode to product, or connect the direction vector to the product, or simply employ Eqn. 103. In any case, σ is then reduced, while the energy generally increases (although it need not if the search has gone up on the shoulder of the transition direction valley) to that of the transition state.

3.5 More on the Meaning of Orbitals

We have previously discussed the meaning of the occupied orbitals and their energies. It is useful to continue that discussion in greater detail, for orbitals are a popular and easy to grasp concept, far more easier to grasp than the physics behind many electron functions.

Koopmans' approximation suggests that the energy difference between a parent molecule and the same system with one less electron should be given by the negative of the orbital energy, *if* the orbitals are unchanged (the frozen orbital approximation). This is provable for RHF and UHF calculations, and either the closed-shell or open-shell orbitals of a ROHF calculation, but not both.[70] The corrections required in the ROHF theory are often quite small.[70] In reality ionization energies are given as the differences between the energies of two many-electron states, and these do contain both orbital relaxation and differing amounts of correlation.

If the orbital energies reflect the energies of ionization processes, the *aufbau* principal pertains, and electrons can be assigned to orbitals in increasing order of energy until all are assigned, generating the ground state SCF wavefunction. As Figure 1 suggests, this may not be the case, especially in open-shell situations. Were this to be the case, however, the symmetry of the frontier orbitals would reflect the symmetry of the states generated from loss or gain of electrons from these orbitals. This is a huge simplification, and underlines such important concepts as the Woodward–Hoffman symmetry rules for concerted reactions.

In very many cases the orbital energies *do* reflect the ionization energies—in those cases where the relaxation energies are small, and tend to cancel the correlation differences. In those cases in which Koopmans' approximation holds (RHF, UHF), the orbital energies are upper-bounds to the ionization energies obtained from ΔSCF calculations, as the energy obtained from a separate calculation on the ion using the same (frozen) orbitals of the neutral can only improve through an SCF procedure that minimizes the energy, thus relaxing those orbitals to best fit the ion. The correlation energy of the ion, however, with one less electron, is somewhat less in the ion, and the correlation energy differences tend to cancel the relaxation energy errors.

Under what conditions is relaxation small? When the orbital is delocalized, and the shock to the system when an electron is lost is distributed over many atoms. This is not the case for localized orbitals: core orbitals (the $1s$ of O_2 for example) and d and f orbitals centered on transition metal nuclei. Here the relaxation can be very large indeed, as we have mentioned in the discussion around Figure 1. The first ionization potential in any of the states of Fe(II) porphyrin is from the d MO's, even though these orbitals are quite low in orbital energy.

Even if the HOMO orbital energy approximates the IP well, this does not imply that LUMO energy yields an accurate electron affinity for the negative ion. The frozen orbital approximation fails with negative ions because negative ions have orbitals that are often quite diffuse, whereas the virtual orbitals of the parent neutral species are more compact. This occurs because negative ions have very small ionization energies and, consequently, the anion LUMOs approximate states lying in or near the continuum.

Models such as crystal field theory, ligand field theory and EHT theory can be parametrized to yield occupied orbitals with energies that produce excellent ionization energies. This is because the diagonal elements of the effective one-electron operators are generally equated to atomic IP's (see Eqns. 69a and 70) and then slightly perturbed in the complex formation. The virtual orbital energies can then be set to reproduce Δ or $10Dq$. For example, this was done in reference 171 for transition metal porphyrins, yielding a parameter $K_{\mu\nu}$ (see Eqn. 69b) of 1.89, only slightly different from the value of 1.75 originally assigned to reproduce the rotational barrier of ethane, a quite different observable. But it is now easy to realize that such a model that produces an IP (generally around 10 eV in a neutral complex)) and a virtual orbital $10Dq$ higher in energy (generally about 1.5 eV higher in energy) cannot possible produce a correct electron affinity (this would yield a virtual orbital energy at $-10 + 1.5 = -8.5$ eV, far lower than most EA's of neutral systems!)

The problem here is that we are dealing with many-electron phenomena, not single electron phenomena. For the discussion below we consider a neutral complex and a correct many electron theory (as Hartree–Fock). If the frontier orbital is delocalized, we might think that the occupied orbital would have an energy reproducing an ionization potential. The LUMO is generally a bad description for an electron affinity because the approximation of no orbital relaxation for negative ions is not good. For argument's sake, let us say that Koopmans' theorem *is* good. Then the use of this idea leads to excitation energies given by

$$\Delta E(i \rightarrow a) = \varepsilon_a - \varepsilon_i - J_{ia} + 2K_{ia}$$

if the ground state is closed-shell (see the A matrix of RPA theory in Section 4.3). In other words, even within the frozen orbital approximation, spectroscopy cannot be given by the differences in orbital energies, if the orbitals represent ionization and electron attachment processes.

To understand the above equation for excitation, the frozen orbital approximation yields ε_a as an electron affinity for a system that already contains N electrons. The empty orbitals see an N electron repulsion, whereas the occupied see $N - 1$ (an electron in an occupied orbital does not interact with itself). For spectroscopy an electron is excited from ϕ_i to ϕ_a, and now an electron in the a'th orbital has one less repulsion then it would have if this were a simple EA, namely with ϕ_i. The quantity $\varepsilon_a - J_{ia}$ is equivalent within these approximations to the EA of an electron in ϕ_a with one less electron in ϕ_i.

Can *ab-initio* models be made to yield good ionization potentials and electron affinities from orbital energies? Attempts have been made to yield occupied orbitals that reproduce ionization potentials, but these have really met with little success. And the energies of the virtual orbitals, the orthogonal complement to the occupied (they are simply what is left over) have very little definition until they are occupied. This is also easy to understand by increasing the basis set to contain many diffuse functions. These diffuse function will begin to build up the continuum, with orbital energies clustered around zero (for neutral complexes). Adding an electron to these orbitals will lead to electron loss. Adding an electron to a more localized valence virtual orbital (often a higher energy anti-bonding orbital) will cause the other occupied orbitals to relax to accommodate it, and, perhaps even to a bound state and a correct EA. (This same observation is important in spectroscopy also. The Coulomb repulsion J_{ia} between two localized orbitals will be considerably larger than one between a localized occupied and delocalized virtual, and lead to lower-lying excited states than those suggested by differences in orbital energies, $\varepsilon_a - \varepsilon_i$).

Can semiempirical models be made to yield good ionization potentials and electron affinities from orbital energies? Yes. Again experimental ionization potentials of atoms enter into the theories, and this builds in atomic relaxation and correlation. What is missing then is molecular relaxation and correlation, and this is much smaller. Numerical experience still produces the types of orbital energies seen in Figure 1, but the relaxation energy is 5 eV, compared to the *ab-initio* values of about 15 eV in these systems.[64] In the INDO/s model electron affinities are built into the Fock operator (see Eqns. 84 and 72b), easily seen by setting the charge $-(P_{AA} - Z_A^c)$ to -1. Spectroscopic transitions, however, must still be given through the CI, in this case CIS, the first approximation being the diagonal of the CI (A) matrix: that is, by the equation above for $\Delta E(i \rightarrow a)$.

Before leaving this Section, it is necessary to stress that excellent ionization potentials of neutral and positive systems can be calculated. These are not obtained from orbital energies directly, but from full many-electron correlated calculations. The calculation of ionization potentials from negative ions (electron affinities for neutrals) is still a challenging problem.

4 ELECTRONIC SPECTROSCOPY

4.1 Introduction

Many different types of spectroscopy are discussed elsewhere in this book, and here we confine ourselves only to the calculation of molecular electronic spectroscopy, as this is particularly difficult in the case of transition metal complexes, and it is particularly—well—rich!

In general the electronic spectroscopy of transition metal complexes can be classified as[114,180]

1. The ligand spectroscopy, $L \rightarrow L$
2. The Laporte $d \rightarrow d^*$ spectroscopy (ligand field excitations)
3. The metal-to-ligand and ligand-to-metal charge transfer spectroscopy, MLCT, $M \rightarrow L^*$ and LMCT, $L \rightarrow M^*$
4. Metal ligand exchange coupling, $ML \rightarrow ML^*$
5. Spin-orbit transitions.

Accurate calculation of the spectroscopy of these systems requires a reasonably accurate treatment of both the ligand orbitals and the d or f orbitals of the transition metal. This is not always easy to ensure, as the former generally participate in delocalized molecular orbitals, whereas the latter are generally localized on the metal center. The first three types of transitions are sketched in Figure 16, and examples from INDO calculations are given in Table 6 for the case of Fe(II) porphyrin,[180] Figure 1.

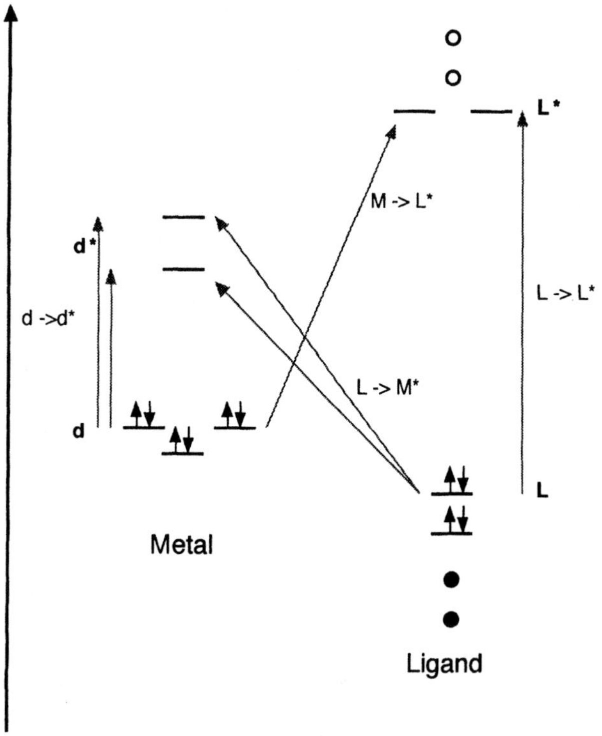

Figure 16. A schematic of the three most common types of spectroscopy associated with transition metal complexes, ligand field $d \rightarrow d^*$, ligand $L \rightarrow L^*$, and charge transfer, $L \rightarrow M^*$ (LMCT) and $M \rightarrow L^*$ (MLCT).

Table 6 Calculated Spectrum of Fe(II) porphyrin Assuming the Ground State is $^3A_{2g}$, in 1000 cm^{-1}. Numbers in Parentheses Are Osc. Strengths, from Reference 64

State	Calculated	Observed	Comments
1. $^3A_{2g}$	0.0	0.0	calc. grd state
2. ^3E_g	2.4	likely thermal[a]	$d_z^2 \rightarrow d_\pi$ ^3E_g is on occasion assigned as ground state
3. ^3E_g	11.0		$d_{xy} \rightarrow d_\pi$
4. ^3E_u	8.5 (0.012)	12.5?	$a_1u(\pi) \rightarrow e_g(\pi^*)$ Trip-Triplet
5. ^3E_u	13.2 (0.002)	15.0?	$a_2u(\pi) \rightarrow e_g(\pi^*)$ Trip-Triplet
6. ^3E_u	15.9 (0.115)	16.6	Q: $(a_1u(\pi) \rightarrow e_g(\pi^*) + a_2u(\pi) \rightarrow e_g(\pi^*))$

[a]Calculate 200 cm^{-1} lower than $^3A_{2g}$ in ^3E_g SCF-CI

Ligand spectroscopy is often much the same as it would be in the absence of the transition metal, with three exceptions. There is often a shift due to inductive effects, that is, the ligand may lose or gain a fraction of an electron (charge transfer) in the complex and this affects the transition energies. Frequently one can readily determine the parentage of such transitions in the isolated ligand. A second effect involves shifts of transitions that involve the σ orbitals of the chelating atom. For example, the $n \rightarrow \pi^*$ excitations in pyridine are shifted greatly to the blue (higher energy), a consequence of the bonding nature of the lone pair of the nitrogen atom. A third effect, sometimes quite subtle, relates to metal–ligand exchange coupling, item 4 in the above list.

In Table 6, the 6th transition is the $\pi \rightarrow \pi^*$ transition calculated at 15,900cm^{-1}. This is a typical porphyrin based transition, and, in this case, inductively shifted to give blood a red color! There is a very low-lying $d_z2 \rightarrow d_\pi$ transition giving rise to a state of ^3E_g symmetry that is believed to be the ground state of the related phthalocyanine compound.[181] Note that this calculation is the same as depicted in Figure 1 in the second panel, and the calculated spectroscopy is accurate in spite of the rather unusual appearance of the MO manifold which appears to have half-occupied d orbitals calculated well below the ligand based MO's, as remarked upon earlier. There are no ligand to metal charge transfer transitions calculated in the low energy region of this spectrum, and there are none observed.

Two bands have been characterized as trip-triplets, and this requires some explanation. These transitions are essentially $\pi \rightarrow \pi^*$ excitations which are spin forbidden (singlet ground state to triplet excited state) in the isolated ligand. Metal-ligand exchange coupling ($ML \rightarrow ML^*$ delineated in the above list)[179,182] produces features in the $L \rightarrow L^*$ spectrum that are not easily observable in the isolated ligand. In a ligand with ground state spin value S, the strongest $L \rightarrow L^*$ transitions arise from transitions from the ground state to excited states of the same multiplicity. In coupling with a transition metal of spin state S', the ground state of the complex is one of $S + S', S + S' - 1, \ldots, |S - S'|$. In the case of Table 5, the ground state of the complex is a triplet. Excited states of the ligand with multiplicities other than S, say S^*, are usually very weak, but also couple with the metal spin yielding states of $(S^* + S'), (S^* + S' - 1), \ldots, |S^* - S'|$. One of these states may now be of the same spin-multiplicity as is the ground state, and be quite observable. The situation is depicted in Figure 17 for the case of Fe(II) porphyrin.

Metal–ligand exchange coupling is often very difficult to calculate using methods that have as their starting point Hartree–Fock SCF calculations, as there is a very strong bias in this method for states of highest multiplicity. The unpairing of electrons is generally accompanied by a relative gain over the closed-shell situation of *about* 1 eV in correlation energy, a relative error corrected only through consideration of a great deal of electron correlation. The parametrization of the INDO/CI scheme includes some of this dynamic correlation through the parameters.

Spin-orbit effects can also be quite large in all complexes that include heavy elements. In general, the importance of these splittings grows with Z^4, where Z is the atomic number of the element considered. An example of very large spin-orbit effects is given in Table 7 for the complex $[Ce(III)(H_2O)_9]^{3+}$. This is essentially the

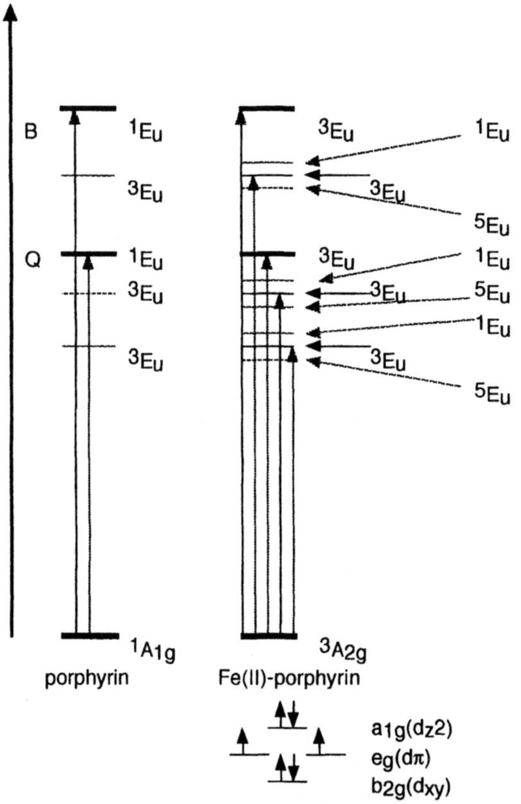

Figure 17. Metal–ligand exchange coupling. In this case the metal electronic configuration gives rise to a triplet, causing the ground state to also become triplet. The formerly forbidden ligand triplets now become signets, triplets, and quintets, see text.

Table 7 Comparison of the Band Maxima in $Ce(H_2O)_9(EtSO_4)_3$ and Solvent Spectrum of Ce(III) with the Calculated Energies in $(cm^{-1})^c$ Using the INDO/s Model. This Spectroscopy is Essentially $4f \rightarrow 5d$ and Dominated by the Spin-Orbit Splitting of the f Orbital Manifold

Ce:crystal[a]	Aqueous[b]	INDO/S-CI
39,060	39,630	40,228
41,950	41,970	42,933
44,740	45,330	45,670
47,400	47,620	48,665
50,130	50,000	50,917

[a]S.P. Sinha, reference 205

[b]C.K. Jørgensen and J.S. Brinen, reference 206

[c]M. Kotzian, N. Rösch and M.C. Zerner, reference 181

metal ion spectroscopy $4f \rightarrow 5d$. What makes this example particularly interesting is that when the ligand field effects are isolated from the spin orbit effects (computationally, not experimentally!) the ligand field account for only 2000 cm^{-1} of the observed 10,000 cm^{-1} total splitting.[183,184]

In addition to being essential in calculating the spectroscopy of complexes containing heavy elements, the inclusion of spin-orbit effects can also be important as a tool in characterizing the excited states, even in those cases where the effects are reasonably modest, but are detected in the experiment. Then not only can the location of the bands and the calculated intensity be used for characterizing the excited states, but also the spin-orbit splitting. The calculated transition energies are but one piece of information available from a good calculation in assigning transitions.

4.2 Comments Regarding Calculations of Molecular Electronic Spectra

Ligand field theory has for years been used to estimate $d \rightarrow d$ spectra, and within certain limits this has been very successful.[114,182,183] By and large this model is qualitative, unless use is made of the positions of one or two transitions to characterize the Slater–Condon or Racah integrals, and these then can be used to estimate the position of other transitions. Ligand field theory is very successful in presenting physical pictures for transitions. Generally the *d* orbitals are the frontier orbitals in such simple pictures of spectroscopy.

Models based on the noniterative Extended Hückel theory are also very successful in yielding a qualitatively easy to understand picture of the important transitions. The ligand orbitals generally lie below the occupied *d*, as is expected on the basis of ionization potentials, but this is not the picture obtained from SCF models, as, for example, that shown in Figure 1. These models, however, have not been made to yield quantitative results, although attempts to parametrize the SCC Extended Hückel model on spectroscopy have been made.[11] Unfortunately, levels of theory that are capable of quantitatively describing the excited states of molecules and their transition energies are often very difficult to describe in terms of one-electron independent particle models. Spectra are inherently two-electron phenomena.

The methods for calculating spectroscopy can be cataloged into three types:

1. *Delta Methods.* Two calculations are performed, and a subtraction is made. These calculations can be of the SCF type, ΔE(SCF), the SCF/CI types, ΔE(SCF-CI), etc. This simplest of methods can only be applied to the lowest state of each given space and spin symmetry, due to *variational collapse*: that is, that the variational principle seeks the lowest energy for a given trial wavefunction without constraint. In its simplest form, only the specified spatial and spin symmetry of the trial wavefunction are constraints. In other words, if there is no spatial symmetry, this model can only be used for the lowest state of a given multiplicity: i.e., lowest singlet state, lowest triplet state, etc. Unfortunately, there have been many errors in the application of this method. It is, however, very simple, and yields a good description of each state, even at

the Hartree–Fock level, as all the one-electron orbitals are relaxed by the SCF procedure.

2. *CI methods.* Here the CI matrix is diagonalized, and the various roots (eigenvalues) of the matrix form upper bounds (the Hyleraas–Undheim theorem) to the various excited states, in order. This does not imply that the energy *differences* between states are bounded, but of course, it is the expectation that as each state approaches a more accurate description, so will the differences. Generally the CI models use a common set of MO's, and thus the calculation of intensities is easy. This is not the case for the delta methods described above.
3. *Direct Methods.* This methodology solves for the excitation energies directly. through the Liouville space equations. The simplest and most common direct method, sketched below, is the RPA model. This formalism also yields intensities directly and, as showed by Bouman and Hansen, very accurately.[183]

These three basic methods appear in *very* many variants and combinations. For example, a CI problem might be generated from RPA-like excitations out of a coupled cluster reference state. We describe a few of these combined methods below.

Ab-Initio calculations of transition metal spectroscopy remain state-of-the-art, especially for open-shell ground states, both because of the size of the CI matrices involved and the difficulty in finding good CI ground state reference functions. As discussed above, the CI matrices grow exponentially with the number of electrons in the system, and current CIF calculations, for example, remain limited to only 14 orbitals. Also as discussed above, variational calculations on large open-shell systems can be quite difficult. Rather than attempt to obtain the best MO's for the ground state for a given CI expansion, the orbitals may be taken from state-averaged calculations (i.e. CAHF, state-averaged MCSCF or CASSCF, improved virtual orbital (IVO) sets[115,121]) which seek a "democratic" set of orbitals equally accurate for the low-lying excited states of interest.

4.3 Direct Methods

When applied to spectroscopy, the response methods, often called Green's function methods or Liouville methods, yield the energies of transitions and the intensities (transition moments or oscillator strengths) *directly* without recourse to either the ground state or the various excited states.

Starting with the generalized Schrödinger equation

$$H|\Psi_\iota\rangle = E_\iota|\Psi_\iota\rangle \tag{104}$$

one examines a Liouville approach defining an excitation operator Ω_ι that when operating on the ground state $|\Psi_0\rangle$ generates an excited state $|\Psi_\iota\rangle$

$$\Omega_\iota|\Psi_0\rangle = |\Psi_\iota\rangle \tag{105a}$$

A formal realization of this excitation operator might be $\Omega_\iota = |\Psi_\iota\rangle\langle\Psi_0|$, an operator which is, of course, unknown as $|\Psi_\iota\rangle$ is not known. Continuing, the Liouvillian operator $\mathbf{L}_\iota$ is formed from the commutator

$$\mathbf{L}_\iota = [H, \Omega_\iota] = H\Omega_\iota - \Omega_\iota H \tag{105b}$$

$$\mathbf{L}_\iota|\Psi_0\rangle = E_\iota\Omega_\iota|\Psi_0\rangle - E_0\Omega_\iota|\Psi_0\rangle = \omega_\iota|\Psi_\iota\rangle \tag{105c}$$

There are many ways to develop these equations. One is by projection as described in Section 2.4. We will pursue another below.

Note that the energy differences ω_ι appear naturally as a consequence of these calculations. Note also that Ψ_0 occurs on the left side of this equation, but Ψ_i on the right. This points out one of the difficulties of this method: it is difficult to extract specific information on the individual states. On the plus side, this methodology generally gives not only excellent transition energies, but also transition moments.

The simplest of these models is the RPA theory, or random phase approximation, and this model uses the SCF wavefunction for Ψ_0 and the single excitation operator for Ω_ι $(i \to a) = [(a_\alpha^+ i_\alpha + a_\beta^+ i_\beta)/\sqrt{2}] + [(a_\alpha i_\alpha^+ + a_\beta i_\beta^+)/\sqrt{2}]$. (Note that the second term is the *de-excitation* operator which, when operating on the reference determinant in the RPA theory, yields zero, as ϕ_i is doubly occupied, but is required in response theory methods to maintain Hermiticity in the subsequent development.) Pursuing projection, the resulting RPA non-Hermitian eigenvalue problem takes the form:[27,143]

$$\begin{vmatrix} \mathcal{A} & \mathcal{B} \\ \mathcal{B} & \mathcal{A} \end{vmatrix} \begin{vmatrix} \mathbf{Z} \\ \mathbf{Y} \end{vmatrix} = \omega_\iota \begin{vmatrix} 1 & 0 \\ \mathbf{0} & -\mathbf{1} \end{vmatrix} \begin{vmatrix} \mathbf{Z} \\ \mathbf{Y} \end{vmatrix} \tag{105d}$$

where the $[\mathbf{Z}, \mathbf{Y}]$ vector denotes the RPA eigenvector solution of this matrix problem. The matrices $\mathcal{A}$ and $\mathcal{B}$, called the A and B matrices, contain four spin-orbital indices, and can be expressed as

$$\begin{aligned} \mathcal{A}_{ai,bj} &= \langle \Psi_{\text{SCF}} | \mathbf{i}^+\mathbf{a}[\mathbf{H}, \mathbf{b}^+\mathbf{j}] | \Psi_{\text{SCF}} \rangle \\ &= \delta_{a,b}\delta_{i,j}(\varepsilon_a - \varepsilon_i) - \langle ai|bj\rangle + 2\langle ai|jb\rangle \\ \mathcal{B}_{ai,bj} &= \langle \Psi_{\text{SCF}} | \mathbf{i}\mathbf{a}^+[\mathbf{H}, \mathbf{b}^+\mathbf{j}] | \Psi_{\text{SCF}} \rangle \\ &= 2\langle ib|aj\rangle - \langle aj|bi\rangle \end{aligned} \tag{105e}$$

where ε_a, ε_i denote SCF orbital energies, and $\langle ab|ij\rangle$ represents standard two-electron integrals (Eqns. 28b,c). The non-Hermitian eigenvalue problem of Eqn. 105g may be readily symmetrized, yielding a pair of generalized eigenvalue problems.[27,142,143]

Although the RPA theory can be developed as a response of the ground state SCF wavefunction to a perturbation, and generally describes SCF (uncorrelated) derivative properties, when applied to the calculation of excited states as outlined above, contains correlation to the ground through the B matrix in an unusual sense. Each excited state has a separate ground state wavefunction to which it is coupled. These ground states have differing amounts of double excitations, included in such a fashion as to minimize the differences between oscillator strengths calculated using the dipole length and dipole velocity operators for a given excitation and basis set.[143,183]

$$f_i^r = 2/3\omega_\iota(\langle\Psi_{HF}|\mu|\Psi_\iota\rangle)^2$$
$$f_i^\Delta = 2/3\omega_\iota^{-1}(\langle\Psi_{HF}|\nabla|\Psi_\iota\rangle)^2$$

(ω_ι, μ and $\nabla = 1/i(\underline{x}\partial/\partial x + \underline{y}\partial/\partial y + \underline{z}\partial/\partial z)$ are in atomic units. f is dimensionless and is the area under an absoption peak, normalized to the $v = 0$ to $v = 1$ transition of the harmonic oscillator. $\underline{x}$, etc. are the normalized unit vectors in the x direction, etc.). These two formulations of oscillator strength should yield identical values should the commutator $[r, \Delta] = ih/2\pi$, seldom true in practical calculations.

More sophisticated theories either employ a better reference function (MCSCF, MBPT, CC) or include more terms in the excitation operator Ω_ι^+, or develop the model consistent to a given order in many-body perturbation theory. A particularly interesting and accurate example of the latter is the SOPPA (Second order polarization propagator approximation) model.[142] Section 4.7. discusses two variants of the coupled cluster linear response (CCLR) approach, the symmetry adapted coupled cluster CI (SAC-CI), and the coupled cluster equations of motion (CC-EOM) approaches.

4.4 The INDO/S Semiempirical Method

The INDO/s model has been described above in some detail, and it has been used now many times for estimating the spectroscopy of transition metal complexes.[156–160,179] Toward this goal, the spectroscopy of model compounds is used to fit the parameters of the theory at the CIS level of theory.[127,128,157,158] There is now quite a bit of experience using this model at the MR-CIS level, and reparametrization has not been necessary. The INDO/s model has also been used at the RPA level without reparametrization, and with considerable improvement in calculated oscillator strengths, as expected from an RPA theory. There is not much experience using this level of theory for transition metal complexes, for the presence of open-shell ground-states complicates matters considerably, and this has not yet been programmed into semiempirical models.

As an example of INDO/S RPA calculations, Table 8 compares INDO/S CIS and RPA results for an Fe(II) porphyrin-CO complex. Recall that the INDO/S method was parametrized at the CIS level of theory for small molecules, but, nevertheless, the CIS and RPA calculations predict very similar transitions, with the RPA method yielding much improved oscillator strengths. The CIS and RPA spectra only differ significantly for the strongly allowed Soret, or B, band, which is observed at about 24,000 cm^{-1}. The oscillator strengths can be calculated using either the dipole length or the dipole velocity formalism. In the ideal case, such as for full CI wavefunctions which preserve the commutation relation between the length and velocity, the dipole length and velocity formalisms should yield the same value. The CIS approaches generally provide quite different values for these two formulations, whereas the RPA formalism yields quite similar results. For example, the CIS values for the intensity in

Table 8 The Calculated Spectrum of Fe(II)-CO Porphyrin, Comparing the RPA Model with CIS, See Text

		Singles CI for singlet			
		Oscillator strength			
INDX	E (cm^{-1})	Length	Velocity	Product	
1	6146.6	0.00024	0.00707	0.00129	
2	6146.6	0.00024	0.00707	0.00129	
3	10941.6	0.00000	0.00000	0.00000	
4	15333.8	0.00000	0.00000	0.00000	
5	15585.1	0.04691	0.00499	0.01530	Q
6	15585.2	0.04691	0.00499	0.01530	
7	24519.1	0.00219	0.00002	0.00021	
8	24519.1	0.00219	0.00002	0.00021	
10	26860.4	1.41171	0.26361	0.61003	B
11	26861.0	1.41181	0.26360	0.61004	
		RPA singlet			
1	5245.9	0.00014	0.00363	0.00072	
2	5245.9	0.00014	0.00363	0.00072	
3	10023.1	0.00000	0.00000	0.00000	
4	14339.7	0.04198	0.05074	0.04615	Q
5	14339.9	0.04199	0.05075	0.04616	
6	15167.0	0.00000	0.00000	0.00000	
7	23783.1	0.75769	0.91865	0.83430	B
8	23783.5	0.75752	0.91839	0.83409	
9	24145.1	0.14841	0.19440	0.16985	
10	24145.2	0.14864	0.19469	0.17011	

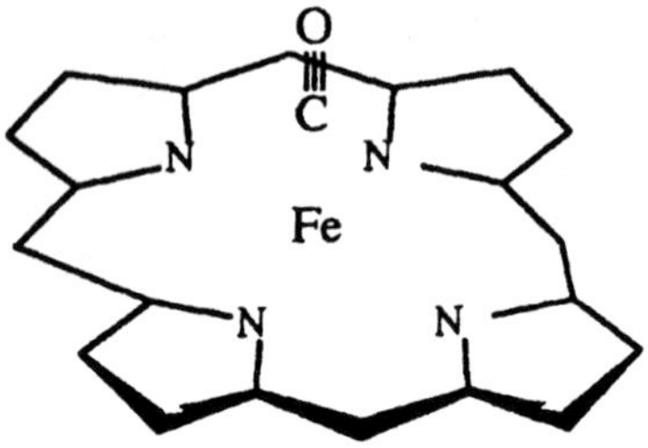

the Q band differ by a factor of 10, 0.0469 vs. 0.00499, while RPA values yield nearly identical values of 0.04198 and 0.05075. Likewise, for the B band, the CIS methods yields 1.4118 and 0.2636, while the RPA approach provides the quite similar 0.7576 and 0.9184 results. The experimental values is quoted between 1.2 and 1.5, the sum of the two degenerate E_u transitions.

The RPA method also characterizes various *Hartree–Fock instabilities*. These situations arise when the SCF calculations converge poorly. For example, assuming the $^1A_{1g}$ state of Fe(II) porphyrin (Figure 1), the RPA calculation immediately indicates that the ground state is a triplet (a *triplet instability*).

4.5 Multireference Singles and Doubles Configuration Interaction (MRSDCI)

The most general, and, perhaps, the most powerful method for obtaining spectra is the MRSDCI procedure,[20,21,32,33] briefly described above in Section 2.4.2. Unfortunately, the MRSDCI approach also requires the most computation resources, and only a few high quality *ab-initio* calculations exist for transition metal complexes. For instance, Fulde has recently performed MRSDCI calculations on a number of sandwich complexes $M(C_8H_8)_2$, M=Ce, Nd, Tb, Yb, and U,[140,184] further described below in Section 4.7. Other recent examples include Pt_3Au and PtAu clusters,[185] GeSe,[186] and a number of dimers, halides, and metal oxides. Because of the computational complexity of MRSDCI calculations, there exists only a few available *ab-initio* packages for performing MRSDCI, including the COLUMBUS package,[20,32] the MOLPRO package,[21,33] and the MELD suite (described in the MOTECC series[187]).

4.6 Complete Active Space Second Order Perturbation Theory(CASPT2)

Recently, there have appeared a number of highly correlated *ab-initio* calculations of transitional metal spectroscopy using the CASPT2 method.[23,47–53] The CASPT2 theory employs a Complete Active Space Self Consistent Field (CASSCF) calculation (more correctly a *full* active space), followed by second order state-selective multiconfigurational perturbation theory. The idea here is to include, if possible, essential correlation in the CAS, a diagonalization, and the dynamic correlation through perturbation theory. This is a successful idea if the CAS is small (similar to a multireference method) and then the effects of the millions of configurations that each add but small amounts of correlation to each state might be included by efficient perturbation theory.

The first step of a CASPT2 calculation is a CASSCF calculation and this involves selecting a set of the metal and ligand molecular orbitals for each electronic state or small set of electronically excited states desired. The current CASSCF technology limits the size of the orbital space to about 14 orbitals, for all excitations possible within this space are included, and thus CASPT2 calculations are limited at present to small molecules in which only a small set of metal–ligand orbitals encapsulates the multireference character of the electronic states of interest. Since each first-row transition metal atom requires at least five 3*d* orbitals, and, additionally, sometimes a sixth 4*s* orbital, this limits CASPT2 calculations to either transition metal dimers, or small organometallic compounds.[47] The most successful CASPT2 transition metal calculations have dealt with small organometallic molecules containing a single transition metal and up to 6 small ligands (such as F, H_2O, NH_3, CN, CO, NO, O), many utilizing only 10 active orbitals. In addition, it may be necessary to include a second 3*d* shell (termed a 3*d* *double-shell*), and this has been found very important, for example, in some Ni complexes (including the Ni atom, NiH, and $Ni(CO)_4$, and in complexes where there is strong ionic charge transfer between the Cr 3*d* orbitals and ligand π^* orbitals, such as in $Cr(NO)_4$, and between the Cr 3*d* and ligand 2π orbitals in CrO_4^- and CrF_6.

The initial CASSCF calculation defines a one-electron operator $\mathbf{H}_0$ to which dynamic correlation can be added using second order perturbation theory. Clearly a one-electron H_0 (Fock-like) operator cannot generate a multi-configuration CASSCF wavefunction, therefore the CASPT2 orbital energies arise from some configuration averaged Fock operator. Frequently, excited state calculations require using state-averaged CASSCF orbitals and appropriate orbital energies. Numerous definitions abound in the literature for both CASPT2 and other implementations of $\mathbf{H}_0$ second order, state-selective, multi-configurational perturbation theory, and research in this area continues. For example, good zero'th order Fock operators, especially for describing potential energy surfaces, should preserve rotational invariance of the orbitals, not break spin symmetry, and not overweight open-shell or closed-shell configurations. Indeed, it is this arbitrariness in the CASPT2 theory which will, we believe, preclude its widespread use by nonexperts for some time.

Once having defined the CASSCF orbitals and $\mathbf{H}_0$, the electronic energy E is computed using the first order wavefunction $|\Psi_1\rangle$ and the second order energy expression E_2. The equations of second order perturbation theory are:[48]

$$\mathbf{H}_0|\Psi_0\rangle = E_0|\Psi_0\rangle, \tag{106a}$$

$$E_1 = \langle\Psi_0|\mathbf{H}_1|\Psi_0\rangle, \tag{106b}$$

$$(\mathbf{H}_0 - E_0)|\Psi_1\rangle = -(\mathbf{H}_1 - E_1)|\Psi_0\rangle, \tag{106c}$$

$$E_2 = \langle\Psi_1|\mathbf{H}_1|\Psi_0\rangle\rangle. \tag{106d}$$

To construct the first order wavefunction, one decomposes $|\Psi_1\rangle$ into four subspaces, $\mathcal{V}_0$, $\mathcal{V}_K$, $\mathcal{V}_{SD}$, $\mathcal{V}_{TQ}$, where $\mathcal{V}_0$ contains only the CASSCF reference function $|\Psi_0\rangle$, $\mathcal{V}_K$ is the orthogonal complement to $|\Psi_0\rangle$ within the CAS reference space, $\mathcal{V}_{SD}$ contains all single and double excitations from $|\Psi_0\rangle$ not included in $\mathcal{V}_0$ and $\mathcal{V}_K$, and $\mathcal{V})_{TQ}$ contains all higher excitations. With these subspaces defined, one may write the $\mathbf{H}_0$ as:

$$\mathbf{H}_0 = \mathcal{P}_0\mathbf{F}\mathcal{P}_0 + \mathcal{P}_K\mathbf{F}\mathcal{P}_K + \mathcal{P}_{SD}\mathbf{F}\mathcal{P}_{SD} + \mathcal{P}_{TQ}\mathbf{F}\mathcal{P}_{TQ}, \tag{107}$$

where $\mathbf{F}$ is a generalized Fock operator, $\mathcal{P}_0$ is the projection operator onto the $\mathcal{V}_0$ subspace, etc.

The first order wavefunction, $|\Psi_1\rangle$, contains only contributions from $\mathcal{V}_{SD}$, and, accordingly, the $\mathcal{V}_{SD}$ subspace is called the *first order interactions* (FOI) space.

The CASPT2 method can produce highly accurate energetics. However, as with many other formulations of multireference perturbation theory, it may also suffer from severe convergence problems related to the *intruder* state problem. Intruder states are low-lying states that are not included in the CAS, but are of lower energy than states that are within the CAS. These arise due to the necessity of choosing a small enough CAS space to be tractable.[47,48,99]

Frequently, intruder state problems can be avoided, once detected, by including them in the CAS reference space if this is possible. Recall that when a single orbital is added to the CAS space, very many excitations are added because of the full CI nature of this space. One might also remove certain low-lying virtual orbitals from the perturbation calculations (such as completely removing some low-lying Rydberg orbitals that are not of interest), or simply energy shifting certain offending core orbitals (level- shifting). Clearly, even choosing an appropriate CAS is far from automatic and level-shifting remains confined again to the experts, and even here it is controversial.

A second type of convergence problem arises due to the fact that CASPT2 is a state selective, or single reference, perturbation theory. Notice that in Eqn. 106d for the second order energy E_2, the electron-electron correlation contributions do not include the interactions between the reference function $\mathcal{V}_0$ and nearly degenerate electronic configurations in the CAS space complement $\mathcal{V}_K$. Consequently, the CASPT2 method can not resolve the near degeneracies between electronically excited states with vastly differing dynamic correlations. Such cases arise when two electronic states of the same symmetry cross, and the multireference character of each state changes rapidly and dramatically (i.e. from covalent to ionic).

4.7 Symmetry Adapted Cluster Configuration Interaction (SAC-CI) and Coupled Cluster Equations of Motion (CCEOM)

The symmetry adapted couple-cluster CI model was introduced by Nakatsuji.[34–46] The starting point is a ground state wavefunction of the coupled cluster doubles type, Eqn. 61c.

The *SAC* wavefunction is given as

$$|\Psi_0\rangle = \Theta_S \exp(\mathbf{S})|0\rangle = \Theta_S|\psi_g\rangle \tag{108a}$$

$$(\mathbf{H} - E)|\Psi_0\rangle = (\mathbf{H} - E)\Theta_S \exp(\mathbf{S})|0\rangle = 0 \tag{108b}$$

where the cluster operator

$$\mathbf{S} = \mathbf{S}_1 + \mathbf{S}_2 + \cdots \tag{108c}$$

$$\mathbf{S}_1 = \sum C_{ia}\mathbf{S}^+_{ia} \tag{108d}$$

$$\mathbf{S}_2 = \sum C_{ijab}\mathbf{S}^+_{ijba} \tag{108e}$$

is spin-symmetry adapted

$$\mathbf{S}^+_{ia} = 1/\sqrt{2}(a^+_\alpha i_\alpha + a^+_\beta i_\beta) \text{ for a singlet} \tag{108f}$$

$$\mathbf{S}^+_{ia} = (a^+_\alpha i_\beta) \text{ for a triplet} \tag{108g}$$

A CI is now generated using a symmetry adapted excitation operator Ω^+_k on the coupled cluster reference wavefunction $|\psi_g\rangle$.

$$|\Phi_k\rangle = N_k \mathbf{Q}\Theta_s\Omega^+_k|\psi_g\rangle \text{ \textit{excited state k}} \tag{109a}$$

where

$$\mathbf{Q} = 1 - |\Psi_0\rangle\langle\Psi_0| \tag{109b}$$

is a projection operator that keeps the excited state $|\Phi_k\rangle$ orthogonal to the reference state, and N_k is a normalization factor. The treatment then proceeds with the assumption of a *CI* (SAC-CI) wavefunction

$$|\psi_k\rangle = |\Phi_k\rangle d_k + |\Psi_0\rangle d_0 \tag{109c}$$

$d_0 = 0$ if the symmetry of the excited state and ground state are different, or if the system of equations (the CI matrix) is diagonalized. Most often further approximations are made:

$$|\Psi_0\rangle \sim \Theta_S \exp(\mathbf{S})|0\rangle \sim \left[1 + \mathbf{S} + (1/2!)\mathbf{S}^2\right]|0\rangle \tag{110a}$$

$$\mathbf{S} \sim \mathbf{S}_1 + \mathbf{S}_2 \tag{110b}$$

$$\mathbf{S}_1 = \sum C_{ia}\mathbf{S}_{ia}^{+} \tag{110c}$$

$$\mathbf{S}_2 = \sum C_{ijab}\mathbf{S}_{ia}^{+}\mathbf{S}_{jb}^{+} \tag{110d}$$

and

$$\Omega_k^{+} = \Omega_{ia}^{+} = \mathbf{S}_{ia} \tag{110e}$$

and then the resulting equations are solved either by projection, in the usual couple cluster fashion,[56,58]

$$\begin{aligned} &\langle 0|(\mathbf{H} - E)|\Psi_0\rangle = 0 \\ &\langle \mathbf{S}_{ia}^{+}0|(\mathbf{H} - E)|\Psi_0\rangle = 0 \\ &\langle \mathbf{S}_{ia}^{+}\mathbf{S}_{jb}^{+}0|(\mathbf{H} - E)|\Psi_0\rangle = 0 \end{aligned} \tag{111}$$

or by matrix diagonalization,

$$H_{ij} = \langle \Phi_i|\mathbf{H}|\Phi_j\rangle \tag{112a}$$

$$\mathbf{Hd} = \mathbf{dE} \tag{112b}$$

The latter methodology has the advantage that it creates wavefunctions that are not only orthogonal, but also non-interacting through the Hamiltonian. This procedure, however, reintroduces *unlinked clusters* into the methodology, and the states are no longer size-extensive, and matrix diagonalization is a more time consuming technology. Nevertheless, the results seem to suggest that diagonalization is the better methodology.

Another related technique has been proposed by Bartlett and his coworkers and is named the Coupled Cluster Equation of Motion model (CCEOM).[28] Starting with the generalized Schrödinger equation

$$\mathbf{H}|\Psi_\iota\rangle = E_\iota|\Psi_\iota\rangle$$

one examines a Liouvillian approach much as in the RPA theory described previously,

$$\Omega_\iota^R|\Psi_0\rangle = |\Psi_\iota\rangle \text{ (i.e., } \Omega_\iota^R = |\Psi_\iota\rangle\langle\Psi_0|) \tag{113a}$$

$$L_\iota = [\mathbf{H}, \Omega_\iota^R] = \mathbf{H}\Omega_\iota^R - \Omega_\iota^R\mathbf{H} \tag{113b}$$

$$L_\iota|\Psi_0\rangle = E_\iota\Omega_\iota^R|\Psi_0\rangle - E_0\Omega_\iota^R|\Psi_0\rangle = \omega_\iota|\Psi_\iota\rangle \tag{113c}$$

The starting point is a reference coupled cluster wavefunction (as was the SAC-CI model)

$$|\Psi_0\rangle = e^{\mathbf{T}}|\Phi_0\rangle$$

with $|\Phi_0\rangle$ the Hartree–Fock SCF wavefunction (although it needn't be).

Then with

$$\mathbf{T} = \mathbf{T}_1 + \mathbf{T}_2 + \mathbf{T}_3 + \cdots$$
$$\mathbf{T}_1 = \sum t_{ai}a^+i$$
$$\mathbf{T}_2 = \tfrac{1}{2!}\sum t_{abij}a^+b^+ji, \text{ etc.}$$

As before, in the RPA model,

$$\left[\mathbf{H}, \Omega_\iota^R\right]\left\{e^{\mathbf{T}}|\Psi_0\rangle\right\} = \omega_\iota\left\{\Omega_\iota^R\right\}\left\{e^{\mathbf{T}}|\Phi_0\rangle\right\} \tag{113d}$$

Multiplying the above by $e^{-\mathbf{T}}$

$$e^{-\mathbf{T}}\left[\mathbf{H}, \Omega_\iota^R\right]e^{\mathbf{T}}|\Phi_0\rangle = \omega_\iota\Omega_\iota^R|\Phi_0\rangle \tag{113e}$$

yields

$$\left[\underline{\mathbf{H}}, \Omega_\iota^R\right]|\Phi_0\rangle = \omega_\iota\Omega_\iota^R|\Phi_0\rangle \tag{114}$$

To obtain the above, $[\Omega_i^R, \mathbf{T}] = 0$ constrains Ω_i^R to contain only excitation operators and not the excitation/de-excitation pair that arises in the standard EOM formalism described above. Although developed as an equation of motion method, this constraint destroys the necessary conditions imposed on a true response model.

Note that

$$\underline{\mathbf{H}} = e^{-\mathbf{T}}\mathbf{H}e^{\mathbf{T}} \tag{115}$$

is not Hermitian.

A big advantage to this method is the generation of a Schrödinger-like equation for the energy differences. A complicating factor is that there exist two Schrödinger equations,

$$H\Omega^R = \Omega^R \omega \tag{116a}$$

$$\Omega^L H = \omega \Omega^L \tag{116b}$$

which is to say that there is a left and a right sided Schrödinger equation. Only one need be solved for the excitation energies, but other properties may require the solution of both. The resulting equations closely resemble those obtained from the SAC-CI model, and therefore both methods should yield similar results for transition metal complexes.

To date, the SAC-CI has been applied to a wide range of small transition metal complexes, including metal oxides, halides, and organometallic complexes.[34–46] The CC-EOM has only been applied, to date, to one transition metal complex, $[FeCl_4]^{2-}$.[188]

4.8 Remaining Difficulties

There remain a number of unresolved problems in using either semiempirical or *ab-initio* methods to study the ground and electronically excited states of transition metal complexes. In treating the ground state, the single reference methods, such as coupled cluster and the more approximate QCI, still face problems with systems that have extensive multi-reference character, or at least require extensive correction to the HF description. For example, because the QCI method only approximates the triples contributions in a full CC calculation, it can fail to properly describe the dissociation energy of even modest systems such as $CuCH_3$.[134] An even more pathological case arises with the insertion of NO into a Co-C bond.[189] The highly multireference character of the transition state precludes studies of this reaction with CCSD(T), QCI, DFT, HF+MP2, or any other single reference method. Unfortunately, the size of such complexes also precludes extensive MRSDCI calculations, therefore making theoretical studies most cumbersome. Multi-reference character is not limited to transition states. For instance, Fulde has pointed out the inherent difficulties in treating the ground state of $Ce(C_8H_8)_2$,[140,184] noting the similarities between problems in treating metallic Ce solids already exist in the simpler Ce sandwich complexes. The formal valence of Ce is 4, however, the Ce $4f$ orbitals mix strongly with the ligand π-orbitals, forcing the ground state into an admixture of the $Ce^{4+}(C_8H_8^{2-})_2\ 4f^0\pi^4$ (20%) and $Ce^{3+}(C_8H_8^{1.5-})_2\ 4f^1\pi^3$ (80%) configurations. Because of the multi-configuration ground state, the DFT-X_α method counts as many as 1.619 f-electrons in the ground state which formally has none. Obtaining fractional occupations is typical for DFT-X_α calculations on strongly correlated systems. In contrast, semiempirical INDO/S calculations predict the ground state to be a $4f^1$ configuration. Only most recently, with the development of accurate relativistic effective core potentials and advances in computer technologies, have *ab-initio* MRSDCI calculations been possible on such a large molecule, and the *ab-initio*

MRSDCI calculations confirm the semiempirical prediction. It should be noted that semiempirical calculations proceed readily on much larger complexes, such as Ce bis(octaethylporphyrinate).[190]

A traditional multireference problem, which today seems more anomalous than ubiquitous, arises in the chromium dimer.[47,145,146] Here, the sextuple metal-metal bond requires an extensive multireference treatment. For instance, the CASPT2 theory can not properly treat the dimer with only 12 orbitals (the two $4s$ and 10 $3d$ AO's of the two Cr atoms) in the active space because of low-lying $3d$-$4d$ excitations required to represent the changing size of the orbitals as one goes from a strong valence bond AF description at longer distances to the importance of the sextuple bonded molecular orbital description.[47] The problem is inherently multireference with important configurations related by as many as 6 electron excitations from the sextuple (or AF) CSF. Problems that exist in the ground state description of this dimer are exasperated for the excited states, where degeneracy effects are worse due to the increasing density of states. The CASPT2 approach, being inherently a state-selective approach, cannot account for the strong coupling of the near degeneracies between states, a particular example being the two low-lying ${}^1\Sigma_u^+$ states. To date, no *ab-initio* method treats the Cr_2 system correctly.

The SAC-CI and CC-EOM approaches, both variants of the same approach, seem the most viable for calculations on transition metal complexes, providing some competition for INDO/S and DFT calculations. For example, recent SAC-CI calculations on metal oxides (i.e., MO_4, M=Ru, Os, Mn, Cr) provide excellent spectroscopic data[34–36] whereas the standard CIS procedure in the INDO/s model perform poorly on many (but not all) oxygen containing complexes.

The accuracy of SAC-CI and CC-EOM calculations depends on the level of correlation in the ground state and the types of excitation operators included. For example, the current CC-EOM formulation, based on a CCSD ground state and single and double excitation operators, does not treat two-photon excited states well (which are important, among other things, for interpreting ultra fast pump-probe laser spectroscopy[109,135,136,191]). Likewise, only recently has the SAC-CI formulation been extended to treat Auger spectra, which require triple excitation operators $\mathbf{S}_3$. It remains to be seen how well the SAC-CI and CC-EOM approaches will treat inherently multireference complexes, such as $Ce(C_8H_8)_2$, especially in the current formulations. There is evidence to suggest that the CC framework can treat many multireference situations; however, the CC ground state may require triples corrections $\mathbf{T}_3$.

Finally, we should like to speculate on the potentially enormous difficulties with treating the spectroscopy of transition metal clusters, which, by the nature of their many open-shell electrons, may exhibit extensive multireference character and degeneracies. Such situations arise, for example, in the iron sulfur complexes ubiquitous in biological electron transfer. With such large numbers of open-shells, one can conceive of a highly degenerate ground state lying just below a number of highly multi-configurational, nearly degenerate excited states. It is difficult to imagine methods such as CASPT2 or SAC-CI/CCEOM treating such highly degenerate, strongly correlated systems, and few other *ab-initio* methods offer any solutions. Methods that average exchange, such as the CAHF model described above, could form the basis of CI calculations that would sort this out, but the size of the CI is likely to be prohibitive. DFT models, since they ignore specific exchange, would yield useful insight but the results would lose the fine resolution of the underlying physics because they *do* ignore detailed (non-local) exchange.

5 RELATIVISTIC EFFECTS

5.1.1 ***Basic Consideration*** As discussed in Section 2.1, relativistic effects become important in heavy atoms. We briefly examine a semiclassical formulation to provide a qualitative feel for the contributions to the RECPs and, additionally, to derive the mathematical framework for the effective spin-orbit operator utilized in current *ab-initio* and semiempirical calculations. The theory of relativistic effects is an area rapidly developing today, and those interested in more detail are referred to recent reviews.[98,100,111,112,193]

Following the discussion of Pyykko,[193] we note that electrons near heavy nuclei have velocities that approach the speed of light. This has a major effect, for example, on electrons in the $1s$ orbital, with sizable probability near the nucleus, that now contracts. Other orbitals with probability near the nucleus also contract.

The speed of light, in atomic units, is given by

$$c = \alpha^{-1} = 137.037 \text{ a.u.},$$

where α is the *fine structure constant* (in cgs units, $\alpha = 2\pi e^2/hc$). The average radial velocity of an electron in a $1s$ orbital in a hydrogen-like atom (with nuclear charge Z) is

$$\langle V_r \rangle = \langle 1s|V_r|1s \rangle = Z.$$

Thus, for a typical transition metal, such as Hg, the average velocity of an electron in a $1s$ orbital is roughly 58% the speed of light ($80/137 \times 100\%$). For such a large velocity, the electron mass m_e increases according to

$$m = m_e\left[1 - (v/c)^2\right]^{-1/2} = 1.23m_e$$

and therefore causes the Bohr radius a_0 to contract by 20%, since (in cgs units)

$$a_0 = h^2/(2\pi e)^2 m.$$

In a similar fashion to the $1s$, the inner s-, p-, and d-shells likewise contract. These contracted inner-orbitals lead to greater shielding of the nucleus, and therefore cause the outer d and f shells, with few or no node, and thus little probability of being near the nucleus, to actually expand (see Figure 8). This qualitative treatment cannot readily explain the origin of the spin-orbit coupling, explained below, which becomes most important for treating the spectroscopy of second and third row transition metal complexes (The importance of spin-orbit coupling grows as Z^4). Fortunately, it is not necessary to delve into Quantum Electrodynamics (QED) to understand the corrections typically employed in today's quantum chemical calculations on systems containing heavy elements (although a QED formulation can be applied to atoms and metal hydrides[100]). A fairly rigorous calculation might employ the relativistic Dirac–Fock Hamiltonian ($\mathbf{H}_{\mathrm{DF}}$) in variational SCF or CI calculations. Simpler and more approximate calculations may use perturbation theory to add relativistic corrections to a nonrelativistic calculation. These perturbative corrections are summarized in the Breit–Pauli Hamiltonian ($\mathbf{H}_{\mathrm{PB}}$), which arise from the Dirac–Fock Hamiltonian via a power series expansion in (v/c). Finally, the most approximate calculations will employ RECPs to model some of $\mathbf{H}_{\mathrm{PB}}$ corrections, and then include some form of a spin-orbit operator into the SCF or, more commonly today, a CI calculation.

5.1.2 ***The Breit–Pauli Hamiltonian*** A complete derivation of the Breit–Pauli Hamiltonian lies outside the scope of this review, therefore we will only consider the origin of the individual correction terms.

Expressing the relativistic corrections to $\mathbf{H}_{\mathrm{nr}}$ (Eqn. 117) as a power series expansion in v/c leads to the following simplified form of the well known Breit–Pauli Hamiltonian:[98,196,207]

$$\mathbf{H}_{\mathrm{BP}} = \mathbf{H}_{\mathrm{nr}} + \mathbf{H}_{\mathrm{MV}} + \mathbf{H}_{\mathrm{D}} + \mathbf{H}_{\mathrm{SO}}, \tag{117a}$$

where the three relativistic corrections, the mass-velocity correction ($\mathbf{H}_{\mathrm{MV}}$), the spin-orbit interaction ($\mathbf{H}_{\mathrm{SO}}$), and the Darwin mass term ($\mathbf{H}_{\mathrm{D}}$) are given by (in cgs units)

$$\mathbf{H}_{\mathrm{MV}} = \left(1/8m^3c^2\right)\mathbf{p}^4 \quad \text{(mass-velocity)} \tag{117b}$$

$$\mathbf{H}_{\mathrm{SO}} = -(\mathcal{A}_c^2/4)(1/R)(\mathrm{d}V(R)/\mathrm{d}R)\mathbf{L}\cdot\mathbf{S} \quad \text{(spin-orbit)} \tag{117c}$$

$$\mathbf{H}_{\mathrm{D}} = (\mathcal{A}_c^2/8)\Delta V(R) \quad \text{(Darwin)} \tag{117d}$$

and $\mathcal{A}_c$ is the Compton wavelength $(h/2\pi m_e c)$. These expressions are slightly more complicated for a many-electron Hamiltonian.[98,193] (Notice that in atomic units, $\mathcal{A}_c = \alpha = 1/c$, where α is the fine structure constant.)

We encountered the mass-velocity correction $\mathbf{H}_{\mathrm{MV}}$ earlier when constructing RECPs. This term may be obtained from the first perturbative correction to the relativistic energy for a classical particle.

To qualitatively understand electric, magnetic and the above relativistic corrections, we start with the exact, nonrelativistic Hamiltonian, $\mathbf{H}_{\mathrm{nr}}$, for a single electron, obtained by replacing the classical momentum $\mathbf{p}$ with the momentum $\pi = \mathbf{p} - e/c\mathcal{A}$, an expression that contains the effect of a particle of charge e and mass m_e moving in a field specified by the vector potential $\mathcal{A}$,[98,135,194–196]

$$\mathbf{H}_{\mathrm{nr}} = \mathbf{p}^2/2m_e + \mathbf{V}(R) \Leftarrow \pi^2/2m_e + \mathbf{V}(R) \tag{118a}$$

$\mathbf{V}(R)$ is again the potential energy. In the approximate formalism presented here, relativistic corrections become important when the electron has a kinetic energy comparable to the speed of light. We therefore express the relativistic corrections to $\mathbf{H}_{\text{nr}}$ as a power series expansion in v/c, leading to the following simplified form

$$\mathbf{H}_{\text{rel}} \cong \mathbf{p}^2/2m_e - \mathbf{p}^4/8m_e^3c^2 + \cdots + V(R) \tag{118b}$$

The $\mathbf{p}^4/8m_ec^2$ term in Eqn. 118b is the mass-velocity correction.

Choosing a *Coloumb gauge*, $\mathbf{p} \cdot \mathcal{A} = \mathcal{A} \cdot \mathbf{p}$, then leads to terms generally assigned to $\mathbf{H}_{\text{nr}}$[135]

$$\mathbf{H}_{\text{f}} = -e/m_ec\mathcal{A} \cdot \mathbf{p} + 1/2(e/m_ec)^2\mathcal{A}^2 \tag{118c}$$
$$\mathbf{H}_{\text{nr}} = \mathbf{H}_{\text{o}} + \mathbf{H}_{\text{f}}$$

where $\mathbf{H}_{\text{o}}$ is the usual field-free Hamiltonian. The terms in $\mathbf{H}_{\text{f}}$ are very important when considering the interaction of atoms or molecules with radiation (the absorption of light, for example).

The spin-orbit interaction $\mathbf{H}_{\text{SO}}$ is obtained from generalizations of the Dirac equation, but it is, perhaps, more informative to consider the effects of special relativity which gives rise to a spinning electron with intrinsic magnetic moment $M_s = (eh/2\pi m_e)\mathbf{S}$.[197] The electron can be thought of, in the classical sense, as moving with velocity $\mathbf{v}$ in an electric field $E = -\mathrm{d}V(R)/\mathrm{d}R$ due to the nucleus (in general, $E = -\nabla\phi - 1/c\partial\mathcal{A}/\partial t$, where ϕ is the scalar potential), and the electron generates a magnetic field, $B = -1/c^2 v \times E$. (Note that in cgs units, $\mathbf{p} = \mathbf{v}/m_e$, but in atomic units, the mass of the electron $m_e = 1$. In general, $B = \nabla \times \mathcal{A}$ or $\mathcal{A} = 1/2\mathbf{B} \times \mathbf{R}$). These two magnetic fields interact, giving rise to an interaction energy $-\mathbf{M}_s \cdot \mathbf{B}$. To first order in v/c:

$$\mathbf{B} = -1/c^2\mathbf{v} \times \mathbf{E} = (1/mc^2)(1/R)(\mathrm{d}V(R)/\mathrm{d}R)(\mathbf{p} \times \mathbf{R}) \tag{119a}$$

Expressing $\mathbf{p} \times \mathbf{R} = -(h/2\pi)\mathbf{L}$ ($\mathbf{L}$, angular momentum, is dimensionless in this development) in Eqn. 119a leads to the spin-orbit interactions as expressed in Eqn. 117d above. For a many-electron molecule, the spin-orbit operator also contains electron-electron interactions, and may be expressed as[98,100]

$$\mathbf{H}_{\text{SO}} = (\mathcal{A}_c e/2)^2 \Big/ \Big[\sum_A \sum_i Z_A (1/R_{iA}^3) \mathbf{L}_{iA} \cdot \mathbf{S}_i - (2\pi/h) \sum_{i \neq j} (1/r_{ij}^3)(\mathbf{r}_{ij} \times \mathbf{p}_i)(\mathbf{S}_i + 2\mathbf{S}_j) \Big] \tag{119b}$$

where (i, j) denote electrons, R_{iA} is the position of electron i relative to nucleus A, and $\mathbf{r}_{ij}$ is the vector between electrons (i, j). Note the Z_A^4 dependence of $\mathbf{H}_{\text{SO}}$, since $\langle(1/R_{iA}^3)\rangle \sim Z_A^3$, an expectation value dominated by the values of the wavefunction closest to the nuclei. Eqn. 119b has been multiplied by 1/2, the *Thomas correction factor*, obtained by comparison of this more classical treatment with the fully relativistic treatment.

The final relativistic correction, the Darwin term $\mathbf{H}_{\mathrm{D}}$, can also be understood in term of a power series expansion in v/c leading to a weakly-relativistic reformulation of the Dirac equation.[195] These effects appear indirectly through the functional form of the Coulomb interaction. In the exact Dirac equation, the Coulomb interaction is completely local and thus of the functional form $\mathbf{V}(\mathbf{r})$. But in the v/c series expansion, the potential becomes nonlocal, and takes on a form

$$\int \mathrm{d}^3\rho f(\rho)\mathbf{V}(\mathbf{r}+\rho), \tag{119c}$$

where $f(\rho)$ is a function normalized to 1, depends only on $|\rho|$, and falls to zero outside the distance of the Compton wavelength, $\mathcal{A}_c$. Replacing $\mathbf{V}(\mathbf{r}+\rho)$ with a Taylor series about $\rho = 0$, the integral above takes on a value, at second order, proportional to $(\mathcal{A}_c^2)\nabla V(R)$. This correction is the mass Darwin term in Eqn. 117e, and is of similar magnitude to the other two corrections, and is incorporated into the RECPs through parameters.

The Breit–Pauli Hamiltonian $\mathbf{H}_{\mathrm{BP}}$ is not bounded from below and therefore unstable for all-orbital variational calculations.[98,100,207] For example, the mass-velocity term diverges in regions close to the atomic nuclei and, consequently, will cause all-electron variational calculations to collapse. Many workers avoid variational collapse in *ab-initio* variational calculations by contracting the high exponent s orbitals, therefore removing the variational freedom near the atomic nuclei. Note that semiempirical calculations do not experience variational collapse because of the limited variational freedom in the minimal basis set. Variational collapse can also be avoided by treating the relativistic effects using first order perturbation theory rather than variationally, but frequently the relativistic effects can be too great to expect accuracy from a low-order perturbative treatment. Finally, another way to avoid variational collapse is to avoid the power series expansion in the electron energy in Eqn. 118b and to, instead, replace the mass-velocity correction with the exact expression for the relativistic kinetic energy E_i. This leads to the SQRD (Square-Root plus Darwin) correction[100,198]

$$\mathbf{H}_{\mathrm{SQRD}} = E_i + \mathbf{H}_{\mathrm{D}}. \tag{119d}$$

5.1.3 ***Relativistic Quantum Electrodynamics*** Because the Breit–Pauli Hamiltonian diverges near the atomic nuclei, one cannot rigorously derive a relativistic Hamiltonian using an expansion in v/c. The proper formulation of a relativistic theory requires relativistic quantum electrodynamics (QED), recently examined and reviewed by Hess.[100] This development lies outside the scope of this chapter. Hess shows that $\mathbf{H}_{\mathrm{PB}}$ actually represents a low energy approximation to a more consistent quantum electrodynamical Hamiltonian ($\mathbf{H}_{\mathrm{QED}}$).

5.2 The Spin-Orbit Interaction in Transition Metal Spectroscopy

The spin-orbit interaction $\mathbf{H}_{\text{SO}}$ becomes important for late first row transition metal complexes, especially when considering open-shell electronic configurations. The spin-orbit interaction both affects electron kinematics and couples otherwise degenerate atomic configurations with the same orbital spin, and thus spin is no longer a quantum number of the system. In a molecule the spin configurations may not only split but also mix, thus greatly complicating electronic spectroscopy, and, in turn, nonradiative decay. On occasion, spin-orbit coupling can affect chemical bonding and reactivity by causing states of different multiplicities to mix. For a well known example in spectroscopy, the lustrous yellow color of gold, as compared to silver, results from a relativistic spin-orbit splitting of the $5d$ shell of gold.[202] Typical spectroscopic shifts arise between electronic states with differing numbers of d-electrons (Laporte $d \rightarrow d$ transitions remain unaffected). Note that d-electron rich states also can require large amounts of electron correlation, such as the $3d$ double shells required in CASPT2 calculations;[47] thus even further complicating the spectroscopy of such states. The spin-orbit effect also results in spin-forbidden radiative transitions, such as in phosphorescence, and radiationless *inter-system* crossing (between different multiplicities caused by the spin-orbit interaction) followed by *internal conversion* (transitions between excited states of the same multiplicity).

As described in Section 2.3, the mass-velocity and Darwin corrections lend themselves to a compact, phenomenalogic description in terms of relativistic effective core potentials (RECPs), often referred to as *scalar* relativistic effects. The RECPs might be of the form described in Section 2.3.3, and the spin-orbit interaction included as a post Hartree–Fock correction, or they can be included directly in the SCF step by using an appropriate RECP for treating the spin-orbit interaction. In this case the RECP is j-dependent, with $j = |l + 1/2|$ and $j = |l - 1/2|$, called *Pauli two-component spinors*. The RECPs may then take the form:

$$\mathbf{V}^{\text{REP}}(\mathbf{r}) = \mathbf{V}_{L,J}(\mathbf{r}) + \sum_{l=0,L} \sum_{j=|l-1/2|,|l+1/2|} \sum_{m=-j,j} \mathbf{V}_{lj}^{\text{REP}}(\mathbf{r}) - \mathbf{V}_{L}^{\text{REP}}(\mathbf{r}) |ljm\rangle\langle ljm| \quad (120)$$

Two sets of MOs then result, one for spin up and one for spin down. The previous RECPs defined in Eqn. 46 may be obtained by averaging $\mathbf{V}^{\text{REP}}(\mathbf{r})$ over the two j values for each l.

As with the mass-velocity and Darwin corrections, the spin-orbit interaction will also affect the charge distribution of the electrons, and, in particular, affect differently orbitals with different angular momentum.

For calculations that go beyond the HF level, it is simpler to evaluate the spin-orbit interaction $\mathbf{H}_{\text{SO}}$ directly between correlated wavefunctions using conventional MO's

$$\langle \Psi | \mathbf{H}_{\text{SO}} | \Psi' \rangle = \langle \Phi \xi | \zeta \mathbf{L} \cdot \mathbf{S} | \Phi' \xi' \rangle = \zeta \langle \Phi | \mathbf{L} | \Phi' \rangle \langle \xi | \mathbf{S} | \xi' \rangle \quad (121)$$

where Φ and ξ represent the spatial and spin parts of the correlated, many-electron wavefunction Ψ, and ζ is the *spin-orbit constant coupling parameter*, often chosen as an empirical parameter. The matrix elements are evaluated using standard formulas for evaluating Hamiltonian matrix elements between many-electron Slater determinants.

Notice that the fully *ab-initio* spin-orbit operator requires both one- and two-electron coupling terms. In the semiempirical INDO/S method, the two-electron components of $\mathbf{H}_{\mathrm{SO}}$ are absorbed into an effective one-electron operator[200]

$$\mathbf{H}_{\mathrm{SO}} = \sum_i \sum_k \zeta_k \mathbf{L}_k(i) \cdot \mathbf{s}(i) \tag{122}$$

where ζ_k represents the effective one-electron atomic *spin-orbit coupling parameter* (essentially $\zeta_k = Z\langle k|1/r^3|k\rangle$ *averaged*, see Eqn. 121).

The INDO/S parametrization for transition metals, coupled with the above form of $\mathbf{H}_{\mathrm{SO}}$, provides the tools for computing spin-orbit effects in a large number of transition metal compounds. In contrast, and in comparison to the CASPT2 and SAC-CI *ab-initio* studies of first row transition metal complexes, few *ab-initio* studies at all consider transition metal compounds with spin-orbit splittings. A number of *ab-initio* works do examine the spin-orbit coupling in transition metal atoms and diatomic compounds, including hydrides, oxides, and dimers.[88,100]

It should be noted that understanding excited state energetics and decay processes continues to challenge theory, especially for systems with both weak and strong spin-orbit couplings, and especially near curve crossings. In the weak coupling cases, such as in first-row transition metal oxides, the spin-orbit coupling will perturb the many low electronic states, but often not enough in order to experimentally observe spin-forbidden transitions. In the strong coupling case, the Born–Oppenheimer approximation might break down, thus giving rise to fast non-radiative decay between near degenerate states. The spin-orbit interaction itself causes intersystem crossing (i.e., between low-lying singlet and triplet states in organic molecules), but it may compete with nonadiabatic internal conversion. This competition hinders computational studies since single-reference methods, such as CASPT2 and CC-EOM, may fail near many curve crossings, where the requisite multireference theories, such as MRCI and MRPT, again require enormous computational resources. Indeed, for complicated excited state decay processes, semiempirical methods, because they resemble approximate formulations of MRPT, may prove to be the methods of choice for many years to come.

6 ON THE FUTURE OF TRANSITION METAL CALCULATIONS

The study of the electronic structure of transition metal systems is developing today at a very fast pace. For the reasons given in this chapter, there are still many unknowns in treating transition metal systems using modern quantum chemical techniques, and there are still no black box methods such as those available today for the treatment of systems that contain only main group elements. This is even more true in the open-shell systems that often characterize transition metal complexes. Nevertheless, huge strides have been made in applying quantum chemical techniques to these systems. And there are new ideas appearing every day. Even if methods for studying transition metal systems are not at the same stage of development as they are for other systems, they can, and are, being applied to a wide variety of very interesting chemical and biochemical problems and yield useful and interesting suggestions, especially when combined creatively with experimental knowledge. For example, if the nature of the ground state of the molecule is known experimentally—its spatial and spin symmetry—geometries can be obtained assuming these symmetries. This is also true when calculating other properties—ESR, NMR, vibrational spectroscopy, electronic spectroscopy, Mössbauer spectroscopy, and a host of others. In fact, it is common to predict the ground state from calculated properties that are more sensitive to spatial and spin symmetry than is the total energy itself. Quantum chemistry has certainly become a strong partner of experiment in interpreting and predicting observation in transition metal chemistry.

This chapter is not complete. It hopefully gives a snapshot of the available techniques and their strengths that are promising in studying transition metal systems based on a Hamiltonian that can be expressed in conventional form (i.e., second quantized form, see Eqn. 30 and Appendix 1). Density functional methods that are described elsewhere in this book have already demonstrated enormous utility in treating transition metal systems. The Kohn–Sham equations that underlie present DFT methods cannot be written in second quantized form, and therefore have formal difficulties (although not always *practical* difficulties) with such properties as spin multiplets (since they start with an averaged local exchange operator), excited states (see, however, the approximate schemes that are available, and the response methods) and negative ions (a consequence of the present exchange functionals).

What about semiempirical quantum chemistry? Are the days numbered? It is certainly true that *ab-initio* methods are becoming more accurate and are applicable to larger systems. Semiempirical models can often yield information on large systems in seconds, rather than months, and even if they cannot be relied upon for great accuracy, they have a great record of offering explanation where there was none, and in making predictions. They are efficient preliminary calculations for more accurate and time consuming *ab-initio* work and they have the great advantage of keeping up with the imagination of the experimentalist. And finally they provide a convenient model for testing approximate methods of calculating the correlation energy, which, if successful, can then be applied to the more computationally demanding *ab-initio* methods.

ACKNOWLEDGMENTS

We are grateful for a thorough reading of this manuscript by Ted O'Brien and by Barry Lever. Much of the work reported that deals with the INDO model was supported in part through grants from the National Science and Engineering Council of

Canada, and from the Office of Naval Research. Original calculations reported in this Chapter were supported, in addition, through an IBM SUR grant.

APPENDIX 1. SECOND QUANTIZATION

In Eqn. 60 the Fermion creation μ^+ and annihilation operator ν appear.[25,27] These are convenient constructs of *second quantization* that, once understood, allow rather simple expressions to replace complicated looking ones. In some of the equations that follow we will express concepts in second quantized form. All of these relate back to rather simple constructs. μ^+ creates an electron in orbital ϕ_μ according to

$$\mu^+|\phi_1\phi_2\ldots\phi_n| = |\phi_\mu\phi_1\phi_2\ldots\phi_n|$$

if ϕ_m should be occupied in the original n particle determinant, then two rows of the determinant (see Eqn. 3) are the same, and this is zero. Otherwise it generates an $n+1$ electron determinant.

The operator ν destroys an electron in orbital ϕ_ν

$$\begin{aligned}
\nu|\phi_1\phi_2\ldots\phi_\nu\ldots\phi_n| &= = \nu(-1)^\nu|\phi_\nu\phi_1\phi_2\ldots\phi_n| \\
&= (-1)^\nu|\phi_1\phi_2\ldots\phi_n| \qquad \text{if } \phi_\nu \text{ was occupied} \\
&= 0 \qquad \text{if it was not}
\end{aligned}$$

To be consistent with the creation operator definition, the annihilated electron and its orbital must be moved to the first position in the determinant before being annihilated. This generates the factor $(-1)^\nu$ since each time an orbital is moved one row in the determinant there is a sign change.

From this

$$\begin{aligned}
\mu\nu|\phi_1\phi_2\ldots\phi_n| &= \mu\nu(-1)^\nu(-1)^\mu|\phi_\nu\phi_\mu\phi_1\phi_2\ldots\phi_n| \\
&= (-1)^\nu(-1)^\mu|\phi_1\phi_2\ldots\phi_n| \quad \text{if } \phi_\mu \text{ is before } \phi_\nu \\
\nu\mu|\phi_1\phi_2\ldots\phi_n| &= \nu\mu(-1)^\mu|\phi_\mu\phi_1\phi_2\ldots\phi_n| \\
&= (-1)^\mu(-1)^{\nu-1}|\phi_1\phi_2\ldots\phi_n| \quad \text{if } \phi_\mu \text{ is before } \phi_\nu \\
&= -(-1)^\nu(-1)^\mu|\phi_1\phi_2\ldots\phi_n|
\end{aligned}$$

and

$$\mu\nu = -\nu\mu$$

The same result is obtained if ϕ_ν is before ϕ_μ in the original determinant, but then it is $\mu\nu|\phi_1\phi_2\ldots\phi_n|$ that yields $-(-1)^\nu(-1)^\mu|\phi_1\phi_2\ldots\phi_n|$. This expression is still true should the product be zero, the case that arises if either or both ϕ_ν or ϕ_μ are not occupied in $|\phi_\nu\phi_1\phi_2\ldots\phi_n|$. Similarly

$$\mu^+\nu^+ = -\nu^+\mu^+$$

Then

$$\begin{aligned}\mu^+\nu|\phi_1\phi_2\ldots\phi_n| &= (-1)^\nu|\phi_\mu\phi_1\phi_2\ldots\phi_n| \quad \text{if } \phi_\nu \text{ is occupied and } \phi_\mu \text{ is not, or}\\ &= 0 \quad \text{otherwise.}\end{aligned}$$

Similarly

$$\begin{aligned}\nu\mu^+|\phi_1\phi_2\ldots\phi_n| &= -(-1)^\nu|\phi_\mu\phi_1\phi_2\ldots\phi_n| \quad \text{if } \phi_\nu \text{ is occupied and } \phi_\mu \text{ is not, or}\\ &= 0 \quad \text{otherwise.}\end{aligned}$$

The extra minus sign in the above arises because ϕ_ν must be moved one additional orbital (the newly created ϕ_ν) before it can be annihilated.

This leads to

$$\mu^+\nu = -\nu\mu^+ \qquad \mu \neq \nu$$

In the special case in which $\mu = \nu$

$$\mu^+\mu + \mu\mu^+ = 1$$

since ϕ_μ is either occupied or it is not! This can be summarized in terms of the *Fermion anti-commutation relationships*

$$\begin{aligned}[\mu^+, \nu]_+ &= \mu^+\nu + \nu\mu^+ = \delta_{\mu\nu}\\ [\mu, \nu]_+ &= [\mu^+, \nu^+]_+ = 0\end{aligned}$$

It is conventional to designate occupied orbitals in the reference determinant by $i, j, k, \ldots$ unoccupied orbitals by $a, b, c, \ldots,$ partially filled orbitals by $m, n, o, \ldots$ and general orbitals by $q, r, s, \ldots$ Thus, if Ψ_0 (is a closed shell ground state, a singly excited singlet state can be generated by

$$\begin{aligned}\Omega_{ia}|\psi_0\rangle &= (a_\alpha^+ i_\alpha + a_\beta^+ i_\beta)\sqrt{2}|\Psi_0\rangle\\ &= \left[|\phi_1\alpha\phi_1\beta\ldots\phi_a\alpha\phi_i\beta\ldots\phi_n| + |\phi_1\phi_2\ldots\phi_i\alpha\phi_a\beta\ldots\phi_n|\right]/\sqrt{2}\end{aligned}$$

Then, for example, $\mathcal{D}_1$ of Eqn. 39 can be expressed as

$$\mathcal{D}_1 = \sum_{i,a} d_{ia}\Omega_{ia}$$

and so forth.

APPENDIX 2. THE RELATIONSHIP BETWEEN *AB-INITIO* AND SEMIEMPIRICAL THEORIES

Semiempirical methods such as INDO/S employ a minimum basis set and empirical parameters and may seem as nothing more than elaborate curve fitting schemes. But the great number of successful semiempirical calculations, including over forty years with the Pariser–Parr–Pople π-electron theory[152–155,161–165,203] and twenty years of experience with INDO/S calculations on transition metal spectroscopy[156–158] indicates that perhaps the parametrization procedure works well for quite subtle reasons. It is, in fact, possible to formulate a highly accurate *ab-initio* theory, based on third order QDPT,[115–121,131] which helps to explain how semiempirical parameters include dynamic correlation.

Consider Eqn. 85, which expresses the ZDO Hamiltonian in second quantized form, and includes the semiempirical parameters $\langle\mu|\mathbf{h}|\nu\rangle$ and $\langle\mu\nu||\sigma\lambda\rangle$. Typically one interprets these parameters as matrix elements of the bare (uncorrelated) *ab-initio* one- and two-electron integrals. Within the framework of *ab-initio* QDPT, however, one also obtains a ZDO-like effective valence shell Hamiltonian $\mathbf{H}^{\mathbf{V}}$ which takes an algebraic form similar to that of Eqn. 8 and includes *ab-initio* effective integrals that explicitly include dynamic correlation:

$$\begin{aligned}\mathbf{H}^V &= \sum_{\mu\nu}\langle\mu|\mathbf{U}_1^{\mathbf{V}}|\nu\rangle\mu^+\nu + \tfrac{1}{4}\sum_{\mu\nu}\sum_{\sigma\lambda}\langle\mu\nu|\mathbf{V}_{1,2}^{\mathbf{V}}|\sigma\lambda\rangle\mu^+\nu^+\lambda\sigma \\ &\quad + \tfrac{1}{6}\sum_{\mu\nu}\sum_{\sigma\lambda}\sum_{\omega\xi}\langle\mu\nu\sigma|\mathbf{W}_3^{\mathbf{V}}|\lambda\omega\xi\rangle\mu^+\nu^+\sigma^+\lambda\omega\xi + \cdots\end{aligned}$$

Notice that the *ab-initio* $\mathbf{H}^{\mathbf{v}}$ contains not only one-electron ($\langle\mu|\mathbf{U}_1^{\mathbf{v}}|\nu\rangle$) and two-electron ($\langle\mu\nu|\mathbf{V}_{1,2}^{\mathbf{v}}|\sigma\lambda\rangle$) effective integrals, but also three-electron effective integrals ($\langle\mu\nu\sigma|\mathbf{W}_3^{\mathbf{v}}|\lambda\omega\xi\rangle$). These effective integrals contain dynamic correlation, as a result of the perturbation expansion, and the essential, or non-dynamical, correlation arises by diagonalizing the $\mathbf{H}^V$ CIF matrix. *Ab-initio* $\mathbf{H}^V$ calculations generally perform as well, and sometimes better than, the best MRSDCI, CASPT2, and SAC-CI *ab-initio* calculations on small molecules.

$\mathbf{H}^V$ may be constrained so that it spans exactly the same minimal basis set of atomic orbitals as does the ZDO Hamiltonian, and the effective integrals become *ab-initio* analogs of the semiempirical parameters. Constrained *ab-initio* $\mathbf{H}^V$ calculations on small polyenes validate many of the fundamental approximations in semiempirical theory, at least for π-electron theory, such as the use of a minimal basis set of atomic orbitals, the ZDO approximation, and the values of the parameters in π-electron theory.

$\mathbf{H}^V$ calculations on systems with transition metal using semiempirical methods, such as INDO/S, do not proceed as cleanly as they did with π-electron systems because of the presence of large three-electron effective integrals, and because the INDO/S method is parametrized using a CIS rather than CIF diagonalization. $\mathbf{H}^V$ calculations on transition metals atoms display large three-electron effective integrals (as large as 4 eV), whereas the INDO/S model only contains one- and two-electron integrals.[204] Fortunately, the $\mathbf{H}^V$ three-electron effective integrals are diagonally dominant, and it can be shown that these large interactions may be averaged into the one- and two-electron effective integrals. The situation resembles the case of the semiempirical spin-orbit operator in INDO/S. The *ab-initio* spin-orbit operator contains both one- and two-electron terms, whereas the semiempirical spin-orbit operator only includes one-electron terms because it averages the diagonally dominant two-electron terms. Likewise, one can view current semiempirical parametrizations as having averaged over the $\mathbf{H}^V$ three-electron effective integrals, a portion of the essential correlation in the $\mathbf{H}^V$ CIF matrix, thus compactly and efficiently approximating a high level *ab-initio* QDPT.

Because INDO/S is parametrized at the CIS level of theory, it is likely that the INDO/S parameters also average over other portions of the nondynamic correlation in the *ab-initio* $\mathbf{H}^V$ CIF matrix. Consequently, it is extremely difficult to directly compare effective integrals with INDO/S parameters. It is interesting to note that the success of the INDO/S CIS theory has motivated new *ab-initio* theories, such as STEOM, a formulation of Fock Space Coupled Cluster Theory, that constructs an effective Hamiltonian over only the CIS subspace.[204]

REFERENCES

1. M.C. Zerner. Appendix C. In A. Szabo and N. Ostlund, editors, *Modern Quantum Chemistry*, New York, 1989. McGraw Hill.
2. D. Head and M.C. Zerner. New ton-based optimization methods for obtaining molecular geometries. *Adv. Quant. Chem.*, 20: 239, 1988.
3. H.B. Schlegel. Optimization of equilibrium geometries and transition states. In K.P. Lawley, editor, *Ab-Initio Methods in Quantum Chemistry I*, page 249, New York. John Wiley and Sons.
4. H.B. Schlegel. Optimization of equilibrium geometries and transition structure. In D.R. Yarkony, editor, *Modern Electronic Structure Theory, Part I*. World Scientific Press, 1995.
5. A. Kormornicki and J.W. McIver. *Chem. Phys. Lett.*, 10: 303, 1971.
6. A. Kormornicki and J.W. McIver. *J. Am. Chem. Soc.*, 94:2625, 1971.
7. J.A. Pople. *Int. J. Quantum Chem.*, 37:349, 1990.
8. ZINDO: M.C. Zerner and coworkers, Quantum Theory Project, The University of Florida, Gainesville, FL 32611: Distributed by Biosym Technologies, 10065 Barnes Canyon Road, San Diego, CA 92121.
9. AMPAC: M. Dewar and co-workers: Now distributed by Andrew Holder, Semichem, Dept. of Chemistry, University of Missori, Kansas City, MO.
10. MOPAC: J.J.P. Stewart and coworkers (1989) QCPE 455.
11. M.C. Zerner. *In Reviews In Computational Chemistry, Volume II*. VCH, New York, 1991.
12. J.P. Stewart. In K.P. Lipkowitz and D.B. Boyd, editors, *Reviews in Computational Chemistry, Vol 1*, New York, 1990. VCH Publishing.
13. J.P. Stewart. In K.P. Lipkowitz and D.B. Boyd, editors, *Reviews in Computational Chemistry, Vol 2*, New York, 1991. VCH Publishing.

14. J. Andzelm and E. Wimmer. *JCP*, 96:1280, 1992.

15. DMon: Distributed by Biosym Technologies, 10065 Barnes Canyon Road, San Diego, CA 92121.

16. DMol: Distributed by Biosym Technologies, 10065 Barnes Canyon Road, San Diego, CA 92121.

17. GAUSSUAN94: M.J. Frisch, G.W. Trucks, H.B. Schlegel, P.M.W. Gill, B.G. Johnson, M.A. Robb, J.R. Cheeseman, T. Keith, G.A. Petersson, J.A. Montgomery, K. Raghavachari, M.A. Al-Laham, V.G. Zakrzewski, J.V. Ortiz, J.B. Foresman, C.Y. Peng, P.Y. Ayala, W. Chen, M.W. Wong, J.L. Andres, E.S. Replogle, R. Comperts, R.L. Martin, D.J. Fox, J.S. Binkley, D.J. Defrees, J. Baker, J.P. Stewart, M. Head-Gordon, C. Gonzalez and J.A. Pople: Gaussian, Inc. Pittsburgh PA.

18. GAMESS: M.W. Schmidt, K.K. Baldridge, J.J. Boatz, S.T. Elbert, M.S. Gordon, J.H. Jensen, S. Koseki, N. Matsunaga, K.A. Nugyen, S. Su, T.L. Windus, M. Dupuis and J.A. Motgomery, Jr: Distributed by M. Schmidt, Dept. of Chemistry, Iowa State Univ., Ames, Iowa.

19. TURBOMOL: Reinhart Ahlrich and coworkers: Distributed by Biosym: Technologies, 10065 Barnes Canyon Road, San Diego, CA 92121.

20. COLUMBUS Program System, R. Shepard, I. Shavitt, R.M. Pitzer, D.C. Comeau, M. Pepper, H. Lischka, P.G. Szalay, R. Ahlrichs, F.B. Brown and J.-G. Zhao, Department of Chemistry, The Ohio State University, Columbus, OH 43210.

21. MOPLRO: written by H.-J. Werner, P.J. Knowles, with contributions from R.D. Amos, A. Berning, D.L. Cooper, M.J.O. Deegan, A.J. Dobbyn, F. Eckert, C. Hampel, T. Leininger, R. Lindh, A.W. Lloyd, W. Meyer, M.E. Mura, A. Nickla, P. Palmieri, K. Peterson, R. Pitzer, P. Pulay, G. Rauhut, M. Schtz, H. Stoll, A.J. Stone and T. Thorsteinsson, School of Chemistry, University of Birmingham, Edgbaston, Birmingham, B15 2TT, UK.

22. ACESII: R. Bartlett and coworkers. Quantum Theory Project, The University of Florida, Gainesville, FL 32611.

23. MOLCAS: K. Andersson, M.R.A. Blomberg, M.P. Fulscher, G. Karlstrom, V. Kello, R. Lindh, P.A.A. Malmqvist, J. Noga, J. Olsen, B.O. Roos, A.J. Sadlej, P.E.M. Siegbahn, M. Urban and P.-O. Widmark, University of Lund, Sweden.

24. D.A. McQuarrie. *Quantum Mechanics*. University Science Books, Mill Valley, CA, 1983.

25. A. Szabo and N. Ostlund. *Modern Quantum Chemistry*. Macmillan Press, New York, 1982.

26. R. McWeeny. *Ab Initio Methods in Quantum Chemistry*. Academic Press, London, 1989.

27. P. Jorgensen and J. Simons. *Second Quantization-Based Methods in Quantum Chemistry*. Academic Press, New York, 1981.

28. R.J. Bartlett. Coupled-cluster theory: An overview of recent developments. In D.R. Yarkony, editor, *Modern Electronic Structure Theory*. World Scientific Press, 1995.

29. R.G. Parr and W. Yang. *Density Functional Theory of Atoms and Molecules*. Oxford University Press, 1989. The X-alpha DFT method is also described elsewhere in this text.

30. B.P. Hay. Methods for molecular mechanics modelling of coordination compounds. *Coor. Chem. Rev.*, 126: 177–236, 1993.

31. C. Landis, D.M. Root, and T. Cleavland. In K.B. Lipkowitz and D.B. Boyd, editors, *Reviews in Computational Chemistry, Vol V*, New York, 1991. VCH Publishing.

32. R. Shepard, I. Shavitt, R.M. Pitzer, D.C. Comeau, M. Pepper, H. Lischka, P.G. Szalay, R. Ahlrichs, F.B. Brown, and J.-G. Zhao, *Int. J. Quantum Chem.*, S2:149, 1988.

33. H.-J. Werner. Matrix-formulated direct multiconfigurational self-consistent field and multiconfigurational reference configuration interaction methods. In K.P. Lawley, editor, *Ab Initio Methods in Quantum Chemistry II*, New York. John Wiley and Sons.

34. T. Yonezawa, H. Nakatsuji, T. Kawamura and H. Kato. *J. Chem. Phys.*, 51:669, 1969.

35. H. Nakatsuji, H. Kato, and T. Yonezawa. *J. Chem. Phys.*, 51: 3175, 1969.

36. N. Matagata and K. Nishimoto. *Z. Physik. Chem.*, 13: 140, 1957.

37. H. Nakatsuji, M. Komari, and T. Yonezawa. *J. Chem. Phys.*, 83:723, 1985.

38. H. Nakatsuji, M. Komari, and T. Yonezawa. *Chem. Phys. Lett.*, 142:446, 1987.

39. H. Nakatsuji and S. Saito. Theoretical study for the excited states of $MoO_{4-n}S_n^{2-}$ ($n = 0, 4$) and $MoSe_4^{2-}$. *J. Chem. Phys.*, 93:1865, 1990.

40. H. Nakatsuji, M. Ehara, M.H. Palmer and M.F. Guest. Theoretical study on the excited and ionized states of titanium tetrachloride. *J. Chem. Phys.*, 97:2651, 1993.

41. H. Nakatsuji and M. Ehara. Symmetry adapted cluster-configuration interaction study on the excited and ionized states of $TiBr_4$ and TiI_4. *J. Chem. Phys.*, 101:7658, 1994.

42. M. Hada, Y. Imai, M. Hidaka, and H. Nakatsuji. Theoretical study on the excitation spectrum and the photofragmentation reaction of $Ni(CO)_4$. *J. Chem. Phys.*, 103:6993, 1995.

43. M. Ehara and H. Nakatsuji. Collision induced absorbtion spectra and line broadening of CsRg system (Rg = Xe, Kr, Ar, Ne) studies by the symmetry adapted cluster-configuration interaction (SAC-CI) methods. *J. Chem. Phys.*, 102:6822, 1995.

44. H. Nakai, Y. Ohmori, and H. Nakatsuji. Theoretical study on the ground state and excited states of MnO_4^-. *J. Chem. Phys.*, 95:8287, 1994.

45. S. Jitsuhiro, H, Nakai, M. Hada, and H. Nakatsuji. Theoretical study on the ground state and excited states of chromate anion CrO_4^{2-}. *J. Chem. Phys.*, 101:1029, 1994.

46. K. Yasuda and H. Nakatsuji. Theoretical study of the visible and ultraviolet spectra of crhomyl chloride (CrO_2Cl_2). *J. Chem. Phys.*, 99:1945, 1993.

47. B.O. Roos, K. Andersson, M.P. Fülsher, P.-A. Malmqvist, L. Serrano-Andrés, K. Peirloot, and M. Merchán. *Adv. Phys. Chem.*, 93:219, 1996.

48. K. Andersson and B.O. Roos. In D.R. Yarkony, editor, *Modern Eectronic Structure Theory*. World Scientific Press, 1995.

49. B.O. Roos, M.P. Fülscher, Per-Åke Malmqvist, M. Merchán, and L. Serrano-Andrés. Theoretical studies of electronic spectra of organic molecules. In S.R. Langhoff, editor, *Quantum Mechanical Electronic Structure Calculations with Chemical Accuracy*, Dordrecht, The Netherlands, 1995. Kluwer Academic Publishers.

50. K. Pierloot, E. Tsokos, and B.O. Roos. $3p-3d$ intershell correlation effects in transition metal ions. *Chem. Phys. Lett.*, 214(6):583, 1993.

51. U. Ryde, M.H.M. Olsson, K. Pierloot, and B.O. Roos. Blue copper proteins are not entatic (preprint).

52. B.J. Persson, B.O. Roos, and K. Pierloot. A theoretical study of the chemical bonding in $M(CO)_x$ (M=Cr, Fe, and Ni). *J. Chem. Phys.*, 101(8):6810–6821, 1994.

53. L. Serrano-Andrés, M. Marchán, I. Nebot-Gil, R. Lindh, and B.O. Roos. *J. Chem. Phys.*, 98:3151, 1993.

54. E.R. Davidson. In D. Salahub and M.C. Zerner, editors, *The Challenge of d and f Electrons*. A.C.S. Symposium Series 394, 1989.

55. R. Shepard. The multiconfigurational self-consistent field method. In K.P. Lawley, editor, *Ab Initio Methods in Quantum Chemistry II*, New York. John Wiley and Sons.

56. B.O. Roos. *Ab Initio Methods in Quantum Chemistry*. Wiley, New York, NY, 1987.

57. J.C. Slater. *Quantum Theory of Atomic Structure*. McGraw-Hill, New York, 1960.

58. R. Pauncz. *Spin Eigenfunctions*. Plenum Press, New York, 1979.

59. C.C.J. Roothaan. *Rev. Mod. Phys.*, 23:69, 1951.

60. G.G. Hall. *Proc. Roy. Soc. Lond.*, A205:541, 1951.

61. M.C. Zerner and M. Hehenberger. A dynamic damping for converging molecular SCF calculations. *Chem. Phys. Lett.*, 62:550–554, 1979.

62. P. Pulay. *J. Comp. Chem.*, 62:556, 1982.

63. H.L. Hsu, E.R. Davidson, and R.M. Pitzer, 1976.

64. W.D. Edwards, B. Weiner, and M.C. Zerner. *J. Phys. Chem.*, 92:6188, 1988.

65. J.A. Pople and R.K. Nebset. Self-consistent orbitals for radicals. *J. Chem. Phys.*, 22:571, 1954.

66. A.T. Amos and G.G. Hall. *Proc. Roy. Soc. Lond.*, A263:483, 1961.

67. A.T. Amos and L.C. Snyder. *J. Chem. Phys.*, 41:1773, 1964.

68. R. Caballol, R. Gallifa, J. Ricra, and R. Carbo. *Int. J. Quantum Chem.*, 8:373, 1974.

69. R. Carbo, L. Domingo, and J. Gregori. *Int. J. Quantum Chem.*, 7:725, 1980.

70. W.D. Edwards and M.C. Zerner. A generalized restricted open-shell Fock operator. *Theoret. Chim. Acta*, 72:347, 1987.

71. A. Veillard. In G.H.F. Diercksen, B.T. Sutcliffe, and A. Veillard, editors, *Computational Techniques in Quantum Chemistry*, Boston, 1975. Nato Adv Series C.

72. J.C. Slater, J.B. Mann, T.M. Wilson, and J.H. Wood. *Phys. Rev.*, 184:672, 1969.

73. J.F. Abdulnar, J. Linderberg, N.Y. Ohrn, and P.W. Thulstrup. *Phys. Rev. A*, p. 889, 1972.

74. P.O. Lowdin. Symposium in molecular physics in Nikko. Tokyo, 1954.

75. R. Ake and M. Gouterman. *Theoret. Chim. Acta*, 15:20.

76. J.E. Harriman, *J. Chem. Phys.*, 40:2827, 1964.

77. M.G. Cory and M.C. Zerner (in preparation).

78. M.G. Cory, K.K. Stavrev, and M.C. Zerner. An examination of the configuration averaged Hartree–Fock proceedure for model ferredoxin and its electronic spectroscopy. *Int. J. Quantum Chem.*, 63:781, 1997.

79. A. Hardison and J.E. Harriman. *J. Chem. Phys.*, 46:3639, 1967.

80. M.C. Zerner. A configuration averaged Hartree–Fock. *Int. J. Quantum Chem.*, 35:567–575, 1989.

81. G. Estiu and M.C. Zerner. *Int. J. Quantum Chem.*, 98:4793, 1994.

82. G. Estiu and M.C. Zerner. Icosahedral symmetry and magnetic properties of small nickel metal clusters. *J. Phys. Chem.*, 98:9972–9978, 1994.

83. G. Estiu and M.C. Zerner. Interplay between geometric and electronic structure and the magnetism of small Pd clusters. *J. Phys. Chem.*, 98:4793–4799, 1994.

84. G. Estiu and M.C. Zerner. Structural, electronic, and magnetic properties of small Ni clusters. *J. Phys. Chem.*, 100(42):16874–16880, 1996.

85. S. Wilson. Basis sets. In K.P. Lawley, editor, *Ab Initio Methods in Quantum Chemistry I*, New York, 1987. John Wiley and Sons.

86. S. Huzinaga. *J. Chem. Phys.*, 42:1293, 1965.

87. T.H. Dunning, Jr. *J. Chem. Phys.*, 53:2823, 1970.

88. T.H. Dunning, Jr. *J. Chem. Phys.*, 90:1007, 1989.

89. P.Å. Malmqvist, P.-O. Widmark, and B.O. Roos. *Theoret. Chim. Acta*, 77:291, 1990.

90. P.-O. Widmark, B.J. Persson, and B.O. Roos. *Theoret. Chim. Acta*, 79:419, 1991.

91. W.C. Ermler, R.B. Ross, and P.A. Christiansen. Spin orbit coupling and other relativistic effects in atoms and molecules. In P.-O. Lowdin, editor, *Advances in Quantum Chemistry, Vol. 19*, San Diego, CA, 1988. Academic Press.

92. W. Stevens, Bausch, and M. Krauss. *J. Chem. Phys.*, 81:6026, 1984.

93. R.L. Martin and P.J. Hay. Relativistic constributions to the low-lying excitation energies and ionization potentials of the transition metals. *J. Chem. Phys.*, 74(9):4539, 1981.

94. P.J. Hay and R.W. Wadt. *J. Chem. Phys.*, 82:270, 1985.

95. R.W. Wadt and P.J. Hay. *J. Chem. Phys.*, 82:284, 1985.

96. P.J. Hay and R.W. Wadt. *J. Chem. Phys.*, 82:299, 1985.

97. M.C. Zerner. Removal of core orbitals in valence orbital only calculations. *Mol. Phys.*, 23:963, 1972.

98. K. Balasubramanium and K.S. Pitzer. Relativistic quantum chemistry. *Adv. Chem. Phys.*, 67:287, 1987.

99. M.R. Hoffman. Quasidegenerate perturbation theory using effective Hamiltonians. In D.R. Yarkony, editor, *Modern Electronic Structure Theory*, World Scietific Press, 1995.

100. B.A. Hess, M. Marian, and S.D. Peyerimoff. Ab Initio calculations of spin-orbit effects in molecules including electron correlation. In D.R. Yarkony, editor, *Modern Electronic Structure Theory*. World Scietific Press, 1995.

101. I. Lindgren. *Atomic Many Body Theory*. Springer, New York, 1986.

102. C. Cohen-Tannoudji, B. Diu, and F. Laloe. *Quantum Mechanics. Volumes I and II*. John Wiley and Sons, New York, 1977.

103. M. Krauss and W.J. Stevens. *Ann. Rev. Phys. Chem.*, 35:357, 1984.

104. C. Nash, B. Bursten, and W.C. Elemer. Ab initio relativistic effective potentials with spin-orbit operators. VII. Am through element 118. *J. Chem. Phys.*, 106(12):5133, 1997.

105. M. Dolg, H. Stoll, and H. Preuss. *J. Chem. Phys.*, 90(3):1730, 1989.

106. F. Basalo and R.C. Johnson. *Coordination Chemistry*. Mid-Country Press, 2nd edition, 1986.

107. F.A. Cotton, *Chemical Applications of Group Theory*. John Wiley and Sons, New York, 1963.

108. Tanabe and Sugano. *J. Phys. Soc. Japan*, 9:753, 1954.

109. A. Zewail. *The Chemical Bond. Structure and Dynamics*. Academic Press, Inc., 1992.

110. J.E. Stevens, R.K. Chaudhuri, and K.F. Freed. Global three-dimensional potential energy surfaces of H_2S from the *ab-initio* effective valence shell Hamiltonian method. *J. Chem. Phys.*, 104(8754), 1996.

111. P. Pykko. Relativistic effects in structural chemistry. *Chem. Rev.*

112. P.E.M. Siegbahn. Electronic structure calculations for molecules containing transition metals. *Adv. Phys. Chem.*, 93:219, 1996.

113. C.K. Jorgensen. *Modern Aspects of Ligand Field Theory*. North Holland, Amsterdam, 1971.

114. A.B.P. Lever. *Inorganic Electronic Spectroscopy*. Elsevier, Amsterdam, 1984.

115. Karl F. Freed. In E. Kayachko and J.-L. Calais, editors, *Conceptual Trends in Quantum Chemistry, Vol II*, Amsterdam, 1995. Kluwer.

116. W. Thiel. Perspectives on semiempirical molecular orbital theory. *Adv. Phys. Chem.*, 93:703, 1996.

117. P.-O. Lowdin. On the connection between semi-empirical and ab-initio methods in quantum theory of molecular electronic structure (in press).

118. C.H. Martin. Highly accurate ab-initio pi-electron Hamiltonians for small protonated schiff bases. *J. Phys. Chem.*, 100(34):14310, 1996.

119. C.H. Martin. Redesigning semiempirical-like pi-electron theory with second order effective valence shell Hamiltonian theory: Application to large protonated schiff bases. *J. Phys. Chem.*, 100(34):14310, 1996.

120. C.H. Martin and R.R. Birge. Reparameterizing mndo for excited state calculations using ab-initio effective Hamiltonian theory: Application to the 2,4-pentadien-1-iminium cation. *J. Phys. Chem. A.*, 102.

121. C.H. Martin and K.F. Freed. Ab-initio computation of semiempirical pi-electron methods. V geometry dependence of pi-electron integrals. *J. Chem. Phys.*, 105:1437, 1996.

122. C.D. Sherrill and H.F. Schaefer III. The configuration interaction method: Advances in highly correlated approaches. *Adv. Quant. Chem.*

123. F.A. Matsen and R. Pauncz. *The Unitary Group in Quantum Chemistry*. Elsevier, Amsterdam, 1986.

124. E.R. Davidson. *Chem. Phys. Lett.*, 21:565, 1973.

125. B. Liu. In numerical algorithms in chemistry: Algebraic methods. Technical report, Lawrence Berkeley Laboratory, 1978. Edited by C. Moler and I. Shavitt, NRCC Report LBL-8158.

126. D.J. Rowe. *Nuclear Collective Morion, Models and Theory*. Methuen and Co., London, 1970.

127. J.B. Foresman, M. Head-Gordan, J.A. Pople, and M.J. Frish. Towards a systematic theory for excited states. *J. Phys. Chem.*, 96:135, 1992.

128. M. Head Gordan, R.J. Rico, M. Oumi, and T.J. Lee. *Chem. Phys. Lett.*, 219:21, 1994.

129. S.R. Langhoff and E.R. Davidson. *Int. J. Quantum Chem.*, 8, 1961.

130. J.C. Cullen and M.C. Zerner. The linked singles and doubles method: An approximate theory of electronic correlation based on the coupled cluster ansatz, 1982.

131. B.H. Brandow. *Int. J. Quantum Chem.*, 15:207, 1979.

132. R.J. Bartlett and J.F. Stanton. Applications of post-Hartree–Fock methods: A tutorial. In K.B. Lipkowitz and D.B. Boyd, editors, *Reviews in Computational Chemistry, Vol V*, New York, 1991. VCH Publishing.

133. J.A. Pople, M. Head-Gordon, and K. Ragacari. *J. Chem. Phys.*, page 5986, 1987.

134. M. Bohme and Frenking. The Cu-C bond dissociation energy of CuC_3 a dramatic failure of the QCISD(T) method. *Chem. Phys. Lett.*, 224:195, 1994.

135. G.C. Schatz and M.A. Ratner. *Quantum Mechanics in Chemistry*. Prentice Hall, Inc., 1993.

136. S. Mukamal. *Principles of Nonlinear Optical Spectroscopy*. Oxford University Press, New York.

137. W.E. Boyce and R.C. Diprima. *Elementary Differential Equations and Boundary Value Problems*. John Wiley and Sons, New York.

138. D. McQuarrie. University Science Books, Mill Valley, CA, 1983.

139. J.P. Hansen and I.R. McDonald. *Theory of Simple Liquids*.

140. Peter Fulde. *Electron Correlations in Molecules and Solids*. Springer-Verlag, Berlin, 1991.

141. M.F. Herman, K.F. Freed, and D.L. Yeager. Analysis and evaluation of ionization potentials, electron affinities, adn excitation energies by the equations of motion-green's function methods.

142. J. Oddershede, P. Jorgensen, and D.L. Yeager. Polarization propoator methods in atomic and molecular physics.

143. J.D. Baker and M.C. Zerner. Characterization of the random phase approximation with the intermediate neglect of differential overlap Hamiltonian for electronic spectroscopy. *J. Phys. Chem.*, 95:8614–8619, 1991.

144. H. Koch and P. Jorgensen. *J. Chem. Phys.*, 93:3333, 1990.

145. C.W. Bauschlicher Jr. and H. Partridge. Cr_2 revisited. *Chem. Phys. Lett.*, 231:277–282, 1994.

146. M.M. Goodgame and W.A. Goddard III. The sextuple bond of Cr_2. *J. Phys. Chem.*, 5:215–217, 1981.

147. M.J.S. Dewar, K.M. Merz Jr., and J.J.P. Stewart. *J. Am. Chem. Soc.*, 106:4040, 1984.

148. J.J.P. Stewart. Optimization of parameters for semiempirical methods. I. Method. *J. Comp. Chem.*, 10:209, 1989.

149. J.J.P. Stewart. Optimization of parameters for semiempirical methods. II. Applications. *J. Comp. Chem.*, 10:221, 1989.

150. W. Anderson, T. Cundari, R. Drago, and M.C. Zerner. On the utility of the semi-empirical INDO/1 method for the calculation of the geometries of second row transition metal species. *Inorg. Chem.*, 29:1–3, 1991.

151. W. Anderson, T. Cundari, and M.C. Zerner. INDO/1 calculations on second row transition metal species. *Int. J. Quantum Chem.*, 39:31–45, 1991.

152. J. Sadlej. *Semiempirical methods in quantum chemistry (translated by I. L. Cooper)*. Elis Horwood, division of Wiley, New York, 1985.

153. R.G. Parr. *Quantum Theory of Molecular Electronic Structure*. Benjamin, Boston, 1963.

154. R. Pariser and R.G. Parr. *J. Chem. Phys.*, 21:767, 1953.

155. R.Pariser. Theory of electronic spectra and structure of the polyacenes and alternant hydrocarbons. *J. Chem. Phys.*, 24:250, 1956.

156. M.C. Zerner. *J. Chem. Phys.*, 62:2788, 1975.

157. J.E. Ridley and M.C. Zerner. An intermediate neglect of differential overlap technique for spectroscopy: Pyrrole and the azines. *Theoret. Chim. Acta*, 32:111, 1973.

158. J.E. Ridley and M.C. Zerner. *Theoret. Chim. Acta*, 42:223, 1976.

159. J.C. Culberson, P. Kanppe, N. Rosch, and M.C. Zerner. An intermediate neglect of differential overlap (INDO) technique for lanthinide complexes: Studies of lanthinide halides. *Theoret. Chim. Acta*, 71:21, 1987.

160. Pople and Beveridge. *Approximate Molecular Orbital Methods*. McGraw Hill, New York, 1970.

161. J.A. Pople, D.P. Santry, and G.A. Segal. Approximate self-consistent molecular orbital theory. I. Invariant procedures. *J. Chem. Phys.*, 43:A129, 1965.

162. J.A. Pople and G.A. Segal. Approximate self-consistent molecular orbital theory. II. Calculations with complete neglect of differential overlap. *J. Chem. Phys.*, 43:S136, 1965.

163. J.A. Pople and G.A. Segal. *J. Chem. Phys.*, 44:3289, 1966.

164. J.A. Pople, D. Beveridge, and P. Dobosch. *J. Chem. Phys.*, 47:2026, 1967.

165. M.B. Hall and R.F. Fenske. Electronic structure and bonding in methyl- and perflouromethyl (pentacarebonyl) manganese. *Inorg. Chem.*, 11:768, 1972.

166. M. Wolfsberg and L. Helmholtz. The spectra and electronic structure of tetrahedral ions MnO_4^-, CrO_4^{2-}, and ClO_4^-. *J. Chem. Phys.*, 20:837, 1952.

167. L.L. Lohr, Jr. and W.N. Lipscomb. Molecular orbital theory of spectra of Cr^{3+} ions in crystals. *J. Chem. Phys.*, 38:1607, 1963.

168. R.B. Woodward and R. Hoffman. Verlag Chemie, Weinham, FDR, 1970.

169. C.J. Ballhausen. *Introduction to Ligand Field Theory*. McGraw Hill, New York, 1962.

170. M.C. Zerner and M. Gouterman. Porphyrins IV: Extended Huckel calculations on transition metal complexes. *Theoret. Chim. Acta*, 4:44–63, 1966.

171. R.S. Mulliken. *J. Chem. Phys.*, 46:497, 1949.

172. P. Coffey. Potential energy integrals in semiempirical MO methods. *Int. J. Quantum Chem.*, 8:263, 1974.

173. K. Rudenberg. Three- and four-center integrals in molecular quantum mechanics. *J. Chem. Phys.*, 19:1433, 1951.

174. J.D. Head and M.C. Zerner. An approximate Hessian for molecular geometry optimization, 1986.

175. J.S. Griffith. *The Theory of Transition Metal Ions*. Cambridge University Press, Cambridge, 1961.

176. P. Pulay. Analytical derivative techniques and the calculations of vibrational spectra. In D.R. Yarkony, editor, *Modern Electronic Structure Theory*. World Scientific Press, 1995.

177. J.D. Baker. An algorithm for the locations of transition states. *J. Phys. Chem.*, 95:8614, 1991.

178. P. Coppens and L. Li. *J. Chem. Phys.*, 81, 1984.

179. K.K. Stavrev and M.C. Zerner. A theoretical treatment of the absorbtion spectrum and emission properties of Cu(II) porphryn. *Chem. Phys. Lett.*, 233:179–184, 1995.

180. M.G. Cory, H. Hirose, and M.C. Zerner. Calculated structure and electronic absorbtion spectroscopy for magnesium pthalocyanine and its anion radical. *Inorg Chem.*, 34:2969–2979, 1995.

181. M.G. Cory and M.C. Zerner. Metal-ligand exchange coupling in transition-metal complexes. *Chem. Rev.*, 91:813, 1991.

182. M. Kotzian, N. Rosch, and M.C. Zerner, *Int. J. Quantum Chem.*, page 545, 1991.

183. M.C. Zerner. Intermediate neglect of differential overlap calculations on the elecronic structure of transition metal complexes. In N. Russo and D.R. Salahub, editors, *Metal Ligand Interactions: Structure and Reactivity NATO ASI series, Volume 474,* Dordrecht, 1995. Kluwer.

184. A.E. Hansen and T.D. Bouman. *Mol. Phys.*, 37:1713, 1979.

185. M. Dolg, P. Fulde, W. Kulch, C.-S. Neuman, And H. Stoll. *J. Chem. Phys.*, 94(4):3011, 1991.

186. D. Dai and K. Balasubramanian.

187. B. Manna and K.K. Das. Low-lying electronic states of $GeSe$: Multiferences singles and doubles configuration interaction study. *J. Phys. Chem. A.*, 102:214, 1998.

188. MOTECC: Modern Techniques in Computational Chemistry. Ab Initio Programs. List and description of IBM-licensed codes in MOTECC: ATOMSCF, HYCOIN, ALCHEMY-II, HONDO-8, KGNMOL, QMDCP, KGNMD, NSCATT, KGNNCC, KGNMCYL, BROWNIAN, PRONET, KGNCA, KGNFLOW, TRB3D, KGNFEM, AIRPOL, KGNGRAF, XWIB. Enrico Clementi, Centro di Ricerca, Sviluppo e Studi Superiori in Sardegna (CRS4), Casella Postale 488, 09100 Cagliari, ITALIA.

189. N. Oliphan and R. Bartlett. Theoretical determination of the charge-transfer and ligand field transition energies for $FeCl_4$ using the EOM-CCSD method. *J. Am. Chem. Soc.*, 116(9):4091–4092, 1994.

190. S. Niu and M. Hall. Comparison of Hartree–Fock, density functional, Möller–Plesset perturbation, coupled cluster, and configuration interactions methods for the migratory insertion of nitric oxide into a cobalt-carbon bond. *J. Chem. Phys.*, 101:1360, 1997.

191. G.L. Estiu, N. Rosch, and M.C. Zerner. Ground state characteristics and optical spectra of Ce and Eu octahethyl bispophyrinate double deckers. 99:13819–13829, 1995.

192. Feng Gai, K.C. Hansen, J. Cooper McDonald, and P.A. Anfinrud. Chemical dynamic in proteins: The photoisomerization of retinal in bacterirhodopsin. *Science*, 279:1886–1891, 1997.

193. R.L. Liboff. *Introductory Quantum Mechanics*. Holden-Day, Inc., 1980.

194. P. Pykko. Relativistic quantum chemistry. *Adv. Quant. Chem.*, 11:354–399, 1978.

195. H.A. Bethe and E.E. Salpeter. *Quantum Theory of One and Two Electron Atoms.* Plenum, New York, 1977.

196. G.E. Uhlenbeck and S. Goudsmith. *Nature*, 117, 1926.

197. R.J. Buenker, P. Chandra, and B.A. Hess. *Chem. Phys.*, 81:1, 1984.

198. P. Pykko and J.P. Desclaux. *Acc. Chem. Res.*, 12: 276, 1979.

199. M. Bockmann, M. Klessinger, and M.C. Zerner. Spin-orbit coupling in organic molecules: A semiempirical configuration interaction approach towards triplet state reactivity. *J. Phys. Chem.*, 100:10570–10579, 1996.

200. M. Kotzian, N. Rosch, R.M. Pitzer, and M.C. Zerner. A spin-orbit interaction enhanced INDO/S-CI technique: Application to main group and transition metal heteronuclear diatomic molecules. *Chem. Phys. Lett.*, 160:168–174, 1989.

201. W.P. Anderson, T. Cundari, and M.C. Zerner. INDO/1 calculation on second row transition metals. *Int. J. Quantum Chem.*, 39:391, 1990.

202. C. Ribbing and C. Daniel. Spin-orbit coupled excited states in transition metal complexes: A configuration interaction treatment of $HCo(CO)_4$. *J. Chem. Phys.*, 100(9):6591, 1994.

203. A. Messiah. *Quantum Mechanics*. North Holland Publishing Company, Amsterdam.

204. K. Jug. *Theoret. Chim. Acta*, 74:91, 1969.

205. M. Nooijen and R.J. Bartlett. Similarlity transformed equation-of-motion coupled-cluster theory: Details, examples, and comparisons. *J. Chem. Phys.*, 107(17):6812–6830, 1997.

206. C.K. Jorgensen and S. Brinen. *Mol. Phys.*, 6:629, 1963.

207. S.P. Sinha. *Malays Sci.*, 6:88, 1971.

11 Electronic Structure Calculations: Density Functional Methods with Applications to Transition Metal Complexes

JIAN LI, LOUIS NOODLEMAN, AND DAVID A. CASE
Department of Molecular Biology, TPC15
The Scripps Research Institute
La Jolla, CA 92037, USA
E-mail: lou@scripps.edu
case@scripps.edu

Inorganic Electronic Structure and Spectroscopy, Volume I: Methodology.
Edited by E. I. Solomon and A. B. P. Lever.
ISBN 0-471-15406-7.

1 INTRODUCTION

Transition metal complexes can have complex and diverse ground and excited states, and no simple description or review can cover this field in any systematic manner. Books by Ballhausen[1] and Lever[2] provide an excellent short introduction to transition-metal spectroscopy, and classic texts on ESR[3,4] cover many aspects of magnetic behavior. All of these deal primarily with systems containing a single metal ion, and were written before many of the modern computational techniques were developed. In this chapter, we provide an overview of density functional studies of coordination complexes of the first transition series. The main part of the discussion is pedagogical, emphasizing general features of density functional theories, interpretations of energy-level diagrams, and approaches to multiple-metal clusters. In addition, we have tried to make a quantitative assessment of the expected level of accuracy of current calculations, using the hexa-aquo species as a particular example. We emphasize that this is not a comprehensive review; other important aspects of inorganic systems are covered in the chapters by Solomon and Lever, Gatteschi, and Zerner in this book, and in other books and review articles.[5–8]

Electronic structure calculations on transition metal complexes face a number difficulties beyond the obvious problem that most interesting systems contain large numbers of atoms and electrons. One principal difficulty is that molecular orbital theory, which can supply a satisfying qualitative account of a great many features of transition metal systems, exhibits a variety of systematic deficiencies, so that quantitative results are often unreliable. It is useful to think in terms of "dynamical" and "non-dynamical" correlation effects that go beyond the Hartree–Fock molecular orbital approach.[8] Roughly speaking, dynamical correlation is a short-ranged effect that decreases electron-electron repulsion within a single configuration; additional non-dynamical effects are often long-ranged, and arise when there is a near-degeneracy of more than one configuration. Transition metal complexes turn out to be more difficult than typical organic molecules on both counts.[8] First, the small spatial extent of the 3*d* shell leads to large dynamical correlation effects, and high angular momentum basis functions are needed to describe these in a CI or perturbation approach. Second, systems with unpaired spins frequently have a number of near-degenerate configurations for both ground and excited states, so that non-dynamical correlation effects are frequently important. Even at a qualitative level, the presence of unpaired electrons can lead to large spin-polarization effects that can complicate or invalidate simple molecular orbital arguments. Since there is no practical single approach that handles all of these problems, a variety of approaches have to be considered; many of these are discussed in M. Zerner's chapter in this book; others are covered in recent reviews.[7,9]

In this chapter, we focus on the use of density functional theory (DFT) as a practical computational tool for understanding a variety of energetic and spectroscopic features of transition metal complexes. The pictures and ideas of DFT can often be used in a qualitative fashion to correlate and rationalize experimental data; this can be especially helpful in systems with high-spin metal sites, where large spin-polarization effects limit the usefulness of conventional ligand-field or extended Huckel models. DFT is also used as a quantitative tool whose expected level of accuracy is becoming increasingly clear. Among the most exciting and promising developments in this area are models to include environmental effects in liquid and solid-state systems, providing for the first time realistic models of phenomena like oxidation/reduction and protonation events. In addition to their quantitative value, calculations of redox potentials and pK_a's of related systems provide insights into the physical effects contributing to shifts in natural and synthetic systems.

2 DENSITY FUNCTIONAL THEORY FOR TRANSITION METAL COMPLEXES

Density functional theory is now becoming widely recognized as a high-level method for carrying out quantum chemistry calculations, particularly for transition-metal clusters, which are difficult to handle by more conventional *ab initio* techniques.[7,10,11] A distinctive feature of density-functional theory is that both the exchange and the correlation part of the electronic energy is approximated by terms that depend only on functionals of the electron density. In principle, there exists a universal and exact functional that yields the total ground state energy, given the electron density. Furthermore, the energy that arises from the true density is lower than that arising from any other density that integrates to the correct number of electrons, so that a variational procedure can be used to optimize the energy and corresponding electron density.[10,11] A computational procedure developed by Kohn and Sham allows this optimization to be carried out (using approximate exchange-correlation functionals) in a manner similar to that of classical molecular orbital theory: given an orbital basis set, one constructs and diagonalizes a Fock-like matrix to determine the Kohn–Sham orbitals and energies. With this approach, one retains much of the conceptual simplicity and appeal of molecular orbital theory, obtaining numerical results that are generally better (and simpler to compute) than Hartee–Fock orbitals and energies. Because approximate density functional methods can scale as a lower power of basis set size than Hartree–Fock or more complex *ab initio* methods, large basis sets and realistic ligand models are often feasible with a density functional approach.

Density functional methods have become enormously popular recently, with a correspondingly large literature that can only be touched upon here. A variety of recent compilations and reviews cover many aspects of the field.[7,12–14] Our goal here is to provide a pedagogical overview of some of the salient points, paying particular attention to systems with significant spin polarization and non-dynamical correlation effects, and on finding energies and properties of excited states as well as ground states.

2.1 The Total Energy in Density Functional Theory

In order to provide a framework for a qualitative understanding of recent trends in density functional theory, we begin with some basic concepts from the theory of reduced density matrices.[15,16] Let a general coordinate $x = (r, s)$ represent both space and spin variables. The total energy is then:

$$\begin{aligned} E &= \int_{x'=x} [-\nabla^2/2]\rho_1(x; x')\, dx + \int_x \rho(x) V_N(x)\, dx \\ &\quad + \frac{1}{2}\int_{x_1,x_2} \rho_2(x_1, x_2) r_{12}^{-1}\, dx_1\, dx_2 + U_{NN} \\ &= T + V_{Ne} + U_{ee} + U_{NN} \end{aligned} \tag{1}$$

Here the first term (T) is the total kinetic energy of the electronic system, the second is the nuclear-electron attraction energy (V_{Ne}), the third is the total electron-electron repulsion energy of the system (U_{ee}), and the last term is the nuclear-nuclear repulsion energy; $\rho_2(x_1, x_2)$ is the diagonal part of the second order density matrix, giving the joint probability of finding electron 1 at x_1 and electron 2 at x_2. It is useful to separate this joint distribution into four terms ($\rho_2^{\alpha\alpha}$, $\rho_2^{\beta\beta}$, $\rho_2^{\alpha\beta}$ and $\rho_2^{\beta\alpha}$), involving only the two space variables (r_1, r_2), indexed by the spin labels α, β.

For a system with N electrons, we have the following conservation equations

$$\int \rho(r) = N \tag{2}$$

$$\int \rho_2(r_1, r_2) dr_1 dr_2 = N(N-1) \tag{3}$$

Since the conservation equation applies equally to α and β spin electrons, $N_\alpha + N_\beta = N$. Eq. (3) can be interpreted as saying that every electron in the system interacts with all other electrons (but not with itself). We can expand $N(N-1)$ in the following illuminating way

$$\begin{aligned} N(N-1) &= (N_\alpha + N_\beta)(N_\alpha + N_\beta - 1) \\ &= N_\alpha(N_\alpha - 1) + N_\beta(N_\beta - 1) + N_\alpha N_\beta + N_\beta N_\alpha \end{aligned} \quad (4)$$

The individual terms on the right-hand side are the integrals of $\rho_2^{\alpha\alpha}(r_1, r_2)$, $\rho_2^{\beta\beta}(r_1, r_2)$, $\rho_2^{\alpha\beta}(r_1, r_2)$, $\rho_2^{\beta\alpha}(r_1, r_2)$. All α electrons interact with all β electrons, and with the other $N_\alpha - 1$ electrons of the same spin.

2.2 Exchange Energy and the Fermi Hole

For $\rho_2^{\alpha\alpha}$, Eq. (3) implies that correct joint probability function deviates substantially from the classical uncorrelated form, since

$$\int \rho_{2,uncorr}^{\alpha\alpha}(r_1, r_2)dr_1dr_2 \equiv \int \rho^\alpha(r_1)\rho^\alpha(r_2)dr_1dr_2 = N_\alpha^2 \quad (5)$$

rather than $N_\alpha(N_\alpha - 1)$. The deviations from an uncorrelated pair distribution can be expressed in the following form:[15]

$$\rho_2^{\alpha\alpha}(r_1, r_2) = \rho^\alpha(r_1)\rho^\alpha(r_2)\left[1 + f^{\alpha\alpha}(r_1, r_2)\right], \quad (6)$$

consisting of an uncorrelated part plus an additional term describing Fermi correlation. Dividing both sides by $\rho^\alpha(r_2)$ yields the conditional probability for finding an electron of spin α at r_1 given that there is another spin α electron at r_2. The difference between this and the average density at r_1 is the Fermi hole density, that is, the density deficit at r_1 due to the presence of an electron at r_2:

$$\left[\frac{\rho_2^{\alpha\alpha}(r_1, r_2)}{\rho^\alpha(r_2)} - \rho^\alpha(r_1)\right] = \rho^\alpha(r_1)f^{\alpha\alpha}(r_1, r_2) \quad (7)$$

The integral of this density over r_1 is -1 for any r_2. Further, the Pauli exclusion principle requires that as $r_1 \to r_2$, $f^{\alpha\alpha}(r_1, r_2) \to -1$, so that no two α spin electrons can be at the same place. Thus, electrons of the same spin exhibit Fermi correlation in their interactions. Since this fundamentally arises from an antisymmetry requirement, it is present as well even in hypothetical non-interacting systems where the electron-electron interaction is turned off.[17,18] In this exchange-only model, the electron-electron repulsion energy becomes

$$\begin{aligned} U_{ee} = \frac{1}{2}\int \rho(r_1)\rho(r_2)r_{12}^{-1}dr_1dr_2 + \frac{1}{2}\int \rho^\alpha(r_1)U_x^\alpha(r_1)dr_1 \\ + \frac{1}{2}\int \rho^\beta(r_1)U_x^\beta(r_1)dr_1 \end{aligned} \quad (8)$$

where

$$U_x^\alpha = \int \rho^\alpha(r_2) f^{\alpha\alpha}(r_1, r_2) r_{12}^{-1} dr_2 \tag{9}$$

with a corresponding equation for U_x^β as a Coulomb integral of $\rho^\beta f^{\beta\beta}$. The first term in Eq. (8) is the classical electron-electron repulsion, and the second two give the total exchange energy. A straightforward and often effective approximation is to replace the r_2 integrals involving $f^{\alpha\alpha}$ and $f^{\beta\beta}$ by the corresponding integrals over the Fermi hole of a uniform electron gas having the same density as that of the real system at point r_1. This gives

$$U_x^\alpha = -(3)[3/4\pi]^{1/3}(\rho^\alpha)^{1/3} \tag{10}$$

This is the local spin density (LSD) approximation, and (since detailed numerical properties of the uniform electron gas are known), it can be extended to include correlation effects as well. It turns out, however, that correlation between electrons of the same spin is much smaller in finite systems than in the uniform electron gas,[19] so that ignoring or reducing the LSD correlation contribution for parallel spins is a common modification to the LSD model.

The study of the implications of the simple exchange model of Eq. (10) for molecular and solid-state system was pioneered by Slater,[17] and a slight variant of this model called $X\alpha$, where Eq. 10 is multipled by $3\alpha/2$ with $\alpha \approx 0.7$, was used for many early studies on transition metal complexes.[20] As discussed below, models that are generally more accurate than LSD for total energies are now available, but it is not clear that these refined models predict better densities than those arising from LSD itself.[21]

2.3 Correlation for Opposite Spins and the Coulomb Hole

Consider now the joint distribution function for electrons of opposite spin. We can write this in a form similar to Eq. 6:[15]

$$\rho_2^{\alpha\beta}(r_1, r_2) = \rho^\alpha(r_1)\rho^\beta(r_2)[1 + f^{\alpha\beta}(r_1, r_2)] \tag{11}$$

but now the Coulomb hole density integrates to zero for any r_2:

$$\int \rho^\alpha(r_1) f^{\alpha\beta}(r_1, r_2) dr_1 = 0 \tag{12}$$

This follows directly from the fact that the integral of $\rho_2^{\alpha\beta}$ is $N_\alpha N_\beta$. The Coulomb hole is not necessarily small, despite the fact that it must integrate to zero over all space. The important physical principle is that as $r_1 \rightarrow r_2$, $\rho_2^{\alpha\beta}(r_1, r_2)$ becomes small so that electrons of opposite spin avoid each other so as to reduce their Coulomb repulsion. The Coulomb hole density is thus negative (representing an electron deficit) at short distances and positive (electron surplus) at larger distances, to give a zero integral over all space.

As with parallel spins, these opposite-spin exchange-correlation effects could be approximated by using an $f^{\alpha\beta}$ function derived for a uniform electron gas. This gives generally reliable results for solid-state systems, but is often not sufficiently accurate to provide reliable bond energies or geometries in molecules. Over the past decade, considerable practical success has been achieved by extending the LSD model to write the exchange-correlation energy in terms of the densities and their local gradients:[22]

$$E_{XC}^{GGA}[\rho^\alpha, \rho^\beta] = \int f(\rho^\alpha, \rho^\beta, \nabla\rho^\alpha, \nabla\rho^\beta)dr \tag{13}$$

Here "GGA" stands for "generalized gradient approximation"; the differences between the LSD and GGA models are commonly called non-local corrections. These functions can be constrained to comply with a variety of asymptotic limits, normalizations and scaling laws, and may include some amount of empirical adjustment as well.[11,22,23] The largest adjustments to the LSD model involve the exchange energies, although non-local contributions to correlation can be important as well.

These non-local theories have had a dramatic effect in quantum organic chemistry, providing for the first time numerical results from density functional methods that systematically improve on Hartree–Fock theory for a variety of energies and properties. Some useful generalizations can be drawn about the performance of various types of density functional theory that generally apply to transition metal complexes as well:

(1) Bond energies are generally overestimated in LSD theory, and are considerably better at the GGA level. Since density gradients tend to be smaller in molecules than in the isolated atoms or fragments, an approximation neglecting gradient information works better for molecules than for atoms: the GGA models energetically stabilize density inhomogeneities more than LSD does. For many purposes, non-local DFT provides a useful level of approximation for considering geometries and chemical energetics. For small organic molecules, non-local density functional methods usually give bond energies accurate to 3–5 kcal/mol, bond lengths accurate to 0.02 Å, and bond angles accurate to a few degrees. In addition, accurate energies have been obtained for a variety of chemical reactions, excitation energies, electron affinities, proton affinities, and ionization potentials.[7,24,25] Transition metal complexes provide greater challenges for all quantum chemical methods. Geometries and vibrational frequencies have been calculated with good accuracy in recent work (average bond length error 0.03 Å).[26] We have calculated bond energies for a variety of small spin polarized transition metal complexes ($MnCl_2$, $FeCl_2$, $FeCl_3$, ScO, TiO, VO, CuO), which gave bond energies with errors in the range 1–17% (0.8 to 11 kcal/mol), in agreement with the findings of other groups for example, on simple metal hydrides.[27] For transition metal carbonyl compounds, the M-CO bond lengths

can be calculated to an accuracy of about 0.03 Å, and the bond energies to an accuracy of about 5 kcal/mol by a GGA model with quasi-relativistic corrections.[28]

(2) Correlation effects often tend to oppose exchange effects, particularly in open shell systems. In the absence of correlation, exchange provides the only direct pathway to reduce U_{ee}, but this must compete with related changes in kinetic and nuclear attraction energies, and with changes in the classical part of U_{ee}. Correlation can provide additional pathways for opposite spin electrons to avoid each other, reducing the need for large exchange effects. In practice, this means that exchange forces that favor localization of parallel-spin electrons (such as in a high-spin transition metal ion) will be reduced. Both LSD and GGA models are thus considerably better than the Hartree–Fock model, which can significantly overstabilize high-spin states.

(3) When correlation is turned on, the hole around a given electron becomes deeper and more short-ranged, so that approximations based on local information become more useful.[29,30] The distinction between exchange and correlation is not always a useful concept, especially in the presence of non-dynamical correlation effects; further, the so-called exchange-correlation energy of density functional theory also contains a kinetic energy component.[10,18] Many GGA models show some cancelation of errors in their treatment of exchange and correlation, so that the sum is more accurate than the individual components. As with LSD, GGA accounts of correlation are expected to be more trustworthy for antiparallel spins than for parallel spins,[31] although the practical consequences of this for open-shell systems are not yet clear.

Density functional methods are now available in a large number of quantum chemistry codes, and it is increasingly common to see systematic evaluations of results with different basis sets and energy functionals.[25,32] The most extensive tests involve organic molecules, where comparisons can be made to a large body of well-calibrated gas-phase experimental data. Here the best overall results, particularly for energetics, currently appear to come from hybrid models in which a fraction of the Hartree–Fock exchange energy is combined with the density functional exchange terms described above.[33] The B3-LYP functional is currently the most popular of these hybrid models.[34] For two reasons, it is harder to draw conclusions about the performance of various functionals in transition metal systems. First, gas-phase geometries and energetics are less well known for transition metal systems, particularly for coordination complexes of greatest interest to us here. Second, several of the programs most widely used by inorganic chemists (such as *deMon*, *DGAUSS* or *ADF*) do not include hybrid functionals, so that these have been less extensively tested. From the available evidence,[35–40] gradient corrections appear to offer a systematic improvement over local functionals, various implementations of gradient corrections (such as BP, B-LYP, or PW91) give approximately equivalent results, and hybrid functionals also look attractive for transition-metal systems. Basis-set quality is likely to be at least as important as the density functional form in influencing the reliability of computed results, and relativistic effects can have a quantitative impact, even for first-row transition metal systems. It is likely that our understanding of these effects will improve rapidly over the next few years, and users will need to continue to keep abreast of current developments.

3 GENERAL FEATURES OF TRANSITION METAL ELECTRONIC STRUCTURES

3.1 The Pauli Principle: Steric Interactions and Hund's Rule

The Pauli principle is fundamental to the electronic structure of all transition metal complexes, and, in fact, to all of chemistry. This states that no two electrons of the same spin index can occupy the same space orbital, and implies that all orbitals having the same spin index must be mutually orthogonal.

One major effect of the Pauli principle is the steric interaction associated with formation of a transition metal complex, arising from orbital repulsion between filled orbitals. Consider bringing a transition metal ion together with a set of ligand molecules. The occupied ligand orbitals must be orthogonalized to the occupied metal spin-orbitals; this will lift the orbital degeneracy of the $M(d)$ orbitals, and push up those $M(d)$ orbitals, such as σ^* anti-bonding orbitals, which interact strongly with the occupied ligand orbitals. This repulsive interaction between filled orbitals (after accounting for the classical electrostatic interaction of the metal ion with the ligands) is called the Pauli repulsion.[41–43] The sum of the classical and Pauli terms gives the total steric interaction.

A useful convention is to consider the steric interaction of the ligands with a high-spin metal ion in the spherically averaged $M(d)$ configuration with equal occupancy of all d orbital components. This convention allows for a reasonable definition of the ligand field stabilization energy, and its separation from the steric energy term. (The same methodology can be used with a low-spin metal ion interacting with ligands for low-spin complexes.) The ligand field stabilization energy (LFSE) is due to splitting of the d orbital degeneracy (for open d shells), the electrons preferentially occupying the lower energy "mainly d" orbitals. LFSE is the gain due to this non-spherical density. After bond formation, orbital polarization and charge transfer can occur in the averaged $M(d)$ configuration, and then orbital re-occupancy gives the LFSE energy.

A further consequence of the Pauli principle was alluded to above: the exchange (or "Fermi") hole functions are generally different for α and β electrons if there are unpaired electrons, since the spin densities are distinct. The resulting potential due to the hole density is more attractive for those spin-orbitals whose spin index has the greater density. Hence, electrons of majority spin avoid one another more effectively than do the minority spin electrons, providing a driving force toward spin localization and high-spin states. Indirectly, this leads to Hund's rule for atoms and ions. Hund's rule is the experimental observation that the ground states of free atoms and ions nearly always have maximum spin; that is, the electrons residing in degenerate (or nearly degenerate) orbitals have parallel spin alignment. We say indirectly[44–49] because, when comparing the ground and excited state spin multiplets, second order effects, especially changes in nuclear-electron attraction, are often larger than direct changes in the electron-electron repulsion energy (including exchange and correlation terms). The change in the kinetic energy is another significant second order effect.

Examples of these second order effects are provided by the excited and ground state multiplets of the Cu atom and Cu^+, where Hartree–Fock results have been compared with two different forms of non-local density functional potentials.[50] Second order effects are associated with changes in the shapes of orbitals and the electron density. For example, the different Cu multiplets have different orbital shapes for all valence orbitals, even where these may not be directly involved (active) in the excitation process.

As an example, consider the Cu atom electronic transition $d^9s^2 \rightarrow d^{10}s^1$. When the excitation energy ($\Delta E_{\mathrm{total}}$) is evaluated with a density functional potential (B-LYP), the individual energy terms for the total one-electron energy term (Δh_1, kinetic energy plus nuclear-electron attraction energy), the Coulomb repulsion energy (ΔE_{Coul}), the exchange energy (ΔE_{exch}), and the correlation energy (ΔE_{corr}) are (-150.16, $+154.01$, -5.74, -0.16 eV) giving $\Delta E_{\mathrm{total}} = -2.05$ eV. So the net energy difference is the resultant of large but opposite energy contributions from one-electron and two-electron terms. Further, these energies differ substantially from those evaluated with the same B-LYP energy expression, but applied to the Hartree–Fock density (-179.88, $+185.18$, -6.99, -0.18 eV). These results are indicative of large electron density changes in several orbitals, which depend also on the exchange-correlation potential used.

3.2 Inverted, Mixed, and Normal Energy Level Schemes

Despite the complicating factors associated with the collective phenomenon of electron relaxation, energy level diagrams can give an overall picture of which orbitals are involved in oxidation, reduction, or excitation, and an assessment of their metal versus ligand character. For example, metal → metal, metal → ligand, ligand → metal, ligand → ligand excitations can be distinguished in calculations and tested against spectroscopic observations. One classification scheme divides energy level diagrams into normal (N), mixed (M), and inverted (I) level schemes, illustrated in Figure 1. Like divisions of the electromagnetic spectrum, this classification divides a continuum of possible diagrams into convenient categories, and there are cases at the boundaries between classes.

3.2.1 ***Normal Level Scheme*** In the normal (N) level scheme, depicted in many textbooks,[51] all occupied metal *d* levels lie above all occupied ligand levels. The lowest unoccupied levels can be either of ligand or, more typically, of metal type. This scheme is often found in low spin systems in low metal oxidation states, as in the bridged metal carbonyls $Fe_2(CO)_9$, $Fe_2(CO)_6(PPh_2)_2^{0,2-}$, and $Co_2(CO)_6(CH)_2$. The strong ligand field forces low spin sites with corresponding metal orbital energy levels above the occupied ligand levels.[52,53] Among high-spin systems, the ferrous-aquo complex $Fe(H_2O)_6^{2+}$ obeys the normal level scheme based on our calculations, as shown in Figure 2(a). Both the five highest occupied majority spin levels and

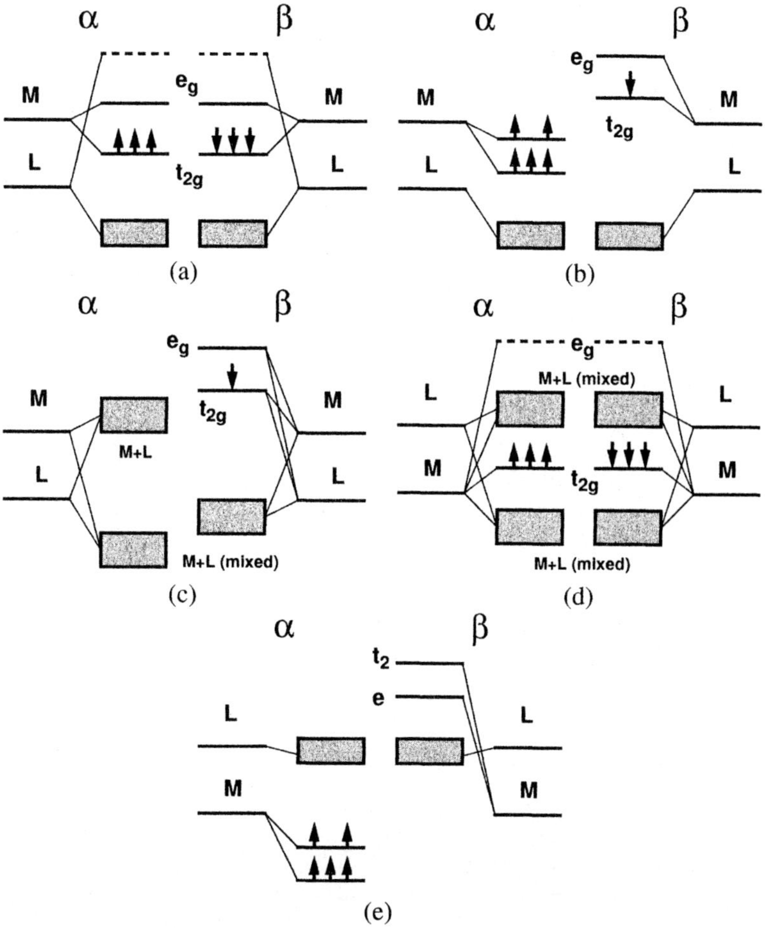

Figure 1. Energy level schemes of transition metal complexes. (a) Normal level scheme for a low spin, octahedral d^6 ML_6 complex; (b) Normal level scheme for a high spin, octahedral d^6 ML_6 complex; (c) Mixed level scheme for a high spin, octahedral d^6 ML_6 complex; (d) Mixed level scheme for a low spin, octahedral d^6 ML_6 complex; (e) Inverted level scheme for a high spin, tetrahedral d^5 ML_4 complex.

the lowest lying minority spin levels (one occupied, four empty) are mainly metal, consistent with the formal Fe(II) high-spin occupation, $d^5\alpha d^1\beta$.

In systems where there is more extensive metal-ligand covalency and/or strong spin-polarization effects, large deviations from the normal level scheme can arise. The most extreme difference is found in the inverted level scheme, and a smaller, but still substantial difference is exhibited by the mixed level structure.

3.2.2 ***Inverted Scheme*** The inverted level scheme is induced by strong spin polarization effects. Here spin polarization is strong enough to drive the majority-spin

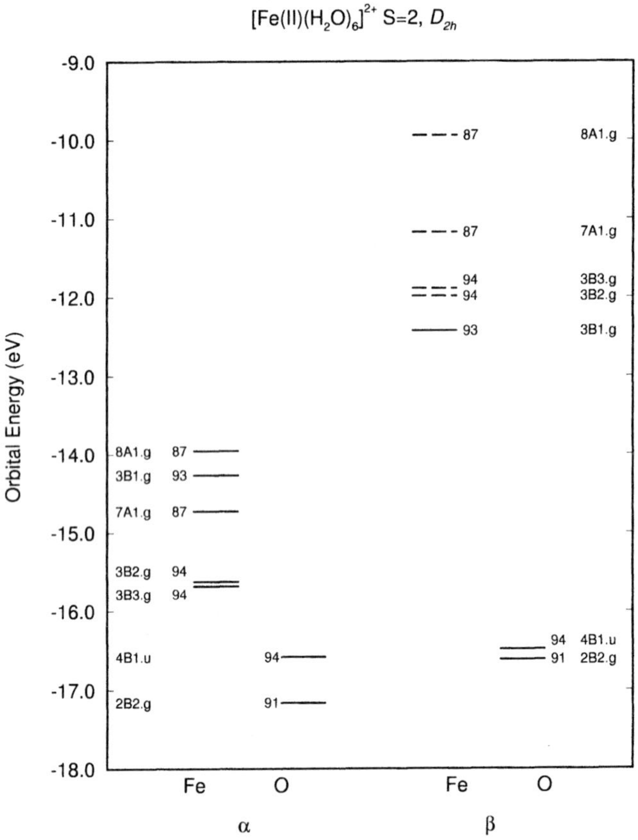

Figure 2a. Energy level diagrams of (a) $[Fe(II)(H_2O)_6]^{2+}$, normal scheme; (b) $[Fe(III)(H_2O)_6]^{3+}$, mixed scheme; (c) High spin $[Co(III)(H_2O)_6]^{3+}$, mixed scheme; (d) Low spin $[Co(III)(H_2O)_6]^{3+}$, mixed scheme; (e) $[Fe(II)Cl_4]^{2-}$, mixed scheme; (f) $[Fe(III)Cl_4]^{1-}$, inverted scheme. Solid lines are filled; dashed empty levels.

mainly-metal levels well below the highest occupied mainly-ligand levels. Above the ligand levels lie the minority spin metal levels, which may be either empty or partially occupied. Overall, the spin polarization energy exceeds the ligand field stabilization energy in a weak ligand field, and the system is high-spin. Examples are provided by iron-sulfur complexes and by iron-chloride complexes, particularly in higher metal oxidation states. The oxidized ferric complex $FeCl_4^{1-}$ is depicted in Figure 2(f). The majority spin Fe(3*d*) band lies below the occupied Cl(3*p*) band in energy. Above this lies the empty minority spin Fe(3*d*) levels, which are ligand field in character (weakly or strongly anti-bonding). Electron levels of both α and β spin are occupied for Cl(3*p*), but the ligand β levels are stabilized by back donation (Cl-Fe covalency) into the empty Fe(3*d*). The inverted level scheme predicted from our calculations and those of earlier workers for ferric chloride and ferric thiolate complexes[54–57] have been confirmed by spectroscopic measurements in $Fe(SR)_4^{1-}$ and

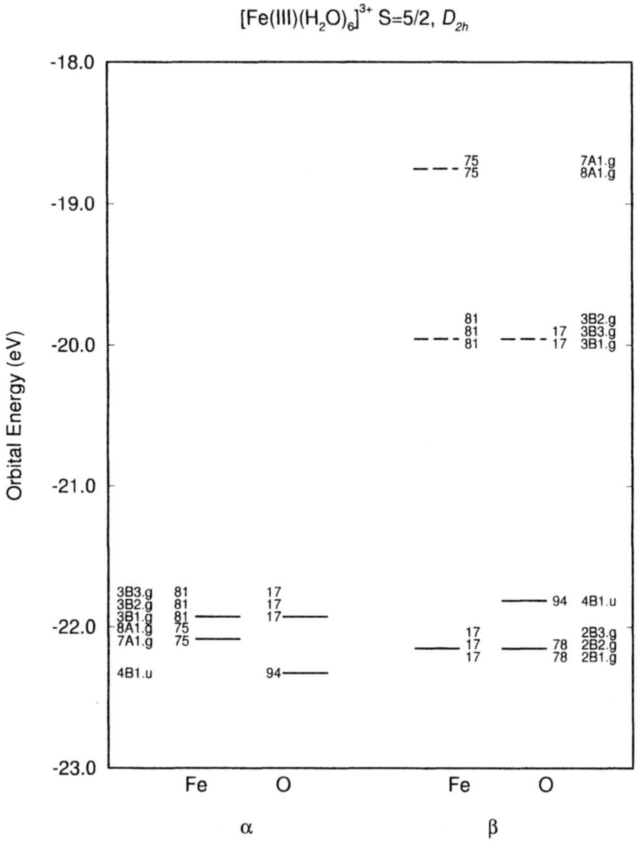

Figure 2b. (*Continued.*)

$FeCl_4^{1-}$ systems.[58–61] For example, spin-forbidden transitions occur at low energy, and the energy shift between $FeCl_4^{1-}$ and $Fe(SR)_4^{1-}$ is predicted well only if these spin-forbidden transitions have substantial ligand → metal charge transfer character as required in the inverted level scheme.[62,63] Further supporting evidence for the inverted level scheme comes from resonance Fe($3p \rightarrow 3d$) absorption edge measurements with variable photon energy photoelectron spectroscopy.[60] Below, we show the energy level scheme for a dinuclear (diferric) iron-sulfur complex,[56] which also shows an inverted level scheme, as do 4Fe4S complexes.[9,64]

3.2.3 ***Mixed Level Scheme*** Adding one electron to the complex above gives $FeCl_4^{2-}$, with formal oxidation state Fe^{2+}, shown in Figure 2(e). The added electron goes into the lowest empty minority spin Fe($3d$) β orbital of the oxidized system $2e\beta$. The metal spin density drops and so does the spin-splitting between the Fe($3d$) minority spin levels and the deeper lying Fe($3d$) majority spin levels. We consider this a mixed level scheme case,[64] since below the highest occupied $2e\beta$ orbital are a

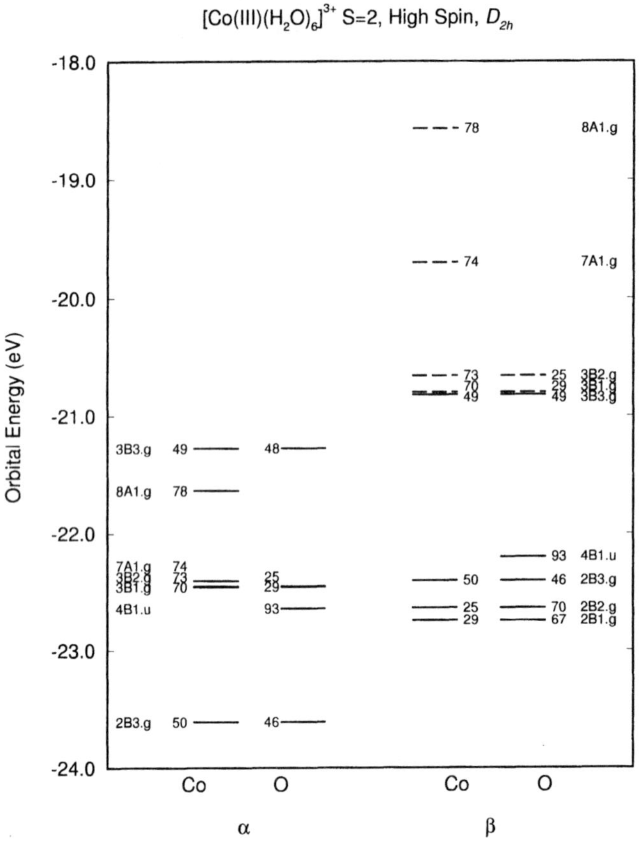

Figure 2c. *(Continued.)*

number of fairly covalent metal-ligand combination orbitals. We also find a mixed-level scheme for related high-spin ferrous $Fe(SR)_4^{2-}$ complexes; these exhibit more metal-ligand covalency than the ferrous chloride complex both in occupied majority and minority spin orbitals and for empty minority spin mainly Fe(3*d*) orbitals.

Solomon and coworkers[59] have found that $FeCl_4^{2-}$ obeys a more normal level scheme based on spectroscopic studies. This assessment preceded our proposal of the mixed level scheme category.[64] The HOMO is a minority spin Fe(3*d*) level in both cases, but our prediction is that the next lower lying orbitals are not simply Fe(3*d*) majority spin, but rather exhibit significant metal-ligand covalency consistent with a mixed level scheme. Substantial metal-ligand covalency is found also from the observations of variable photon energy photoelectron spectroscopy on the $FeCl_4^{2-}$ system.[60]

In general, we define a system as obeying a mixed level scheme[64] if either the highest occupied orbitals or the lowest unoccupied orbitals are highly covalent (i.e., there is thorough mixing of metal-ligand character in these molecular orbitals, or

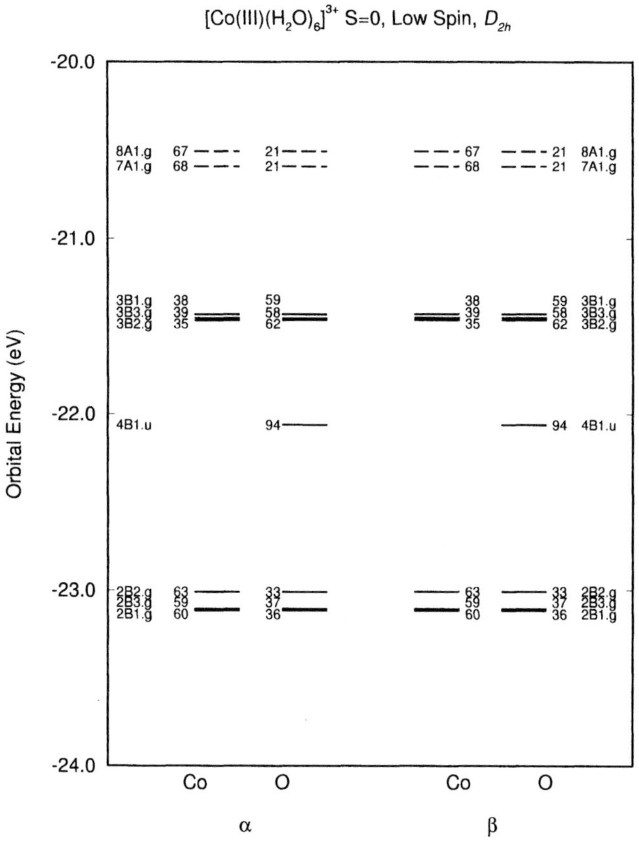

Figure 2d. *(Continued.)*

metal or ligand levels are closely spaced in this energy region). We include in the category of highest occupied levels those within about 2eV of the HOMO, or 3eV of the LUMO, so that the energy range is relevant for optical and near-UV transitions. The mixed level scheme provides an additional useful category beyond the previous inverted *vs.* normal scheme dichotomy, and a large number of systems fit into the mixed level category.

Switching to octahedral coordination, the low-spin $Co(H_2O)_6^{3+}$ shown in Figure 2(d) has lowest empty orbitals which are mainly metal type (anti-bonding e_g in idealized octahedral symmetry, and $e_g \rightarrow (a_{1g}, a_{1g})$ in the lower D_{2h} group) whereas the highest occupied orbitals have strong metal-ligand covalency, being about 40% metal and 60% ligand. The mainly-metal occupied orbitals are considerably deeper, and the level structure has all the features expected of a mixed level scheme. The high metal-ligand covalency in the highest occupied orbitals is associated with the high formal metal oxidation state, since there is considerable $L \rightarrow M$ back-donation in the higher lying occupied orbitals. There is not enough ligand char-

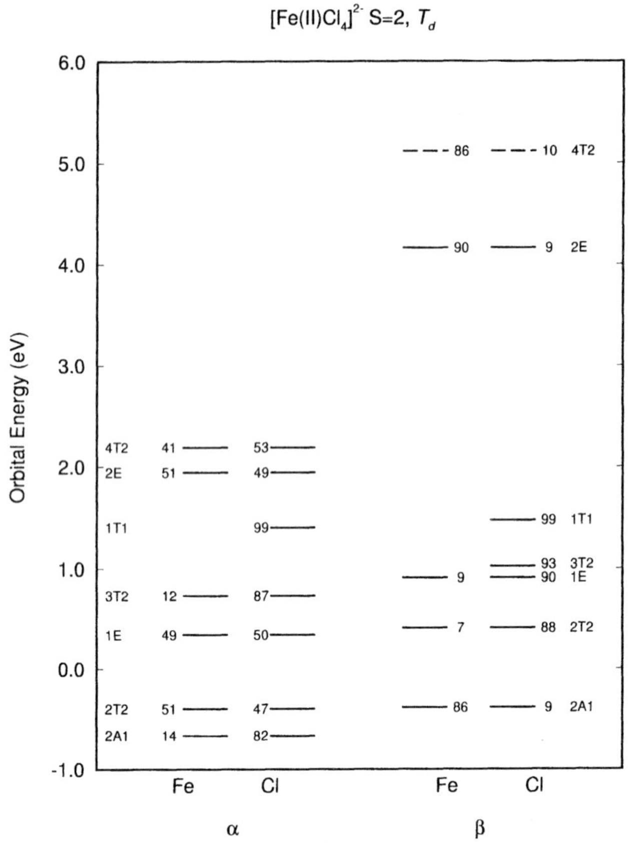

Figure 2e. (*Continued.*)

acter in the highest occupied orbitals to give an inverted level pattern, nor are the highest occupied orbitals mainly metal, as exhibited by the normal level scheme.

Perhaps more typical of mixed level schemes is the corresponding high-spin $Co(H_2O)_6^{3+}$ complex (formal d occupancy $= d^5\alpha d^1\beta$) [Figure 2(c)]. Here there is strong competition between spin polarization, giving enhanced spin splitting in high spin sites, and ligand field strength increasing metal orbital splitting, particularly the $t_{2g} \rightarrow e_g$ gap in approximately octahedral complexes.[53,65,66] The large ligand field strength favors a low-spin occupation, as in the low-spin $Co(H_2O)_6^{3+}$ observed as the ground state experimentally.[2,67] However, the low-spin state occurs with total loss of the spin polarization energy, and in this energy balance, the low-spin is only slightly favored over the high-spin form in our calculations. In the high-spin state, the HOMO is of minority spin $b_{3g}(t_{2g})\beta$ type, but with extensive metal-ligand covalency. The symmetry orbital correlation is $t_{2g} \rightarrow (b_{1g}, b_{2g}, b_{3g})$ giving the HOMO and the two lowest unoccupied orbitals. The deeper orbitals, both of majority and minority spin, shows extensive intermixing of metal and ligand based orbitals, and

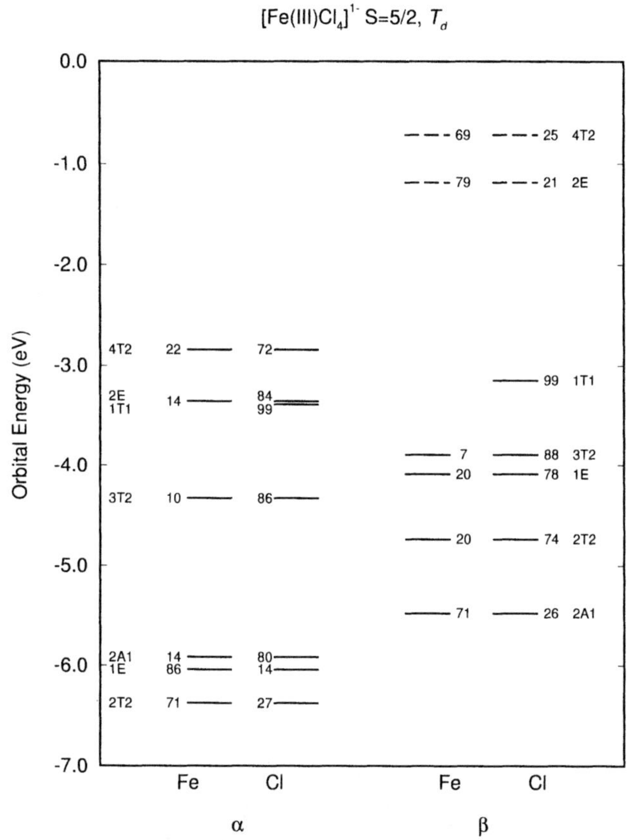

Figure 2f. *(Continued.)*

also some strongly ligand levels near mainly metal levels. Similarly, the high spin ferric aquo complex $Fe(H_2O)_6^{3+}$ displays a mixed level structure (Figure 2(b)).

Overall, the higher oxidation state complexes show more metal-ligand covalency. The other important influence is the presence of large spin polarization splitting, associated with weak ligand fields. Both of these effects lead to mixed or inverted level schemes, and are in sharp contrast to the normal level scheme found in most textbooks. The normal scheme is a consequence of both strong field ligands and low oxidation states in some metal complexes.

3.3 Electron Relaxation Effects

Electron relaxation can be defined as the remaining change in electronic structure associated with any change in orbital occupation after accounting for differences in the active orbitals alone. The active orbitals are defined as those where the orbital occupation numbers change, and the passive orbitals are those where the occupation numbers remain the same.[56,68–70] We can then consider electron relaxation effects on oxidation or reduction, optical excitation, spin state interconversion, or on metal $\rightarrow$ ligand or ligand $\rightarrow$ metal electron transfer.

In Hartree–Fock theory, when ionization is treated accounting only for changes in occupancy of a single active orbital, one has the frozen orbital approximation, and the total energy difference in this approximation equals the one-electron energy for that orbital; this is referred to as Koopmans' theorem.[69,70] This Koopmans' energy or frozen orbital energy difference is known to give only a rough approximation to the ionization energy, and to be less satisfactory for electron affinities.[71] In principle, the highest occupied Kohn–Sham orbital energy in density functional theory gives the exact ionization potential,[21] but this equivalence is not well satisfied in current approximate calculations. Frozen orbital total energy differences can be defined even for DFT, and have considerable utility for analysis of ionization phenomena, separating frozen orbital from relaxation effects. Further, ionization and corresponding relaxation energies will differ substantially depending on whether the ionization produces a localized or delocalized hole state.[72–76]

We can contrast the frozen orbital energy difference with the result of performing full self-consistent-field SCF calculations on both the initial and final many-electron states. The resulting energy difference is called the ΔSCF energy, and this is typically what is used to evaluate excitation or ionization energies, or electron affinities in density functional theory.[71] Within density functional theory, there is an alternative one-electron method, called the Slater transition state method,[17,77] where ionization, one-electron excitation, or electron affinities are computed from DF one-electron eigenvalues, but where half an electron is removed from the active orbital for ionization (added for electron affinities), and excitations are described by one-electron energy differences, where half an electron is removed from an initial orbital and placed in the final orbital. It is important to realize that the Slater transition state method is simply a convenient approximation to the ΔSCF method, built in such a way that relaxation is effectively included in the modified orbital energy difference equation.[17,77]

Electron relaxation is a consequence of the strongly interacting character of the electron-electron Coulomb repulsion and the nonlinearity of the SCF equations. Consequently, all changes in spin state or orbital occupancy lead to changes in all orbitals. Electron relaxation produces particularly strong effects in compact shells as in the $M(3d)$ shell of the first row transition metal series.[68,78,79]

Relaxation effects can often be readily seen in comparisons of atomic charges. There are a number of different ways of partitioning electrons between different atoms (finding partial charges). The two different methods referred to in this review are Mulliken charge partitioning,[69] and electrostatic potential (ESP) charges[80] implemented into density functional codes.[81,82] Mulliken charges are obtained from the quantum mechanical electron density, expressed as a density matrix in a given basis set, by allocating the diagonal element density terms to the corresponding atomic center, and the overlap density half-and-half to each of the two centers in the density matrix term. The electron populations (with negative sign) are then added to the nuclear charges to obtain the Mulliken partial charges. Although the half-and-half allocation of the overlap density is somewhat arbitrary, Mulliken population analysis has the advantage that it can be applied also to particular populations, such as M(3*d*) in the same way as to the total electron density.

In contrast, the ESP charges are determined from a least-squares fit of point charges to the electrostatic potential from the actual electron density. The electrostatic potential is calculated for a particular region of space, usually outside a Van der Waals envelope around the molecule or complex. The main purpose of the ESP charges is to provide a compact and fairly transferable representation of the electrostatic potential at long or medium range; the purpose is to describe electrostatic interactions between molecules, or of a molecule with its environment where the electron clouds of the interacting molecules do not overlap strongly. This is a classical electrostatic description. The advantage of ESP charges is that these are tied to a specific physical property. There are two limitations: the fit depends somewhat on the region where the electrostatic potential is considered; also, unlike Mulliken analyses, there is no corresponding scheme for allocating particular subpopulations, like M(3*d*) populations. Both Mulliken and ESP charges are useful for analyzing charge re-distributions with changes in electronic structures due to ionization or excitation.

The total charge redistribution upon oxidation or reduction of a metal site, including electron relaxation, differs substantially from that associated with changes of occupation in the active orbital alone. The final electron density change is typically greater on the ligands and smaller on the metal ion when electron relaxation is included. One way of thinking about a one-electron oxidation is to evaluate the hole introduced by emptying the active orbital, and then to examine the backflow of negative charge onto the metal ion associated with electron relaxation. Conversely, on one-electron reduction, after filling an empty metal-based orbital, there is an outflow of electronic (negative) charge to the ligands. Figure 3 provides an example of this for the Fe(III) $\rightarrow$ Fe(II) reduction process in $Fe(H_2O)_6^{3+,2+}$. Figure 3(a,b) shows the electron densities of the oxidized complex LUMO, and the very similar density of the reduced HOMO. Notice that there is only a single type of shading (light), representing the electron density in that orbital when an electron is added to the system. By contrast, in the total electron density difference map (Figure 3(c)), there are regions of electron density increase both on Fe and on all the water oxygens (light shading; positive change in electron density on reduction); further, there is a significant electron depletion region on the Fe atom (dark shading; negative density change) which has no counterpart in the frozen orbital density picture. The net change in Fe charge is only about 0.15 from electrostatic potential, ESP, charges, and 0.21 from Mulliken charges (total density difference), compared to 0.81 (LUMO density) and 0.93 (HOMO density) based on Mulliken populations. The difference is even larger when compared with the change in metal charge (1.0) predicted from the formal oxidation states of the metal ion and its ligands. This applies even more strongly in more covalent systems.[78,79,81] A proper assessment of metal-ligand covalency includes both active orbital effects and relaxation effects, since the collective effect of the passive orbitals can be very significant.

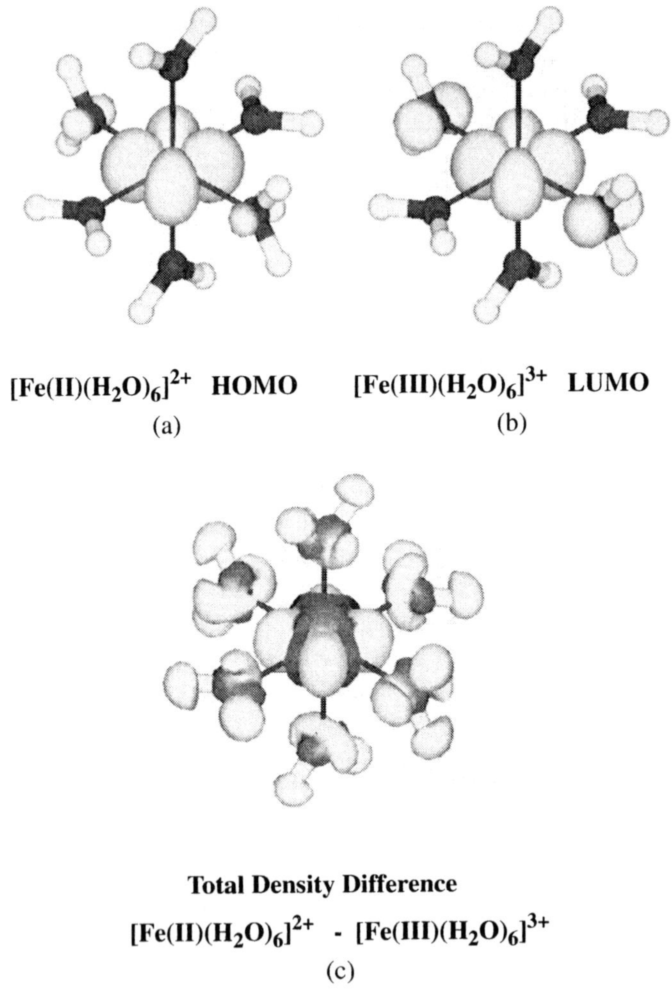

Figure 3. (a) Charge density of HOMO of $[Fe(II)(H_2O)_6]^{2+}$; (b) Charge density of LUMO of $[Fe(III)(H_2O)_6]^{3+}$; (c) Total density difference $[Fe(II)(H_2O)_6]^{2+}$–$[Fe(III)(H_2O)_6]^{3+}$. Density contour is 0.0035|e|.

Electron relaxation has two principal chemical consequences in redox processes: (1) the geometry change in metal-ligand bond lengths is diminished by electron relaxation with implications for reorganization energies.[78,79,81] The reorganizational energy for a redox process from the theory of Marcus and Hush (with a net negative free energy change) is defined as the energy required for a transition from the reactant (B) to the product (A) surface, but where this occurs with a fixed initial state geometry and with no rearrangement in the solvent environment (vertical transition in geometry/solvent space).[83,84] These are called inner sphere or vibronic and outer sphere or solvent reorganization terms, respectively; both types of relaxation occur on a fast time scale comparing to isothermal processes at normal temperatures (with liquid phase solvents). The reorganization energy is an essential parameter in determining the barrier for thermal electron transfer. More directly, this is also the relevant energy for the main peak in internal optical electron transfer (giving the vertical energy of the inter-valence charge transfer band).[85,86] (2) The ligand charge distribution is altered becoming more negative on reduction (or more positive on oxidation) with implications for properties including geometry, hydrogen bonding strength[9,87–89] to the ligands, proton transfer[90] and ligand pK_a shifts.[78] In Figure 4, we show the predicted change in mean metal-ligand bond lengths on M(III) → M(II) reduction *vs.* the calculated change in ESP charge for a series of first row high-spin transition metal hexa-aquo complexes. There is a rough but distinct correspondence between the differences in calculated ESP charges and the predicted bond length increases on reduction. In contrast, one might have expected a horizontal line if only the formal difference in oxidation state on the metal ion, +1 for all [M(III)-M(II)] systems, were important; of course, the differences in ESP charges are much less than this for all the first row metal ions and these also roughly correlate with electronegativity (comparing Sc with Fe, Co). The points for Cr and Mn, which show the greatest deviation from the line, also are the systems which display the largest Jahn–Teller distortion (bond length asymmetry), for Cr(II) and Mn(III). The reductions of the Cr(III) → Cr(II) and Mn(III) → Mn(II) hexa-aquo complexes both correspond to addition of an antibonding e_g electron, $t_{2g}^3 \rightarrow t_{2g}^3 e_g^1$ and $t_{2g}^3 e_g^1 \rightarrow t_{2g}^3 e_g^2$ for the chromium and manganese complexes, respectively.

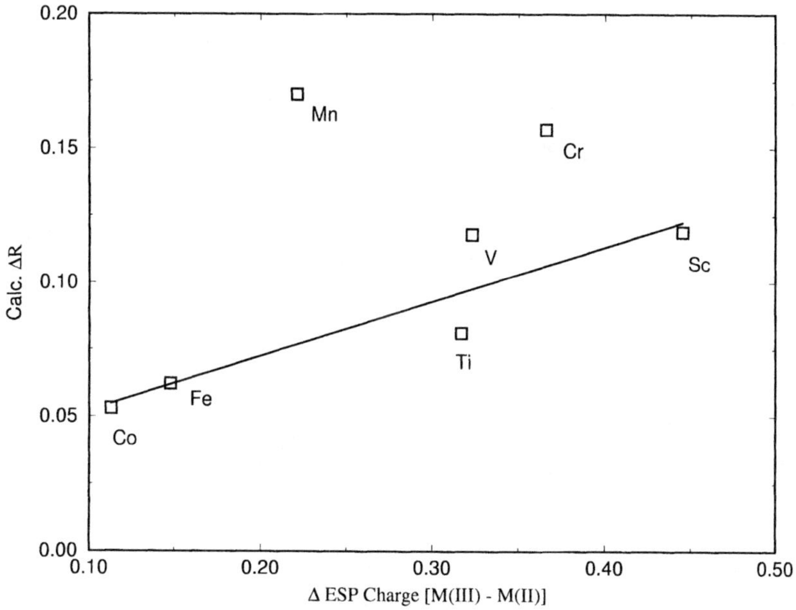

Figure 4. Correlation between calculated distance change (in Ångstroms) and ESP charge changes. The calculation were done for high spin $[M(II)(H_2O)_6]^{2+}$ and $[M(III)(H_2O)_6]^{3+}$. Mn and Cr are excluded from regression for the correlation line.

4 ENERGETICS OF METAL IONS AND METAL-HEXA-AQUO COMPLEXES

The energetics of redox processes for metal ions in solution can be partitioned into two contributions: the gas phase ionization potential of the reduced species, and the solvation energy difference between the oxidized and reduced states. The importance of ionization and excitation processes is even broader, since these are fundamental to both bonding and net electron transfer processes. We begin with an assessment of the performance of density functional calculations for ionization and excitation energies of the first row transition metal ions. Following this, we examine the energetics of formation of the metal ion-hexa-aquo complexes.

Figure 5 shows the calculated *vs.* experimental IP's for the $M^{2+,3+}$ (IP_{III}) and $M^{1+,2+}$ (IP_{II}) processes. Despite some quantitative differences, the calculated IP's track the observed IP's quite well, with errors of about 0.5–1 eV. The IP_{III} plot shows a relative maximum at $Mn^{2+,3+}$, where removing a Mn(3d) electron corresponds to losing the enhanced stability of the half-filled 3d shell $d^5 \rightarrow d^4$, and the relative minimum at $Fe^{2+,3+}$ which corresponds to having the half-filled shell for the ox-

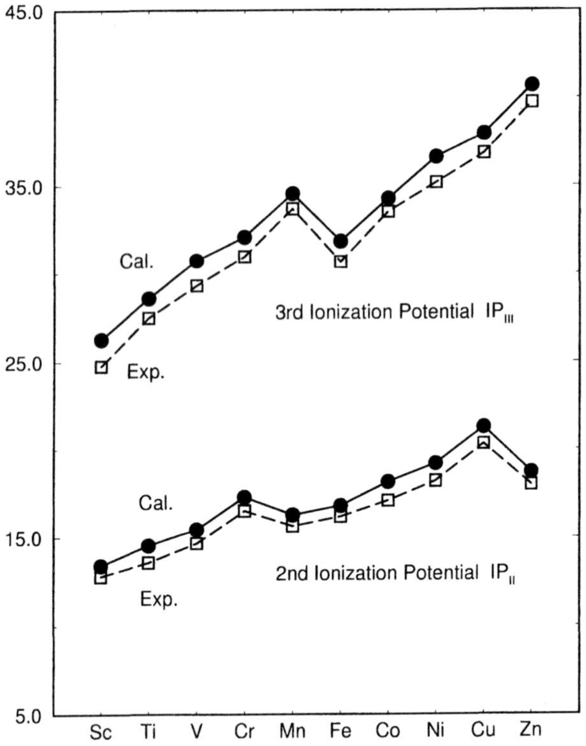

Figure 5. Calculated (filled circle) and experimental (open square) second and third ionization potentials (eV) of first-row transition metals.

idized state $d^6 \rightarrow d^5$. A similar argument can be made for the relative maxima of $Cr^{1+,2+}$ and $Cu^{1+,2+}$, losing a half-filled shell and filled shell for the oxidized species. The standard ligand-field argument has been made that these maxima are due to the stronger exchange term for the half-filled or filled $3d$ shell,[91] but indirect terms are also very important. These include changes particularly in nuclear-electron attraction and Coulomb repulsion in addition to exchange effects. The gradual increase shown across the series Sc → Zn is a consequence of the increasing nuclear charge across the series, and that electron screening of this charge is only partially effective. The other notable observation is that the IP's are very large, particularly for the $M^{2+,3+}$ couple. Comparing this with the experimental first IP of water, 12.6 eV, we see that the IP_{III} values are much larger for the entire series, and the very large electron relaxation effects observed for the M(III)-hexa-aquo ions becomes very plausible.

Figure 6 shows the calculated *vs.* experimental $3d^{(n-1)}4s^1 \rightarrow 3d^n$ promotion energies of the first row transition metal cations M^+, which is a model for the $d \rightarrow s$ charge polarization (or orbital hybridization) contribution to bonding in transition metal complexes. Here the agreement of the calculations with experiment is closer than for the IP's. There are minima for the excitation energies where the $d^5(Cr^+)$ and $d^{10}(Cu^+)$ final states have half-filled and filled $3d$ shells, again corresponding to the

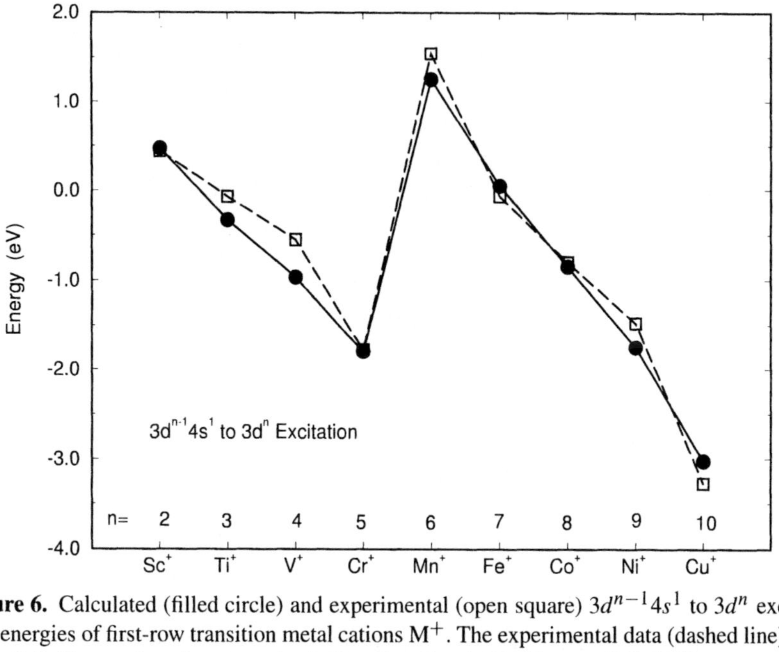

Figure 6. Calculated (filled circle) and experimental (open square) $3d^{n-1}4s^1$ to $3d^n$ excitation energies of first-row transition metal cations M^+. The experimental data (dashed line) are from C.E. Moore, Atomic energy level, Nat. Bur. Stand. Washington, DC, 1958, corrected to nonrelativistic limits according to T. Russo, R.L. Martin, P.J. Hay, *J. Chem. Phys.*, 101:7729, 1994, and K. Raghavachari, G.W. Trucks, *J. Chem. Phys.*, **91**, 1062 (1989) and *ibid.*, **91**, 2457 (1989).

enhanced stability of these configurations. It is clear that current density functional models provide a well-balanced account of the relative stabilities of many low-lying atomic configurations.

Transition metal-hexa-aquo complexes provide good examples of the comparative influences of ligand field stabilization and spin polarization energies in determining the ground electronic state. These energies are often in competition, depending on the metal d electron count. This is particularly evident when comparing the energies of high spin vs. low spin complexes.

In Figure 7 we compare the calculated binding energies for $M(H_2O)_6^{2+,3+}$ formation from the ion and separated water molecules with the experimental hydration enthalpies for the first row transition metal series. The calculated binding energies are defined for the isolated $M(H_2O)_6^{2+,3+}$ complexes. The experimental hydration energies are much larger because of the presence of the second solvation shell and the extended solvent environment. We will examine such effects later when we discuss condensed phase results. Despite the large overall energy shift, both the binding and hydration energies display very similar double bowl shapes for both oxidation states. All the ground state complexes are high-spin both experimentally and from the calculations with the single exception of $Co(H_2O)_6^{3+}$, which is observed to be low spin by UV/visible spectroscopy.[2] The effects of spin polarization combined with the increasing nuclear charge across the series can be seen in the downward sloping dashed line which interpolates between the empty (d^0), half-filled ($d^5\alpha$) and filled ($d^5\alpha d^5\beta = d^{10}$) $3d$ shell. The double bowl shape (open circles for calculated high-spin states) is then associated with the additional ligand field stabilization energy, which is strongest at $d^3\alpha = t_{2g}^3\alpha$ (Cr(III) and V(II)) and at $d^5\alpha d^3\beta = t_{2g}^3\alpha e_g^2\alpha t_{2g}^3\beta$ (Ni(II)). The low spin states are usually higher in energy since the loss of spin polarization is not sufficiently compensated by LFSE energy. The low spin state energies (triangles) are plotted for the Mn, Fe, and Co complexes. The oxidizing power of the M(III) ions is so large that the Ni(III), Cu(III), and Zn(III) ions are unstable in water, leading to rapid oxidation of H_2O and reduction of the metal to M(II). Co(III) is the last metal ion in the series for which the hydration energy and aqueous redox potential are measurable.[67,91,92] Consequently, Ni(III), Cu(III), and Zn(III) hydration energies were omitted from the figure.

Only for Co(III) is the low spin complex the ground state. The predicted low spin complex is nearly degenerate with the high spin $Co(H_2O)_6^{3+}$. The LS state is only 0.02 eV higher in energy than HS; however, this ordering reverses (LS becoming 0.14 eV lower than HS) when a smaller frozen core is used (comparing non-relativistic calculations, freezing only Co($1s, 2s, 2p$) instead of Co($1s, 2s, 2p, 3s, 3p$)). Even rather large ab initio multireference CI (MRCI) calculations do not yield the experimental HS vs. LS state ordering, with the predicted E(LS)-E(HS) energy difference being 10 kcal/mol.[67]

The active sites of metalloproteins are more complicated and of lower symmetry, but they also share a number of features of basic coordination geometries and electronic structure with the simpler aqueous ions above. Electron relaxation upon

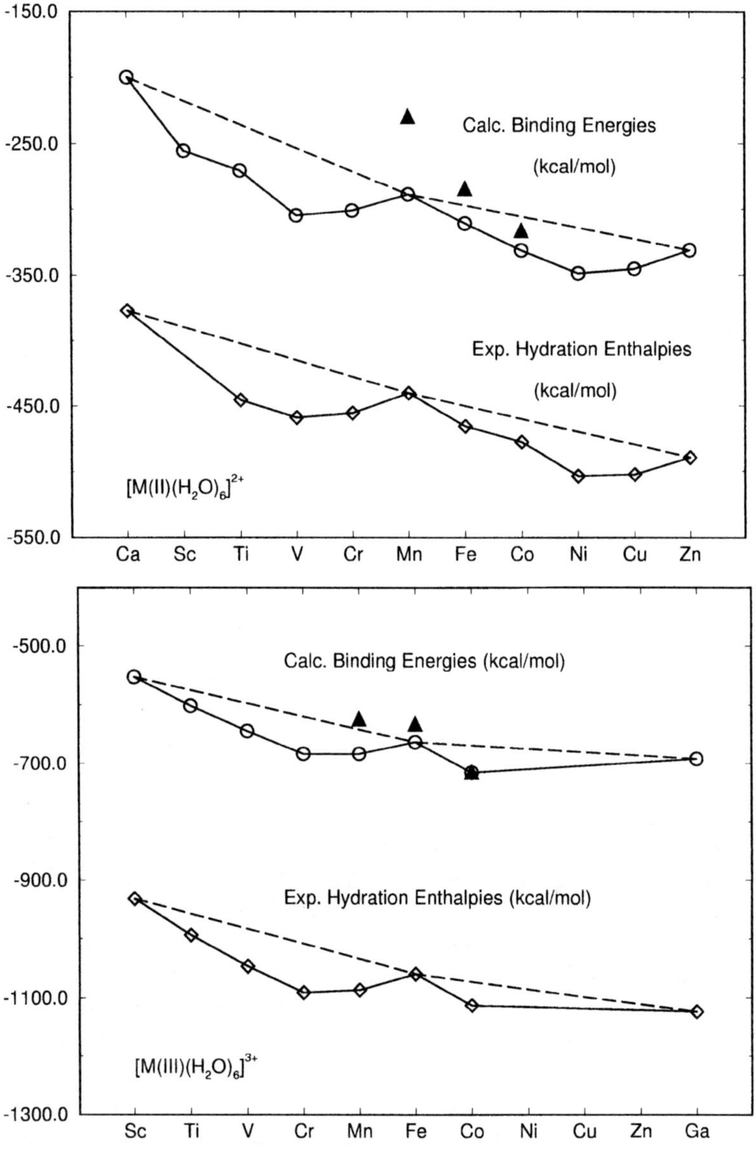

Figure 7. Calculated binding energies and experimental hydration energies of $[M(II)(H_2O)_6]^{2+}$ and $[M(III)(H_2O)_6]^{3+}$. For comparison, several low spin results are marked by filled triangles. The experimental hydration enthalpies are from D.W. Smith, *J. Chem. Edu.*, **54**, 540 (1977).

redox[78,79] and the associated charge transfer from ligands to metal upon oxidation leads to much lower pK_a's in higher metal oxidation states, and smaller than expected geometry changes (that is, expected from changes in formal oxidation states).

5 FEATURES OF MULTIPLE-METAL COMPLEXES

Complexes with more than one metal ion can be difficult to describe by quantum-mechanical methods. Except in the presence of significant metal-metal bonding, spins are often fairly well-localized in the vicinity of the metal ions, with relatively weak (magnetic) interactions between sites. A valence-bond is often more appropriate than molecular orbital models for qualitative descriptions, yet most feasible computations are based on an MO starting point.

The Kohn–Sham procedure uses orbitals from a single-determinant picture to construct the densities, and builds in correlation effects through an approximate density functional. This procedure works well for systems whose wavefunction is dominated by a single configuration, and whose correlation is therefore primarily a dynamic one; the description of short-range, dynamical correlation in terms of a transferable function of the density and its gradients is accurate enough to be quantitatively useful in a wide range of systems. For systems with substantial mixing of configurations, however, where non-dynamical correlation is important, currently available Kohn–Sham procedures run into both conceptual and practical difficulties. Basically, these non-dynamical effects are longer ranged,[18] and can have a wide variety of consequences, since the nature of the near-degenerate configurations that mix can be completely different from one system to another. No simple approximate density functional, especially if based on local information, can be expected to describe such diverse behavior. These features are especially important for excited states and for complexes with more than one metal atom.

5.1 Broken Symmetry Model for Non-Dynamical Correlation

One approach to this problem breaks the underlying symmetry of the spin density distribution.[9] Unlike the Pauli principle, which is a precise symmetry requirement for fermions, the Coulomb hole is essentially a means by which a wavefunction can lower its electron-electron repulsion, and also its total energy. One way to achieve this is to remove whatever constraints are keeping the e-e repulsions too high. In a standard spin-restricted framework, $\rho^{\alpha}(r) = \rho^{\beta}(r) = \rho/2$, and the electron repulsion between α and β electron densities is high. The broken symmetry model changes the form of ρ^{α}, ρ^{β}, allowing the electron densities for α and β spin to differ in space; how they differ precisely will be determined by the self-consistent field of the system in the density functional calculation. For the individual orbitals, this means removing the constraint that the orbitals be members of the irreducible representations of the molecular point group defined by the nuclear geometry. The orbitals are instead calculated in a subgroup compatible with the symmetries of ρ^{α} and ρ^{β}, although it is the localization properties more than the point group symmetry that is essential. By using unrestricted spin-orbitals, all of the energy terms (kinetic energy, nuclear electron attraction, and e-e repulsion) will change in a variationally optimal way. Both the classical electron-electron repulsion and the exchange-correlation energy terms in U_{ee} are altered when passing from the spin restricted to the spin unrestricted case. In general, the α and β densities become more compact, and occupy different regions in space. These effects are quite strong when there are spin polarized transition metal atoms, because the energetic cost on the kinetic energy of having a compact α and β densities is outweighed by the advantageous lowering of some combination of the classical electron-electron repulsion, the exchange part of the spin polarization, and the nuclear-electron attraction energies. Further, the broken symmetry state and its energy will differ significantly from the high spin state and its energy, even though both will experience spin polarization, because the two states have different covalency and delocalization, affecting all the energy terms.

The broken symmetry approach for spin-polarized and spin-coupled systems, in combination with spin projection methods, provides a intuitive picture of energetic interactions that contains the main physics involved. The basic idea is to construct broken-symmetry single determinant wavefunctions, which are not pure-spin states, and therefore are not expected to represent the true wavefunctions of the system. Rather, spin-projection techniques are used to estimate the energies and properties of the correct pure-spin states. This represents a synthesis of density functional ideas with the ideas of broken space and spin symmetry perhaps best exemplified by Lowdin's symmetry dilemma from the 1950's–60's and his DODS (different orbitals for different spin) methods.[93,94] A simple model is that of H_2 at large internuclear separations,[95] where the overlap of atomic orbitals between the two sites becomes small, so that a broken symmetry function, with one electron localized on one atom, and a second localized on the other atom, is more stable than the delocalized molecular orbital description. Transition metal clusters often satisfy similar overlap criterion, even at equilibrium distances. For example, in systems with multiple metal-metal bonds, the weaker δ and δ^* components can break space and spin symmetry to form an overlapping valence bond pair.[96–98]

We can represent an antiferromagnetic spin-coupled state (broken symmetry state) within density functional theory as analogous to a spin-unrestricted determinant in which spin-up electrons are predominantly located on some metal sites, and spin-down electrons on others.[9,99] The broken symmetry method has been applied to a variety of transition metal complexes.[100–102] Although the spin localization provides important stabilization for the system, and leads to single-determinant models that more faithfully represent many aspects of the underlying (multi-configurational) electronic wavefunction, there are important and difficult questions about how the energies and properties of the broken symmetry states are to be interpreted, for example, whether and how the energies of multi-configurational states can be estimated from such density functional studies. We start with a discussion of a very simple two-electron system.

5.2 Singlet-Triplet Splittings for Two Unpaired Electrons

The theory for a system with only two unpaired electrons is especially simple.[15,16] Consider an open shell composed of two orthogonal space orbitals a, b with opposite spins

$$\Psi_B = A(ab\alpha\beta) \equiv |ab\alpha\beta| \tag{14}$$

Following common practice, the coordinate labels can be omitted, with the convention that the labels increase from left to right. The function Ψ_B is not an eigenfunction of S. Proper spin eigenfunctions $\Psi(S, M_S)$ are given by[15]

$$\begin{aligned}\Psi(1,1) &= |ab\alpha\alpha|, \quad \Psi(1,-1) = |ab\beta\beta|,\\ \Psi(1,0) &= 1/\sqrt{2}(|ab\alpha\beta| + |ab\beta\alpha|),\\ \Psi(0,0) &= 1/\sqrt{2}(|ab\alpha\beta| - |ab\beta\alpha|)\end{aligned} \tag{15}$$

from elementary spin algebra. Therefore,

$$\Psi_B = 1/\sqrt{2}[\Psi(1,0) + \Psi(0,0)] \tag{16}$$

The state Ψ_B has $M_S = 0$, but it is not an eigenstate of spin. Rather, it is a state of mixed spin (with both $S = 1$ and $S = 0$ contributing), and will be of broken spatial symmetry if $a(r) \neq b(r)$.

In general, a mixed spin, broken symmetry state takes the form[9,16,97]

$$\Psi_B = \sum_S C(S)\Psi_S \tag{17}$$

(This is trivially true for the two-electron system in Eq. (16), but this idea also holds more larger numbers of unpaired electrons, and the $C(S)$ coefficients can be explicitly calculated, as we discuss below.) Since the Hamiltonian cannot mix eigenfunctions of different spin (in the absence of spin-orbit coupling), we have an energy equation for the broken symmetry state:[16,97]

$$E_B = \langle\Psi_B|H|\Psi_B\rangle = \sum_S [C(S)]^2\langle\Psi_S|H|\Psi_S\rangle \tag{18}$$

There are no cross terms here between states of different S.

To evaluate the energy of the singlet ($S = 0$) and triplet ($S = 1$) states, we need also a separate equation for $S = 1$. Keeping always with energy evaluations involving only single configuration wavefunctions:

$$E(S=1) = \langle\Psi(1,1)|H|\Psi(1,1)\rangle = \langle\Psi(1,-1)|H|\Psi(1,-1)\rangle \tag{19}$$

Note that we have chosen not to make use of $\Psi(1,0)$ since this has more than one configuration. There is nothing wrong with multi-configurational wavefunctions, but local density functional theory is most easily given a concrete and well defined form in terms of single configuration wavefunctions.

The energy of the broken symmetry wavefunction in the present case is now easily found

$$E_B = (1/2)[E(S=0) + E(S=1)] \tag{20}$$

Of more interest is the singlet-triplet splitting:

$$E(S=1)-(S=0)=2[E(S=1)-E_B]\equiv J \tag{21}$$

where J is the Heisenberg parameter in a Hamiltonian of the form $H = J\mathbf{S}_1 \cdot \mathbf{S}_2$. Eq. (21) has been used for many years to estimate singlet-triplet splittings in molecular excited states using $X\alpha$ or local density functional theory.[7,16] The above derivation is of course a very simple one, but we will see that the basic concepts carry through to more complex situations; only the algebra becomes more complicated.

5.3 Nonorthogonal Orbitals

So far, we have assumed that the space orbitals $a(r)$ and $b(r)$ are orthogonal, that is

$$\int a(r)b(r)dr \equiv S_{ab} = 0 \tag{22}$$

This will be true in many molecular excited states where the orbitals may be orthogonal by symmetry. For the case of coupling of two transition metal ions, it may be approximately true if the coupling is weak. Hund's rule implies that the coupling will be ferromagnetic ($J < 0$) when a and b are orthogonal, but it can be either ferromagnetic or antiferromagnetic when a and b are non-orthogonal. We now consider the more general situation for non-orthogonal orbitals (e.g. valence-bond orbitals) a', b' with overlap $S_{a'b'}$. Then it may be shown that the corresponding broken symmetry wavefunction is:[97,103]

$$\Psi_B = \left[\left(1+S_{a'b'}^2\right)/2\right]^{1/2}\Psi(0,0)+\left[\left(1-S_{a'b'}^2\right)/2\right]^{1/2}\Psi(1,0) \tag{23}$$

The corresponding energy equation is

$$E(S=1)-E(S=0)\equiv J=\frac{2[E(S=1)-E_B]}{1+S_{a'b'}^2} \tag{24}$$

For small $S_{a'b'}$, this reduces to Eq. (21); as $S_{a'b'} \to 1$, we have the strong bonding limit, in which the broken symmetry state approaches the singlet state. The two-electron valence bond problem is thus solvable throughout the entire range of $S_{a'b'}$.

The question of how best to estimate singlet-triplet energy differences arises repeatedly, particularly in conjunction with biradical species in organic chemistry, and in binuclear transition metal complexes. A straightforward approach uses the configuration interaction or MCSCF methods to construct separate variational estimates of the singlet and triplet species; this approach can be applied in the context of density functional theory as well as in conventional quantum chemistry.[104] A more common approach has exploited that fact that broken-symmetry solutions are often an effective way to build in non-dynamical correlation effects; these approaches use Eqs. (21) and (24), or developments of them. One approach involves analysis of spin contamination (from higher spin states) in unrestricted states. For organic radicals, this contamination is generally smaller than for corresponding unrestricted Hartree–Fock wavefunctions,[105] and spin-projection or annihilation can offer one approach to the extraction of pure spin-state energies.[106,107] The general problem of symmetry restrictions and breaking in density functional theories is an important and unsolved problem,[108–111] but the approximate approaches outlined here give useful answers for a wide variety of organic diradicals.

In transition metal systems, a classic example has involved dinuclear copper(II) complexes, with varying strengths of direct metal-metal bonds, and ligand-bridged metal-metal interactions. An early model that concentrated on orbital contributions to variations in magnetic coupling was proposed by Hay, Thibeault, and Hoffman,[112] where

$$J = 2K_{ab} - \frac{(\varepsilon_1 - \varepsilon_2)^2}{J_{aa} - J_{ab}} \tag{25}$$

Here ε_1 and ε_2 are the orbital energies of the singly occupied molecular orbitals, and K_{ab}, J_{aa}, and J_{ab} are two electron integrals. This is based on open shell spin-restricted calculations, and in practice ascribes the dependence of coupling constant on geometry primarily to variations in one-electron energies, so that the two-electron integrals need not be evaluated. This can give a useful account of the effects of geometry variation within a family of structures.

More quantitative calculations have been based on *ab initio* CI[113–117] or MC-SCF[118] methods. The broken symmetry density functional approach (essentially with Eq. (21) or Eq. (24) in either the weak or strong bonding limits) has been applied to Cu(II) dimers by a number of workers.[101,119–121] While there is a broad general agreement with experimental values, the splittings involved are generally small (on the order of a few hundred wavenumbers) and difficult to evaluate reliably. Different density functionals can give significantly different results, depending also on whether the strong bonding[121] or weak bonding limits apply (or intermediate bonding),[120] and questions of geometry optimization and basis set convergence are still difficult to resolve for ligands that are commonly used. Nevertheless, geometric variations appear to be predicted well. For hydroxo-bridged copper dimers, the geometric variations predicted by Eq. (25) closely match those obtained from broken symmetry DFT calculations,[121] with the DFT method offering an approach to absolute, as well as relative, splittings.

5.4 Interactions of More Than Two Electrons

When more than two unpaired electrons are weakly coupled, the complete problem is not solvable, but substantial progress can be made through recourse to perturbation theory.[99,114,122,123] Consider a single configuration wavefunction (determinant) in which both spin and space symmetry restrictions have been lifted. For a dinuclear metal cluster, this would have non-orthogonal magnetic orbitals d_l', d_r', mainly on the metal centers (where l and r denote the left and right transition metal centers), and non-orthogonal ligand orbitals $l^{k'}\alpha$, $l^{k''}\beta$, including both the bridging and terminal ligands. Typically, this single configuration wavefunction would be variationally optimized as a broken symmetry UHF or local density functional wavefunction. This wavefunction can formally be expanded in terms of determinants[99,122]

$$\begin{aligned}\Psi_B &= |[d_l'][\alpha\alpha\alpha\ldots][l^{k'}l^{k''}][\alpha\beta][d_r'][\beta\beta\beta\ldots]| \\ &= D_0 + \sum_u a_u(B)D_u \end{aligned} \tag{26}$$

The excited configurations D_u are created by excitations from the principal determinant D_0, and the $a_u(B)$ are mixing coefficients in the expansion of the broken symmetry state Ψ_B.

To treat the states of pure spin, we recognize that each determinant in Eq. (26) is like an outer product of monomer spins. The Clebsch–Gordan algebra can be used to express this in terms of the spin eigenstates of the coupled system:[124]

$$|S_1M_1\rangle|S_2M_2\rangle = \sum_S C(S_1S_2S;\, M_1M_2)|SM\rangle \tag{27}$$

where C is a Clebsch–Gordan coefficient. In the principal determinant D_0, there are n unpaired electrons on the left and right, and the two spin vectors $S_1 = S_2 = n/2$ are coupled to total spin S. If the mainly-metal orbitals have significant overlap, there will be a direct exchange term favoring parallel spin alignment. For contributions to super-exchange, the relevant excited state configurations D_u arise from $d_l \rightarrow d_r$ or $d_r \rightarrow d_l$ excitations that leave $n-1$ unpaired electrons per site, so that $S_1 = S_2 = \frac{1}{2}(n-1)$; for contributions to ligand spin polarization, the excited configurations arise from $l^k \rightarrow d_l$ or $l^k \rightarrow d_r$ excitations, again with $S_1 = S_2 = n/2$. Hence, there are two important types of coupling coefficients, $C_1(S) \equiv [C(\frac{1}{2}n\frac{1}{2}nS;\, \frac{1}{2}n - \frac{1}{2}n)]$ and $C_2(S) \equiv [C(\frac{1}{2}(n-1)\frac{1}{2}(n-1)S;\, \frac{1}{2}(n-1) - \frac{1}{2}(n-1))]$.

Combining Eqs. (26) and (27) then yields an expression for the broken symmetry state in terms of spin eigenstates:

$$\begin{aligned}\Psi_B &= \sum_S C_1(S)\Phi_0(S) + \sum_S\sum_u a_u(B)C_u(S)\Phi_u(S) \\ &= \sum_S C_1(S)\left\{\Phi_0(S) + \sum_u [a_u(B)C_u(S)/C_1(S)]\Phi_u(S)\right\}\end{aligned} \tag{28}$$

where Φ represents the appropriate spin-coupled state $|SM\rangle$, and $C_u(S)$ will be $C_2(S)$ for determinants of the super-exchange type and $C_1(S)$ for ligand spin polarization excitations. Eqs. (17) and (18) then hold with $C_1(S)$ replacing $C(S)$. Hence, even for multiple unpaired electrons, the broken symmetry wavefunction and its energy can be connected in a simple way to pure spin states. This forms the foundation for the calculation of spin properties and spin-state energetics.

The second main ingredient in our approach to spin coupling involves the development of an expression for E_S. It may seem natural to expect E_S to be proportional to $S(S+1)$, as in a Heisenberg Hamiltonian. This turns out to be approximately correct; a perturbation theory justification for the Heisenberg form for the pure spin states has been given in several places,[9,56,99,122] and will not be repeated here.

We thus adopt the *ansatz* that the E_S values are of Heisenberg form with $E_S = S(S+1)J/2$. The Clebsch–Gordan algebra also allows one to directly show that:[99]

$$\sum_{S=0}^{S_{\max}} C_1(S)^2 S(S+1) = n = S_{\max} \tag{29}$$

Combining Eqs. (18) and (29) yields an expression for the energy difference between two single configuration states, that of $S_{\max}$ and the broken symmetry state:

$$E(S_{\max}) - E_B = (S_{\max}^2 J/2) \tag{30}$$

Note that Eq. (21) is a special case of this relation, with $S_{\max} = 1$.

5.5 Analysis of a Mixed-Valence Dimer

When an electron is added to a symmetric dimer (say Fe^{+3}-Fe^{+3}) to form a mixed-valence system, the resulting energy terms are of two types. First, there are Heisenberg terms which originate from direct exchange, super-exchange, and ligand spin polarization, analogous to those described above. In addition, there are resonance delocalization interactions resulting from the mixing of the degenerate configurations created by the (arbitrary) assignment of the extra electron to the left or right side of the system.[56,125]

It is easiest to characterize the principal and excited functions by their vector coupling structure. Consider an outer product spin state of the form $|S_1M_1\rangle_{\mathrm{L}}|S_2M_2\rangle_{\mathrm{R}}$, where L, R are the left and right metal centers. The principal spin configuration is $|5/2\ 5/2\rangle_{\mathrm{L}}|2\ -2\rangle_{\mathrm{R}}$ with the electron added to side R. Excited configurations of super-exchange type can be included by methods analogous to those in the previous section. These terms lead to a Heisenberg Hamiltonian for the pure spin states and to the corresponding energy difference equation:[56,99]

$$E(S_{\max}) - E_B = n(n-1)J/2 \tag{31}$$

According to the theory of double exchange (also called resonance delocalization coupling), developed by Anderson and Hasegawa,[126] the energy of delocalization for a mixed-valence pair of transition metal sites depends on the alignment of the spins of the sites (i.e., on the pair spin quantum number S_{ij}) in a linear fashion: $E_{res} = \pm B(S_{ij} + 1/2)$, with the plus/minus sign representing the anti-bonding/bonding combination of orbitals associated with the delocalized electron. (For an alternative proof for the E_{res} equation using only Clebsch–Gordan, rather than Racah algebra,[126] see Noodleman and Baerends.[56]) Hence, for an example of an Fe^{2+}/Fe^{3+} dimer, the bonding states will have energies $JS(S+1)/2 - B(S+\frac{1}{2})$. For a bridged system,

J will be positive (antiferromagnetic), and the parallel-spin ($S = 9/2$) state will be lowest when $|B|/J > 4.5$, and the antiferromagnetic ($S = 1/2$) state lowest with $|B|/J < 1.5$. Clusters of the type $Fe_2S_2(SR)_4^{3-}$ in model compounds and proteins are antiferromagnetic. Recently, completely delocalized (Robin–Day Class III) mixed-valence dimers have been found (with total spin $S = 9/2$) both in synthetic complexes based on $[Fe_2(OH)_3L_2]^{2+}$,[127] and in mutant (cysteine → serine) ferredoxin proteins based on a $Fe_2S_2(SR)_3(OR)^{3-}$ structure.[128,129] This now appears to be part of a growing class of dinuclear, trinuclear, and tetranuclear complexes containing various metals where there is a delocalized mixed-valence pair.[130]

Nuclear motion and geometric relaxation are also important components of electronic structure in mixed-valence systems, since this can determine whether the extra electron is localized or delocalized. Both the shape of the potential well, and the dependence of the potential energy surface on the alignment of the site spins are important. A good overview of these subjects is found in the review by Blondin and Girerd.[86]

Consider a mixed-valence pair with spin coupled Fe^{2+}-Fe^{3+} ions, where each site is internally high spin ($S_i = 2$ or $5/2$). The presence of two possible sites for the extra sixth Fe(3d) electron suggests a simplified model Hamiltonian illustrated in Figure 8. This is a slight generalization of Figure 3 of Blondin and Girerd,[86] and we follow their notation. In the absence of a resonance interaction between the two sites, this model has two "diabatic" surfaces, labeled A and B, that correspond to adding the extra electron to one of the two sites. The vibrational distortion parameter q^- represents changes that take place in the site environment on going from oxidized to reduced localized wavefunctions; these include changes in bond lengths

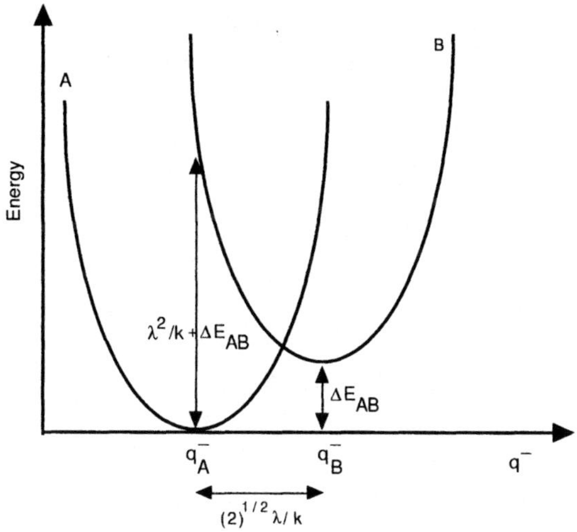

Figure 8. Schematic vibrational model for a mixed-valence dimer. Curves A and B are the diabatic energies (in the absence of resonance delocalization) for localized states, as a function of a nuclear distortion parameter q^-.

and other rearrangements of the ligand/solvent environment. This model is characterized by three parameters: the curvature k of the curves about the oxidized and reduced minima, the separation of these minima along the q^- coordinate (which is $(2)^{1/2}\lambda/k$ in this notation), and a shifting of the relative energies of curves A and B by ΔE_{AB}, reflecting a static preference (arising from an asymmetric environment) for the extra electron to reside on one site rather than the other. At the minimum of the A curve (point $q^-_{\min}(A)$ in Figure 8), state A has energy 0 and state B has energy $\Delta E \equiv \lambda^2/k + \Delta E_{AB}$, where the first term can be considered a vibronic contribution and the second a static asymmetric contribution to a total localization energy.

We now consider a resonance interaction between these localized states. If we focus on the point $q^-_{\min}(A)$ (where the the extra electron resides on site A), the energy matrix can be written:

$$\begin{bmatrix} 0 & B(S+\frac{1}{2}) \\ B(S+\frac{1}{2}) & \Delta E \end{bmatrix} \tag{32}$$

The first row or column of the matrix refers to the state with the electron localized on the left, and the second to the state where the electron is localized on the right, both at the geometry of $q^-_{\min}(A)$.

The properties of this simple resonance Hamiltonian model have been discussed in many places. Generally, valence trapping will occur unless $2|B(S_{ij}+1/2)| > \Delta E$; as delocalization becomes important, the minimum in the ground state energy moves to $q^- = 0$ (see Figure 3 of Ref. 86). Asymmetries in site properties are governed by the ratio $R_{loc} = \Delta E/|B(S+\frac{1}{2})|$. For small values of this ratio, the system is almost completely delocalized: $R_{loc} \leqslant 0.4$ corresponds to $0.4 \leqslant c_A^2, c_B^2 \leqslant 0.6$, where c_A^2, c_B^2 are the weights of the local states. Large values of $R_{loc} (> 6)$ corresponding to $c_A^2 > 0.9$, give nearly complete localization.

Overall, this model provides a simple description of why low-spin Fe(II)–Fe(III) dimers in iron-sulfur systems (for example) appear as distinct valences, whereas high-spin analogues (which often form parts of larger clusters) are delocalized in an effective +2.5 oxidation state. For the antiferromagnetic dimer ($S = \frac{1}{2}$), the resonance splitting is quite small, so that minor static or dynamic distortions that break the strict equivalence of the iron sites will lead to trapped valence states with large values of R_{loc}. In the ferromagnetic ($S = 9/2$) case, however, the resonance interaction is much larger, and delocalized states with small values of R_{loc} are obtained. In a simple language, it is easier for the final d-electron to hop between iron sites when the spins are parallel than when they are opposed, since no net change in exchange interactions is involved in the former shift. A detailed analysis along these lines for Fe_2S_2 dimers has been given by Noodleman and Baerends.[56] For these clusters J is antiferromagnetic (positive), favoring low S, while the resonance interaction stabilizes states of high S, so that the final spin state often represents a balance of opposing forces.

5.6 Analysis of a Three-Spin System

The ideas presented can readily be extended to systems with more than two metal ions, and the important ideas can be illustrated by considering the simplest such model, a three-spin system with equivalent metal sites. Consider a cubane 3Fe cluster in the oxidized state, where the three iron sites are geometrically equivalent and are all formally in the +3 oxidation state.[131] The crux of our computational approach arises from the recognition that broken symmetry wavefunctions (in which otherwise equivalent metal sites have different spin populations) are relatively easy to compute and interpret. By contrast, the correct wavefunctions (which are eigenfunctions of $\hat{S}^2$) are generally multi-configuration states that are considerably more difficult to approximate and understand. Hence, we choose to fit an (assumed) spin Hamiltonian to energies computed from broken symmetry wavefunctions, and use the resulting parameters to estimate the locations of the pure spin states, including the pure-spin ground state.

We assume that the true electrostatic interactions that couple iron spins together can be replaced by an interaction of the Heisenberg type:

$$\hat{H} = J(\hat{\mathbf{S}}_1 \cdot \hat{\mathbf{S}}_2 + \hat{\mathbf{S}}_1 \cdot \hat{\mathbf{S}}_3 + \hat{\mathbf{S}}_2 \cdot \hat{\mathbf{S}}_3) \tag{33}$$

Griffith[132] has worked out the eigenstates of this Hamiltonian, and related methods may also be used to estimate the energies of broken symmetry states in which each iron atom is either spin-up (with $S = M_S = 5/2$) or spin-down (with $S = 5/2$, $M_S = -5/2$). If for convenience we denote these monomer states as $|\alpha\rangle$ and $|\beta\rangle$, respectively, then broken symmetry kets will have forms such as

$$|\alpha\alpha\beta\rangle \equiv |\alpha_1\rangle|\alpha_2\rangle|\beta_3\rangle \tag{34}$$

which represents a state with $M_S = 5/2$ which can be approximately identified with a broken symmetry molecular orbital wavefunction that places five unpaired spin-up d-electrons on centers 1 and 2, and five spin-down d-electrons on center 3. These broken symmetry kets are not intended to approximate eigenstates of the Hamiltonian; rather, they represent mixed states whose energies can be computed by both spin-unrestricted molecular orbital and by spin Hamiltonian methods, so that the two may be compared.

The energies of kets like that in Eq. (34) are relatively straightforward to evaluate since each term in the Hamiltonian couples only two centers at a time. Hence, with $S' \equiv S_1 + S_2$,

$$\begin{aligned}\langle\alpha\alpha\beta|\hat{\mathbf{S}}_1 \cdot \hat{\mathbf{S}}_2|\alpha\alpha\beta\rangle &= \langle\beta|\beta\rangle\langle\alpha\alpha|\hat{\mathbf{S}}_1 \cdot \hat{\mathbf{S}}_2|\alpha\alpha\rangle \\ &= (1/2)[S'(S'+1) - S_1(S_1+1) - S_2(S_2+1)] = (25/4)\end{aligned} \tag{35}$$

since $|\alpha\alpha\rangle$ is a pure spin state with $S' = 5$. The mixed-spin states are slightly more complicated to evaluate, but also yield simple results:

$$\langle\alpha\alpha\beta|\hat{\mathbf{S}}_2 \cdot \hat{\mathbf{S}}_3|\alpha\alpha\beta\rangle = \langle\alpha\alpha\beta|\hat{\mathbf{S}}_1 \cdot \hat{\mathbf{S}}_3|\alpha\alpha\beta\rangle = -(25/4) \tag{36}$$

For a three-iron system we can define a high-spin ket $|\alpha\alpha\alpha\rangle$ (which is a pure spin state with $S = S_{\max} = 15/2$), and three equivalent broken symmetry states, $|\alpha\alpha\beta\rangle$, $|\beta\alpha\alpha\rangle$, and $|\alpha\beta\alpha\rangle$, all with $M_S = 5/2$. Their energies are $E(S_{\max}) = 75J/4$ and $E(B) = -25J/4$. J can thus be estimated by comparing the energy differences arising from these formulas with those computed from a broken symmetry molecular orbital approach, and estimates of the pure spin state energies are then made from the resulting parameterized spin Hamiltonian. By comparing the energies of the high spin and broken symmetry states, we can generate the entire Heisenberg spin state ladder.

It is worth noting that the ground state of this cluster would be very difficult to describe in pure density functional terms. A determinant of doubly-occupied Kohn–Sham orbitals describes a system with low-spin irons, and in any event has a much higher computed total energy than the broken symmetry state. The broken symmetry state itself has $M_S = 5/2$, and hence has zero overlap with the ground state wavefunction that has $S = 1/2$ with $M_S = 1/2$, or $M_S = -1/2$. It is the construction of the spin Hamiltonian intermediate model that allows connections to be made between single configuration density functional results and the (properly symmetrized) approximate ground state.

For the Fe(III) sites in these clusters, the five parallel-spin d-electrons have fairly low orbital energies, and there is little delocalization among metal sites, although there is still substantial iron-sulfur covalency. When one of these sites is reduced, however, the extra electron can be distributed over more than one site, and this delocalization can interact strongly with spin coupling effects.[133,134] In the high spin configurations, the first five d-electrons on each site are aligned in a parallel fashion, say spin-up. While there are general formal solutions to the double exchange problem,[135] it is useful to outline the simplest approach first.[136] We form three basis configurations by allowing the final d-electron (which must be spin-down) to reside in turn on each of the three sites. The Heisenberg matrix elements are computed as we outlined above; for the delocalization terms we assume that a single parameter B' characterizes resonance interactions between each pair of sites. Hence, the spin Hamiltonian matrix for the high-spin state becomes:

$$\mathbf{H}_{hs} = \begin{bmatrix} (65/4)J & -5B' & -5B' \\ -5B' & (65/4)J & -5B' \\ -5B' & -5B' & (65/4)J \end{bmatrix} \tag{37}$$

Here and below, the diagonal elements represent the system energy in the absence of delocalization, and the off-diagonal elements give the specific resonance delocalization effects, recognizing that $(S' + \frac{1}{2}) = 5$ for parallel spin Fe(II)/Fe(III) dimers. The eigenvalues are $E_1 = (65/4)J - 10B'$, and $E_{2,3} = (65/4)J + 5B'$ (doubly degenerate). For these clusters we find $B' > 0$, and hence E_1 lies lowest.

For the broken symmetry state, the first five d-electrons of one of the iron atoms (which we call a) is of opposite spin to that of an equivalent pair b. There are still three basis configurations, corresponding to the three possible locations of the last d-electron. We will adopt the simplest delocalization hypothesis, that resonance interaction is important only between the two irons of the same spin, pair b. The spin Hamiltonian matrix becomes:

$$\mathbf{H}_{bs} = \begin{bmatrix} -(25/4)J & 0 & -5B \\ 0 & -(15/4)J & 0 \\ -5B & 0 & -(25/4)J \end{bmatrix} \quad (38)$$

Here we have allowed the delocalization parameter in the broken symmetry state, B, to differ from that in the high spin state. Eigenvalues for the broken symmetry case are $E_{1,2} = -(25/4)J \pm 5B$ and $E_3 = -(15/4)J$. The splitting between E_1 and E_2 (which is $10B$ in this model) reflects the difference between bonding and antibonding orbitals delocalized over the b sites, and is the same as that in an isolated high-spin iron dimer.

The J's and B's can thus be estimated by comparing the energy differences arising from these formulas with those computed from a broken symmetry molecular orbital approach, and estimates of the pure spin state energies (including the ground state energy and its spin value) are then made from the resulting parameterized spin Hamiltonian. For the simple model used here, the eigenstates for various values of B and J are given by Papaefthymiou et al.,[133] and numerical estimates are reported by Sontum et al.[136]

In the next two sections, we illustrate the application of many of these general ideas to bridged complexes with several iron or manganese ions.

5.7 Iron-Sulfur Dimer and Tetramer Complexes

Iron-sulfur dimer and tetramer complexes lie at the active sites of many iron-sulfur proteins.[137,138] These proteins are often single-electron transfer agents, as in the electron-transport chain of mitochondria,[139] in bacterial electron transfer, and in photosynthetic electron transfer in plants and cyanobacteria.[140] Other functions are known for related clusters. There are regulatory functions based on cluster decomposition/synthesis in response to iron or oxygen concentration (sensors).[141] In aconitase, the presence of an open coordination site (not present in the iron-sulfur complexes examined here) is associated with hydrolytic chemistry.[142] More complex polynuclear clusters (the P and MoFe clusters) are used for multi-electron redox catalysis (conversion of molecular nitrogen to ammonia) in nitrogenase.[143–145]

Synthetic analogues of the simpler iron-sulfur dimer and tetramer systems have been constructed involving organic ligands and having the same redox couples as observed in the protein active sites.[137] In this section, we examine the fundamental electronic structure features of these complexes with emphasis on energy level structure, localization versus resonance delocalization of electrons in mixed-valence systems (particularly delocalization over mixed-valence pairs of Fe sites), and spin-coupling parameters from Heisenberg exchange and resonance delocalization (double exchange). These aspects of electronic structure are valuable for analyzing both the simple iron-sulfur clusters discussed below, and, by a building block approach,[146,147] for examining the less-well understood polynuclear and mixed-metal clusters in nitrogenase, aconitase, and other systems.

Figure 9 shows schematically the structures and alignment of site spin vectors for $Fe(SR)_4$, $Fe_2S_2(SR)_4$, and $Fe_4S_4(SR)_4$ clusters found in the electron transfer proteins rubredoxin (1Fe), and 2Fe and 4Fe ferredoxins. In these proteins, SR is cysteine, and these residues are connected to the protein polypeptide chain. In smaller synthetic analogues, organic groups are connected to the various terminal sulfurs (SR). We show antiferromagnetic alignment of the various site spin vectors (each $S_i = 5/2$). For the reduced $Fe_2S_2(SR)_4$ systems, we depict the final minority spin electron added as a smaller spin oppositely aligned to the main spin vector, giving a net ferrous site with spin $S_i = 2$. A trapped valence system is shown as is typically seen experimentally. Because the AF Heisenberg coupling competes with resonance delocalization, and in the presence of environmental forces favors localization, the ground state of the reduced 2Fe complex is trapped valence $S = 1/2$. When considering the 4Fe4S complex in the $Fe_4S_4(SR)_4^{2-}$ oxidation state, with an average Fe oxidation state of 2.5+, the globular structure of the 4Fe sites allows for parallel alignment of some site spins and anti-parallel alignment of others (in fact, in a 2:4 ratio). This arrangement allows easy delocalization of two minority spin electrons over two opposite cube faces, consistent with a delocalized mixed-valence system. These results are also borne out by the observed pattern of delocalization of the one-electron oxidized (high potential oxidized) and one-electron reduced (reduced ferredoxin) systems from Mossbauer and ENDOR spectroscopies.[148–151] This simple qualitative picture persists when quantum mechanical spin coupling among the multiple sites is analyzed.

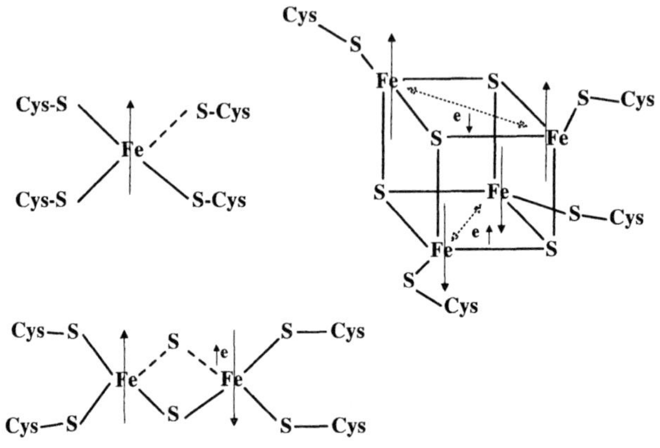

Figure 9. Schematic structures and spin alignments for iron-sulfur complexes $[Fe(SR)_4]^{1-}$, $[Fe_2S_2(SR)_4]^{3-}$ and $[Fe_4S_4(SR)_4]^{2-}$ from rubredoxins and ferredoxins. The long arrows represent high-spin $s_i = 5/2$ spin vectors in Fe_i sites and the short arrows represent additional electrons of opposite spin $s_i = 1/2$.

Returning to the oxidized (diferric) form of the Fe_2S_2 complex, we can examine the energy level structure found for the broken symmetry state. Spin polarization between majority and minority spin electrons leads to an inverted level scheme[9,56] (found also in $Fe(SR)_4^{1-}$ and $FeCl_4^{1-}$ systems as discussed above).[58,60,61] From the 2Fe2S energy level diagram (Figure 10), the occupied majority spin Fe 3*d* levels lie below the mainly bridging sulfur (S*) and terminal S 3*p* levels, with the minority spin Fe 3*d* levels above S*, S(3*p*). The majority spin Fe *d*-electrons are oppositely aligned on the two sites, but with nonorthogonal valence-bond type overlap of the spatial orbitals of opposite spin. Addition of one electron to the $Fe_2S_2(SR)_4^{2-}$ oxidized cluster yields the reduced system with one minority spin mainly $Fe(3d_{z^2})$ orbital occupied, $16a_1\alpha$ (assuming reduction on the right, see Figures 9 and 10).

The compactness of the Fe(3*d*) shell leads to considerable electron relaxation, so that the net charge on the reduced Fe atom after reduction is not much different from that in the oxidized system. The electron density on the reduced Fe atom is far more asymmetric after reduction, however. The main changes in the net electron populations are increases on all sulfurs, particularly those directly neighboring the reduced Fe site, based on calculated ESP charges. An electron density difference map (reduced minus oxidized) is shown in Figure 11.[64]

A direct comparison has been made between the calculated ligand-to-metal charge transfer transitions in $Fe_2S_2(SR)_4^{2-}$ and the bands observed experimentally in both proteins and synthetic analogues.[56,152–155] The calculations used an $X\alpha$ potential

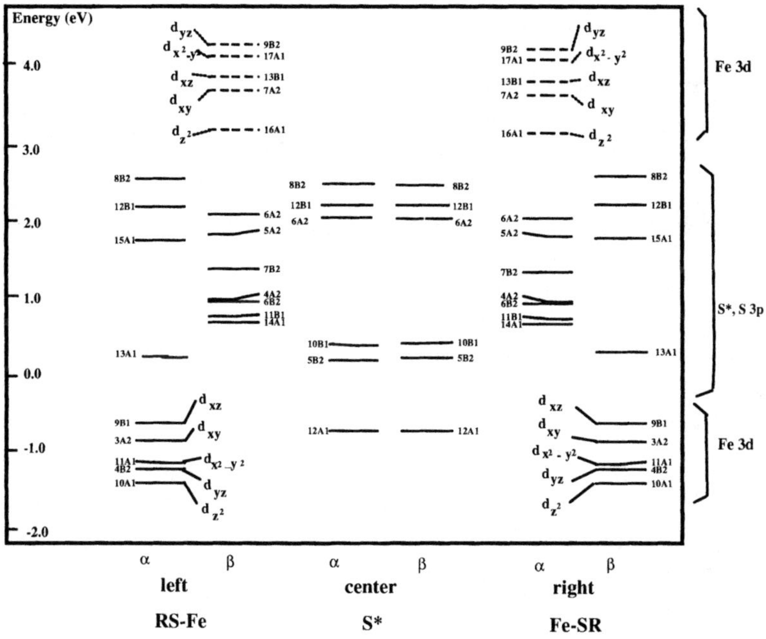

Figure 10. MO energy level diagram for the broken symmetry state of oxidized ferredoxin $[Fe_2S_2(SCH_3)_4]^{2-}$.

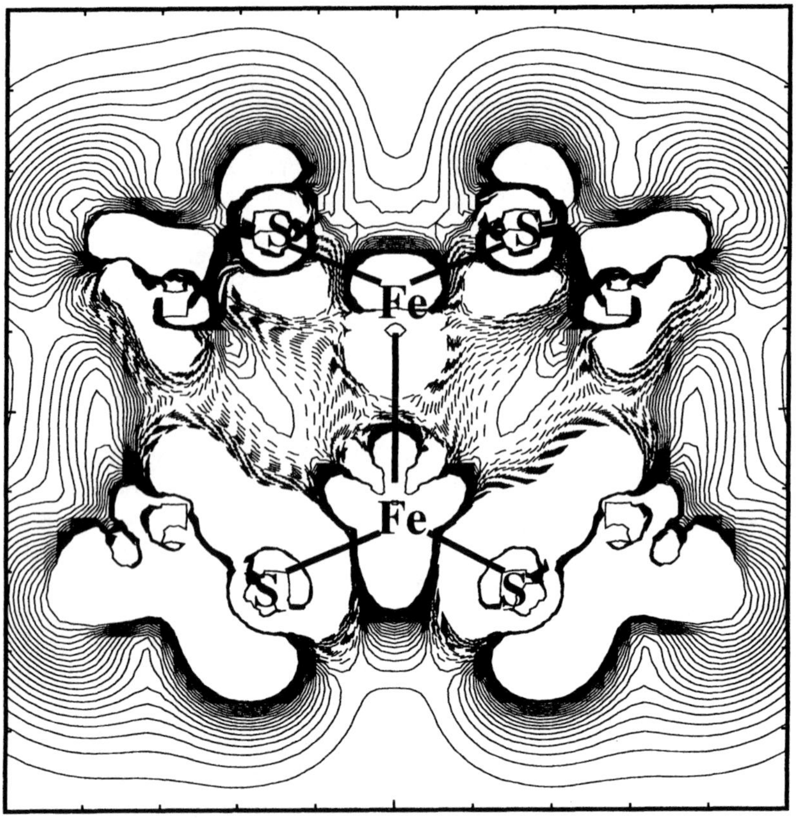

Figure 11. Electron density difference map for the reduced $[Fe_2S_2(SR)_4]^{3-}$ minus oxidized $[Fe_2S_2(SR)_4]^{2-}$ cluster. The reduced half of the complex is the lower part of the plot. The solid contour lines represent increase of the electron density.

on the simplified model SR = SH system and the Slater transition state method to account for orbital relaxation.[56] Calculated and observed spectra are in good agreement, although the experimental spectra are quite broad with increasing band intensities at higher energies. Further, both experiment and excited state broken symmetry calculations show low lying transitions in the energy range 10700–13900 cm^{-1} attributed to spin- and dipole-allowed ligand-to-metal charge transfer (LMCT), mainly into the LUMO $16a_1(Fe(d_{z^2}))$. These low-energy transitions are specific to the dimer systems, and since the corresponding low-lying transitions are higher in the 1Fe protein rubredoxin, at 15000–18000 cm^{-1}, it appears likely that there is a moderate excited state antiferromagnetic coupling effect[100] in the ferredoxins.

In Table 1, we report calculated J and B parameters for iron-sulfur dimers and tetramers.[81,156–160] The theoretical J coupling parameters typically are larger than experiment (with $VS + B > X\alpha$), but the experimental trend $J_{ox} > J_{red}$[161] is properly portrayed. The size of the calculated B parameters are substantial, and values obtained for lower-symmetry geometries more representative of proteins are typically

Table 1 *J* and *B* Coupling Parameters

	$X\alpha$	$VS+B$	*Exp.*
	$Fe_2S_2(SCH_3)_4^{2-,3-}$		
J_{ox}	527	763	298,366
J_{red}	340	514	196,220
B	486	394	—
	$Fe_4S_4(SCH_3)_4^{1-,2-,3-}$		
State (2−)			
$J(Fe^{2.5+}-Fe^{2.5+})$	454	645	340
B	740	795	<570
State (1−)			
$J(Fe^{3+}-Fe^{3+})$		833	797
$J(Fe^{3+}-Fe^{2.5+})$		673	652
B		878	592
B'		286	
State (3−)			
$J(Fe^{2+}-Fe^{2.5+})$		519	
B		753	
B'		618	

Units are cm^{-1}; Heisenberg Hamiltonian convention $H_{spin} = JS_1 \cdot S_2$; Symmetries for the calculations were C_{2v} for all systems except the 1− state of the four-iron clusters, which were C_2. B is the delocalization parameter within a mixed-valence pair (or pairs); B' is the delocalization parameter between a mixed-valence and a ferrous pair (3−), or between a mixed-valence and a ferric pair (1−). Citations for all values are given in Ref. 160

higher still. The B term is not usually observed in reduced $Fe_2S_2(SR)_4^{3-}$ systems, because a combination of trapping forces and Heisenberg coupling usually leads to a trapped $S = 1/2$ ground state. Some mutant proteins do have delocalized cores, as discussed above.[128,129] In $Fe_4S_4(SR)_4^{1-,2-}$ systems, there have been experimental studies where both J and B parameters were used in fits of magnetic susceptibility for a 1– system[158] and solid state NMR-MAS spectroscopy for the 2– systems.[162] The J parameters increase as the cluster oxidation state becomes more positive, and for the higher oxidation state Fe pairs within a cluster (1−) both experimentally and computationally. The calculated B parameters are fairly large, (in the range 700–900 cm^{-1}) as are those found experimentally from fits of the magnetic data,[158] and this is also consistent with delocalization within one (1− or 3−) or two (2−) mixed-valence pairs, as seen by Mossbauer and ENDOR spectroscopies.[134,149–151] Further discussions of iron-sulfur protein electronic structures are given elsewhere.[9,64]

5.8 Manganese-Oxo Dimer and Tetramer Complexes

Multi-electron transfer catalysis is exhibited by the tetranuclear Mn cluster in the water-oxidation-complex (WOC) of photosystem II.[163] In a light-driven process involving the chromophore P680 and a redox active tyrosine as electron acceptors, successively higher oxidation states are generated in this Mn cluster, eventually removing electrons and protons from bound waters. These electrons and protons are used in photosynthesis, and molecular oxygen is generated from water and released.[164] The WOC is critical to photosynthesis in plants and cyanobacteria, and is the major source of oxygen in the atmosphere. While the detailed coordination geometry of the WOC is not known, the chemistry of high oxidation state Mn dimers and tetramers with bridging oxo, carboxylate and with terminal nitrogen or oxygen donors is relevant.

We have recently carried out calculations on the electronic structures of a series of five manganese-oxo complexes with additional bridging carboxylates (or bridging peroxo- or additional oxo-)ligands.[66] The terminal ligands were TACN = triazacyclononane. The formal oxidation states varied from Mn(III)Mn(III), Mn(III)Mn(IV), to Mn(IV)Mn(IV). The structures of these complexes are depicted in Figure 12. These systems show extensive metal-ligand covalency. Figure 13 shows the calculated ESP charges on the manganese ions across the series. The Mn ion charges are considerably less than those indicated by the formal Mn oxidation states, and slowly decrease with increasing formal oxidation number. This trend is due to the extensive donation from the oxo, peroxo, and TACN ligands; the expected trend of increasing charge with increasing oxidation state can be recovered when the two TACN ligands are removed, giving weaker electron donation (labeled "core" in Figure 13). The ligand atom charge transfer order (calculated per donor atom to a single metal ion) is peroxo $\geqslant$ oxo $\gg$ TACN $>$ acetate. While these synthetic systems are probably not close analogues of the tetranuclear Mn cluster in the WOC, this charge transfer ordering may be significant. Water oxidation requires that the oxidizing power of the cluster be directed toward particular ligands, $2(H_2O)$, O_2H^-, and O_2^{2-}, which are oxidized and/or acidified, resulting in the final product O_2.

Returning to issues of spin coupling, the exchange coupling and ligand field effects in these synthetic systems are interrelated. The strongest exchange coupling pathways involve the bridging oxo or peroxo ligands. The ligands surrounding each Mn ion produce $t_{2g} - e_g$ splittings in both minority spin and majority spin orbitals, while the spin polarization splitting of the high spin (or intermediate spin) Mn sites raises the minority spin levels compared to the majority spin levels. These two effects are competitive in their influence on spin coupling energies. Super-exchange pathways can involve metal $\rightarrow$ metal donation either into minority spin orbitals, leading to antiferromagnetic coupling contributions, or into the higher lying majority spin ligand orbitals (of e_g type), leading to a crossed-interaction, and making a ferromagnetic coupling contribution. A good way of assessing the energetics of spin coupling across this series of metal-oxidation states and ligand environments is to plot the $\Delta E_{HS,BS} \equiv E(S_{\max}) - E_B$ for the calculations vs. experiment (where the experimental high spin minus broken symmetry energy difference is extracted from the measured J value). This approach has the advantage of showing how sensitive the calculation needs to be for quantitative agreement with experiment, since the procedure precisely parallels that employed in the J calculation (cf. Eq. 30). The results of this approach are displayed in Figure 14. When the experimental J values are also translated into an experimental $\Delta E_{HS,BS}$, the experimental range is

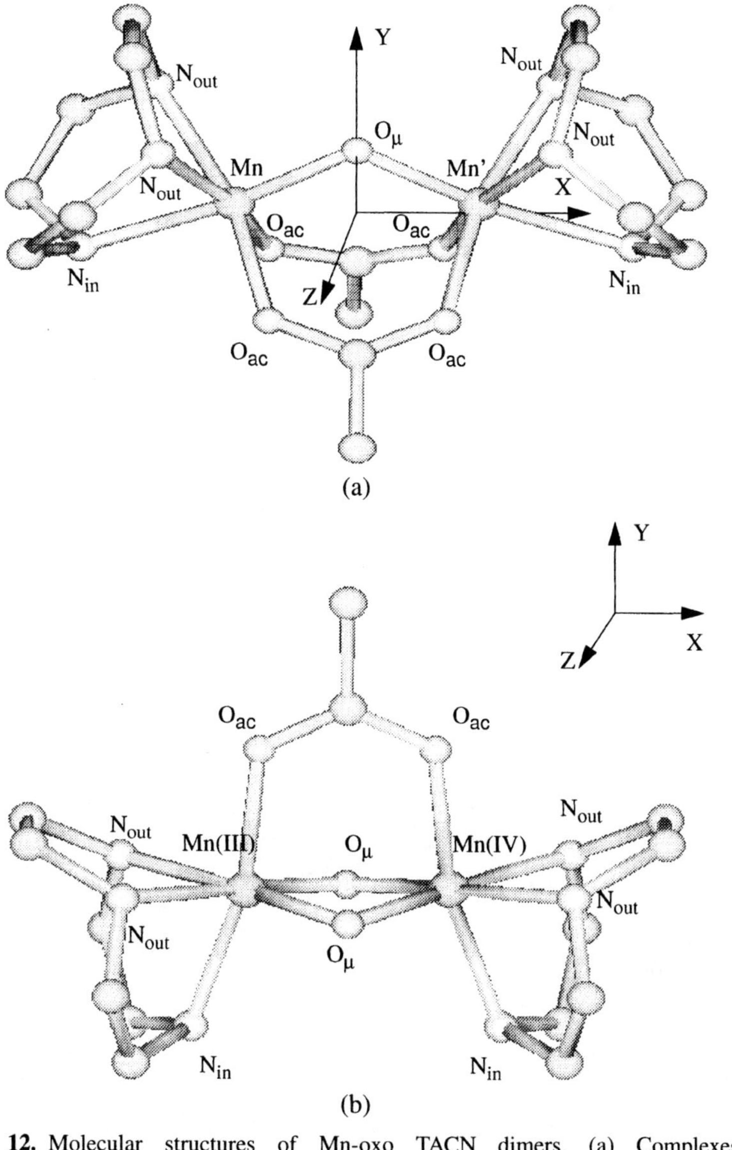

Figure 12. Molecular structures of Mn-oxo TACN dimers. (a) Complexes A1 $[Mn_2^{III}(\mu\text{-}O)(OAc)_2(TACN)_2]^{2+}$ and A2 $[Mn^{III}Mn^{IV}(\mu\text{-}O)(OAc)_2(TACN)_2]^{3+}$; (b) Complex B $[Mn^{III}Mn^{IV}(\mu\text{-}O)_2(OAc)(TACN)_2]^{2+}$; (c) Complex C $[Mn_2^{IV}(\mu\text{-}O)_2(\mu\text{-}O_2^{2-})(TACN)_2]^{2+}$; (d) Complex D $[Mn_2^{IV}(\mu\text{-}O)_3(TACN)_2]^{2+}$.

−0.02 eV to 0.44 eV, while the calculated range is from 0.06 eV to 0.77 eV. The RMS error of 0.23 eV is comparable to that found for ionization and excitation energies by density functional methods,[35] and to the error in calculated J values for the $Na_2Fe_2(\mu\text{-}O)Cl_6$ complex studied by Hart, Rappe, and coworkers.[165] Despite

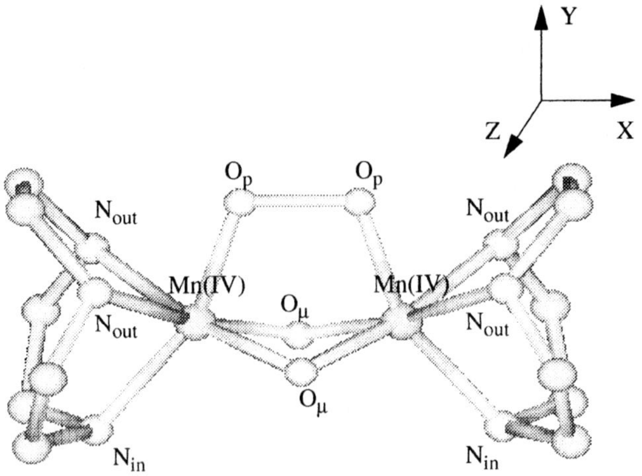

(c)

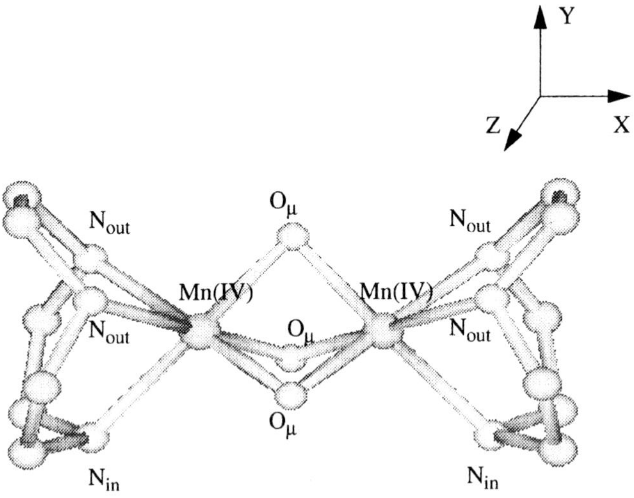

(d)

Figure 12. (*Continued.*)

quantitative deficiencies, the pattern of the calculated results is similar to the experimental trend, ranging from weakly AF coupled for Mn(III)Mn(III) to moderately or strongly AF coupled for Mn(III)Mn(IV)$(O)_2$(OAc) with a relative minimum at Mn(IV)Mn(IV)(μ-O_2)(μ-O$)_2$, and maximum at Mn(IV)Mn(IV)(μ-O$)_3$. The relative minimum at Mn(IV)Mn(IV)(μ-O_2)(μ-O$)_2$ is probably due to the favorable energy of triplet O_2 compared with singlet O_2, which will lead to enhanced μ-O_2^{2-} to Mn charge transfer in the excited $S = 3$ spin state.

The strong predicted metal-ligand covalency, the competition seen between spin-polarization and ligand field splitting, and the ability of the calculations to represent

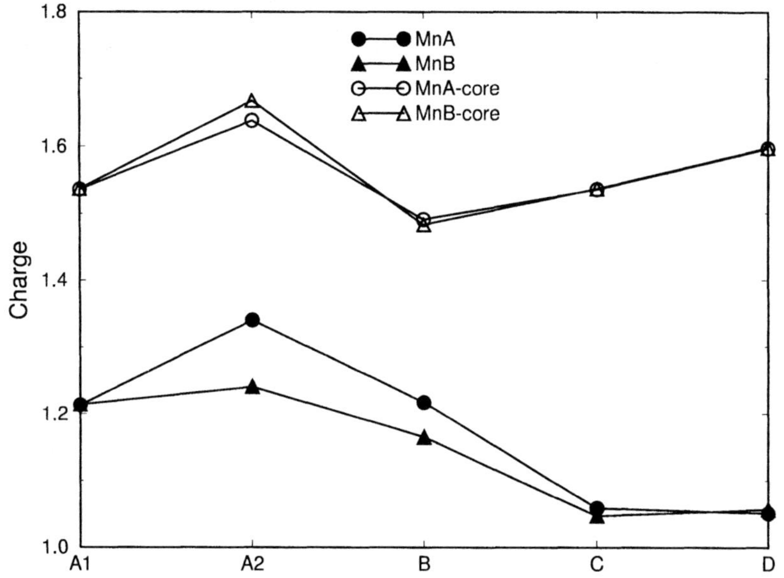

Figure 13. Calculated ESP charges on Mn sites for full and core-only Mn-oxo dimers.

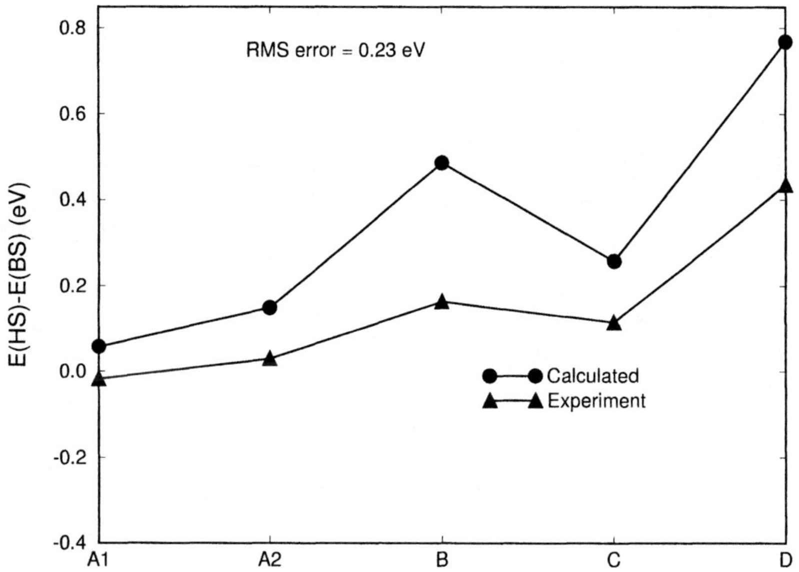

Figure 14. Calculated energy difference, E(HS)–E(BS), *vs* "experimental" data. The latter were evaluated from $E(HS)–E(BS) = -4J'S_1S_2$ with assumed monomer spin states, corresponding to a Heisenberg Hamiltonian convention $H_{spin} = -2J'S_1 \cdot S_2$ commonly used for Mn dimer complexes. The convention throughout the text is $H_{spin} = JS_1 \cdot S_2$, and $J = -2J'$.

trends in spin-coupling energies with reasonable accuracy are all important consequences of this study which show promise for understanding electron and proton removal from oxygen-containing species (including water and peroxide or hydroperoxide anions) by high-oxidation state dinuclear and tetranuclear manganese complexes.

6 ELECTRONIC STRUCTURE IN CONDENSED PHASES

Although the application of quantum chemistry to structural and energetic problems in gas-phase molecules has reached a high level of sophistication and maturity, many interesting problems remain for molecules in more complex condensed-phase environments. Among the simplest such environments is a molecule in dilute solution, but it is also of interest to consider more complex surroundings, such as might be encountered at the active site of a protein. The electrostatic aspects of such surroundings can often be usefully described with continuum dielectric theory; for example, the protein interior might be described with a low local dielectric that contains embedded charges (to represent charged side chains and polar groups within the protein) surrounded by a high dielectric solvent. Even for complex geometries, the resulting electrostatic interactions can be determined by numerical solutions to the underlying differential equations, and such theories have come into widespread use in the past few years.[166,167]

The idea of modeling the electrostatic portion of solvation by placing the solute in a cavity immersed in a dielectric continuum dates back to the Born model for ionic solvation[168] and to work by Kirkwood[169] and Onsager[170] on solvation effects in polar molecules, and the subsequent history of continuum-model calculations has recently been reviewed.[171,172] The simplest models use spherical cavities and consider just the coupling of the molecular dipole to the reaction field arising from polarization of the solvent dielectric.[173] This model involves straightforward computations, adding a term proportional to the dipole-moment operator to the gas-phase Hamiltonian. It provides reasonable results for simple solutes, but has obvious limitations for systems that are not very spherical, and it is known that both the shape of the cavity and higher-order moments of the charge distribution have important effects on the results.[174,175] Analytic expressions based on multipole expansions in ellipsoidal cavities are also feasible,[174,176] but these are still restricted to relatively compact systems, and the multipole expansions involved often converge very slowly.

A more general approach has been developed by Tomasi and coworkers[177] that permits the boundary between the solute and the solvent to take on an arbitrary shape. In this approach the reaction potential is considered to arise from surface charge densities at the boundaries between different dielectric regions. A boundary-element method is used to discretize this charge density, turning the electrostatic integral equations into linear systems that can be solved by iterative or matrix-inversion methods.[177,178] Because the reaction potential arises from discrete charges (at each boundary element), integrals of this potential over the quantum-mechanical basis functions can be accomplished in the same manner as nuclear attraction integrals, and no multipole expansion of the charge distribution is required. This model, when combined with careful quantum calculations, has been shown to give an excellent account of many aspects of aqueous solvation of small organic molecules.[171,179,180]

An alternative approach computes the reaction potential from a finite-difference solution to the Poisson–Boltzmann equation. Although the underlying physical models for the reaction potential are similar in the finite-difference and boundary-element methods, the finite-difference approach has two principal advantages: First, the effects of mobile counterions (salt effects) can easily be incorporated at the Debye–Huckel level; this can be done with some difficulty in boundary-element methods,[181–183] but important elaborations like the inclusion of a counterion exclusion region are not readily incorporated. Second, the incorporation of nonhomogeneous environments, such as a protein-solvent environment consisting of multiple dielectric regions with embedded charges, is straightforward and imposes little additional computational cost in the finite-difference approach but adds significantly to the number of boundary elements and the complexity of the relations between them in the boundary-element method.

6.1 Basic Equations for Continuum Solvation Models

Within the density functional approximation, the electronic energy of a molecule can be written as,[82]

$$E = \sum_i \langle \psi_i | h(1) | \psi_i \rangle + \frac{1}{2} \int \rho_{tot}(\mathbf{r}) G(\mathbf{r}, \mathbf{r}') \rho_{tot}(\mathbf{r}') d\mathbf{r} d\mathbf{r}' + E_{xc}[\rho_{el}(\mathbf{r})] \tag{39}$$

where the one-electron Hamiltonian, $h(1)$, contains the kinetic energy integrals, the ψ_i are the Kohn–Sham orbitals, ρ_{tot} is the total charge density due to both electrons and nuclei, and $E_{xc}[\rho_{el}(\mathbf{r})]$ represents the exchange and correlation energy for the electron density $\rho_{el}(\mathbf{r})$. (For consistency in what follows, we have deviated from the usual formalism by removing nuclear attraction integrals from $h(1)$ and including nuclear charges in ρ_{tot} instead.) $G(\mathbf{r}, \mathbf{r}')$ is the Coulomb interaction operator, which will just be $1/|\mathbf{r}-\mathbf{r}'|$ in the gas-phase, but will have a more complex form in solution, as we discuss below.

Applying the variational principle, $\delta E/\delta\rho_{el}(\mathbf{r}) = 0$ to Eq. (39) yields the Kohn–Sham equations for the one-electron orbitals:

$$\left[h(1) + \int G(\mathbf{r}, \mathbf{r}') \rho_{tot}(\mathbf{r}') d\mathbf{r}' + v_{xc}[\rho_{el}(\mathbf{r})] \right] \psi_i(\mathbf{r}) = \varepsilon_i \psi_\varepsilon(\mathbf{r}) \tag{40}$$

where

$$v_{xc} = \frac{\delta E_{xc}[\rho_{el}(\mathbf{r})]}{\delta \rho_{el}(\mathbf{r})} \tag{41}$$

and

$$\rho_{el}(\mathbf{r}) = \sum_i |\psi_i(\mathbf{r})|^2 \tag{42}$$

The crux of this approach to solvation is to take the effective electron-electron interaction, $G(\mathbf{r}, \mathbf{r}')$ to be the Green function of the Poisson equation,

$$\nabla\varepsilon(\mathbf{r})\nabla G(\mathbf{r}, \mathbf{r}') = -4\pi\delta(\mathbf{r}, \mathbf{r}') \tag{43}$$

where $\varepsilon(\mathbf{r})$ is the (nonuniform) dielectric constant, and $\delta(\mathbf{r}, \mathbf{r}')$ is a Dirac's delta function representing a unit charge at $\mathbf{r}'$. We set $\varepsilon(\mathbf{r})$ to the solvent static dielectric constant for $\mathbf{r}$ in the exterior (solvent) region, although it could be set in a variety of ways to reflect more complex environments, such as a protein in solution. For the molecular interior in which most of the electron density is found, we set $\varepsilon(\mathbf{r}) = 1$. This makes it convenient to divide the Green function into a gas-phase-like Coulomb part and a remainder which we define as the reaction potential contribution, G^*, representing the effect of solvent:

$$G(\mathbf{r}, \mathbf{r}') = \frac{1}{|\mathbf{r} - \mathbf{r}'|} + G^*(\mathbf{r}, \mathbf{r}') \tag{44}$$

As long as $\mathbf{r}$ and $\mathbf{r}'$ are both in the interior (the most relevant case since most of the charge density is in the interior), $G^*(\mathbf{r}, \mathbf{r}')$ can be thought of as the potential due to the surface charge induced on the dielectric interface by a unit charge at $\mathbf{r}'$. In the exterior, $G^*(\mathbf{r}, \mathbf{r}')$ can be thought of as the amount by which the Coulomb potential is reduced due to dielectric screening. With this separation, and using the fact that $\frac{1}{2}\int \rho\phi d\mathbf{r}$ remains a valid expression for the electrostatic work of charging in a linear dielectric medium,[184] Eq. (39) can be rewritten:

$$\begin{aligned} E = & \sum_i \langle\psi_i|h(1)|\psi_i\rangle + \frac{1}{2}\int \frac{\rho_{tot}(\mathbf{r})\rho_{tot}(\mathbf{r}')}{|\mathbf{r} - \mathbf{r}'|} d\mathbf{r}d\mathbf{r}' \\ & + E_{xc}[\rho_{el}(\mathbf{r})] + \frac{1}{2}\int \rho_{tot}(\mathbf{r})G^*(\mathbf{r}, \mathbf{r}')\rho_{tot}(\mathbf{r}')d\mathbf{r}d\mathbf{r}' \end{aligned} \tag{45}$$

where the solvent dependence enters only in the last term which, because of the factor of $1/2$, includes the work of polarizing the solvent in the variational formulation. The other terms correspond to the gas-phase problem and are handled in usual way in the quantum mechanics program while the solvent reaction term is calculated by a finite difference method and an effective point charge approximation described below. The gas-phase and solvent parts of the calculation are brought to self-consistency by an iterative procedure. This division makes good numerical sense because the Coulomb integrals of the gas-phase part can be performed analytically (or semi-analytically, depending upon the code), and much work has gone into ensuring high accuracy for the estimation of these interactions. The reaction potential contribution, on the other hand, is much smaller and is a relatively smooth function of position, and more straightforward numerical integration techniques can be used for it.

There has been a revival of interest in these sorts of methods in quantum organic chemistry over the past few years, and there is not space to review these advances here, and other overviews are available.[185,186] Most approaches that allow an arbitrary cavity shape use boundary element methods to compute the solvent response, rather than the finite-difference approach outlined above. Among the many trends that might be mentioned, we would point out the increasing availability of gradients with respect to nuclear position (and hence, of automatic geometry optimization),[187] and the use of a simplified model called COSMO, which starts from the infinite dielectric (conductor) limit for the solvent, allowing simplified equations and derivatives.[188] Warshel has proposed models based on similar ideas, but using a grid of Langevin dipoles, rather than a continuum dielectric, to represent the solvent response.[189]

Much of the work in the organic chemistry area has concentrated on neutral species, which are certainly of interest but are often easier to treat than the highly charged moieties characteristic of many transition metal complexes. As an illustration of work on organic ions, we present in Figure 15 results of absolute pK_a's for models of ionizable groups in proteins.[190] These were determined using the density functional methods outlined above and the thermodynamic cycle shown in Figure 16.

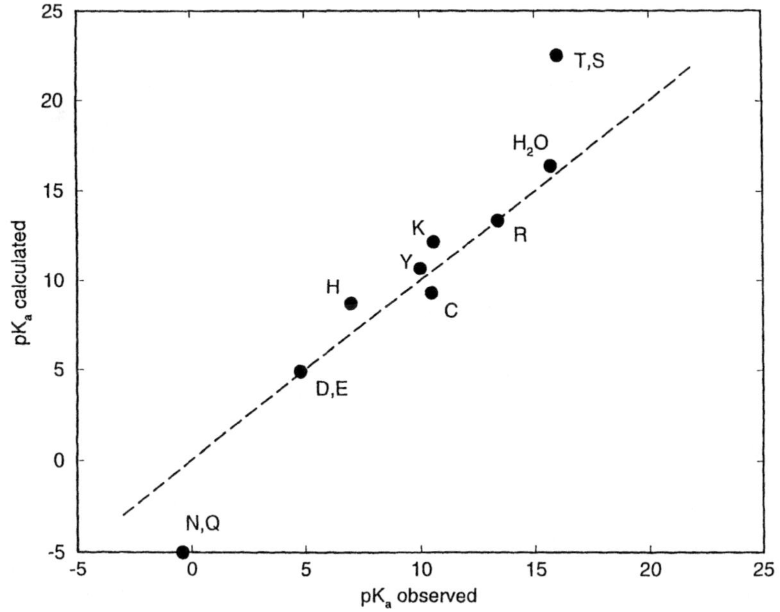

Figure 15. Calculated and experimental pK_a values for model complexes resembling the side chains of amino acids: T,S is methanol; K is methylamine; R is N-methylguanidine; Y is phenol; D,E is acetic acid; Q,N is acetamidium; S is mercaptoethanol; H is imidazole.

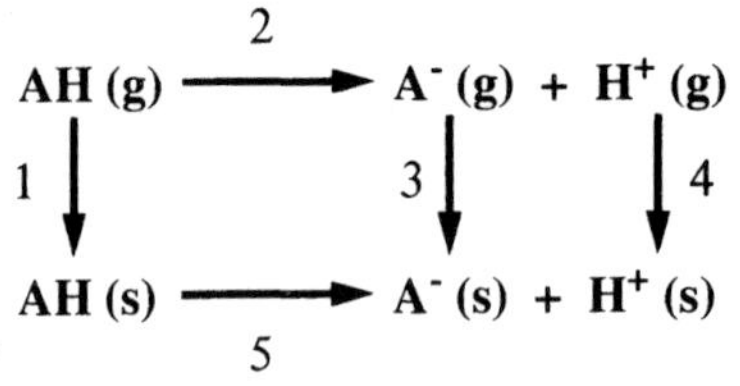

$$1.37\text{p}K_a = \Delta G_5 = -\Delta G_1 + \Delta G_2 + \Delta G_3 + \Delta G_4$$

Figure 16. Thermodynamic cycle for determining absolute pK_a values.

Although very strong and very weak acids are not predicted reliably, all groups titrating in the normal pH range show excellent agreement with observed values. An analysis of the individual contributions to these values[190] suggests that the gas-phase proton affinities are fairly accurate, with most of the error arising from estimates of the solvation free energies for the charged species; nevertheless, the general level of agreement is impressive, and provides a useful foundation for studies of general acid-base reaction mechanisms.

6.2 Redox Potentials for Aqueous Transition-Metal Ions

Of greater interest to the present chapter are similar calculations on oxidation/reduction and deprotonation of hydrated transition-metal cations.[78] Models for Mn^{2+}, Mn^{3+}, Fe^{2+} and Fe^{3+} have been constructed with either one or two shells of water molecules (six or eighteen molecules) treated explicitly in the density functional calculation, and the effects of the remaining water approximated by a continuum solvent. These are extremely challenging calculations: in the gas phase, the third ionization potentials of iron and manganese are about 750 kcal/mol, and hydration enthalpies are around 450 kcal/mol for the divalent cations, and over 1000 kcal/mol for the trivalent species. Table 2 shows results for hydration enthalpies, redox potentials, and the pK_a values for water in the first hydration shell. Although the pK_a values of waters bound to Fe^{3+} or Mn^{3+} are underestimated (i.e. the stability of the deprotonated forms is overestimated) overall the results are in reasonable agreement with experiment, and suggest that there are no major deficiencies in either the density functional or the continuum solvent models being used.

6.3 Redox Potentials and pK_a's of the Manganese Superoxide Dismutase Active Site

As an example of more complex systems, consider superoxide dismutases (SOD's), a diverse group of metalloenzymes crucial to the existence of oxygen-tolerant organisms.[191,192] Among these, MnSOD's are found both in prokaryotes[193] and in the mitochondria of higher organisms,[194] and contain a single metal active site per monomeric subunit of the tetrameric enzyme. In MnSOD, the rapid catalytic dismutation of two superoxide radical anions to molecular oxygen and hydrogen per-

Table 2 Properties of Iron and Manganese in Water

		Calc.	*Exp.*
Hydration enthalpies kcal/mol	Fe^{2+}	431	465
	Fe^{3+}	1046	1060
	Mn^{2+}	409	440
	Mn^{3+}	1073	1087
Redox potentials eV	$Fe^{3+} \rightarrow Fe^{2+}$	1.06	0.77
	$Mn^{3+} \rightarrow Mn^{2+}$	1.59	1.56
pK_a for bound water	Fe^{2+}	9.0	9.5
	Mn^{2+}	14.0	10.6
	Fe^{3+}	−4.0	2.2
	Mn^{3+}	−6.5	0.1

Adapted from Ref. 78, using 18 explicit waters and continuum solvent.

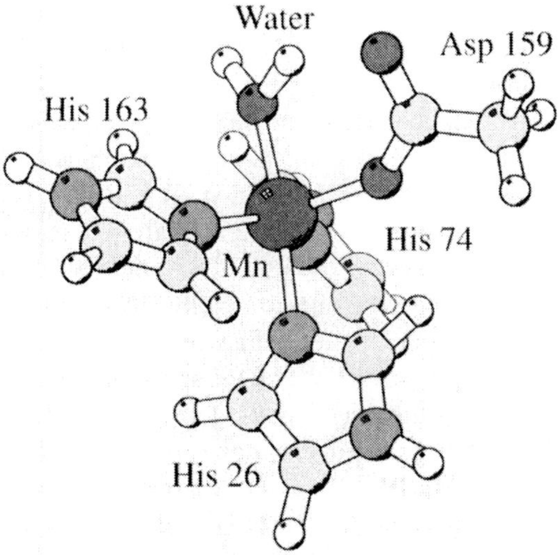

Figure 17. Active-site model of human manganese superoxide dismutase used for the calculations. Histidines 26, 74, and 163 are represented by imidazoles and aspartate 159 by an acetate ion. The water-bound form is depicted. The large central sphere is the Mn ion, and the medium gray spheres are nitrogen, light gray are carbon, and white hydrogen atoms.

oxide proceeds via an alternating reduction of the Mn(III) enzyme and following re-oxidation of the Mn(II) enzyme:[193–195]

$$\mathrm{E - Mn(III) + O_2^- \rightarrow E - Mn(II) + O_2} \tag{46}$$

$$\mathrm{E - Mn(II) + O_2^- + 2H^+ \rightarrow E - Mn(III) + H_2O_2} \tag{47}$$

The active site model of manganese superoxide dismutase is depicted in Figure 17, as used for our active site density functional calculations.[79] There are three histidine ligands (replaced by imidazole in the calculations) and one aspartate (replaced by acetate) plus one water molecule (or hydroxyl). The overall site is five coordinate and the incoming superoxide likely occupies a sixth site.

The enzyme active site has a net positive charge which, along with the greater protein environment, facilitates the encounter with the superoxide anion by electrostatic attraction. Further, the extensive electron relaxation predicted by the density functional calculations allows the second superoxide anion encountered to have an electrostatic attraction to the active site comparable to that of the first superoxide.[79] One critical aspect of the dismutation process is the energetics of reduction and reoxidation of the Mn site. It is very likely that the resting enzyme E-Mn(III) has a bound hydroxyl in place of water at any reasonable pH.[79,193,195] For example, the pK_a for simple aqueous Mn^{3+} is 0.1 experimentally, and -6.5 from our Poisson–Boltzmann self-consistent-reaction-field (SCRF) calculations.[78] Upon reduction, the hydroxyl is very likely protonated, again by analogy to aqueous Mn^{2+} ions, with p$K_a = 10.6$ (experiment), and 14.0 (SCRF MnSOD calculation). In the re-oxidation of E-Mn(II), the peroxide anion is probably protonated to HO_2^- from the bound water, regenerating hydroxyl.

We have calculated redox potentials and pK_a values for the model active sites by solvating the system using a continuum dielectric with an aqueous dielectric constant ($\varepsilon = 80$).[78] When the coupled protonation is taken into account in the calculated redox potential, the calculated value is too negative -1.09 V (versus a standard hydrogen electrode) compared with 0.26, 0.31 V from experiment for related bacterial enzymes. (However, the experiments are difficult and there are serious problems of reversibility and variability depending on experimental conditions.[196–198]) Overall, the very negative pK_a calculated for the active site (AS) model AS-Mn(III)-($H_2O \rightarrow OH^-$) is a major contributor (calculated p$K_a = -13.4$) to the very negative calculated redox potential since when the redox potential is calculated with water as the ligand throughout AS-Mn(III$\rightarrow$II)-(H_2O), we find $+0.17$ V; accounting for the expected pK_a of about 0 for AS-Mn(III)-($H_2O \rightarrow OH^-$), we would predict a redox potential at pH $= 7$ with coupled protonation of $0.17 - 0.42 = -0.25$ V (where -0.42 eV is the free energy of 7 pH units for the oxidized AS-Mn(III)-($H_2O \rightarrow OH^-$ state); this is much better than -1.09 V, but a considerable error still remains. Further examination of the Mn(SOD) active site reveals an extensive network of protein residues hydrogen-bonded to the charged active site. Charged hydrogen bonds are known to display large stabilization energies; a nearby glutamate, a glutamine which can potentially H-bond to water and/or hydroxyl depending on side-chain conformation (glutamine flip), and a nearby tyrosine need to be considered along with other H-bonding residues. Extended models need to be constructed including the entire protein framework and solvent electrostatically, and with the closest charged H-bonds treated quantum mechanically. Geometric changes at the active site on reduction/oxidation and in the protein may also prove important. This strategy is analogous to including the second water layer explicitly in the hydrated metal ions calculations.

6.4 Absolute Redox Potentials of Iron-Sulfur Clusters

Since many iron-sulfur proteins are electron transfer agents, it is important to understand the underlying basis for the general range of redox potentials observed in these systems. As a first step toward this goal, we have calculated the redox potentials of simpler active site model clusters in organic and aqueous solvents,[81] using the continuum dielectric representation, and comparing these with the redox potentials measured for related synthetic iron-sulfur clusters. Synthetic clusters display trends in redox potentials which are similar to those in analogous proteins with respect to both cluster size and redox couple. Overall, the proteins show a positive redox shift with respect to the synthetic systems, and a narrower redox range over cluster types and redox couples than the synthetic systems.

Redox potentials were calculated for iron-sulfur clusters with 1Fe, 2Fe and 4Fe sites,[81] and with SCH_3^- in place of organic thiols for the synthetic clusters, and cysteine ligands for the proteins. The standard redox potential (E^0) can be calculated as the sum of the gas phase ionization potential for the reduced species (IP(red)) and the solvation energy difference for the oxidized minus the reduced states (ΔE_{PB}) adding a known constant (−4.5 V) to reference the result to a standard hydrogen electrode.

Density functional calculations were performed for the model 1Fe, 2Fe, and 4Fe complexes, $Fe(SR)_4^{1-,2-}$, $Fe_2S_2(SR)_4^{2-,3-}$ and $Fe_4S_4(SR)_4^{1-,2-,3-}$ (two different redox couples), using gradient-corrected density functional theory. For the dinuclear and tetranuclear systems, the broken symmetry and high spin state energies were obtained, and the spin-coupled ground state energies determined. This includes the Heisenberg exchange coupling energy difference between the broken symmetry state and the corresponding pure spin ground state. In this process, the spin Hamiltonian parameters for Heisenberg coupling (J), resonance delocalization within a mixed-valence pair (B), and resonance delocalization between pairs of a 4Fe system (B') were calculated. The solvation energies were computed from the broken symmetry state ESP charges for the continuum solvent model. There is a good correlation between predicted and measured redox potentials, as shown in Figure 18. The more negative redox potentials occur for the systems with the most negative charges for the redox couple (3−/2− versus 2−/1−) and for the smaller systems (1Fe, 2−/1− versus 4Fe, 2−/1−) as would be expected from considerations of electron-electron repulsion alone. However, the solvation energy contribution to the redox potential is large in all clusters and for all redox couples, and compensates for a large part of the differences in electron-electron repulsion energies in vacuum. Further analysis shows that the Heisenberg spin coupling is responsible for a large part of the negative potential shift for the 2Fe redox couple compared to the 1Fe. Both Heisenberg spin coupling and resonance delocalization contribute to the much higher redox potential for the 4Fe high potential ($HP_{ox,red}$) compared to the reduced ferredoxin couple ($Fd_{ox,red}$) in synthetic analogues, and probably in proteins as well. The higher redox potential for $HP_{ox,red}$ vs. $Fd_{ox,red}$ is consistent with the number of delocalized mixed-valence Fe pairs (1, 2, 1) for the 1−/2−/3− oxidation states; the maximum energy stabilization then occurs for the 2− state, which is the oxidized state of the 3−/2− couple, but the reduced state of the 2−/1− couple.

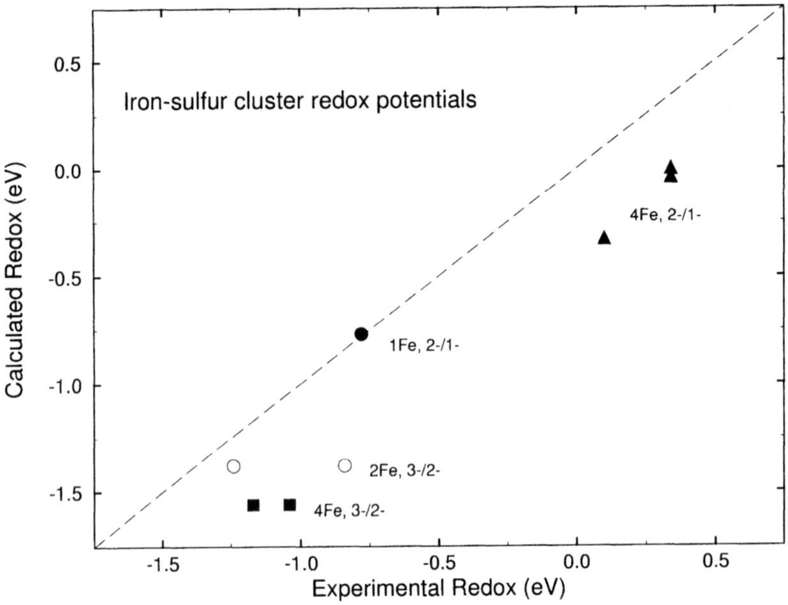

Figure 18. Calculated vs. experimental redox potentials for 1Fe, 2Fe, and 4Fe iron-sulfur clusters.

Experimentally, the redox potentials of these synthetic clusters spans a total range of about 1.6 V from about −1.24 to +0.34 V. The reduced 2Fe, 3−/2− and 4Fe, 3−/2− ferredoxin redox couples have the most negative redox potentials, covering overlapping ranges, followed by the 1Fe, 1−/2− and finally by the 4Fe 2−/1− high potential couple at most positive potential. The theoretical calculations (with $\varepsilon = 37$ as in DMF) give a similar total span of about 1.67 V, with similar ordering. The maximum error in the calculations is about −0.5 V compared to experiment, or generally less than 10% of the solvation energy difference term for that redox couple. The total span of redox potentials in related proteins is smaller, about 0.85 V, and positively shifted by about +0.5 V with respect to synthetic clusters, but with very similar redox potential ordering for different clusters and oxidation states.

The extension of these ideas to protein environments appears straightforward, but there are a number of subtleties, such as where the quantum region can be terminated, and the simpler classical electrostatic representation of protein/solvent begun, or where a discrete molecular representation of solvent is preferable to a continuum representation, or how protein conformational change should be treated. The relative redox potentials of related iron-sulfur proteins have been studied by a number of groups, often with charge sets derived from our work.[199–202] We are currently working on extensions of these ideas to absolute redox potential calculations for proteins.

7 CONCLUSIONS

It is worth reiterating that no single review can cover more than a fraction of the topics of potential interest in inorganic structure and spectroscopy. We have emphasized here applications of density functional theory to discrete entities involving mostly first-row transition metal ions, and where the metal ions for the most part are in high-spin configurations. We have tried to emphasize both qualitative ideas that arise from the calculations (in terms of orbital energy schemes, the importance of spin polarization and electronic relaxation, and the nature of spin Hamiltonians involved in multi-metal systems) and the expected reliability of quantitative results (for geometries, bond energies, spin coupling parameters, proton affinities and redox potentials). We have addressed the near-degeneracy correlation problem particularly as it is exhibited in dinuclear and polynuclear transition metal complexes. We have shown the value of broken symmetry and spin projection methods for spin-coupled systems, including both Heisenberg spin-coupling terms and spin dependent electron delocalization (also called double exchange or resonance delocalization). Further applications of the broken symmetry method to spin coupled transition metal dimer and tetramer complexes including Fe, Cu, Mn, Mo, and Cr dimers is summarized in a recent review paper, including comparisons with alternative methodologies.[160] The prospects for doing condensed phase quantum chemistry, and hence making more direct comparisons to many experiments, appear quite promising, but applications of these ideas to transition-metal systems are still in their infancy. The continued development of computers and quantum chemistry algorithms will provide access to increasingly realistic models of inorganic systems. The ideas outlined here should help to provide a foundation for the design and interpretation of such calculations.

ACKNOWLEDGMENTS

This work was supported by NIH grants GM39914, GM43278, and GM50154. We thank Bernard Lamotte and Don Bashford for many helpful discussions. We thank our coworkers Jean-Marie Mouesca, Jun L. Chen, Xin-Gui Zhao, Cindy Fisher, William H. Richardson, and Chun-Yang Peng for their contributions to the work reviewed here.

APPENDIX: COMPUTATIONAL DETAILS

The calculations reported in this chapter, if not referred elsewhere, were carried out using Amsterdam Density Functional package ADF2.01.[203] These used the local spin density approximation (LSD) for exchange and correlation, and the Becke88 and Perdew86 gradient terms for exchange and correlation. A set of uncontracted triple-zeta Slater-type orbitals (STO) was employed for the $4s$, $4p$, and $3d$ valence orbitals of the transition metal atoms. For the $2s$ and $2p$ orbitals of oxygen and carbon, the $3s$ and $3p$ orbitals of sulfur and chlorine, and the $1s$ orbital of hydrogen, the same quality basis sets was used. The inner core shells were treated by the frozen core approximation, through $(3s, 3p)$ of transition metals, $(2s, 2p)$ of sulfur and chlorine, $(1s)$ of carbon and oxygen. The basis, core and fit sets used correspond to Basis Set IV of ADF201. All calculations were done with a spin-unrestricted scheme.

Atomic Calculations: The second and third ionization potentials (IP_{II} and IP_{III}) of first-row transition metals are defined as $IP_{II} = E_{GS}(M^{2+}) - E_{GS}(M^{+})$ and $IP_{III} = E_{GS}(M^{3+}) - E_{GS}(M^{2+})$. The experimental ground states chosen for the cations are as follows:

	Sc	Ti	V	Cr	Mn	Fe	Co	Ni	Cu	Zn
M^+	s^1d^1	s^1d^2	d^4	d^5	s^1d^5	s^1d^6	d^8	d^9	d^{10}	s^1d^{10}
	^3D	^4F	^5D	^6S	^6S	^6D	^3F	^2D	^1S	^2S
M^{2+}	d^1	d^2	d^3	d^4	d^5	s^1d^5	d^7	d^8	d^9	d^{10}
	^3D	^3F	^4F	^5D	^6S	^7S	^4F	^3F	^2D	^1S
M^{3+}	d^0	d^1	d^2	d^3	d^4	d^5	d^6	d^7	d^8	d^9
	^1S	^2D	^3F	^4F	^5D	^6S	^5D	^4F	^3F	^2D

Since each of these states is the highest spin state of corresponding electronic configuration and meanwhile M_L of such state adopts the maximum possible value, it can be represented by a single Slater determinant. Therefore, the state energy can be evaluated using a non-spherical density with integer occupation numbers. For example, the ^{3}F state of $Ti^{2+}(3d^2)$ and the ^{4}F state of $V^{2+}(3d^3)$ are described by the state functions $\Phi(2^+1^+) = |3d_2^+3d_1^+|$ and $\Phi(2^+1^+0^+) = |3d_2^+3d_1^+3d_0^+|$, respectively. Because only real atomic orbitals instead of complex orbitals were used in this calculation, a D_{2h} symmetry constraint was applied to specify the occupation of distinct $3d$ orbitals belonged to different irreducible representations. Therefore, for ^{3}F state of $Ti^{2+}(3d^2)$, the state we calculated is $|3d_{xy}^+3d_{xz}^+|$ by assigning one electron in b_{1g} and another in b_{2g}. Since $3d_{xy}$ is a linear combination of $3d_2$ and $3d_{-2}$ and $3d_{xz}$ is a linear combination of $3d_1$ and $3d_{-1}$, the energy of $|3d_{xy}^+3d_{xz}^+|$ is equal to that of $|3d_2^+3d_1^+|$, the energy of ^{3}F state of $Ti^{2+}(3d^2)$. The other states can be calculated in the same way and Slater-sum rule or other multiplet method are not necessary in these special cases. If the $4s$ orbital has to be taken into account, O_h symmetry can be used to prevent mixture of $4s$ and $3dz^2$ orbitals. It is always important to check the occupations after the self-consistent field iterations converge and make sure that the electrons occupy correct orbitals.

The calculated ionization energies are in good agreement with experimental data with errors of about 0.5–1 eV. Further improvement may be achieved by including relativistic effects and unfreezing the outermost $(3s, 3p)$ core shell of metals. Inclusion of $(3s, 3p)$ shell into valence space reduces the errors of IP_{III} by about 0.2 eV. We have also carried out a quasi-relativistic density functional calculations to evaluate the influence of relativistic effects on IP_{III}. The relativistic correction amounts to about 0.2 eV and improves the overall agreement with experiment. The quasi-relativistic density functional method takes into account of the mass-velocity term and Darwin term in the first-order Hamiltonian, along with the density change induced by these terms in a self-consistent way.[204] In this scheme, the frozen core orbitals are replaced by the relativistic ones which are obtained by numerical solution of the atomic Dirac equation.

Molecular Calculations: For hexa-aquo transition metal complexes $[M(H_2O)_6]^{m+}$, geometry optimizations of the complex structures were done using analytic energy gradient methods.[205,206] The optimization used the Newton–Raphson method and the Hessian was updated with Broyden–Fletcher–Goldfarb–Shanno strategy. Convergence is achieved when changes in coordinate values are less than 0.005 Å and the norm of all gradient vectors is smaller than 0.01. D_{2h} symmetry was used for all complexes which allows Jahn–Teller distortion along all axial directions. The optimized M-OH_2 distances in general compare well with available X-ray crystallography data with an average error of 0.046 Å for 17 complexes. The energies of the optimized structures were used to calculate the binding energy of the complex: $BE = E([M(H_2O)_6]^{m+}) - 6E(H_2O) - E(M^{m+})$. $E(H_2O)$ was calculated from the optimized structure of H_2O.

REFERENCES

1. Ballhausen, C.J., *Molecular Electronic Structures of Transition Metal Complexes*, McGraw-Hill, Chatham, U.K. (1979).
2. Lever, A.B.P., *Inorganic Electronic Spectroscopy, 2nd Ed.*, Elsevier, Amsterdam, Netherlands (1984).
3. Abragam, A.; Bleaney, B., *Electron Paramagnetic Resonance of Transition Ions*, Dover, New York (1986).
4. Pilbrow, J.R., *Transition Ion Electron Paramagnetic Resonance*, Clarendon Press, Oxford (1990).
5. Kahn, O., *Molecular Magnetism*, VCH Publishers, New York (1993).
6. Cory, M.G.; Zerner, M.C. *Chem. Rev.* **1991**, *91*, 813–822.
7. Ziegler, T. *Chem. Rev.* **1991**, *91*, 651–667.
8. Siegbahn, P.E.M. *Adv. Chem. Phys.* **1996**, *93*, 333–387.
9. Noodleman, L.; Case, D.A. *Adv. Inorg. Chem.* **1992**, *38*, 423–470.
10. Parr, R.G.; Yang, W., *Density-Functional Theory of Atoms and Molecules*, Oxford University Press, New York (1989).
11. Kohn, W.; Becke, A.D.; Parr, R.G. *J. Phys. Chem.* **1996**, *100*, 12974–12980.
12. Seminario, J.M.; Politzer, P., eds., *Modern Density Functional Theory: A Tool for Chemistry*, Elsevier, Amsterdam (1995).
13. Seminario, J.M., ed., *Recent Developments and Applications of Modern Density Functional Theory*, Elsevier, Amsterdam (1996).
14. Laird, B.B.; Ross, R.B.; Ziegler, T., eds., *Chemical Applications of Density Functional Theory*, American Chemical Society, Washington, DC (1996).
15. McWeeney, R.; Sutcliffe, B.T. *Methods of Molecular Quantum Mechanics*, Academic Press, San Diego, CA (1976).
16. Ziegler, T.; Rauk, A.; Baerends, E.J. *Theor. Chim. Acta* **1977**, *43*, 261.
17. Slater, J.C., in *The Self-Consistent Field for Molecules and Solids: Quantum Theory of Molecules and Solids*, McGraw-Hill, New York (1974).

18. Ernzerof, M.; Burke, K.; Perdew, J.P., in *Recent Developments and Applications of Modern Density Functional Theory*, ed. J.M. Seminario, Elsevier, Amsterdam (1996).
19. Stoll, H.; Pavlidou, C.M.E.; Preuss, H. *Theoret. Chim. Acta* **1978**, *149*, 143–149.
20. Case, D.A. *Annu. Rev. Phys. Chem.* **1982**, *33*, 151–171.
21. Baerends, E.J.; Gritsenko, O.V.; van Leeuwen, R., in *Chemical Applications of Density Functional Theory*, ed. B.B. Laird; R.B. Ross; T. Ziegler, American Chemical Society, Washington, DC (1996), pp. 20–41.
22. Perdew, J.P.; Burke, K.; Ernzerhof, M. *Phys. Rev. Lett.* **1996**, *77*, 3865–3868.
23. Becke, A.D. *J. Chem. Phys.* **1986**, *84*, 4524–4529.
24. Baker, J.; Muir, M.; Andzelm, J. *J. Chem. Phys.* **1995**, *102*, 2063–2079.
25. Scheiner, A.C.; Baker, J.; Andzelm, J.W. *J. Computat. Chem.* **1997**, *18*, 775–795.
26. Sosa, C.; Andzelm, J.; Elkin, B.C.; Wimmer, E.; Dobbs, K.D.; Dixon, D.A. *J. Phys. Chem.* **1992**, *96*, 6630–6636.
27. Ziegler, T.; Li, J. *Can. J. Chem.* **1994**, *72*, 783–789.
28. Li, J.; Schreckenbach, G.; Ziegler, T. *J. Am. Chem. Soc.* **1995**, *117*, 486–494.
29. Becke, A.D. *J. Chem. Phys.* **1993**, *98*, 1372–1377.
30. Burke, K.; Perdew, J.P.; Levy, M. *Phys Rev A* **1996**, *53*, R2915.
31. Burke, K.; Perdew, J.P. *Int. J. Quantum Chem.* **1995**, *56*, 199.
32. Curtiss, L.A.; Raghavachari, K.; Redfern, P.C.; Pople, J.A. *J. Chem. Phys.* **1997**, *106*, 1063–1079.
33. Becke, A.D. *J. Chem. Phys.* **1993**, *98*, 5648–5652.
34. Bauschlicher, C.W., Jr. *Chem. Phys. Lett.* **1995**, *246*, 40–44.
35. Ziegler, T. *Can. J. Chem.* **1995**, *73*, 743–761.
36. Holthausen, M.C.; Heinemann, C.; Conrehl, H.H.; Koch, W.; Schwarz, H. *J. Chem. Phys.* **1995**, *102*, 4931–4941.
37. Eriksson, L.A.; Pettersson, L.G.M.; Siegbahn, P.E.M.; Wahlgren, U. *J. Chem. Phys.* **1995**, *102*, 872–878.
38. Russo, R.V.; Martin, R.L.; Hay, P.J. *J. Chem. Phys.* **1995**, *102*, 8023–8028.
39. Andrews, L.; Chertihin, G.V.; Ricca, A.; Bauschlicher, C.W., Jr. *J. Am. Chem. Soc.* **1996**, *118*, 467–470.
40. Glukhovtsev, M.N.; Bach, R.D.; Nagel, C.J. *J. Phys. Chem. A* **1997**, *101*, 316–323.
41. Ziegler, T.; Rauk, A. *Theor. Chim. Acta* **1977**, *46*, 1–10.
42. Bickelhaupt, F.M.; Nibbering, N.M.M.; van Wezenbeek, E.M.; Baerends, E.J. *J. Phys. Chem.* **1992**, *96*, 4864–4873.
43. Bickelhaupt, F.M.; Baerends, E.J.; Ravenek, W. *Inorg. Chem.* **1990**, *29*, 350–354.
44. Lowdin, P.O. *J. Mol. Spectr.* **1959**, *3*, 46–66.
45. Colpa, J.P.; Brown, R.E. *Mol. Phys.* **1973**, *26*, 1453–1463.
46. Colpa, J.P.; Islip, M.F.P. *Theoret. Chim. Acta* **1972**, *27*, 25–32.
47. Katriel, J. *Theoret. Chim. Acta* **1972**, *23*, 309–315.
48. Messmer, R.P.; Birss, F.W. *J. Phys. Chem.* **1969**, *73*, 2085.
49. Boyd, R.J. *Nature* **1984**, *310*, 480.
50. Russo, R.V.; Martin, R.L.; Hay, P.J. *J. Chem. Phys.* **1994**, *101*, 7729–7737.

51. Huheey, J.E. *Inorganic Chemistry, 2nd. Ed.*, Harper and Row International Edition, New York (1979). pp. 839–851.
52. Rosa, A.; Baerends, E.J. *New J. Chem.* **1991**, *15*, 815–829.
53. DeKock, R.L.; Baerends, E.J.; Hengelmolen, R. *Organomet.* **1984**, *3*, 289–292.
54. Norman, J.G., Jr.; Jackels, S.C. *J. Am. Chem. Soc.* **1975**, *97*, 3833–3835.
55. Bair, B.A.; Goddard, W.A. *J. Am. Chem. Soc.* **1978**, *100*, 5669.
56. Noodleman, L.; Baerends, E.J. *J. Am. Chem. Soc.* **1984**, *106*, 2316–2327.
57. Noodleman, L.; Norman, J.G., Jr.; Osborne, J.H.; Aizman, A.; Case, D.A. *J. Am. Chem. Soc.* **1985**, *107*, 3418–3426.
58. Butcher, K.D.; Gebhard, M.S.; Solomon, E.I. *Inorg. Chem.* **1990**, *29*, 2067–2074.
59. Butcher, K.D.; Didziulis, S.V.; Briat, B.; Solomon, E.I. *Inorg. Chem.* **1990**, *29*, 1626–1637.
60. Butcher, K.D.; Didziulis, S.V.; Briat, B.; Solomon, E.I. *J. Am. Chem. Soc.* **1990**, *112*, 2231–2242.
61. Shadle, S.E.; Hedman, B.; Hodgson, K.O.; Solomon, E.I. *J. Am. Chem. Soc.* **1995**, *117*, 2259–2272.
62. Gebhard, M.S.; Koch, S.A.; Millar, M.; Devlin, F.J.; Stephens, P.J.; Solomon, E.I. *J. Am. Chem. Soc.* **1991**, *113*, 1640–1649.
63. Gebhard, M.S.; Deaton, J.C.; Koch, S.A.; Millar, M.; Solomon, E.I. *J. Am. Chem. Soc.* **1990**, *112*, 2217.
64. Noodleman, L.; Peng, C.Y.; Case, D.A.; Mouesca, J.-M. *Coord. Chem. Rev.* **1995**, *144*, 199–244.
65. Schmitt, E.A.; Noodleman, L.; Baerends, E.J.; Hendrickson, D.N. *J. Am. Chem. Soc.* **1992**, *114*, 6109–6119.
66. Zhao, X.G.; Richardson, W.H.; Chen, J.-L.; Li, J.; Noodleman, L.; Tsai, H.-L.; Hendrickson, D.N. *Inorg. Chem.* **1997**, *36*, 1198–1217.
67. Åkesson, R.; Pettersson, L.G.M.; Sandström, M.; Wahlgren, U. *J. Am. Chem. Soc.* **1994**, *116*, 8691–8704.
68. Ziegler, T.; Rauk, A.; Baerends, E.J. *Chem. Phys.* **1976**, *16*, 209–217.
69. Szabo, A.; Ostlund, N.S., *Modern Quantum Chemistry*, Dover, Mineola, N.Y. (1996).
70. Wittel, K.; McGlynn, S.P. *Chem. Rev.* **1977**, *77*, 745–771.
71. Ziegler, T.; Gutsev, G.L. *J. Comput. Chem.* **1992**, *13*, 70–75.
72. Banna, M.S.; Frost, D.C.; McDowell, C.A.; Noodleman, L.; Wallbank, B. *Chem. Phys. Lett.* **1977**, *49*, 213–217.
73. Wood, J.H. *Chem. Phys. Lett.* **1977**, *51*, 582.
74. Noodleman, L.; Post, D.; Baerends, E.J. *Chem. Phys.* **1982**, *64*, 159–166.
75. Jonkers, G.; Lange, C.A. de; Noodleman, L.; Baerends, E.J. *Mol. Phys.* **1982**, *46*, 609–620.
76. Post, D.; Baerends, E.J. *Chem. Phys. Lett.* **1982**, *86*, 176–180.
77. Slater, J.C. *Adv. Quantum Chem.* **1972**, *6*, 1.
78. Li, J.; Fisher, C.L.; Chen, J.L.; Bashford, D.; Noodleman, L. *Inorg. Chem.* **1996**, *35*, 4694–4702.
79. Fisher, C.L.; Chen, J.-L.; Li, J.; Bashford, D.; Noodleman, L. *J. Phys. Chem.* **1996**, *100*, 13498–13505.

80. Breneman, C.M.; Wiberg, K.B. *J. Comp. Chem.* **1990**, *11*, 361–373.
81. Mouesca, J.-M.; Chen, J.L.; Noodleman, L.; Bashford, D.; Case, D.A. *J. Am. Chem. Soc.* **1994**, *116*, 11898–11914.
82. Chen, J.L.; Noodleman, L.; Case, D.A.; Bashford, D. *J. Phys. Chem.* **1994**, *98*, 11059–11068.
83. Marcus, R.A. *Angew. Chem. Int. Ed. Engl.* **1993**, *32*, 1111–1121.
84. Suppan, P., *Chemistry and Light*, The Royal Society of Chemistry, Cambridge, UK (1994).
85. Hupp, J.T.; Dong, Y. *J. Am. Chem. Soc.* **1993**, *115*, 6428–6429.
86. Blondin, G.; Girerd, J.J. *Chem. Rev.* **1990**, *90*, 1359–1376.
87. Carter, C.W., in *Iron-Sulfur Proteins*, ed. W. Lovenberg, Academic, New York (1977). pp. 157–204.
88. Adman, E.; Watenpaugh, K.D.; Jensen, L.H. *Proc. Natl. Acad. Sci. U.S.A.* **1975**, *72*, 4854–4858.
89. Backes, G.; Mino, Y.; Loehr, T.M.; Meyer, T.E.; Cusanovich, M.A.; Sweeny, W.V.; Adman, E.T.; Sanders-Loehr, J. *J. Am. Chem. Soc.* **1991**, *113*, 2055–2064.
90. Butt, J.N.; Sucheta, A.; Martin, L.L.; Shen, B.; Burgess, B.K.; Armstrong, F.A. *J. Am. Chem. Soc.* **1993**, *115*, 12587–12588.
91. Johnson, D.A., *Some Thermodynamic Aspects of Inorganic Chemistry, 2nd Ed.*, Cambridge University Press, Cambridge, UK (1982).
92. Shriver, D.F.; Atkins, P.W.; Langford, C.H., *Inorganic Chemistry*, W.H. Freeman, New York (1990).
93. Lowdin, P.O. *Phys. Rev.* **1955**, *97*, 1509.
94. Lowdin, P.O. *Rev. Mod. Phys.* **1962**, *34*, 80.
95. Fukatome, H. *Prog. Theor. Phys.* **1972**, *47*, 1156.
96. Hay, P.J. *J. Am. Chem. Soc.* **1978**, *100*, 2897.
97. Noodleman, L.; Norman, J.G., Jr. *J. Chem. Phys.* **1979**, *70*, 4903–4906.
98. Benard, M. *J. Chem. Phys.* **1979**, *71*, 2546.
99. Noodleman, L. *J. Chem. Phys.* **1981**, *74*, 5737–5743.
100. Solomon, E.I.; Tuczek, F.; Root, D.E.; Brown, C.A. *Chem. Rev.* **1994**, *94*, 827–856.
101. Bencini, A.; Gatteschi, D. *J. Am. Chem. Soc.* **1986**, *108*, 5763–5771.
102. Albonico, C.; Bencini, A. *Inorg. Chem.* **1988**, *27*, 1934.
103. Dunlap, B.I. *Phys. Rev. A* **1984**, *29*, 2902–2905.
104. Grimme, S. *Chem. Phys. Lett.* **1996**, *259*, 128–137.
105. Baker, J.; Scheiner, A.; Andzelm, J. *Chem. Phys. Lett.* **1993**, *216*, 380–388.
106. Cramer, C.J.; Dulles, F.J.; Giesen, D.J.; Amlöf, J. *Chem. Phys. Lett.* **1995**, *245*, 165–170.
107. Lim, M.H.; Worthington, S.E.; Dulles, F.J.; Cramer, C.J., in *Chemical Applications of Density Functional Theory*, ed. B.B. Laird; R.B. Ross; T. Ziegler, American Chemical Society, Washington, DC (1996). pp. 402–422.
108. Dunlap, B.I. *Adv. Chem. Phys.* **1987**, *87*, 287.
109. Goursot, A.; Malrieu, J.P.; Salahub, D.R. *Theor. Chim. Acta* **1995**, *91*, 225.
110. Wang, S.G.; Schwarz, W.H.E. *J. Chem. Phys.* **1996**, *105*, 4641–4648.
111. Goldstein, E.; Beno, B.; Houk, K.N. *J. Am. Chem. Soc.* **1996**, *118*, 6036–6043.

112. Hay, P.J.; Thibeault, J.C.; Hoffman, R. *J. Am. Chem. Soc.* **1975**, *97*, 4884.
113. Loth, P. de; Cassoux, P.; Daudey, J.P.; Malrieu, J.P. *J. Am. Chem. Soc.* **1981**, *103*, 4007.
114. Loth, P. de; Karafiloglou, P.; Daudey, J.P.; Kahn, O. *J. Am. Chem. Soc.* **1988**, *110*, 5676.
115. Astheimer, H.; Haase, W. *J. Chem. Phys.* **1986**, *85*, 1427.
116. Miralles, J.; Castell, O.; Caballoll, R.; Malrieu, J.-P. *Chem. Phys.* **1993**, *172*, 33.
117. Castell, O.; Miralles, J.; Caballoll, R. *Chem. Phys.* **1994**, *179*, 377.
118. Fink, K.; Fink, R.; Staemmler, V. *Inorg. Chem.* **1994**, *33*, 6219.
119. Bencini, A.; Midollini, S. *Coor. Chem. Rev.* **1992**, *120*, 87–136.
120. Ross, P.K.; Solomon, E.I. *J. Am. Chem. Soc.* **1991**, *113*, 3246–3259.
121. Ruiz, E.; Alemany, P.; Alvarez, S.; Cano, J. *J. Am. Chem. Soc.* **1997**, *119*, 1297–1303.
122. Noodleman, L.; Davidson, E.R. *Chem. Phys.* **1986**, *109*, 131–143.
123. Kollmar, C.; Kahn, O. *J. Chem. Phys.* **1993**, *98*, 453–472.
124. Rose, M.E., in *Elementary Theory of Angular Momentum*, Wiley, New York (1957), p. 46.
125. Anderson, P.W., in *Magnetism*, ed. G.T. Rado; H. Suhl, Academic, New York (1963). pp. 25–83.
126. Anderson, P.W.; Hasegawa, H. *Phys. Rev.* **1955**, *100*, 675–681.
127. Gamelin, D.R.; Bominaar, E.L.; Kirk, M.L.; Wieghardt, K.; Solomon, E.I. *J. Am. Chem. Soc.* **1996**, *118*, 8085–8097.
128. Crouse, B.R.; Meyer, J.; Johnson, M.K. *J. Am. Chem. Soc.* **1995**, *117*, 9612–9613.
129. Achim, C.; Golinelli, M.-P.; Bominaar, E.L.; Meyer, J.; Münck, E. *J. Am. Chem. Soc.* **1996**, *118*, 8168–8169.
130. Beissel, T.; Brikenbach, F.; Hill, E.; Glaser, T.; Kesting, F.; Krebs, C.; Weyhermüller, T.; Wieghardt, K.; Butzlaff, C.; Trautwein, A.X. *J. Am. Chem. Soc.* **1996**, *118*, 12376–12390.
131. Noodleman, L.; Case, D.A.; Aizman, A.J. *J. Am. Chem. Soc.* **1988**, *110*, 1001–1005.
132. Griffith, J.S. *Struct. Bonding* **1972**, *10*, 87.
133. Papaefthymiou, V.; Girerd, J.J.; Moura, I.; Moura, J.J.G.; Münck, E. *J. Am. Chem. Soc.* **1987**, *109*, 4703–4710.
134. Münck, E.; Papaefthymiou, V.; Surerus, K.K.; Girerd, J.J., in *Metal Clusters in Proteins, ACS Symposium Series 372*, ed. L. Que, Jr., American Chemical Society, Washington, DC (1988). pp. 302–325.
135. Borrás-Almenar, J.J.; Clemente, J.M.; Coronado, E.; Palli, A.V.; Tsukerblat, B.S.; Georges, R. *J. Chem. Phys.* **1996**, *105*, 6892–6909.
136. Sontum, S.F.; Noodleman, L.; Case, D.A., in *The Challenge of d and f Electrons: Theory and Computation*, eds. D.R. Salahub; M.C. Zerner, American Chemical Society, Washington, DC (1989). pp. 366–377.
137. Ibers, J.A.; Holm, R.H. *Science* **1980**, *209*, 223–235.
138. Beinert, H. *FASEB J.* **1990**, *4*, 2483–2491.
139. Beinert, H.; Albracht, S.P.J. *Biochim. Biophys. Acta* **1982**, *683*, 245–277.
140. Golbeck, J.H. *Curr. Opin. Struct. Biol.* **1993**, *3*, 508–514.
141. Rouault, T.A.; Klausner, R.D. *T.I.B.S.* **1996**, *21*, 174–177.
142. Beinert, H.; Kennedy, M.C.; Stout, C.D. *Chem. Rev.* **1996**, *96*, 2335–2373.
143. Burgess, B.K.; Lowe, D.J. *Chem. Rev.* **1996**, *96*, 2983–3011.

144. Howard, J.B.; Rees, D.C. *Chem. Rev.* **1996**, *96*, 2965–2982.

145. Bolin, J.T.; Campobasso, N.; Muchmore, S.W.; Morgan, T.V.; Mortenson, L.E., in *Molybdenum Enzymes, Cofactors, and Model Systems*, eds. E.I. Steifel; D. Coucoucanis; W.E. Newton, American Chemical Society, Washington, DC (1993). pp. 186–195.

146. Mouesca, J.-M.; Noodleman, L.; Case, D.A. *Inorg. Chem.* **1994**, *33*, 4819–4830.

147. Mouesca, J.-M.; Noodleman, L.; Case, D.A.; Lamotte, B. *Inorg. Chem.* **1995**, *34*, 4347–4359.

148. Papaefthymiou, V.; Millar, M.M.; Münck, E. *Inorg. Chem.* **1986**, *25*, 3010–3014.

149. Rius, G.; Lamotte, B. *J. Am. Chem. Soc.* **1989**, *111*, 2464–2469.

150. Middleton, P.; Dickson, D.P.E.; Johnson, C.E.; Rush, J.D. *Eur. J. Biochem.* **1978**, *88*, 135–141.

151. Middleton, P.; Dickson, D.P.E.; Johnson, C.E.; Rush, J.D. *Eur. J. Biochem.* **1980**, *104*, 289–296.

152. Eaton, W.A.; Palmer, G.; Fee, J.A.; Kimura, T.; Lovenberg, W. *Proc. Natl. Acad. Sci. U.S.A.* **1971**, *68*, 3015.

153. Gray, H.B.; Siiman, O.; Rawlings, J. *Proc. Natl. Acad. Sci. U.S.A.* **1974**, *71*, 125.

154. Mayerle, J.J.; Denmark, S.E.; Pamphilis, B.V. De; Ibers, J.A.; Holm, R.H. *J. Am. Chem. Soc.* **1975**, *97*, 1032–1045.

155. Blum, H.; Adar, F.; Salerno, J.C.; Leigh, J.S. *Biochem. Biophys. Res. Commun.* **1977**, *77*, 650.

156. Mouesca, J.-M.; Noodleman, L.; Case, D.A. *Int. J. Quant. Chem.: Quantum Biol. Symp.* **1995**, *22*, 95–102.

157. Noodleman, L.; Case, D.A.; Mouesca, J.-M.; Lamotte, B.M *J. Biol. Inorg. Chem.* **1996**, *1*, 177–182.

158. Jordanov, J.; Roth, E.K.H.; Fries, P.H.; Noodleman, L. *Inorg. Chem.* **1990**, *29*, 4288–4292.

159. Noodleman, L.; Case, D.A.; Baerends, E.J., in *Density Functional Methods in Chemistry*, eds. J.K. Labanowski; J.W. Andzelm, Springer-Verlag, New York (1991). pp. 109–123.

160. Noodleman, L.; Li, J.; Zhao, X.G.; Richardson, W.H., in *Density Functional Methods: Applications in Chemistry and Materials Science*, ed. M. Springborg, John Wiley and Sons, New York (1997). pp. 149–188.

161. Petersson, L.; Cammack, R.; Rao, K.K. *Biochim. Biophys. Acta* **1980**, *622*, 18–24.

162. Crozet, M.; Bardet, M.; Emsley, L.; Lamotte, B.; Mouesca, J.-M. *unpublished.*

163. Sauer, K. *Acc. Chem. Res.* **1980**, *13*, 249.

164. Wieghardt, K. *Angew. Chem. Int. Ed. Engl.* **1994**, *33*, 725–728.

165. Hart, J.R; Rappe, A.K.; Gorun, S.M.; Upton, T.H. *Inorg. Chem.* **1992**, *31*, 5254–5259.

166. Honig, B.; Sharp, K.; Yang, A.-S. *J. Phys. Chem.* **1993**, *97*, 1101–1109.

167. Davis, M.E.; McCammon, J.A. *Chem. Rev.* **1993**, 509–521.

168. Born, M. *Z. Phys.* **1920**, *1*, 45–48.

169. Kirkwood, J.G. *J. Chem. Phys.* **1939**, *7*, 911.

170. Onsager, L. *J. Am. Chem. Soc.* **1936**, *58*, 1486.

171. Tomasi, J.; Bonaccorsi, R.; Cammi, R.; Valle, F.J. Olivares del *J. Mol. Struct. (Theochem)* **1991**, *234*, 401–424.

172. Karelson, M.M.; Zerner, M.C. *J. Phys. Chem.* **1992**, *96*, 6949–6957.

173. Wong, M.W.; Frisch, M.J.; Wiberg, K.B. *J. Am. Chem. Soc.* **1991**, *113*, 4776–4782.

174. Rivail, J.L.; Terryn, B.; Rinaldi, D.; Ruiz-Lopez, M.F. *J. Mol. Struct. (Theochem)* **1985**, *120*, 387–400.

175. Rinaldi, D.; Rivail, J.-L.; Rguini, N. *J. Computat. Chem.* **1992**, *13*, 675–680.

176. Ford, G.P.; Wang, B. *J. Computat. Chem.* **1992**, *13*, 229–239.

177. Miertus, S.; Scrocco, E.; Tomasi, J. *Chem. Phys.* **1981**, *55*, 117–129.

178. Grant, J.A.; Williams, R.L.; Scheraga, H.A. *Biopolymers* **1990**, *30*, 929–949.

179. Tuñón, I.; Silla, E.; Tomasi, J. *J. Phys. Chem.* **1992**, *96*, 9043–9048.

180. Tuñón, I.; Silla, E.; Pascual-Ahuir, J.-L. *J. Am. Chem. Soc.* **1993**, *115*, 2226–2230.

181. Yoon, B.J.; Lenhoff, A.M. *J. Comp. Chem.* **1990**, *11*, 1080–1086.

182. Zauhar, R.J. *J. Computat. Chem.* **1991**, *12*, 575–583.

183. Juffer, A.H.; Botta, E.F.F.; van Keulen, B.A.M.; Ploeg, A.van der; Berendsen, H.J.C. *J. Computat. Phys.* **1991**, *97*, 144–171.

184. Jackson, J.D., *Classical Electrodynamics*, Wiley, New York (1975). pp. 158–159.

185. Tomasi, J.; Persico, M. *Chem. Rev.* **1994**, *94*, 2027–2094.

186. Tawa, G.J.; Martin, R.L.; Pratt, L.R.; Russo, T.V. *J. Phys. Chem.* **1996**, *100*, 1515–1523.

187. Cossi, M.; Tomasi, J.; Cammi, R. *Int. J. Quant. Chem. Quant. Chem. Symp.* **1995**, *29*, 695–702.

188. Andzelm, J.; Kölmel, C.; Klammt, A. *J. Chem. Phys.* **1995**, *103*, 9312–9320.

189. Warshel, A., *Computer Modelling of Chemical Reactions in Enzymes and Solutions*, Wiley, New York (1991).

190. Richardson, W.H.; Peng, C.; Bashford, D.; Noodleman, L.; Case, D.A. *Int. J. Quantum Chem.* **1997**, *61*, 207–217.

191. Halliwell, B.; Gutteridge, J.M.C.; Blake, D. *Phil. Trans. R. Soc. London B* **1985**, *311*, 659–671.

192. Halliwell, B.; Gutteridge, J.M.C., in *Free Radicals in Biology and Medicine, 2nd Ed.*, Clarendon, Oxford, UK (1989).

193. Ludwig, M.L.; Metzger, A.L.; Pattridge, K.A.; Stallings, W.C. *J. Mol. Biol.* **1991**, *219*, 335–358.

194. Borgstahl, G.E.O.; Parge, H.E.; Hickey, M.J.; Breyer, W.F., Jr.; Halliwell, R.A.; Tainer, J.A. *Cell* **1992**, *71*, 107–118.

195. Lah, M.S.; Dixon, M.M.; Pattridge, K.A.; Stallings, W.C.; Fee, J.A.; Ludwig, M.L. *Biochemistry* **1995**, *34*, 1646–1660.

196. Whittaker, J.W.; Whittaker, M.M. *J. Am. Chem. Soc.* **1991**, *113*, 5528–5540.

197. Verhagen, M.F.J.M.; Meussen, E.T.M.; Hagen, W.R. *Biochim. Biophys. Acta* **1995**, *1244*, 99–103.

198. Clair, C.S. St.; Gray, H.B.; Valentine, J.S. *Inorg. Chem.* **1992**, *31*, 925–927.

199. Smith, E.T.; Tomich, J.M.; Iwamoto, T.; Richards, J.H.; Mao, Y.; Feinberg, B.A. *Biochemistry* **1991**, *30*, 11669–11676.

200. Correll, C.C.; Ludwig, M.L.; Bruns, C.M.; Karplus, P.A. *Prot. Sci.* **1993**, *2*, 2112–2133.

201. Langen, R.; Jensen, G.M.; Jacob, U.; Stephens, P.J.; Warshel, A. *J. Biol. Chem.* **1992**, *267*, 25625–25627.

202. Jensen, G.M.; Warshel, A.; Stephens, P.J. *Biochemistry* **1994**, *33*, 10911–10924.

203. te Velde, G.; Baerends, E.J. *J. Comp. Phys.* **1992**, *99*, 84–98.

204. Ziegler, T.; Tschinke, V.; Baerends, E.J.; Snijders, J.G.; Ravenek, W.J. *J. Phys. Chem.* **1989**, *93*, 2050.

205. Versluis, L.; Ziegler, T. *J. Chem. Phys.* **1988**, *88*, 322.

206. Fan, L.; Ziegler, T. *J. Chem. Phys.* **1991**, *95*, 7401–7408.

INDEX